Wetlands: Ecology, Conservation and Management

Volume 1

Series Editor:

Max Finlayson
International Water Management Institute
Colombo, Sri Lanka
email: m.finlayson@cgiar.org

Aims & Scope:

The recognition that wetlands provide many values for people and are important foci for conservation worldwide has led to an increasing amount of research and management activity. This has resulted in an increased demand for high quality publications that outline both the value of wetlands and the many management steps necessary to ensure that they are maintained and even restored. Recent research and management activities in support of conservation and sustainable development provide a strong basis for the book series. The series presents current analyses of the many problems afflicting wetlands as well as assessments of their conservation status. Current research is described by leading academics and scientists from the biological and social sciences. Leading practitioners and managers provide analyses based on their vast experience.

The series provides an avenue for describing and explaning the functioning and processes that support the many wonderful and valuable wetland habitats, such as swamps, lagoons and marshes, and their species, such as waterbirds, plants and fish, as well as the most recent research directions. Proposals cover current research, conservation and management issues from around the world and provide the reader with new and relevant perspectives on wetland issues.

The titles published in this series are listed at the end of this volume.

The Becher Wetlands – A Ramsar Site

Evolution of Wetland Habitats and Vegetation Associations on a Holocene Coastal Plain, South-Western Australia

by

Christine Semeniuk

Wetlands Research Association,
Perth, WA, Australia

Springer

A C.I.P. Catalogue record for this book is available from the Library of Congress.

ISBN-10 1-4020-4671-5 (HB)
ISBN-13 978-1-4020-4671-1 (HB)
ISBN-10 1-4020-4672-3 (e-book)
ISBN-13 978-1-4020-4672-8 (e-book)

Published by Springer,
P.O. Box 17, 3300 AA Dordrecht, The Netherlands.

www.springer.com

Cover image:
Prominent peripheral ring of grass trees (*Xanthorrhoea preissii*), blackened by fire, ringing a small wetland basin in the Becher Suite inhabited by sedge.

Printed on acid-free paper

CONTENTS

PREFACE

Listed as a Wetland of International Importance as site 1048 under the Ramsar convention in January 2001, the Becher Point wetlands are an important and unusual wetland system in Western Australia. The Becher Point Wetlands are an example of shrub swamps and seasonal marshes that have formed in an extensive sequence of inter-dunal depressions that have arisen from seaward advancement of the coastline over recent millennia. This type of wetland system is rare in southwestern Australia, and examples of this type of geomorphological sequence in equally good condition and within a protected area are rare worldwide. Knowledge and understanding about the wetlands derived from the research which is reported in this book made possible the effective nomination of this wetland as a Wetland of International Importance.

The Ramsar site includes a substantial part of the suite of approximately 200 discrete, very small wetlands located between Becher Point (Indian Ocean coast) and the Perth-Mandurah Road. I first stumbled upon this system in 1978, while working in Western Australia, and was immediately entranced by its complexity, as well as its beauty – and importance as a key landscape element. The site's wetlands are within 0.2-1.5 km of the Indian Ocean and they comprise chains of micro-scale linear, ovoid or irregular swamps arranged in about 10 groups roughly parallel to the coast, separated by sand ridges. There is usually no surface water late summer to autumn. The fresh surface water of winter, derived primarily from groundwater flow and direct precipitation, is generally less than 0.3 m deep.

Some of the vegetation types are included in the national list of threatened ecological communities. Key swamp vegetation dominated by sedges and rushes has mixtures of different species/genera dominant, including *Baumea articulata*, *B. juncea* and *Typha* spp. Wooded areas are dominated by *Melaleuca rhaphiophylla*. At least 21 reptile and four amphibian species have been recorded.

The ephemeral nature of the wetlands is an important element in ensuring viability of the whole hydrological cycle of the coastal dunes systems in the Perth Region. This monograph represents a considerable volume of time effort and funding which has been placed into an investigation of these systems over the last 11 years, using monthly measurements for over 100 sites. This kind of information is rarer even than the wetland types, but is essential if we are to effectively understand, and thus manage, our wetlands and associated ecosystems.

The Ramsar convention now has over 1400 wetlands of International importance listed, and very few have the amount of detailed science background available that the Becher Suite wetlands now do. While it is too much to hope we will have all the Wetlands of International Importance documented to the extent the Becher wetlands now are, this effort sets the standard and the pace we need to have in the convention if we are to stem, and reverse, the continued decline of the world's wetlands.

The author is to be congratulated on her efforts and perseverance in undertaking this research, and Springer for making the material available in this book format. I commend the information to all who live close to or around the Becher Point wetlands and the whole study as a model for wetland science globally.

Peter Bridgewater
Secretary General
Ramsar Convention, Gland, Switzerland
December 2005.

ACKNOWLEDGEMENTS

In a piece of work with wide ranging subject matter and extensive duration, such as this, there are many people to acknowledge and to thank for their contribution and perseverance. Many people followed through to completion tasks which incorporated several stages and several years. They expressed a real commitment to the project and a love for the wetlands in the area. It has been uplifting to work in their company. Together, we have amassed a repertoire of stories that are funny, disappointing, bizarre, dangerous and inspirational.

I wish to begin by thanking the many people who assisted with fieldwork because this is where the story really started. They are: Ben Asper, Derek Bazen, Theo Bazen, Anthony Bougher, Gary Dietrich, Martine Desbureaux, Toby Nisbet, Kaylene Parker, Julie Pech, Karen Semeniuk, Trudi Semeniuk, Tony Smith, and Joy Unno. Regional surveying of sites was carried out by Ric Stephenson. The people who assisted in preparing samples for laboratory analyses were Penny Clifford, Toby Nisbet, and Joy Unno. The people responsible for final electronic drafting of the diagrams were Craig Miskell, Vic Semeniuk and Glynn Kernick.

Radiometric analyses (^{14}C) were undertaken at CSIRO laboratories in South Australia and supplementary analyses (AMS) at the University of Sydney. While most chemical analyses were undertaken by the author, some cation analyses in groundwater and all extracted soil waters were carried out by AMDEL P/L Perth, analyses of total phosphorus were undertaken by SGS Laboratories, Perth, mixed acid digests, and all analyses of mineral samples and vegetation for determination of elements were carried out by UltraTrace Analytical Laboratories, Perth. XRD analyses were carried out by AMDEL, South Australia. SEM work was carried out at the CSIRO Laboratories, Bentley (W.A.) and EMPA Laboratories at Zurich. Pollen preparation was undertaken by the Geography Department, University of Western Australia, and pollen identification by Dr Lynne Milne.

Funding for this project over the 11-year period, and publication costs came from VCSRG Pty Ltd as part of their R&D endeavour, registered as AusIndustry Project #3. A grant from Lotterywest (Western Australia), mediated by the Wetlands Research Association, went towards the costs of the coloured plates in Chapter 12.

Margaret Brocx, Dr Don Glassford, Dr Philip Ladd, and Dr Vic Semeniuk read the manuscript and made constructive comments - the manuscript benefited from this thorough and careful reading and correction. For their time and patience I am most grateful. I wish to thank Dr Max Finlayson for the support he provided in publishing this book. Finally, I wish also to personally and professionally thank Dr Vic Semeniuk for taking on the role of mentor, which, in this instance, involved equal discipline in silence, advice, and encouragement.

1. INTRODUCTION

1.1 General introduction

In 2001, a suite of relatively small natural basins, known as the Becher Point wetlands, were designated as Wetlands of International Importance, and nominated for protection under the Ramsar Convention. An internationally important wetland very often has attributes which make it immediately apparent to the observer that this habitat is special. The scenery is breathtaking, the wetland is usually large and filled with water, it is a habitat for rare flora, and international migratory birds or rare fauna inhabit or regularly visit the site. These attributes are readily observable, and ones, with which we, as humans, strongly identify. In stark contrast, the subjects of this study are a group of very small basin wetlands, comprising the Becher Suite (C. A. Semeniuk 1988), which are seasonally inundated or waterlogged by groundwater rise, colonised by herbs, sedges and shrubs which comprise no rare taxa, and support a local population of marsupials and reptiles. The wetlands were nominated for their outstanding scientific values, values that became apparent after the landscape in which the wetlands reside was identified as a rare coastal type in Western Australia (Woods 1984; Searle and Semeniuk 1985; Semeniuk *et al.* 1989; Sanderson *et al.* 1999), and that the wetlands themselves were unusual, globally. In 1990, a holistic research programme was commenced to detail their developmental history, functions and biotic components. The findings of this scientific study endorsed the nomination of the Becher Suite as wetlands of international importance with respect to their inter-beachridge setting, their underlying carbonate rich sands, which, in this particular climate setting, influenced hydrochemical patterns, the occurrence of carbonate muds as basin fills, the archival information contained in the Holocene sediments, the hydrological responses to a varied stratigraphy, and the resulting diversity of plant communities. The results stand to inspire similar endeavours elsewhere.

The wetlands are located in beachridge swales on a vegetated coastal plain which forms the Holocene surface of an accretionary cuspate foreland, the Becher cuspate foreland, in southwest Western Australia (Searle *et al.* 1988) (Fig. 1-1). The distinct chains of wetlands within the swales mirror the orientation of the beach ridges and the changing asymmetry of the cusp, suggesting that a continuous record of wetland development for the region could be obtained along an axis from the eastern landward boundary of the cuspate foreland to its western shore. This continuum in wetland development is a unique occurrence in Western Australia, and is rare globally (Gulliver 1896, 1897; Lewis 1932; Moslow and Heron 1981; Rosengren 1981; Lubke and Avis 1982; Thom 1984; Lees 1987; Coakley 1989; Penland and Suter 1989; Ying Wang 1989; Anthony 1989, 1991; Isla *et al.* 1996; Rasch *et al.* 1997; Bonorino *et al.* 1999; Fontolan and Simeoni 1999; Saito *et al.* 2000; Sanderson 2000; Huh 2001).

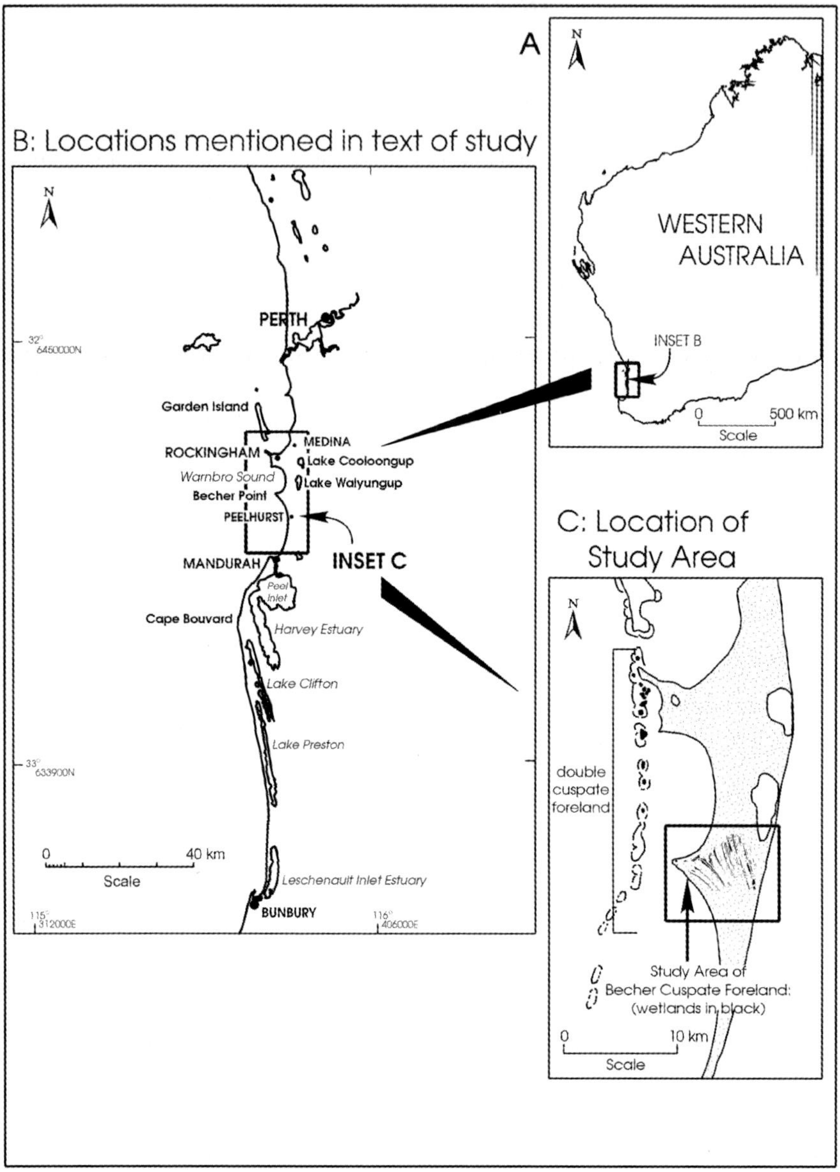

Figure 1-1. *Location of study area and key geographic locations in southwestern Australia.*

An important feature of this beachridge plain, which distinguishes it from coastal plains and cuspate forelands in eastern Australia and globally, is that it is underlain by a relatively carbonate rich calcareous quartzose sand, which has implications for wetland evolution and hydrochemistry. The interplay between calcium carbonate rich parent material and the natural *in situ* wetland production of organic material created a series of stratigraphic sedimentary sequences in the Becher Suite wetlands which contain a detailed and reliable sedimentary and vegetation record of the late Holocene period. In addition, this varied stratigraphy was discovered to influence the hydrological responses from regional to pellicular scale, and therefore, the resulting distribution and composition of plant communities.

It is against this coastal beachridge plain historical background that wetland initiation, evolution, and response to climatic variability were examined. This monograph documents the features, processes and developmental history of a number of these wetlands, from the youngest to the oldest, demonstrating the considerable scientific information inherent in the individual wetlands, and the extent to which that could be enriched by drawing upon the cumulative studies to provide context for the entire wetland suite.

1.1.1 This study

The relative homogeneity of regional geomorphology, in that beachridges and swales are the recurring pattern on this landform, the relative consistency of the underlying sediments, the discrete location of the suite of wetlands within the (current) climatic setting, the relatively young age of the Becher Cuspate Foreland, and the range of ages of the wetlands from *circa* 5000 years to <1000 years old, provided an ideal setting for attempting to isolate variable wetland processes from regional processes, to delineate wetland intra-basin complexity using inter-basin comparison, and to trace wetland evolution.

This type of study requires a multi-disciplinary approach appropriate to the study of complex systems and necessitates drawing upon a variety of concepts, methods, measuring techniques, and terminology. It incorporates the following components: the geomorphic and stratigraphic evolution of the wetlands; description, synthesis and geohistorical unravelling of their hydrology and hydrochemistry; description of their vegetation and its relationship to the physical, stratigraphic, hydrological and hydrochemical attributes of the wetlands; and a reconstruction of the vegetation history and physical evolution of the wetlands using pollen and stratigraphic patterns.

In the considerable body of detailed information on wetlands and wetland processes, there are few studies that employ a multi-disciplinary approach, either for a single wetland or a suite of genetically similar wetlands. There are even fewer studies that attempt to relate the evolutionary history of a given wetland to its current setting. As a relatively new discipline, wetland science has been the arena for two fundamental types of investigations: 1) ecological investigations of wetland vegetation distributions, zonation, and dynamics (van der Laan 1979; Andrus *et al.* 1982; Vitt and Chee 1990; Robertson *et al.* 1991; Wheeler 1999; and references therein), and 2) investigation of traditional scientific subjects in a wetland habitat, *e.g.*, well established methodologies in hydrology and geochemistry applied within a variety of wetland types, *e.g.*, riverine, estuarine, tidal, non-tidal, lakes and seasonal wetlands (Hite and Cheng 1996; Fisher *et al.* 1998; Kehew *et al.* 1998). The studies of mires, coastal areas and floodplains, by Malmer (1986); Hayden *et al.* (1995); Mertes *et al.* (1995); and Gurnell (1997), serve as exceptions to this general pattern, in that the authors attempted to synthesise a range of physical, hydrological, and chemical processes into an integrated picture of wetland evolution and functioning. This study also steps outside established boundaries in order to highlight the considerable scientific lessons which lie beneath the wetlands' surface.

1.2 Location of study area

The Becher Cuspate Foreland is part of a double cusp system located along the coast in subhumid southwestern Australia (Fig. 1-1; Searle *et al.* 1988). The wetlands, referred to as the Becher Suite (C. A. Semeniuk 1988), are located mainly along the axis of progradation of the cuspate foreland (Fig. 1-1), but the study area also incorporates some of the sub-regional aspects of the terrain, in that it extends south to Mandurah to capture the southern limits of the cuspate foreland, east to the Spearwood Ridge of Pleistocene limestone that forms a natural boundary to the beachridge plain, and north to Rockingham to incorporate the northern cuspate foreland in the system.

1.3 Objectives

No systematic investigation of the Becher Suite wetlands has ever been undertaken with the exception of the work on wetland classification (C. A. Semeniuk 1987, 1988). This study aims to describe sedimentary, pedogenic, hydrologic, hydrochemical and vegetation features of the wetlands in order to characterise them, and to interpret causal processes and progressive changes in those processes which are relevant to wetland development. These aims can be more fully stated as follows:

1. document the sedimentary stratigraphic sequences in order to reconstruct the history of wetland development and the history of deposition within the wetland basins relating sequences to possible changes in regional and local conditions;

2. document the hydrological functioning of the wetlands in terms of present sources, processes, pathways, influences, and patterns in relation to sedimentary stratigraphic sequences;

3. document some basic components of the hydrochemical environment as an adjunct to inferring hydrological processes;

4. relate present distribution of wetland plants to hydrological, hydrochemical and sedimentological factors, and use these data to create, in retrospect, a model for wetland habitat evolution;

5. relate pollen abundances in the sedimentary profile to radiocarbon dates of wetland commencement and present vegetation patterns, and generate models of vegetation change in this setting;

6. construct a model of wetland evolution on the Becher cuspate foreland for the period 5000 years BP to the present.

1.4 Nature and scope of study

The unravelling of the evolution of wetland development, its habitats and plant associations [from 5000 years BP - present] is a holistic study, in the sense that it uses present day cause/effect relationships to help interpret past change, and uses chronosequence studies to describe past general temporal trends. It is an attempt to revitalise an older, more traditional custom in scientific enquiry, that of the pursuit of natural history in its complexity, to deviate from the straight line and single correlation and to enter the messy world of natural action and numerous reactions. Inherent in any holistic study of this type are numerous hypotheses relating to the nature of each wetland attribute and process, the links between processes and effects, the relationships between ancestral and current wetland processes, as well as the interplay between all these aspects. The development and continual unfolding of the research occurs in response to the questions generated by data collection, rather than from the directed enquiry and empirical testing associated with hypothetical constructs. Specific hypotheses were formulated and tested, where necessary, otherwise the substance of the investigations was defined by the objectives outlined at the beginning of each chapter, culminating in a final interpretation and synthesis.

In order to pursue these enquiries it was necessary to systematically address various scales of study: sub-regional, local, the basin and bedding scale. In this study, sub-regional studies encompassed the broadest view, that of the cuspate foreland of the Rockingham-Becher Plain, incorporating the setting and characterisation of the Becher Suite wetlands, comparative analysis between wetlands, and establishment of relationships between sub-regional variables. Investigations which encompassed beachridges and adjacent wetlands were termed local scale. Examples of local scale processes and patterns are local precipitation and local groundwater gradients. Other studies contained within the dimensions of a single basin or sedimentary layer were termed respectively “basin” and “bedding” scales. Processes and patterns which can be observed at basin scale are wetland fills, seepages, and vegetation assemblages, while examples of processes and patterns which can be observed at bedding scale are layered sequences, rain infiltration through the profile, interstitial water content down profile, chemical characteristics, and pollen profiles. These scales of reference are alluded to in the description of methods, and throughout.

A holistic study requires a well-defined framework. There are four physical frameworks upon which this work rests. At the largest scale, the physical framework is defined by the cuspate foreland setting. At a finer scale, a single suite of wetlands, the Becher Suite, forms the second framework, in that from a regional perspective, it is a relatively consistent set of wetland types embedded in a similar geomorphic setting, with initially similar parent substrates and general hyrologic processes. Thus, selection of a group of wetlands in the same consanguineous suite ensures a similarity of wetland size and shape, a recurring pattern of similar genetic types, similar origin, and at the broad scale, similar hydrological recharge and discharge mechanisms and patterns. The stratigraphic framework is the next scale of reference in that wetland evolutionary history, detailed hydrological and hydrochemical patterns, and palynology rest firmly on this foundation. Finally, the entire system of the Becher Suite wetlands have formed wholly within the middle to very late Holocene period, and this means that the various wetlands, once dated, can be used to determine wetland evolution within relatively small time frames. Selection of wetlands whose origin and history is contained within the late Holocene time frame, decreases the probability of loss of data that is often associated with longer time periods and a higher number of cataclysmic events.

The conceptual framework as defined by the objectives of the study is rooted in the perspective that wetlands which are complex interactive systems can be viewed as comprising 1) fundamental attributes which are water and land, and 2) biological attributes which result from the responses by flora and fauna to the interplay of fundamental attributes, species competition, and over time, their own modification of the habitat. This perspective dictates both the central elements of the study and the order of data presentation depicted in Table 1.1, that is, wetland geomorphic setting, stratigraphy, and hydrology, followed by their interactions through hydrochemistry and vegetation (Chapters 5, 6, 7, 8, 9, 10). These aspects are then integrated in discussion sections.

1.5 History of work in similar areas

Very few studies of similar wetland habitats to the wetlands in the Rockingham-Becher Plain have been documented globally (van der Laan 1979). The Becher Suite wetlands, being expressions of groundwater in interdune depressions on coastal plains, have affinities with dune slacks. However, there are some important similarities and some equally important differences. Dune slacks are defined as low lying areas within a coastal dune terrain where seasonally the water table inundates, or is close to, the surface (Van Dieran 1934 cited in Lammerts *et al.* 2001; Ranwell 1959; De Raeve 1987; Grootjans *et al.* 1998). The hydrological system is freshwater and the underlying parent material is generally nutrient poor mineral sediment comprising a mixture of calcareous and quartz sand (Willis *et al.* 1959; Jones and Etherington 1971; Noest 1992; Van Dijk and Grootjans 1993; Mc Lachlan *et al.* 1996). In these respects the Becher Suite wetlands are similar, however, the term "dune slack" was originally applied either to beach plains, partly or completely secluded from marine influence by a new beachridge or foredune, or areas within the mobile dune field deflated to the water table surface. In this latter respect the Becher Suite wetlands are fundamentally different. They are basins within the swales of a stable beachridge plain under which the groundwater table has risen concomitant with beachridge plain progradation. Several elements and developmental processes inherent in the formation of dune slacks, *i.e.*, salt spray, sand mobilisation and incipient soil development (Lubke and Avis 1982), are either absent or minor in the Becher Suite wetlands.

Table 1.1 *Titles of Chapters and order of presentation*

Chapter sequence	*Title*
Chapter 1	Introduction
Chapter 2	Methods and Terminology
Chapter 3	Regional Setting
Chapter 4	Description of wetlands
Chapter 5	Development of proto-wetland type
Chapter 6	Wetland sedimentology and stratigraphy
Chapter 7	Linkage between stratigraphy and wetland hydrology
Chapter 8	Wetland hydrology
Chapter 9	Wetland hydrochemistry
Chapter 10	Vegetation assemblages
Chapter 11	Vegetation history
Chapter 12	Synthesis

The second important difference between the Becher Suite wetlands and dune slacks is the nature and composition of the wetland fills. The Becher Suite wetland fills can be viewed as patches of sedimentation on, and intermixing with, the highly calcareous sediments underlying the undulating Holocene surface of a beachridge plain. Most dune slacks (Tansley 1949) contain organic material overlying quartzose/calcareous sand, however, the calcareous component of the sand is often minimal and sometimes absent (Ovington 1951; Ranwell 1959; Crawford and Wishart 1966; Moreno-Casasola 1986,1988; Robertson *et al.* 1991; Van Dijk and Grootjans 1993; Munoz Reinoso 2001). In this respect, the Becher Suite wetland fills also differ from other coastal inter-dune swamps or lagoons which contain carbonate mud precipitated under mesosaline (marine) or hypersaline conditions (Moreno-Casasola 1988).

The types of studies carried out in the dune slack habitat have much in common with this study. Plant distribution and zonation description and analysis have been the focus of numerous studies because the dune slacks appeared to be habitats with recognisable boundaries, with plant diversity clearly underscored by environmental attributes and gradients (Godwin and Bharucha 1932; Ranwell 1959; Willis *et al.* 1959; Jones and Etherington 1971, 1987; van der Laan 1979; Olff *et al.* 1993; Dickinson and Mark 1994; McLachlan *et al.* 1996; Hellemaa 1998; Lammerts *et al.* 2001). More recently, specific attributes of dune slacks have been the main focus of investigation, such as nutrients, hydrogeochemistry, the role of algae in seed germination (van Dijk and Grootjans 1993; Grootjans *et al.* 1996; Vazquez *et al.* 1998; Malcolm and Soulsby 2001). The studies have mirrored two general trends in ecology, the first toward conducting experiments and field measurements, which are designed to be integrated into the body of information on the interplay of hydrological, hydrochemical, geomorphic and geological factors, and the second towards experiments and field measurements on the physiological and anatomical responses of an individual plant species to change in a single environmental attribute. Both of these ecological aspects are considered in the present study.

In addition, dune slacks have been the sites around which theoretical and experimental debates about succession and community structure have centred. Succession of plant communities is one aspect of evolution. There have been several studies indicating that there is a transition from seasonally inundated dune slacks to permanently submerged slacks to shrub and forest swamp vegetation based on the following attributes: gradation of wetland types in the dune terrain; the gradient from coast to inland of depth to water table; the increase in organic matter and the increase in nutrients (Lubke and Avis 1982; De Raeve 1987). In this study, the pollen has been used to deduce wetland vegetation history.

A third aspect of dune slack evolution which has attracted researchers has been the cumulative chemical effects of groundwater upon the sediments. In this case the dissolution of carbonate has been a primary finding (Grootjans *et al.* 1996; Sival and Grootjans 1996) which is discussed further in Chapter 5. However, other aspects of the dune slacks, such as their short period of existence or their simple wetland stratigraphy, do not obviously embody evolutionary principles. In contrast, the Becher Suite wetland stratigraphic sequences and their interaction with wetland hydrology, hydrochemistry, and vegetation, provide a rich archive of wetland evolutionary history from the middle to late Holocene period on the southwest coast of Western Australia.

A considerable body of research on the Rockingham-Becher Plain and its nearshore and offshore surroundings had been carried out prior to this study, and this information provided detailed baseline and historical data about regional geologic and geomorphic processes which indirectly and directly related to the current study. Information on regional coastal sediment dynamics was drawn from descriptions of nearshore bathymetry, summarised regional coastal wind and wave dynamics, and history of coastal sedimentation and erosion, cuspate foreland geometry, stratigraphy and origin (Steedman and Craig 1979, 1983; Collins 1983; Woods and Searle 1983; Searle 1984; Woods 1984; Searle and Semeniuk 1985a; Searle *et al.* 1988; Semeniuk *et al.* 1988) to derive a history of intra-swale basin evolution and geometry. Reference was made to petrological descriptions and diagnostic characteristics for lithofacies in wave dominated sandy shores (Semeniuk and Johnson 1982; Searle and Semeniuk 1988) to characterise the wetland basal sheets, and determine the environmental depositional setting of the landforms which comprised the template for the proto-wetland. Reference was also made to definitive descriptions and stratigraphic relationships of new formations within the sequence of Holocene sediments (Semeniuk and Searle 1985a, 1987; Searle *et al.* 1988). Of direct relevance to this study also were the descriptions of sheets of groundwater calcrete in the stratigraphic sequences of Holocene beach/dune units (Semeniuk and Meagher 1981; Semeniuk and Searle 1985b; Semeniuk 1986a, 1986b), in that, according to Semeniuk (1986), it provides a proxy record of climate change, and at the local scale, it affects hydrological patterns. Other pertinent information, related to Holocene sealevel records 7,400 yrs BP to the present (Semeniuk 1985; Semeniuk and Searle 1986), which were incorporated in the final synthesis of wetland data, led to a theory for wetland habitat evolution on the Becher cuspate foreland.

1.6 Selection of photographs of the Point Becher area

A series of photographs illustrating the setting and some of the landscape and vegetation features of the Becher Suite wetlands are provided in Figures 1-2 to 1-5. These illustrations transcend a range of scales: from the large scale aerial view showing the nature of the cuspate foreland that is the setting for the wetlands, to a typical view at the medium scale of a moderately high beach ridge that borders many wetlands on their eastern side, to views at smaller scales of the shape of the wetlands and the nature of the vegetation that inhabits them. As will be described in later chapters, *Stipa* grassland and various scrub and heath inhabit the beachridges, and paperbark scrub, sedges, rushes and herb formations inhabit the wetlands.

Figure 1-2. *Aerial view looking east from the coast. showing the apex of the Becher Cuspate Foreland and the beachridge nature of this accretionary promontory.*

Figure 1-3. *View to east across a linear wetland. A moderate relief beachridge borders the wetland. Mottled wetland vegetation comprises sedgelands (Lepidospermum gladiatum and Typha spp. are light green; living Baumea juncea is medium green, senescent is light grey; isolated clumps are Juncea kraussii). Seacouch Sporobolus virginicus and a band of Melaleuca teretifolia heath are in the foreground along the wetland shore.*

Figure 1-4. *Scrub of the paperbark Melaleuca rhaphiophylla fringing a wetland, with the herb Centella asiatica, various sedges, and patches of open water in the wetland interior.*

Figure 1-5. *Small round wetland, typical of the Becher Suite, with the sedge Baumea articulata in the wetland centre, and a circumferential ring of grass trees Xanthorrhoea preissii along the wetland margin. Flowering Acacia and various low heath species inhabit the surrounding beachridges.*

2. METHODS AND TERMINOLOGY

2.1 General introduction

This chapter is divided into three parts. The first part explains the classification and terminology used herein for wetlands, their vegetation and their sediments. The second part is allocated to general terminology, and in the third part, the various methods used in the investigation of the Becher Suite wetlands are described. Extensive fieldwork, (incorporating long term monitoring and sampling, single sampling episodes, and several field experiments), sample preparation and laboratory analyses, and statistical software programmes were components in this study. After 12 months of laboratory work, the ongoing laboratory analyses were dispersed to commercial laboratories. A summary of the scale and scope of the methods is presented in Figure 2-1.

2.2 Local scale wetland classification systems

The local scale classification of wetlands and their vegetation follows C. A. Semeniuk (1987), Semeniuk and Semeniuk (1995, 1999), and C. A. Semeniuk *et al.* (1990). While details are provided in these papers (*op cit.*), a summary is presented below.

2.2.1 Local scale wetland classification

The classification used herein for individual wetlands is a standard classification scheme which utilizes the two primary components of land based wetlands, "wetness" and "landform" (C. A. Semeniuk 1987; Semeniuk and Semeniuk 1995; Semeniuk and Semeniuk 1999). The water component is the major feature that distinguishes the wetland habitat from other terrestrial habitats, and the component which influences biological response by its presence, depth, chemistry and movement. The landform is essentially the water container and thus it determines size, shape, and depth of a wetland.

Landform can be categorised on the basis of cross-sectional wetland geometry, scale and plan geometry. In the local classification of wetlands, a fundamental attribute of landforms is cross-sectional geometry. Descriptive terms are provided that further classify landform as to size and shape (Fig. 2-2).

Water within wetlands can be categorised on the basis of its persistence or longevity, its quality, its constancy of water quality and the mechanism by which it maintains the wetland. In the classification, the fundamental attribute is water permanence or longevity. Similarly, descriptive terms are provided for each of the characteristics which may be used to develop secondary wetland categories (Fig. 2-2).

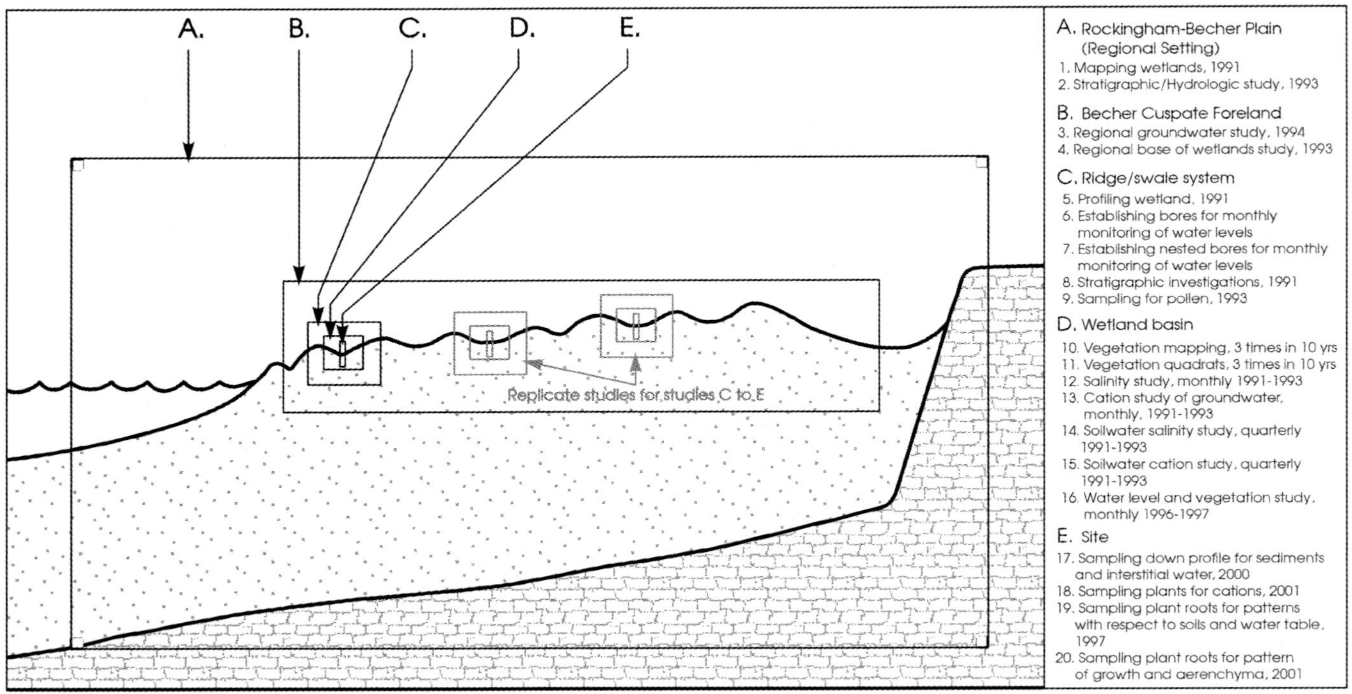

Figure 2-1. *Scope and time frames of field studies over 10 years.*

Wetland types are delineated from a matrix constructed using combinations of primary subdivisions of cross-sectional geometry and categories of water persistence or longevity. The following geomorphic forms which host wetlands are: hills, slopes, flats, channels and basins. Basins are the only wetland geomorphic form in the study area. The length of time water resides in a wetland is subdivided into permanent inundation, seasonal inundation and seasonal waterlogging. When wetness and landform attributes are combined, 13 categories of common wetlands result (Table 2.1):

Table 2.1 *Geomorphic classification of wetlands*

Water Longevity	*Cross-sectional Geomorphology*				
	basin	*channel*	*flat*	*slope*	*hill*
permanent inundation	lake	river	-	-	-
seasonal inundation	sumpland	creek	floodplain	-	-
seasonal waterlogging	dampland	trough	palusplain	paluslope	palusmont
intermittent inundation	playa	wadi	balkarra	-	-

The wetlands occurring in the study area are lakes, sumplands, and damplands, and for the Becher Suite of wetlands, sumplands, and damplands. In this study, wetlands are differentiated to the level of one of these categories. Water and landform descriptors (Fig. 2-2) are used to further augment the nomenclature of the primary categories and discriminate individual wetlands. One of the most important descriptors is that of scale. The terms of scale referred to herein are as follows.

- Macroscale: large scale wetlands encompassed by a frame of reference 1000 m x 1000 m to 10 km x 10 km
- Mesoscale: medium scale wetlands encompassed by a frame of reference 500 m x 500 m to 1000 m x 1000 m
- Microscale:small scale wetlands encompassed by a frame of reference 100 m x 100 m to 500 m x 500 m
- Leptoscale: fine scale wetlands encompassed by a frame of reference several metres by tens of metres

2.2.2 Local scale wetland vegetation classification system

A classification for wetland basin vegetation was designed based on the scale of wetland vegetation complexes, extent of vegetation cover over the wetland, internal organisation of vegetation in plan, vegetation structure, and details of the floristic/ structural components (Fig. 2-3) (Semeniuk *et al.* 1990). Only emergent, perennial, woody or herbaceous macrophytes are considered in this classification.

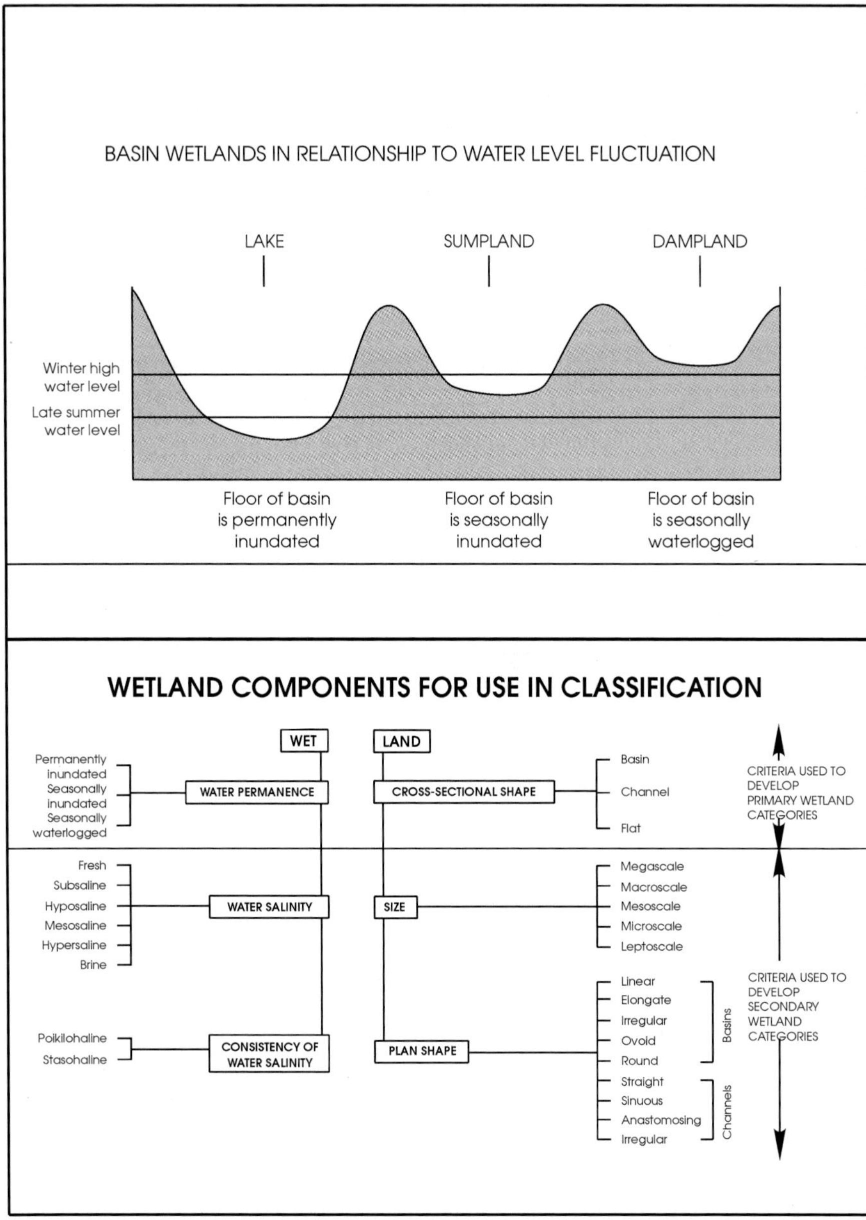

Figure 2-2. *Classification of wetland basins, and terminology for components of wetlands used in the local wetland classification (after C. A. Semeniuk 1987).*

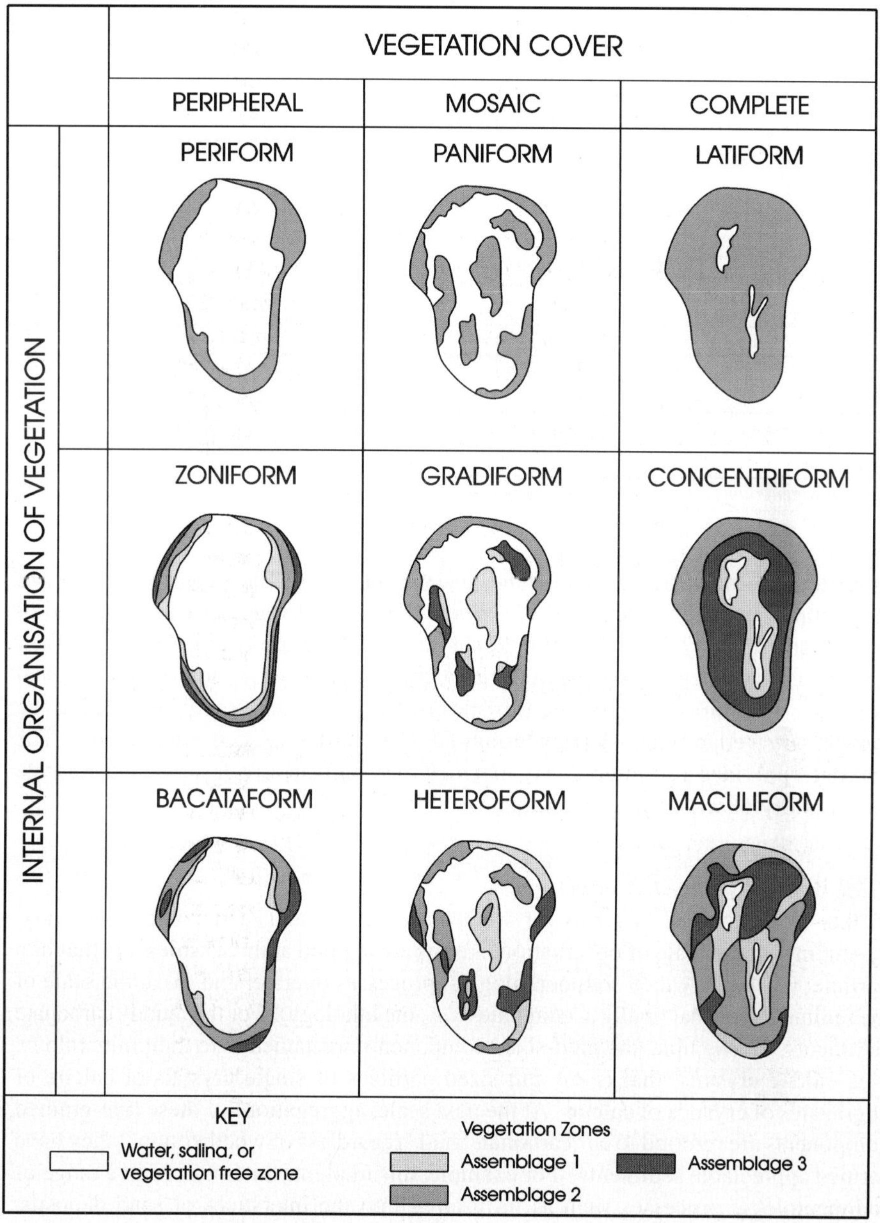

Figure 2-3. *Classification of vegetation organisation in wetland basins (after Semeniuk et al. 1990).*

Wetland vegetation cover is described in two ways, firstly, by extent of cover, and secondly, by the complexity of organisation. The vegetation cover is classified as peripheral, mosaic, or complete. The classes grade into each other. Wetland pattern is described as homogeneous, zoned, or heterogeneous. The combination of areal extent and internal organisation of the wetland vegetation results in 9 basic wetland categories (Table 2.2).

Table 2.2 *Classification of wetland basin vegetation*

Organisation of Vegetation	*Areal extent of Vegetation Cover*		
	Peripheral	*Mosaic*	*Complete (>90%)*
Homogeneous	periform	paniform	latiform
Zoned	zoniform	gradiform	concentriform
Heterogeneous	bacataform	heteroform	maculiform

These terms form the primary part of the binary terminology. The second part comprises structural terms after Specht (1981). Where the wetland vegetation is composed of several structural types arranged in zonal pattern, these are listed in order of their occurrence from margin to centre of wetland, essentially mirroring some environmental gradient. The information on the floristics of the assemblages may be added to the main binary wetland classification terminology as a suffix, or secondary adjunct. The approach provides a structured way in which to systematically describe and compile an inventory of wetland vegetation units.

2.2.3 Wetland sediment terminology

In this study, the use of terms for wetland sediments and their components vary, depending on the scale of observation. Terms are applied at three scales: (1) that of a particle; (2) that at which sedimentological processes operate; and (3) at the scale of the sediment type that finally accumulates (*i.e.*, the lithology). For the muddy carbonate sediments, for example, the mud-sized components are named as to their mineralogy, *e.g.*, calcite crystals; that is <4 μm sized particles of single crystals of calcite or aggregates of crystals of calcite. At the next scale, aggregations of these fine-grained components are referred to as "carbonate mud" regardless of whether or not they have formed appreciable sediments. For example, this mud may be involved in a range of sedimentologic processes such as infiltrating into the interstices of sand deposits, coating sand grains as pellicular film, accumulating as a sheet-like fine film on the wetland basin floor, or accumulating to form relatively thick sediment deposits. When carbonate mud has accumulated to sufficient thickness to be termed a carbonate mud sediment, a lithologic term is applied, *viz.*, calcilutite (a term already established in the literature for carbonate mud deposits *cf.* Semeniuk and Semeniuk 2004).

The same principles of nomenclature can be applied to organic matter. The term "organic matter" is applied to the particles and to material that is involved in sedimentologic processes (*e.g.*, coating of sand grains, or infiltrating into sand deposits). Where organic matter has accumulated within sandy sediment or carbonate mud sediment to the extent that the resulting material is dark, the descriptor "organic matter enriched" is applied. Where organic matter has accumulated to form a distinct sediment type, the term "peat" is applied.

A classification of fine grained biogenic wetland sediments and mixtures between these fine-grained sediments and sand has been constructed by Semeniuk and Semeniuk (2004). In the first instance, a ternary diagram with carbonate mud, organic matter, and diatoms at the apices of the triangle is used to identify and name end-member sediments and their mixtures (Fig. 2-4A). In this study, the majority of fine-grained sediments are located between the carbonate mud and organic matter compositional fields, and hence sediment terms such as calcilutite, organic matter enriched calcilutite, calcilutaceous peat, and peat are applicable.

For sedimentary mixtures between sand and biogenic mud-sized components, a simplified classification and nomenclature of the fabric classes and hence sediment classes is used (Figure 2-4B, after Semeniuk and Semeniuk 2004). Fabric rather than percentage boundaries are used to separate the classes of muddy sand and sandy mud because the category of "grain-support" will have different sizes of interstitial space, and hence different sand to mud ratio, dependent on grain shape and sphericity (Dunham 1962; Semeniuk and Semeniuk 2004). In this classification system, while the descriptor terms "peaty" and "calcilutaceous" carry implication that these sediment types are muddy sands, Semeniuk and Semeniuk (2004) suggest that the term "muddy" be inserted between the descriptors referring to the mud fraction and the term "sand", *e.g.*, calcilutaceous muddy sand. When the mud-sized fraction is left undifferentiated as to its particle types, the sediments may be termed "muddy sand" or "sandy mud". If the composition of the muddy component of the sediment is known and has been classified as to its position on the ternary diagram, the category of the "mud" in the muddy sand can be adfixed to the sediment name *e.g.*, organic-matter-enriched calcilutaceous muddy sand.

2.3 Terminology

Some of the terms not readily located in the literature, or those used in a specific sense in this study, are defined below.

Basal sheet: Thin or thick layer of muddy sand at the base of wetland fill.

Becher cuspate foreland: The asymmetric deltoid shaped cuspate foreland coastal feature whose apex is located at Point Becher.

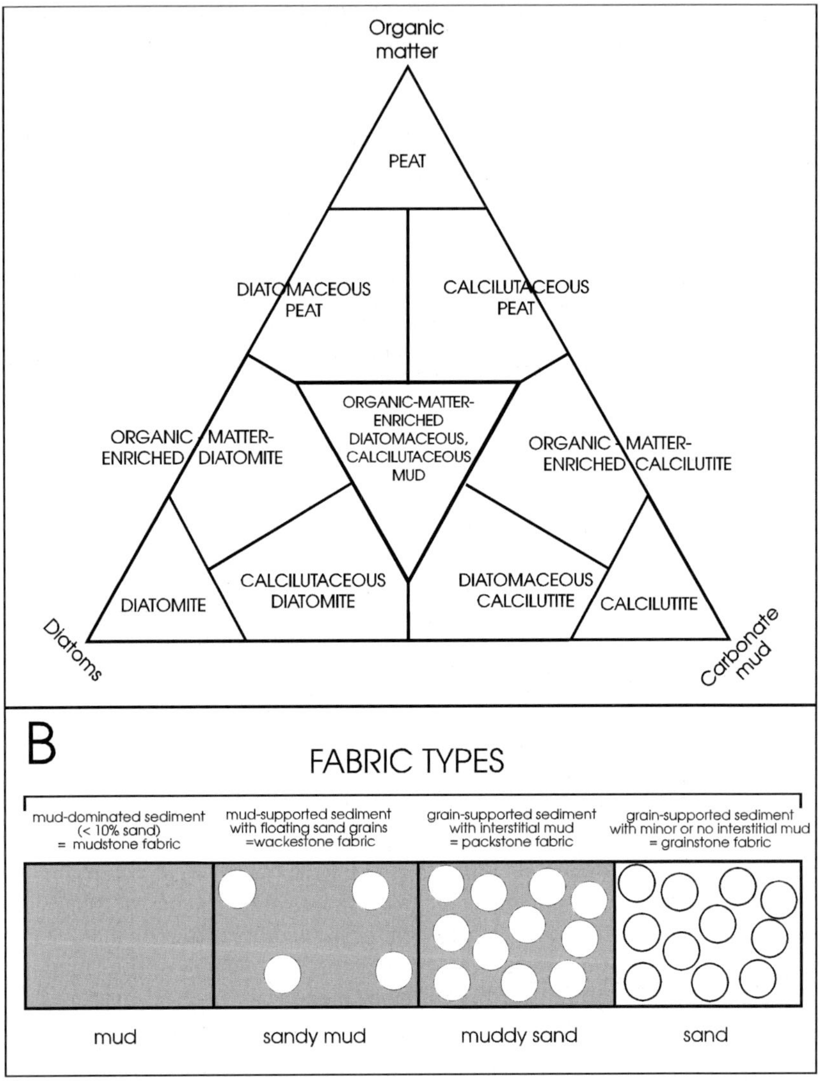

Figure 2-4. *Sediment classes and terminology.*

Becher Cusp: Informal term used to denote the cuspate form of the coast and coastal plain at Point Becher.

Becher Suite: The group of related wetlands in the parallel swales of the beachridge plain on the Becher cuspate foreland (C. A. Semeniuk 1988).

Consanguineous suite: In a physiographic setting, a group of wetlands fundamentally related because of similar geometry, size, geomorphology, stratigraphy, hydrology, and origin (C. A. Semeniuk 1988).

Grain sizes: Gravel, sand, and mud, and the subdivisions of sand into very fine, fine, medium, coarse, and very coarse follow the Udden-Wentworth scale (Wentworth 1922, cited in Folk 1974).

Local scale: The area of study encompassing the area around specific wetlands, and generally involves an area some 100 m x 100 m.

mM/L: Unit of molality used for comparisons of cation concentrations in waters (Fetter 1994).

$$\textit{Molality} = \frac{\textit{milligrams per litre} \times 10^{-3}}{\textit{formula weight in grams}}$$

mM/Kg: Unit of molality used for comparisons of cation concentrations in plants and sediments.

$$\textit{Molality} = \frac{\textit{milligrams per kilogram} \times 10^{-3}}{\textit{formula weight in grams}}$$

Peat: The accumulated remains of dead plants. Originally, the term was applied to organic matter containing <20% of "unburnable inorganic matter" and measured as loss of carbon (Clymo 1983); the inorganic compounds embodied in living organic matter were later combined with other inorganic materials in the analysis of the ash. In this study, sediments contain variable organic matter, from 0-100%, and "peat" is the class of sediment with organic matter content ranging from 75-100% (Semeniuk and Semeniuk 2004). "Peat" refers to sediments grading from accumulations of decomposing wet organic matter to organic mud, incorporating the terms peat and muck as defined by the Soil Science Society of America (1997). Peat may accumulate *in situ* or form from organic matter transported to a deep water environment (Semeniuk and Semeniuk 2004).

Poikilohaline: (after C. A. Semeniuk 1987) A term used to denote variable water salinity. Water quality that markedly fluctuates throughout the year such that several salinity fields are encompassed, is termed poikilohaline.

Regional scale: The area of study encompassing the region for 100 km x 100 km.

Stasohaline: (after C. A. Semeniuk 1987) A term used to denote constant water salinity. Water quality that is consistent throughout the year, *i.e.*, remains totally within a given salinity field, is termed stasohaline.

Sub-regional scale: The area of study encompassing the Becher cuspate foreland: 10 km x 10 km.

TDS: Total dissolved solids, a measure of water quality based on the total amount of solids in milligrams per litre that remain when a water sample is evaporated to dryness (Fetter 1994).

2.4 Methods

2.4.1 Introduction

Field and laboratory investigations in this study involved a wide range and large number of methods. These are described below within a framework largely following the organisation of the chapters, *viz.*, wetland mapping and selections of wetlands for study, stratigraphy, hydrology, hydrochemistry, and vegetation. Each of these sections are subdivided according to the decreasing scale categories described in the introductory chapter, *i.e.*, regional to subregional scale, local scale (*i.e.*, at the scale of wetland basins and their encompassing beachridges), basin scale (*i.e.*, the scale of individual wetland basins), and bedding scale (*i.e.*, at the scale of individual beds or layers of sedimentary fill within a wetland).

2.4.2 Wetland mapping, selection of wetlands for study, and description

The wetlands of the sub-region were mapped as natural groupings of consanguineous suites, *viz.*, Becher Suite, Coolongup Suite, and Peelhurst Suite by C. A. Semeniuk (1988). The delineation of this mapping formed the basis for the study area. In the early stages of this study, as most of the 275 wetland basins in the Becher Suite were unnamed, a numbering system was used to distinguish between them. Numbering began at the coast and progressed eastwards towards the mainland Spearwood Dune ridge. Some of the wetlands were informally named as WAWA (since that wetland was located on land vested in the former Water Authority of Western Australia) and as basins (i), (ii) and (iii) within the southwestern part of swale number 1, hence wetlands swi, swii, swiii. All the individual wetland basins of the Becher Suite were located, mapped and classified as lake, sumpland or dampland using various coloured and black and white vertical and oblique aerial photography at scales of 1:25,000 and 1:4000, followed by extensive field verification.

The physical, hydrochemical, and vegetation attributes of individual wetland basins within the Becher Suite also were described on a preliminary basis as part of a study for the National Heritage Trust (V and C Semeniuk Research Group 1991). This information is presented in Chapter 4, the Description of Wetlands.

Eighteen wetland basins within the Becher cuspate foreland were selected for detailed investigation as part of this study. The rationale for their selection was as follows:

1. wetlands were selected from the chain of wetlands located to the far east of the Becher Suite, where the basins were sited on beachridge isochrons of *circa* 5000

years; some were selected from the central regions of the Becher Suite, *i.e.*, where the basins were sited on beachridge isochrons of *circa* 2000-3000 years old; and some were selected from the chain of wetlands located to the far west of the Becher Suite, where the basins were sited on beachridge isochrons of *circa* 1000 years;

2. wetlands represented a range of sumplands and damplands, varying from linear to oval to circular in shape; and

3. wetlands were selected to encompass a range of vegetation types; generally three basins of a given vegetation type were selected, but if there was only one example of a vegetation type then that basin was included.

The wetlands selected for detailed study are (from east to west) wetlands 161, 162, 163, WAWA, 142, 135, 136, 63, 72, 45, 35, 9-3, 9-6, 9-14, swi, swii, swiii, and 1-N. These wetlands capture a wide range of ages, wetland types, and vegetation types across the Becher Suite (Table 2.3).

2.4.3 Wetland stratigraphy

Various investigations were undertaken in the field and laboratories to determine aspects of wetland stratigraphy at different scales. The general objectives behind these investigations were:

1. to describe the Holocene stratigraphy of the Becher Cuspate foreland
2. to define the freshwater aquifer underlying the Becher Cuspate foreland
3. to compare the basal sediments of the wetlands to those of dune and beach samples collected from the present geomorphic units within the Rockingham-Becher Plain, using the univariate statistical parameters mean, mode, standard deviation, skewness and kurtosis
4. to compare sedimentary structures in cores of beach, dune and swale with those in the basal sheet of the wetland sequence
5. to describe the sediments comprising the wetland fill
6. to determine the wetland stratigraphy for each wetland
7. to quantitatively assess the proportion of quartz under the wetland and in the corresponding layer beneath the beachridge/dune
8. to produce a 3-dimensional picture of the land surface in an area of modern beachridge/swale construction on the Becher Cusp and demonstrate the relationship between the land surface and the groundwater table
9. to collect *in situ* cores of wetland sediments for analyses of sedimentary microstructures and for fine scale sampling

Table 2.3 Wetlands on the Becher cuspate foreland selected for the Study

Wetland	*Age Setting*	*Wetland type*	*Vegetation (dominant species)*
161	eastern part of suite	sumpland	*Baumea articulata*
162	eastern part of suite	sumpland	*Melaleuca teretifolia*
163	eastern part of suite	sumpland	*Juncus kraussii*
WAWA	eastern part of suite	sumpland	*B. articulata, Typha orientalis*
142	middle part of suite	dampland	*M. teretifolia*
135	middle part of suite	sumpland	*M. rhaphiophylla, Centella asiatica*
136	middle part of suite	sumpland	*M. rhaphiophylla, C. asiatica*
63	middle part of suite	dampland	*C. asiatica*
72	middle part of suite	dampland	*B. juncea, C. asiatica*
45	middle part of suite	sumpland	*M. rhaphiophylla, C. asiatica*
35	middle part of suite	sumpland	*M. rhaphiophylla, J. kraussii, C. asiatica*
9-3	western part of suite	sumpland	*B. juncea*
9-6	western part of suite	sumpland	*B. juncea*
9-14	western part of suite	sumpland	*J. kraussii*
swi	western part of suite	dampland	*Lepidosperma gladiatum*
swii	western part of suite	dampland	*L. gladiatum*
swiii	western part of suite	sumpland	*Schoenoplectus validus*
1-N	western part of suite	dampland	*B. juncea*

10. to determine the composition of the gravel, sand and mud fraction of wetland sediments down the stratigraphic profile in terms of carbonate minerals, organic matter, and siliciclastic grains
11. to determine the nature and origin of the carbonate mud particles using photomicrographs (scanning electron microscope) and X-ray diffraction
12. to date the carbonate mud and/or peat using radiocarbon radiometric methods for samples at the surface, at the base of the mud, and at the lowest level in the stratigraphic sequences in selected wetlands, and to determine the age structure of selected wetlands
13. to determine the nature of the geomorphic and sedimentary surface that formed the base of the proto-wetlands

These objectives are embedded in the Methods described below. While both fieldwork and laboratory work were undertaken in this study, emphasis was placed on field investigations, which were viewed as the foundation of the research. As described below, in the field many methods were designed to provide information which could be used to achieve the objectives (*cf.* Carter 1993).

Regional and sub-regional scale

The Holocene stratigraphy of the Becher Cuspate Foreland was investigated using reverse air circulation coring to depths of 12-40 m, with continuous sample recovery. In addition to information obtained from the literature (Searle *et al.* 1988), and regional drill sites from the Research and Development endeavour of the V and C Semeniuk Research Group (Semeniuk and Semeniuk 2002), 7 further sites were drilled for this study. Drill bore information was used to construct a regional to sub-regional setting for the Holocene stratigraphy, to confirm and augment stratigraphic profiles presented in the literature, and to provide a context for the surface wetland and beachridge sediments (Fig. 2-5). Sediment samples were taken at 1 m intervals, and if considered necessary, at 0.5 m intervals. Water samples for TDS and cation analyses were collected at 3 m intervals.

To provide a perspective of the variability of beach sand granulometrics, sediments from the surface and the profile of beaches (winter and summer) and dunes within the Rockingham-Becher Plain region were collected for sieving analysis (Fig. 2-5). To provide a sedimentary structural and granulometric comparison between buried (fossil) beach, dune and swale sediments under wetlands and modern beach, dune, and swale sediments, short cores (70 cm) of modern beach, dune and swales were also collected, frozen and cut longitudinally. The sedimentary structures in the vertical sections were described as standards for comparison with cores of the wetland basal sheet.

Local scale (wetland and adjacent beachridges)

Fieldwork to determine stratigraphy at the local scale mainly involved manual auger drilling to the minimum position of the water table (<6 m), with samples retrieved, structurally disturbed, but in sequence in 5 cm increments. Drill sites were located along a west/east transect through each wetland basin, to include the centre of each wetland, the wetland margins and the adjacent east and west beachridge dunes. This work provided the stratigraphic foundation to the wetland studies. Initially, a total of 87 sites were drilled. These are notated on Figure 2-6. Later, 50 additional intermediate intra-wetland sites, and extra beachridge/dune sites were drilled to resolve lithological boundaries, stratigraphic complexities, the geometry of lenses, and any textural anomalies. All drill sites were surveyed in relation to Australian Height Datum (AHD) and stratigraphically correlated. Sediments sampled at the 5 cm intervals were described under stereoscopic microscope in terms of colour, fabric, texture, and general composition.

To determine whether sediments under the wetland fills were depleted of carbonate by acidic waters deriving from the wetland groundwaters, multiple sampling of the upper layer of the beach horizon, a distinct sedimentary marker, was undertaken.

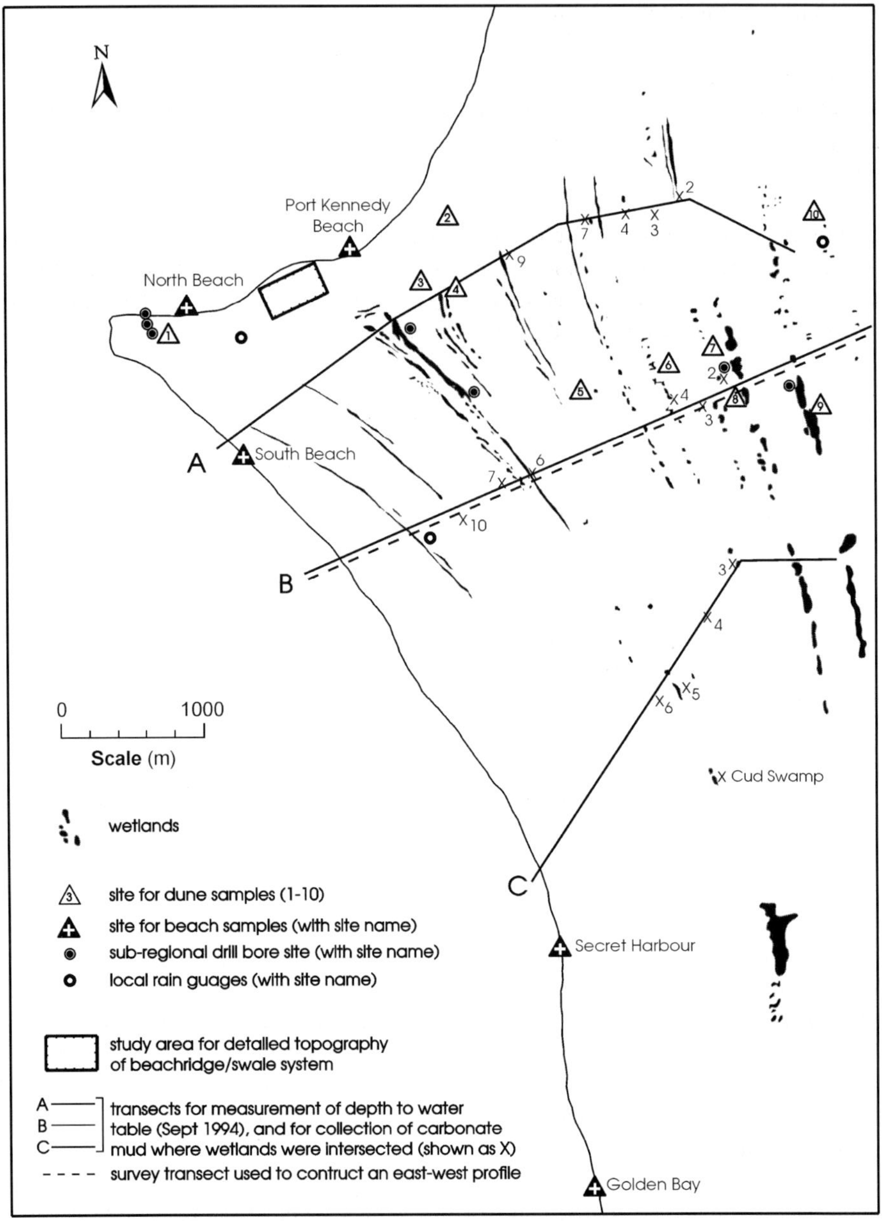

Figure 2-5. *Location of sites and transects used for sub-regional data collection.*

Buried beach sediments were sampled under the base of the ridge and wetland centre at 4 of the older wetlands and 2 of the younger wetlands, *viz.*, wetlands 161, 162, 163, WAWA, 35 and swii. Five to ten replicate samples were collected to address the potential natural lateral and vertical variation in grain composition of beach sediments. The proportions, by weight, of quartz to carbonate grains, were determined for each site.

To provide a context for the development of a proto-wetland by water tables rising into beachridge swales, an area of modern beachridge/swale construction on the northern shoreline of the Becher Cusp was surveyed at a grid of 2 m x 5 m intervals, over an area of 100 m x 40 m, to produce a 3-dimensional picture of the land surface (Fig. 2-5). The levels of the groundwater were also measured to provide a picture of the relationship between the land surface and the groundwater table. Data were imported into image processing and 3-dimensional visualisation programme "ERmapper" to produce the 3-dimensional diagram of topography and the water table.

Basin scale

Using PVC pipes 10 cm in diameter, intact cores of wetland sediment, were collected from the centre of each of the study wetlands to provide an *in situ* sequence of the wetland sedimentary fill (Fig. 2-6). Each core varied in length according to the depth of wetland sediments developed (30-110 cm). The cores were sealed, frozen, and cut in half longitudinally while frozen, exposing structures and horizons. The sedimentary formations were identified, *e.g.*, Safety Bay Sand, Becher Sand, (Semeniuk and Searle 1985; Searle *et al.* 1988) and the various lithologies were numbered, logged as to their thickness, and described in terms of colour, structure, fabric, texture, general composition and pedogenic effects. These cores were also used for pollen and for radiocarbon sampling.

To determine the composition of the gravel, sand and mud fraction of wetland sediments down the stratigraphic profile in terms of carbonate minerals, organic matter, and siliciclastic grains (mainly quartz), the sediments from the centre of each wetland (Fig. 2-6), collected at 10 cm intervals down profile, were separated by sieving into gravel fraction (>2000 μm), sand fraction (2000-63 μm), and mud fractions (<63 μm), and then each size fraction was further subdivided into three basic compositional categories: carbonate, organic carbon, and quartz (though the siliciclastic fraction is mainly quartz, felspar generally comprises <2% of grains). Hydrochloric acid digest was used to dissolve and remove carbonate, and this was followed by rinsing to remove solutes remobilised by the acid leaching, drying, and combustion at 550°C to remove organic carbon, leaving quartz and other minor mineral compounds as residue. This sampling was carried out only once during the study period, and was not replicated in the majority of cases so that analysis of variability is not presented.

To assist in interpreting the nature and origin of carbonate mud particles, and to determine the extent of surface grain corrosion by groundwaters, a range of mud samples were processed using a Scanning Electron Microscope (SEM) and an Environmental Scanning Electron Microscope (ESEM) at the CSIRO laboratories in Perth, and the EMPA Laboratories at Zurich. With the former, mud samples from Cooloongup A1, Cooloongup A2, 161-3, 162-3, 9-6, and 1-N (Fig. 2-6) were first carbon coated (or carbon and gold coated for better image quality) using a Dynavac MiniLab coater to minimise charging during analysis. The SEM was a JOEL 5800LV. The samples were loaded onto the stage and a vacuum of approximately 10E-5 Torr attained in the analysis chamber. Samples were analysed at an appropriate accelerating voltage for the sample type, image contrast and topography. At optimum performance, the spatial resolution of the instrument is approximately 4nm. Samples of interstitial mud from the surface of 1-N were also processed using the ESEM, with analysis of the back-scatter of electrons using EMPA Laboratories to determine particle composition at the 10 μm size and larger. The equipment involved an ESEM, XL30, FEG (= Field Emission Gun), with a BSE (back-scattered electron) detector. Acceleration voltage was 15 kV. The ESEM is equipped with an EDAX (TM) energy dispersive X-ray (EDX) analysis system. Analyses were used for elemental identification, and no quantification was attempted.

The mud samples used for the SEM, *i.e.*, samples Cooloongup A1, Cooloongup A2, 161-3, 162-3, 9-6, and 1-N, were also processed to determine mud size particle distribution using a laser beam scattering device at the CSIRO laboratories. The instrument used was a Malvern Instruments Mastersizer MS2000. Material >2000 mm in diameter was manually sieved out of the sample prior to analysis. The dry samples were subsampled (0.25-2.00 g of the <2000 μm fraction, by rotary riffler), and dispersed in approximately 75 ml of 1000 ppm sodium hexametaphosphate (Calgon) using a Cole-Palmer 8851 (or equivalent) ultrasonic bath for 20 minutes. This 'dispersed' subsample was then homogenised before being dispensed into the sample presentation chamber of the Malvern Mastersizer. The optical properties (refractive index, absorption coefficient) of the particulate material was set at an appropriate value for the sample. The sample was analysed in the diffraction system using a single optical lens with two laser sources. This provided two size ranges of data which were blended to give the reported size distribution of 0.02 mm to 2000 mm. For the Becher samples, this procedure provided estimates of the volume percentage of particles in intervals of 1 μm between the ranges 1-10 μm, in intervals of 10 μm between 10-100 μm, and in intervals of 100 μm between 100-1000 μm.

To determine the nature of the geomorphic and sedimentary surface that formed the base of the proto-wetlands, basal sediments underlying each wetland sedimentary fill from each wetland study site were sampled for granulometric analysis. Basal sediments

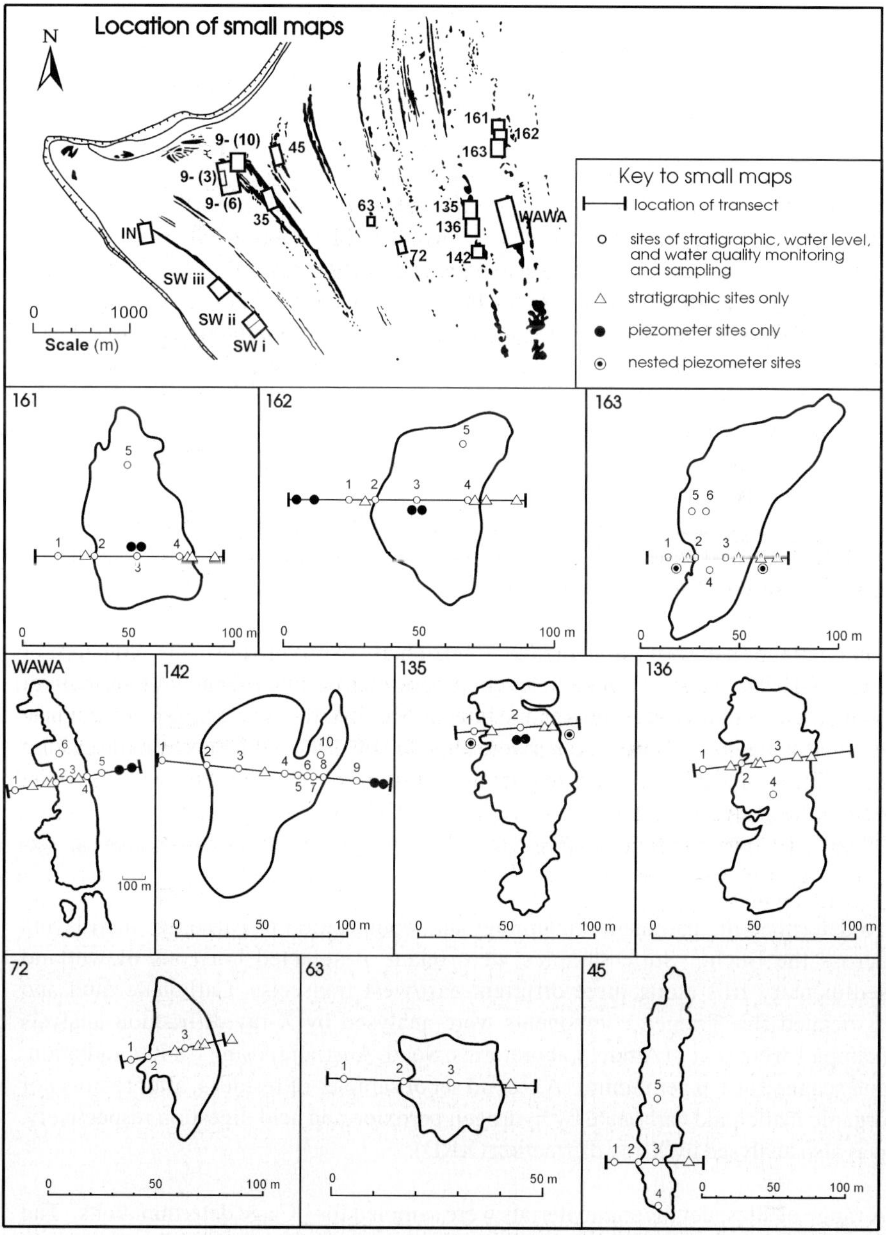

Figure 2-6. Location of study wetlands in Becher wetland suite. Plan view of each wetland showing monitoring and sampling sites.

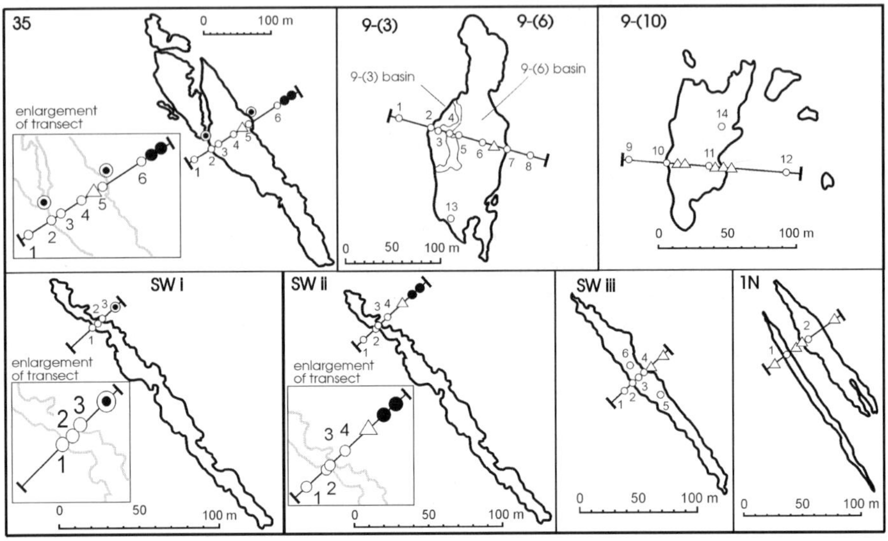

Figure 2-6 (cont). *Plan view of each study wetland showing monitoring and sampling sites.*

were defined as the muddy sand layer immediately overlying the unaltered beach or dune material, regardless of the relevant proportions of mud to sand. A minimum of 30 g of sediment was wet sieved at half phi intervals into fractions >63 μm, to eliminate the mud fraction. Carbonate grains were removed from the sediments by hydrochloric acid digestion, and frequency histograms were produced from the weights of the remaining quartz fractions.

Bedding scale

To determine the carbonate mineralogy and composition of carbonate mud layers across the Becher Suite samples were taken of selected horizons of wetland sedimentary fills along three different east/west transects. Carbonate mud and associated fine grained components were analysed by X-ray diffraction analysis (samples were sent to Amdel Laboratories, South Australia) using Cu Kα- radiation, and scanned at 1 θ per minute. A second set of samples of residues, after removal of organic matter and carbonate by hydrogen peroxide and acid digestion respectively, was also analysed by X-ray diffraction (XRD).

A range of sites, depths and materials were sampled for ^{14}C age determinations. The sampling sites, in relation to the wetlands, are shown in Figure 2-7. The location of the samples down the stratigraphic profile are given in Chapter 4, the description of the database. The materials used for ^{14}C dating were carbonate mud, organic matter, and freshwater snails. The rationale for the sampling was manifold: 1) the base of the wetland fill sediments was dated to determine when a given wetland was initiated (wetlands 161, 162, 163, 229, Cud Swamp, WAWA, 142, 135, 136, 72, 63, 45, 35, 9-14, 9-6, A2, A3, A4, A7, A9, C5) (Fig. 2-6); 2) horizons throughout the stratigraphic sequence in selected cores were dated to determine the rate of accumulation of the carbonate muds, and the rate of formation of the basal sheet (wetlands 161, 162, 163, WAWA, 135, 35, 9-6, 9-14); 3) dated horizons also were used to provide the temporal framework for the palynological study (wetlands 161, 162, 163, WAWA, 135, 9-14); 4) selected horizons that were organic matter-enriched carbonate mud units were separated into carbonate and organic matter fractions, and dated separately to determine whether there was an age difference between carbonate mud and organic mud accumulation, *i.e.*, whether these sediment types were accumulating contemporaneously, or as alternating episodes (wetlands 161, 161, 163, and 135); and 5) where sufficiently abundant, freshwater snails were dated from some units to ascertain their age relative to the enclosing sediment (wetlands WAWA, Cud Swamp).

Samples were collected either in the field by excavating to the selected horizon and retrieving the sample from *in situ* deposits, or from cores. Where the carbonate mud formed a minor part of the stratigraphic sequence (*e.g.*, the muddy sands at the base of the sequence), the sample was obtained from some 10 cm of profile to obtain sufficient mud. Generally, the samples were obtained from 5 cm or less of profile. Because there is rarely a sharp boundary between the layer of mud infiltration and the underlying parent material (beach sand, humic beach sand, dune sand, humic dune sand), the base of the infiltration layer was identified using the following methods:

1. collection and analysis of cores of parent materials, *viz.*, beach, dune crest, swale, seagrass
2. collection of cores containing the contact between wetland sediments and underlying parent materials
3. identification of parent material sediments through textural and compositional grain size analysis
4. identification of buried humic horizons (interpreted to be the sediments in the swales)
5. identification of mud component of muddy sand horizons

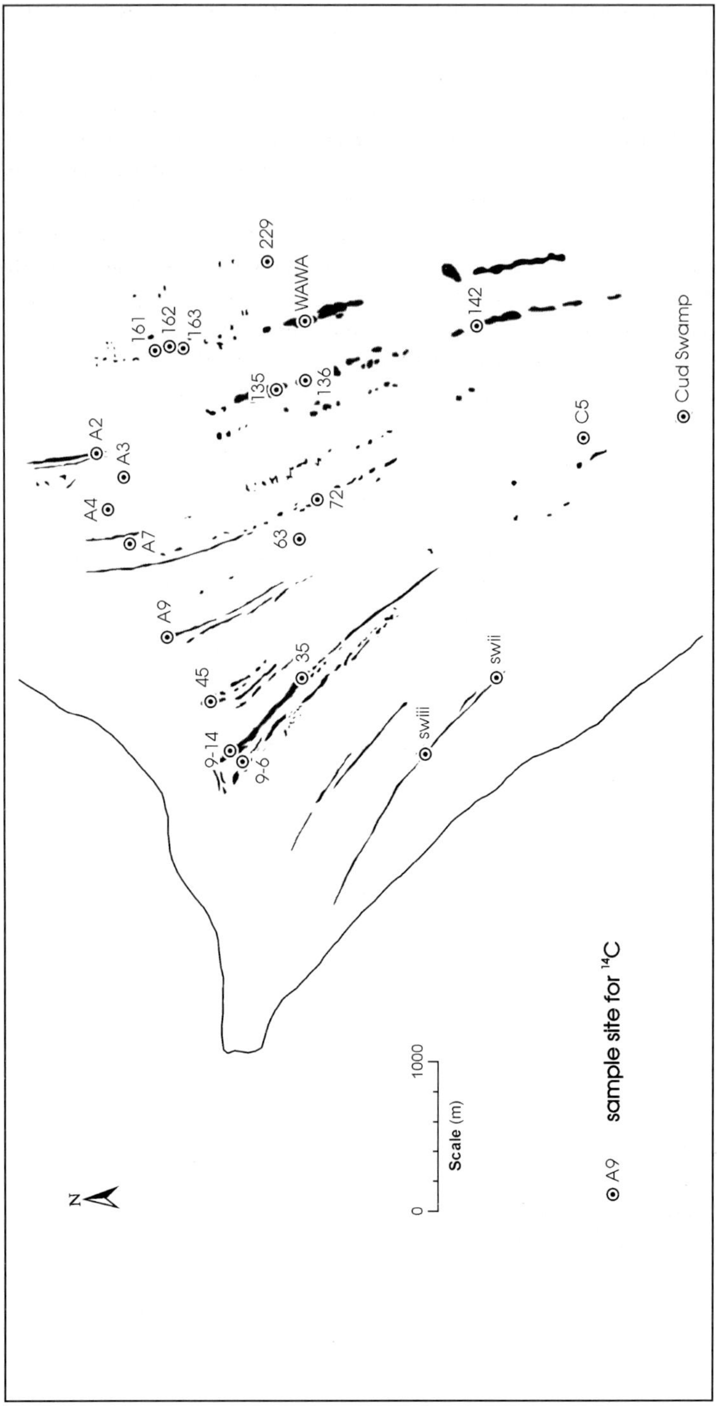

Figure 2-7. *Location of sampling sites for carbonate mud, peat or shell used in ^{14}C dating.*

In the laboratory, the sand fractions that contained earlier Holocene grains, or even Pleistocene grains, were eliminated by wet sieving. Sieving separated all mud fractions (particles <63 μm,) from biogenic, or exogenic sand and rootlet material of size >63 μm. The sieved mud was collected in a large clean container. Because it was important that only wetland mud was used for radiocarbon analyses, slurry samples of mud after the first sieving were re-sieved through a 63 μm mesh to ensure no sand grains which might add older carbon to the mud sample accidentally by-passed the first sieving process. After drying this second sieved sample, the base of the dried mud cake was examined with stereoscopic microscope to ensure that there were no sand grains that would have preferentially settled to the base of the slurry. The dried mud was weighed to ensure the minimum required weight (18-20 g). As this mud was sometimes a mix between organic material and carbonate mud, the organic and carbonate components of the muds were combusted to obtain a pure carbonate sample. When both organic carbon and carbonate mud components of a given sample were required to be analysed, sufficient sample size was forwarded to the laboratories to ensure sufficient carbon was present in the organic mud fraction and in the carbonate mud fraction. The CSIRO laboratories acid-digested the mud to collect the CO_2 for radiocarbon analysis of the carbonate mud fraction, and later combusted the organic carbon and collected the CO_2 for radiocarbon analysis of the organic carbon mud fraction. Analysis of carbon-14 was undertaken at CSIRO laboratories in South Australia, using a method for direct absorption of CO_2 for counting. A separate carbon-13 analysis of CO_2 was also undertaken for all samples.

Freshwater snails were broken apart to ensure that they were free of sediment internal to the chambers so that only shell wall material was submitted for analysis.

2.4.4 Wetland hydrology

Investigations were undertaken in the field with the following objectives:

1. to describe the variability in local rainfall on the Becher Cuspate foreland
2. to produce an empirical set of groundwater contours for the Becher Cuspate foreland
3. to define the inland limit of tidal effects on groundwater in the wetlands
4. to produce hydrographs for wetlands and beachridge/dunes
5. to examine the morphology of the water table beneath high beachridge/dunes
6. to investigate and quantify the effect of upward discharge
7. to quantify the effect of surface water perching
8. to quantify soil moisture variability in wetlands
9. to demonstrate the effect of small scale structures on vertical water movement

The approach and designs of this part of the study were similar to those described in Kutilek and Nielsen (1994) and Butler (1998).

Again, the objectives are embedded in the various investigations described below in decreasing scale, from regional and sub-regional to bedding scale.

Regional and sub-regional scale

Rainfall gauges were established at three sites on the Becher Cuspate Foreland at varying distance from the coast: near the apex of the cusp, in the most easterly swale, and at the southern end of Lake Walyungup (Fig. 2-5). Monthly measurements were recorded to quantify temporal and spatial rainfall variability on a sub-regional scale.

To obtain a picture of the sub-regional groundwater table in relation to the coast and to the wetlands, during September 1994, the period of maximum water levels, depth to groundwater was measured and levelled to AHD in all wetlands and selected adjoining beachridge/dunes intersected by three transects: 1) normal to the north shore of the Becher Cuspate foreland, 2) along the main axis of the cusp, and 3) normal to the southern shore of the Becher Cuspate foreland (Fig. 2-5). Maps of water table contours for the sub-region were produced from the data points.

To define the inland limit of tidal effects on groundwater in the wetlands, groundwater levels in bores were monitored over the same diurnal interval in several coastal wetland sites nearest the coast, as well as under the foredune and the beach (Fig. 2-8). A control site was set up in one of the wetlands distal from the coast. Monitoring was undertaken at the commencement of winter during intense atmospheric activity which coincided with the period when groundwater levels at the control site were falling. The procedure was repeated during the period when groundwater levels at the control site were rising.

Local scale (wetland and adjacent beachridges)

For the eighteen wetlands studied in detail (Fig. 2-6), shallow piezometers (<6m) were installed for the purposes of measuring the depth to the water table and collecting water samples, in each wetland vegetation type intercepted by the west/east transect, at the wetland margins, adjacent beachridges, and in additional vegetation types not intersected by the transect. These sites were manually augered to the position of minimum water table during summer, and a 40 mm diameter PVC pipe, capped at the lower end, and slotted for 1 m from the base, was positioned in the auger hole, which was then packed with the original sediments in the reverse order of extraction. The cap at the base of the pipe was slotted to allow water to drain from the pipe if the water table ever fell below its base. The top of the pipe had a removable cap. Water levels at each site were monitored monthly over a period of 5 years (August 1991-July 1996). A modified measuring programme incorporating only two or three sites per wetland (in the centre and on adjacent beachridges) was undertaken for an additional 5 years (1996-2001), providing 10 years of data for each wetland monitored.

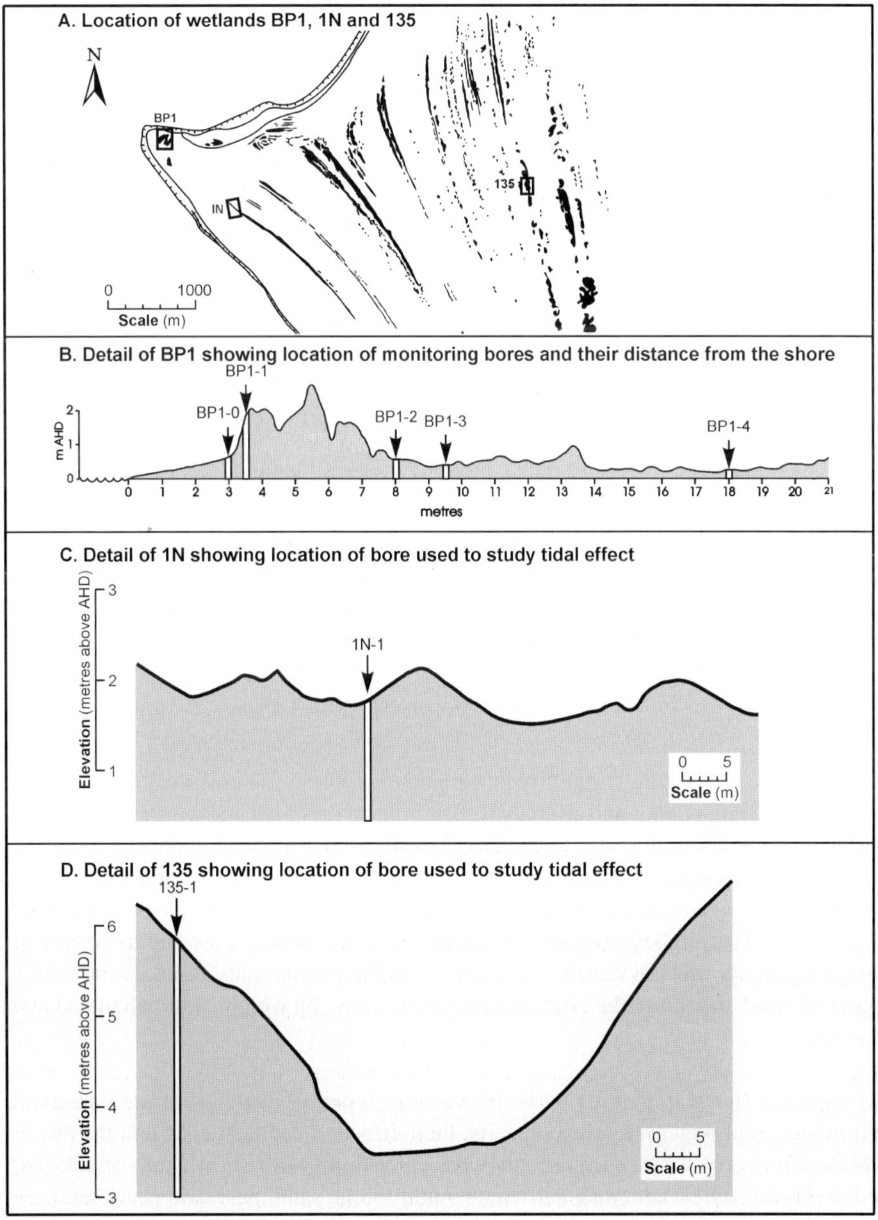

Figure 2-8. *Locations for the study of tidal effects on groundwater.*

To explore the morphology and lateral relationship of the groundwater table between high beachridges/dunes and adjoining wetlands, a piezometer was installed at the crest of the highest beachridge/dune adjacent to each of five selected wetlands (Fig. 2-6). Water levels at each site were monitored monthly from August 1995-July 2001.

To ascertain whether there was vertical water movement within the aquifer and whether there was upward or downward pressure in the aquifer over the seasons, sets of three piezometers (with slotting over an interval of 30 cm at depths 3 m, 9 m, and 18 m respectively) to simulate nested piezometers were located at both the east and west margins of 4 wetlands (Fig. 2-6). The stratigraphic intervals intersected were the Safety Bay Sand (at 3 m depth), and the Becher Sand (at 9 m and 18 m). The piezometers were within 1 m radius, but separated by 40-50 cm laterally so as to avoid water infiltration between the shallow and deeper sites. Water level measurements from these piezometers, for the period 1998-2001, were recorded monthly.

Basin scale

At each site where perching had been observed, two piezometers were installed above and below the impermeable layer (*e.g.*, calcrete or carbonate mud). Where the impermeable layer extended from 10 cm depth to the basal sheet, the shallow piezometer was inserted 10 cm into it. Measurements were taken daily for up to two weeks following significant rainfall events.

Bedding scale

To quantify soil moisture variability in wetlands, samples were collected down profile under 3 beachridge/dunes and all study wetlands during summer and winter. In addition, surface sediment samples were collected quarterly from wetland sites during the period 1991-1994. All samples were treated in the same way. Samples were cooled immediately in the field, refrigerated overnight, halved for replication, weighed, oven dried at 100° C, cooled and reweighed. Because of the number of sampling sites, the frequency of sampling events, and the variable bulk density of the soil samples, (comprising humic material, sand and mud), the ratio of water to 50 g soil, by weight, was calculated and graphed.

To measure the water content of the various types of water-saturated wetland sediments, their water retention capacity, their effect on capillary rise, and the rate at which they freely drain, experiments were carried out with short cores of selected sediment, *viz.*, carbonate mud, carbonate muddy sand, sand, peat, and peaty sand. *In situ* cores of sediment from the surface of each wetland were collected in PVC containers (10 cm diameter and 10 cm high). The cores, with sediment intact, were capped at the top to retard any evaporation during the experiment. Some small holes in the lid allowed water to escape once the experimentally induced rising water table level reached the top of the sediment core. The cores were then progressively saturated by standing them with their bases in a bath of cooled boiled water 1 cm deep. The level of the water in the bath was raised 1 cm every 24 hrs to simulate natural groundwater rise. Saturation was determined by water escaping from a hole in the lid of the core. Saturated sediments were weighed, oven dried, then reweighed. From these measurements, porosity was calculated.

In order to demonstrate the effect of small scale structures on vertical water movement through Becher Suite wetland sediments, a set of seven PVC pipes, 20 cm in length, arranged in a circle around a central one were inserted into the peat in wetland WAWA for 10 cm at two levels: ground surface and into the floor of an excavated flat pit 20-30 cm below the surface (Fig. 2-9). The sediment at the surface initially was dry, whereas the sediment at 20-30 cm was moist. Each PVC pipe at the surface was sited on *in situ* root boles of *Typha orientalis* and *Schoenoplectus validus*. The PVC pipes at 20-30 cm contained no obvious roots or burrows. The PVC pipes were filled to the top with water and the rate of fall for each centimetre (1-7 cm) was recorded.

2.4.5 Wetland hydrochemistry

Groundwater samples were collected from the field and returned to the laboratory for various types of analyses. The pH of groundwater, however, was undertaken directly in the field. The samples returned to the laboratory were processed for salinity determinations, cation content, and nutrients. The majority of analyses were carried out by the writer, but some were carried out by commercial chemical laboratories. Information on instrumental methods of analysis were derived from the following: Franson (1985); Broberg and Petterson (1988); Willard *et al.* (1988); Clementson and Wayte (1992); Rayment and Higginson (1992); and Mudroch *et al.* (1997). Methods are briefly described below in relation to the specific objectives, which were:

1. to determine the extent and nature of the freshwater prism that underlies the Becher Suite wetlands, and its relationship vertically and laterally, to the saline aquifers
2. to analyse groundwater samples for fundamental chemical characteristics: pH, salinity, cation concentration
3. to analyse groundwater samples for nutrient concentrations (orthophosphate, nitrate, ammonium)
4. to analyse sediments for nutrient concentrations
5. to analyse plant material for nutrient and cation concentrations
6. to analyse down profile sediment samples for cation concentration
7. to analyse down profile interstitial water samples for cation concentration
8. to analyse interstitial water samples in the sediment layer 0-10 cm below the surface for fundamental chemical characteristics: salinity, cation concentration

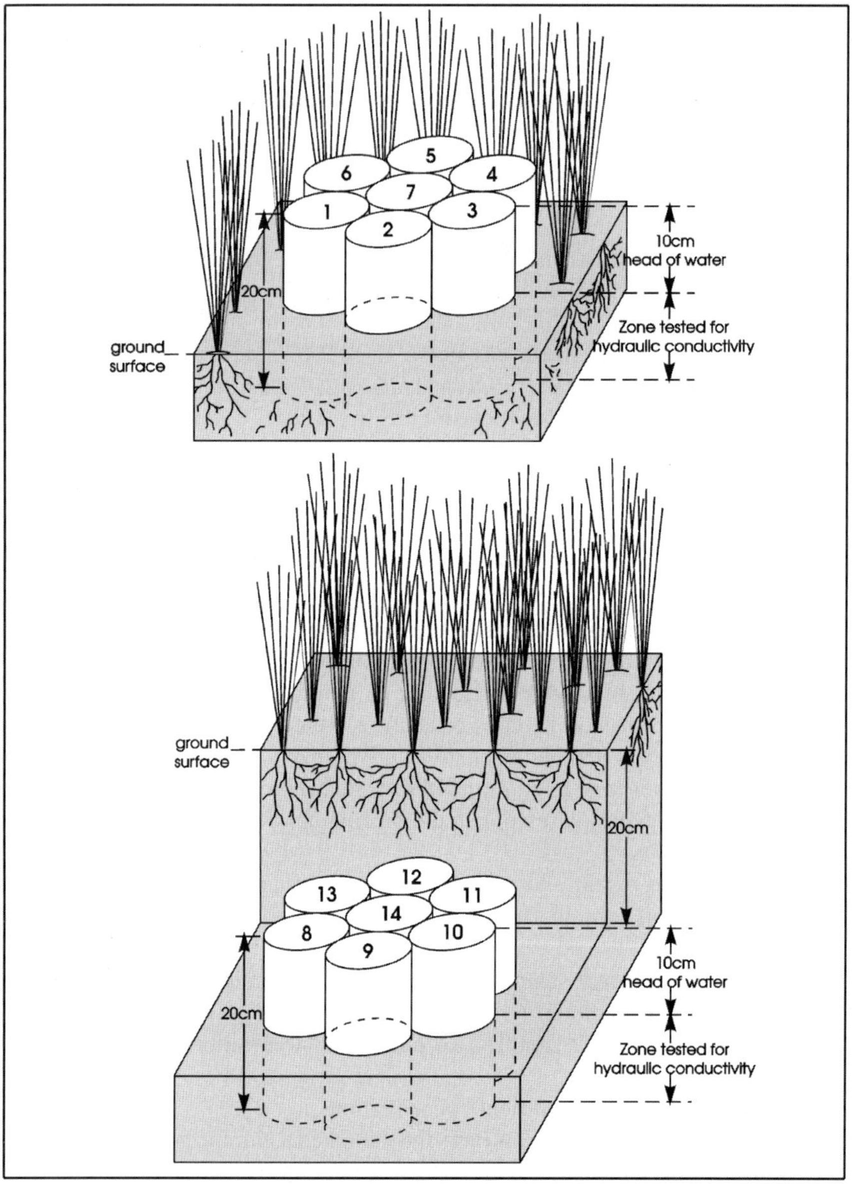

Figure 2-9. *Design of experiment to compare hydraulic conductivity through root and non-root zones.*

Regional to sub-regional scale
To characterise the groundwater hydrochemical fields, water samples were obtained at 3 m intervals during the various drilling exercises described earlier (Fig. 2-5). In particular, the freshwater/seawater interface near the coast was explored by drilling and collection of water samples at 3 m intervals (Fig. 2-5).

Local scale (wetland and adjacent beachridges)
To characterise the groundwater under each wetland and beachridge/dune site, samples were collected monthly in clean plastic vials from capped piezometers, after extracting the stagnant water. pH was measured in the field on unfiltered samples using a standard pHScan 1/2 microprocessor, calibrated to buffer solutions pH 4, 7, and 9. The pHScan has a resolution of 0.1 pH units. In the laboratory, water samples were filtered through a 0.45 μm millipore silica fibre filter prior to further processing. TDS of water samples were measured using a conductivity meter (an LC81 meter). Micro-siemens were converted to total dissolved solids using a calibration graph based on NaCl concentrations at 22° C (Schlumberger 1985). The water samples were classified according to the water salinity classification of Hammer (1986) in Table 2.4.

Table 2.4 *Classification of water salinity based on total dissolved solids*

Salinity mg/l	*Water category*
<1,000	Fresh
1,000-3,000	Subhaline
3,000-20,000	Hyposaline
20,000-50,000	Mesosaline
50,000-100,000	Hypersaline

Cation concentration in groundwater was analysed in order to investigate whether one or more cations could be used as tracers of water movement across the wetland or between the wetland and the adjacent dune ridges. The following cation concentrations in groundwater were measured: Na^+, K^+, Ca^{++}, Mg^{++}. Samples were collected monthly from each site along an east/west transect through each wetland. The samples were frozen for up to two years, then thawed, injected with 1-3 ml (depending on volume of water sample) of 10 M reagent grade hydrochloric acid to dissolve any precipitation of calcium carbonate which may have taken place during storage, and diluted by 10, 100, and occasionally 1000, depending on TDS, to bring all samples to the same order of magnitude of concentration for spectrometer analysis. A flame spectrometer was used to analyse the absorbance of Na^+, and K^+. For the relevant range, absorbance was compared to standard samples of known concentrations of sodium chloride and potassium chloride respectively. A blank of acidified, de-ionised water was used to ensure that the HCl did not contaminate the water samples with cations. Calcium ion absorbance was determined using the ionic coupling plasma mass spectrometer (ICP-MS) with standard calcium carbonate solutions, and magnesium ion absorbance was measured using the atomic absorption spectrometer and magnesium sulphate standards with Lanthanum solution added. Twelve out of 24 sampling runs were dispatched to a NATA accredited commercial laboratory (Amdel Laboratories) for determination of cation concentrations using atomic absorption spectrometry. Although replicate samples for each site were not used (due to the large number of samples, and the level of precision required), accuracy was assessed by repeated monthly sampling from the same sites.

Water samples were collected for nutrients, *viz.*, orthophosphate, nitrate, ammonia. Prior to nutrient analysis, several experiments with the method of sampling and storage were carried out. Three methods of water collection were tried: 1) direct collection from the piezometer, 2) collection after bailing and 3) collection from freshly augered holes. Orthophosphate, nitrate and ammonia were particularly sensitive to the method used, and showed variability in the order from high to low for each site tested. The method selected was to auger the holes as required and collect muddy samples and filter them on site when required. In a second trial, filtered and unfiltered samples were collected and analysed for ammonia. Concentration increased by a factor of 100 when unfiltered. In a third trial to test frozen and unfrozen samples, there was no significant difference between the mean concentrations for ammonia in the two groups.

Preliminary sampling in autumn of 1994 showed that Kjeldahl-N (organic nitrogen and ammonia) in the groundwater exceeded, equalled, and was less than nitrate-nitrite N at different sites. Further experimentation showed that concentrations of each of these forms in groundwater, collected at the same time and in the same way, under sites less than one metre distant also varied. It was felt that the complexity evident from these trials required more resources and more time than were available for this aspect of the study. A decision was made to determine orthophosphate concentrations only to characterise the groundwater and the sediments in the wetlands. For this purpose, event based groundwater sampling was used rather than monthly sampling. Events which could impact on orthophosphate concentrations were identified as 1) early winter first flush of rain, 2) late winter dilution in maximum water table position, 3) spring conditions coinciding with commencement of falling water levels, plant growth and evapo-transpiration, and 4) late autumn minimum water table position (Good *et al.* 1978; Campbell and Capece 1999; Kadlec 1999).

The procedure for analysis of orthophosphate content in groundwater was as follows. Groundwater samples were collected in the field from freshly augered holes and filtered as pre-treatment through a 0.45 μm pore diameter glass fibre membrane to separate dissolved from suspended forms of phosphorus, immediately after collection, using a mobile vacuum pump unit. Samples were frozen in plastic containers and stored for up to one month. All glass receptacles used in the preparation of standards were cleaned using phosphate free detergent and rinsed in distilled water. Colorimetric determination

of dissolved orthophosphate according to the methods of Murphy and Riley (1962), and Major *et al.* (1972), was made using a Varian spectrophotometer. Limit of detection on spectrometer was 2 μg/L. Sediment samples were collected from 5-15 cm and 40-50 cm in selected wetland and beachridge sites in order to compare variation in total phosphorus between beachridge sites and wetlands, between wetlands, and down profile in relation to organic matter content. The procedure for determining total phosphorus in sediments involved two steps:

1. conversion of phosphorus to dissolved orthophosphate by nitric and perchloric acid
2. colorimetric determination of dissolved orthophosphate

The procedure was similar to that described above with the omission of the filtration and with the additional steps of oxidation of organic phosphorus by perchloric acid, and dissolution in distilled water and sodium hydroxide prior to preparation for colorimetric determination.

Basin scale

Plant material was analysed for cation and orthophosphate concentrations in order to investigate the relationship between plant uptake and groundwater and pore water chemistry. Plant materials were prepared prior to sending to laboratory where they were digested in a mixture of nitric and perchloric acids and then analysed using ICP optical emission spectrometry and photospectrometry respectively. Detection limit was 10 ppm. Preparation included field collection (5-10 replicate samples) of living leaves, stems, flowers, live and dead roots, bark, and woody fruits of the following species of dicotyledons: *Melaleuca rhaphiophylla, M. teretifolia, M. cuticularis,* and *Centella asiatica*, and the following species of monocotyledons: *Baumea articulata, B. juncea, Juncus kraussii* and *Typha orientalis,* and from kangaroo scats. All plant components were separated and sediment carefully removed from the material, even between sheets of the paperbark.

Further sediment samples and interstitial waters were collected from the central site of each wetland every 10 cm down profile to the water table. Sampling occurred in autumn when water table position is usually at minimum (*viz*., March 2000). Samples were leached of interstitial water and soluble salts by immersing them in 100 ml of distilled water and stirring vigorously for 5-10 minutes until no clods remained. The supernatant was decanted and filtered. The liquid was analysed for Ca^{++}, Na^{+}, K^{+} and Mg^{++}, using an ICP optical emission spectrometer. The sediment was dried and digested using nitric, perchloric, hydrochloric and hydrofluoric acids, the residues dissolved in hydrochloric acid and then separately analysed for Ca^{++}, Na^{+}, K^{+} and Mg^{++} using an ICP optical emission spectrometer.

Bedding scale
To investigate temporal variability in salinity and cation concentrations in the interstitial water in the 0-10 cm sediment layer, samples of approximately 40 g of surface sediment were collected from wetland sites at quarterly intervals coinciding with the seasons (*i.e.*, March, June, September and December) over three years (1991-1994). The samples were halved, one half was stored frozen and the other half weighed, dissolved in 50 ml of distilled water and the slurry stirred on a mechanical shaker for a prescribed length of time, ensuring that no clods remained. The solution was filtered and retained for analysis. The conductivity of the solution was measured, and cations determined by atomic absorption spectrometry. The sediment in the jar and on the filter paper was dried and weighed. Concentrations were calculated for the original interstitial water content in each sediment for use in calculations of TDS and cation content.

2.4.6 Wetland vegetation (including pollen)
Wetland vegetation mapping, classification and quantitative analysis were undertaken. A variety of methods was used, the most important being observation and documentation in the field 1991-2001. The methods are briefly described below within the framework of the specific objectives, which were:

1. to classify the wetland vegetation into assemblages based on species composition and structure
2. to map the distribution of wetland vegetation assemblages
3. to quantify plant assemblage compositions, and classify them using numerical classification techniques
4. to relate spatial variation in cover abundance of species to environmental attributes through ordination in order to develop hypotheses about vegetation distribution patterns
5. to test hypotheses about plant distribution under field experimental conditions
6. to investigate plants for anatomical features which could be related to seasonal inundation or waterlogging

In the electromicroscopic study of wetland vegetation, the objectives were:

1. to determine the composition of pollen in the surface sediments of the wetlands and to classify it according to the location of derivation
2. to identify and quantify the pollen taxa in the sedimentary profiles of selected wetlands
3. to date the sediments at selected intervals of the sedimentary profiles in order to compare pollen composition and abundance between wetlands

Regional and sub-regional scale
A sub-sample of vegetated wetlands was selected for classification of wetland vegetation into assemblages. This selection was based on the following rationale.

- Wetlands should lie along a line, transverse to the orientation of ridges and swales, concomitant with the direction of cusp progradation *i.e.*, east-west, to encompass the complete ridge/swale complex time series (Searle *et al.* 1988).
- Wetlands should include sumplands and damplands.
- Wetlands selected should encompass the range of wetland plant assemblages in the Becher Suite.

The best location for a given transect approximated a locus central to the cuspate foreland. Fortuitously, the wetlands were also more frequent and less disturbed in this location. Exclusion of disturbed wetlands and wetlands difficult to access resulted in a final selection of 9 wetland swales incorporating 18 wetland basins.

Local scale (wetland and adjacent beachridges)
Pollen was collected from plant species presently colonising the wetlands and ridges as well as species, cited in the literature, which occur in the region. Pollen samples were collected from herbarium sheets of the W A Herbarium, as well as from living plants collected in the field during this study. Samples were acetolysed following the standard technique of Erdtman (1952), as outlined by Phipps and Playford (1984), with minor modifications.

For each preparation four to eight (depending on availability) of the most mature anthers still containing pollen (*i.e.*, on the point of dehiscing), and one or two less mature anthers were selected from a single plant specimen. Anthers were placed in a clean watch glass, held with fine forceps, and dissected with a scalpel. Viewing the contents of the watch glass with a binocular dissecting microscope, pollen was released from anthers and unwanted plant debris carefully removed. The procedure for acetolysis of the pollen grains is outlined below.

1. Pollen was washed-centrifuged three times in glacial acetic acid (CH_3COOH) to remove all H_2O before acetolysis.
2. Samples were boiled for approximately 10 minutes in an acetolysis mixture of 1 part concentrated sulphuric acid (H_2SO_4): 9 parts acetic anhydrite [$(CH_3CO)_2O$].
3. When the acetolysis process was complete, *i.e.*, all pollen contents were removed from specimens, residues were centrifuged then washed-centrifuged three times in CH_3COOH.
4. Alkali treatment followed by acid treatment was carried out as for the sediment samples.
5. To remove large extraneous plant material, the residues were routinely sieved with either 100 μm or 150 μm mesh. Sievings were examined before discarding.

Basin scale

Vegetation assemblages in 18 wetlands were surveyed and classified according to the procedures outlined in Specht (1981), Semeniuk *et al.* (1990), and Kent and Coker (1992). Wetlands were classified according to vegetation cover, pattern, dominant physiognomies and corresponding dominant species (Specht 1981; Semeniuk *et al.* 1990). Coloured and black and white aerial photography at scales 1:4000 and 1:2000, respectively, was used to map internal wetland boundaries, and every zone was inspected in the field. The wetland vegetation was then classified into assemblages using floristic and structural characteristics. Specht's classes were adopted because much of the wetland vegetation was structurally homogeneous with minimal multi-layering. The emphasis in floristic classification was on abundance, emphasising plant frequency and constancy, over rarity, opportunism and invasion.

Several methods were attempted in the collection of phytosociological abundance data.

1. Species numbers of wetland plants were counted in nested quadrats of increasing size from 10 cm x 10 cm up to 500 cm x 500 cm in order to determine appropriate quadrat size (Barbour *et al.* 1987).
2. Plant density data were collected in quadrats varying from 10 cm x 10 cm up to 100 cm x 100 cm.
3. Presence/absence of species was recorded in quadrats from 10 cm x 10 cm up to 100 cm x 100 cm.
4. East-west transects were established in each of the wetlands from dune crest to dune crest, along which, percent cover of wetland species in 1 m^2 quadrats was recorded (species not rooted in the quadrat were included).

Method 1 was employed at the commencement of the study to determine the most appropriate quadrat size for the range of physiognomic forms present, but also served as baseline data for temporal comparisons. Method 2 was eliminated because plant density in the herblands and sedgelands was not always a true indicator of fluctuations in short term biomass production in response to seasonal environmental stress. Method 3 was also eliminated because, in seasonal wetlands, dynamic changes in vegetation response expressed in terms of presence/absence would require quadrat coverage approaching that of the wetland itself to ensure that seasonal expansion and contraction of individuals were captured. Method 4 was selected because it made possible the inclusion of some of the dynamic attributes of the vegetation; it incorporated the transitional zones, and minimised the effect of clustered, regular, and random patterns within the vegetation. This method is conducive to settings where there are rapid changes in vegetation and marked environmental gradients.

Measurement of cover using a cover pin frame and measurement of biomass were not feasible, given the range of vegetation structural types and the need to conserve them. It was noted that in many vegetation types there were several dominant species

with high frequency and cover, therefore subjective estimates of both total cover and cover of individual species were jointly undertaken by two operators. Although the transect method of quadrat selection is somewhat problematical for independent sample analysis because of spatial auto-correlation between quadrats (Legendre and Fortin 1989; Cressie 1991; Kent and Coker 1992; Palmer 1995), the placement of the transect within the wetland, its beginning and end were independent of vegetation associations.

A 1 m^2 quadrat was selected based on consideration of the following factors:

1. size of quadrat relative to size of wetland;
2. size of wetland vegetation units;
3. dominance of plant lifeform (tree *vs* sedge); and
4. minimal area and species area concepts (Dietvorst *et al.* 1982)

Nested quadrats of 1 m^2 and 25 m^2 were used in wetlands 135, 136, 142 to account for different life forms, *i.e.*, ground covers *vs* trees.

The classes of Domin were applied (Kent and Coker 1992):

<1% cover
1-4% cover
4-10% cover
11-25% cover
26-33% cover
34-50% cover
51-75% cover
76-90% cover
90-100% cover

Cluster analysis was applied to the untransformed transect quadrat data without the use of the adjacency constraints option. An hierarchical-agglomerative-polythetic clustering method (flexible UPGMA), which gives equal weight to objects rather than pair groups, was used in conjunction with a dissimilarity type association measure, (*i.e.*, smaller values = increased similarity), asymmetric in respect of 0-0 matches (Austin 1976; Belbin 1992). The Bray and Curtis association measure was used. This resulted in 23 groups of vegetation types throughout the Becher area. Eleven of these assemblages contained only wetland species, and sites were selected such that each of the vegetation types were replicated in order to investigate environmental parameters.

The ordination procedure used was semi strong hybrid multidimensional scaling (SSH) (Belbin1991), which uses Guttman and Lingoes monotone regression and a single symmetric matrix. It breaks tied input values but will not equalise different input values. Once the regression is determined, the aim is to minimise stress which is defined as the disparity between distances between all points in the current configuration and the best estimates of the same values. The iterations stop when the ratio of stress to last stress is minimal. Hybrid scaling is a combination of interval/ratio-ordinal scaling which is suited to ecological data. Environmental attribute data values were not transformed but were standardised according to the recommended procedure for equal weighting of attributes. Three dimensions were used and the maximum number of iterations selected was 50.

Subsequent to the scaling ordination, a multiple linear regression programme was run to determine how the set of attributes could best be fitted to the ordination space. As the assumptions associated with linear regression have not been tested, the correlation coefficients were not used and the resulting patterns were viewed as a preliminary step in hypothesis generation.

To determine the composition of pollen in the surface sediments of the wetlands and to classify it according to the location of derivation, samples were collected from the centre of each study wetland. In wetlands with a thick cover of leaf litter, samples from the surface of the litter were also collected. A volume of 1 cm^3 of sediment was processed for each site. Sediment samples were processed to extract organic material following the general procedures of Phipps and Playford (1984). Depending on the sample type, processing involved all or several of the procedures listed below, *i.e.*, some samples did not require silicate and fluoride removal (steps 2 and 3) or heavy liquid separation (Step 7). With the exception of the acetolysis procedure, residues were washed-centrifuged in distilled H_2O three times after each step involving chemical treatment.

1. *Pre-HF treatment*: Calcium and magnesium carbonates were removed by heating the sediments in 10% or 50% hydrochloric acid (HCl).
2. *Silicate removal:* Samples were left in hydrofluoric acid (HF) for several days to remove silicates.
3. *Fluoride removal:* The majority of fluoride precipitates that resulted from the HF treatment were removed by heating to boiling point in 50% HCl to release the organic matter from the fluoride gel.
4. *Acetolysis:* To remove pollen contents and lipids; see acetolysis procedure for modern pollen preparation above.
5. *Alkali treatment:* Humic acids were cleared by warming the residues in 5% potassium hydroxide.
6. *Acid treatment:* Residues were warmed in 5% HCl to ensure a slightly acid environment to enhance preservation of pollen.
7. *Heavy-liquid separation:* Large mineral particles and heavy organic material were removed using zinc bromide with a specific gravity (SG) of 2.0 initially, and increasingly lower SG until palynomorphs appeared in the sinks (SG 1.8-1.75).

8. *Sieving*: To remove large extraneous plant material the residues were routinely sieved with either 100 μm or 150 μm mesh. Sievings were examined before discarding.
9. Residue strews were mounted on glass microscope slides using 'Eukitt', a plastic medium.

Pollen identification and counting were undertaken by Dr L Milne (University of Western Australia). Two hundred pollen grains were counted for all surface samples. Pollen identification was referred to modern pollen standards, additional pollen was either referenced to publications or assigned to the undifferentiated category. Samples were seeded with 0.02 ml *Betula* grains which is approximately 2,200 ± 400/cm^3.

To identify and quantify the pollen taxa in the sedimentary profile of selected wetlands, sediment samples at 10 cm intervals, beginning at 3-5 cm below the ground surface, were extracted from *in situ* cores of wetland muds and muddy sands from selected wetlands (161, 162, 163, 135, 9-14). These wetlands were selected for the following reasons. Wetland 161 is the oldest of the study wetlands and has the greatest thickness of wetland sediment. Wetland 162 is the second oldest of the study wetlands and has the thickest layer of carbonate mud, free from organic material. In addition, wetland 162 supports very different vegetation assemblages from the two wetlands adjacent to it (161, 163). Wetland 163 is younger than the other two wetlands, although adjacent to them, and has different vegetation assemblages. Wetland 135 is a wetland of medium age and supports a different vegetation assemblage to those described above. Wetland 9-14 is a relatively young wetland, but has sufficient fill to sample for pollen changes.

Anatomical or physiological features of plants which could be adapted to hydrological or sedimentological conditions were investigated, specifically rhizome, root, tiller and culm attributes. These components were measured in three species of monocotyledons: *Baumea articulata, B. juncea* and *Juncus kraussii*. A continuous series of ramets from current shooting tip to oldest tiller were collected spanning 4 years' growth. For three replicate plants, measurements were made between consecutive tillers of the following attributes: the rhizome depth from the surface, rhizome length, rhizome diameter, rhizome internode lengths, and number of internodes. As each tiller supported culms from various years, the maximum number of culms in any tiller was recorded and categorised. It was assumed that recognisable stages in the culm (new shoot, flowering, fruiting, and senescence) corresponded with 2001, 2000, 1999 and pre-1999 growing seasons. An attempt to correlate measurements of each attribute with the time span of the culms supported by a select interval of rhizome, was undertaken to investigate whether changes in these measurements reflected the hydrological conditions recorded for a particular season.

In addition, the presence/absence of aerenchyma was recorded along the length of rhizomes and roots to see if variability in size or type correlated with identifiable years of growth. Specimens were held in the laboratory under controlled conditions of permanent inundation for comparison of aerenchyma with plants collected from the field.

Radiocarbon dating of the sedimentary profile, described earlier, also was used as a basis to provide an age structure for the fossil pollen studies. In order to compare pollen composition and abundance between wetlands, samples of both carbonate mud and carbon were collected and prepared for radiocarbon dating. The pollen study, to some extent, determined the sampling frequency for radiocarbon samples. Included in the selection were the base and surface layer (3-5 cm) of each core. In between these two layers, samples were collected at approximately 20 cm intervals. Dates for pollen sampled in intervals between horizons with actual radiocarbon dates were inferred.

Pollen identification was again referred to the modern pollen standards, referenced to publications or assigned to the undifferentiated category. In these samples 100 pollen grains were counted; in two samples with very low pollen frequency, (wetland 161-85, 95), 50 pollen grains were counted. Pollen grains were counted along evenly spaced transects over each slide to account for preferential distribution of different sized particles in the plastic medium.

A new procedure was used in transforming pollen counts to pollen numbers per standard volume of sediment (volume containing 15 *Betula* grains). A mathematical problem potentially occurs for pollen taxa with zero frequency in the volume of sample prepared. In multiplying the actual numbers of pollen recorded by the ratio of *seeded pollen recorded* to the *number added per cubic centimetre*, zero still remains zero in contrast to all other taxa which increase proportionately, whereas, in fact, pollen recorded as zero in a low volume of sediment may be detected in larger volumes. To avoid this problem, all numbers of pollen were reduced to the lowest common denominator, and then for diagrammatic presentation, were multiplied by 1000 to remove any decimal places and plotted using a logarithmic value. In addition to pollen counting, notes were made about the state of preservation and degradation of pollen in order to link the pollen with oxidation processes. Algae abundance and type were also noted to be used in cases of low pollen numbers.

2.4.7 Experiments

Experiment 1

The hypotheses generated as a result of ordination highlighted water level, soil moisture, groundwater salinity, and interstitial water chemistry as environmental factors that affected plant distribution. In order to test these hypotheses, a field trial was set up in which these environmental parameters were measured in 5 quadrats containing increasing cover abundance of a single species, *e.g.*, 0, 25, 50, 75, 100 percent cover. The trial was set up in four separate wetlands along a single swale to minimise other variables such as age of wetland, distance from coast and geographic location with respect to the shallow aquifer. The species included the sedge *Baumea articulata*, the heaths *Melaleuca teretifolia* and *M. viminea*, and the rush *Juncus kraussii*. The herb *Centella asiatica* constituted the cover in the 0% quadrat in each wetland. Measurements of depth to water, groundwater salinity, groundwater pH, and soil moisture content were collected monthly from August 1995 to October 1996, as well as single measures of the groundwater concentrations of orthophosphate, ammonia and nitrate.

Experiment 2

To confirm a relationship between the hydrochemistry of interstitial waters and sediment type developed through leaching of cations by groundwaters and interstitial waters, a series of elution experiments were undertaken. Approximately 40 g of humic, peaty and carbonate mud surface sediments and kangaroo scats from selected wetlands were placed in 100 ml of de-ionised water for several minutes, and the water decanted to rinse them of any precipitated solutes such as NaCl, KCl, and $MgSO_4$. The rinsed samples were then placed in a fresh 100 ml aliquot of de-ionised water for a day to allow elution (and leaching) of cations from the samples to take place, and another set of rinsed samples were left to stand in 100 ml of de-ionised water for a week. The entire experiment was repeated with slightly acidified de-ionised water of pH 6.5 (using HCl acid to mimic the pH of rainwater), and weakly acidified de-ionised water of pH 6.5 (using carbon dioxide charged water). All water samples from the experiment were analysed for Na^+, K^+, Ca^{++}, Mg^{++}, and phosphorous.

Experiment 3

A second leaching experiment was undertaken where five samples of various dune sands were thoroughly rinsed several times with de-ionised water to free the sand of any labile cations present in pellicular water, or as precipitated salts. The sand was then left to drain until only the film of pellicular water remained to allow leaching of cations from the sand grains. The pellicular waters collected after one day and one week were analysed for cations to determine the extent that they could leach Na^+ and K^+ from the felspars, and Ca^{++} and Mg^{++} from the carbonate grains as a basis to interpreting natural patterns in wetland hydrochemistry.

Experiment 4

Similar experiments using de-ionised water were carried out on dried, comminuted wetland plant material to determine how rapidly and to what extent cations could be leached from such matter. Water samples were analysed for Na^+, K^+, Ca^{++}, and Mg^{++}.

3. REGIONAL SETTING

3.1 Introduction

This chapter describes the environmental features and natural processes of the larger scale physiographic and climatic region which is the context for the study area, and briefly reviews some of the previous work which provides the foundation to the study.

3.2 The Swan Coastal Plain

The Rockingham-Becher Plain is located in a larger physiographic unit - the Swan Coastal Plain. The latter is a ribbon shaped coastal feature, extending from Geraldton in the north to Busselton in the south (Gentilli and Fairbridge 1952). Its eastern margin is demarcated by the Darling Scarp, a landform related to the Darling Fault (Warner 1977).

3.2.1 Climate

The Rockingham-Becher Plain lies within the temperate belt of warm climate, with a cool wet winter and hot dry summer (Gentilli 1972). Climatic data used in this section are based on climatic averages for the station most representative of the Rockingham-Becher Plain locality, *i.e.*, Kwinana Station (averaged over 37 years), except for data pertaining to evaporation which was available only from the Perth Meteorological Station (averaged over 26 years). These data are presented in Table 3.1.

Table 3.1 *Regional climatic data*

Summer		*Winter*	
Mean maximum summer temperature	28.3°C	Mean maximum winter temperature	18°C
No. days above 35°C	11		
Hottest month	February		
Mean monthly summer rainfall	12 mm	Mean monthly winter rainfall	146 mm
Mean monthly summer evaporation	192 mm	Mean monthly winter evaporation	67 mm

At a regional scale, climate in this region is controlled by the belt of anticyclones moving from west to east in the middle latitudes during November to April. In the winter, this belt moves northward and is replaced by low pressure systems from the south bringing lower temperatures. At the zone of contact, fronts develop which bring rainfall. This seasonality is particularly striking in the rainfall and evaporation distribution. Ninety percent of rainfall occurs between April and October, and precipitation exceeds evaporation only during the months of May to August.

Regional wind patterns are also controlled by the seasonal north-south movement of the pressure systems. In summer, the dominant winds derive from the south and south-west with subsidiary winds from the north-east as the system progresses eastward. In autumn and winter, winds derive from the west. Local wind patterns, generated daily, particularly on the coast, are also very important. In summer, the landbreeze-seabreeze system reinforces prevailing winds, and is more consistent than regional wind occurrences. Another local coastal wind pattern is that related to storms. Steedman and Craig (1979) analysed the frequencies, direction, duration and velocity of the wind systems, and identify a long term periodicity of 7-10 years for storms.

3.2.2 Geology

The Swan Coastal Plain (Fig. 3-1C), is the Quaternary surface expression of the Phanerozoic Perth Basin (Playford *et al.* 1976). The Quaternary section of the Perth Basin, which underlies the Swan Coastal Plain, consists largely of unconsolidated sediments and sedimentary rocks, *i.e.*, sands, limestone, and clays. The deposits form a narrow linear body juxtaposed against the Yilgarn Craton, an extensive body of Precambrian rock.

3.2.3 Geomorphology

The surface morphology of the Swan Coastal Plain consists of various dune forms and alluvial flats (McArthur and Bettenay 1960). It has been subdivided into five regional scale units (Fig. 3-1D) orientated roughly parallel to the coastline. From east to west they are: 1) Ridge Hill Shelf - a shelf formed by coalescing alluvial fans, 2) Pinjarra Plain - alluvial plains, 3) Bassendean Dunes - Pleistocene aeolian dune fields, 4) Spearwood Dunes - parallel Pleistocene quartz sand covered limestone ridges, and 5) Quindalup Dunes - Holocene aeolian and beach deposits (McArthur and Bettenay 1960).

The Spearwood Dunes are relevant to this study in that they form the foundation and palaeo-shoreline to the accretionary Holocene deposits, and because they have influenced the development of the cuspate forelands. The Spearwood Dunes comprise a series of medium to high (20-60 m) parallel dune ridges, composed of limestone, semi-indurated aeolian sand, and yellow sand (Semeniuk and Glassford 1989). They are oriented NNW to SSE. Some of these occur offshore and form islands (Searle *et al.* 1988). These ridges mark the positions of the Pleistocene shorelines. Juxtaposed against these Pleistocene ridges is the most western geomorphic unit, the Quindalup Dunes (McArthur and Bettenay 1960), a discontinuous zone of Holocene dunes and beaches, comprising quartzose calcareous sand. The Rockingham-Becher Plain and the study area are located in this geomorphic unit (Fig. 3-1E, F).

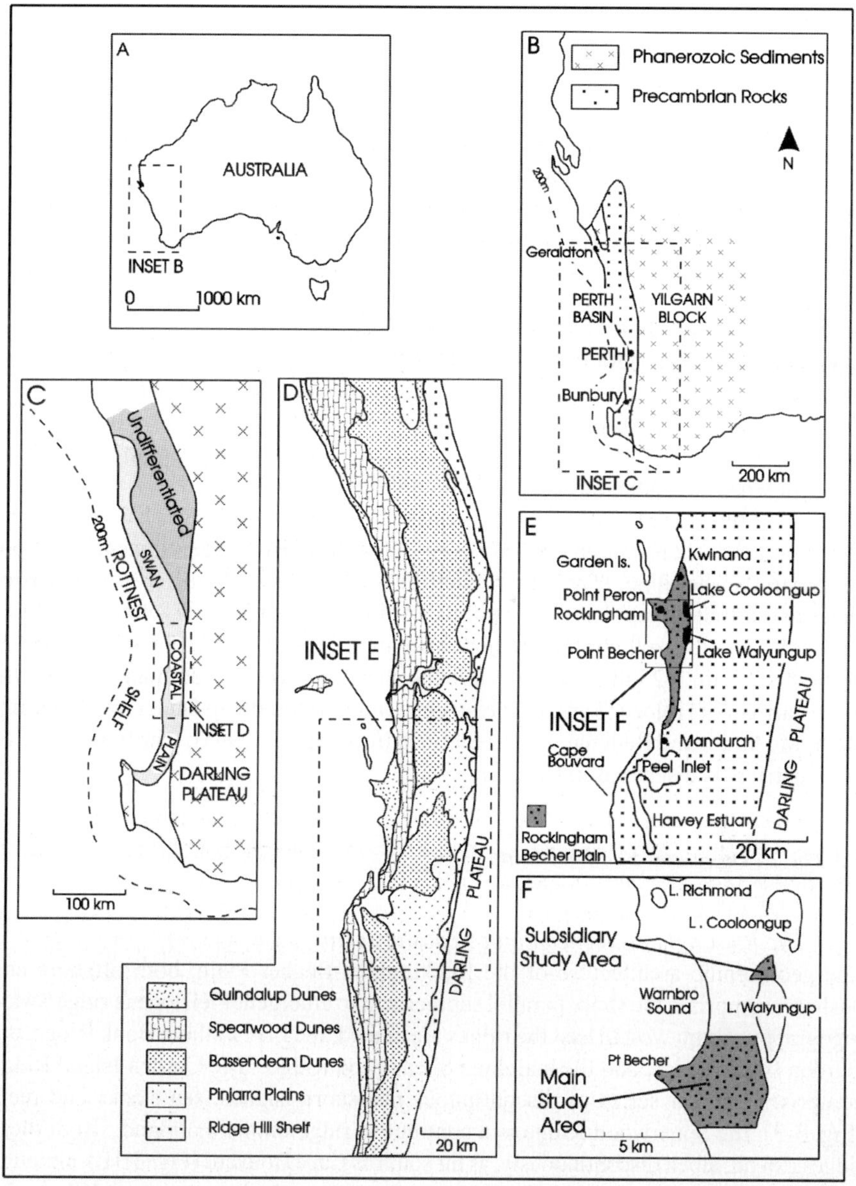

Figure 3-1. *Location and setting of Becher cuspate foreland. A. Location. B. Geological setting. C. Geomorphic setting. D. Geomorphic units of Swan Coastal Plain (after McArthur & Bettenay, 1960). E. Rockingham-Becher Plain. F. Study area.*

3.2.4 Hydrology

The Swan Coastal Plain is underlain by shallow unconfined groundwater. Surface water, in channel flows, finds expression only in local shore normal river/creek systems spaced intermittently from north to south across the Swan Coastal Plain. The study area lies between the southern two largest fluvial networks, the Swan and Canning Rivers discharging at Fremantle, and the Serpentine and Murray Rivers discharging at Peel-Harvey Estuary, but is not directly affected by either. The study area itself, is located in a zone of unconfined groundwater.

3.2.5 Coastal sectors and nearshore morphology

Searle and Semeniuk (1985) categorised the inner Rottnest Shelf coast adjoining the Swan Coastal Plain into distinct sectors, each of which can be recognised by its combined geomorphology, sedimentation processes and stratigraphic evolution (Fig. 3-2). Two of these sectors are relevant to this study: the Cape-Bouvard to Trigg Island sector in which the study area is located; and the Leschenault-Preston sector to the south of the study area. The Cape-Bouvard to Trigg Island sector is characterised by 1) complex near shore bathymetry, and 2) discrete cells of Holocene sediment accretion, normal to the shoreline. The Leschenault-Preston sector is characterised by barrier dunes and estuarine lagoons, and exhibits a simple submarine bathymetry. South of Cape Bouvard, the near coastal offshore bathymetry slopes seaward to merge with an inner shelf plain, 1-2 km offshore, while the shoreline becomes dominated by north/south barriers of Holocene and/or Pleistocene ridges (Quindalup Dunes and Yalgorup Plain- McArthur and Bettenay 1960, redefined by Semeniuk 1995). The importance of these barriers is discussed in Chapter 5.

3.3 The Rockingham-Becher Plain

3.3.1 The Rockingham-Becher Plain - coastal sector

The geomorphic architecture of the Rockingham-Becher Plain, both offshore and onshore, comprises the shore parallel submergent to emergent Pleistocene ridge/swale topography. From west to east the ridges are named, the Five Fathom Bank Ridge, the Garden Island Ridge, and the mainland Spearwood Ridge. The Garden Island Ridge comprises a linear series of discontinuous nearshore islands, sea stacks and reefs (Fig. 3-3); the Spearwood Ridge is a continuous ridge on the mainland. Both these ridges extend, albeit discontinuously, as far south as Cape Bouvard (Fig. 3-1E), although they are submerged south of Peel-Harvey Estuary exchange channel. Where the ridges intersect the coast, they form headlands.

Discrete cells of Holocene sediment have accreted in the inter-ridge depression (Fig. 3-3). The major accretionary part of these cells comprises submarine banks and beachridge plains. The Rockingham-Becher Plain, submergent to emergent banks,

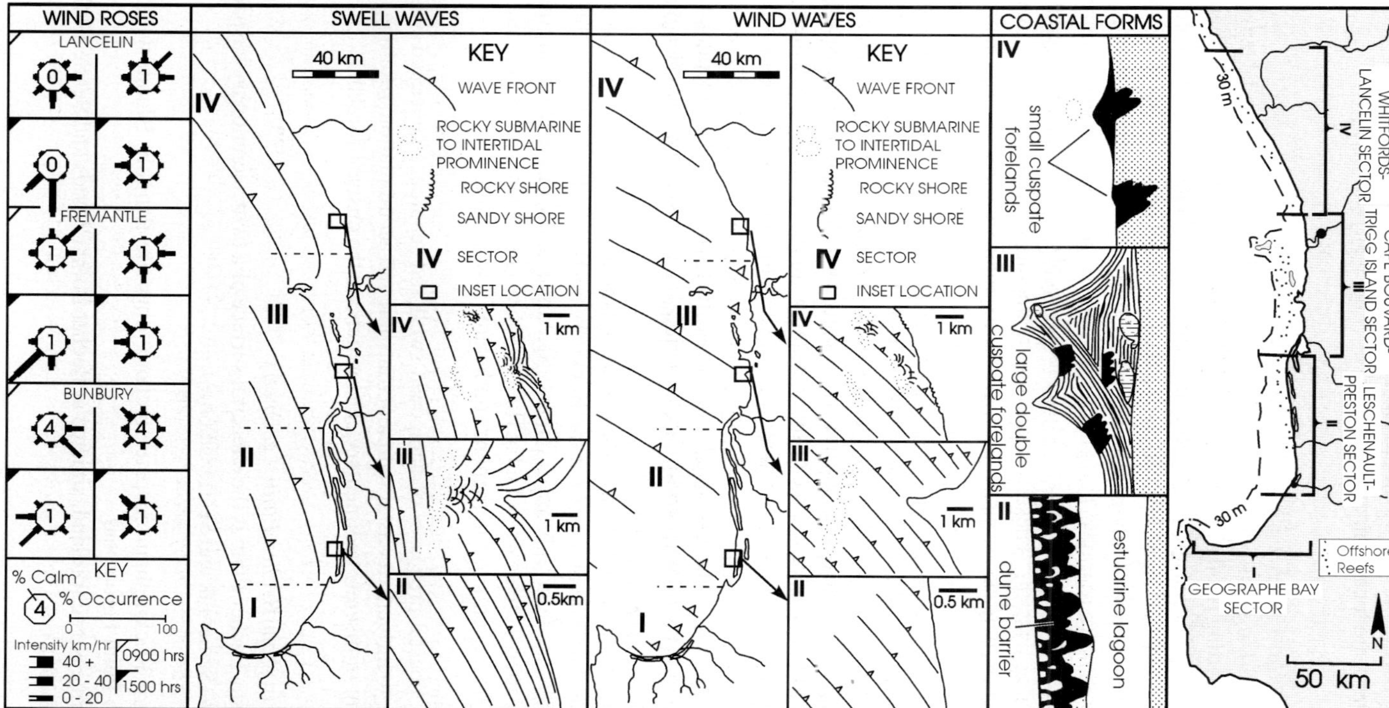

Figure 3-2. *Summary of wind and wave processes and resultant coastal forms for coastal sectors II, III, and IV (after Searle & Semeniuk 1985) A. Wind Rose Diagrams, B. Swell Wave Patterns, C. Wind Wave Patterns, D. Coastal Forms, E. Location of sectors.*

and several submarine basins: Cockburn Basin, Warnbro Basin, and Madora Basin (Searle *et al.* 1988), are located in one of these cells (Fig. 3-3).

3.3.2 The Rockingham-Becher Plain - offshore oceanography

The coast of the Rockingham-Becher Plain is located in a microtidal (0.4 m tidal range), wave-dominated setting (Hodgkin and Dilollo 1957; Steedman and Craig 1983; Searle and Semeniuk 1985). The location of the Rockingham-Becher Plain within a Pleistocene dune ridge topography has significant effects on the nature and impact of the shoreline processes. In the absence of any nearshore and offshore ridges, the coast would be dominated by two wave regimes: 1) regional swell waves generated in the Southern Ocean and southern part of the Indian Ocean, and 2) local wind waves generated by the land/sea breeze system. These two pressure/temperature driven processes may occur independently, may occasionally function so as to amplify effects, or may function to dampen effects.

In the study area, prevailing swell waves derive from west to south-west all year round (Fig. 3-2). In summer, they are supplemented by south-west wind waves generated by sea breezes (Fig. 3-2). In winter, storm driven wind waves approach from west to north-west. However, the occurrence of submergent to emergent islands, rocky reefs, and stacks causes refraction and diffraction of wave orthogonals, resulting in wave convergence, divergence, and reduced energy (Searle and Semeniuk 1985). Swell waves with longer periods (10-14 seconds) are more affected by submarine topography than locally generated wind waves with shorter period (<10 seconds) and wavelength.

Regional circulation is located west of the Garden Island Ridge, and therefore has no influence on the shoreline processes of the Rockingham-Becher Plain. In nearshore waters, wind generated currents are dominant (Steedman and Craig 1979). Shifting seasonal wind patterns create a reversal in flow directions in seasonal littoral currents, *i.e.*, northward flow occurs in summer and southward flow in winter.

In this context, coastal processes of dampened swell and seabreeze generated wind waves combine to influence coastal sediment transport and accumulation, resulting in the development of discrete sedimentary cells which are the cuspate forelands.

3.3.3 The Rockingham-Becher Plain - geometry

The Rockingham-Becher Plain is the subaerial surface of a twin accretionary cuspate foreland system (Semeniuk and Searle 1986b):

- the northern Rockingham cuspate foreland, opposite Penguin Island, with a tombolo to Point Peron; and
- the southern cuspate foreland, situated at Becher Point (Woods and Searle 1983; Searle *et al.* 1988).

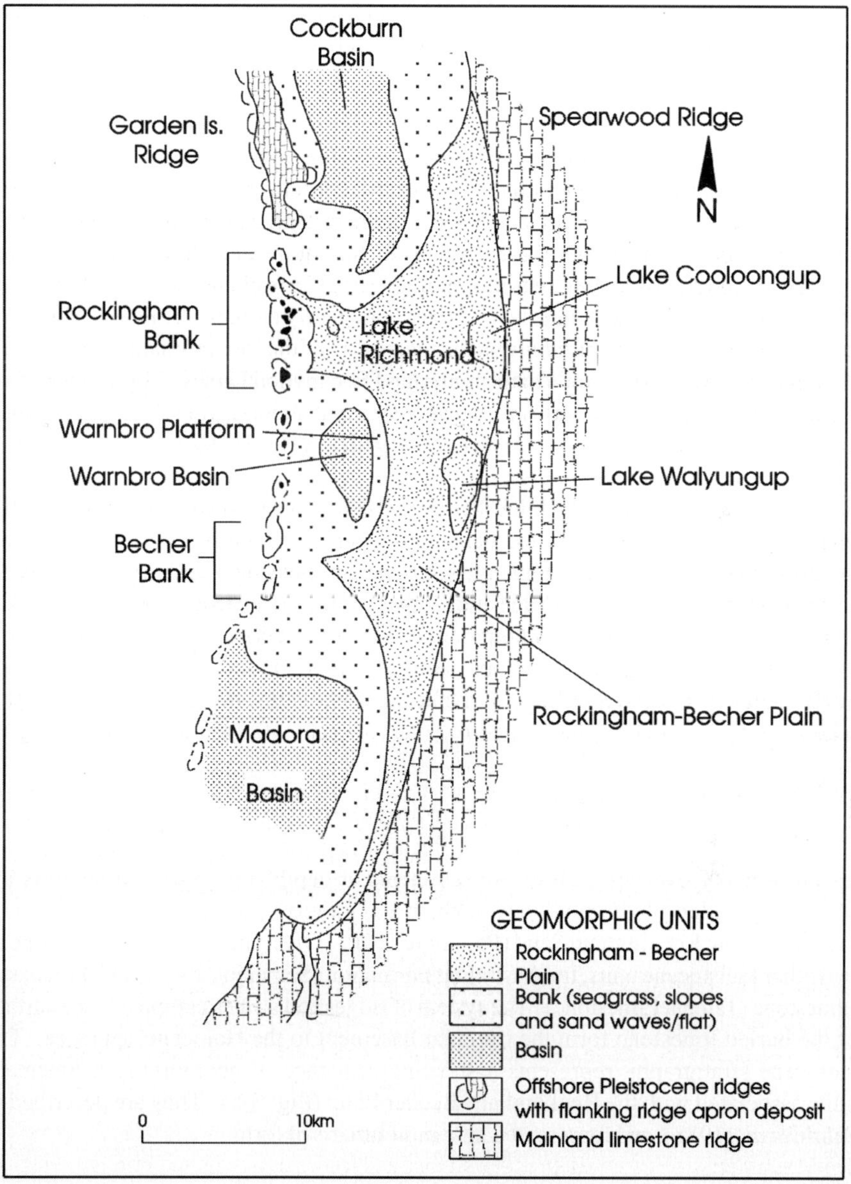

Figure 3-3. *Geomorphic units of the Rockingham-Becher Plain and the near shore zone (after Searle et al. 1988).*

The Plain is 10 km at its maximum width, and extends approximately 40 km from Cockburn Sound to the north to the Peel-Harvey Estuary exchange channel to the south where it is restricted to a narrow linear shaped body (Figs. 3-1E, 3-3).

3.3.4 The Rockingham-Becher Plain - geomorphology

The Rockingham-Becher Plain has been subdivided into the following geomorphic units (Searle *et al.* 1988): beach zone, and beachridge and dune plain. The beach zone comprises the subtidal shoreface, the beach foreshore and the beach backshore (Semeniuk and Johnson 1982). Its morphology varies in response to seasonal shoreline processes, but in general is narrow (5-30 m), and gently sloping. The beachridge and dune plain, including the foredune, is the major unit of the Rockingham-Becher Plain. It extends from the beach zone to the hinterland Spearwood Dunes. The surface of the plain is characterised by a series of linear ridges and swales, whose elevation ranges from 2-12 metres above mean sea level (MSL), with a mean elevation of 5 metres. At the coastal margin, the ridges are disrupted by blowouts associated with the formation of small parabolic dunes. At the hinterland margin, *i.e.*, where the Rockingham-Becher Plain abuts the Spearwood Dunes, there are two types of contact: 1) that characterised by the abutment of NW/SE oriented small scale parallel ridges and the medium scale NE/SW oriented Spearwood Ridge, and 2) that characterised by two macroscale lakes (Lake Cooloongup and Lake Walyungup), enclosed by a series of ridges which were recurved spits. The main ridge (spit) is oriented NW/SE and splays at its distal end into a series of ENE/WSW fingers. Dunes and blowouts are superimposed on this ridge.

3.3.5 The Rockingham-Becher Plain - stratigraphy

The stratigraphy under the Rockingham-Becher Plain consists of Quaternary sediments unconformably overlying Mesozoic sediments at depth (Fig. 3-4). At depth is the Mesozoic Leederville Formation (Cockbain and Playford 1973), which is overlain by the Pleistocene Rockingham Sand (Passmore 1970). Overlying this formation is one of two other Pleistocene units, the Australind Formation (Semeniuk 1983), and the Coastal Limestone (Tamala Limestone). The system of ridges and swales comprising the surface of the buried limestone form the irregular basement to the Holocene sequence. The Holocene stratigraphy represents a shoaling sequence of geomorphic/sedimentary units associated with the Rockingham-Becher Plain (Fig. 3-5). They are described in Searle *et al.* (1988), and represented here in summarised form:

- sands of the beachridge/dune plain (Safety Bay Sand)
- sands of the beach zone (Safety Bay Sand)
- sands of the submergent to emergent banks (Becher Sand)
- sand, muddy sand, and mud of former estuaries and lagoons (Leschenault Formation)
- muds of the submarine basins (Bridport Calcilutite)

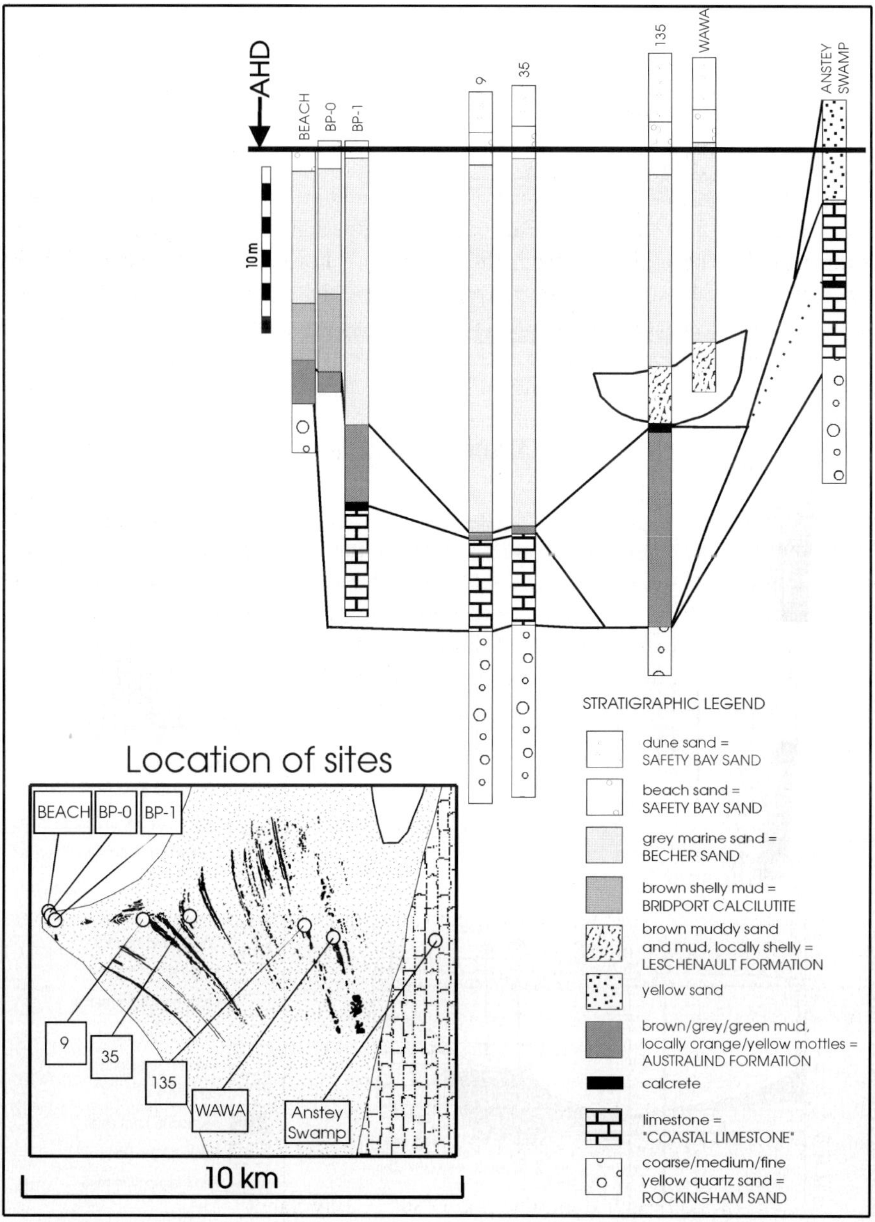

Figure 3-4. *Stratigraphic profile across the Becher beachridge plain showing the geohydrologic framework.*

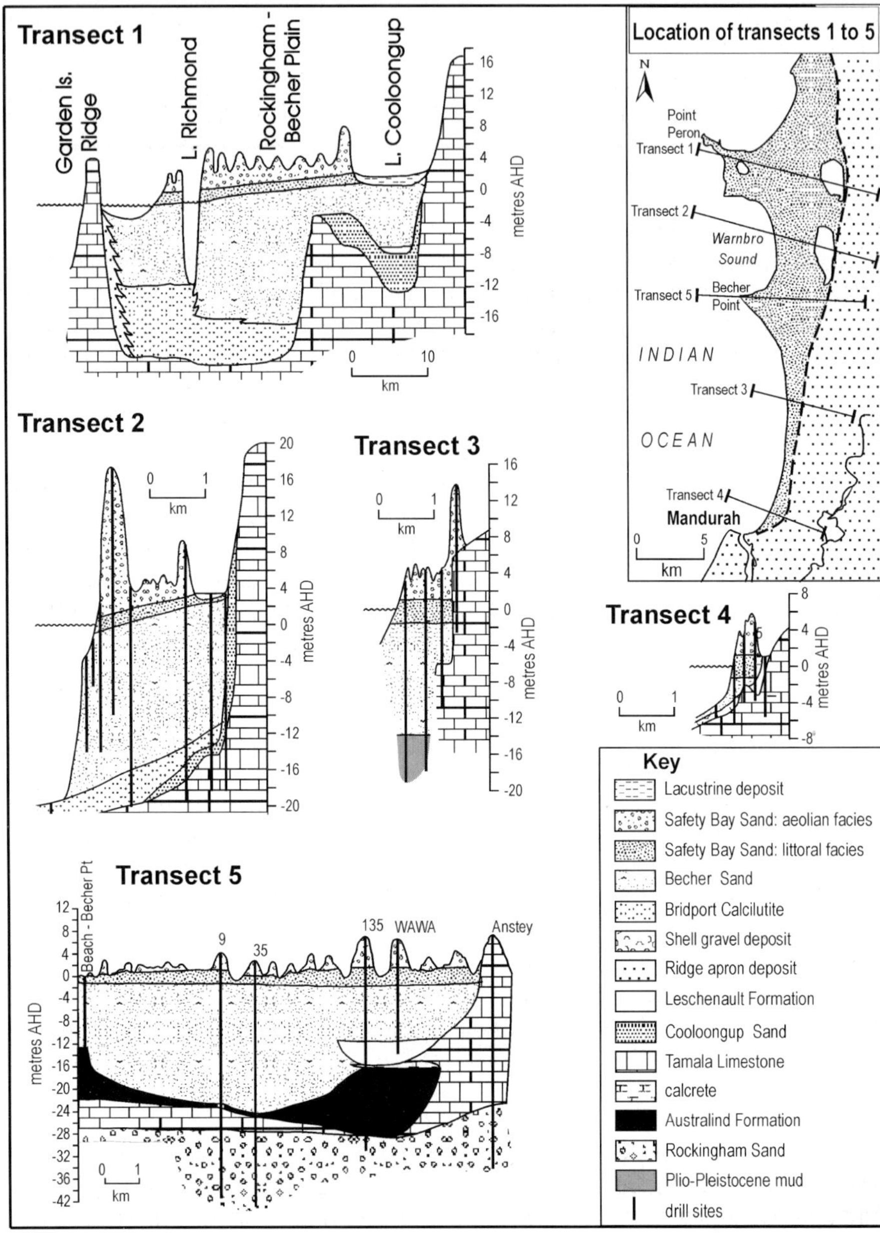

Figure 3-5. *Stratigraphic profiles of five transects showing base and margins of Safety Bay Sand and Becher Sand aquifers. Transects 1-4 amalgamated from Searle et al. 1988 and this study. Transect 5 from this study.*

The Holocene sedimentary units have been related to basin, beachridge and dune facies within the geomorphic units, providing an environmental context for each sediment type (Searle *et al.* 1988). The stratigraphic relationship, between the three Holocene sand sediments, is one of simple superposition without the disruption of any major eustatic or isostatic events.

3.3.6 The Rockingham-Becher Plain - groundwater hydrology

There are several water bearing aquifers at varying depths in the Holocene and Pleistocene sediments under the Rockingham-Becher Plain. At the base is the Quaternary Rockingham Sand aquifer which is a channel shaped body of medium to coarse grained sand, unconformably overlying the early Cretaceous Leederville Formation (Figure 5 in Passmore 1970). Overlying the Rockingham Sand are several formations which have been amalgamated, for water supply purposes, into a group informally termed the "superficial formations" (Allen 1976). The "superficial formations" unconformably overlie both the South Perth Formation and the Rockingham Sand. Detailed investigation or revision of the "superficial formations" for stratigraphic purposes, has resulted in five formations: 1) the Tamala Limestone of Pleistocene age, (Playford 1976), and 2) the Bridport Calcilutite (Semeniuk and Searle 1987), 3) the Cooloongup Sand (Passmore 1970), 4) the Becher Sand (Semeniuk and Searle 1985), and 5) the Safety Bay Sand (Passmore 1970, redefined by Semeniuk and Searle 1985), of Holocene age.

The Tamala Limestone is variable in composition (Klenowski 1976), comprising cemented rock, calcretised limestone, cavernous and vugular limestone, quartz sand, and thin layers of calcareous silt and clay, and so its properties as an aquifer vary from unconfined to semi-confined. The properties of the aquifer range from lenses and sheets which are highly porous to sheets which act as aquitards. The Bridport Calcilutite (Semeniuk and Searle 1987), unconformably overlies the Tamala Limestone. It is a sequence of structureless to bioturbated calcareous mud and shelly mud. This Formation may be that described by Passmore as a thin layer of clay which is an effective barrier to vertical mixing of water between the Rockingham Sand and the Becher Sand. In the Cooloongup Sand, Becher Sand and the Safety Bay Sand, the water is unconfined and although lithologically distinct, the three formations act as a single aquifer directly recharged by infiltrating meteoric waters.

The water in the surficial aquifers of the northern cuspate foreland is fresh, varying slightly in different formations (Passmore 1970). It ranges from subhaline to hyposaline in the western part of the Tamala Limestone, and from fresh to hyposaline in the Becher Sand, Safety Bay Sand, and Cooloongup Sand, depending on the local occurrences of clay beds in the latter. Its geometry approximates an elongate mound, sloping downward to the west, north, and east (Passmore 1970).

There are two low groundwater mounds in the region (WAWA 1988), the Rockingham Mound which is located southeast of the northern cuspate foreland centred

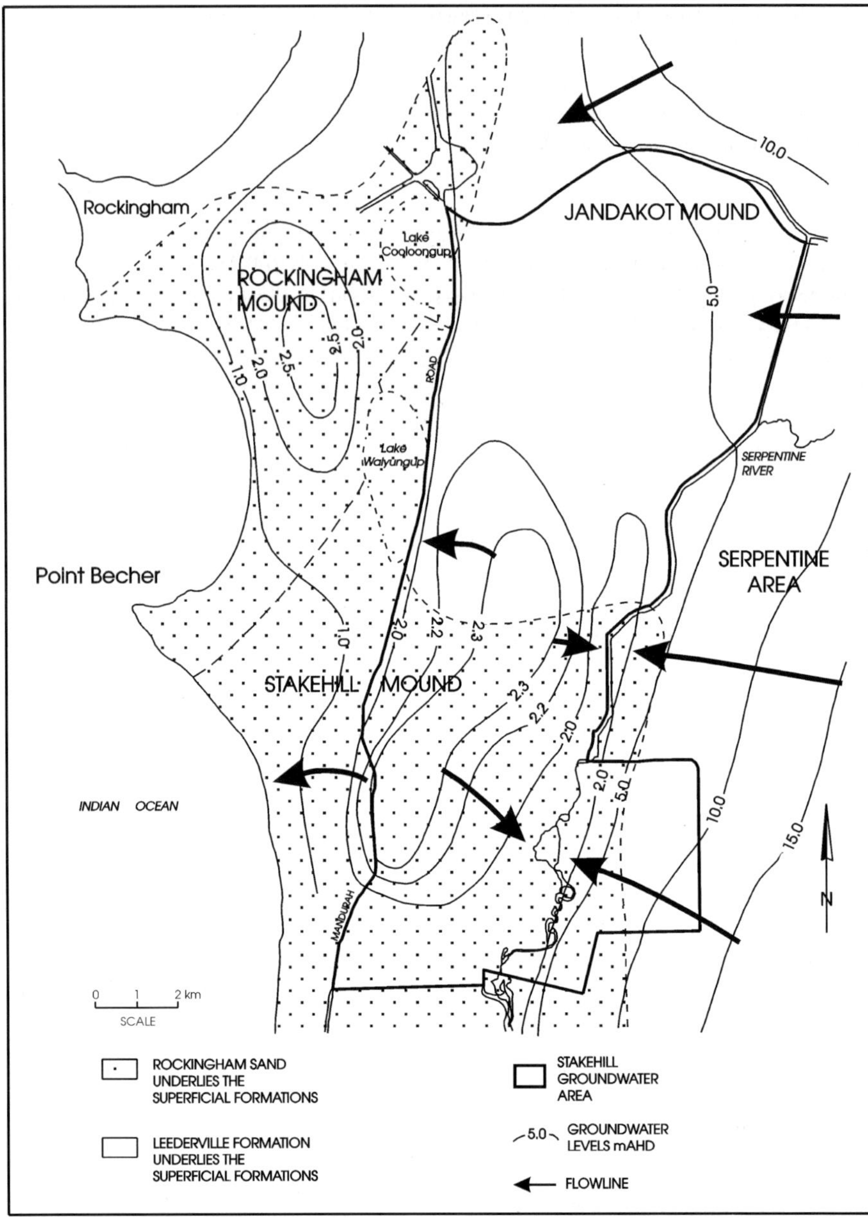

Figure 3-6. *Groundwater flow systems in the subregion. (after Water Authority of WA 1988).*

on Safety Bay, and the Stakehill Mound, situated east of the southern cuspate foreland centred on Stakehill (Fig. 3-6). The maximum height of the Rockingham Mound is approximately 2.5 m; its steepest hydraulic gradient (approximately 1:570) is to the southwest towards Warnbro Bay. The maximum height of the Stakehill Mound is approximately 2.3 m; its steepest hydraulic gradient is approximately 1:3000 to the west towards Becher Point, and approximately 1:2200 to the nearest wetland. The groundwater flows westward with a general hydraulic conductivity throughout the Safety Bay Sand aquifer of 40 m/day (Passmore 1970). At the seaward edge, the fresh groundwater depresses the seawater below MSL. Passmore (1970) describes the mean depth of the zone of diffusion between fresh and saline water as being approximately 1-8 metres in the vicinity of Point Peron.

3.3.7 The Rockingham-Becher Plain - wetlands

The majority of wetlands within this region have formed as a result of groundwater intersecting the topography. Different settings, origins, and developmental histories have resulted in three regional suites of wetlands in the Rockingham-Becher Plain (Fig. 3-7): Cooloongup Suite, Becher Suite and the Peelhurst Suite (C. A. Semeniuk 1988). In the Cooloongup Suite there are three lakes and two seasonal wetlands (sumplands) behind barrier spits. Lakes Cooloongup and Walyungup are macroscale and hyposaline to hypersaline, and Lake Richmond is mesoscale and subhaline. Lakes Cooloongup and Walyungup are recharged partly by groundwater seepage from the Tamala Limestone and Safety Bay Sand aquifers to the east, and partly by precipitation which is perched (C. A. Semeniuk 1988). Locally, groundwater from beneath the basins is partly to completely confined. Lake Richmond is recharged by precipitation and groundwater flow. The two sumplands at Point Becher itself are microscale to lepto-scale, and fresh to hypersaline, recharged by precipitation and groundwater rise (C. A. Semeniuk 1988). In the Becher and Peelhurst Suites there are numerous leptoscale, seasonal wetlands (sumplands and damplands) in the inter-ridge depressions of the beachridges and in the bowls of the parabolic dunes, respectively. These wetlands are fresh to subhaline, recharged by precipitation and groundwater rise.

3.3.8 The Rockingham-Becher Plain - evolutionary environmental history relating to beachridge and swale development

Radiocarbon dating of shells from swash and inshore zones, and peat from specific stratigraphically determined levels at 35 sites on the Rockingham-Becher Plain enabled Woods and Searle (1983), Searle and Woods (1986), and Searle *et al.* (1988), to reconstruct the age structure of the Rockingham-Becher Plain, its Quaternary history, and its palaeogeography (Fig. 3-8).

Progradation commenced at around 6645 ^{14}C yrs BP with the establishment of beach conditions adjacent to the Spearwood Ridge. Sediment, transported predominantly by shore drift along the Spearwood Ridge mainland, but supplemented by easterly

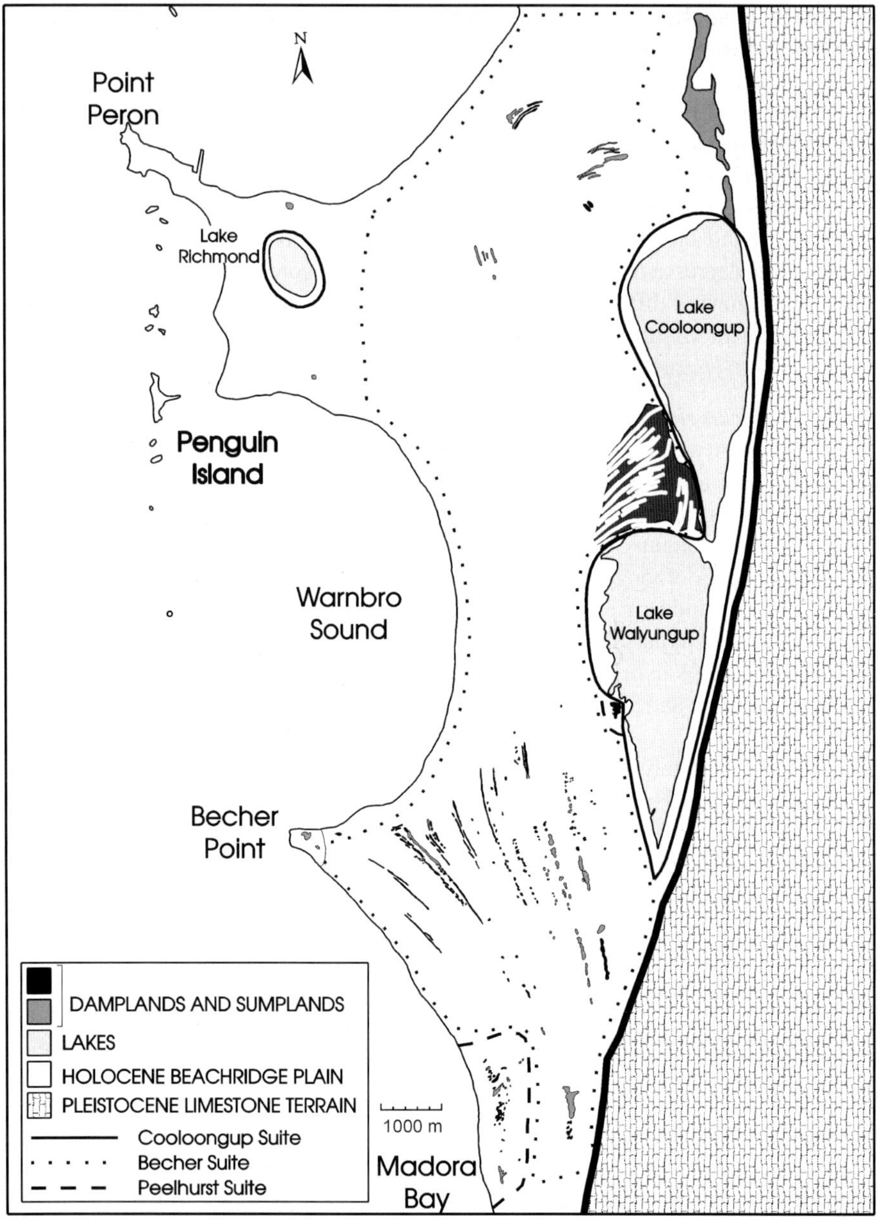

Figure 3-7. *Location of consanguineous wetland suites on Rockingham-Becher Plain and distribution of wetland types according to Semeniuk (1987).*

entrainment of northerly shore sediment drift debris from marine erosion of the Garden Island Ridge (Searle *et al.* 1988), and skeletal material from seagrass banks, accumulated in the low energy environment southeast of Garden Island itself. Here it formed a submarine bank, topped by a series of recurved spits whose morphology is readily recognisable today as the barrier between the Becher Cusp and the Becher sandplain.

From 6000-5000 ^{14}C yrs BP the major aggradation of sediment was still occurring in the area southeast of Garden Island. An asymmetrical cuspate foreland was developing in response to the prevailing southwesterly swell/wind wave system. The apex of this cuspate foreland was near what is known today as Point Peron. The southern extremity of this cuspate foreland was very narrow, probably as a result of localised littoral drift northwards to the cusp apex. As a result, in this area, the initial barrier spit was overprinted by aeolian sediment increasing its height to 20 m which is significantly higher than the rest of the beachridge plain.

From 5000-4000 ^{14}C yrs BP, substantial progradation continued through the development of submarine banks and the subaerial beachridge plain, along both the primary locus at Point Peron, and along a secondary locus in the vicinity of Point Becher. This region of the Becher beachridge plain exhibits the greatest number of high ridges, suggesting periods of rapid progradation interspersed with periods of coastal stability. Between 4000 and 2000 C^{14} yrs BP, the growth of the Becher Bank, and later, the beachridge plain into a cuspate body, began to exert influence on nearshore dynamics. Longshore sediment transport was arrested by the presence of the protruding tip, resulting in the development of beach ridges along the southern shore, the development of planar areas along the axis of the tip, and static shore positions to the immediate north. There is some evidence of periodic interruptions to this pattern in the presence of an occasional high ridge with a deep swale seaward of it. This morphology has been observed in other locations where high energy waves are active (*e.g.*, Windy Harbour, WA), and may therefore be related to patterns of storminess rather than a change in sediment supply.

From 2000 ^{14}C yrs BP to the present, development of the beachridge plain continued along the axis of Pt. Becher; the apex of the cuspate foreland became more prominent through rapid planar progradation, and the beachridges to the south became numerous, low (1-2 m), and closely spaced (3 ridges over 100 m). The low height and relatively close spacing of the beachridges indicate formation in a micro-tidal, low wave energy environment, (in this context low energy relates to intense wave refraction), with an abundant supply of sediment. These conditions favour a process of rapid deposition (Davies 1958). Coastal areas to the north were depleted of sediment, and coastal erosion became more prominent.

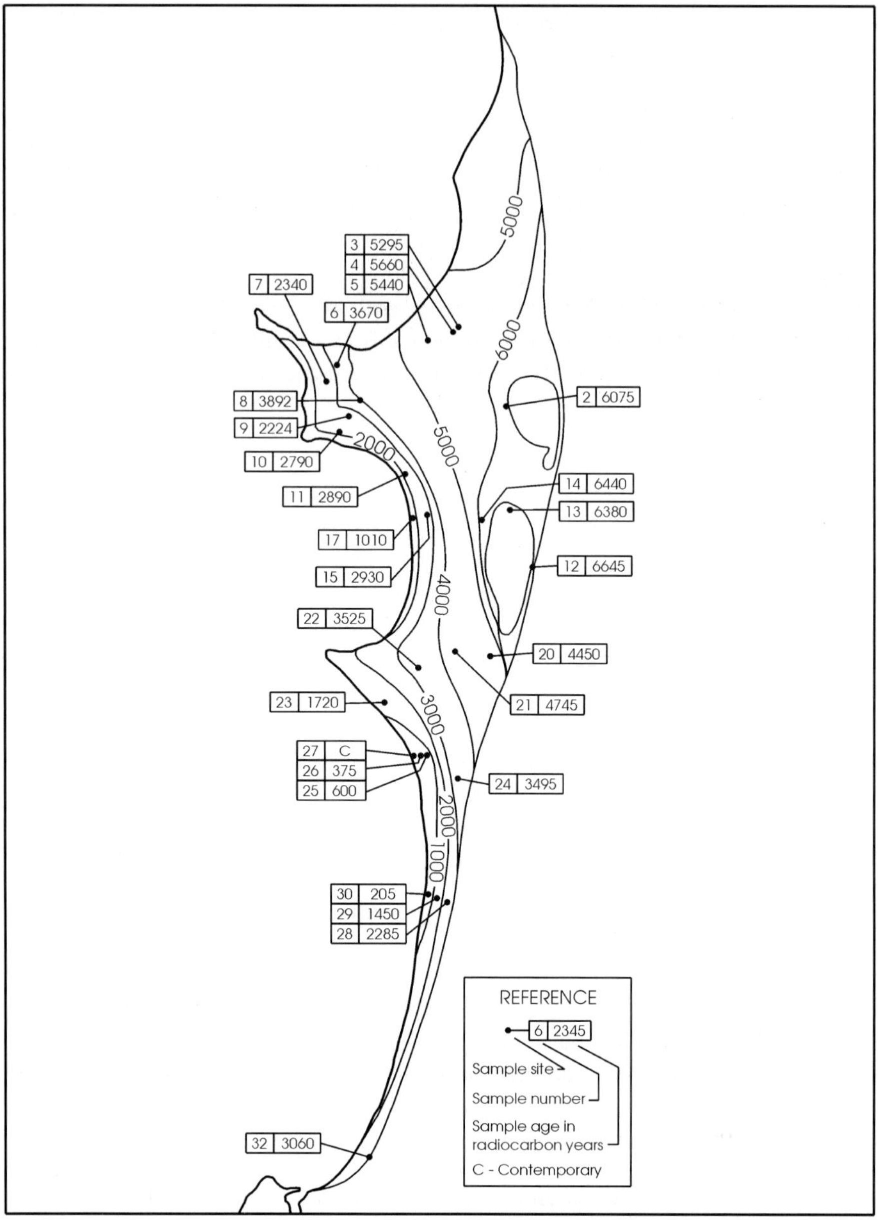

Figure 3-8. *Age structure of beachridge plain. (after Searle et al. 1988).*

3.4 The Becher Cusp

3.4.1 The Becher Cusp - geometry and terminology

The study area is centred on the southern cusp of the Rockingham-Becher Plain, referred to herein as the Becher Cusp, with its apex at Becher Point. It is typical of simple accretionary cusps (Gulliver 1896), *i.e.*, a seaward projecting accumulation of sediment, bounded on both sides by active wave dominated shorelines, and located in a wave shadow area (Zenkovich 1959) behind a chain of offshore islands, reefs and pinnacles. In plan view, it is triangular and asymmetrical, the longer southern shore facing the shore-normal incident southwesterly swell. It is approximately 5 km long north to south, and 5 km from the apex to the hinterland margin.

In more recent literature on beaches, the term "cusp" has been used to refer to ephemeral crescent shaped features on the beach, caused by wave action on steep shore profiles (Snead 1982; Schwartz 1986). The term "cuspate foreland" has been adopted for the landform resulting from progradation of the shore seaward through the process of long continued shore drifting of sediment (Snead 1982; Schwartz 1986). Therefore, in accord with this current practice, usage of the term "cuspate foreland" will be adopted herein in a geomorphic context, while the term "Becher Cusp" will be used as a notation, or geographic location reference.

3.4.2 The Becher Cusp - geomorphology

The surface morphology of the Becher cuspate foreland comprises a variety of forms: shore-parallel ridges and swales, complete and partial parabolic dunes (blowouts), chaots, inter-ridge flats, and basin wetlands which occur singly and in chains along the swales (Fig. 3-9).

3.4.3 The Becher Cusp - stratigraphy and soils

The sedimentary stratigraphic sequence is summarised from Searle *et al.* (1988). The oldest unit of the Holocene deposits is the Bridport Calcilutite (Semeniuk and Searle 1987), which forms the contemporary surface in Warnbro Sound. This unit consists of structureless, grey/brown carbonate mud. Overlying this is the Becher Sand (Semeniuk and Searle 1985b), which is composed of sediment formed in the bank facies. It is mostly grey bioturbated sand (shell, quartz grains), and muddy quartzose calcareous sand with local seagrass fibre. The top unit and the wetland basal material is Safety Bay Sand (Passmore 1970, redefined by Semeniuk and Searle 1985b). This unit is divided into two sedimentary facies: beach and beachridge/dune. The beach facies may be further subdivided into several zones based on sedimentary structures (Semeniuk and Johnson 1982). The texture of the beach sediments ranges from medium to coarse sand. The composition comprises variable proportions of skeletal fragments, carbonate lithoclasts, and quartz. The texture and composition of the beachridge/

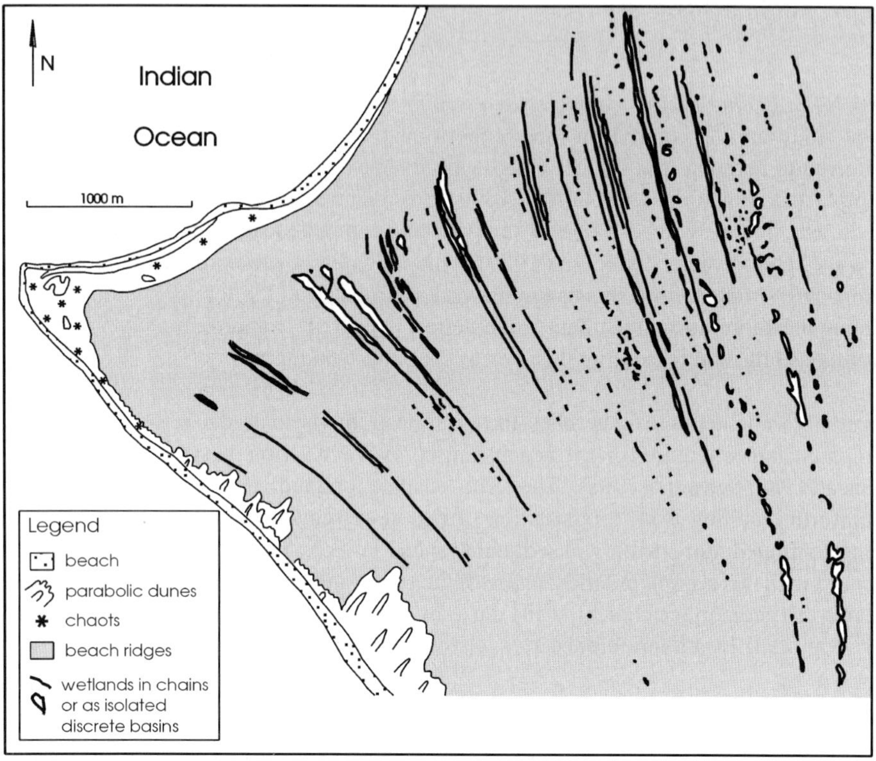

Figure 3-9. *Surface morphology of the Becher cuspate foreland.*

dune facies comprise well to poorly sorted, medium and fine sand composed of rounded skeletal fragments, carbonate lithoclasts, and minor well-rounded quartz.

Soils

The following is summarised from a study of the evolution and development of soils on the Holocene beachridge sequences by Woods (1984). The parent material to soils in this area is sand predominantly composed of calcium carbonate, mainly as skeletal grains and fragments, with lesser amounts of quartz, felspar, mica and heavy minerals, but the concentration of calcium carbonate varies from 30-80% across the beachridge sequence (Woods 1984). Organic horizons are present at the surface where organic material constitutes approximately 4% of the soil (by weight), but there is little organic matter below 0.8 metres. There is very little clay content in the soils (maximum = 0.75%). This content increases with the age of the soil where it tends to concentrate in the surface horizons.

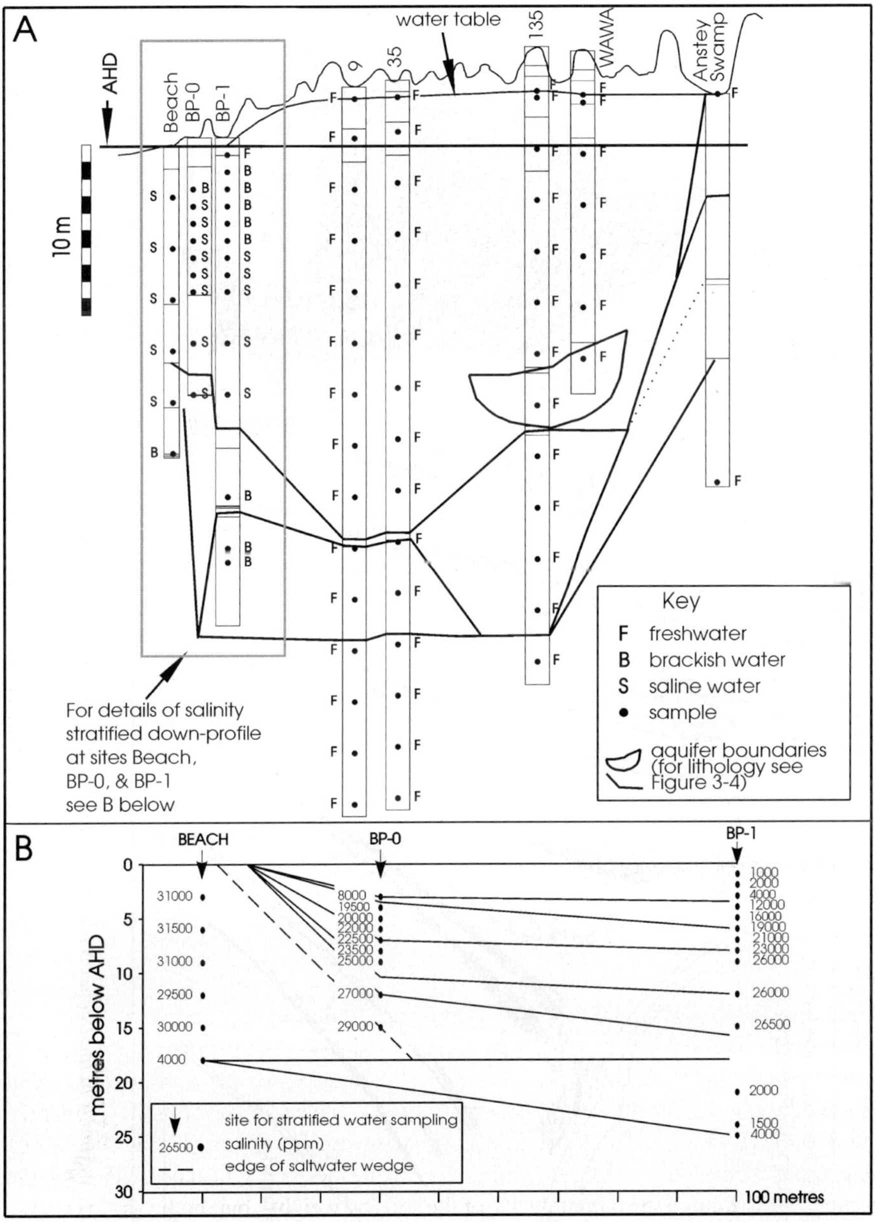

Figure 3-10. *A. Salinity of water body in various aquifers under Becher cuspate foreland. B. Details of ground-water salinities between coast and 100 m inland.*

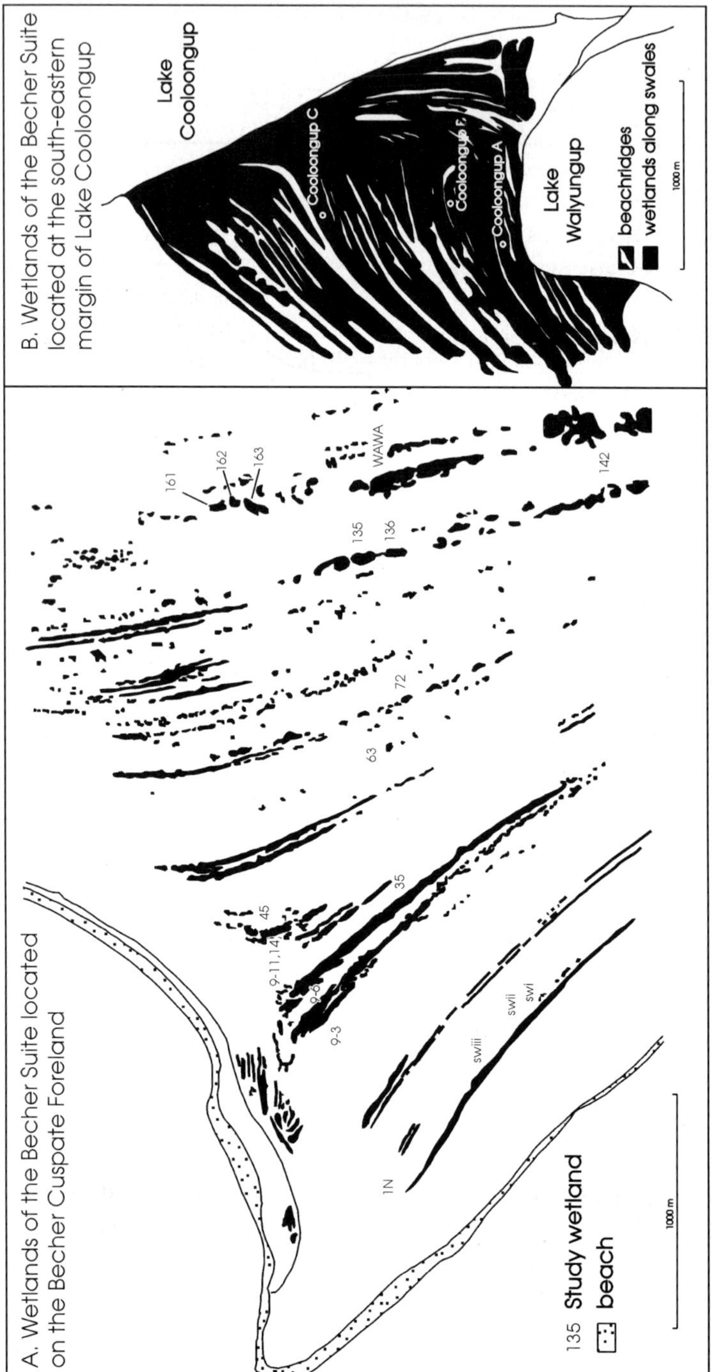

Figure 3-11. *Wetlands in the Becher Suite: on the Becher cuspate foreland, and at the south-eastern margin of Lake Cooloongup.*

Of special interest in this study are the carbonate mineral species and the related cations, Ca and Mg. Woods (1984) found that the parent sands of the beachridges were composed of calcite, Mg calcite and aragonite, but the surface soils became depleted in Mg calcite and aragonite. Down profile patterns in Mg calcite and calcium carbonate content and Mg and Ca cation concentrations were similar, in that they increased to a depth of approximately 80-100 cm (lower limit of pedogenic effects) and then became relatively constant. There was also a marked decline in their relative proportions from young to old sediments. The progressive loss of Ca and Mg from surface sediments was interpreted to reflect dissolution of Mg calcite (Woods 1984). Most of the Ca and Mg leached out of the soil profile, passed out of the system. The value of the Ca/Mg ratio was greater at the surface than at depth, indicating a loss of Mg relative to Ca in the top 1 m. The relevance of these data to the study will be discussed later.

3.4.4 The Becher Cusp - hydrology

The Becher Cusp, as for the whole of the Rockingham-Becher Plain, is a fresh groundwater hydrological system, recharged by seasonal precipitation and discharged primarily through evapo-transpiration. The inter-ridge wetlands are windows to the water table, though locally perched for part of the year. The saturated zone containing freshwater is >40 metres thick and extends to within 100 metres of the shoreline at Point Becher. At the coast its thickness decreases as it is replaced by an underlying brackish to saline water wedge that increases in salinity from 8000-30,000 ppm between depths of 3-15 metres (Fig. 3-10). At 18 metres, within the Rockingham Sand aquifer, there is a body of water separated from the marine water phreatic zone by a calcrete sheet and the underlying calcareous muddy Australind Formation, which together probably act as semi-confining to confining beds. This body of subhaline water derives from the Rockingham Sand and Tamala Limestone aquifers further inland, and has a salinity of 1400 - 4300 ppm.

This pattern of coastal hydrology is at variance with the pattern of the coastal aquifers documented for the northern Rockingham Cusp (Passmore 1970), in that the freshwater/ seawater interface does not extend far inland, and the phreatic zone under the beach face at depth is not marine water but relatively freshwater.

Wetlands

The Becher Suite (Fig. 3-11) includes numerous wetlands (approximately 275), but only two wetland types, leptoscale to microscale sumplands, and leptoscale to mesoscale damplands. The majority are steep sided with undulating floors. In the sumplands, inundation is shallow (up to 35 cm) and lasts approximately 3-5 months. In the damplands, waterlogging varies between the ground surface and a depth of 50 cm below, with a similar hydroperiod. The water is poikilohaline and ranges from freshwater to hyposaline. The recharge mechanisms are direct input through precipitation, concomitant with groundwater rise.

The mesoscale wetlands in the inter-ridge depressions relating to the marginal spits around Lakes Cooloongup and Walyungup (Figs. 3-1, 3-7) are transitional between basins and flats, *i.e.*, they are flat floored and shallow. Although intermittently inundated, recently they have only experienced a rise in the water table to a level of 50 cm below ground surface. The water is poikilohaline and ranges from freshwater to hyposaline. The recharge mechanisms are similar to those in the basins of the Becher beachridge plain, *i.e.*, direct input through precipitation and groundwater rise. Because of their similarities, the wetlands have been included in the study. Three examples of these inter-ridge wetlands have been included in this study, and have been named Cooloongup A, B, and C (Fig. 3-11).

3.4.5 The Becher Cusp - vegetation

Previous work on the vegetation has been carried out at the regional scale by Heddle *et al.* (1978), and Gibson *et al.* (1994), and at the local scale by Trudgen (1988). At the regional scale the vegetation is classified as Quindalup Dune Complex, the term "complex" in this sense being used to define vegetation referred to a specific landform unit, in this case that of Churchward and McArthur (1978). Quindalup Dune Complex is described as:

> "Coastal Dune Complex consisting mainly of two alliances: the strand and foredune alliance, and the mobile and stable dune alliance."

At the local scale, vegetation for part of the Becher cuspate foreland was mapped (Trudgen 1988), using the small scale landform and soil units of McArthur and Bartle (1979), and Wells *et al.* (1985). Again, the term complex was used. For this study, upland vegetation assemblages for each of the dune types on the Becher cuspate foreland (as defined in Semeniuk *et al.* 1989) were described using structural terms of Specht (1981) and species dominance, abundance, and absence/presence (Syrinx Env. 2000). This approach was used because there is a relationship between the age of the dune and the species composition and abundance. Although dune forms are gradational, the dune types may be ordered as follows, from oldest to youngest:

- older beach ridges
- middle beachridges
- young beachridges
- chaots, conical hill residuals and parabolic dunes
- currently forming beachridges and foredune

A summary of the vegetation associated with each broad landform unit is presented below (Table 3.2), together with previous vegetation mapping of a portion of the study area. In Table 3.3, a more detailed description of the upland vegetation undertaken for this study is presented.

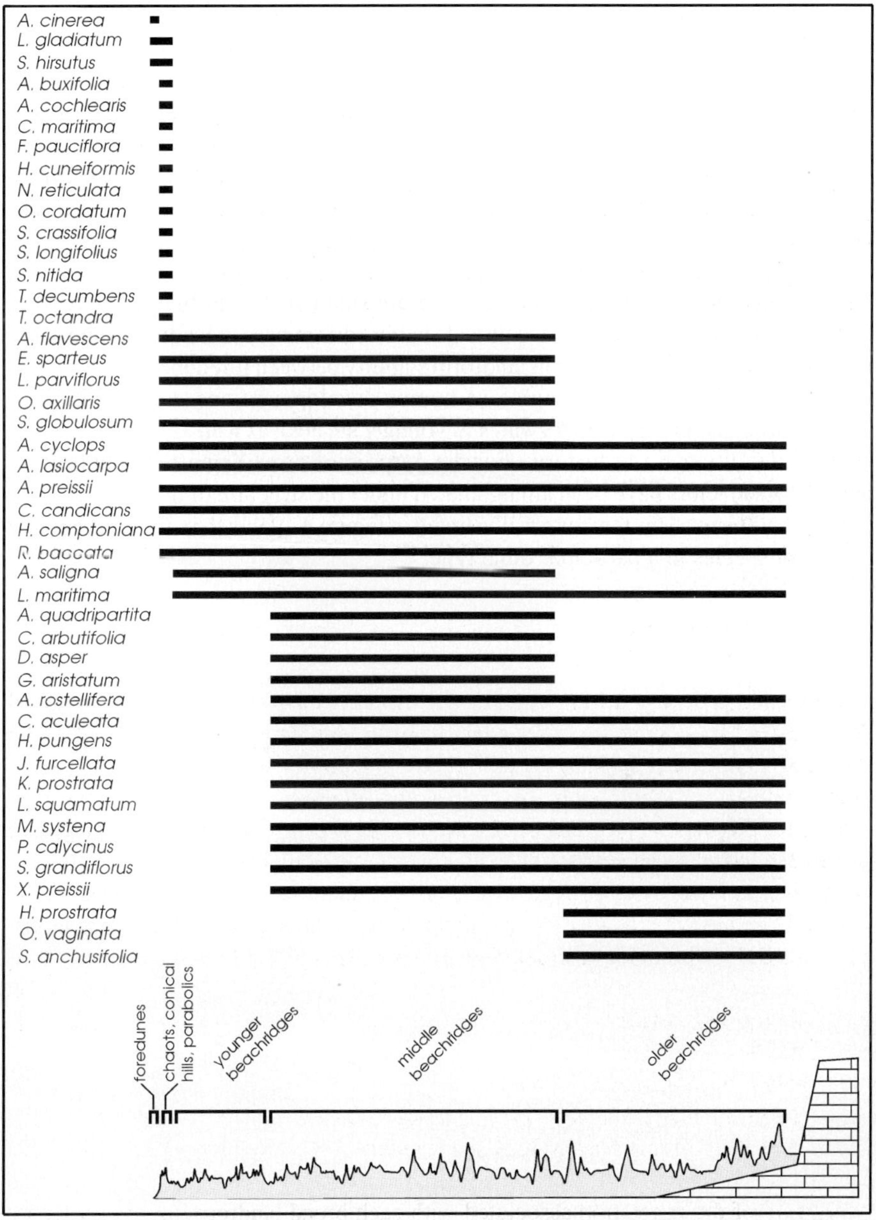

Figure 3-12. *Distribution of main dune/beachridge plant species across the beachridge plain, ordered in terms of relationship to antiquity of Holocene landforms.*

Generally the foredunes are colonised by *Spinifex hirsutis, Tetragonia decumbens* and *Cakile maritima*, the parabolic dunes by *Spyridium globulosum*, *Alyxia buxifolia* and *Acanthocarpus preissii*, the chaots by *S. globulosum*, *Rhagodia baccata*, and *A. preissii,* the more recent beachridges by *Olearia axillaris*, *Acacia cyclops* and *Spinifex longifolius*, the older beachridges by *Jacksonia furcellata*, *Melaleuca systena* and *Hakea prostrata*, and areas disturbed by fire by *A. cyclops* and *Acacia rostellifera*. Wetlands present a distinct suite of habitats and their vegetation assemblages are rich.

Vegetation assemblages of the Becher Cusp are small scale and the distribution is complex. The variability reflects small scale habitat differences in landforms and soils; dune crest, dune swale, variation in landform stability between foredunes, beachridges, chaots, and parabolic dunes, degree of humus development and concentration of calcium carbonate in soils (older swales *vs* younger swales), as well as wind exposure, water availability and fire history. For the purpose of providing a regional setting, many of these factors have been amalgamated under the structure of increasing age of the dune landforms. This structure is illustrated in Figure 3-12 which indicates presence/ absence of species in a particular dune type.

Table 3.2 *A comparison with a previous study of the relationships between landform and vegetation associations*

Soil-Landform Unit (Wells et al. 1985)	*Vegetation assemblages (Trudgen 1988)*	*Geomorphic term Semeniuk et al. 1989 (this study)*	*Vegetation assemblages (this study) structure and dominant species*
Active foredunes	Open herbland:- *Arctotheca calendula, Cakile maritima*	Foredunes	Open herbland:- *A. calendula, C. maritima*
Contemporary relict foredune plain	Low open to closed heath to open shrubland:- *Scaevola crassifolia* Open shrubland to open heath:-*Olearia axillaris* Hummock grassland:- *Spinifex longifolius, Tetragonia decumbens* Closed heath to closed scrub:-*O. axillaris*	Low relief chaots	Low dune heath:-*S. longifolius, O. axllaris* Thickets:-*Acacia cyclops* Low dune heath:- *T. decumbens, Acanthocarpus preissii, Lomandra maritima*
Relict foredunes forming a plain	Complex:-*Acacia rostellifera*	Low disrupted shore parallel ridge systems	Scrub:-*A. rostellifera*
Relict foredunes forming a plain	Complex:- *Jacksonia furcellata, Acacia saligna* Complex:- *J. furcellata, O. axillaris* Complex:- *A. rostellifera* Complex:- *O. axillaris, Melaleuca systena* Complex:- *Acacia lasiocarpa*	Low continuous shore parallel ridge systems	Open to shrubland to low heath:- *J. furcellata, M. systena, Rhagodia baccata, A. lasiocarpa, A. cyclops, A. saligna, Exocarpus sparteus, Spyridium globulosum, Conostylis aculeata, Lepidosperma squamatum, L. maritima, Schoenus grandiflorus, Desmocladus asper, A. preissii, Adriana quadripartita., Phyllanthus calycinus*

Table 3.2 (cont.)

Table 3.2 (cont.)

Soil-Landform Unit (Wells et al. 1985)	*Vegetation assemblages (Trudgen 1988)*	*Geomorphic term Semeniuk et al. 1989 (this study)*	*Vegetation assemblages (this study) structure and dominant species*
Active foredunes, nested parabolic dune complex, deflation basin, active sand sheet		Attenuated and fretted parabolic dunes: elements include ridges, deflation basins, high relief chaots, undifferentiated blowouts	Open low shrubland over dune heath:-*S. globulosum* Mosaic of scattered open low shrubland and dune heath:- *A. preissii, Alyxia buxifolia*
Wetlands	Variable wetland assemblages	Flat	Low forest:-*Melaleuca rhaphiophylla* Heath:-*Melaleuca teretifolia* Sedgelands:- *Isolepis nodosa, Juncus kraussii, Baumea juncea, B. articulata, L. gladiatum*

Table 3.3 *Description of vegetation in relation to dune type*

Landform	*Vegetation Structure*	*Dominant species*	*Type of dominance*	*Common understorey components*	*Components specific to the assemblage*
Warnbro foredune	Open low shrubland and sedgeland	*Spinifex hirsutus, Lepidosperma gladiatum*	consistent	*Olearia axillaris* *Acacia cyclops* *Trachyandra divaricata*	*Atriplex cinerea*
Warnbro chaots - dune crest	Mosaic of open low shrubland, closed sedgeland and low dune heath	*Acacia cochlearis* *Scaevola crassifolia* *Cakile maritima* *Spinifex longifolius*	variable	*Alyxia buxifolia* *O. axillaris, L. gladiatum*	*S. crassifolia* *C. maritima*
eastern slope		*Acanthocarpus preissii* *Spyridium globulosum* *Rhagodia baccata*	consistent	*Austrostipa flavescens, Frankenia pauciflora, Hardenbergia comptoniana, Acacia lasiocarpa*	*F. pauciflora*
Warnbro beachridges	Open dune shrubland	*A. cyclops,* *A. cochlearis* *S. globulosum,* *A. buxifolia*	consistent	*A. preissii* *Lomandra maritima* *A. flavescens*	*Desmocladus asper* *Leptosperma squamatum* *Hemiandra pungens, Leucopogon parviflorus, Phyllanthus calycinus, Gompholobium aristatum, Cryptandra arbutiflora, Conostylis aculeata*

Table 3.3 (cont.)

Table 3.3 (cont.)

Landform	*Vegetation Structure*	*Dominant species*	*Type of dominance*	*Common understorey components*	*Components specific to the assemblage*
Becher older beachridges	Thickets or low open shrubland	*A. cyclops, A. rostellifera Jacksonia furcellata Melaleuca systena, Hakea prostrata Xanthorrhoea preissii*	consistent	*Conostylis candicans C. aculeata, A. flavescens L. squamatum, L. maritima, H. comptoniana, H. pungens D. asper, A. preissii*	*C. candicans, X. preissii C. aculeata, H. prostrata A. rostellifera, Scaevola anchusifolia Opercularia vaginata*
Becher middle beachridges	Open to very open shrubland over low heath	*J. furcellata M. systena, R. baccata A. lasiocarpa A. cyclops, A. saligna Exocarpus sparteus S. globulosum*	consistent	*Trachyandra divericata Pelargonium capitatum*	*E. sparteus, M. systena*
		C. aculeata, L. maritima L. squamatum Schoenus grandiflorus D. asper, A. preissii Adriana quadripartita P. calycinus Kennedia prostrata	consistent		*C. aculeata, L. squamatum S. grandiflorus, A. quadripartita K. prostrata*
Becher chaots - south	Low dune heath	*Tetragonia decumbens A. preissii, L. maritima A. cyclops*	consistent		

Table 3.3 (cont.)

Table 3.3 (cont.)

Landform	*Vegetation Structure*	*Dominant species*	*Type of dominance*	*Common understorey components*	*Components specific to the assemblage*
Becher chaots eastern slopes	Low dune heath	*A. preissii, R. baccata* *S. globulosum* *L. maritima,* *L. gladiatum*	consistent	*H. comptoniana* *C. maritima*	
Becher chaots - north	Low dune heath	*S. longifolius* *O. axillaris, A. cyclops*	consistent		
Becher foredune	Low dune heath	*S. longifolius,* *S. hirsutus*	consistent	*O. axillaris, A. cyclops*	
Secret Harbour chaots	Low dune heath	*T. decumbens* *S. hirsutis,* *C. maritima*	constant		*C. maritima*
50 m	low open shrubland and low dune heath	*O. axillaris, A. cyclops* *T. decumbens* *S. longifolius*			
100 m	low open shrubland, low heath, sedgeland	*R. baccata, A. preissii* *L. gladiatum* *S. globulosum*			*S. crassifolia*

Table 3.3 (cont.)

Table 3.3 (cont.)

Landform	*Vegetation Structure*	*Dominant species*	*Type of dominance*	*Common understorey components*	*Components specific to the assemblage*
Secret Harbour chaots 200 m	low open shrubland	*S. globulosum*	constant	*A. cyclops* *A. buxifolia*	*S. crassifolia*
300 m	low open shrubland and low heath	*A. rostellifera, A. preissii A. cyclops, C. candicans*			
Secret Harbour parabolic dunes - bowl	Open low shrubland over heath	*S. globulosum*	constant	*R. baccata, O. axillaris* *L. gladiatum, S. crassifolia* *A. buxifolia, A. preissii* *L. parviflorus, C. candicans* *H. comptoniana* *A. cyclops, E. sparteus* *A. flavescens*	*Hibbertia cuneiformis* *Nemcia reticulata* *Tetraria octandra*
Secret Harbour parabolic dunes - ridges	Mosaic of scattered open low shrubland and dune heath	*A. preissii* *A. buxifolia*	variable	*Ozothamnus cordatum* *Scaevola nitida* *S. crassifolia, C. maritima* *L. parviflorus, E. sparteus*	

4. WETLAND DESCRIPTIONS

4.1 General introduction

While the main body of this book describes and interprets the various geomorphic, stratigraphic, hydrologic, hydrochemical and botanical features, spatial and temporal patterns, and processes of the Becher Suite wetlands, this chapter provides a basic description of these wetlands, their simplified stratigraphy, and their radiocarbon ages. The wetlands, in terms of wetland type, geometry and scale, underlying stratigraphy, and aspects of hydrology and vegetation are described to a level sufficient to differentiate each one and to provide for the reader a systematic overview of wetland attributes (Figs. 4-1 to 4-21). Aspects of wetlands that relate to temporal changes (*e.g.*, seasonal variation in water levels, salinity and cation content), and mathematical treatment of vegetation units are treated in more detailed later chapters.

Eighteen wetlands of the Becher Suite formed the core of the research; 3 wetlands in the inter-ridge depressions of the Cooloongup/Walyungup spit barrier, and one of the barrier wetlands at the tip of Becher Cusp were used as adjuncts to the study. Study wetlands in the Becher Suite are listed below; numbers on a single line refer to wetlands within the same (sub-regional) swale:

- 161, 162, 163, WAWA
- 135, 136, 142
- 72
- 63
- 45
- 9-3, 9-6, 9-14, 35
- 1N, swi, swii, swiii

The four supplementary wetlands are:

- Cooloongup A
- Cooloongup B
- Cooloongup C
- BP-1

Site 161

Wetland Type: Sumpland

GEOMORPHIC SETTING
Concave gently undulating inter-ridge depression whose surface is at 3.6 m AHD.

WETLAND GEOMETRY AND SCALE
Microscale, triangular, steep sided basin

Location of wetland

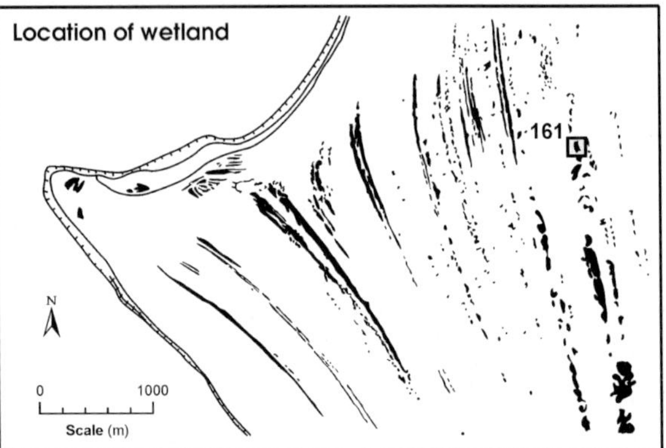

STRATIGRAPHY
Adjoining beachridges are underlain by medium shelly and quartz sand grading into coarse, medium, and fine shelly and quartz sand. The wetland is underlain by root structured and burrow mottled sandy mud, mud and muddy sand grading into coarse, medium, and fine shelly and quartz sand. Granulometric and chemical analyses show the shelly component of the sand in the wetland to be constant down profile but fairly low i.e., from 12-19%, and the mud, to vary from a ratio of 1:3 to 1:7 (by weight) carbon to calcium carbonate.

HYDROLOGY
The groundwater table moves between the mud aquifer to approximately 0.5 m above the surface of the wetland i.e., between 2.7 - 4.1 m AHD. The longer term hydrograph showed winter levels were at their lowest in 1993, 1995 and 1997 while the water table reached its overall minimum positions regularly in autumn between 1995, 1996, and 1998. Groundwater levels under the adjacent beachridge are not always synchronous with those in the wetland, at the monthly incremental measuring period. The salinity of the groundwater is fresh. The soil water content in the top 10 cm varies seasonally, ranging from low to high 5-80%. Down profile, it appears to be relatively constant until it reaches the muddy sand when it falls and restabilises at a lower level in the coarse sand. Soil water salinity ranges from 600-7,500 ppm, with most of the samples registering subhaline values.

VEGETATION
Vegetation pattern and form are gradiform. There are four discontinuous zones each containing one assemblage. From the centre outwards, they are sedgeland, herbland with clumps of mixed sedge, open heath, and heath. The sedge comprises *Baumea articulata*, the herbland *Centella asiatica* with *Baumea juncea*, *Lepidosperma gracile* and *Isolepis nodosa*, (grass trees), the open heath *Acacia saligna*, and the heath *Xanthorrhoea preissii*.

Figure 4-1. *Description of wetland 161.*

Site 162

Wetland Type: Sumpland

GEOMORPHIC SETTING
Concave gently undulating inter-ridge depression whose surface is at 3.8 m AHD.

WETLAND GEOMETRY AND SCALE
Microscale, crescentic, steep sided, basin

Location of wetland

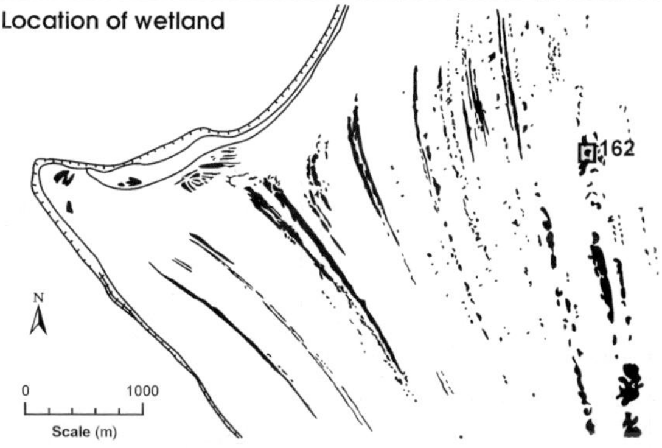

STRATIGRAPHY

Adjoining beachridges are underlain by medium shelly and quartz sand grading into coarse, medium, and fine shelly and quartz sand. The wetland is underlain by texture mottled mud and muddy sand grading into coarse, medium, and fine shelly and quartz sand.

Granulometric and chemical analyses show the shelly component of the sand in the wetland to be increasing down profile but to be fairly low i.e., from 21-32%, and the mud, to vary from a ratio of 1:6 to 1:11 (by weight) carbon to calcium carbonate.

HYDROLOGY

The groundwater table moves between the coarse sand aquifer to the mud, to 0.4 m above the surface of the wetland i.e., between 2.7 - 4.3 m AHD. The longer term hydrograph showed winter levels were at their lowest in 1993, 1997 and 1998 while the water table reached its overall minimum positions regularly in autumn between 1995, 1996, and 1998. Groundwater levels under the adjacent beachridge are not synchronous with those in the wetland, at the monthly incremental measuring period.

The salinity of the groundwater is fresh to subhaline.

The soil water content in the top 10 cm varies seasonally, ranging from low to high 5-90%.

Soil water salinity ranges from 700-12,500 ppm, with most of the samples registering subhaline values. The maximum concentrations are in December and September/October.

VEGETATION

Vegetation pattern and form are concentriform. From the centre outwards, it comprises scrub, herbland, open heath and heath. The scrub is *Melaleuca teretifolia*, the herbland is *Centella asiatica*, the open heath is *Acacia saligna* and the outer heath is *Xanthorrhoea preissii* (grass trees).

Figure 4-2. *Description of wetland 162.*

Site 163

Wetland Type: Dampland

GEOMORPHIC SETTING

Concave gently undulating inter-ridge depression whose surface is at 4.0 m AHD.

WETLAND GEOMETRY AND SCALE

Microscale, irregular, shallow basin

Location of wetland

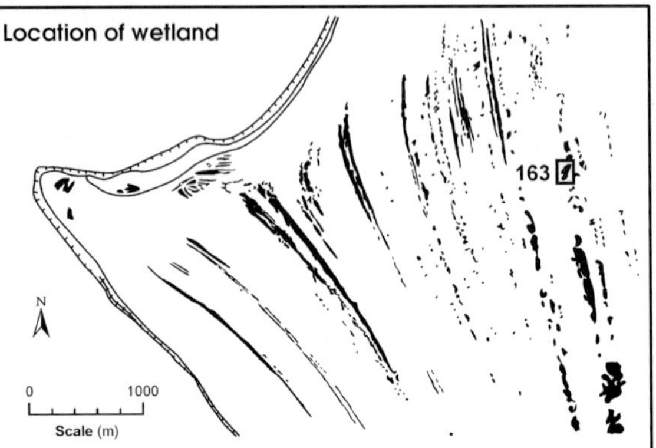

STRATIGRAPHY

Adjoining beachridges are underlain by medium shelly and quartz sand grading into coarse, medium, and fine shelly and quartz sand. The wetland is underlain by root structured and burrow mottled mud and muddy sand grading into coarse, medium, and fine shelly and quartz sand.

Granulometric and chemical analyses show the shelly component of the sand in the wetland to be increasing down profile but to be fairly low i.e., from 4-21%, and the mud, to vary from a ratio of 1:4 to 1:8 (by weight) carbon to calcium carbonate.

HYDROLOGY

The groundwater table moves between the coarse sand aquifer to the wetland surface i.e., between 2.7 - 4.0 m AHD. The longer term hydrograph showed winter levels were at their lowest in 1993 and 1997 while the water table reached its overall minimum positions regularly in autumn between 1994 and 1996. Groundwater levels under the adjacent beachridge are synchronous with those in the wetland, at the monthly incremental measuring period.

The salinity of the groundwater is fresh, to weakly subhaline.

The soil water content in the top 10 cm varies seasonally, ranging from low to medium 5-70%.

Soil water salinity ranges from 650-29,000 ppm, with many of the samples registering subhaline to hyposaline values. The maximum concentrations are in December, March and occasionally in September.

VEGETATION

Vegetation pattern and form are maculiform. There are four assemblages: rushland, open sedgeland, herbland, and open heath. The rush is *Juncus kraussii*, the sedge is *Baumea juncea*. The herb is *Centella asiatica* and the open heath comprises *Xanthorrhoea preissii* and *Acacia saligna*.

Figure 4-3. *Description of wetland 163.*

Site WAWA

Wetland Type: Sumpland

GEOMORPHIC SETTING

Concave gently undulating inter-ridge depression whose surface is at 3.2 m with a western bench at 3.7 m AHD.

WETLAND GEOMETRY AND SCALE

Mesoscale, irregularly elongate, basin

Location of wetland

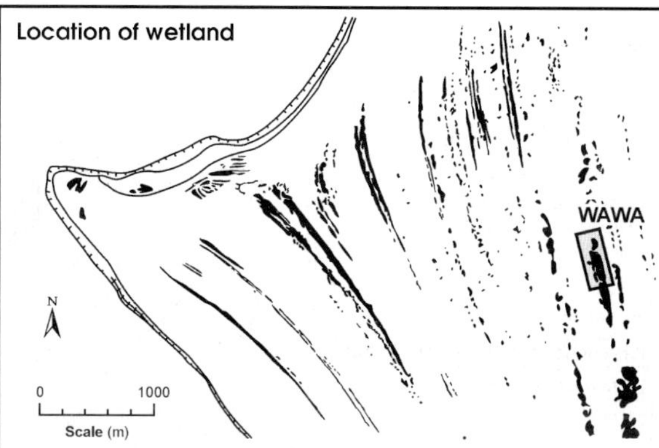

STRATIGRAPHY

Adjoining beachridges are underlain by medium shelly and quartz sand grading into coarse, medium, and fine shelly and quartz sand. The wetland is underlain by homogeneous peat, and peaty sand grading into coarse, medium, and fine shelly and quartz sand.

Granulometric and chemical analyses show the composition of the sand in the wetland to contain virtually no shell other than fragments of Gyraulus sp., but to contain 42%-82% quartz. The mud composition has a 6:1 ratio of carbon to calcium carbonate which reverses at a

HYDROLOGY

The groundwater table moves between the peat aquifer to approximately 0.3 m above the surface of the wetland i.e., between 2.3 - 3.5 m AHD. The longer term hydrograph showed winter levels were at their lowest in 1997 and 1998 while the water table reached its overall minimum positions regularly in autumn between 1998 and 1999. Groundwater levels under the adjacent beachridges are generally synchronous with those in the wetland, at the monthly incremental measuring period.

The salinity of the groundwater is fresh.

The soil water content in the top 10 cm varies seasonally, ranging from low to high 20-80%. Down profile, it appears to be relatively constant until reaching the peaty sand where it begins to fluctuate. In the carbonate muddy sand, it drops sharply and then stabilises at a lower level in the coarse, medium sand.

Soil water salinity ranges from 900-16,000 ppm, with many of the samples registering subhaline to weakly hyposaline values. The maximum concentrations are in March.

VEGETATION

Vegetation pattern and form are maculiform. There are seven assemblages: three sedgelands, herbland, open to closed heath, scrub and low forest. The dominant sedgeland comprises *Baumea articulata* with a minor component of *Schoenoplectus vallidus*. The other sedgelands comprise *Lepidosperma gladiatum* and *Baumea juncea*. The herb is *Centella asiatica*, the closed heath is *Melaleuca viminea*, the shrubland and the low forest are composed of *M. teretifolia* and *M. rhaphiophylla* respectively.

Figure 4-4. *Description of wetland WAWA.*

Site 135

Wetland Type: Dampland

GEOMORPHIC SETTING

Concave to gently undulating inter-ridge depression whose surface is at 3.44 m AHD.

WETLAND GEOMETRY AND SCALE

Microscale, nine shaped, steep sided basin

Location of wetland

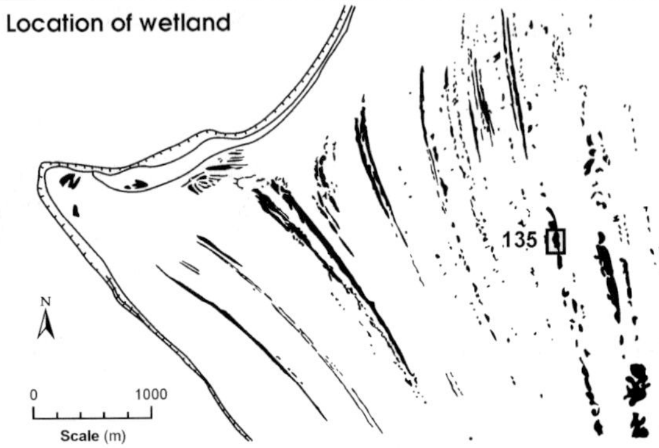

STRATIGRAPHY

Adjoining beachridges are underlain by medium shelly and quartz sand grading into coarse, medium, and fine shelly and quartz sand. The wetland is underlain by root structured and colour mottled mud and muddy sand grading into coarse, medium, and fine shelly and quartz sand.

Granulometric and chemical analyses show the shelly component of the sand in the wetland to increase down profile from 12-44%, and the mud, to range from a ratio of 1:11 to 1:50 (by

HYDROLOGY

The groundwater table moves between the coarse sand aquifer and the wetland surface i.e., between 2.4 - 3.8 m AHD. The longer term hydrograph showed winter levels were at their lowest in 1993 while the water table reached its overall minimum positions in autumn 1995, 1996 and 1998. Groundwater levels under the adjacent beachridge are generally slightly above those in the wetland, at the monthly incremental measuring period.

The salinity of the groundwater ranges from fresh to weakly hyposaline.

The soil water content in the top 10 cm varies seasonally, ranging from low to medium 25-55%. Down profile, it falls in the muddy sand layer and then stabilises in the coarse sand.

Soil water salinity ranges from 1,800-61,000 ppm, with most of the samples registering hyposaline values.

VEGETATION

Vegetation pattern and form are gradiform. There are two assemblages: an inner zone of layered woodland to low forest of *Melaleuca rhaphiophylla* with an understorey of *Juncus kraussii* rushland and *Centella asiatica* herbland and an outer discontinuous zone of *Xanthorrhoea preissii* heath.

Figure 4-5. *Description of wetland 135.*

Site 136

Wetland Type: Dampland

GEOMORPHIC SETTING
Concave gently undulating inter-ridge depression whose surface is at 3.7 m AHD.

WETLAND GEOMETRY AND SCALE
Microscale, hourglass shaped basin

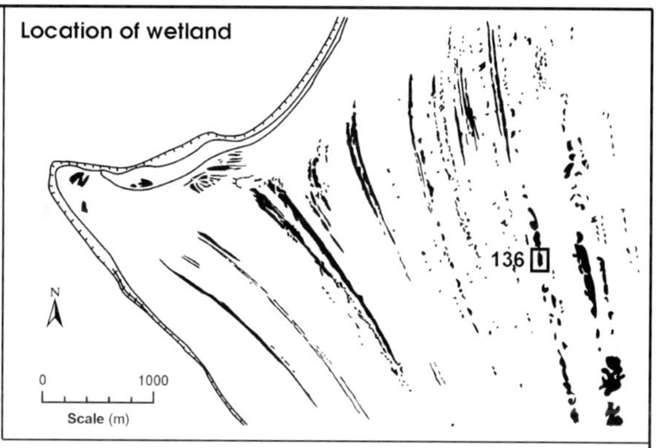

STRATIGRAPHY
Adjoining beachridges are underlain by medium shelly and quartz sand grading into coarse, medium, and fine shelly and quartz sand. The wetland is underlain by colour mottled and texture mottled mud and muddy sand grading into medium and fine shelly and quartz sand.

HYDROLOGY

The groundwater table moves between the coarse sand aquifer and the top of the muddy sand i.e., between 2.3 - 3.7 m AHD. The longer term hydrograph showed winter levels were at their lowest in 1997-8 while the water table reached its overall minimum positions in autumn 1995 and 1998. Groundwater levels under the adjacent beachridge are synchronous with those in the wetland, at the monthly incremental measuring period.

The salinity of the groundwater is fresh to subhaline.

The soil water content in the top 10 cm varies seasonally, ranging from low to medium 15-60%.

Soil water salinity ranges from 3,000-28,000 ppm, with most of the samples registering hyposaline values.

VEGETATION

Vegetation pattern and form are gradiform. There are two assemblages: an inner zone of low forest of *Melaleuca rhaphiophylla* with an understorey of *Juncus kraussii* rushland and *Centella asiatica* herbland with an outer discontinuous zone of *Xanthorrhoea preissii* heath.

Figure 4-6. *Description of wetland 136.*

Site 63

Wetland Type: Dampland

GEOMORPHIC SETTING
Concave inter-ridge hollow whose surface is at 3.27 m

WETLAND GEOMETRY AND SCALE
Leptoscale, circular, steep sided basin

Location of wetland

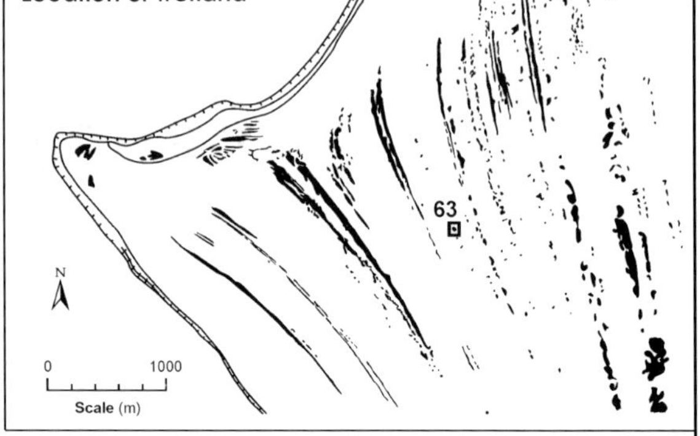

STRATIGRAPHY

Adjoining beachridges are underlain by medium shelly and quartz sand. The wetland is underlain by texture mottled sandy mud and muddy sand grading into medium and fine shelly and quartz sand.

Granulometric and chemical analyses show the shelly component of the sand in the wetland to increase down profile from 23-50%, and the mud, to change from a ratio of 1:2 to 1:7 (by weight) carbon to calcium carbonate.

HYDROLOGY

The groundwater table moves between the coarse sand aquifer to the base of the sandy mud i.e., between 2.0 - 3.0 m AHD. The longer term hydrograph showed winter levels were at their lowest in 1993 while the water table reached its overall minimum position in autumn 1995. The longer term hydrograph is incomplete due to the destruction of the wetland in 1998. Groundwater levels under the adjacent beachridges are synchronous with those in the wetlands, at the monthly incremental measuring period.

The salinity of the groundwater is fresh to weakly subhaline.

The soil water content in the top 10 cm varies seasonally, ranging from low to medium 5-55%. Down profile, it falls significantly in the first 20 cm, and thereafter oscillates around a lower level of 30%.

Soil water salinity ranges from 1,300-36,000 ppm, with most of the samples registering subhaline to hyposaline values.

VEGETATION

Vegetation pattern and form are concentriform. There are four zones. From the centre outwards, they are herbland, sedgeland, and heaths. The herbland is composed of *Centella asiatica*, the sedgeland is *Baumea juncea*, with minor *Isolepis nodosa*, and the heaths are composed of *Xanthorrhoea preissii*, or species of *Acacia* (*A. pulchella* and *A. saligna*).

Figure 4-9. *Description of wetland 63.*

Site 45

Wetland Type: Sumpland

GEOMORPHIC SETTING
Concave inter-ridge depression whose surface is at 1.97 m AHD.

WETLAND GEOMETRY AND SCALE
Leptoscale, narrow, crescentic, steep sided basin

Location of wetland

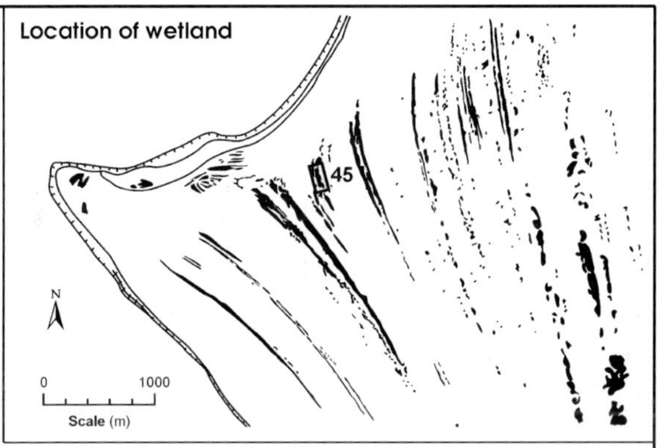

STRATIGRAPHY

Adjoining beachridges are underlain by medium shelly and quartz sand. The wetland is underlain by root structured sandy mud and homogeneous muddy sand grading into medium and fine shelly and quartz sand.

Granulometric and chemical analyses show the composition of the sand in the wetland to change from shell dominated to approximately 50% shell, 50% quartz at the base of the muddy sand. The composition of the mud is carbon and calcium carbonate in a ratio of 1:3 which increases between 40-60 cm depth to 1: 2.5.

HYDROLOGY

The groundwater table moves between the coarse sand aquifer through the muddy sand to the top of the mud i.e., between 0.7 - 2.0 m AHD. Winter levels were at their lowest in 1993. No longer term hydrograph is available due to the destruction of the wetland in 1996. Groundwater levels under the adjacent beachridges are not always synchronous with those in the wetlands, at the monthly incremental measuring period.

The salinity of the groundwater is fresh to weakly hyposaline.

The soil water content in the top 10 cm varies seasonally, ranging from low to high 10-80%. Down profile, it appears to be relatively constant.

Soil water salinity ranges from 760-33,000 ppm, with most of the samples registering hyposaline values.

VEGETATION

Vegetation pattern and form are latiform. There is one assemblage comprising layered woodland to low forest with an understorey of patches of low heath, rushland, sedgeland and herbland. The woodland consists of *Melaleuca rhaphiophylla* the low heath *M. viminea*, the rushland *Juncus kraussii*, the sedge *I. nodosa*, and the herbland is *Centella asiatica*.

Figure 4-10. *Description of wetland 45.*

Site 35

Wetland Type: Sumpland

GEOMORPHIC SETTING
Slightly concave gently undulating inter-ridge depression whose surface lies between 1.90 m and 2.12 m AHD.

WETLAND GEOMETRY AND SCALE
Leptoscale, narrow, linear, shallow basin

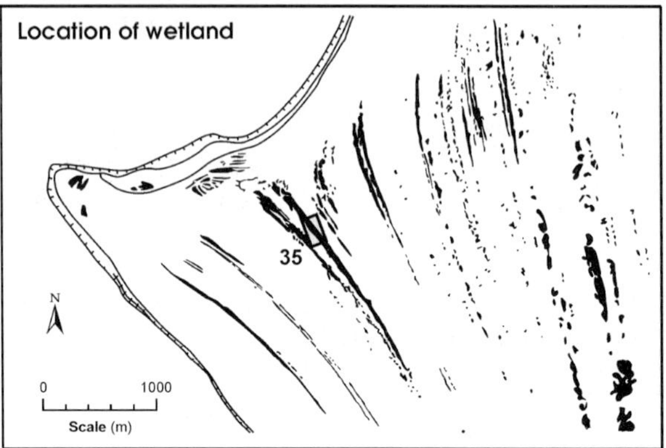

STRATIGRAPHY
Adjoining beachridges are underlain by medium shelly and quartz sand. The wetland is underlain by mud and colour mottled and texture mottled muddy sand. Granulometric and chemical analyses show the composition of the sandy mud and muddy sand in the wetland to increase down profile from 15% to 61% shell material and the mud, and to vary between a 1:3 and 1:5 ratio of carbon to calcium carbonate.

HYDROLOGY
The groundwater table moves between the coarse sand aquifer through the muddy sand and mud to approximately 10 cm above the surface, i.e., between 0.6 - 2.0 m AHD. The longer term hydrograph showed winter levels were at their lowest in 1993 while the water table reached its overall minimum position in autumn 1995. Since 1995 there has been an upturn in water levels. Groundwater levels under the adjacent beachridges are synchronous with those in the wetlands, at the monthly incremental measuring period. The salinity of the groundwater is fresh to subhaline. The soil water content in the top 10 cm varies seasonally, ranging from low to high 5-70%. Down profile, it decreases to 40% with marked changes occurring at the boundaries of the sedimentary layers. Soil water salinity ranges from 1,800-73,000 ppm, with many of the samples registering hyposaline or hypersaline values.

VEGETATION
Vegetation pattern and form are maculiform. There are six assemblages: low forest, scrublands, low heath, rushland and herbland. The low forest contains the freshwater paperbark *Melaleuca rhaphiophylla* and saltwater paperbark *M. cuticularis*, with an understory of *I. nodosa* or *C. asiatica*, the scrublands are composed of either *M. teretifolia* or *Acacia cyclops*, the low heath is *M. viminea*, the rushland is *Juncus kraussii*, and the herbland is *Centella asiatica*.

Figure 4-11. *Description of wetland 35.*

Site 9-1,2,3

Wetland Type: Sumpland

GEOMORPHIC SETTING

Slightly concave to gently undulating inter-ridge depression whose surface is at 1.45 m AHD.

WETLAND GEOMETRY AND SCALE

Leptoscale, narrow, meandering, shallow basin

Location of wetland

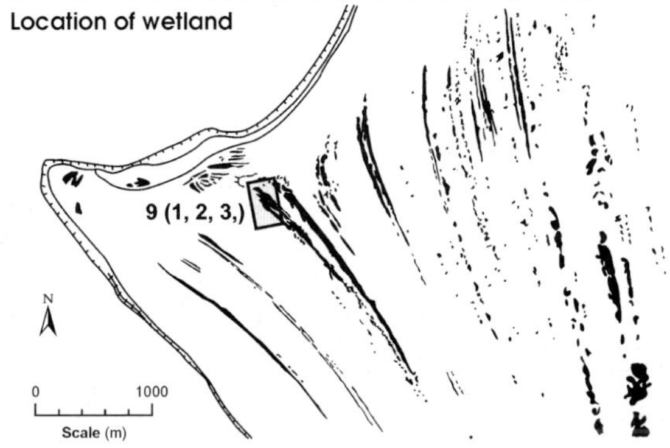

STRATIGRAPHY

Adjoining beachridges are underlain by medium shelly and quartz sand grading into coarse, medium, and fine shelly and quartz sand. The wetland is underlain by root structured and burrow mottled muddy sand with embedded calcrete lenses.

HYDROLOGY

The groundwater table moves between the medium sand aquifer through the muddy sand to approximately 10 cm above the surface i.e., between 0.4 - 1.5 m AHD. The longer term hydrograph showed winter levels were at their lowest in 1993 while the water table reached its overall minimum positions in autumn 1994, and 1995. No clear trend in water levels has occurred since these troughs. Short term perching of winter rains occurs above the muddy sands and infiltration of rain is retarded by the thin sheet of calcrete.

The salinities of the surface water and groundwater are fresh, the salinity of the perched water is subhaline to hyposaline.

The soil water content in the top 10 cm varies seasonally, ranging from low to medium 3-50%.

Soil water salinity ranges from 1,100-418,000 ppm, with many of the samples registering mesosaline and hypersaline values.

VEGETATION

Vegetation pattern and form are latiform. It comprises sedgeland with one emergent low tree (*M. rhaphiophylla*). The dominant species is *Baumea juncea*, with minor *Isolepis nodosa*, and *Sporobilis virginicus*.

Figure 4-12. *Description of wetland 9-1,2,3.*

Site 9-4,5,6,7

Wetland Type: Sumpland

GEOMORPHIC SETTING
Concave gently undulating inter-ridge depression whose surface is at 1.22 m AHD.

WETLAND GEOMETRY AND SCALE
Leptoscale, narrow, ellipsoid, steep sided basin

Location of wetland

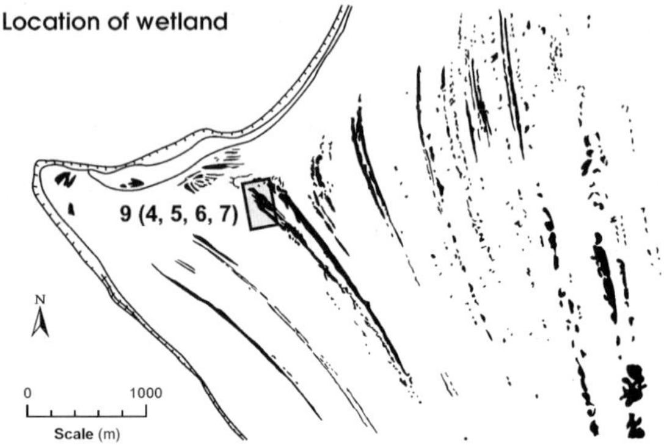

STRATIGRAPHY
Adjoining beachridges are underlain by medium shelly and quartz sand grading into coarse, medium, and fine shelly and quartz sand. The wetland is underlain by colour mottled mud, and sandy mud with Gyraulus sp. shell, grading into a thin layer of muddy sand, calcrete nodules and coarse, medium, and fine shelly and quartz sand. Granulometric and chemical analyses show the composition of the sand in the wetland to be 20% (by weight) Gyraulus sp. shell material and the mud, to comprise a 1:2 alternating with a 1:10 ratio of carbon to calcium carbonate.

HYDROLOGY
The groundwater table moves between the coarse sand aquifer through the mud to approximately 35 cm above the surface i.e., between 0.4 - 1.5 m AHD. The longer term hydrograph showed winter levels were at their lowest in 1993 while the water table reached its overall minimum positions in autumn 1994 and 1995. A slight upturn in water levels is evident. Groundwater levels under the adjacent beachridges are synchronous with those in the wetlands, at the monthly incremental measuring period. The salinity of the surface water and groundwater is fresh, becoming very slightly subhaline in the hottest months. The soil water content in the top 10 cm varies seasonally, ranging from medium to high 20-75% . Down profile, it falls quite sharply within the first 10 cm and thereafter tends to stabilise around 35%. Soil water salinity ranges from 1,000-33,000 ppm, with most of the samples registering subhaline to hyposaline values.

VEGETATION
Vegetation pattern and form are maculiform to concentriform. There are three zones. From the centre outwards, they are herbland, sedgeland, and heath. The herbland is composed of *Centella asiatica*, the sedgeland is *Baumea juncea*, with minor *Isolepis nodosa*, and the heath is composed of *Melaleuca viminea*.

Figure 4-13. *Description of wetland 9-4,5,6,7.*

Site 9-9,10,11,12,14

Wetland Type: Dampland

GEOMORPHIC SETTING

Slightly concave to undulating inter-ridge depression whose surface lies between 1.33 m and 1.66 m AHD.

WETLAND GEOMETRY AND SCALE

Leptoscale, ellipsoid, shallow basin

Location of wetland

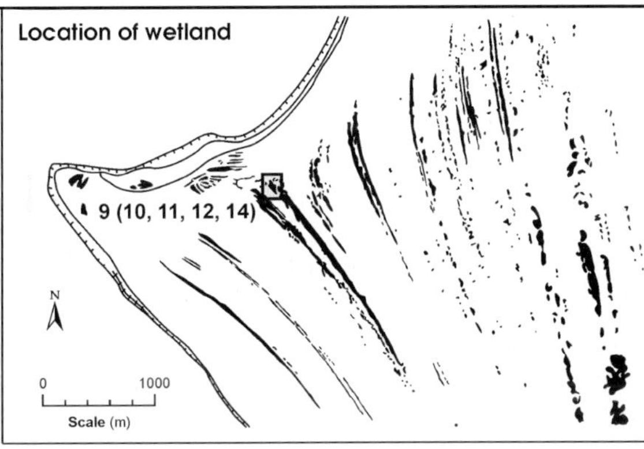

STRATIGRAPHY

Adjoining beachridges are underlain by medium shelly and quartz sand. The wetland is underlain by root structured and burrow mottled mud and muddy sand, grading into coarse, medium, and fine shelly and quartz sand.

Granulometric and chemical analyses show the composition of the sand in the wetland to be approximately 50% shell, 25% quartz, and the mud, to comprise a 1:9 ratio of carbon to calcium carbonate.

HYDROLOGY

The groundwater table moves between the coarse sand aquifer through the muddy sand to the base of the mud, or in places, to approximately 20 cm above the surface i.e., between 0.4 - 1.7 m AHD. The longer term hydrograph showed winter levels were at their lowest in 1993 while the water table reached its overall minimum position in autumn 1995. Since 1995 there has been an upturn in water levels. Groundwater levels under the adjacent beachridges are synchronous with those in the wetlands, at the monthly incremental measuring period.

The salinity of the groundwater is fresh to weakly subhaline.

The soil water content in the top 10 cm varies seasonally, ranging from low to medium 10-60%. Down profile, there is a change at the upper boundary of the muddy sand and thereafter, the content oscillates within 10%.

Soil water salinity ranges from 1,000-18,000 ppm, with most of the samples registering subhaline to hyposaline values.

VEGETATION

Vegetation pattern and form are maculiform. There are three assemblages. They are herbland, rushland and sedgeland. The herb is *Centella asiatica*, the rush *Juncus kraussii* and the sedge is *Baumea juncea*.

Figure 4-14. *Description of wetland 9-9,10,11,12,14.*

Site swi

Wetland Type: Dampland

GEOMORPHIC SETTING
Concave inter-ridge depression with eastern bench. The basin floor lies at 1.49 m AHD, and the bench at 1.59m AHD.

WETLAND GEOMETRY AND SCALE
Leptoscale, narrow, linear, shallow basin

Location of wetland

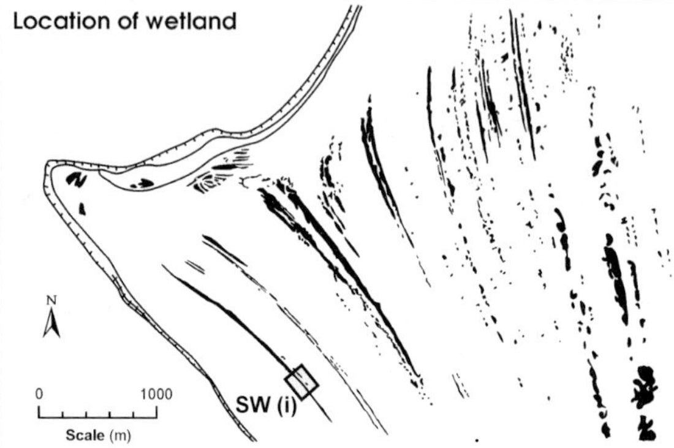

STRATIGRAPHY
Adjoining beachridges are underlain by medium shelly and quartz sand. The wetland is underlain by root structured and burrow mottled muddy sand grading into medium and fine shelly and quartz sand. Granulometric and chemical analyses show the composition of the sand in the wetland to be 49%-54% shelly material and the mud, to vary between a ratio (by weight) of 1:2 to 1:5 carbon to calcium carbonate.

HYDROLOGY
The groundwater table moves between the medium sand aquifer and the muddy sand, 0.6 - 1.5 m AHD. The longer term hydrograph showed winter levels were at their lowest in 1993 and 1997 while the water table reached its overall minimum position in autumn 1996. Since 1996 there has been a slight upturn. Groundwater levels under the adjacent beachridges are synchronous with those in the wetlands, at the monthly incremental measuring period. The salinity of the groundwater is fresh, except after the first rains at the beginning of winter. The groundwater then becomes sub-haline. The soil water content in the top 10 cm varies seasonally, ranging from medium to high i.e., 50-80% . Down profile, it is lower (30%), but constant. Soil water salinity ranges from 880-177,000 ppm, with most of the samples registering subhaline values.

VEGETATION
Vegetation pattern and form are maculiform, comprising sedgelands and rushland. The four assemblages are *Isolepis nodosa* sedgeland, *Baumea juncea* sedgeland, *Juncus kraussii* rushland, and *Lepidosperma gladiatum* sedgeland.

Figure 4-15. *Description of wetland swi.*

Site swii

Wetland Type: Dampland

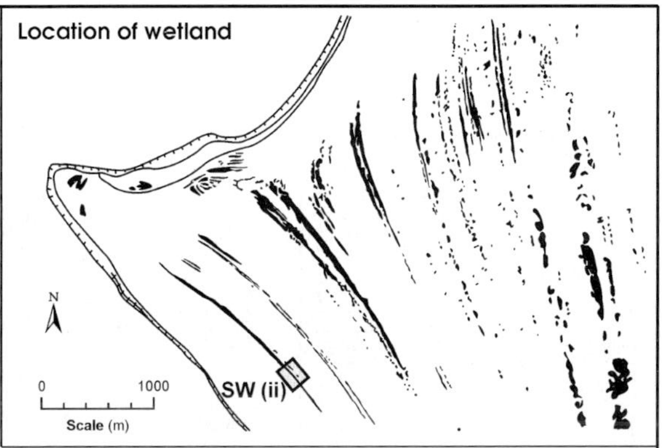

GEOMORPHIC SETTING
Concave inter-ridge depression with western bench. The basin floor lies at 1.59 m AHD, and the bench at 1.84 m AHD.

WETLAND GEOMETRY AND SCALE
Leptoscale, narrow, linear, shallow basin

STRATIGRAPHY

Adjoining beachridges are underlain by medium shelly and quartz sand with lenses of coarse, medium, and fine shelly, quartz sand. The wetland is underlain by root structured and burrow mottled mud and muddy sand grading into medium, and fine shelly and quartz sand.

Granulometric and chemical analyses show the composition of the muddy sand in the wetland to be 44%-61% shelly material and the mud, to vary between a ratio (by weight) of 1:1 to 1:3 carbon to calcium carbonate.

HYDROLOGY

The groundwater table moves between the medium sand aquifer and the muddy sand i.e., between 0.5 - 1.5 m AHD. The longer term hydrograph showed winter levels were at their lowest in 1997 while the water table reached its overall minimum positions in autumn 1995, 1996, and 1998. No upturn in water levels is evident. Groundwater levels under the adjacent beachridges are synchronous with those in the wetlands, at the monthly incremental measuring period.

The groundwater is fresh.

The soil water content in the top 10 cm varies seasonally, ranging from low to medium 5-70%. Down profile, it drops slightly to a depth of 30 cm and then stabilises at around 30%.

Soil water salinity ranges from 1,500-154,000 ppm, with most of the samples registering subhaline values.

VEGETATION

Vegetation pattern and form are maculiform, comprising rushland and sedgelands. The four assemblages are *Juncus kraussii* rushland, *Baumea juncea* sedgeland, *Lepidosperma gladiatum* sedgeland and *Isolepis nodosa* sedgeland.

Figure 4-16. *Description of wetland swii.*

Site swiii

Wetland Type: Sumpland

GEOMORPHIC SETTING
Concave inter-ridge depression with low eastern bench. The basin floor lies at 1.11 m AHD, and the bench at 1.28 m AHD.

WETLAND GEOMETRY AND SCALE
Leptoscale, ovoid, steep sided basin

Location of wetland

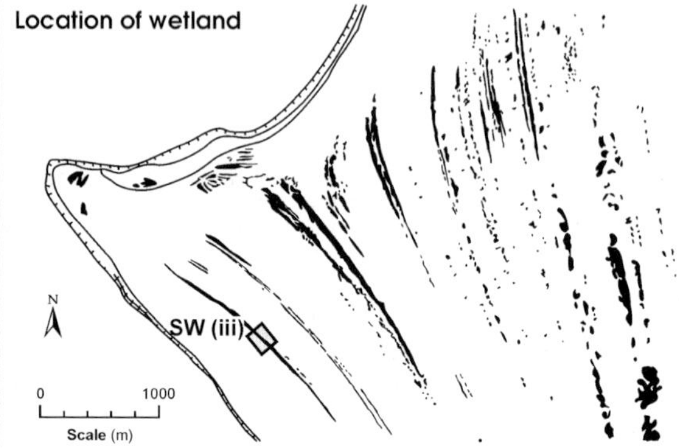

STRATIGRAPHY
Adjoining beachridges are underlain by medium shelly and quartz sand. The wetland is underlain by root structured and burrow mottled muddy sand grading into medium and fine shelly and quartz sand. Granulometric and chemical analyses show the composition of the muddy sand in the wetland to be 58%-66% shelly material and the mud, to comprise a 1:3 ratio of carbon to calcium carbonate.

HYDROLOGY
The groundwater table moves between the medium sand aquifer through the muddy sand to approximately 20 cm above the surface i.e., between 0.5 - 1.4 m AHD. The longer term hydrograph showed winter levels were at their lowest in 1993 and 1997 while the water table reached its overall minimum positions in autumn 1994, 1995, 1996, and 1998. Groundwater levels under the adjacent beachridges are synchronous with those in the wetlands, at the monthly incremental measuring period. The salinity of the groundwater ranges from fresh to hyposaline. The soil water content in the top 10 cm varies seasonally, ranging from medium to high 40-85%. It appears to be decline down profile until reaching a depth approximately 50 cm when there is a slight rise which remains stable thereafter. Soil water salinity ranges from 1,000-72,000 ppm, with most of the samples registering subhaline values.

VEGETATION
Vegetation pattern and form are maculiform. There are four assemblages. They are sedgelands and herbland. The sedgelands are *Schoenoplectus validus* and *Typha orientalis*, *Lepidosperma gladiatum*, and *Baumea juncea*. The herbland is composed of *Centella asiatica*.

Figure 4-17. *Description of wetland swiii.*

Site 1N

Wetland Type: Dampland

GEOMORPHIC SETTING

Slightly concave to gently undulating inter-ridge depression whose surface is at 1.49 m AHD.

WETLAND GEOMETRY AND SCALE

Leptoscale, narrow, linear, shallow basin

Location of wetland

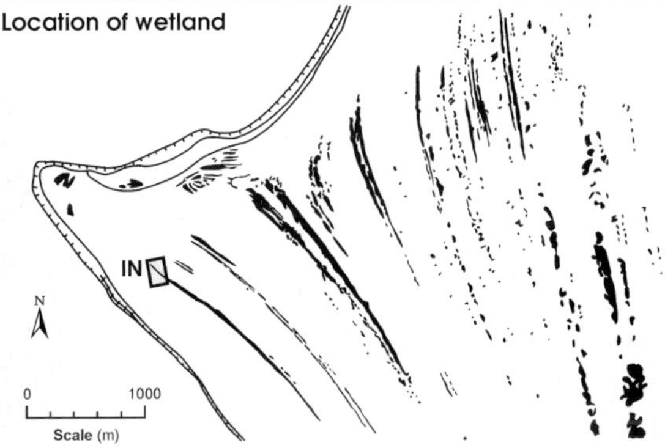

STRATIGRAPHY

Adjoining beachridges are underlain by medium shelly/quartz sand and coarse, medium, and fine shelly/quartz sand. The wetland is underlain by root structured and burrow mottled humic sand grading into coarse, medium, and fine shelly/quartz sand.

Granulometric and chemical analyses show the composition of the sand in the wetland to be 70%-85% shelly material and the mud, in the top 20 cm to vary between ratios (by weight) of 1:1 and 1:5 humus to calcium carbonate.

HYDROLOGY

The groundwater table occurs in the coarse sand aquifer, and fluctuates seasonally between sealevel and 1 m AHD. The longer term hydrograph showed winter levels were at their lowest in 1993 while the water table reached its overall minimum position in 1997. Subsequent to the occurrence of these troughs there was a slight upturn, discernible in higher positions of the annual maximum water table.

The salinity of the groundwater is fresh, except when groundwater levels lie close to sealevel, where they become sub-haline.

The soil water content in the top 10 cm varies seasonally, ranging from low to medium 5-40%. Down profile, soil moisture appears to be relatively constant.

Soil water salinity ranges from 1,000-100,000 ppm, with many of the samples registering mesosaline values.

VEGETATION

Vegetation pattern and form are latiform. There is one assemblage comprising sedgeland with scattered low open heath. The dominant sedge is *Baumea juncea*, with minor *Isolepis nodosa*. The scattered heath comprises *Pelargonium capitatum*, and *Acacia saligna*.

***Figure 4-18.** Description of wetland 1N.*

Site Cooloongup A

Wetland Type: Dampland

GEOMORPHIC SETTING
Slightly undulating inter-ridge depression whose surface is at 2.5 m AHD.

WETLAND GEOMETRY AND SCALE
Microscale, narrow, linear, flat

Location of wetland

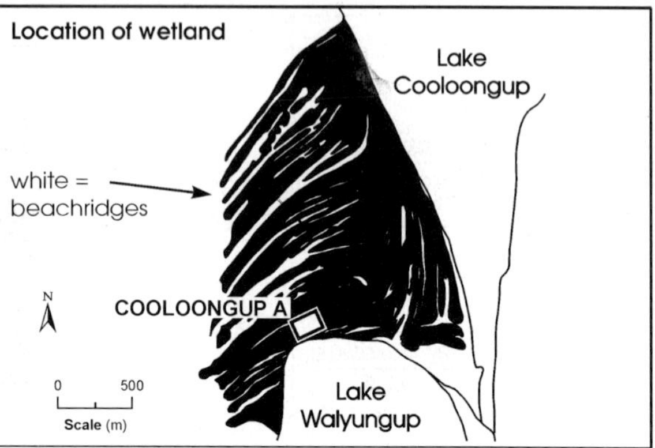

STRATIGRAPHY
Adjoining beachridges are underlain by medium shelly and quartz sand. The wetland is underlain by colour mottled mud, calcrete cemented muddy sand and colour mottled muddy sand grading into medium and fine shelly and quartz sand.

HYDROLOGY
The groundwater table moves between the medium sand aquifer and the base of the cemented layer i.e., between 1.0 - 2.2 m AHD. Groundwater levels under the adjacent beachridges are synchronous with those in the wetlands, at the monthly incremental measuring period. The salinity of the groundwater is subhaline.

VEGETATION
Vegetation cover is complete and zoned. There are four assemblages. They are two sedgelands, layered open shrubland and low closed heath. The dominant sedge is *Baumea juncea*, the secondary one is *Gahnia trifida*. The shrubland comprises *Melaleuca rhaphiophylla* with an understorey of *Baumea juncea* and *Centella asiatica*, the low heath is *M. viminea*.

Figure 4-19. *Description of wetland Cooloongup A.*

Site Cooloongup B

Wetland Type: Dampland

GEOMORPHIC SETTING
Slightly undulating inter-ridge depression whose surface is at 2.7 m AHD

WETLAND GEOMETRY AND SCALE
Microscale, narrow, linear, flat intersected by a low ridge (15 cm)

Location of wetland

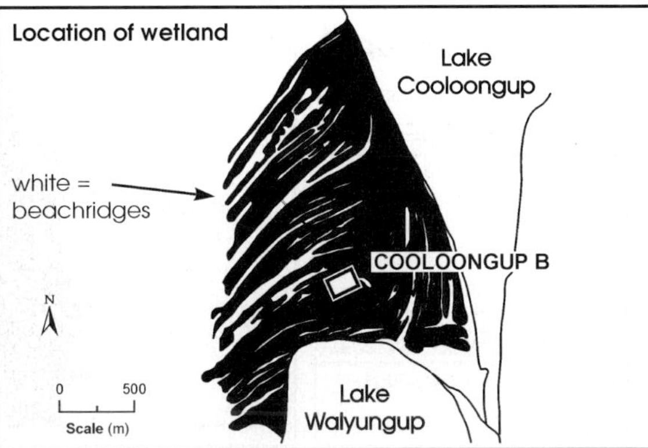

STRATIGRAPHY
Adjoining beachridges are underlain by medium, shelly and quartz sand. The wetland is underlain by root structured mud, colour mottled sandy mud, a narrow calcrete cemented band, and homogeneous and colour mottled muddy sand grading into medium and fine shelly and quartz sand .

HYDROLOGY
The groundwater table moves between the medium sand aquifer and the lower part of the mud horizon i.e., between 1.0 - 2.2 m AHD. Groundwater levels under the adjacent beachridges are not always synchronous with those in the wetland, at the monthly incremental measuring period. The salinity of the groundwater ranges from subhaline to hyposaline.

VEGETATION
Vegetation cover is complete and zoned. There are two assemblages. They are open sedgeland of *Gahnia trifida* with minor *Baumea juncea*, and the same open sedgeland with an overstorey of low closed to open heath of *Melaleuca viminea*.

Figure 4-20. *Description of wetland Cooloongup B.*

Site Cooloongup C

Wetland Type: Dampland

GEOMORPHIC SETTING
Flat inter-ridge depression whose surface is at 2.7 m AHD

WETLAND GEOMETRY AND SCALE
Microscale, narrow, linear, flat

Location of wetland

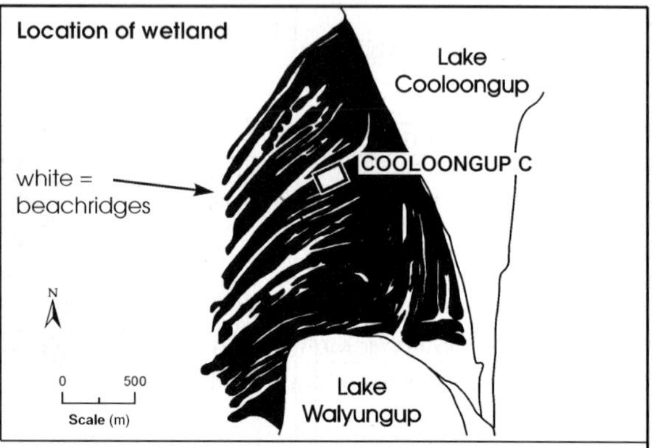

STRATIGRAPHY
Adjoining beachridges are underlain by medium shelly and quartz sand. The wetland is underlain by homogeneous mud and colour mottled muddy sand grading into medium and fine shelly and quartz sand.

HYDROLOGY
The groundwater table moves between the medium sand aquifer and the base of the mud i.e., between 1.0 - 2.2 m AHD. Groundwater levels under the adjacent beachridges are not synchronous with those in the wetland, at the monthly incremental measuring period. The salinity of the groundwater ranges from subhaline to hyposaline.

VEGETATION
Vegetation cover is complete and zoned with respect to understorey. There is one assemblage. It comprises open shrubland to woodland of *Melaleuca rhaphiophylla* with a variable sedge understorey of either *Baumea juncea* or *Gahnia trifida.*

Figure 4-21. *Description of wetland Cooloongup C.*

4.2 Radiocarbon dates

This section provides a summary of radiometric analyses undertaken in this study. Because this information was used in a number of ways in the project (dating wetland commencement, ascertaining rates of accumulation, determing relative ages of components of sedimentary mud, and relating pollen series to periods of wetland development or to sedimentary types), the data were compiled as a database. While the detailed stratigraphy of each wetland is described in Chapter 6, the lithological summary of all cores, together with the location of radiocarbon dating within the profiles, and the types of materials dated, are presented in Figure 4.22 and Table 4.1. The results are also briefly described below prior to their more specific and detailed treatment in later chapters.

All ^{14}C dates derived from the wetlands of this study were middle to late Holocene in age. The oldest date from the base of a wetland located to the far east of the study area was 5740 ^{14}C yrs BP, and the youngest dates were "contemporary" or "modern". The remainder of the dates were spread between, and relate to topgraphic setting and geographic position from the present coast. All dates conform to the fact that the wetlands reside on and have formed wholly within a Holocene landscape.

Dates specifically derived from the base of wetland deposits show that initiation of wetland conditions varied across the beachridge plain (see Section 5.3.4), and continued to at least 680 ^{14}C yrs BP. The dates also show that mud-sized organic matter and carbonate mud from the same horizon appear to provide slightly different ages. The rates of accumulation also varied in different wetlands and in the two components of the mud. These issues are treated in more detail in Chapter 6. Finally, the ^{14}C dates form a framework for use in the palynological studies in that, for the cores studied in detail, a reasonable age structure has been derived within which the pollen sampling took place.

There is scope, however, for further radiocarbon analyses of the wetland sequences, particularly in precise dating of some of the events *within* the wetlands, for example, the age of cliffing within a given wetland, or the beginning of subsidence, or the period of sand incursion along a given wetland margin. However, this more detailed dating of geomorphic events and sedimentologic events was beyond the scope of this study.

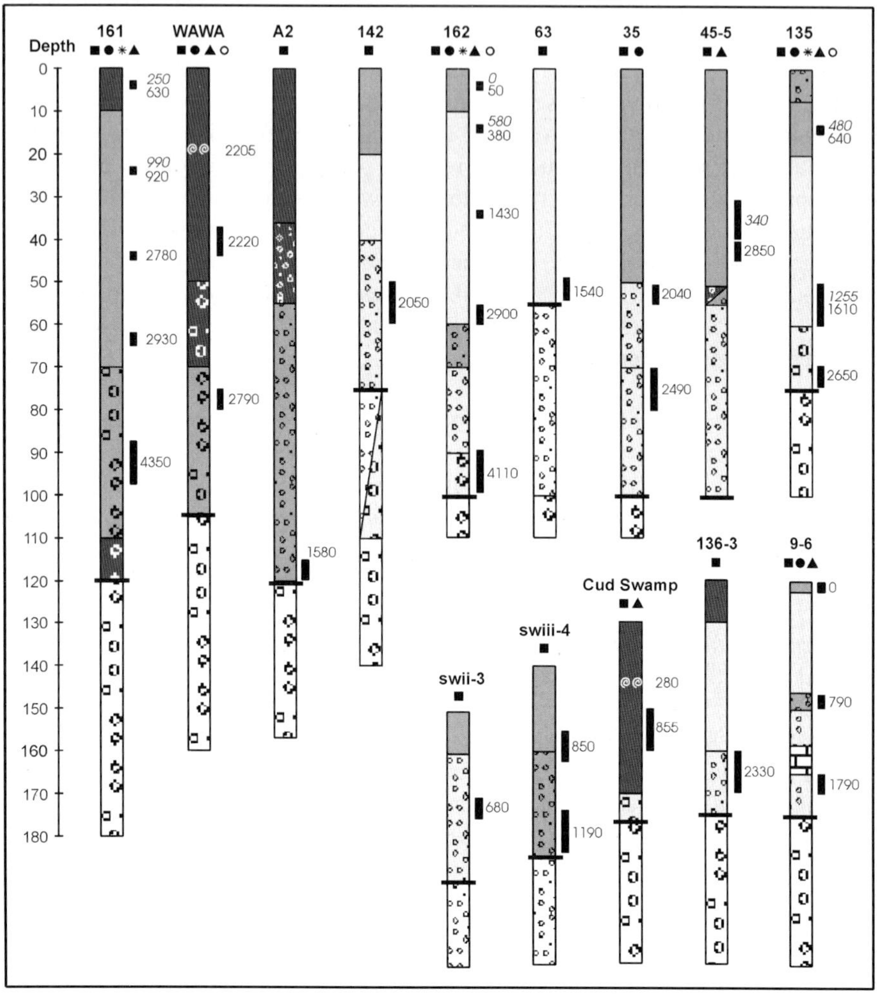

Figure 4-22. *Simplified stratigraphy, showing lithology, location and type of samples, rationale for samples, and ^{14}C dates. Sampling sites are shown in Figure 2-6. The ^{14}C results for base of sequences and pollen are in Figures 5-16, 11-4. Detailed stratigraphy is in Figures 6-3 to 23, and rates of accretion are in Figure 6-64.*

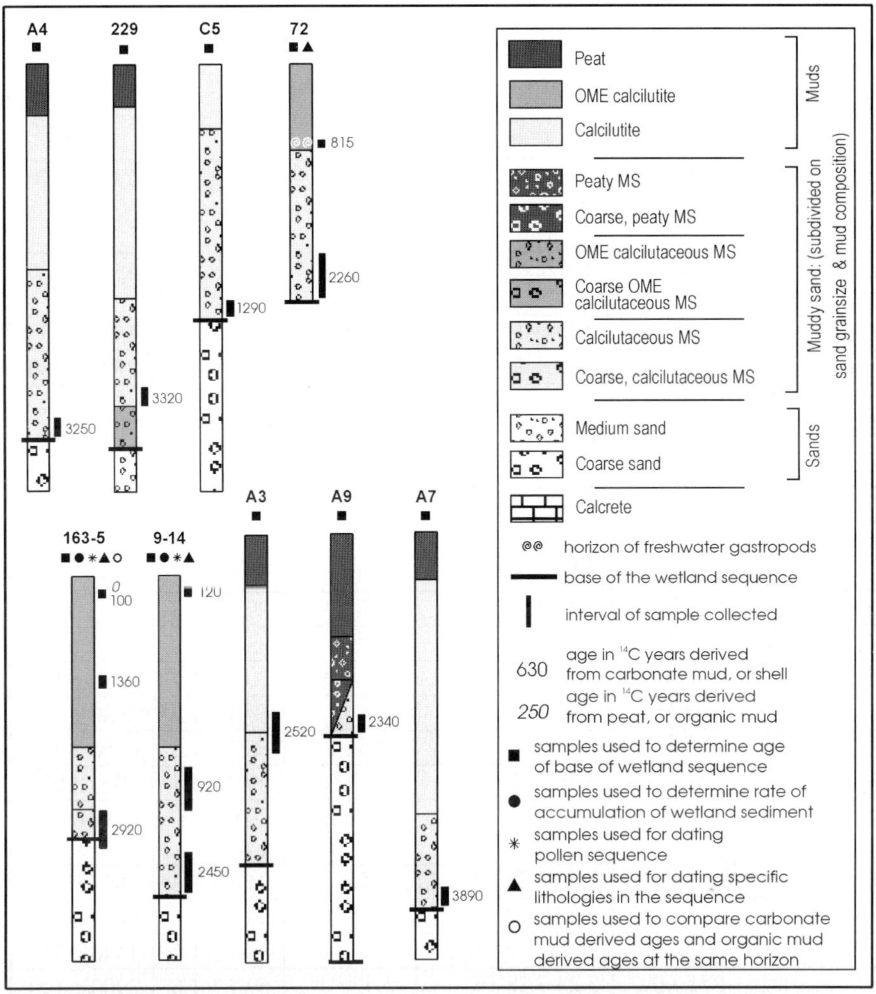

***Figure 4-22 (cont.)**. Simplified stratigraphy, showing lithology, location and type of samples, rationale for samples, and ^{14}C dates. Sampling sites are shown in Figure 2-6. The ^{14}C results for base of sequences and pollen are in Figures 5-16, 11-4. Detailed stratigraphy is in Figures 6-3 to 23, and rates of accretion are in Figure 6-64.*

Table 4.1 *Radiocarbon ages of wetland sediments and freshwater shells*

Wetland	*Sample depth (cm)*	*AHD (m)*	*Mud type*	$\delta^{13}C$ *% PDB*	^{14}C *pMC±1σ*	^{14}C *yr BP ±1σ*
Cooloongup East N/S spits	60-70		C	-4.8	48.9 ± 1.1	5740 ± 170
Cooloongup A2	80-90	2.59-2.69	C	-3.1	56.5 ± 1.1	4590 ± 160
A2	115-120	3.35-3.4	C	-6.0		1580 ± 65
A3	46	3.6	C	-6.2	73.1 ± 1.6	2520 ± 180
A4	88	3.67	C	-6.5	66.7 ± 1.6	3250 ± 190
A7	85-90	2.3-2.35	C	-3.4 ± .1%		3890 ± 120
A9	44-47	1.46-1.49	C	-7.3 ± .1%		2340 ± 70
229	70-80	4.01-4.11	C	-7.6	66.2 ± 1.6	3320 ± 200
161-3	3-5	3.54-3.56	C	-6.4	92.5 ± 1.3	630 ± 110
	3-5	3.54-3.56	peat	-28.6	96.9 ± 1.3	250 ± 110
	23-25	3.34-3.36	C	-7.2	89.2 ± 1.3	920 ± 110
	23-25	3.34-3.36	peat	-28.5	88.4 ± 1.2	990 ± 110
	43-45	3.14-3.12	C	-2.5	70.7 ± 1.2	2780 ± 130
	62-65	2.94-2.96	C	-5.3	69.4 ± 1.2	2930 ± 140
	90-100	2.59-2.69	C	+0.3	58.2 ± 1.5	4350 ± 210
162-3	3-5	3.78-3.80	C	-10.2	99.3 ± 1.4	50 ± 110
	3-5	3.78-3.80	peat	-27.0	105.1 ± 1.4	modern
	13-15	3.68-3.70	C	-9.0	95.3 ± 1.3	380 ± 110
	13-15	3.68-3.70	peat	-26.5	93.0 ± 1.3	580 ± 110
	33-35	3.48-3.50	C	-7.0	83.7 ± 1.2	1430 ± 120
	55-60	3.23-3.28	C	-5.4	69.7 ± 1.6	2900 ± 190
	90-100	2.83-2.93	C	-4.7	59.9 ± 1.1	4110 ± 150
163-5	3-5	3.76-3.78	C	-6.5	98.7 ± 1.3	100 ± 110
	3-5	3.76-3.78	peat	-27.1	107.7 ± 1.4	modern
	23-26	3.55-3.58	C	-9.0	84.4 ± 1.2	1360 ± 120
	60-70	3.11-3.21	C	-5.9	69.5 ± 1.6	2920 ± 190
163-3	40-50	3.48-3.58	C	-6.4	69.2 ± 1.2	2960 ± 140
163 20 m	30-40	4.35-4.45	C	-7.0	84.1 ± 1.3	1400 ± 120

Table 4.1 (cont.)

Table 4.1 (cont.)

Wetland	*Sample depth (cm)*	*AHD (m)*	*Mud type*	*$\delta^{13}C$ % PDB*	*^{14}C pMC±1σ*	*^{14}C yr BP ±1σ*
WAWA 3	50-70	2.47-2.67	peat	-28.5	75.9 ± 2.8	2220 ± 300
WAWA teret	75-80	2.44-2.49	C	-7.2	70.7 ± 1.2	2790 ± 140
WAWA 1/2w	60-70	3.45-3.55	C	-6.4	72.3 ± 1.2	2600 ± 140
136-3	40-50	3.21-3.31	C	-7.5	74.8 ± 1.7	2330 ± 180
135-2	13-15	3.29-3.31	C	-8.0	92.5 ± 1.3	640 ± 110
	13-15	3.29-3.31	peat	-27.9	94.2 ± 1.3	480 ± 110
	50-60	2.84-2.94	C	-6.4	81.9 ± 1.7	1610 ± 170
	75-80	2.64-2.69	C	-7.9	71.9 ± 1.2	2650 ± 130
72-3	45-55	2.93-3.03	C	-6.5 ± .1%		2260 ± 65
63-3	50-60	2.67-2.77	C	-8.2	82.5 ± 1.2	1540 ± 120
45-5	40-45	1.24-1.29	C	-3.3 ± .1%		2850 ± 130
45-5	30-40	1.29-1.39	peat	-28	95 ± 3.0	340 ± 250
35-4	55-65	1.23-1.33	C	-6.5	77.6	2040 ± 180
	70-80	1.08-1.18	C	-6.2	73.4 ± 1.2	2490 ± 130
9-14	3-5	1.38-1.3	C	-5.5	98.5 ± 1.3	120 ± 110
	45-55	0.78-0.88	C	-8.6	89.2	920 ± 160
	65-75	0.58-0.68	C	-2.8	73.7 ± 1.2	2450 ± 130
9-6	0-10	1.12-1.22	C	-6.2	101.7 ± 1.4	modern
	30-40	0.82-0.92	C	-8.3	90.7 ± 1.8	790 ± 160
	50-60	0.62-0.72	C	-8.3	80.0 ± 1.2	1790 ± 160
swiii-4	15-25	1.34-1.44	C	-3.0	90.0 ± 1.8	850 ± 160
	50-60	0.99-1.09	C	-2.8	86.3 ± 1.2	1190 ± 120
swii-3	20-25	1.34-1.39	C	-6.8 ± .1%		680 ± 80
142-3	50-60	2.58-2.68	C	-7.8	77.5 ± 1.7	2050 ± 180
C5	55-60	1.98-2.03	C	-8.4 ± 0.1%		1290 ± 90
Cud Swamp	20-30	1.98-2.08	peat			855 ± 70
WAWA	20-30	2.87-1.97	shell			2205 ± 190
Cud Swamp	10-20	2.08-2.18	shell			280 ± 180

C is the symbol for calcilutite
The shell is freshwater

4.3 General notes on biota

There are descriptions of the effects of fauna in the sections on wetland sedimentary sequences, hydrochemistry, and vegetation. Although wetland fauna have not been specifically addressed in the study because they are itinerant (terrestrial) or intermittent (aquatic), some mention about the general composition of fauna in the region is required prior to discussion of their impacts. The roles fauna play in the wetlands, in terms of bioturbation, nutrient contribution, or skeletal production, are summarised in Table 4.2. The ecological functions detailed in Table 4.2 denote important sedimentological, stratigraphic, pedologic and nutrient contributions. The trophic position of the fauna as primary consumers, or higher order predators, is not directly relevant here, but consumers of detritus are noted because these fauna affect soil structures and small scale nutrient recycling.

The effects of fauna within the wetlands which are relevant to this study are burrowing, comminuting plant material, and recycling soil, detritus, and nutrients. In Table 4.2, the fauna are noted only if they import nutrients to the wetlands (*i.e.*, they are not resident, they forage, or predate within the wetland or elsewhere, and a proportion of their scats contributes allochthonous nutrients). Ants, and other resident invertebrates, which forage or predate wholly within the wetlands, tend to recycle the existing nutrient pool of vegetation and other invertebrates rather than contribute to its accumulation. All fauna that reside or opportunistically utilise the wetlands may contribute skeletons to the deposits. Skeletons of mammals and reptiles have little chance of preservation, and have not been detected in the stratigraphic record nor on the sediment surface beyond a decade. Calcareous invertebrates, on the other hand, contribute enough skeletal material to be detected stratigraphically (*e.g.*, shells of freshwater snails, and carapaces of arthropods).

Endemic macro vertebrate fauna include: 1) temporary resident mammals such as the Black-gloved wallaby (*Macropus irma*), and permanent resident mammals such as the Southern Brown bandicoot (*Isoodon obesulus fusciventer*); 2) locally resident reptiles such as the Dugite snake (*Pseudonaja affinis*), the Tiger snake (*Notechis scutatus*), the Bardick snake (*Echiopsis curta*), Jan's banded snake (*Simoselaps bertholdi*), legless lizards (Pygopodidae), the Western bearded dragon (*Pogona minor*), the Western blue tongue (*Tiliqua occipitalis*), and the Bobtail (*Tiliqua rugosa*); and 3) permanent resident amphibians such as the Slender tree frog (*Litoria adelaidensis*), Western green tree frog (*L. moorei*), and the Moaning frog (*Helioporus eyrei*). Introduced fauna include the rabbit (*Oryctolagus cuniculus*), the fox (*Vulpes vulpes*), and the feral cat (*Felis catus*). Avifauna that predate within and around the wetlands, or that utilise the wetlands as herbivores, fructivores, insectivores, or carnivores include various species of raptor (Falconides), owl (Strigiformes), duck (Anatidae), and passerine birds, noticeably the Black-faced cuckoo shrike (*Coracina novaehollandiae*). Terrestrial invertebrates within the wetlands include, spiders such as the Christmas Spider

(*Gasteracantha minax*), Golden Orb-weaving Spider (*Nephila edulis*), and St. Andrew's Cross Spider (*Argiope aetherea*), ticks (Ixodidae), soil mites (Acari), springtails (Collembola), earthworms (Oligochaeta), butterflies (Lepidoptera), ants (Formicidae), beetles (Coleoptera), bugs (Hemiptera), crickets (Orthoptera) and cockroaches (Blattodea). Aquatic invertebrates include mayflies (Ephemeroptera), dragonflies (Anisoptera), damselflies (Zygoptera), lacewings (Neuroptera), caddis flies (Trichoptera), mosquitoes (Culicudae), midges (Chironomidae), sandflies (Ceratopogonidae), water bugs such as water boatmen (Corixidae) and backswimmers (Notonectidae), diving beetles (Dytiscidae), water mites (Hydracarina), flatworms (Platyhelminthes), roundworms (Nematoda), leeches (Hirudinea) and seed shrimps (Ostracoda).

Table 4.2 *Effects of vertebrate and invertebrate fauna, and plants*
(-) denotes negligible or no contribution

	Effect within wetland	*Nutrient contribution*	*Skeletal contribution*
	Vertebrates		
Wallabies	foraging	scats	-
Southern Brown bandicoot	burrowing	scats	-
Snakes	burrowing on margins	scats	-
Lizards	burrowing on margins	scats	-
Frogs	burrowing on margins	scats	-
Rabbits	burrowing on margins, foraging	scats	-
Ducks	foraging	scats	-
	Invertebrates		
Ants	burrowing	-	-
Spiders	burrowing	-	-
Beetles	burrowing	-	-
Hemiptera larvae	burrowing, foraging	-	-
Lepidoptera larvae	burrowing	-	-
Cockroaches	burrowing, detritus consumer	-	-
Crickets	burrowing	-	-
Snails	-	-	shells
Earthworms	burrowing, detritus consumer	-	-
Roundworms	burrowing, detritus consumer	-	-
Soil mites	burrowing, detritus consumer	-	-
Springtails	burrowing, detritus consumer	-	
Ostracods	-		skeletons

5. DEVELOPMENT OF WETLAND PROTO-TYPE: GEOMORPHOLOGY, BASAL SHEET, HYDROLOGY

5.1 General introduction

As described in the previous chapter, the Becher cuspate foreland comprises a dune and beachridge plain, and a beach zone (Searle *et al.* 1988). Each of these geomorphic units has been subdivided into small scale features (Semeniuk and Johnson 1982; Searle *et al.* 1988; C. A. Semeniuk 1988; Semeniuk *et al.* 1989). The small scale features important to this study are the series of well-defined, continuous, shore-parallel Holocene ridges, which form the beachridge plain, and the leptoscale wetlands, which have evolved within the swales. The two features are related in that the beachridge plain is the template for proto-wetland development.

In order to provide an overview of the physical framework or setting for wetland origin, this chapter begins with a description of the ridge/swale complex and a discussion of the causal processes underpinning its development. This is followed by a description of wetland types and a reconstruction of wetland development.

For the beachridges that frame the wetlands, the specific objectives are:

- to describe the geometry and forms of the ridge/swale complex
- to describe the coastal processes which have resulted in its development
- to discuss the origin of discordant sets of beachridges: the higher than normal beachridges and the modern beachridges
- to describe the elements in the landscape which are proto-wetland forms

For the wetlands, the specific objectives are:

- to describe the basal sheets underlying the wetlands
- to develop models for wetland initiation across the Becher cuspate foreland
- to discuss the age structure of the wetlands and their distribution in relation to that structure

The geomorphic description of the beachridge plain is based on plan view aerial photography; ridge geometry and orientation with respect to the modern shoreline as derived from contour maps (Fig. 5-1A), and field surveys. Interpretation of coastal processes is based on this description, review of the literature, and on basic granulometric analysis of beach and dune sediments collected throughout the cuspate foreland. The data on wetlands were obtained from aerial photographs, contour maps, and laboratory techniques and field investigations, including mapping, classification, coring, augering, land profile surveys, sediment sampling, granulometric analysis and radiocarbon dating.

5.2 Beachridges and swales

This section describes the beachridges and associated swales of the Becher cuspate foreland and discusses their origins. This will form the foundation to interpreting the origin and evolution of the wetlands within, since the beachridge system of ridges and swales forms the framework for the wetlands. In particular, the formation of the various types of swales is explored, be they inter-dune depressions or stranded beach berms within which the wetlands are located.

5.2.1 Definition of shore parallel ridges

The ridges which frame the swales, are linear shoestring sand deposits, convex in cross-section, with elevation above mean high tide and base near to low mean tide. The ridges accord with the definition of "beachridge" (Stapor 1982). Although some authors (Bigarella 1965; Psuty 1965; Stapor 1975; Short *et al.* 1989) restrict beachridge nomenclature to wholly marine derived ridges, other authors (Johnson 1919; Davies 1957; King and Barnes 1964; Hesp 1984; Woods 1984; Searle *et al.* 1988) include ridges with both marine and aeolian components, as long as the fundamental morphology and internal structures mirror beachridge rather than dune development. This perspective accords with the most recent argument that beach ridges should include all relict strandplain ridges, whether dominated by wave/swash-built or by aeolian lithosomes (Otvos 2000). Granulometric analyses show that beachridges have textures associated with both beaches and coastal dunes (Stapor and Tanner 1975).

All of the sediment samples in the study area from the beaches, swales, and some beachridges, showed a negatively skewed distribution with coarse and very coarse sand present. The exceptions occurred in samples taken at, or near, the crest of the beachridges which exhibited very slight positive skewness and contained medium to fine sand. The ridges of the Becher Plain have textural affinities with both beach and dune sediments, that denote an origin pertaining to several construction agents (swell waves, storm waves, wind). Many of the ridges exhibit layering of medium/fine sands, and coarse/medium sands with occasional layers of shell gravel or very coarse sand.

5.2.2 Beachridges and swales of the Becher cuspate foreland: morphology

The beachridges of the Becher cuspate foreland, (as for all of the Rockingham Becher Plain), are parallel to the configuration of the shoreline (Fig. 5-1B), and represent former beaches and ridges stranded by progradation (Stapor 1982; Woods and Searle 1983; Woods 1984). On the northern shore, they are truncated by the development of a narrow zone of modern shoreline ridges (comprising a foredune, chaots, hollows, parabolic dunes, and blowouts), which are related to the Warnbro Embayment. On the southern shore, immediately south of Becher Point, they have been truncated by coastal erosion. Further south, the beachridges have been overprinted by parabolic dunes, blowouts, and a sheet of dust and fine sand; these landforms have been separated from the present coastline by another zone of modern shoreline ridges, related this time to the Madora Embayment.

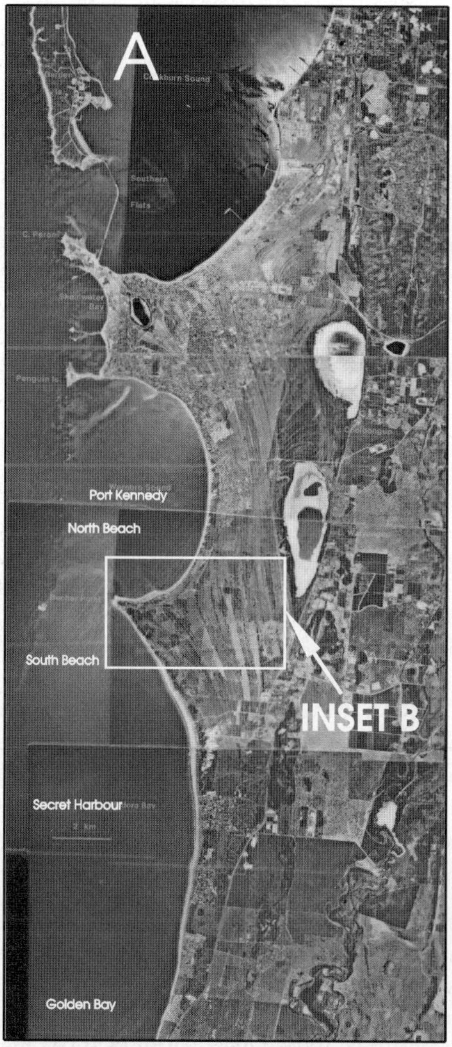

Figure 5-1. A. *Aerial photograph of beachridges on the Rockingham-Becher Plain.*

The beachridges are microscale, commonly standing 1-3 m above the intervening swales and 30-60 m apart (Fig. 5-2). Interspersed within this series of low ridges, there is a second series of beachridges 3-8 m high, at intervals of approximately 400 m. Heights of individual beach-ridges increase along any ridge in a southerly direction.

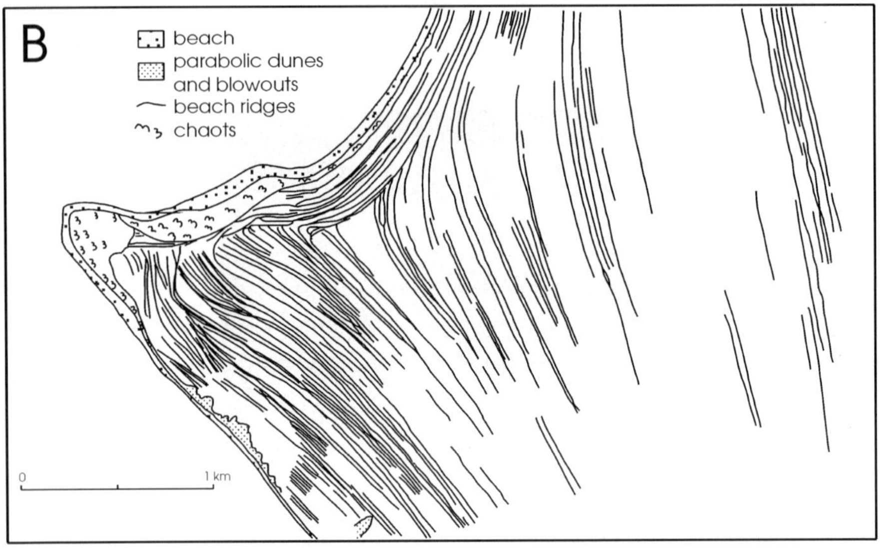

Figure 5-1 (cont.) *B. Map of beachridges on the Becher cuspate foreland showing orientation with respect to shoreline.*

Along the length of some ridges, there is bifurcation. The swales are variable in morphology along their length, grading from continuous linear troughs with slightly undulating floors, to linear troughs bisected by low, transverse ridges, to linear chains of discrete basins. Heights of swales above AHD are variable as are heights of individual basins along a single swale, however, there is a general increase in swale elevation in a southerly direction. All types of swale may be host to wetlands. Narrow flats (approximately 100 m wide and of variable heights above MSL) also occur (Fig. 5-2). Similar features have been observed in the beachridge plains in northern Spencer Gulf (Burne cited in Short *et al.* 1989).

5.2.3 Processes for constructing beachridges

The important coastal agents for constructing beachridges in this setting are 1) swell waves, 2) wind waves, 3) littoral currents, 4) wind, and 5) rainfall. The exact process of beachridge formation has been and continues to be vigorously debated (Johnson 1919; Davies 1957; Tanner 1995, 1996; Taylor and Stone 1996), however, there are some observations which seem to incur general agreement. Firstly, an abundant supply of sediment is required (Wright 1970; Stapor 1975; Roy *et al.* 1980; Anthony 1995). Secondly, initiation of beachridge building is dependent on the offshore profile becoming less steep (Johnson 1919; Davies 1958; Tanner and Stapor 1971). Thirdly, beachridges accumulate on an existing mound, either a persistent wave built berm as nucleus (Bird 1960, 1985), or backshore terrace colonised by vegetation (Hesp 1984;

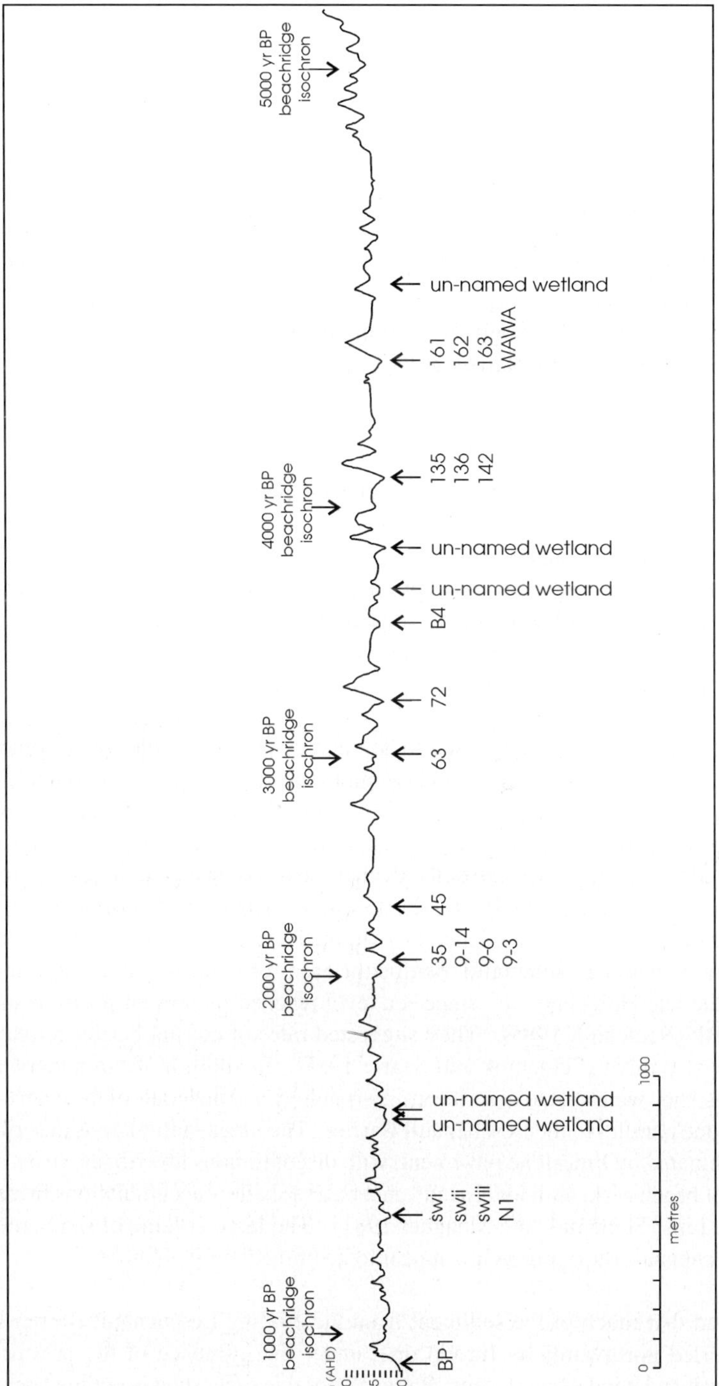

Figure 5-2. *Cross-section along Transect B (see Fig 2-4) showing ridge & swale development, and location of wetlands relative to the beachridge isochrons.*

Bird 1985). Fourthly, a repetitive, although not necessarily cyclic, formational agent is required (Thom 1963; Psuty 1965; Tanner 1995). Each of these conditions is examined with respect to the Becher beachridges.

Sediment source and supply

Three sources of sediment for accretion of the beachridge plain in the Becher area have been suggested (Semeniuk and Meagher 1981; Searle 1984; Semeniuk 1985; Searle *et al.* 1988): 1) sediment from erosion of islands, pinnacles and reefs of the offshore ridges (Five Fathom Bank Ridge and Garden Island Ridge), 2) sediment generated *in situ* in the seagrass banks (Rockingham Bank and Becher Bank), and 3) sediment from the erosion of the Leschenault Barrier to the south.

Stacks, collapsed remnants of platforms and pinnacles on the seafloor, and cliffed primary dunes along the shoreline of emergent islands in the Pleistocene ridges, testify to ongoing erosion of weakly cemented to uncemented Pleistocene limestone and sand, producing sediment in the Cockburn-Warnbro Depression (Searle 1984; Searle *et al.* 1988). Submarine banks extend across the width of the Cockburn-Warnbro Depression (Fig. 3-3) and seagrass meadows colonise the more sheltered parts. These habitats are a source of allochthonous calcareous materials. The seagrass banks, through trapping and baffling processes, also store a large nearshore supply of sediment for potential onshore transfer.

It was questionable whether these two local sources could provide the volume of sediment necessary to build the Becher Cusp, and so the third source from outside the system was investigated. Semeniuk and Meagher (1981), Semeniuk (1985) and Semeniuk and Searle (1987), document large scale erosion in the Holocene of the Leschenault Barrier, located 50-80 km to the south of the Becher Cusp. Using long term photographic analysis of coastal retreat (1941-1971), short term investigation of shoreline erosion during storms (1976-79), and evidence of exhumed stratigraphic units, the authors concluded that in the Leschenault Peninsula area, erosion has been dominant throughout the late Holocene, *i.e.*, since sea level reached its present position some 3000 years BP (Semeniuk 1985). They suggested rates of coastal barrier retreat of between 0.05-1.0 m/yr. (Semeniuk and Searle 1987). In addition, from underwater investigations, they were able to describe the morphology and materials of the nearshore submarine zone parallel to the Leschenault Barrier. The inner shelf plain is described as being dominated by limestone pavement, with discontinuous low ridges, strips and linear slabs of beach rock, with local sand veneers, and shallow accumulations in inter-ridge depressions (Semeniuk and Meagher 1981). The large volume of sand eroded from the Leschenault Barrier does not appear to lie immediately offshore.

It is suggested that much of the sediment from the eroding Leschenault Barrier has been transported northwards by littoral drift under the influence of the prevailing southwest swell and wind wave system. Rationale for this suggestion is outlined below.

- Some 35 km to the north of the Leschenault Barrier is the Peel-Harvey Estuary tidal exchange channel. The tidal exchange channel is the first break in an otherwise uninterrupted shoreline. It is regularly blocked by sediment accumulating from north directed shore drift.

- The nearshore inter-ridge geomorphology and bathymetry adjacent to the Becher Cusp extends as far south as Cape Bouvard. This depression potentially acts as a sediment trap. Sediment, once entrained into this depression, is subject to the prevailing wind and swell wave effects, wave energy levels and sediment drift dynamics which determine nearshore deposition along the length of this coastal sector.

- The description of the textural and compositional characteristics of sand from the Leschenault Barrier is consistent with that collected throughout the Becher beachridge plain, *i.e.*, medium and fine sand with quartz and calcareous components (Semeniuk and Meagher 1981)

- The beachridges on the Becher Cuspate foreland are not evenly spaced, reflecting changes to the supply of sediment. This unevenness of supply is consistent with the pattern of evolution and the sea level history for both the Becher area (Searle and Woods 1986; Semeniuk and Searle 1986b) and the history of erosion in the Leschenault area (Semeniuk 1985).

With reference to beachridge plains elsewhere in Australia, the consensus is that sediment sources are to be found on the continental shelf, and that sediment transport is generally shore-normal with longshore transport being minimal (Roy *et al.* 1980; Short and Hesp 1982; Bowman and Harvey 1986; Short *et al.* 1989). The majority of examples cited in the literature pertain to beachridge systems located along segmented shorelines where the littoral drift cells are localised, and the major source of sediment for beachridge construction is offshore, but there are some examples in the literature of considerable sand movement along coasts. Anthony (1995) cites an example of beachridge development for 200 km along the Bight of Benin, attributed to sediment transported by longshore drift from the Volta Delta. What is particularly noteworthy is the estimated volume of sand being transported in this manner *viz*., 1.5 x 10^6 m^3 $yr.^{-1}$. These are similar to the volumes of sand estimated by Semeniuk (1985) annually transported northwards from the Leschenault Barrier, *i.e.*, 1.0 x 10^6 m^3 $yr.^{-1}$.

Nearshore profile

With an abundant sediment supply, the next requirement for beachridge building is that of a gentle nearshore profile. The extensive submarine banks constitute a gentle shallow base for approaching wave trains. In addition, prevailing low energy swell conditions (0.1-0.2 m^2/s), resulting from wave convergence and divergence relating to the nearshore Five Fathom Bank Ridge and the Garden Island Ridge, have been documented in the coastal area of the Rockingham Becher Plain (Collins 1983; Woods 1984).

The sediments from beachridges throughout the Rockingham Becher Plain exhibit a grain size distribution consistent with a low energy regime (mean = 2.4-2.8 phi). Many of the samples, although negatively skewed, did not show a high degree of sorting, suggesting upper and lower beach facies. A representative collection of sediments was made from a number of locations throughout the Becher Cusp to characterise the sediment types. Summer and winter samples were obtained from the northern and southern facing Becher beaches (in several locations), from Warnbro Sound Beach, Secret Harbour Beach, Golden Bay Beach, and single samples from beachridge swales, from older to younger beachridges, and from different levels within a single beachridge (Fig. 2.5). These samples were sieved at half phi intervals and examples of the grain size distributions are illustrated in Figures 5-3, 5-4. The calcium carbonate grains, having a considerably higher surface area per unit weight than the quartz, complicated the recognition and interpretation of hydrodynamic environment and process, and were therefore excluded from the analysis.

The beach sediments in the summer were found to have a mean size of between 2.5-3 phi, with a standard deviation of 0.5-1 phi class/interval (medium sand size with moderate sorting). In contrast, the end of winter beach sediments had a mean of 1-2.7 phi class/interval with a standard deviation of 0.4-1.0 (coarse to medium sand size with poor sorting). Sediments from the ridges and swales tended to lie somewhere between the two.

These results support the idea that generally the nearshore profile is gently sloping, with prevailing low energy conditions, alternating with periodic storm events.

Mound nuclei

The third requirement is easily satisfied as berm building, and sediment trapping by plants on the beach backshore are common and ubiquitous processes in the Rockingham-Becher Plain. Several colonising plants are present in the Becher area: *Cakile maritima, Spinifex longifolia*, and *Tetragonia decumbens.*

Repetitive formational agent

In contrast to the regularity of swash built beachridge plains described from the north of Denmark and Mexico (Tanner 1993, 1995), the pattern of beachridges on the Becher Cusp suggests that there are a number of ridge building processes which are important, and which occur repetitively or cyclically. The primary repetitive mechanism is the modification of the energy and orientation of the prevailing southwesterly swell and wind wave systems opposite Becher Point by the nearshore bathymetry. The reduction in wave energy induced by localised wave refraction and convergence has continued

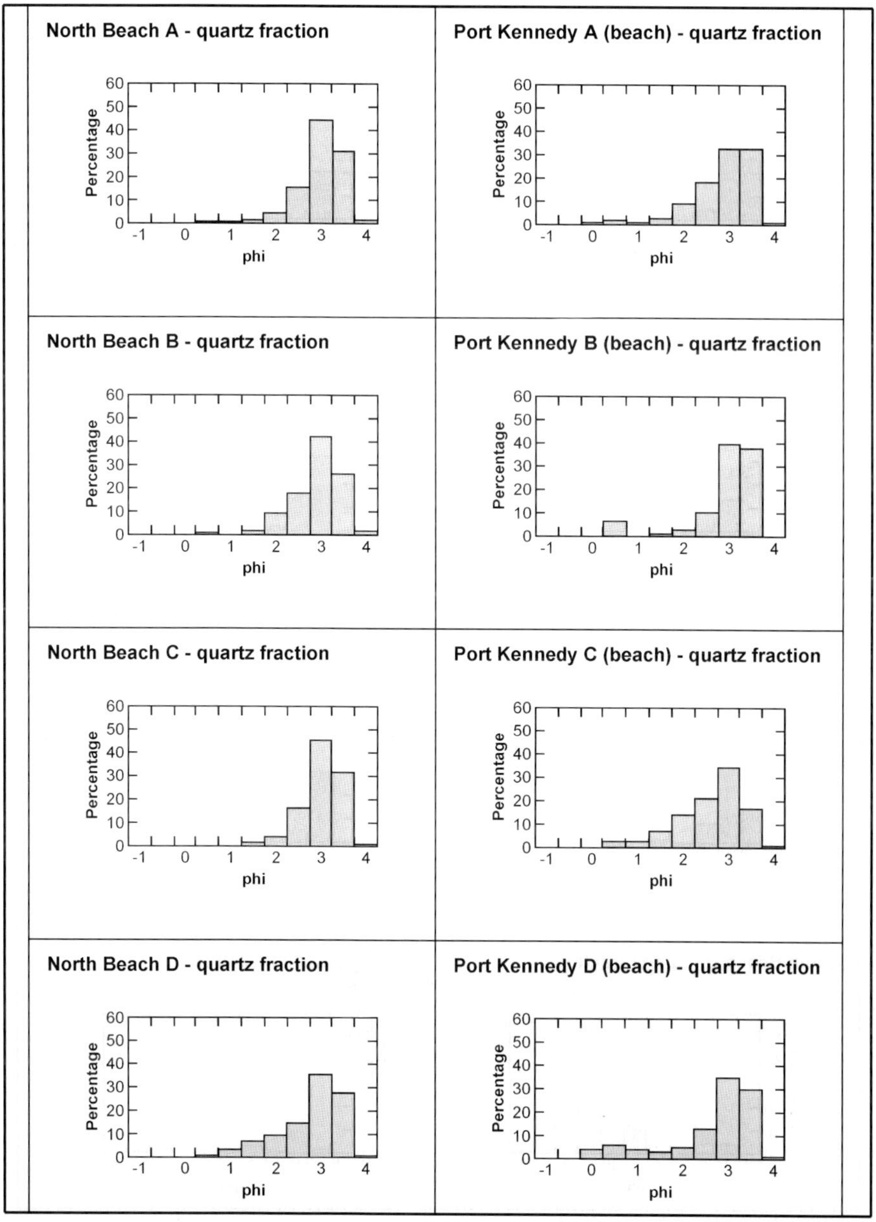

Figure 5-3. *Granulometry of the quartz fraction of parent material (beach and dune sediment)—Summer 1996.*

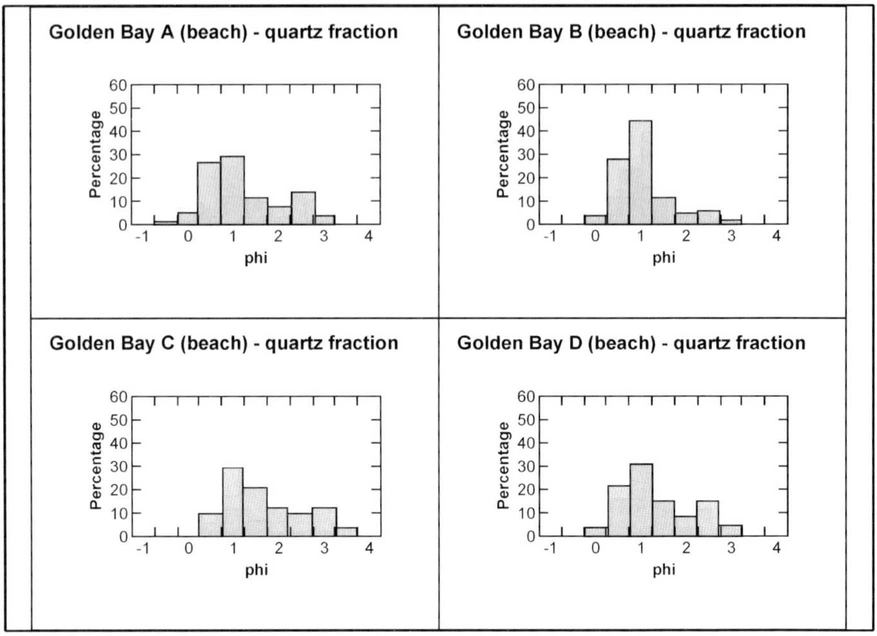

Figure 5-4 (cont.). *Granulometry of the quartz fraction of parent material (beach and dune sediment)—Winter 1996.*

from 8000 yrs BP to the present. Although the rate of erosion of the offshore reefs and the rate of sealevel fall may have been out of phase for some periods, there is no evidence of significant change to the pattern of concentric arcuate ridge lines at Becher Point and their convergence to the north and south in Warnbro Sound and Madora Bay.

The morphology, spacing, and internal structure of the ridges are the result of interruptions to the prevailing low wave energy conditions of the nearshore environment by regular cycles of increased storminess calculated to occur at an interval of *circa* 250 years (Semeniuk 1995). Storms and cyclones from the northwest, with resultant local wind waves, higher rainfall, and elevated shoreline water tables, result in erosion and reworking of shore sediments (Clarke and Eliot 1987). Associated increases in wind velocity and frequency result in mobilisation of sand into parabolic dunes disparate to the shore parallel ridges.

An additional repetitive formational agent, the local seabreeze system (responsible for the aeolian component of the beachridges), also derives from the southwest, thus enhancing the height and configuration of the ridges rather than obscuring them.

5.2.4 Evolutionary environmental history relating to beachridge and swale development

The pattern of accretion outlined in the evolutionary history of Chapter 3 partly explains the distribution of low and high ridges. Along the locus of most rapid accretion (east from the apex), beachridges tend to be low, interspersed with narrow flats lying close to the position of the relative sea level. Beachridges subject to average rates of uninterrupted accretion, tend to be low to medium height, separated by narrow swales of variable height above sea level. The beachridges, formed at times when slow accretion was interspersed with periods of erosion, tend to be high or low, (depending on sediment supply), and closely spaced, with the intervening swale surface above average relief (shore of Warnbro Sound and Madora Bay).

Rate of beachridge development

Analysis of beachridge development in relation to the evolutionary history and position of isochrons, shows considerable variability in the rate and type of beachridge development:

- the development of the larger ridges occurs in the earlier part of the cuspate foreland evolution (5000-3000 ^{14}C yrs BP)
- the number of small ridges is greater in the period of post apex development (2000-present ^{14}C yrs BP)
- the most rapid periods of progradation occurred between 5000-4000 ^{14}C yrs BP and 2000-present ^{14}C yrs BP
- periods of rapid accretion alternated with periods of static shoreline development

As there is significant evidence of coastal erosion and reworking of shoreline features over the last 1000 years, the rate of beachridge construction has been approximated for the period 5000-1000 ^{14}C yrs BP. The pattern of low beachridges suggests formation of a beachridge on a fairly regular basis every *circa* 50 yrs. This finding concurs with that of Semeniuk (1995) who related this frequency to the "Double Hale " 45-year cycle described by Fairbridge and Hillaire-Marcel (1977). During this same period, ten of the ridges constructed were higher than average. However, these larger ridges are not evenly distributed through the beachridge plain (Fig. 5-2): their occurrence is discussed below.

5.2.5 The higher set of beachridges

The higher set of beachridges are indicative of changes in the rate of beach-ridge/ dune development *i.e.*, slower rates of deposition brought about by changes in conditions such as: 1) an increase in wave energy, 2) reduced volumes of sediment supply, 3) a change in refraction intensity of the swell waves, and/or 4) a change in the in-shore beach profile (usually brought about by changing sea levels) (Johnson 1919; Davies 1958; Hails 1969; Wright 1970; Stapor 1975). There is evidence of all of these phenomena at Becher Cusp.

Cyclic storm activity and increased wave energy
An increase in wave energy may result from extensive periods of increased storm activity. Many features testify to the periodic storm activity inherent in the development of the beach ridge plain: truncation of the beachridge plain at the shoreline, erosion of the cuspate tip and erosion of the current beachridges. Relic geomorphic features, similar to those present today, are evident within the beachridge system, *e.g.*, areas of local parabolic dune incursion (now vegetated and stabilised), and, near Becher Point itself, a number of locations showing the former positions of tip structures (Fig. 5-5). These former positions show some lateral movement north and south of the major east west axis, but all are south of the present position. These relic features indicate a possible continuity between the types of processes operating currently and those operating in the earlier part of the Holocene.

Evidence of erosion is also found in the southern part of the Rockingham-Becher Plain (Secret Harbour). Mobile and stabilised parabolic dunes oriented southwest, together with a zone of carbonate dust fallout, unconformably overlie the southern portion of the Becher beachridge sandplain. The parabolic dunes are evidence of an erosional phase in the history of the cuspate foreland. During periods of increased storminess (increased wave energy) or reduced sediment supply, erosion of the coastal dunes takes place through the following process. Onshore winds increase wave height/ steepness and energy, and consequently the steepness of the inshore profile (Wright 1970). Waves reach higher levels on the beach. Wave abrasion creates exposed faces of dunes by removing the vegetation which normally stabilises them. Exposed dune slopes, if not rebuilt in a subsequent depositional phase, become weak points in the ridge, and nodes for wind abrasion (Short and Hesp 1982). In time, at these loci, the dune or beach-ridge is breached or blown out. A series of parabolic dunes develop transverse to the beachridge system.

Sand shadow formation is also strongly linked to storm activity (Clemmensen 1986). Two types of sand bodies dissecting the swales, which are not part of a parabolic dune, have been identified: 1) blowouts preceding parabolic dune development, and 2) aeolian sand shadows behind the chaots within the beachridge system closest to the shore.

Changes to sediment supply
There is evidence of geomorphic change in response to historic fluctuations in sediment supply within the Rockingham Becher Plain: 1) the cessation of progradation in the northern cusps centred on Safety Bay and Pt. Peron, due to the interruption and capture of northerly drift by the advancing Becher Cusp, 2) the erosion of beaches north of Warnbro Sound (Searle *et al.* 1988), 3) the deceleration in progradation of the shore of Warnbro Sound, and 4) the erosion of the southern shore of the Becher Cusp. On Becher Cusp itself, the following effects could also be attributed to variation in sediment supply: 1) a distinct pattern of ridge building interspersed with areas of flats,

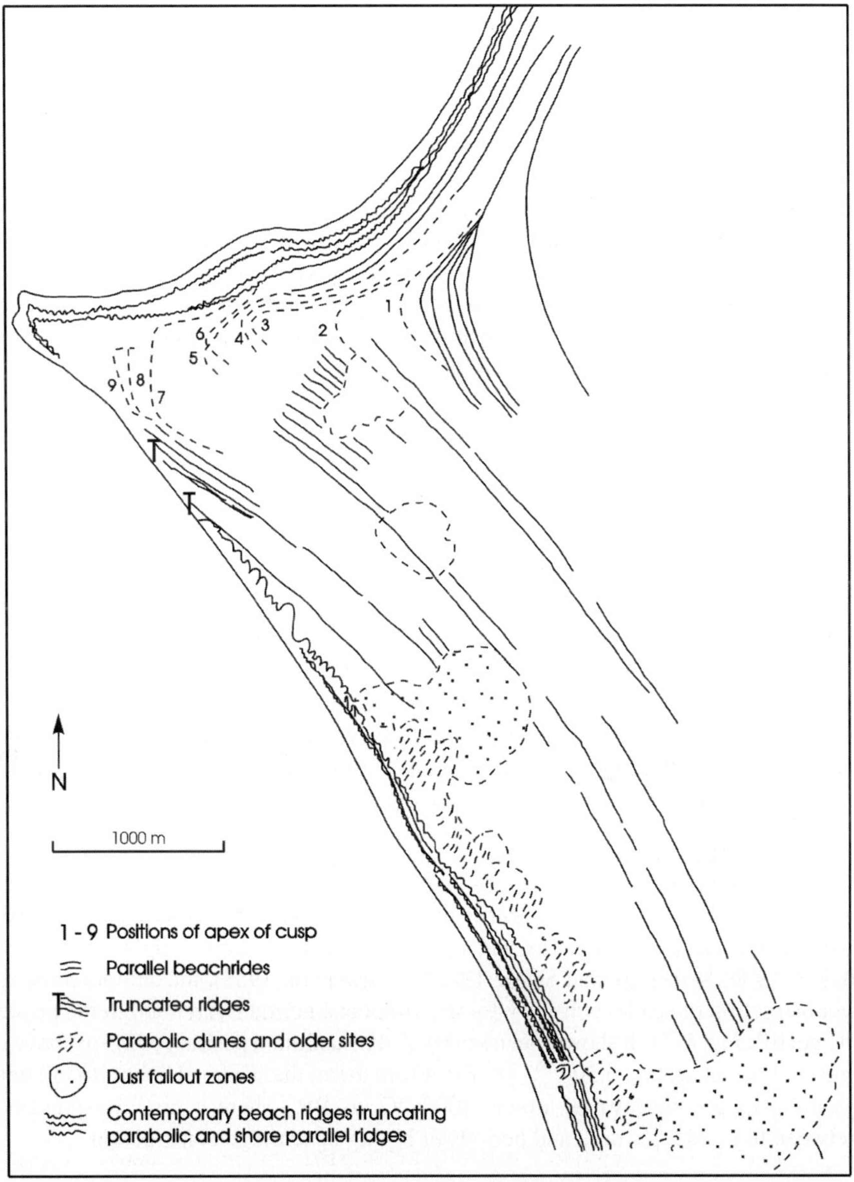

Figure 5-5. *Location of former positions of the apex of the cuspate foreland, areas of local parabolic dune incursion and truncation of beachridges.*

and 2) subsequent reworking of sediments into higher rather than wider shoreline features, resulting in the higher ridges. Woods (1984) notes that the average rate of advance of the shoreline south of Becher Point of about 0.6 m/year does not compare with the rate measured over the last 50 years of 4 m/year (Secret Harbour, 1981). A variable supply of sediment is strongly implied by this anomaly.

Woods (1984), however, states that the mode of growth of the beachridge plain during the Holocene and on the contemporary coast, *i.e.*, rapid progradation followed by a stable shoreline, is dependent on a steady supply of sediment, implicating a repetitive if not cyclic phenomenon, which he suggests may be a series of small scale oscillations in sea level (±50 cm). This idea is echoed by other authors (Stapor 1975; Fairbridge and Hillaire Marcel 1977), however, if the sediment supply is assumed to be inconsistent (fluctuating), as would be the case in the erosion of the Leschenault Barrier, or even the offshore islands, then changes to sediment supply would affect the bathymetry of the Madora Basin and thus, the wave energy, producing the same results as oscillating sea level.

A change in refraction intensity

A change in refraction intensity of the swell waves would be expected to have the most pronounced effect at the cuspate tip. Concordant with normal cuspate foreland development, spits would begin to develop as the cuspate foreland became more pronounced, (*circa* 3000 ^{14}C yrs BP). As a result, wave dynamics and longshore drift patterns would be altered in this local area. However, this effect would be concentrated along the spit itself, as evidenced in the contour map showing the higher elevation of barrier ridges (Fig. 5-6), and is not linked to general high ridge construction.

Sea level changes

A change in the in-shore beach profile may be brought about by changing sea levels. Using both sea level indicators and radiocarbon age of shell samples (Searle and Woods 1986; Semeniuk and Searle 1986b; Searle *et al.* 1988), the authors present a reconstruction of sea level history for the Holocene period *circa* 8000 to the present ^{14}C yrs BP (Fig. 5-7). It shows mean sea level at a maximum position of 2.5 m above its present level at around 6,500 ^{14}C yrs BP. From there, the sea level begins to fall until reaching its present level at around 1000 ^{14}C yrs BP. There is no correspondence between the sealevel curve and periods of higher beachridge construction.

5.2.6 The modern beachridges

The contemporaneous beachridges comprise the presently forming beachridge, two older beachridges and a zone of chaots, hollows and blowouts on the shores of Warnbro Sound and Madora Bay. The beachridges are aligned normal to the north-westerly trajectories of the refracted swell/wind wave system. The source of sand

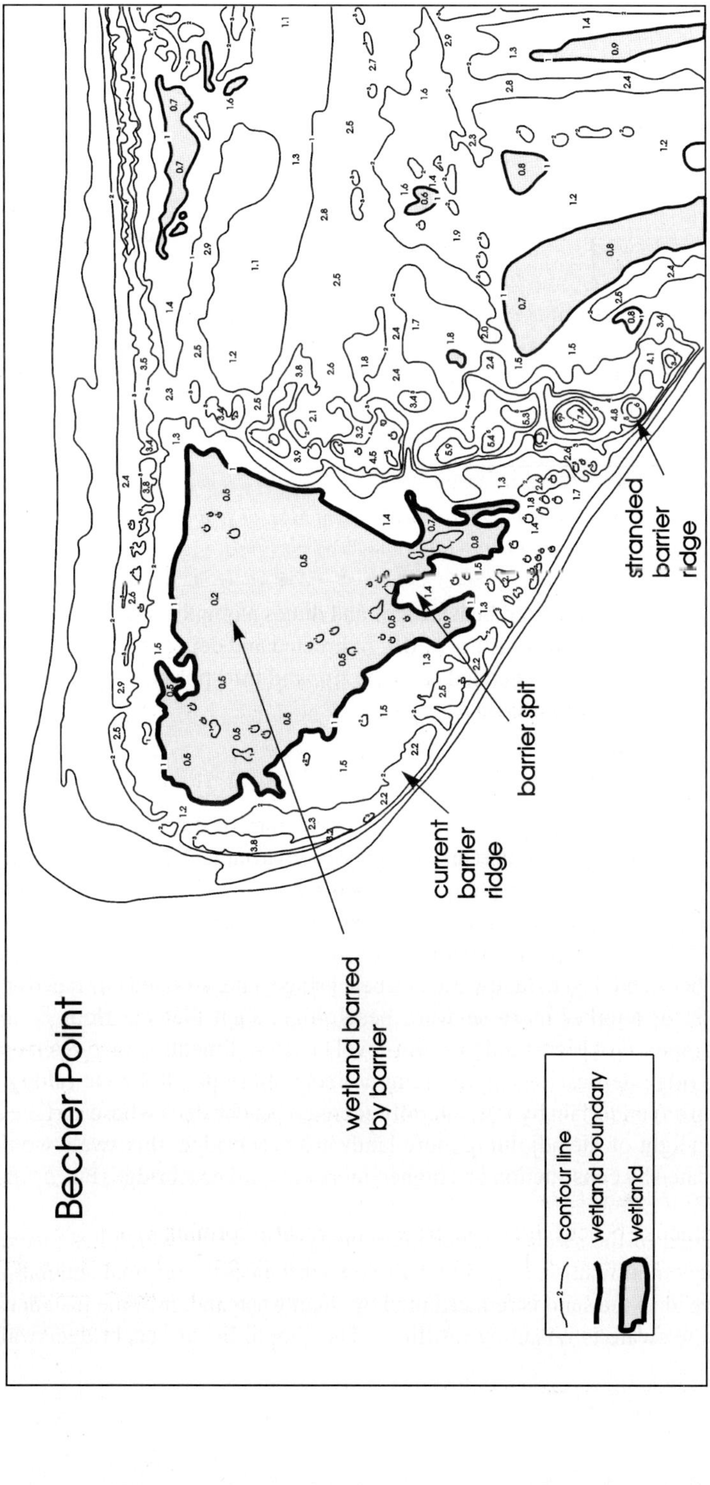

Figure 5-6. *Morphology of current Point Becher tip showing relatively higher barrier ridges, stranded former barrier ridges and spits barring wetlands.*

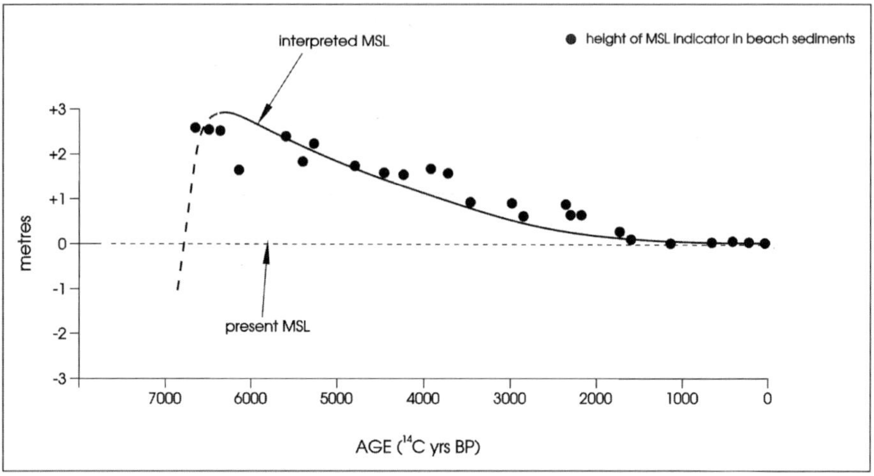

Figure 5-7. *Radiocarbon ages and height above present MSL of sealevel indicators in the Rockingham area (after Searle et al. 1988).*

appears to be material eroded from beachridges and dunes along the southern cuspate shoreline, transported north by longshore drift, redirected and deposited by refracting waves and complex currents associated with the tip, and transported shorewards by the swell waves entering Warnbro Bay.

5.2.7 The development of beachridge swales

The landscape of the beachridge plain, built fundamentally through accretionary processes in a relatively low to moderate wave-energy environment, has been modified by processes associated with the repetitive short term but intense storm activity. The swales of the beachridge plain have formed in one of two settings:

1. as a beach-berm, backed to landward by a beachridge, later isolated (stranded) by construction of another more seaward beachridge, such that the floor of the swale is composed of high-tidal to storm-level beach sediments (Fig. 5-8A); or
2. as an inter-ridge depression, representing a linear shore-parallel zone (along a given isochron) underlain by low, laterally accreted beachridges whose surface is below the height of the adjoining more landward beachridge; this swale would later be isolated by construction of a higher, more seaward beachridge (Fig. 5-8B).

The relative height of the swale depends upon the type of swale which has formed and its subsequent modification (Fig. 5-8C). In some instances the original undulating surface is flattened as the sand is redistributed by sheet wash and, in some instances, the surface of the swale is raised by infilling. The simple linear beachridge/swale

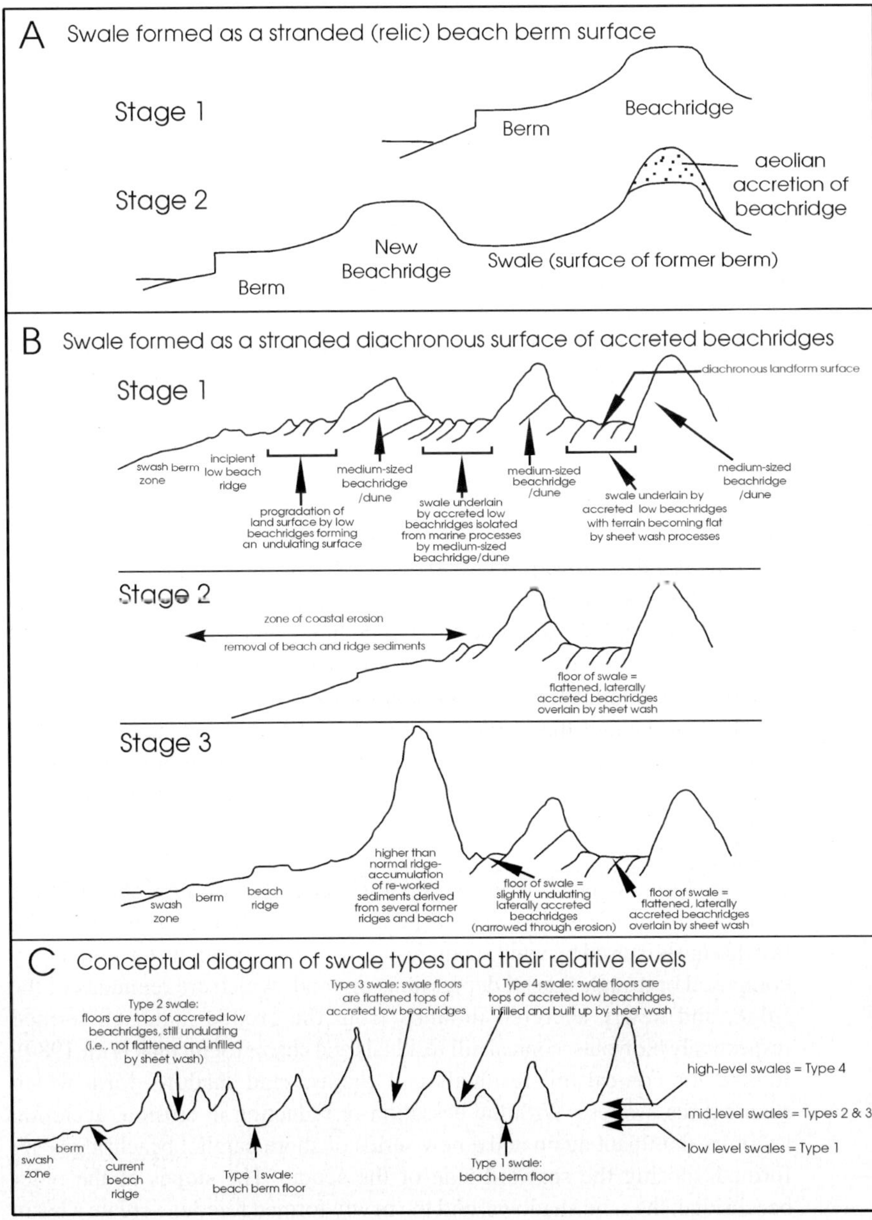

Figure 5-8. *Development of beachridge swales - as stranded berms, and as accreted beachridges - and their relative levels.*

morphology is eventually overprinted in places, by other dune forms such as parabolic dunes, chaots, and sand shadows, resulting in a segmented swale system.

5.2.8 Development of proto-wetland basins

Simple linear swales, through various geomorphic processes of construction and destruction, have been segmented, creating a more complex topography. There are three ways that linear swales have been modified.

1. During the phase of beachridge construction, local variability in the energy of the constructive element, and in the source and supply of material along the cuspate shoreline, can result in variability of ridge construction in adjacent segments or cells. In some cases two ridges form in one cell contemporaneously with a single ridge in the adjacent cell. When these ridges coalesce, the result is an inter-ridge basin between the bifurcated arms of a single ridge (Figs. 5-9A, 10).

2. Subsequent to beachridge development, during periods of storminess, erosion and reworking of coastal ridges takes place. The most common outcome is the development of parabolic dunes oriented in the same direction as the prevailing wind, *viz.*, south-west. These dunes may extend across several beachridges if conditions persist and sand supply is plentiful. However, when supply diminishes, it is common for the advancing edge to be blown out, leaving behind the remnant arms of the parabolic dunes which now become shore-transverse ridges (Semeniuk *et al.* 1989) within a system of disrupted low ridges. These transverse ridges segment the swales, forming basins (Figs. 5-9B, 10).

3. Between formation and stabilisation, the beach ridges are subject to erosion, particularly in areas that lack vegetation. Common disruption of the beachridge is caused by wind scour, resulting in the formation of a topography composed of trough shaped depressions, mounds which are remnants of the ridge, and newly accreted mounds from the eroded material, termed respectively blowouts, conical hill residuals and chaots (Semeniuk *et al.* 1989). Behind the conical hill residuals and chaots, sand shadows form which segment the swales. With the cessation or reduction in erosion, accretion becomes dominant again and a new series of shore parallel beachridges are formed, closing the seaward side of the scour. The slopes of the older beachridge, the sand shadows, and the newly formed foredune create a basin within the swale (Figs. 5-9C, 10).

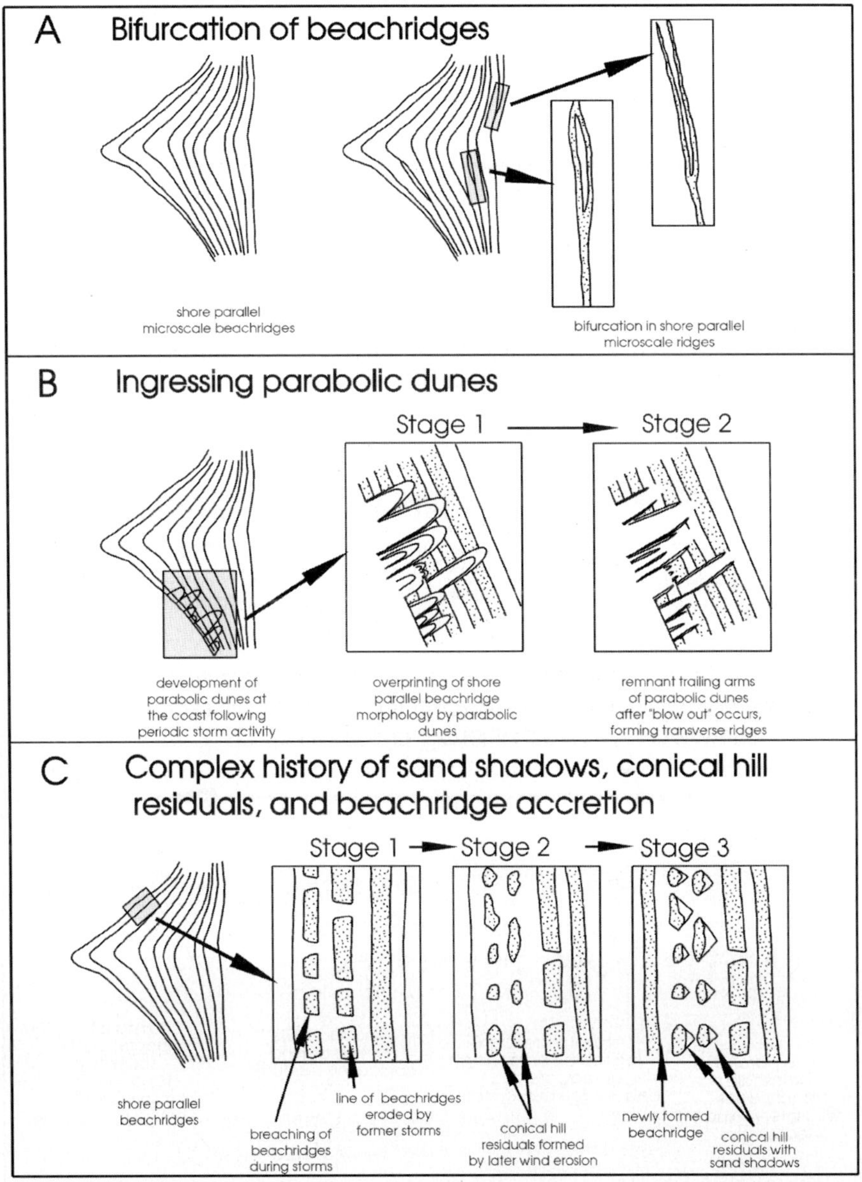

Figure 5-9. *Sedimentologic and geomorphic processes of construction and destruction which result in partitioning of beachridge swales, and the resulting development of proto-wetland basins.*

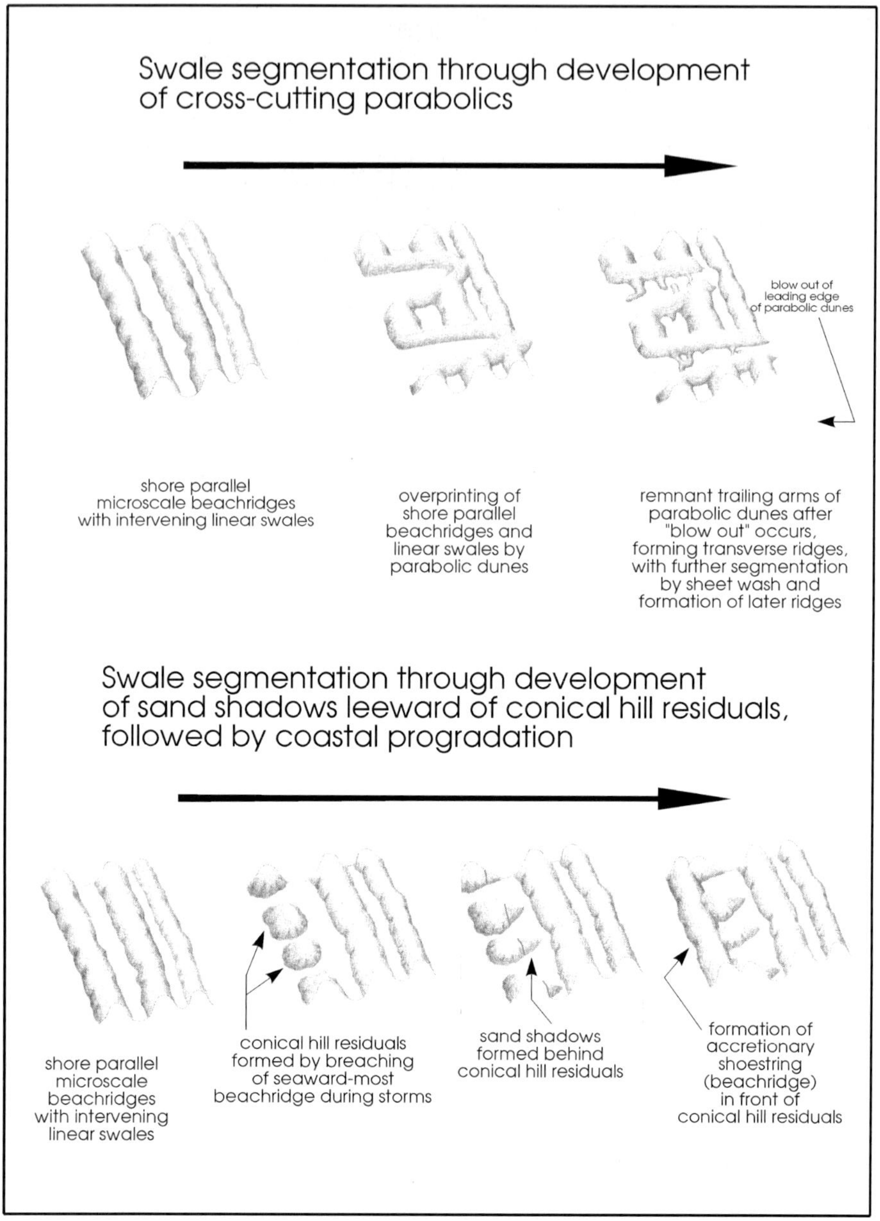

Figure 5-10. *Segmentation of swales through two processes. 1. inland migration of parabolic dunes, and 2. development of sand shadows behind conical hill residuals.*

Thus the various dune forms, through segmentation of some of the linear swales, have created separate basins (Fig. 5-10), the geometry of which has been determined by the same sedimentary processes that created the dune forms. As a result, there are two types of swale:

- a swale which is a simple linear, relatively homogeneous feature; and
- a swale which is a linear feature exhibiting complex internal configuration.

Within the latter swale type, the basins, separated by low ridges, are shallow and steep sided, but vary in size and shape. Size varies from leptoscale to microscale and shape varies from linear to irregular, from ovoid to circular, and from crescentic to irregular. These basins are the proto wetlands.

5.3 Wetlands

5.3.1 Introduction

This section describes the origin of the wetlands on the Becher beachridge plain. The objective is to reconstruct how wetlands were initiated, concentrating on the wetland basal sediments and the parent sediments that were the floor of the proto-wetland, the age structure of the wetlands, and wetland distribution in relation to the age structure.

5.3.2 Basal sediments

Basal sediments (basal sheet and basement sands) are the materials at the base of the sedimentary fill of the wetlands (Fig. 5-11). Analysis of these sediments was undertaken to identify the nature of the parent material and geomorphic setting host to the proto-wetlands, *i.e.*, whether the proto-wetland was founded on beach-berm sediment, or a dune swale. In order to do this it was necessary to identify the formative environment (*i.e.*, beach *vs* dune) of the parent material underlying the wetland basin. Three approaches were taken:

1. Histograms of the granulometry of the basal sediments were compared to those of current dune and beach samples collected from the corresponding geomorphic units within the Rockingham-Becher Plain. Statistical parameters were compared such as mean, mode, standard deviation, skewness and kurtosis, in order to identify beach and dune sands.

2. Quartz sand was separated from the basal wetland sediments by acid-digestion, and the occurrence of a coarse quartz sand fraction was used as an indicator of swash zone hydrodynamic processes. Sands with > 10% coarse quartz sand were categorised as beach sands.

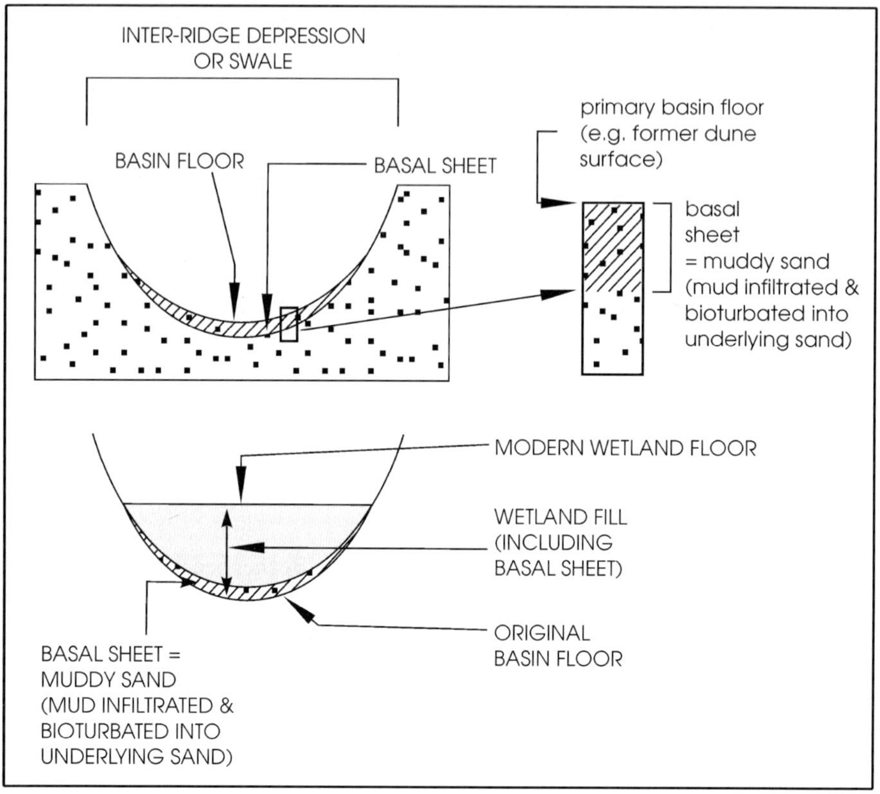

Figure 5-11. *Terminology for geomorphic and stratigraphic components of wetland basins and wetland fills.*

3. Cores of beach, dune, and swale were collected and described as standards for recognition of sedimentary structures reflecting these environments in the cores of the basal sheet.

Descriptions of histograms

Histograms of basal sheet sediments in all wetlands and (for comparison as analogues), histograms of beach and dune sediments collected from the Rockingham-Becher Plain, are presented in Figures 5-3, 5-4 and 5-12. In the histograms of the basal sheet sediments (Fig. 5-12), several of the distributions are bi-modal (wetlands 162, 135, 45, 9-6, 9-14). Distributions are negatively or positively skewed. The sediments are moderately well to poorly sorted. Statistical parameters are tabled below.

Table 5.1 *Statistical parameters of basal sheet sediments underlying study wetlands*

Site	*Mean (ϕ)*	*Mode(ϕ)*	*St. Dev*	*Skew*	*Kurtosis*
161	1.65	1	0.8	0.13	-0.89
162	1.79	1	0.72	0.03	-0.95
163	1.83	2	0.64	0.11	-0.63
WAWA	1.32	1	0.59	0.58	-0.14
142	1.02	1	0.74	0.46	-0.18
135	1.98	2.5	1.04	-0.09	-0.65
136	1.77	1.5	0.67	0.16	-0.63
72	2.15	2.5	0.64	-0.59	0.58
63	2.36	2.5	0.47	-0.22	0.08
45	2.20	3	0.92	-0.39	-1.15
35	2.35	2.5	0.65	-0.8	0.06
9-6	1.56	1	0.91	0.38	-1.09
9-14	1.75	1	0.95	0.18	-1.32
swii	2.58	3	0.62	-0.88	0.70
swiii	2.58	3	0.73	-1.11	1.0

Basal sheet sediments may be subdivided into two categories based on the modal distribution exhibited by the histograms: uni-modal and bi-modal sediment distributions. Within the uni-modal distribution category, further subdivision can be made using values of the mean and mode: those whose mean and mode lie in the coarse to medium sand fraction (161, 163, WAWA, 142, 136), and those whose mean and mode lie in the fine sand fraction (72, 63, 35, swii, swiii). The standard deviation in the uni-modal sediments indicates wetland 63 to be well sorted, and the other wetlands to be moderately well sorted. Wetlands 72, 35, swii and swiii are strongly coarse skewed and wetlands WAWA and 142, are strongly fine skewed, with prominent partitioning of grain sizes in the tails.

Within the bi-modal distribution category, all distributions exhibit one mode in the coarse sand fraction and one mode in the fine sand fraction (162, 135, 45, 9-6, 9-14).

Grain size distribution in wetlands 72, 63, 35, swii, and swiii is consistent with sediment transport by aerodynamic processes. Grain size distribution in wetlands WAWA and 142 are most consistent with sediment transport in the beach swash zone environment. Other sediments which exhibit a mode in the coarse sand fraction and mean in the medium sand fraction and which are moderately well to poorly sorted (wetlands 161, 162, 136, 9-6 and 9-14), may also indicate beach sediments (Stapor 1975). Wetlands 162, 135, 45, 9-6, 9-14, and swiii, exhibit distributions where more than one population is identifiable (Glassford 1980). Wetlands 163 and 136 are examples in which each population is masked in the entire distribution (Glassford 1980).

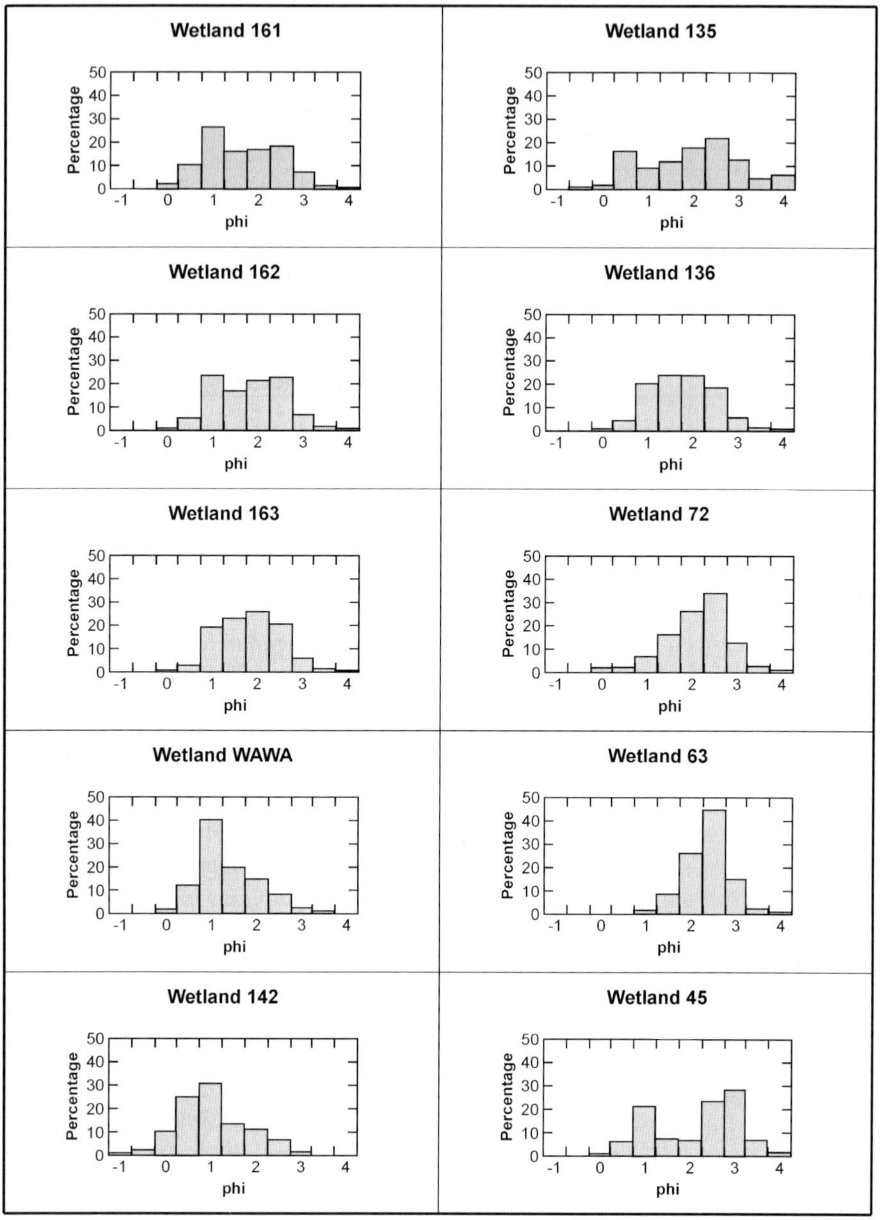

__Figure 5-12.__ Histograms of basal sheet sand granulometry using only the quartz fraction.

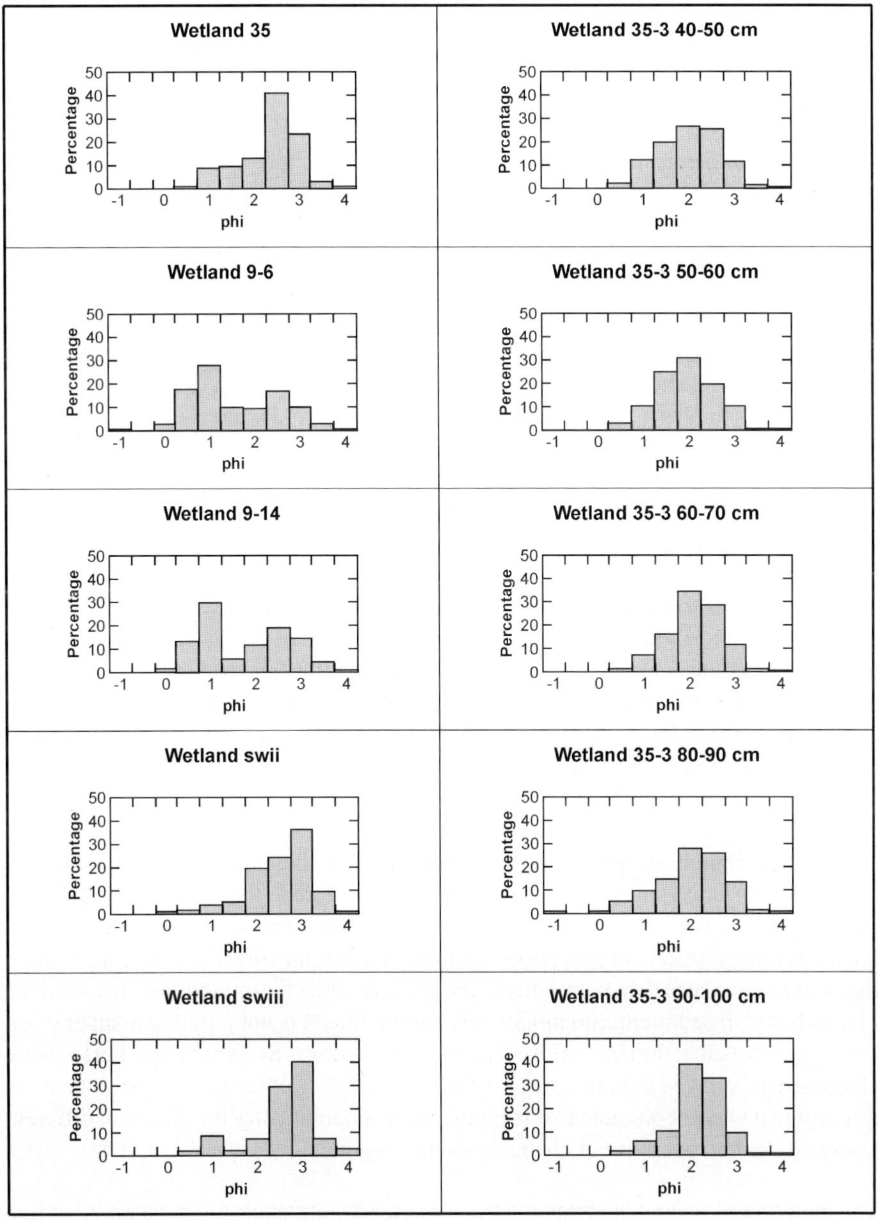

Figure 5-12 (cont.). *Histograms of basal sheet sand granulometry using only the quartz fraction.*

For comparison with wetland basal sheet sediments, statistical parameters of grain size distributions from beach and dune sands in the Rockingham-Becher Plain are presented in Table 5.2.

Description of grain size distributions using modern analogues

Samples of beach sand were collected across the high tide to swash zone in summer and winter from a range of sites because of the dynamic nature of the beach environment. Since the dunes and beachridges are not mobile but fixed by vegetation, samples of dune sand were collected once only. The results of granulometric analyses of the beach samples from summer sampling show that the mode and mean lie in the range of fine sand. The source of this medium/fine sediment is most likely the dunes along the southern shore of the cuspate foreland which have been undergoing erosion. North Beach samples A, B, C and D are well sorted (Fig. 5-3), and all other samples are moderately well sorted with the exception of Port Kennedy (PK) Beach D (backshore). All samples are coarse skewed. Kurtosis values indicate poor sorting in the tails of samples near the Becher Cusp apex, and poor sorting in the central portions in samples from north and south beach.

The granulometric analyses of the beach samples from winter sampling show that the mode and mean of most of the samples lie in the range of coarse sand, but there are several sites composed of medium and fine sand. Port Kennedy Beach is well sorted but all other sites exhibit moderately well sorted grains. Although several samples are coarse skewed (Sth Beach A, D and SH-A, B, C), most other samples are finely skewed (Fig. 5-4). Most samples exhibit lack of sorting in the central portion.

The results show that beach sediments can lie in the range of both coarse and fine sands in this region, depending on the local source of the material and the energy of the waves. During winter, when storm waves are prevalent, the coarser material from the sea grass banks is transported onshore together with fine sand reworked from the local foredune erosion. In the course of subsequent winnowing, the coarse lag is more likely to be preserved than the medium/fine fraction, reflecting conditions characteristic of beaches. All sediments are moderately sorted, thus it is not possible to distinguish dune from beach sediments on the criterion of sorting. Skewness and kurtosis are characteristics linked to both transport mode, and energy of the transporting agent. In this region it was not possible to distinguish in the granulometry the effects of different transport mode from those of changing energy regimes.

The dune samples show that the mean and mode of dune sands lie in the range of fine sand. The samples are all moderately to well sorted, three sites indicate poor sorting in the tails and four in the central portion. All samples are coarse skewed.

Table 5.2 *Statistical parameters of regional beach and dune sands, summer and winter*

SUMMER SAMPLING						*WINTER SAMPLING*					
Site	*Mean (Phi)*	*Mode (Phi)*	*St. Dev*	*Skew*	*Kurtosis*	*Site*	*Mean (Phi)*	*Mode (Phi)*	*St. Dev*	*Skew*	*Kurtosis*
Beaches						*Beaches*					
Nth Beach A	3.04	3	0.44	-0.8	0.8	Nth Beach A	1.73	1.5	0.74	0.62	-0.41
Nth Beach B	2.95	3	0.49	-0.61	-0.65	Nth Beach B	2.0	1.5	0.7	0.22	-0.97
Nth Beach C	3.03	3	0.42	-0.87	0.92	Nth Beach C	2.33	2.5	0.63	0.25	-0.2
Nth Beach D	2.79	3	0.71	-1.14	0.71	Nth Beach D	1.76	1.5	0.58	0.25	0.03
PK Beach A	2.92	3	0.61	-1.26	2.04	PK Beach A	1.52	1.5	0.49	0.01	-0.35
PK Beach B	2.95	3	0.74	-2.27	4.84	PK Beach B	1.69	1.5	0.47	0.04	0.06
PK Beach C	2.65	3	0.69	-0.96	0.69	PK Beach C	1.92	2	0.42	0.19	-0.26
PK Beach D	2.7	3	0.94	-1.6	1.56	PK Beach D	1.95	2	0.42	0.31	-0.10
Sth Beach A	2.7	3	0.54	-0.19	-0.79	Sth Beach A	2.02	3	1.02	-0.3	-1.3
Sth Beach B	2.87	3	0.61	-1.23	1.78	Sth Beach B	1.53	1	0.92	0.58	-0.84
Sth Beach C	2.92	3	0.47	-0.39	-0.75	Sth Beach C	1.8	2.5	0.78	0.10	-0.90
Sth Beach D	2.63	3	0.61	-0.30	-0.84	Sth Beach D	2.7	3	0.61	-0.85	0.15
Dunes						SH-A beach	2.24	3	0.88	-0.88	-0.48
TRN 45	3.08	3	0.55	-1.5	3.8	SH-B beach	2.34	2.5	0.78	-1.05	0.06
WAWA D	1.93	2.5	0.69	-0.20	-0.73	SH-C beach	2.25	2.5	0.74	-0.76	-0.31
Warnbro	2.48	2.5	0.38	-0.07	0.5	SH-D beach	1.72	1	0.87	0.28	-1.32
Opp. 63	2.09	2	0.5	-0.12	-0.11	GB-A beach	1.22	1	0.79	0.64	-0.64
RN 45	2.85	3	0.62	-1.9	3.87	GB-B beach	1.02	1	0.59	1.14	1.51
TBTCA	2.62	2.5	0.44	-0.86	1.51	GB-C beach	1.66	1	0.84	0.55	-0.82
TBTCB	2.6	2.5	0.39	-0.34	0.33	GB-D beach	1.31	1	0.79	0.53	-0.73

Comparison between basal sediments and modern beach/dune sands

Granulometry of beach/dune samples collected throughout the Rockingham Becher Plain was used to assist in the interpretation of the depositional environments of the basal sediments. There was considerable overlap and gradation in the range of mean grain sizes, modes and standard deviations in the beach and dune samples. The granulometric parameters of known beach sediments in the local region showed a greater variation from summer to winter and swash zone to beach backshore than those of beach *vs* dune in a single season. These samples served to illustrate the range of grain size distributions from beach environments within the study area. In attempting to identify beach sediments in the stratigraphic cores, they acted as a caveat with regard to over interpretation.

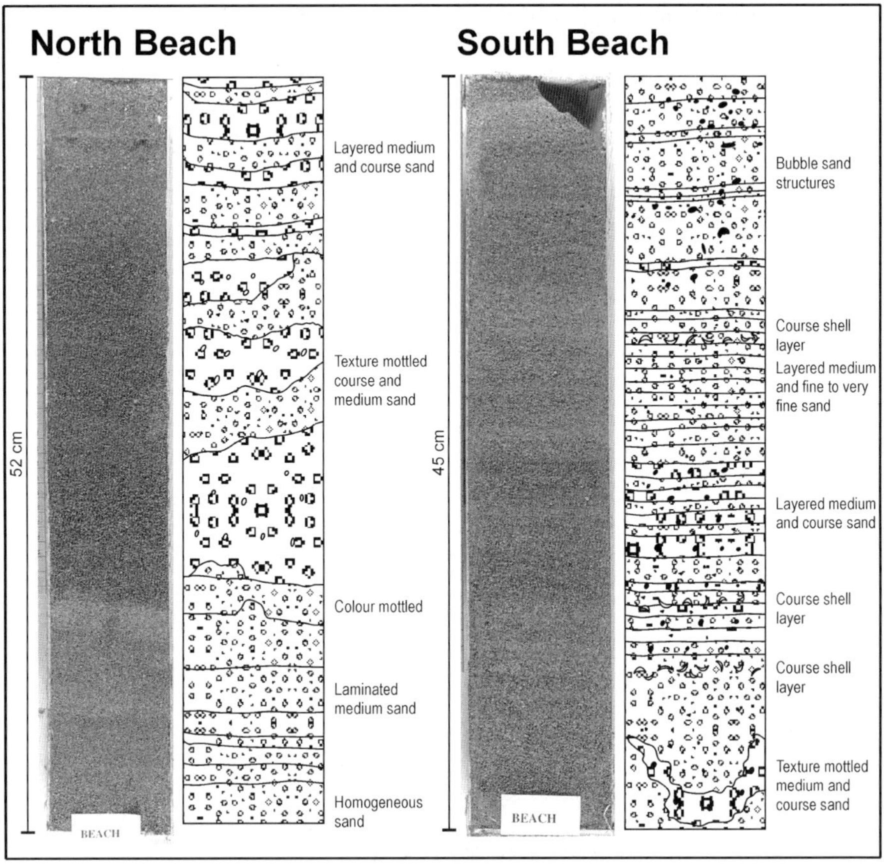

Figure 5-13. *Sedimentary structures in beach sediments.*

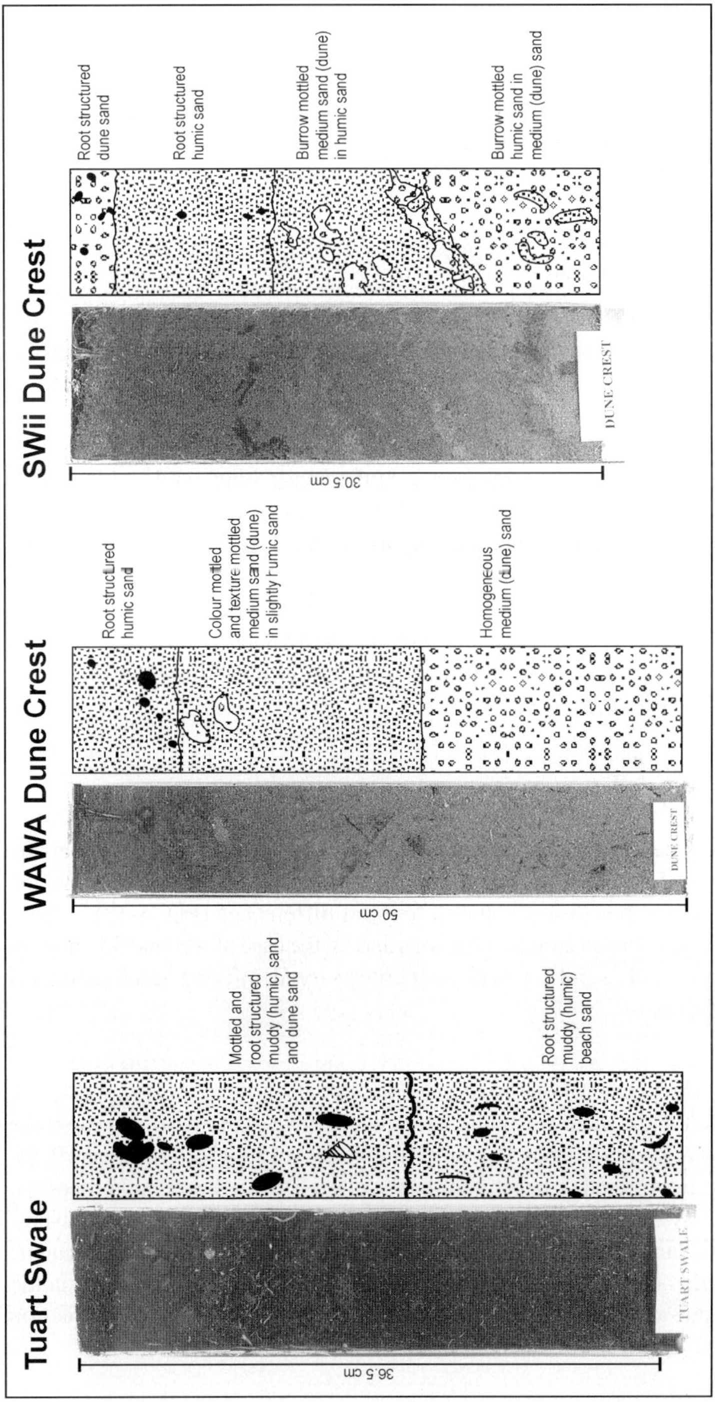

***Figure 5-14**. Sedimentary structures in sediments underlying a swale and dune crests.*

Although mean grain size is an indicator of local availability of sand supply, the degree of skewness and the categories in the coarse tail of the distribution may be used as indicators of lower beach, beach ridges and coastal dunes (Doeglas 1946, Stapor and Tanner 1975). In this study, the categories in the coarse tail of the grain size distributions were used to support the argument for beach origin of many of the basal sands, *i.e.*, the presence of coarse and very coarse sand.

Granulometry of quartz sand as an indicator of beach and dune sediments

Direct comparison of granulometry of modern beach and dune sediments with the sand component of the basal sheet carries with it the problem that a proportion of the grains from wetlands have been generated *within* the wetland, *i.e.*, shell gravel and shell comminuted to very coarse and coarse sand grades will bias the sand analysis of the basal sheet towards the coarse fraction. To circumvent this problem, the carbonate grains of all beach and dune samples and from the basal sheet were removed by acid digestion, prior to granulometric comparisons of the quartz sand grain residue. This approach also circumvents the potential problems in beach and dune environments with carbonate grains of different shapes and density, that all grains are not hydraulically equivalent.

In this context, sediments with dominant coarse quartz sand were categorised as beach, those with medium to fine sand as beachridge/dune, and those with mixed populations as beach-berm.

Description of beach and dune in situ cores

In addition to the surface sampling of sediments, cores were extracted from beach and dune sites to provide information on structures which may be present under the wetland fills (Figs. 5-13, 14). The most important structures identified were the laminations in the beach sediments due to textural differences (Fig. 5-13). These laminations were identified in stratigraphic logs and, in the case of wetland 35, through granulometric analysis (Fig. 5-12) where other criteria for identifying beach sediments proved to be ambiguous.

Interpretation of the results of the three approaches

In many of the wetlands, there was good agreement between the results of the three approaches (Table 5.3). In wetlands 161, 162, 163, WAWA, 135, 136, 142, 35 and 9, the basal sheets were interpreted as beach sediments, altered through wetland processes. In wetlands 72, 63, swi, swii, and 1N, the basal sheets were interpreted as dune sediments, altered through wetland processes. In wetlands swiii and 45 the results are ambiguous. In the first case, this is the result of mixed populations in the basal sheet, and in the second case it is due to incomplete sampling, the consequence of wetland destruction early in the study.

Table 5.3 *Depositional environment of basal sediment based on three criteria*

Wetland	*Depth of basal sands*	*Texture of basal sands*	*Identification based on granulometric distribution histograms*	*Identification based on coarse quartz component*	*Identification based on structures*
161	100 cm	vc, c sand	beach	beach	beach
162	100 cm	vc, c sand	more than 1 population	beach	beach
163	50 cm	c, m sand	inconclusive	beach	beach
WAWA	75 cm	c, m sand	beach	beach	beach
135	75 cm	bi-modal vc, f sand	more than 1 population	beach	beach
136	55 cm	c, m, f sand	beach	beach	
142	100 cm	c sand	beach	beach	beach
72	55 cm	f sand	dune	dune	
63	55 cm	f sand	dune	dune	dune
45	40 cm	bi-modal vc, f sand	inconclusive; more than 1 population	*dune	not sampled
35	100 cm	f sand	dune	dune	beach
9-14	75 cm	bi-modal vc, f sand	more than 1 population	beach	
9-6	55 cm	bi-modal vc, f sand	more than 1 population	beach	
swi	20 cm	f sand	dune	dune	dune
swii	40 cm	f sand	dune	dune	dune
swiii	45 cm	f sand	more than 1 population	dune	
1N	10 cm	f sand	dune	dune	dune

vc very coarse sand; c coarse sand; m medium sand; f fine sand
*Sampled in a different basin within the swale containing wetland 45 which had been cleared for recreation.

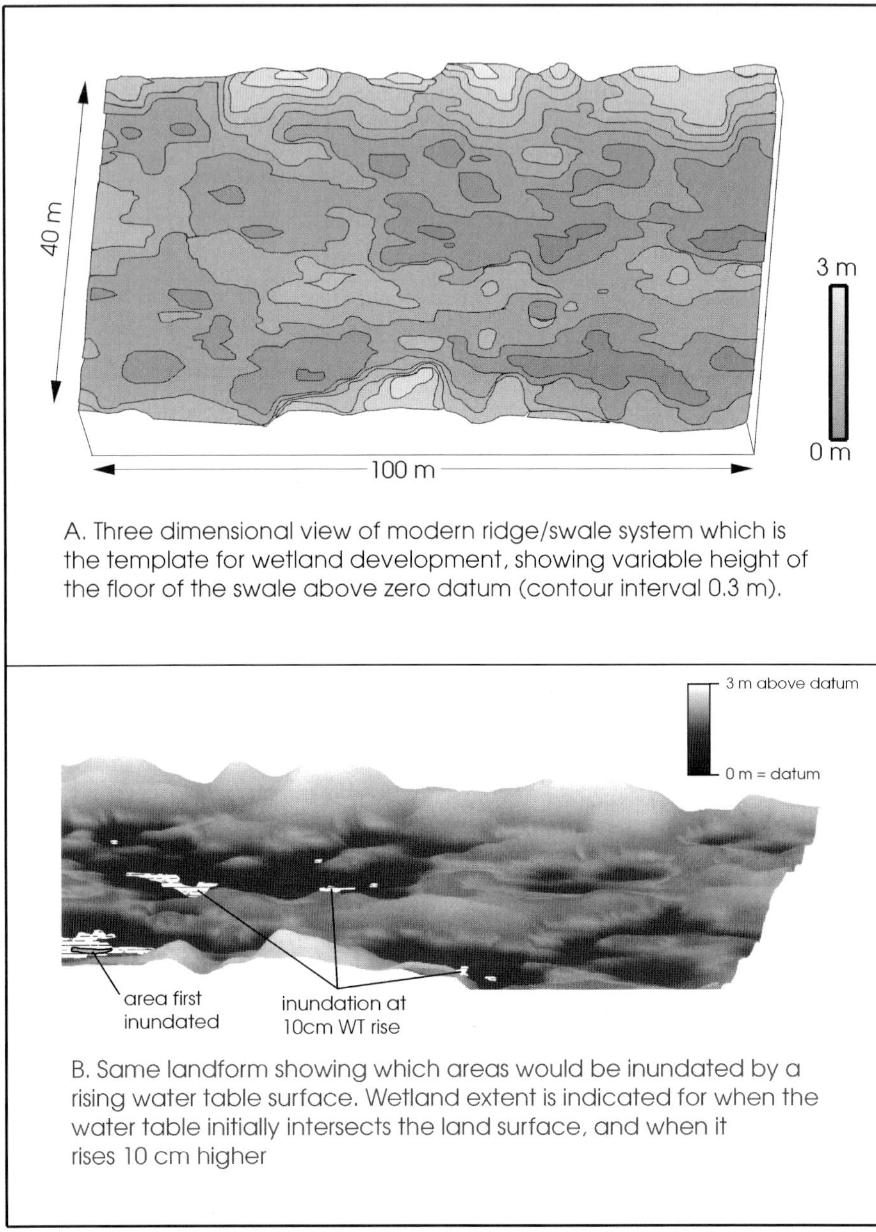

Figure 5-15. *Irregular pattern of wetlands developed by a rising water table intersecting a swale with uneven surface.*

In the majority of cases, the unaltered parent material underlying the wetland sediments is interpreted to be beach sand on the basis of sedimentary structure and grain size. In the older wetlands, located in eastern parts of the Becher Cusp, this sediment is a coarse, medium, and fine sand. In the younger wetlands, located in western parts of the Becher Cusp, this sediment is a laminated, fine to medium sand, or medium to coarse sand.

5.3.3 A model for wetland initiation

Having examined in some detail the "land" component of the term "wetland", a brief discussion of the "wet" or water component is presented. Underlying the Becher beachridge terrain is an aquifer (Safety Bay Sand and Becher Sand), in which the body of unconfined fresh groundwater is seasonally recharged. The groundwater surface is semi-convex, and its morphology, slope, and height are a function of net recharge, sea level, and geographic position within the landscape. During progradation of the land surface seawards, at a fixed point, distance from the sea and relative height of the water table increased. As a result, low topographic areas within the swales were progressively intersected by groundwater. At the same time, near the coast, a combination of falling sea level, rapid progradation, and possibly lower volumes of sediment resulted in the development of swales not much higher than the present mean sea level. In this location also, low topographic areas within the swales were intersected by groundwater. With this process, initiation of the Becher Suite wetlands commenced.

To illustrate the type of processes taking place in the development of the Becher Suite wetlands, an area of modern beachridge/swale construction on the northern shoreline of the Becher Cusp was surveyed at a grid of 2 x 5 m intervals, together with the groundwater levels (Fig. 5-15A). This diagram provides a template for wetland development. The area, near the apex of the Becher Cusp, is located in the midst of presently forming beachridges, chaots, swales dissected by sand shadow features, and basins, typical of the proto-wetland landscape. With further progradation of the cuspate foreland, the groundwater level will rise. A simulated water table rise was programmed for the template resulting in some parts of the swale being inundated (Fig. 5-15B). Because of the undulating nature of the land surface, both across the beachridge plain and along a given swale, the pattern and order of wetland formation does not reflect either spatial or temporal gradients, *viz.*, distance from the coast, nor age of the swale.

5.3.4 Dates for wetland commencement

Radiocarbon dating of base of wetlands

Radiocarbon analysis of the carbonate mud infiltrated into the basal sediments of each wetland was used to procure the date of wetland initiation, based on the assumption that the deepest occurrence of carbonate mud indicates the initiation of wetland processes (Fig. 5-16). Because there is rarely a sharp boundary between the layer of mud infiltration and the underlying parent material (beach sand, humic beach sand, dune sand, humic dune sand), the base of the infiltration layer was identified using the following methods:

1. collection and analysis of cores of parent materials, *viz*., beach, dune crest, swale, seagrass (Fig. 5-14)
2. collection of cores containing the contact between wetland sediments and underlying parent materials
3. identification of parent material sediments through textural and compositional grain size analysis
4. identification of buried humic horizons (interpreted to be the sediments in the swales)
5. determination of mud component of muddy sand horizons

Radiocarbon dates of basal muds are presented in Table 5.4 and Figures 5-16, 17. From the radiocarbon dates it would appear that the wetlands formed in the following order: Cooloongup east wetlands, Cooloongup west wetlands, 161, 162, A7, 229, 163, 45, WAWA, 135, 35, 9-14, A9, 136, 72, 142, 9-6, 63, C5, swiii, swii (Fig. 5-17B).

Evolutionary model for wetland development

It can be seen from Table 5.4 and Figure 5-16 that the relative levels at which the carbonate mud occurs are variable, and that the order of relative levels from deepest to shallowest does not match the order of wetlands in relation to age structure from oldest to youngest. Thus, there is no simple relationship between wetland location, topographic height and age.

In order to see how the concept of wetland initiation described above applied at the regional scale, the relative levels of the deepest carbonate muds in each wetland were plotted against the interpolated position of the palaeo water table surface along 3 transects (Fig. 2-5). Information on age structure of the prograded beachridge plain (Woods and Searle 1983; Searle *et al.* 1988) was used to locate the distance of the palaeo shorelines from the current strandline, and information on Holocene sea levels (Semeniuk and Searle 1986b; Searle *et al.* 1988) was used to position the shore relative to the current AHD. A water table profile was derived from a plot of groundwater contours based on empirical data collected during September 1994 (annual maximum water table position). The water table progressively rises in height with distance from the ocean. This empirical data can be used to determine the height of the groundwater table at any given location on the beachridge plain when its distance from the sea is known. The graphs plot the relative levels of the wetland basal sheets against the position of the (interpolated) water table profile determined by each sealevel datum, and distance of a given site from the "fossil" shoreline for the periods 4000, 3000, and 2000 years BP (Fig. 5-18A, B, C).

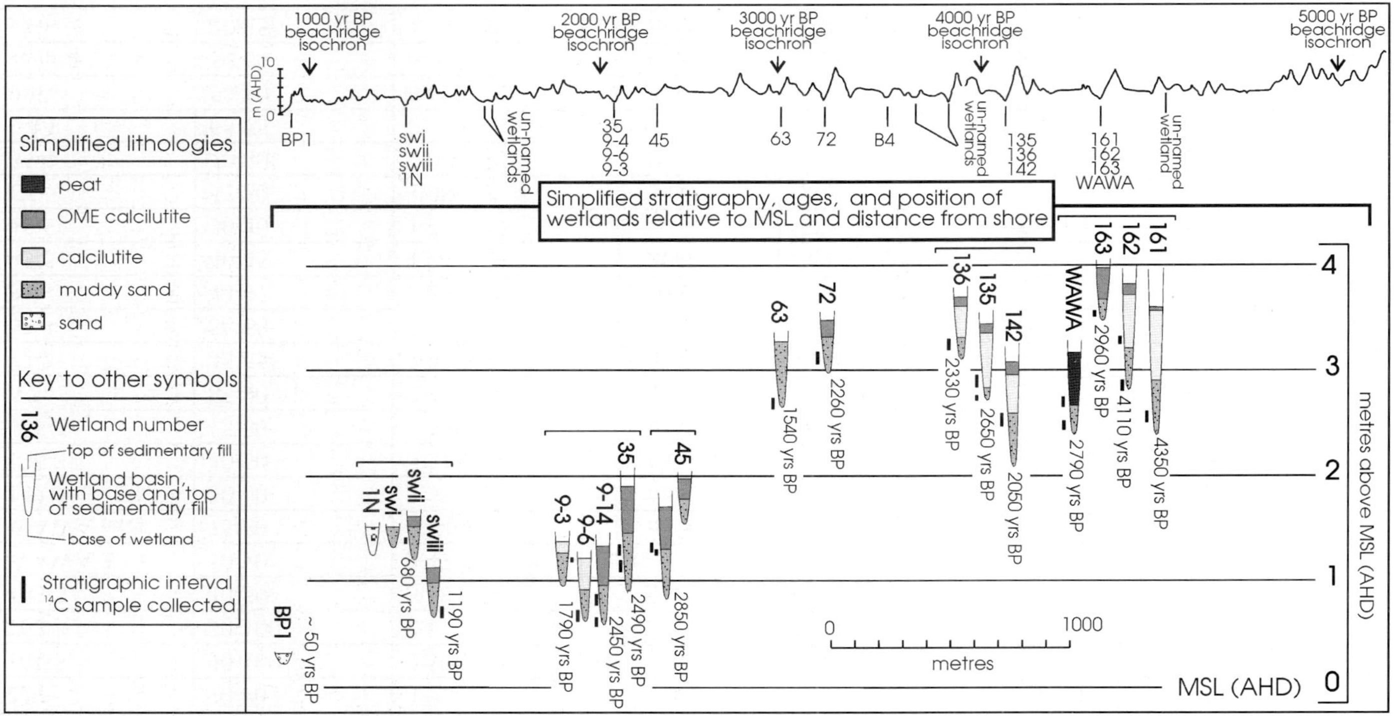

Figure 5-16. *Broad cross-section showing ridge/swale development, location of wetlands with their base and top relative to MSL, their location relative to the beachridge isochrons, the oldest ^{14}C age within the wetland fill, and the stratigraphic intervals of the samples used for the ^{14}C analysis.*

Table 5.4 Age structure (*in order of greatest distance from the present coast to least distance from the coast*)

Wetland	*Depth (cm)*	*AHD (m)*	*Sample type*	*$\delta^{13}C$ % PDB*	*^{14}C pMC±1σ*	*^{14}C yr BP ±1σ*
Cooloongup E	60-70		CO_3 mud	-4.8	48.9 ± 1.1	5740 ± 170
Cooloongup A2	80-90	2.59-2.69	CO_3 mud	-3.1	56.5 ± 1.1	4590 ± 160
229	70-80	4.01-4.11	CO_3 mud	-7.6	66.2 ± 1.6	3320 ± 200
161-3	90-100	2.59-2.69	CO_3 mud	+0.3	58.2 ± 1.5	4350 ± 210
162-3	90-100	2.83-2.93	CO_3 mud	-4.7	59.9 ± 1.1	4110 ± 150
163-3	40-50	3.48-3.58	CO_3 mud	-6.4	69.2 ± 1.2	2960 ± 140
WAWA 3	50-70	2.47-2.67	peat	-28.5	75.9 ± 2.8	2220 ± 300
WAWA teret	75-80	2.44-2.49	CO_3 mud	-7.2	70.7 ± 1.2	2790 ± 140
142-3	50-60	2.58-2.68	CO_3 mud	-7.8	77.5 ± 1.7	2050 ± 180
136-3	40-50	3.21-3.31	CO_3 mud	-7.5	74.8 ± 1.7	2330 ± 180
135-2	75-80	2.69-2.74	CO_3 mud	-7.9	71.9 ± 1.2	2650 ± 130
A7	85-90	2.3-2.35	CO_3 mud	-3.4 ± 0.1		3890 ± 120
72-3	45-55	2.93-3.03	CO_3 mud	-6.5 ± 0.1		2260 ± 65
63-3	50-60	2.67-2.77	CO_3 mud	-8.2	82.5 ± 1.2	1540 ± 120
A9	44-47	1.46-1.49	CO_3 mud	-7.3 ± 0.1		2340 ± 70
45-5	40-45	1.24-1.29	CO_3 mud	-3.3 ± 0.1		2850 ± 130
45-5	30-40	1.29-1.39	peat	-28	95 ± 3.0	340 ± 250
C5	57-60	1.98	CO_3 mud	-8.4 ± 0.1		1290 ± 90
35-4	70-80	1.08-1.18	CO_3 mud	-6.2	73.4 ± 1.2	2490 ± 130
9-14	65-75	0.58-0.68	CO_3 mud	-2.8	73.7 ± 1.2	2450 ± 130
9-6	50-60	0.62-0.72	CO_3 mud	-8.3	80.0 ± 1.2	1790 ± 160
swiii-4	50-60	0.99-1.09	CO_3 mud	-2.8	86.3 ± 1.2	1190 ± 120
swii-3	20-25	1.34-1.39	CO_3 mud	-6.8 ± 0.1		680 ± 80

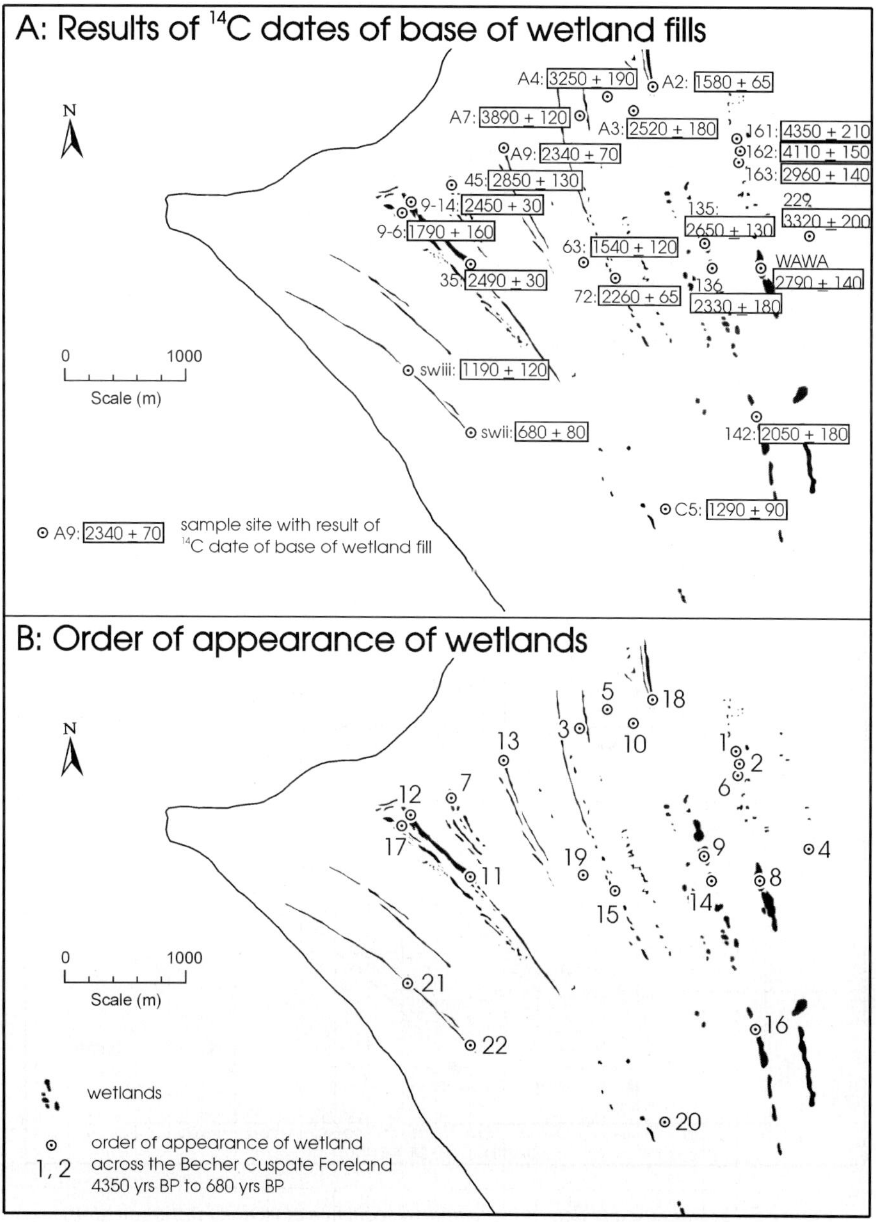

Figure 5-17. *Location, and results of ¹⁴C dates of base of wetland fills, and order of appearance of wetlands based on the ¹⁴C dates.*

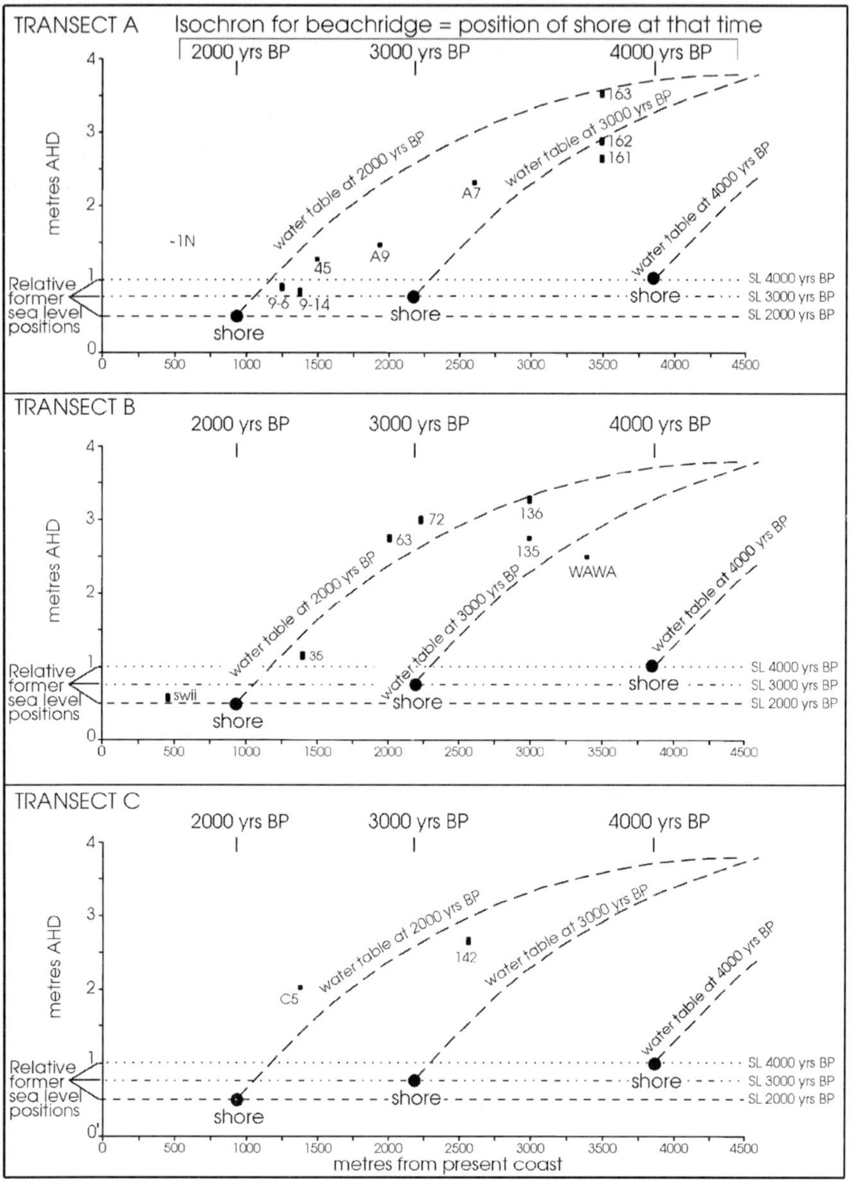

Figure 5-18. Data used to determine temporal sequence of wetland initiation along Transects A, B, & C: isochrons for beachridges, relative former sea level positions, water table configuration relative to a former sealevel, the wetland within the beach ridge age structure, position of lowest carbonate layers relative to AHD, and sampled interval.

This method was used to produce a theoretical sequence of wetland formation on the Becher cuspate foreland based on water table configuration in response to a falling sea level, and the positions of the former strandline. Using the commencement of carbonate mud deposition as the indicator of wetland initiation, two situations were recognised: the commencement of waterlogging (damplands); and the commencement of inundation (sumplands). This theoretical sequence was compared with the order of wetland commencement as determined by radiocarbon dating (Table 5.5).

Table 5.5 *Transect A*

Order of commencement of waterlogging and inundation in wetlands according to the model

Period	*Waterlogging*	*Inundation*
5000-4000 yrs BP		Cooloongup E Cooloongup W
4000-3000 yrs BP	A7 (55 cm) 163(55 cm)	161 162
3000-2000 yrs BP		A7 A9 9-14 163 45 9-6
<2000 yrs BP	1N (80 cm)	

Order of wetland initiation according to radiocarbon dating

Wetland	^{14}C *yr BP* $\pm 1\sigma$
Cooloongup E	5740 ± 170
Cooloongup W	4590 ± 160
161	4350 ± 150
162	4110 ± 150
A7	3890 ± 120
163	2960 ± 140
45	2850 ± 130
9-14	2450 ± 130
A9	2340 ± 70
9-6	1790 ± 120

Table 5.5 *Transect B*

Order of commencement of waterlogging and inundation in wetlands according to the model

Period	*Waterlogging*	*Inundation*
5000-4000 yrs BP		
4000-3000 yrs BP	135 (50 cm)	WAWA
3000-2000 yrs BP	136 (0 cm) 72 and 63 (30 cm)	135 35
<2000 yrs BP	swii (60 cm) swi (70 cm)	136 63 72 swiii

Order of wetland initiation according to radiocarbon dating

Wetland	^{14}C *yr BP* $\pm 1\sigma$
229	3320 ± 200
WAWA	2790 ± 140
135	2650 ± 130
35	2490 ± 130
136	2330 ± 180
72	2260 ± 65
63	1540 ± 120
swiii	1190 ± 120

Table 5.5 *Transect C*

Order of commencement of waterlogging and inundation in wetlands according to the model			*Order of wetland initiation according to radiocarbon dating*	
Period	*Waterlogging*	*Inundation*	*Wetland*	*^{14}C yr BP $\pm 1\sigma$*
5000-4000 yrs BP				
4000-3000 yrs BP				
3000-2000 yrs BP	C5 (55 cm)	142	142	2050 ± 180
<2000 yrs BP		C5	C5	1290 ± 90

There is generally good agreement between the order of wetland commencement determined by radiocarbon dating and the order determined by the hypothetical model. Variation between the two may be explained by a number of factors, the accuracy of the dates, the precision of sampling, and the divergence of a wetland from the transect.

5.3.5 Conclusions

Several conclusions derive from this study. At the largest scale, the information in this chapter shows that along the major east west axis of the cusp, progradation was most rapid; as a result, much of the present land surface is not very much higher than present mean sea level (1-2 m). Given a homogeneous stratigraphy, height of the regional water table will be a function of distance from the present shoreline. These two factors in combination explain why in any particular swale, intersection of the land surface by groundwater will have a higher probability along this axis. As a consequence, there exists the general pattern of diminishing numbers of wetlands along any swale as distance from the major axis of the cusp increases, and there is a random spatial distribution of incipient wetlands.

Patterns at the smaller scales are:

1. granulometric evidence and sedimentary structures indicate that wetland basins are founded on swales that are stranded beach-berms or inter-dune depressions

2. a rising water table, associated with coastal progradation and falling sealevel, progressively waterlogged or inundated these swales to develop proto-wetlands and then wetlands

3. the initiation of wetlands began with the development of a basal sheet (a foundation sheet of parent sand on the basin floor that was impregnated with wetland

sediment), and this basal sheet carries information about the timing of wetland development and the type of material that commenced filling the wetland

4. the initiation of wetland development across the Becher Cusp beachridge plain is related to several interacting factors, *viz.*, the relative height of a given basin within a swale above former sealevel, the location of the basin with respect to its distance from the former coast (and hence the location of the former water table), and the history of sealevel fall;

5. although in a regional view, wetlands are oldest to the east where the terrain is the oldest, and youngest to the west, in detail, the initiation of wetlands within a isochron band is more random and related to the nature of ridge-and-swale topography; thus wetlands within the same swale alignment, even in close proximity, may have different basal ages.

6. WETLAND SEDIMENTOLOGY AND STRATIGRAPHY

6.1 Introduction

Wetland formation and development on the Becher cuspate foreland commenced with the intersection of the topographically low basins within the beachridge swales by a rising regional groundwater table, induced by coastline progradation. These wetlands owed their origin to a combination of regional climatic patterns and other regional processes such as land progradation through beachridge and swale development, groundwater movement, and groundwater rise and fall, but, with regular waterlogging and inundation, soon began to develop site-specific wetland features and processes, such as colonisation by different plant species, and accumulation of incipient *in situ* sediments. Through gradual accumulation of wetland sediments, fills and stratigraphic sequences extant in the modern wetlands were produced, which exhibit characteristics that distinguish them from the surrounding beachridge landscape and soils.

The range of sediments that infill the Becher Suite wetlands, their stratigraphy, and their palaeo-sedimentology are the subjects of this chapter. In a geo-historical context the wetland sediments and stratigraphic sequences provide proxy information about the evolution and history of fill in the wetland. In a hydrological context, they act as small scale aquifers and play an important role in the hydrologic functions of wetlands. From an ecological viewpoint, they provide the foundation to understanding the soils that may determine vegetation distribution and maintenance.

Stratigraphy is concerned with the succession, form, distribution, lithologic composition, fossil content, geophysical and geochemical properties of sedimentary layers, and their interpretation, in terms of environment or mode of origin, and geologic history. Thus, stratigraphy is the foundation to reconstructing wetland origin and wetland history. Wetland basin fills provide the accretionary record of sedimentation (*e.g.*, peat *vs* carbonate mud accumulation), style of fill (*e.g.*, direct vertical accretion, cut and fill, lateral delivery to fill), rate of fill, processes of wetland deepening and shallowing, possible proxy indication of surrounding environmental conditions (*e.g.*, vegetation assemblages, climate, hydrochemistry), and vegetation succession as determined by the occurrence of pollen.

The extant physical, chemical, and biological processes involved in sedimentation within the wetlands are referred to herein as sedimentology. In the context of the Becher Suite wetlands, these processes are those of formation (accumulation), movement (into or out of the wetland basin), alteration, and burial. Sedimentology is used as an adjunct to aid in the interpretation of stratigraphic features.

There are three overall objectives of this chapter:

- to characterise the wetland fills in terms of sediment types and sequences;
- to describe the current sedimentological processes within the wetlands; and
- using stratigraphic sequences, sediment types, sedimentary structures, and diagenetic products, to reconstruct palaeo-environmental and palaeo-sedimentologic processes instrumental to wetland development.

The types, thickness, and stratigraphic relationships of sedimentary units are described for 20 wetlands, including sumplands and damplands, within the Becher Suite.

Terminology for wetland fills and basins is shown in Figure 5-11. The term "basal" is only applied when the layer is relatively thin with respect to the wetland fill.

6.2 Stratigraphic framework to wetland basins

The landscape that is host to the Becher Suite wetlands is the Becher cuspate foreland. Its Holocene stratigraphy is that of a simple shoaling sequence from deep water marine basin facies to beachridge and dune facies (Searle *et al.* 1988) (Fig. 3-4). This sequence, or part thereof, comprising dune or humic dune sand, overlying beach sand, occurs under the ridges and inter-ridge depressions. The wetland sediments (also referred to herein as wetland fill), filling the inter-ridge depressions, are relatively shallow ($D < 1$ m). They overlie or are admixed with the upper parts of the littoral sediments in the inter-ridge depressions. Wetland sediments form two different contacts with the underlying parent sediments: a gradational contact, usually with the dune facies, and a sharper contact, usually with the beach facies.

6.3 Characterisation of wetland basin fills

In terms of physical/chemical characteristics, the wetland basin fills are described as follows:

1. occurrence of sedimentary bodies
2. geometry and thickness of sediment
3. types of sediment
4. vertical stratigraphic relationships
5. lateral stratigraphic relationships
6. small scale structures
7. granulometry
8. sediment composition of grain fractions
9. biota
10. pedogenic and synsedimentary diagenetic overprints
11. age structure

6.3.1 Occurrence of sedimentary bodies
The distribution of wetland sedimentary fill on the Becher cuspate foreland is as follows (Fig. 3-11):

- wetland sediments occur in inter-ridge depressions (or swales), but not every inter-ridge depression
- wetland deposits preferentially accumulate in the swales westward of large ridges
- continuous bodies of wetland sediments form within some swales and discontinuous bodies form within others
- wetland sedimentary deposits are thickest in the older wetlands and in those nearest the major axis of accretion of the cuspate foreland

In general, wetland deposits preferentially accumulate in the swales westward (seaward) of large ridges, *e.g.*, wetlands WAWA, 135, 136, 142, although several examples occur either in depressions between bifurcating arms of a ridge, *e.g.*, wetlands 161, 162, 163, or in hollows bounded by close parallel ridges of a transgressive parabolic dune or sand shadow, *e.g.*, wetlands in swales between 135 and 72. Between the 2000 and 3000 year isochrons on the beachridge plain (Woods and Searle 1983, Searle and Woods 1986, Searle *et al.* 1988), when the occurrence of large ridges begins to decrease, the wetland deposits accumulate westward of approximately every 7th ridge (Fig. 5-16). There is a relationship between the topographic height of the wetland fill relative to MSL, and the position of each 1000 year isochron as determined for the beach ridges. For example, older wetland sediments occur below 4 m AHD, but are restricted to <3.5 m AHD at approximately the position of the 3000 year isochron (Woods and Searle 1983, 1986). The complete listing is presented in Table 6.1 and illustrated in Figure 5-16.

Table 6.1 *Height of basin floor in relation to geographic location*

Location relative to beachridge isochron	*Height of basin floor above modern sea level (m) AHD*
3000-4000 years BP	<4 m
2000-3000 years BP	<3.5 m
1000-2000 years BP	<2 m
<1000 years BP	<1 m

Wetland basin sediment fills may be either continuous or discontinuous along a swale. Continuous linear wetlands lie in close proximity to lines or chains of discrete wetland basins (Fig. 6-1A). Continuous linear wetlands have two types of fill: homogeneous and multiple. Basin sediment fill which is similar longitudinally, occurs in relatively flat floored linear wetlands. Multiple fills occur in continuous linear wetlands with

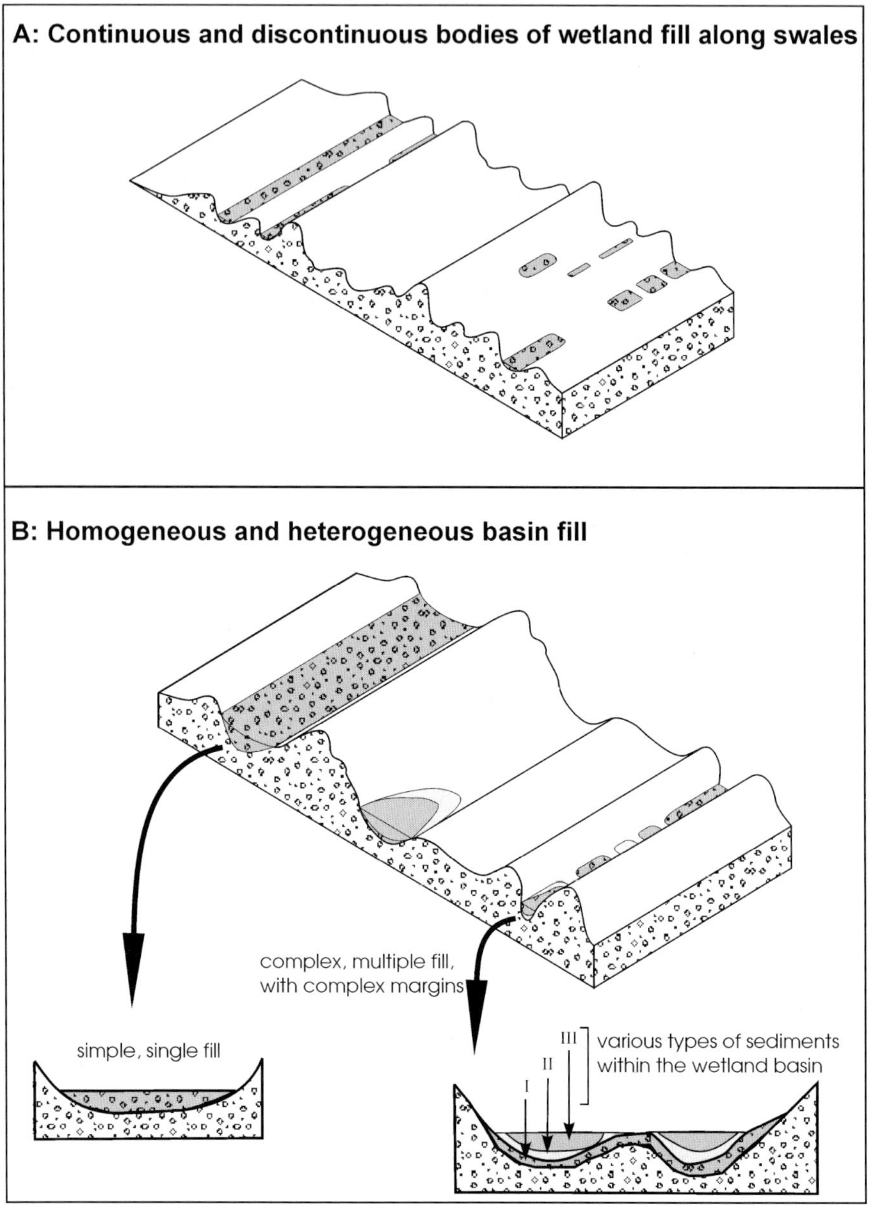

Figure 6-1. *Idealised diagram showing geometry of wetland fills, their distribution along the length of swales, and their nature of their homogeneity or heterogeneity.*

undulating floors (Fig. 6-1B), where partitioning of the long basin provides scope for each sub-basin to have a slightly different stratigraphic history.

Wetland sedimentary deposits are best developed in the older wetlands and in those nearest the axis of cuspate foreland accretion. This pattern reflects and is explained by the model of wetland evolution described in the previous chapter.

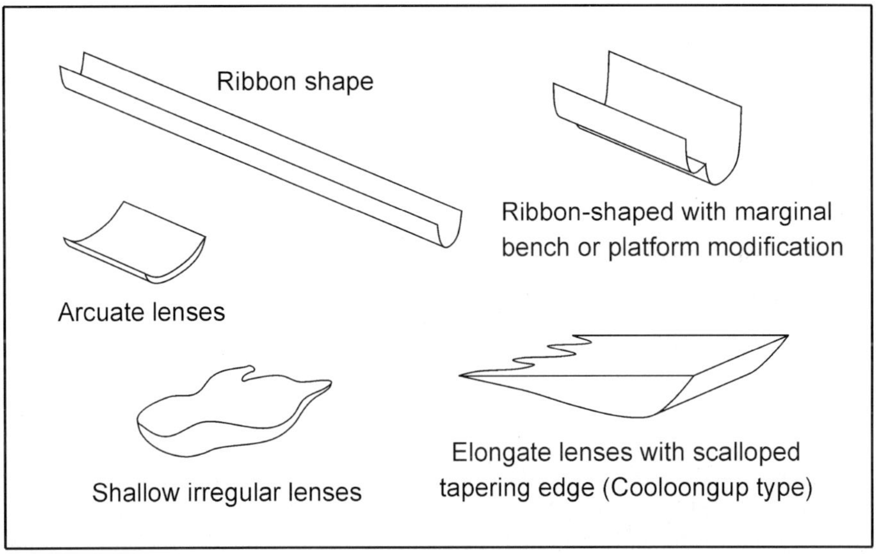

Figure 6-2. *The various three dimensional shapes of wetland fills.*

6.3.2 Geometry and thickness of sediment

The geometry of the wetland fills mirrors the morphology of the basins within the inter-ridge depression. Three types of swale morphology are host to wetlands and these produce similar suites of basins, each with similar dimensions. These are:

1. uninterrupted, relatively shallow swales encompassed by a 2000 m x 10 m x 0.3 m (L x W x D) frame;
2. small, shallow, discrete basins encompassed by the 10 m x 8 m x 0.5 m frame; and
3. well defined deeper basins encompassed by the 160 m x 25 m x 1 m frame.

The three dimensional geometry of the wetland fills is most often ribbon shaped, orientated north/south or northwest/southeast, following the swale, and with relatively steep, almost vertical east and west margins. Variations to this geometry result from undulations of the basin floor, (hence an undulation to the ribbon), and from the addition of a bench or narrow platform extending from one side of the wetland towards the centre, resulting in shallow irregular lenses, arcuate lenses, and circular lenses of wetland fill, all with relatively steep sides (Fig. 6-2).

The wetlands in the Becher Suite in the Cooloongup area are located in the southwestern portion of the Lake Cooloongup system, in the inter-ridge depressions formed by lateral spits. As such, the wetland fill of the Becher Suite wetlands in the Cooloongup area forms finger-like extensions from the main calcilutite mud fill of Lake Cooloongup itself (Fig. 6-2). The entire sedimentary body of Lake Cooloongup is a lens, oriented north-south, with the apex to the south. The wetland fill has a planar surface. In general, these planar bodies are mesoscale, being wider (70 m) than the wetlands on the Becher Cusp.

6.3.3 Types of sediments

Sediments, which have accumulated in the wetland basins of the Becher Suite, consist of sand and mud, mixtures of which generate muddy sand. There are two end member compositional components to the mud: calcium carbonate mud and fine grained organic matter. Various admixtures of sand, calcium carbonate mud, and fine grained organic matter, result in 7 main sedimentary types common to the study area.

These are:

1. peat
2. peaty sand
3. organic matter enriched calcilutite and sandy organic matter enriched calcilutite
4. organic matter enriched calcilutaceous muddy sand
5. calcilutite
6. calcilutaceous muddy sand
7. humic sand

Organic matter enriched calcilutite and sandy organic matter enriched calcilutite are aggregated in a single lithology and termed herein OME calcilutite. Descriptions of each sediment type, together with their distribution in each study wetland, are presented in Table 6.2 and Figures 6-3 to 6-23 A-E.

***Table 6.2** Sediment descriptions*

Sediment type	*Description*	*Site*	*Thickness*
peat	*colour*: black *structure*: root-structured *fabric:* wackestone *texture*: mud (80%), sand (11-17%), gravel (2-8%) *composition*: mud (peat), sand (seeds, shell, quartz) gravel (plant material and pulmonate snails - *Glytophysa* sp. and *Gyraulus* sp.)	161 WAWA	10 cm 50 cm
peaty sand	*colour*: black to dark grey *structure*: root-structured or homogeneous *fabric:* packstone *texture*: mud, medium to fine sand *composition*: mud (peat), sand (quartz grains)	161 162 WAWA 136 45	10 cm, at depth 25-50 cm 20 cm 5 cm, at depth
OME calcilutite and sandy OME calcilutite	*colour*: dark grey *structure*: root-structured, colour mottled, burrow mottled, texture mottled *fabric:* mudstone or wackestone *texture*: mud, medium to fine sand, gravel *composition*: mud (calcite, organic matter), sand (quartz), gravel (roots and pulmonate snails - *Glytophysa* sp. and *Gyraulus* sp.)	161 163 142 135 136 72 45 35 swii swiii Cool A Cool C	60 cm 40 cm 20 cm 15 cm 10 cm 20 cm 50 cm 30 cm 10 cm 20 cm 5 cm 10 cm
OME calcilutaceous muddy sand	*colour*: dark grey *structure*: root structured, homogeneous *fabric:* packstone *texture*: mud, coarse, medium, fine sand *composition*: mud (calcite, organic matter), sand (shell, quartz)	161 162 WAWA 45 swi swiii	40 cm 10 cm 15 cm 5 cm 5 cm 25 cm

Table 6.2 (cont.)

Table 6.2 (cont.)

Sediment type	*Description*	*Site*	*Thickness*
calcilutite	*colour*: light grey *structure*: homogeneous, colour mottled and burrow mottled *fabric:* mudstone to wackestone *texture*: mud (78-94%), sand (6-22%), gravel (0.2-2%) *composition*: mud (calcite with minor aragonite and magnesian calcite), sand (shell, quartz), gravel (pulmonate snails - *Glytophysa* sp. and *Gyraulus* sp.)	161 162 135 136 142 63 9 Cool A Cool B Cool C	30 cm 50 cm 40 cm 30 cm 30 cm 10 cm 30 cm 75 cm 40-60 cm 100 cm
calcilutaceous muddy sand	*colour*: grey *structure*: homogeneous *fabric:* packstone *texture*: mud, coarse, medium, fine sand *composition*: mud (calcite, aragonite and magnesian calcite), sand (shell, quartz)	162 163 142 135 136 72 63 45 35 9 swiii	30 cm 20 cm 50 cm 15 cm 15 cm 30 cm 45 cm 35 cm 50 cm 25 cm 25 cm
humic sand	*colour*: grey to black *structure*: root-structured *fabric:* packstone *texture*: mud, medium to fine sand, gravel *composition*: mud (humus), sand (quartz grains), gravel (roots)	in all wetlands	10-25 cm

What is referred to as peat herein ranges from true peat to muck, a highly organic matter enriched sediment (Collins and Kuehl 2001). It comprises black mud sized to gravel sized particles of organic carbon and decayed plant remains (roots, seeds, leaves and stems). The development of pure peat horizons is restricted both in distribution and accumulation, commonly 10 cm thick, with the thickest accumulation of 50 cm at wetland WAWA. The peat is root structured and has 10-15% fibre content. It accumulates under the present established wetland communities of sedgelands and herblands, *e.g.*, wetlands WAWA and 161.

The organic matter enriched calcilutite, most commonly approximates a homogeneous mixture, albeit, with varying proportions of organic matter and carbonate mud, although mottling does occur (wetlands 161, 135, Cooloongup B4). This deposit occurs in most wetlands, with the exception of the northern basin of the most recent swale and the wetlands located at the tip of the Becher Cusp, and always at the surface. The layers are usually about 20 cm thick, but in wetlands 163 and 45, this sediment type dominates the stratigraphic profile.

The calcilutite deposits are thin, ranging from 20-100 cm. Calcilutite occurs in nearly all of the wetlands, with the exception of N2, swi, swii, 9-3, which are all damplands, and wetland WAWA, which is peat filled. Scanning by the electron microscope showed the grains to be predominantly skeletal, comprising remnants of charophytes, ostracods, and other undifferentiated grains.

The "muddy" sands are intermediate sediment types between sands and biogenic muds and usually form where mud accumulations are interspersed with the influx of sand from wetland basin margins, or at the basal transitional infiltrational zone where the fine-grained wetland sediment fill stratigraphically rests on the underlying basement sand. In the cases of peaty sand and calcilutaceous sand, the mud-sized components are interstitial to the grain-support sand framework. In the cases of sandy peat and sandy calcilutite, sand is dispersed in the mud-support matrix.

As described in Table 6.2, there are four types of matrix interstitial to muddy sand, the most common type being calcilutite. Within this category, more than any of the other types of muddy sand, there is a noticeable gradation from a sediment with only slight mud content to a sediment with interstices fully packed with mud. All these are termed muddy sand. Calcilutaceous muddy sands occur in both a texture mottled (wetlands 63, 72, 136, swii, swiii-4) and homogeneous structure (wetlands 9-3, 35-5).

Peaty sand occurs in a number of wetlands (161, 162, WAWA, 136, 45). It occurs as a thin horizon (10-20 cm) in three different settings: underlying the peat in the centre of the wetland, as a buried horizon at the base of the wetland fill, and at the surface of some wetland margins.

Calcilutaceous muddy sand that is organic matter enriched is both associated with and independent of similarly composed mud horizons and may be root structured or homogeneous.

The humic sands are composed of quartz and shell grains with interstitial organic matter, and occur in the vegetated swales of the beachridge plain under sedge, or *Xanthorrhoea* or *Acacia* heaths. The amount of humus in the sediment is dependent on the species composition of the vegetation cover and its density within the swale. Sediments under *Xanthorrhoea preissii* exhibit various development of humic horizons ranging in depth from 20-100 cm (Fig. 6-24) and organic content from 8-11% (Woods 1984).

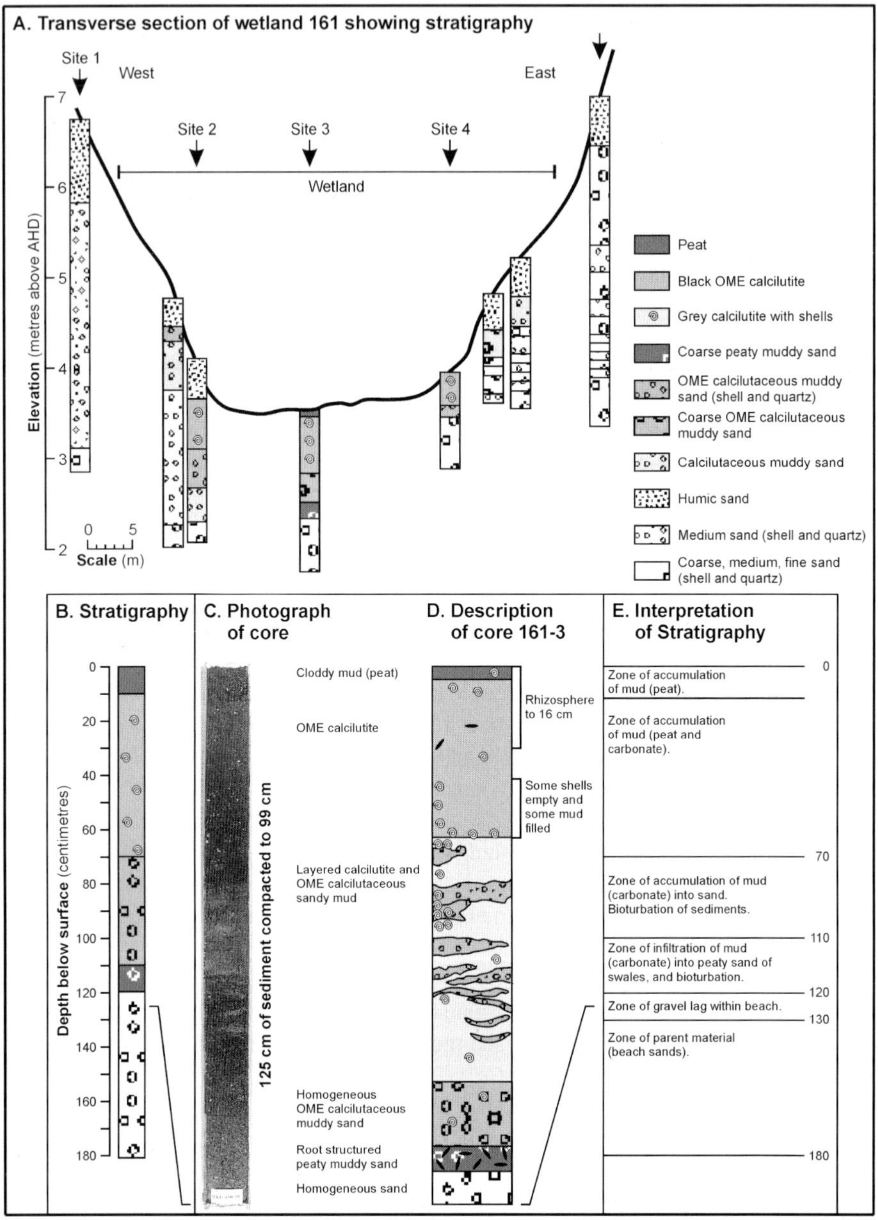

Figure 6-3. *Description and interpretation of sedimentary stratigraphic sequences in wetland 161.*

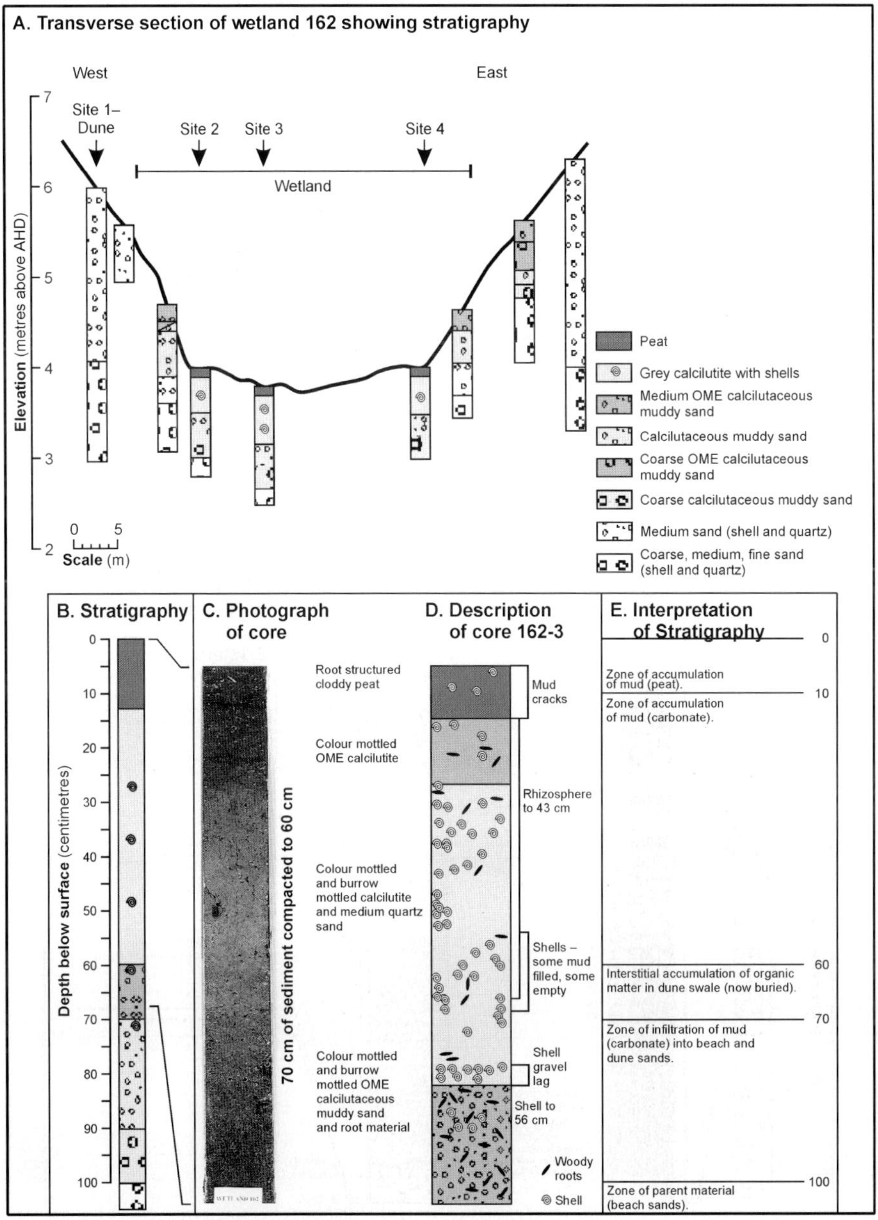

Figure 6-4. *Description and interpretation of sedimentary stratigraphic sequences in wetland 162.*

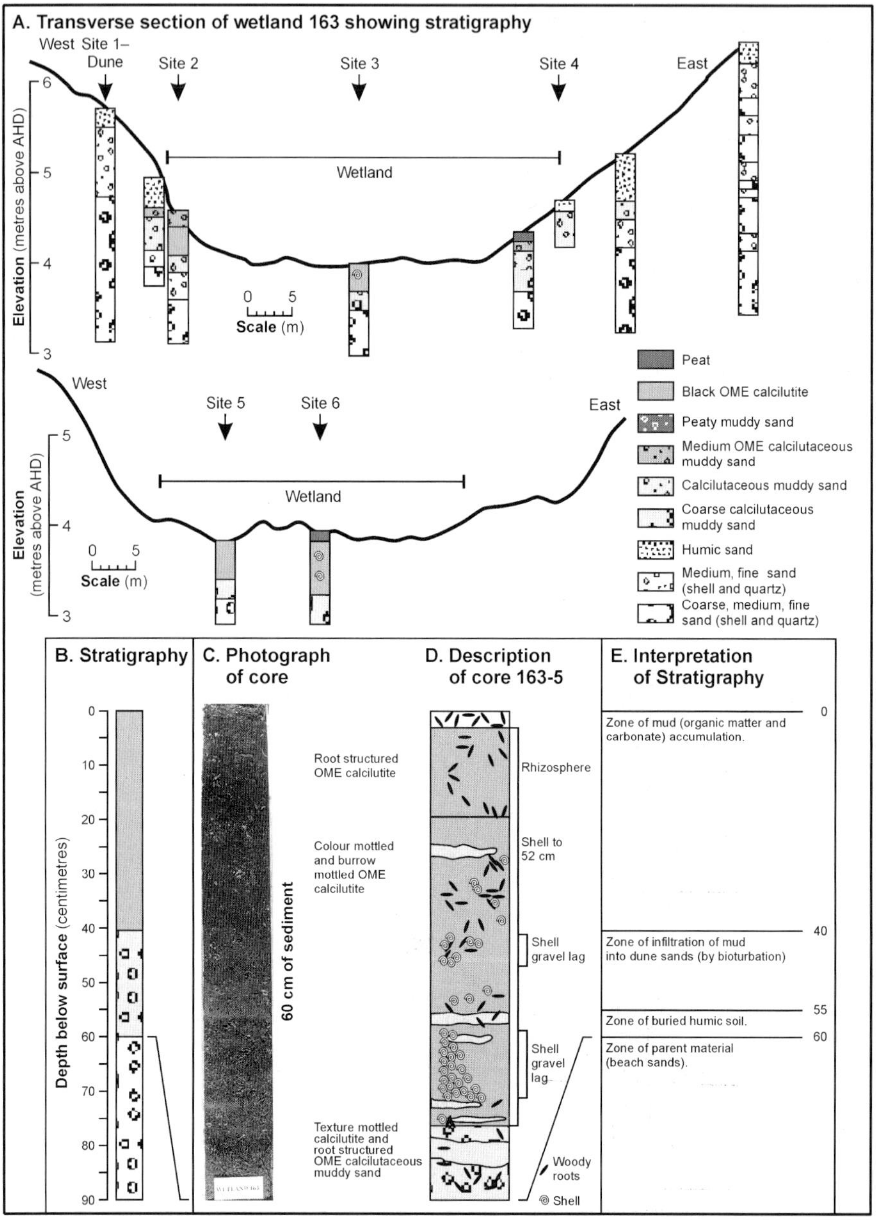

Figure 6-5. Description and interpretation of sedimentary stratigraphic sequences in wetland 163.

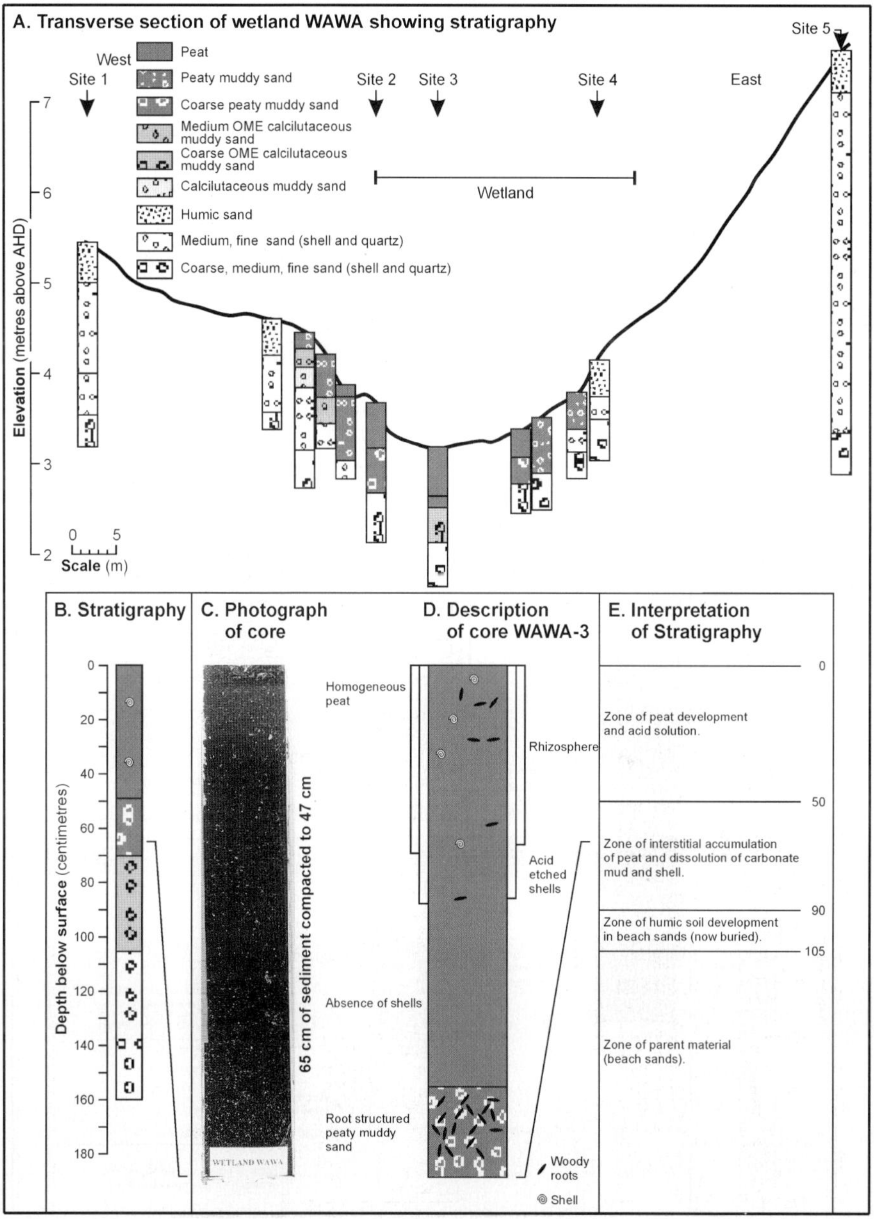

Figure 6-6. *Description and interpretation of sedimentary stratigraphic sequences in wetland WAWA.*

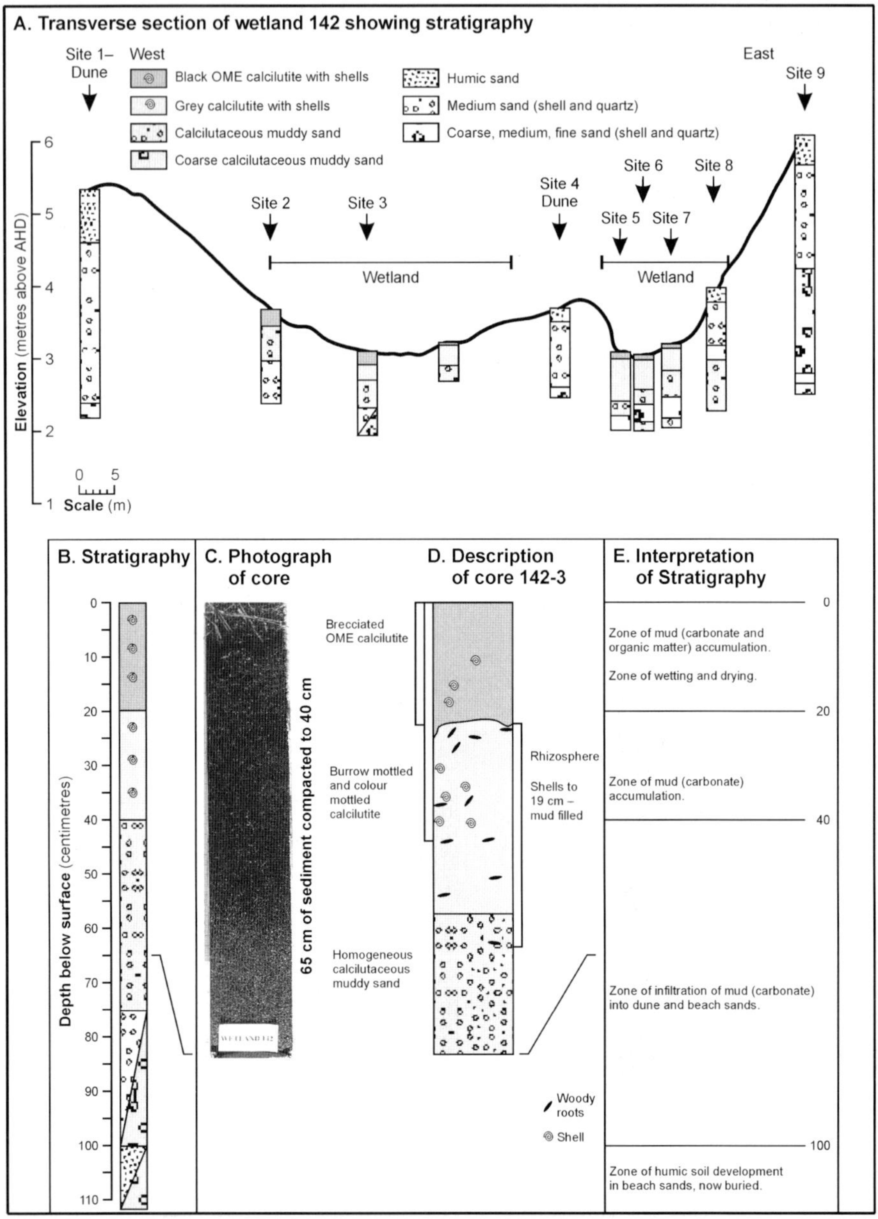

Figure 6-7. *Description and interpretation of sedimentary stratigraphic sequences in wetland 142.*

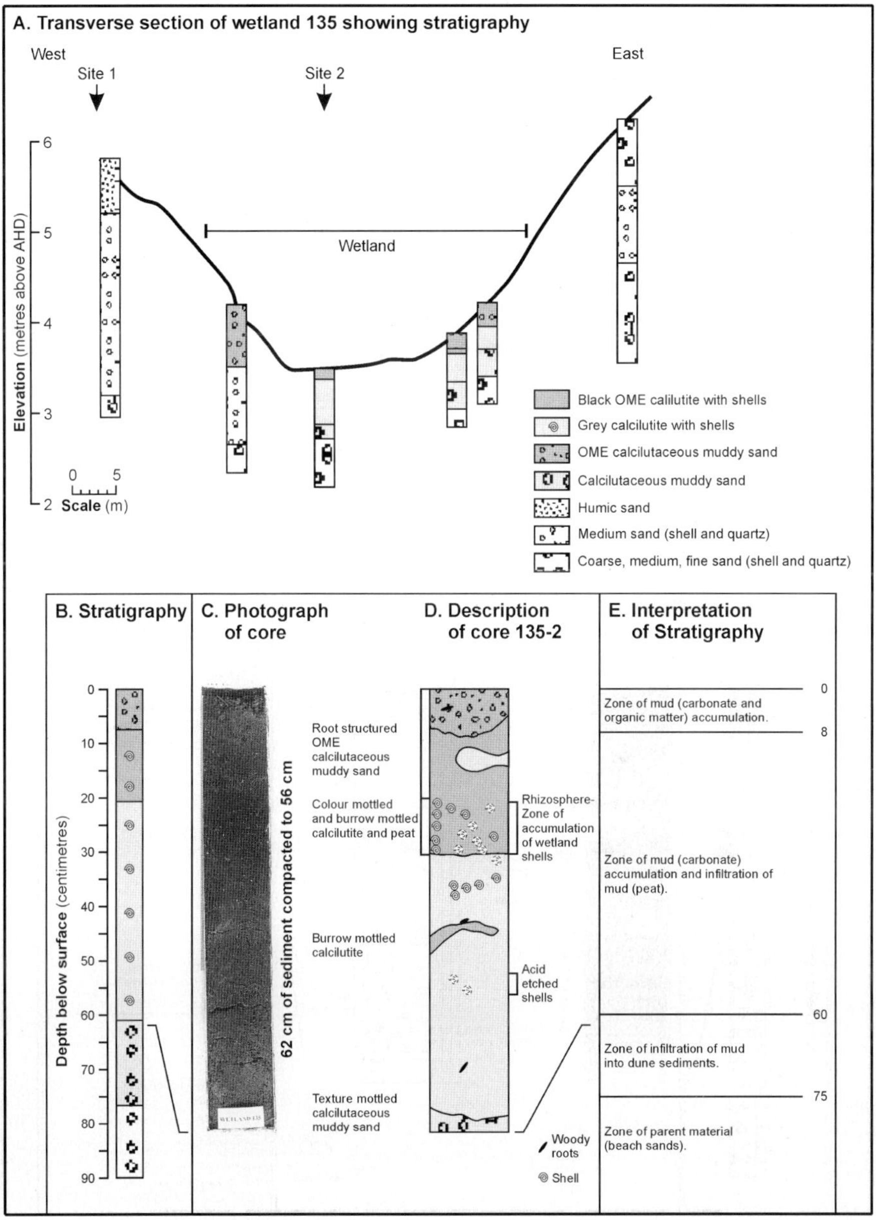

Figure 6-8*. Description and interpretation of sedimentary stratigraphic sequences in wetland 135.*

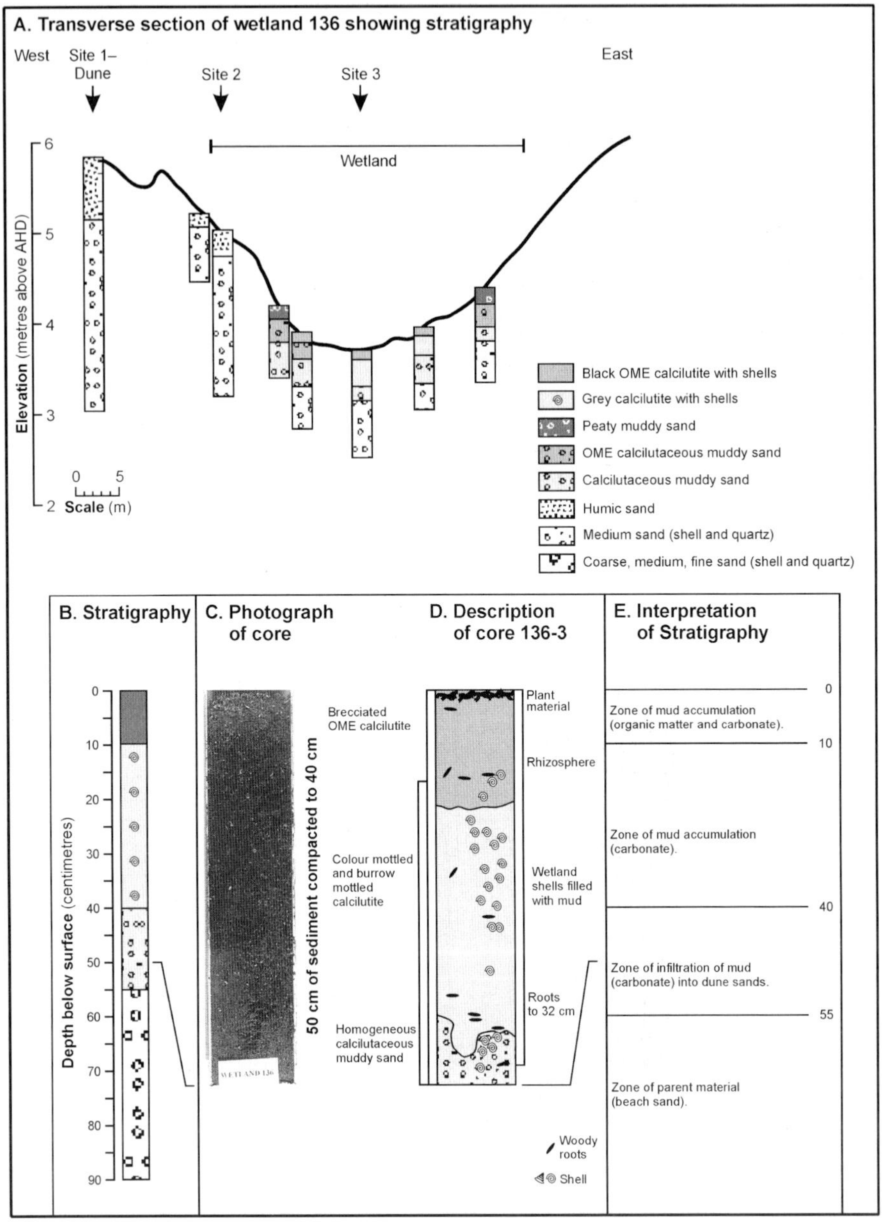

Figure 6-9. *Description and interpretation of sedimentary stratigraphic sequences in wetland 136.*

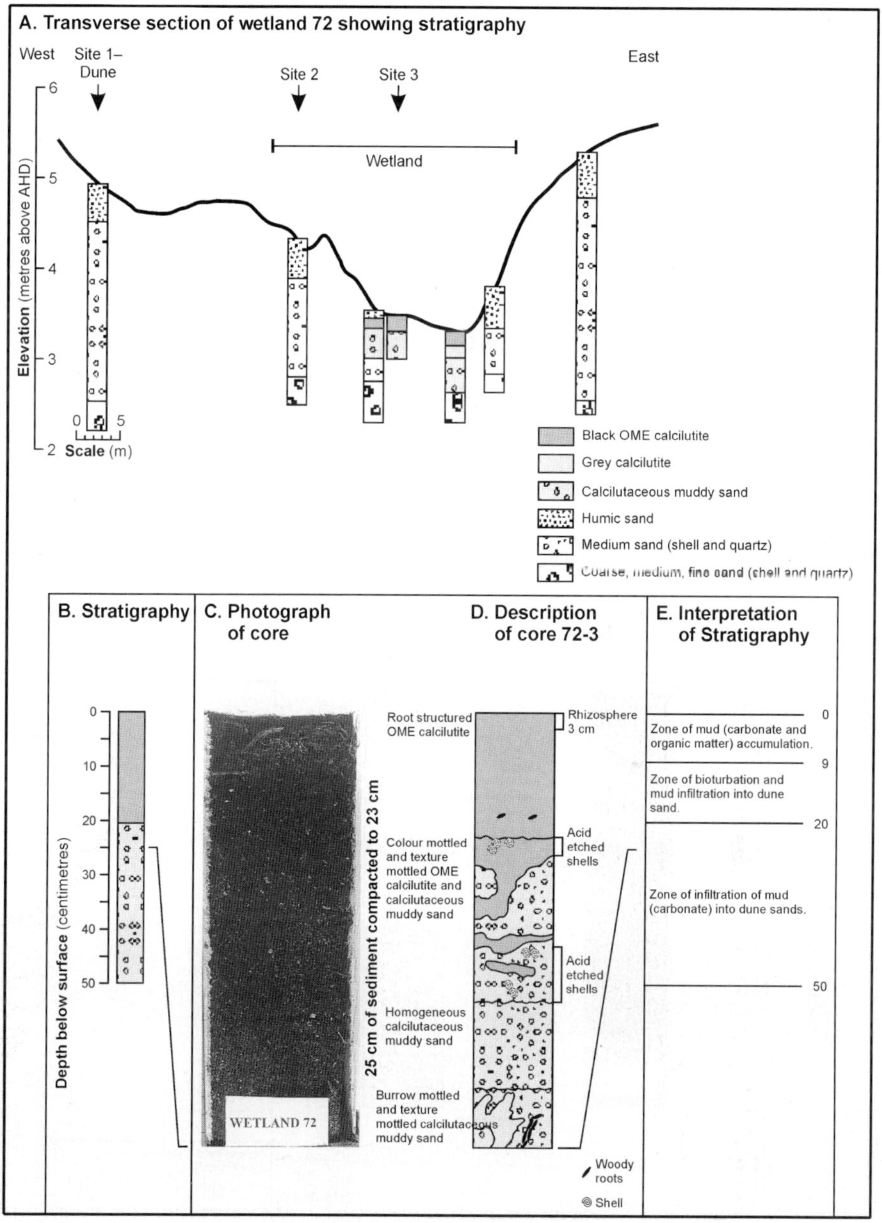

Figure 6-10. Description and interpretation of sedimentary stratigraphic sequences in wetland 72.

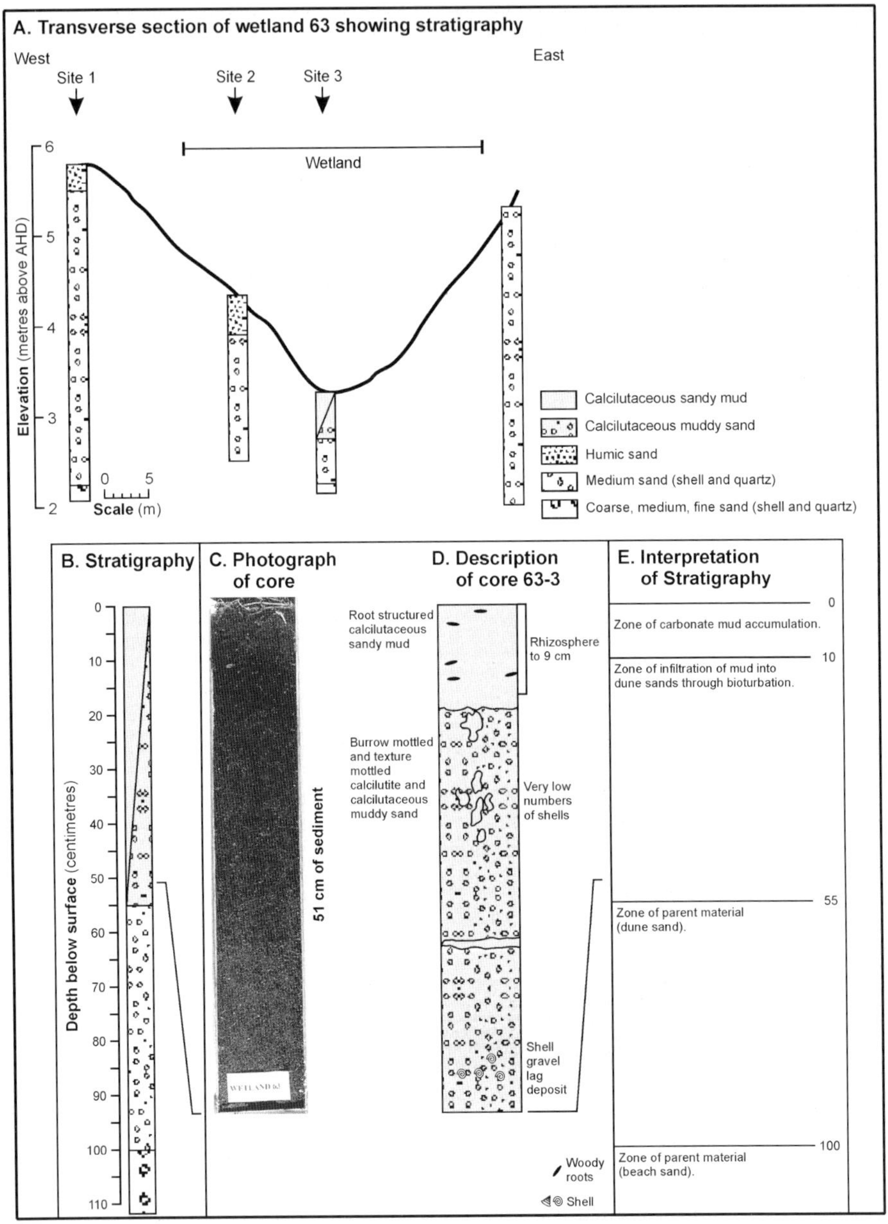

Figure 6-11. *Description and interpretation of sedimentary stratigraphic sequences in wetland 63.*

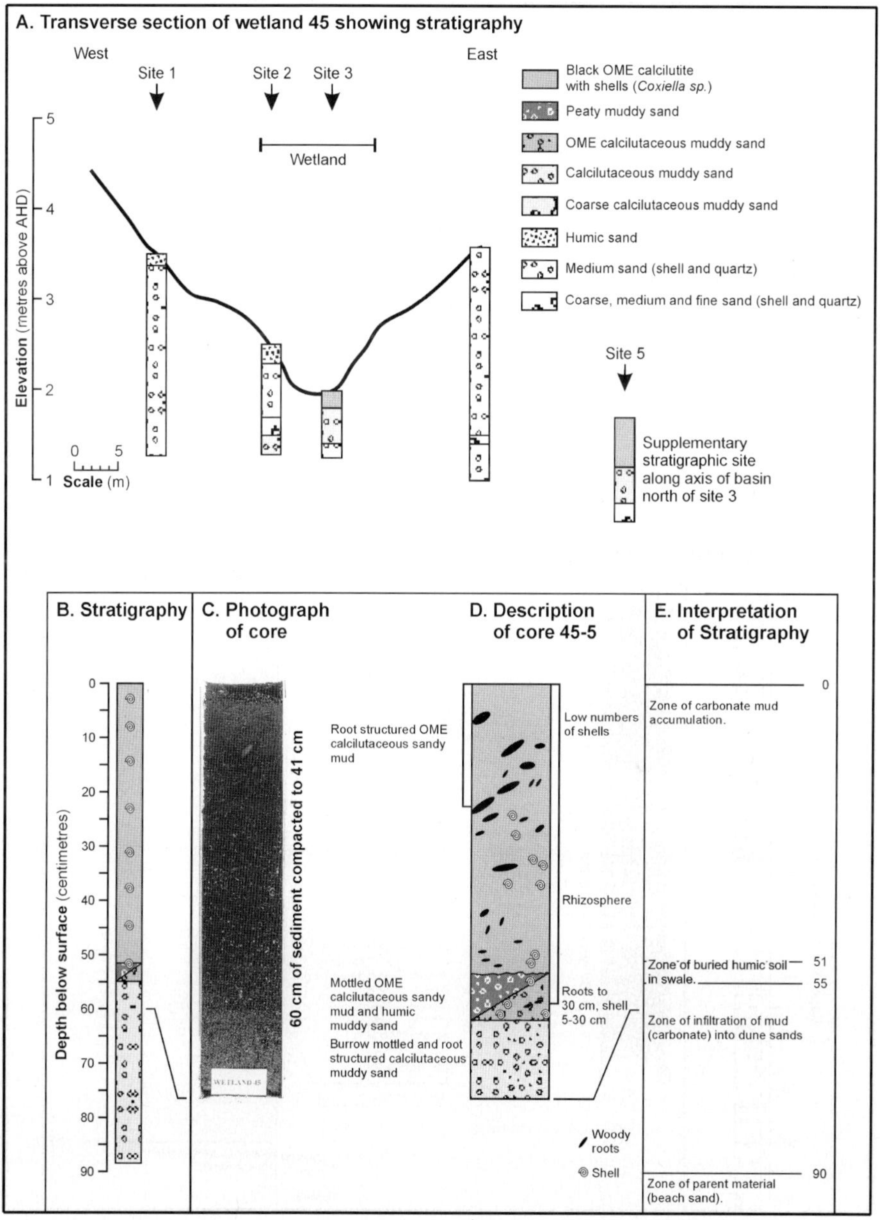

Figure 6-12. Description and interpretation of sedimentary stratigraphic sequences in wetland 45.

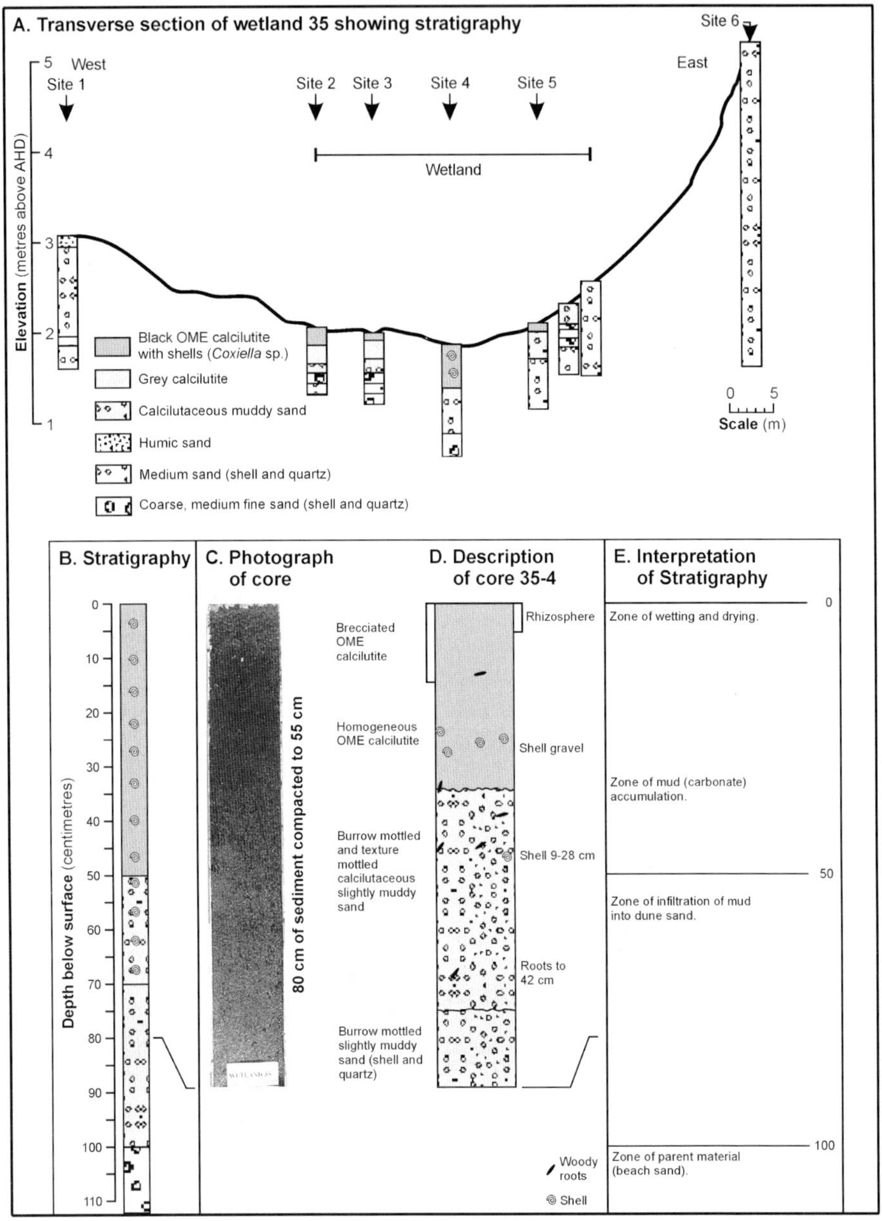

Figure 6-13. Description and interpretation of sedimentary stratigraphic sequences in wetland 35.

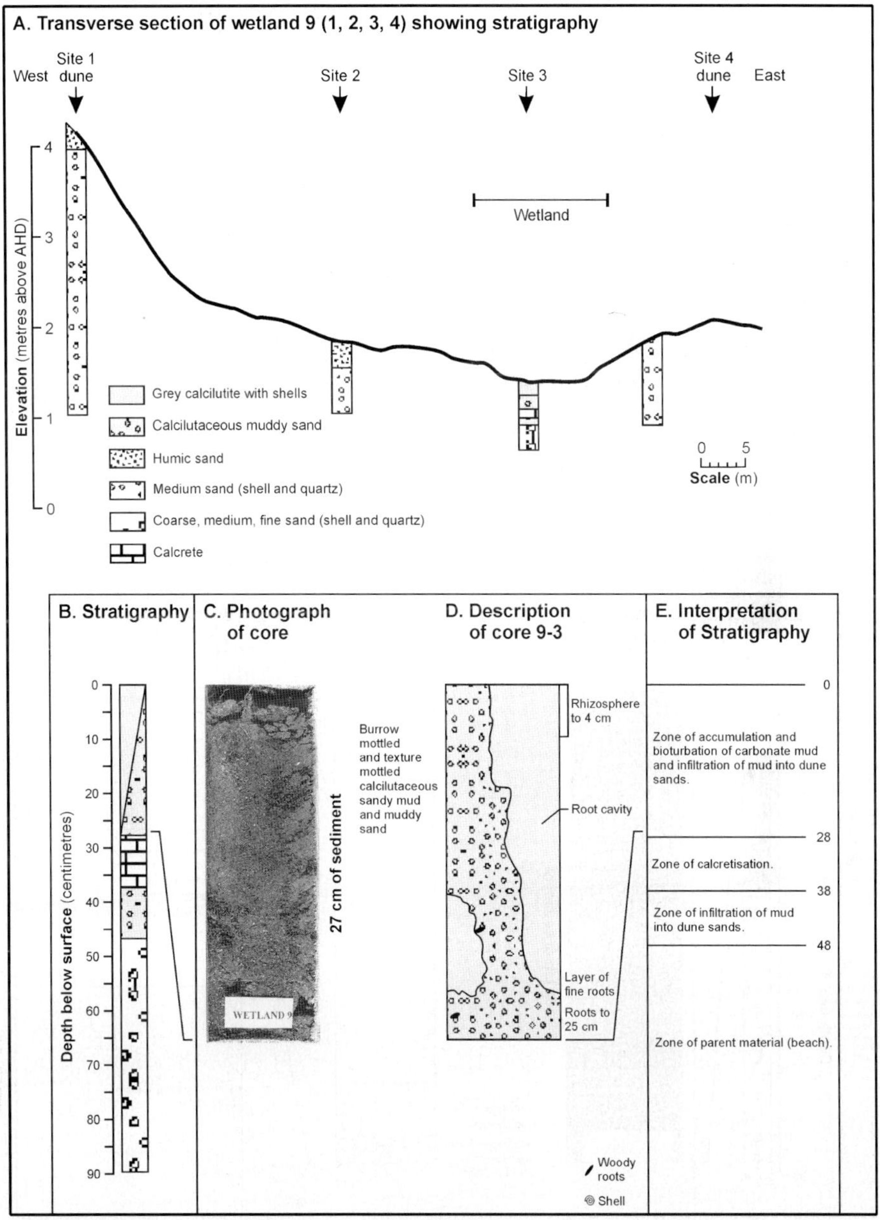

Figure 6-14. *Description and interpretation of sedimentary stratigraphic sequences in wetland 9-3.*

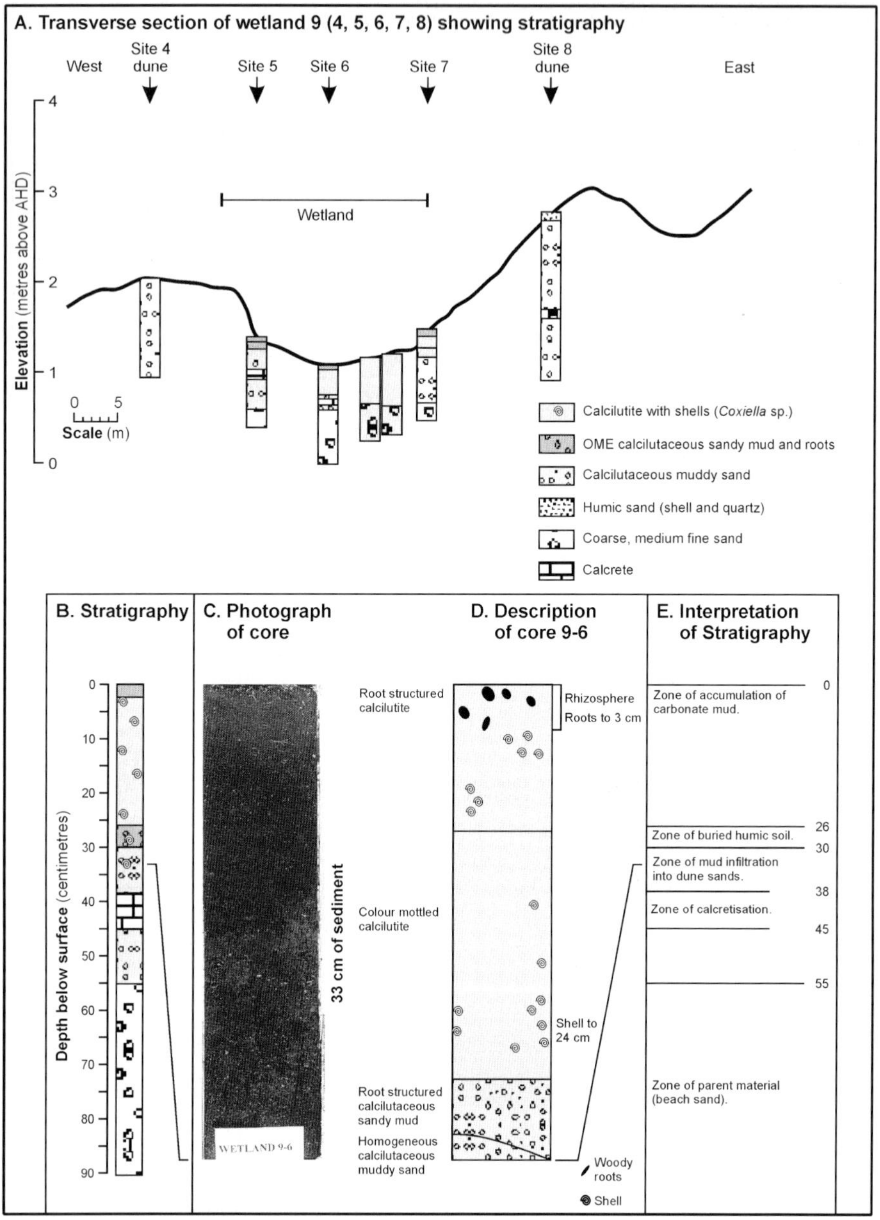

Figure 6-15. *Description and interpretation of sedimentary stratigraphic sequences in wetland 9-6.*

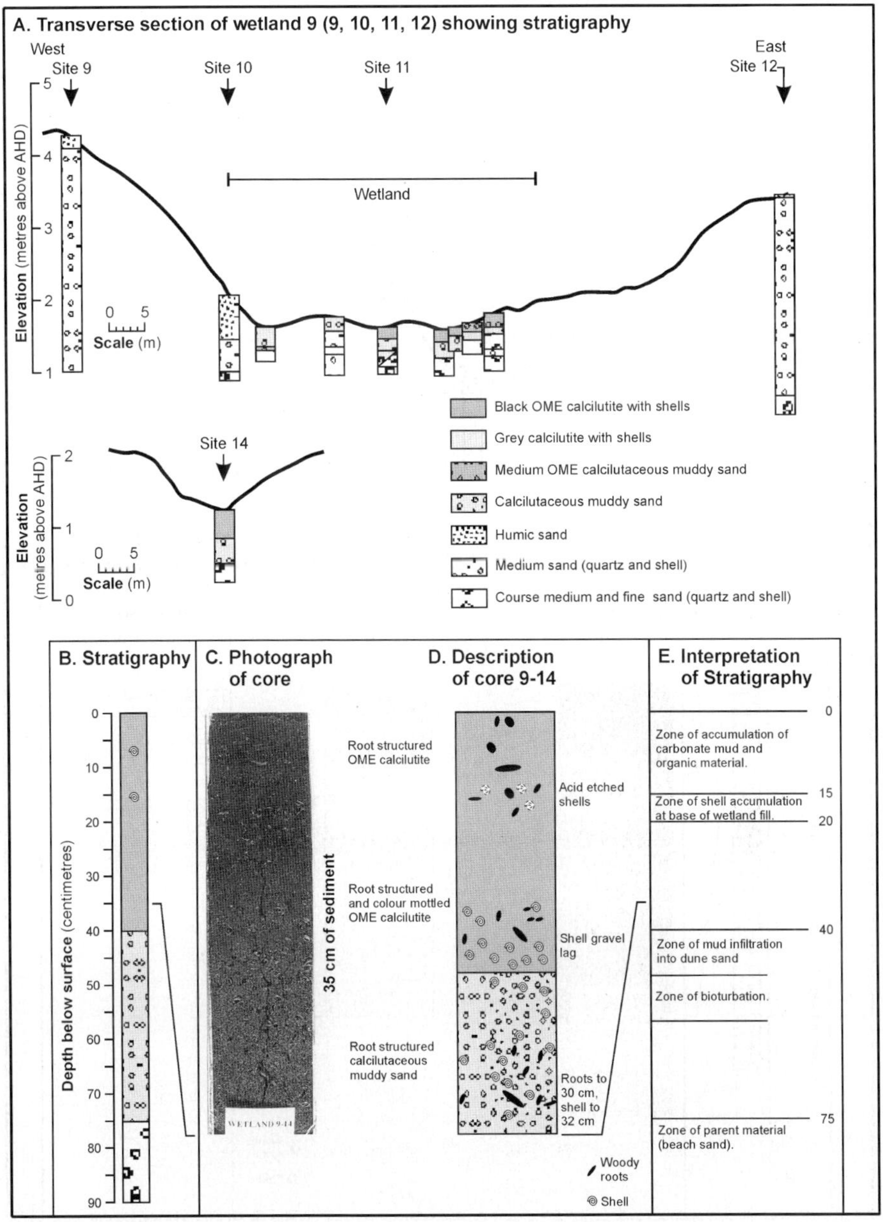

Figure 6-16. Description and interpretation of sedimentary stratigraphic sequences in wetland 9-10.

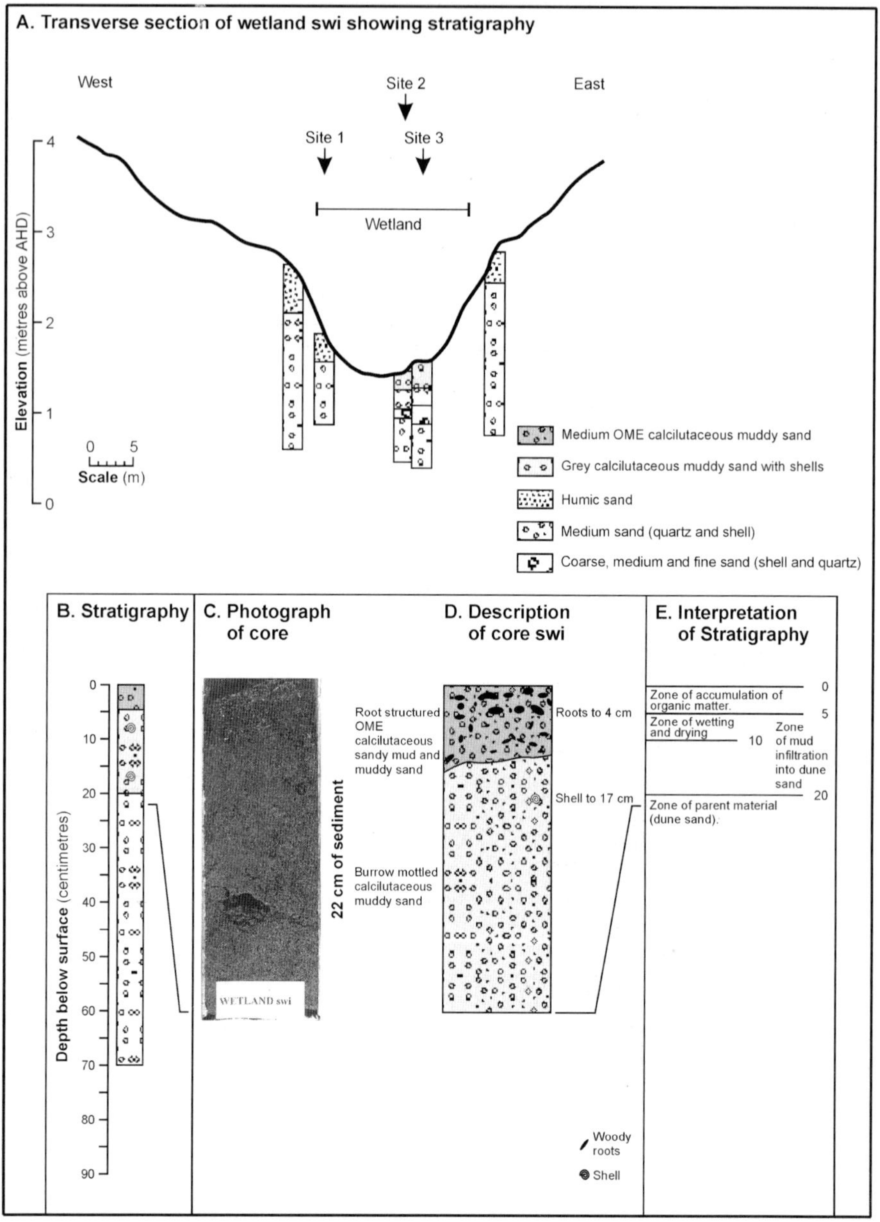

Figure 6-17. *Description and interpretation of sedimentary stratigraphic sequences in wetland swi.*

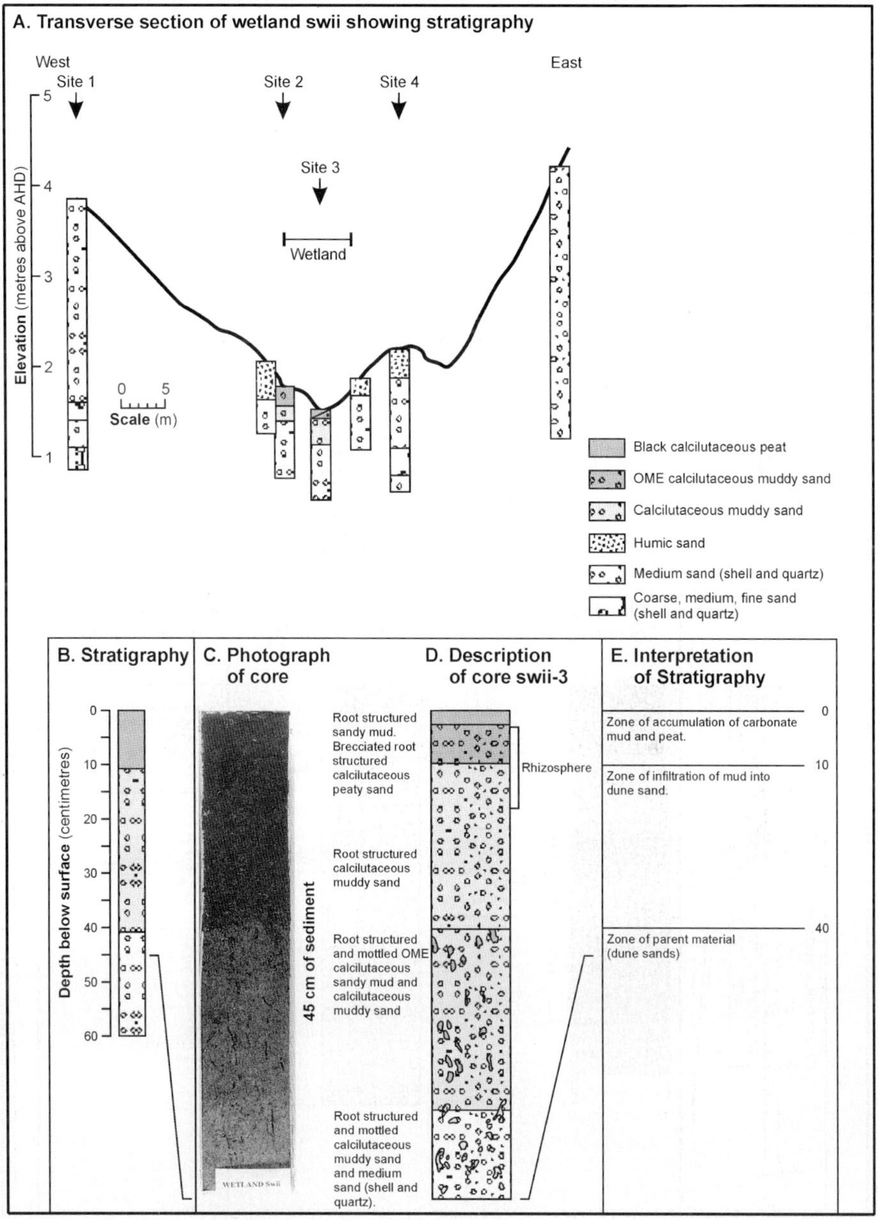

Figure 6-18. Description and interpretation of sedimentary stratigraphic sequences in wetland swii.

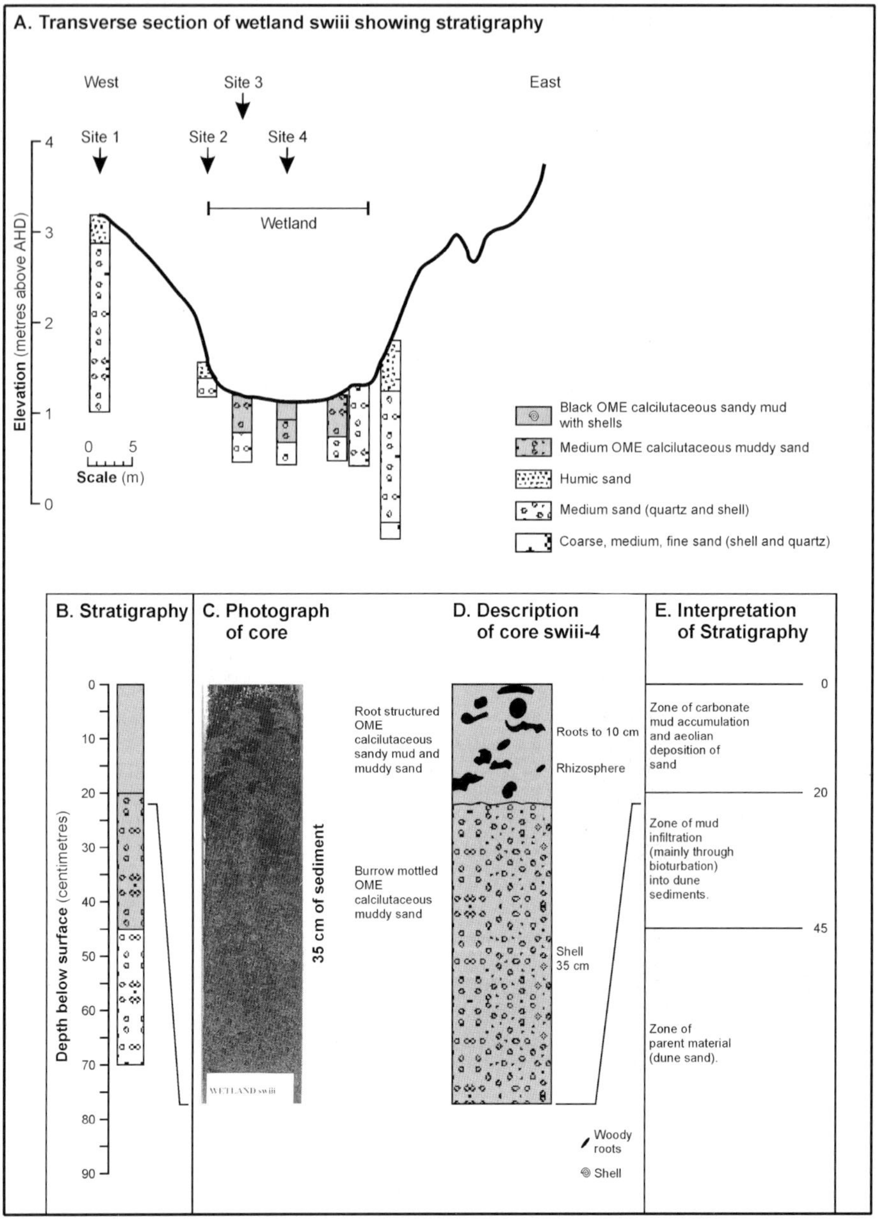

Figure 6-19. *Description and interpretation of sedimentary stratigraphic sequences in wetland swiii.*

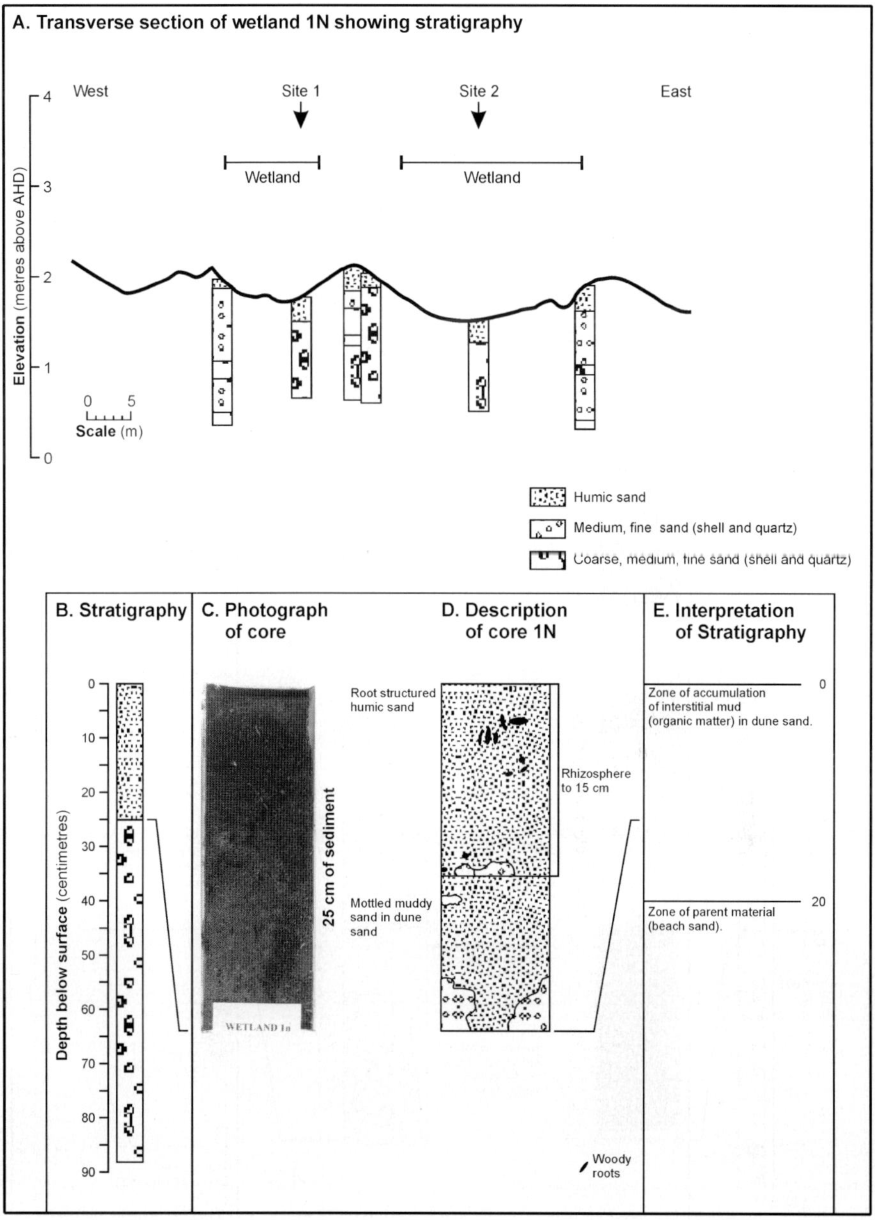

Figure 6-20. *Description and interpretation of sedimentary stratigraphic sequences in wetland 1N.*

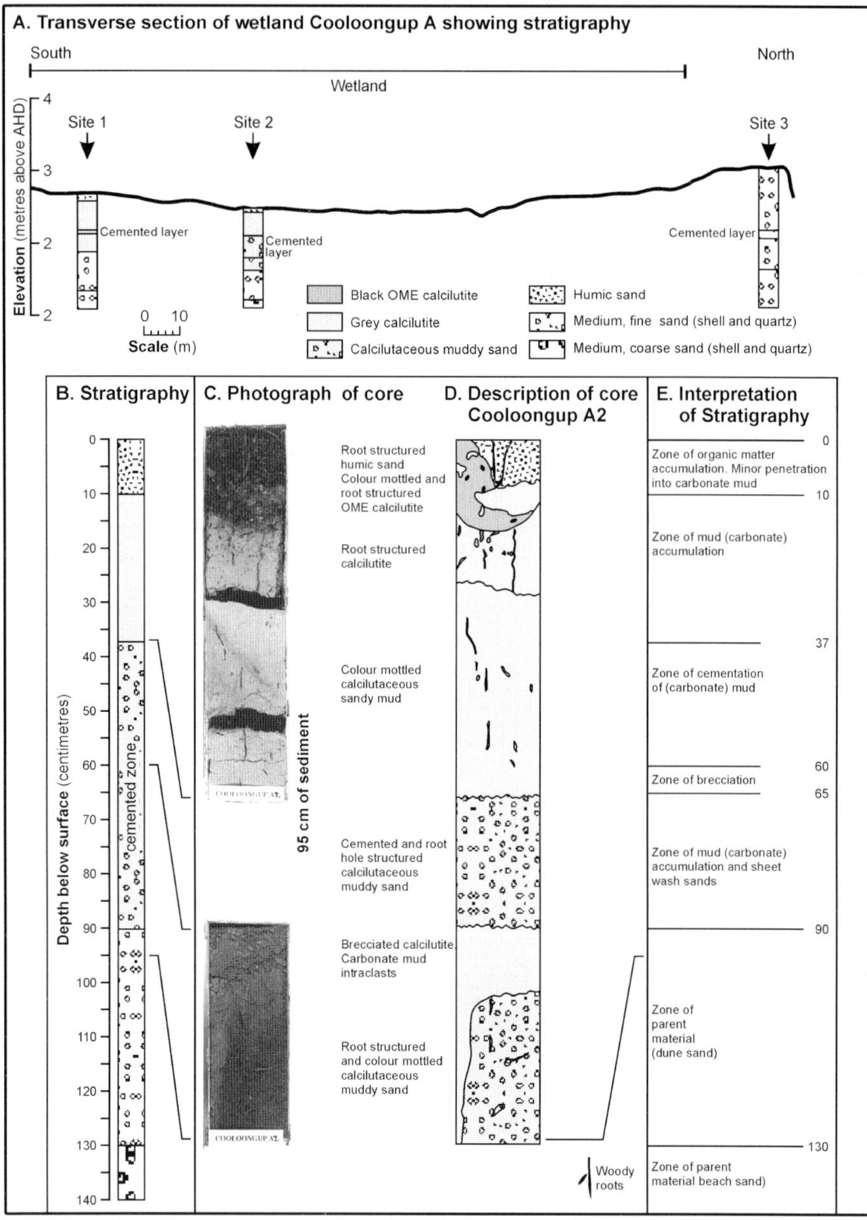

Figure 6-21. *Description and interpretation of sedimentary stratigraphic sequences in wetland Cooloongup A.*

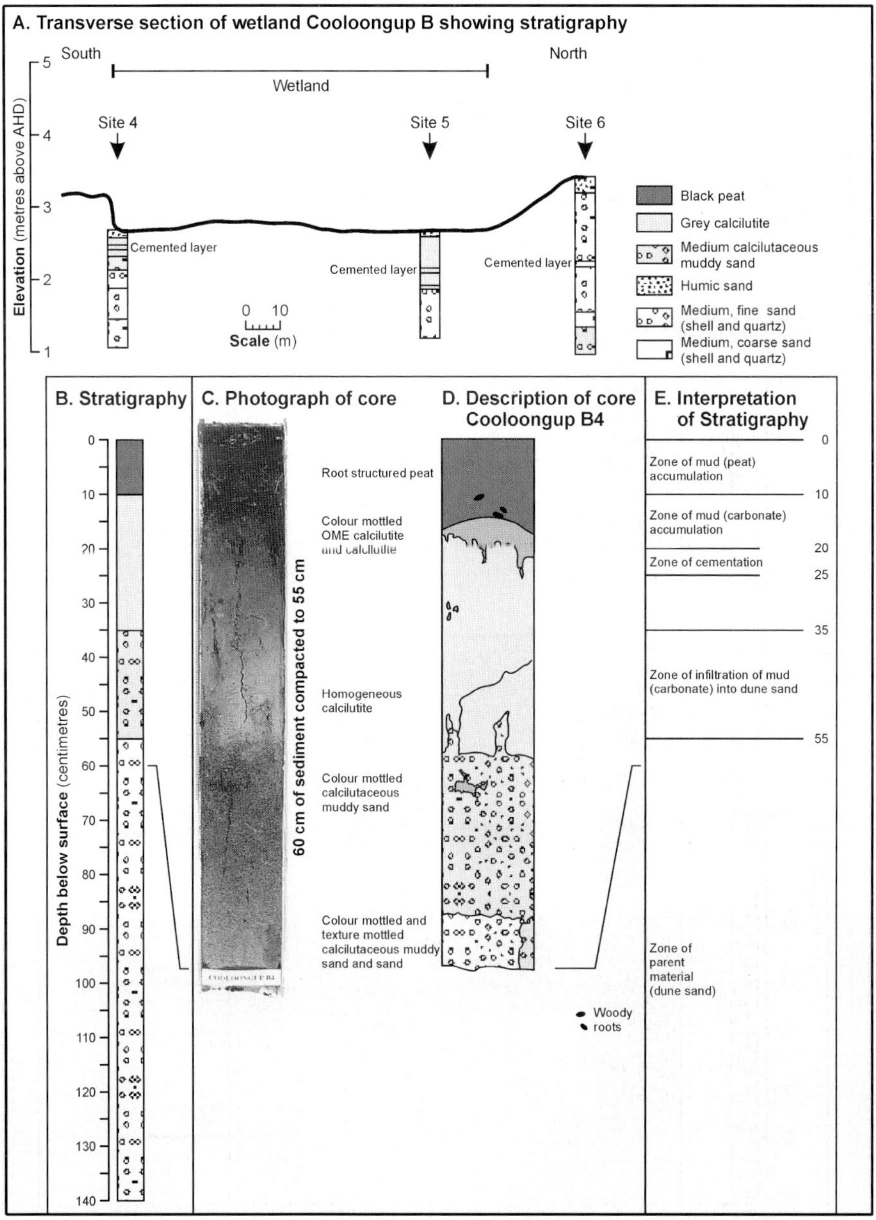

Figure 6-22. Description and interpretation of sedimentary stratigraphic sequences in wetland Cooloongup B.

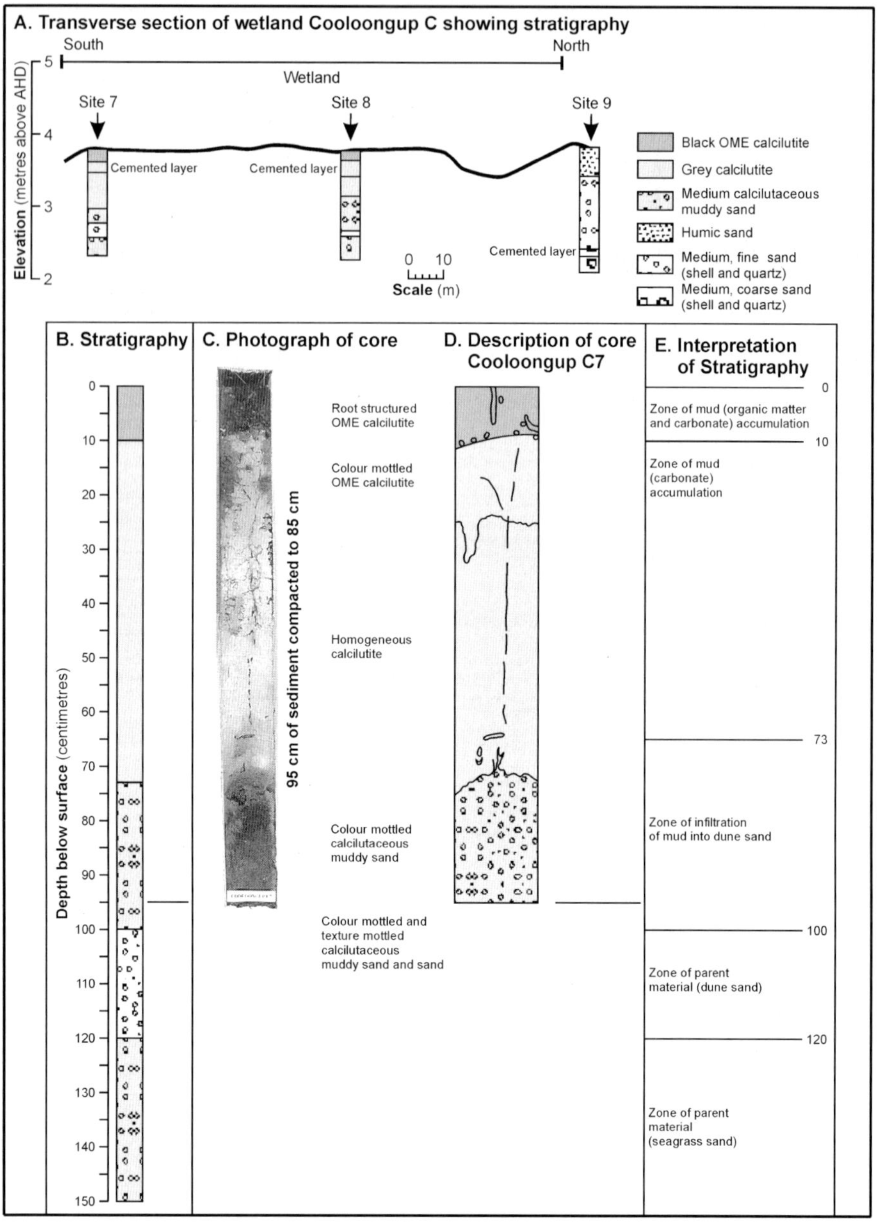

Figure 6-23. *Description and interpretation of sedimentary stratigraphic sequences in wetland Cooloongup C.*

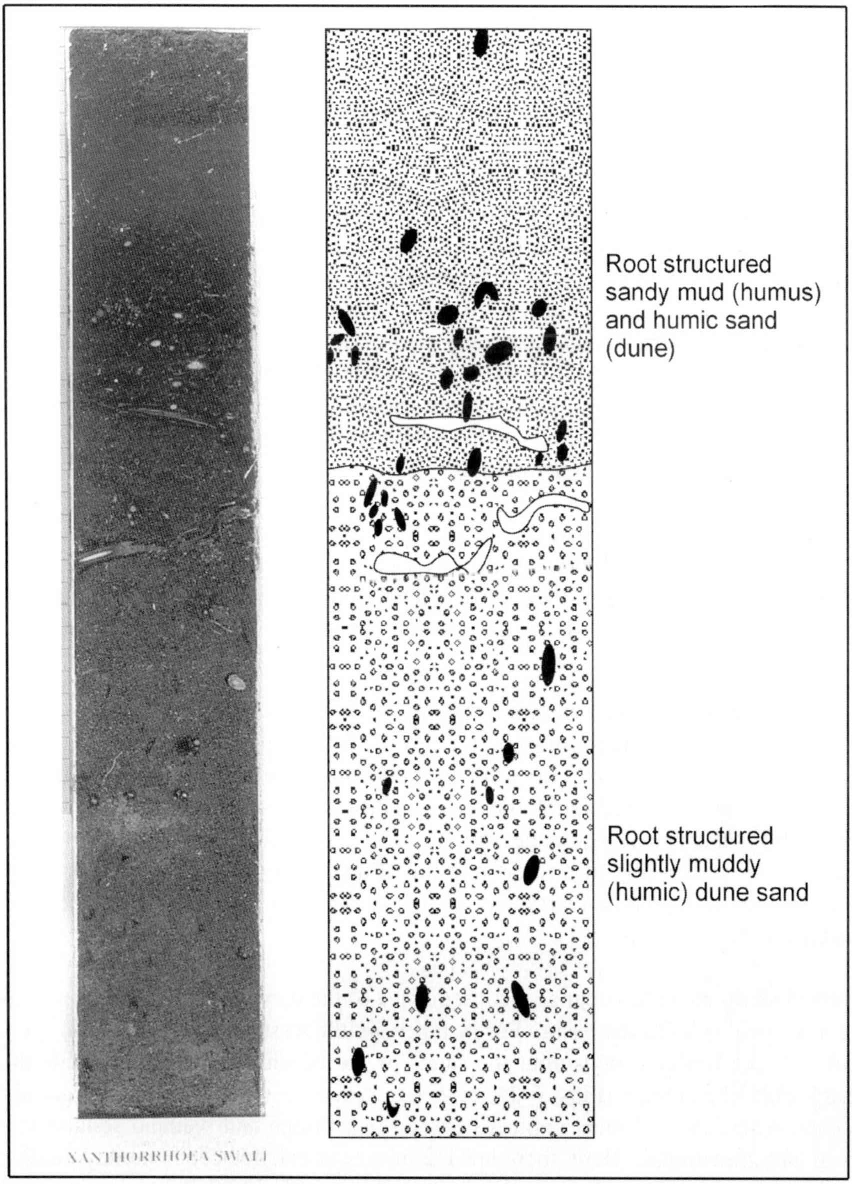

Figure 6-24. *Sediments, structures and textures in a vegetated swale colonised by grass trees (Xanthorrhoea preissii).*

6.3.4 Typical vertical stratigraphic sequences

As noted previously, the general sequence filling the wetland basins (incorporating the mud and muddy sand) has been categorised as wetland fill. The bottom layer of the fill is herein termed the basal sheet and the underlying sediments, the basement sands.

Wetland fills can be categorised into three common stratigraphic sequences (Fig. 6-25). The contact between the various sediments is commonly gradational, but locally irregular due to vertical burrows. The contact between the OME calcilutite and the underlying calcilutite in Sequence 1 is fairly sharp. Muddy sand usually forms the basal sheet, however, this term can only be applied when the layer is relatively thin with respect to the wetland fill. If the layer becomes thicker because of continual sand input through sedimentation, then it forms a wetland filling sediment in its own right. Thus, there are thin basal sheets with overlying calcilutite, thick basal sheets with overlying calcilutite, and thick muddy sand.

Transverse profiles of each study wetland, generally west to east, are illustrated in Figures 6-26 to 6-43A showing the cross-section geometry of the wetland fill, the variation of sediment types, their vertical and lateral stratigraphic relationships, and the relationship to the underlying host parent material.

6.3.5 Lateral stratigraphic relationships

Lateral stratigraphic relationships in these wetland fills are important because 1) they shed light on the evolution of the wetland basin, and 2) they feature in the hydrologic exchange between wetland and upland. Lateral stratigraphic relationships between wetland fills and beachridge/dunes are of four types: vertical and sharp; vertical and gradational; onlapping; and interfingering (Figs. 6-26 to 6-43A). These lateral relationships are variably distributed in the area, and can differ on either side of the wetland.

Vertical sharp contacts occur as a consequence of the downward continuation of the steep vertical sides to the basin. For instance, on the western margin of wetlands 35 and 161, the wetland muds abut the sands of the beachridge/dune sediments in a sharp, cliff like contact (Figs. 6-26, 6-36 A). A second type of cliffed margin also occurs, where the sediment contact between beachridge and wetland sediments is steep and gradational. Here, there are 1-2 metres of calcilutaceous muddy medium sand forming a transitional contact between medium sand and calcilutite *e.g.*, wetlands 162, 136, 9-5, and swiii (Figs. 6-27, 32, 38, 42 A). Onlapping relationships occur on both the eastern and western margin of some wetlands, where wetland sediments have been onlapped by dune sand *e.g.*, wetlands 161, 163, WAWA, 135, 142, 72 (Figs. 6-26, 28, 29, 30, 31, 33 A). In wetlands swi, and swiii, these dune sands have partially encroached into the wetland, either mixing with autochthonous sediments or becoming subject to wetland processes. Interfingering of wetland and beachridge sediments occurs on the eastern margin of some wetlands, *e.g.*, 161 and 163 (Figs. 6-26, 28 A).

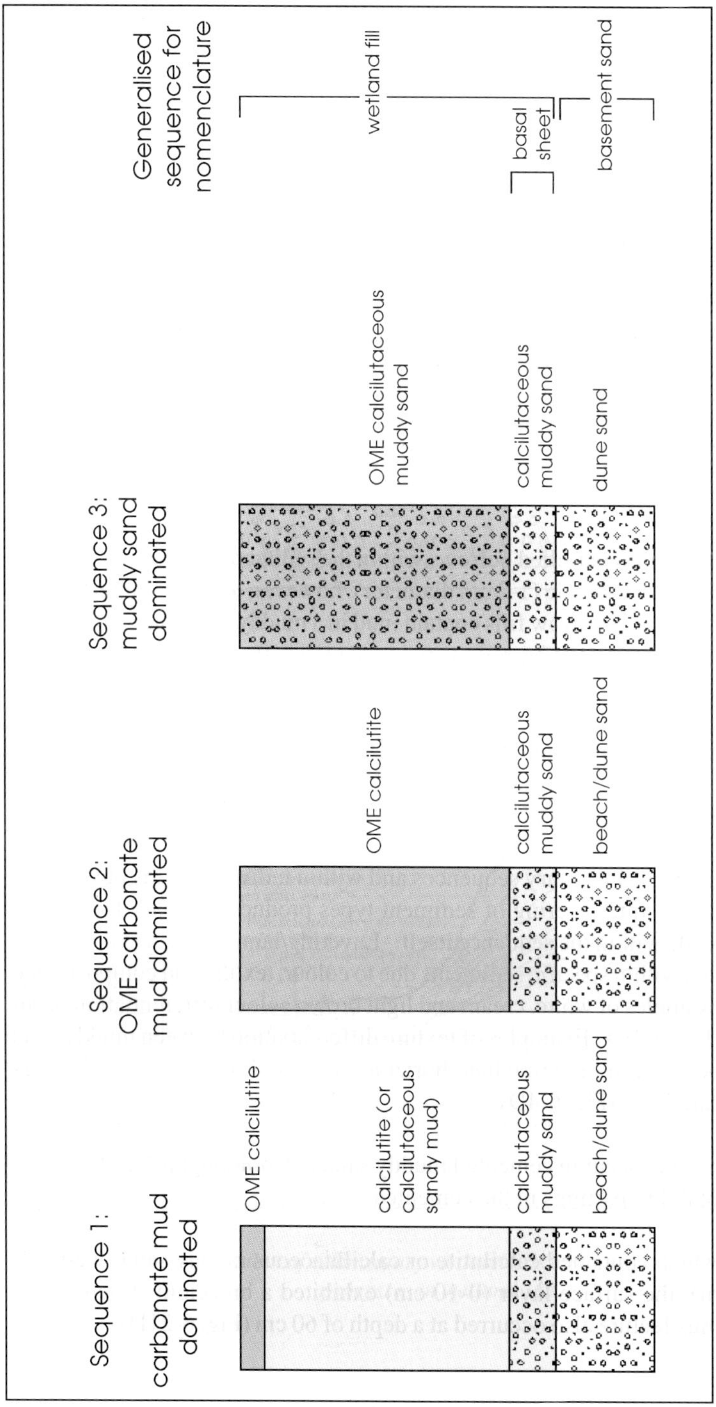

***Figure 6-25**. The three common sequences in wetland fills.*

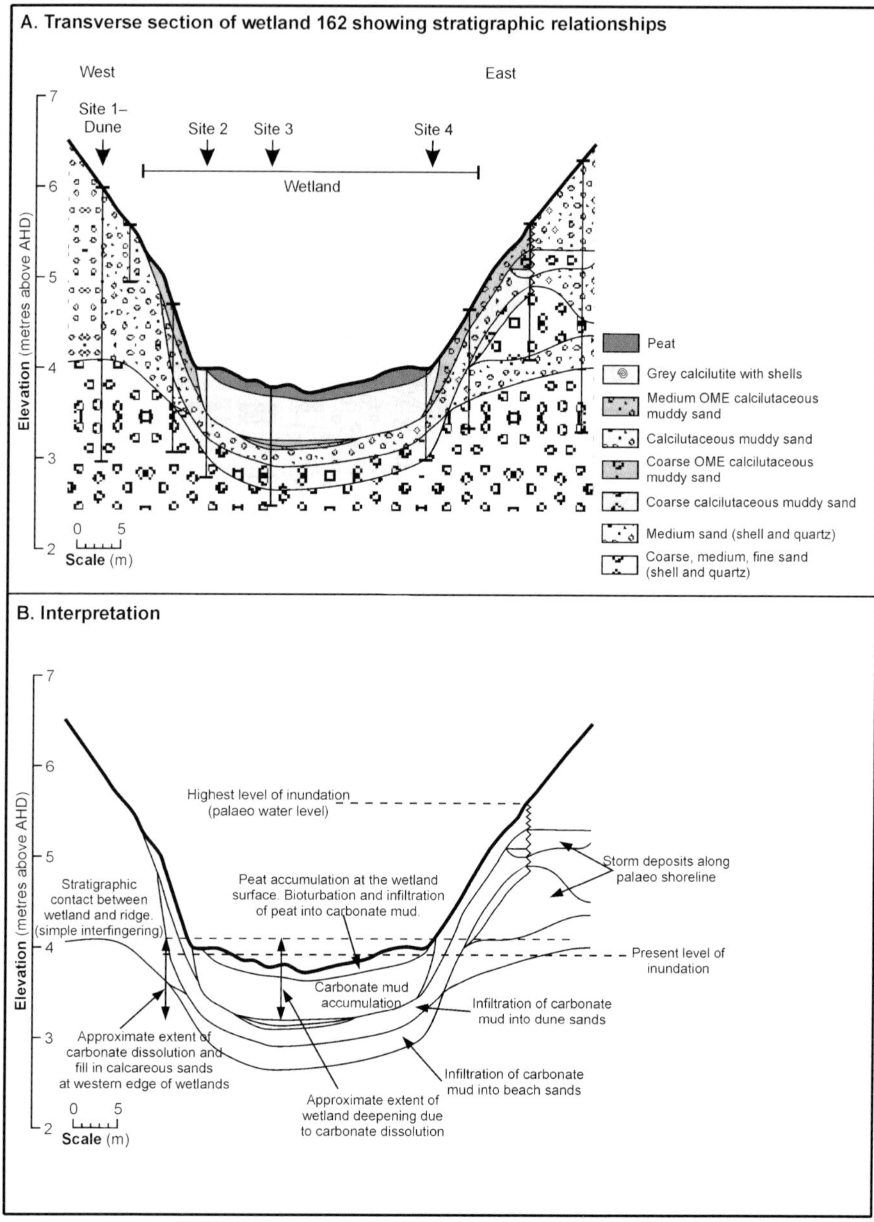

Figure 6-27. *Description of sedimentary wetland fills and interpretation of formative processes in wetland 162.*

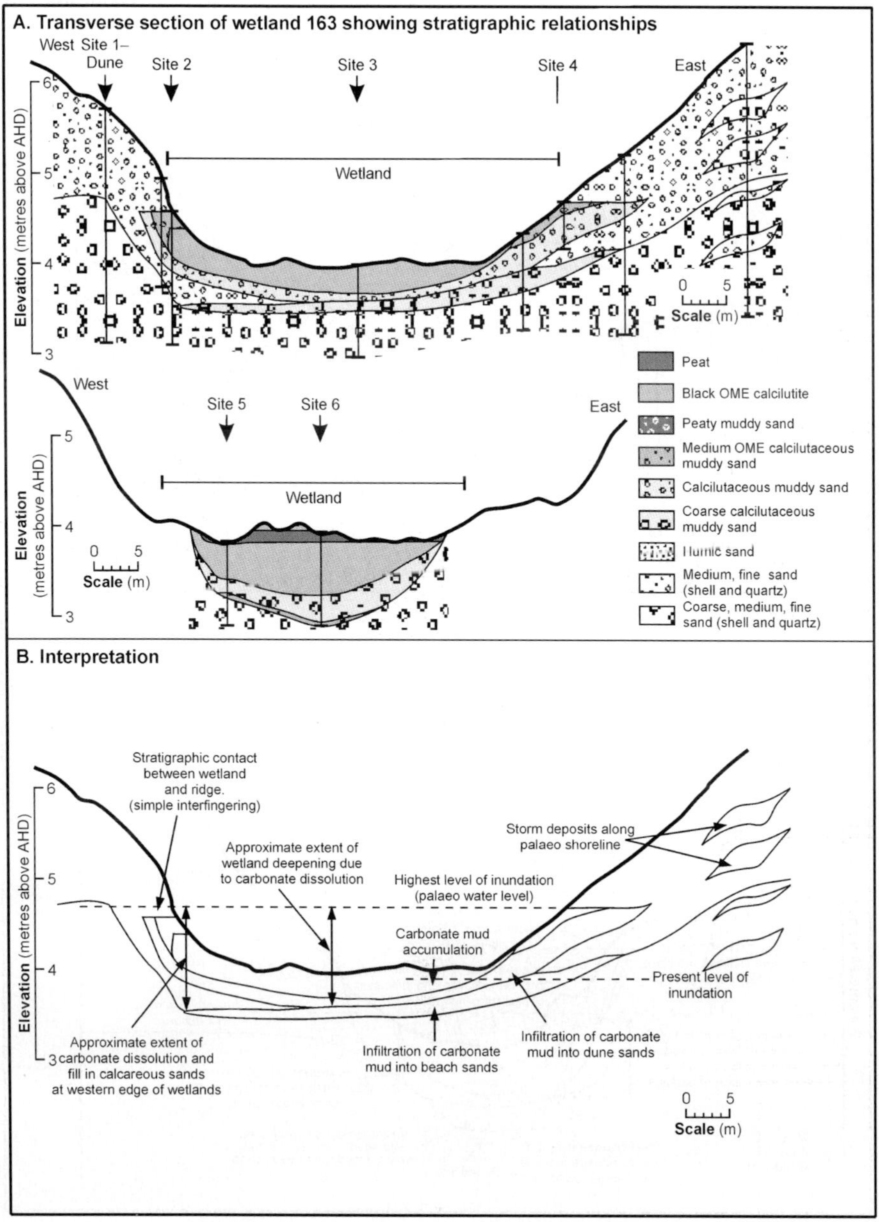

Figure 6-28. *Description of sedimentary wetland fills and interpretation of formative processes in wetland 163.*

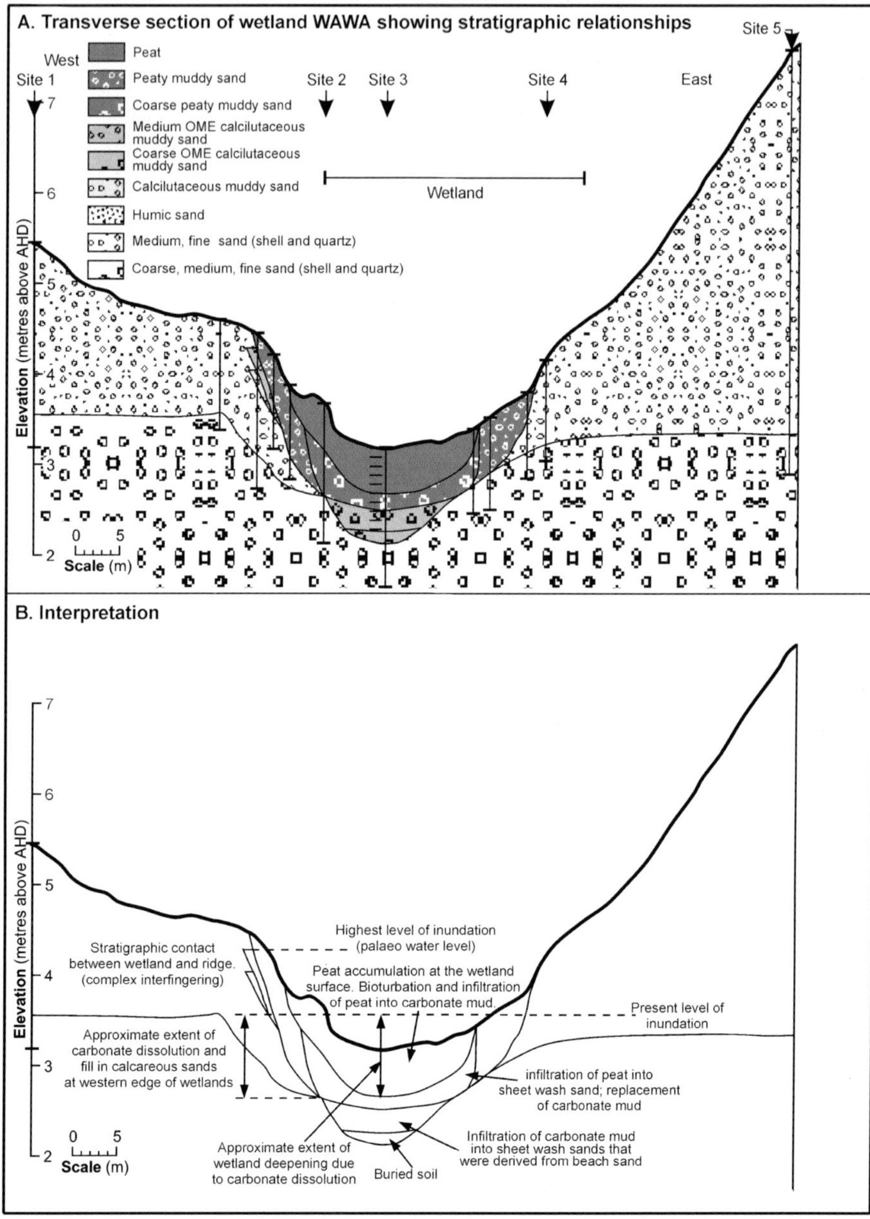

Figure 6-29. *Description of sedimentary wetland fills and interpretation of formative processes in wetland WAWA.*

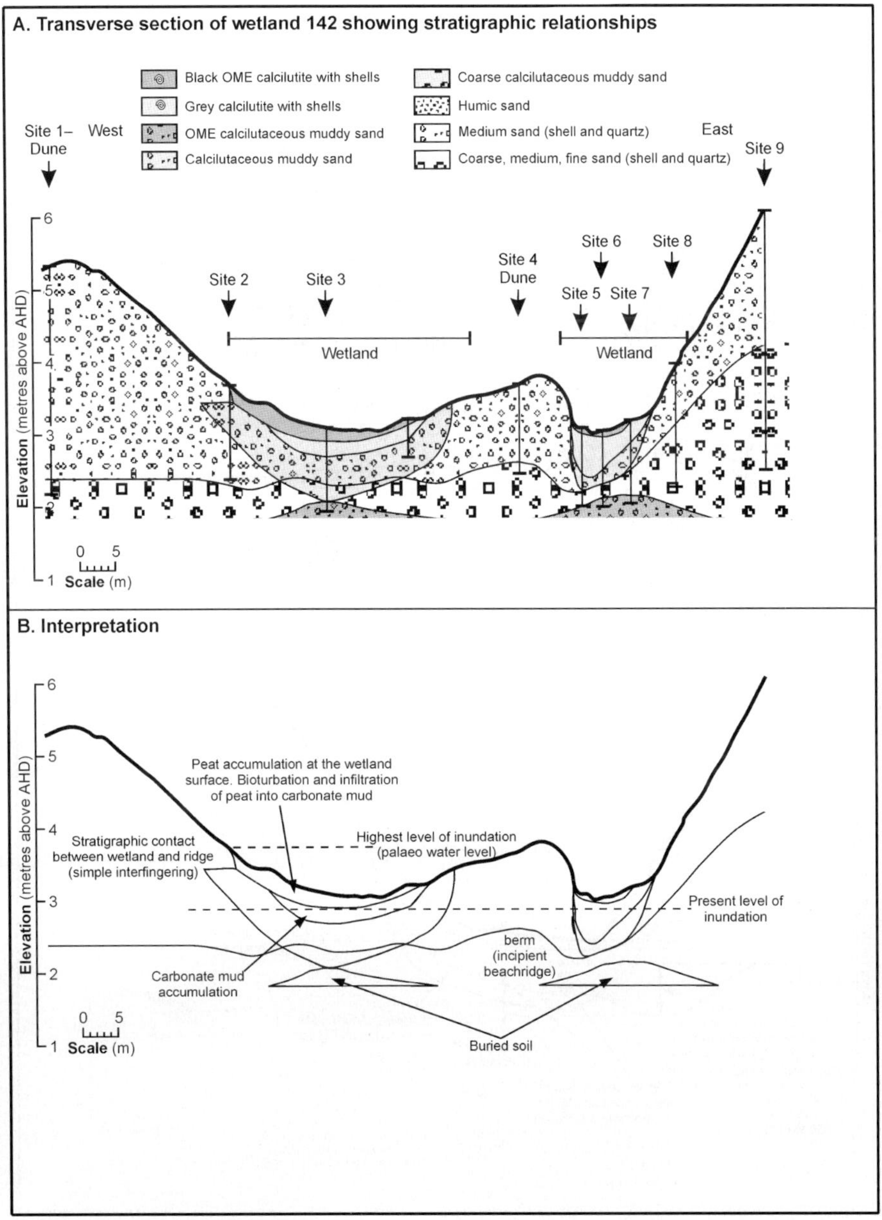

Figure 6-30. *Description of sedimentary wetland fills and interpretation of formative processes in wetland 142.*

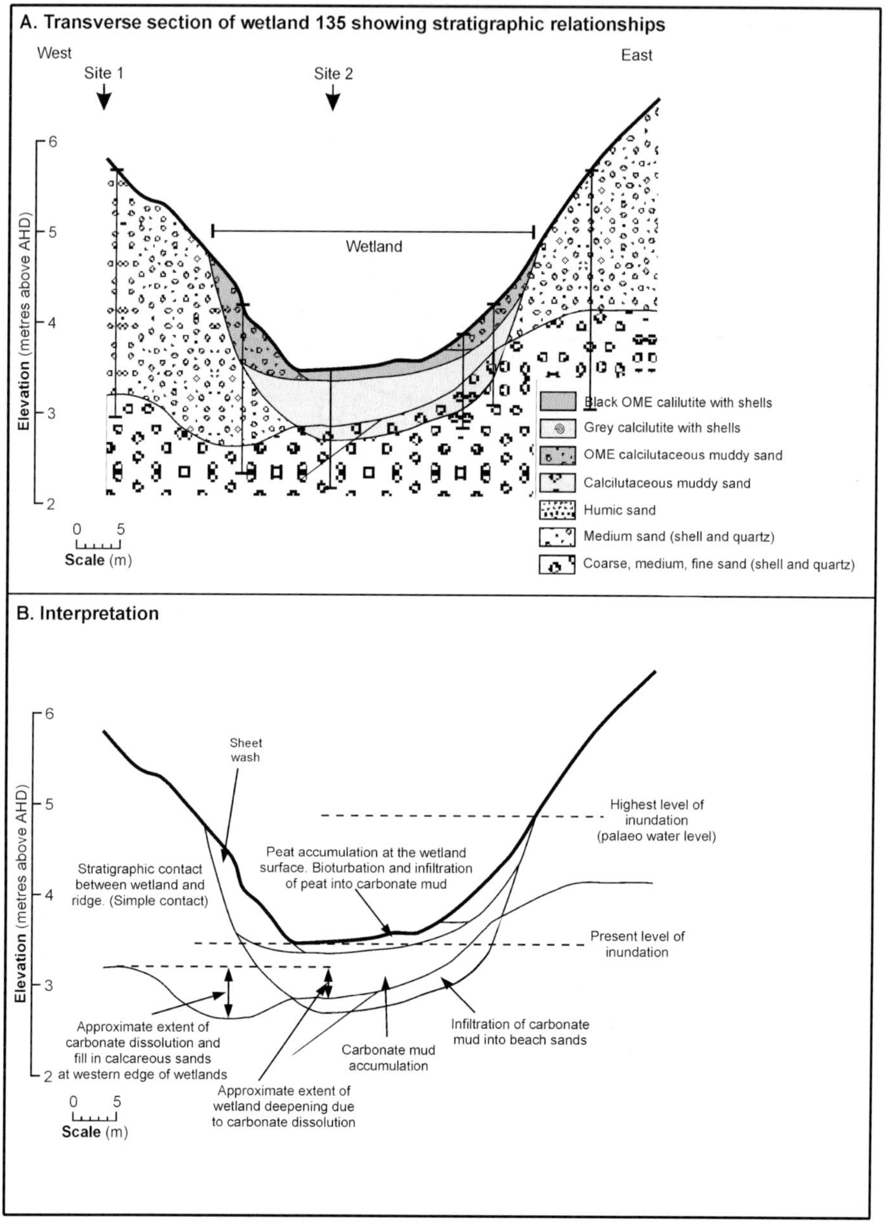

Figure 6-31. *Description of sedimentary wetland fills and interpretation of formative processes in wetland 135.*

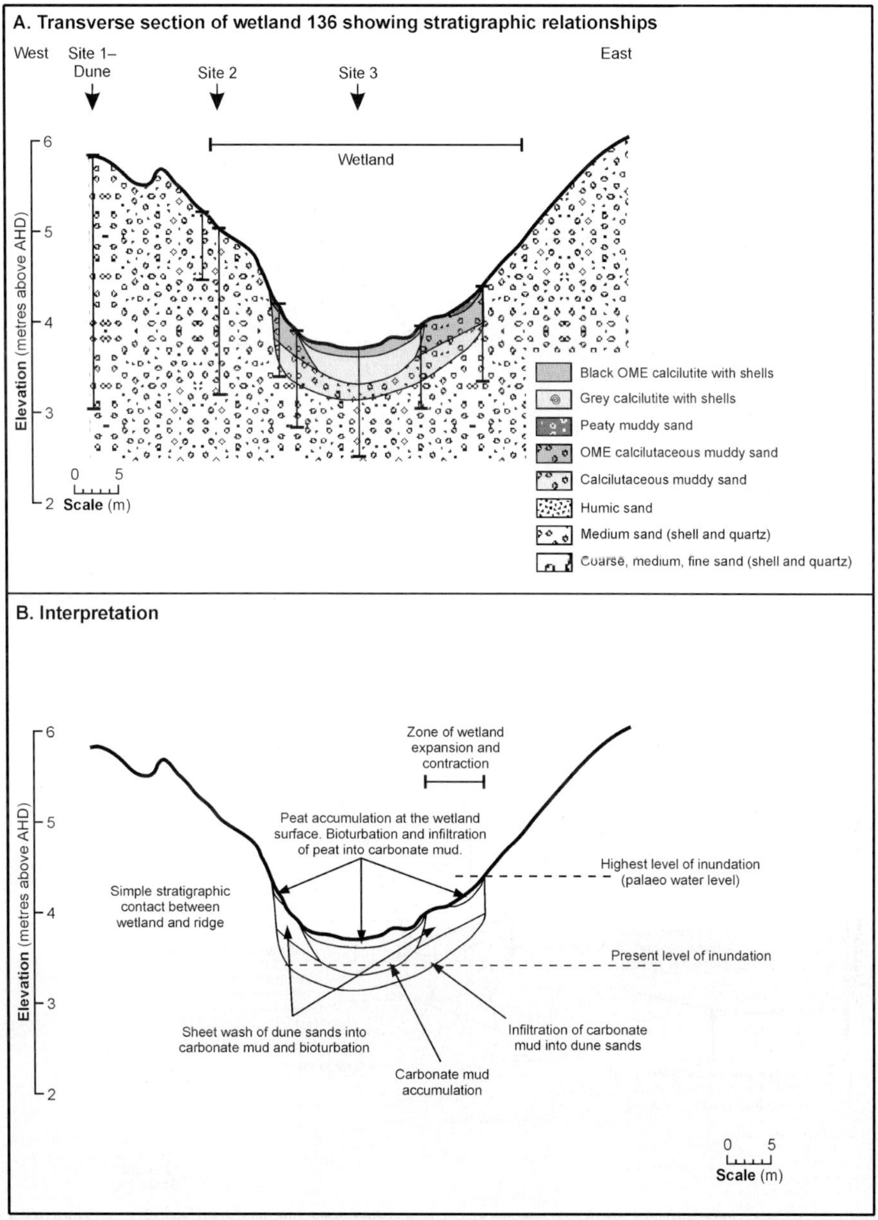

Figure 6-32. *Description of sedimentary wetland fills and interpretation of formative processes in wetland 136.*

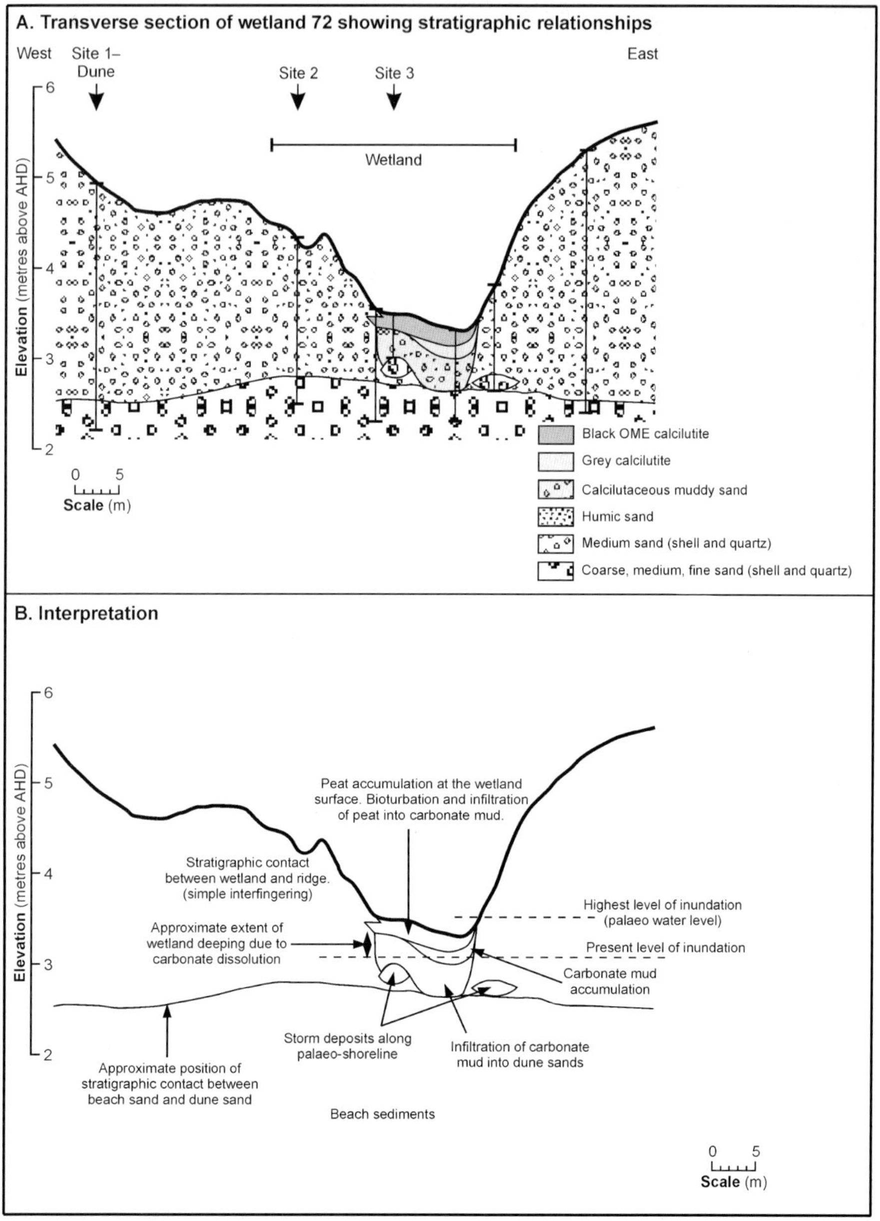

Figure 6-33. *Description of sedimentary wetland fills and interpretation of formative processes in wetland 72.*

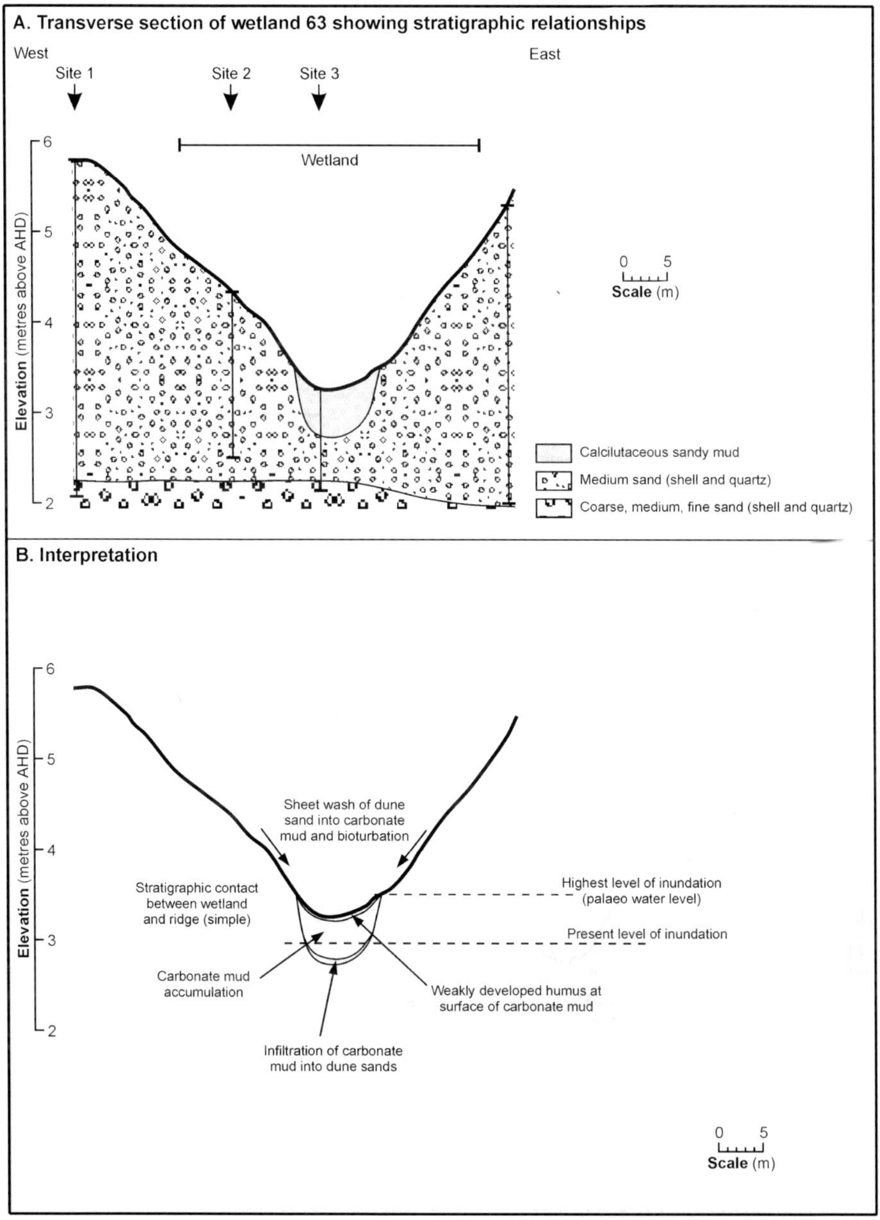

Figure 6-34. *Description of sedimentary wetland fills and interpretation of formative processes in wetland 63.*

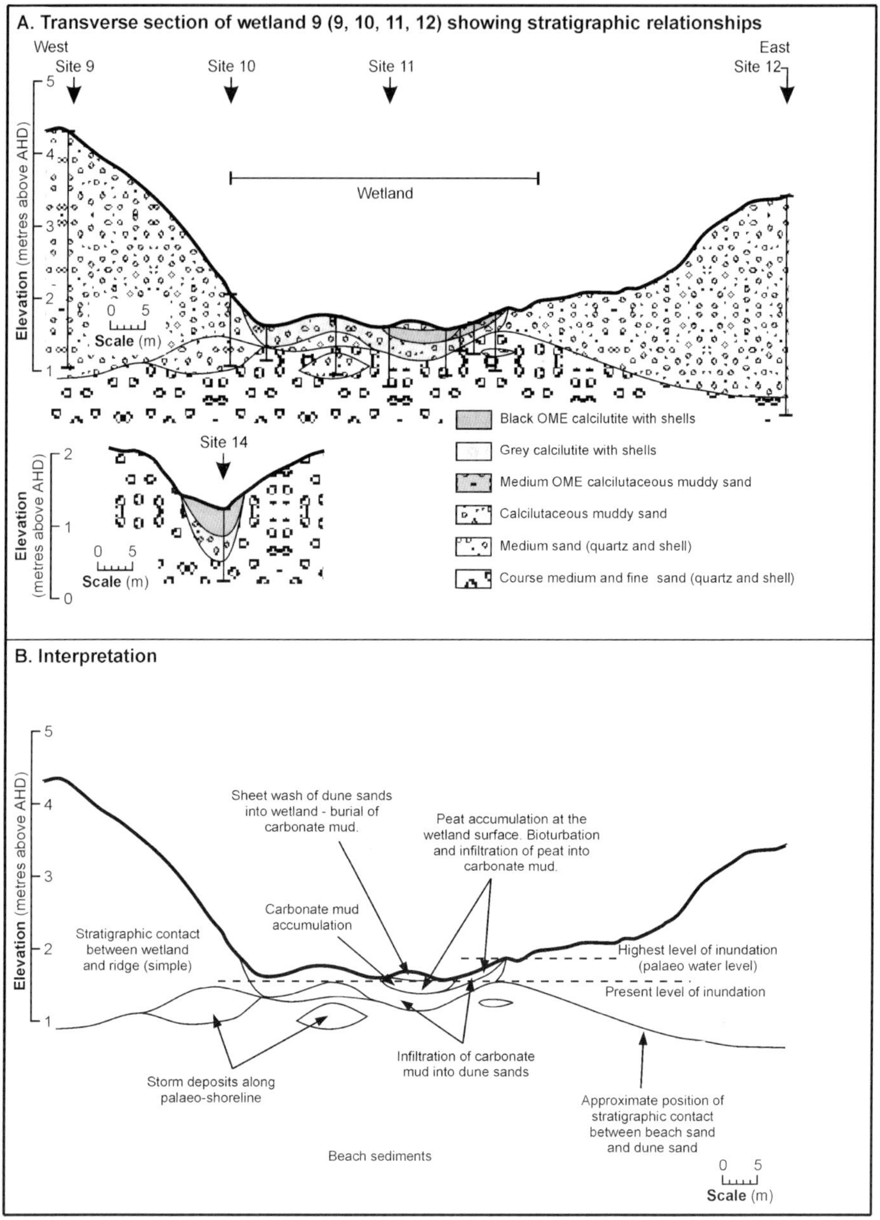

Figure 6-39. *Description of sedimentary wetland fills and interpretation of formative processes in wetland 9-10.*

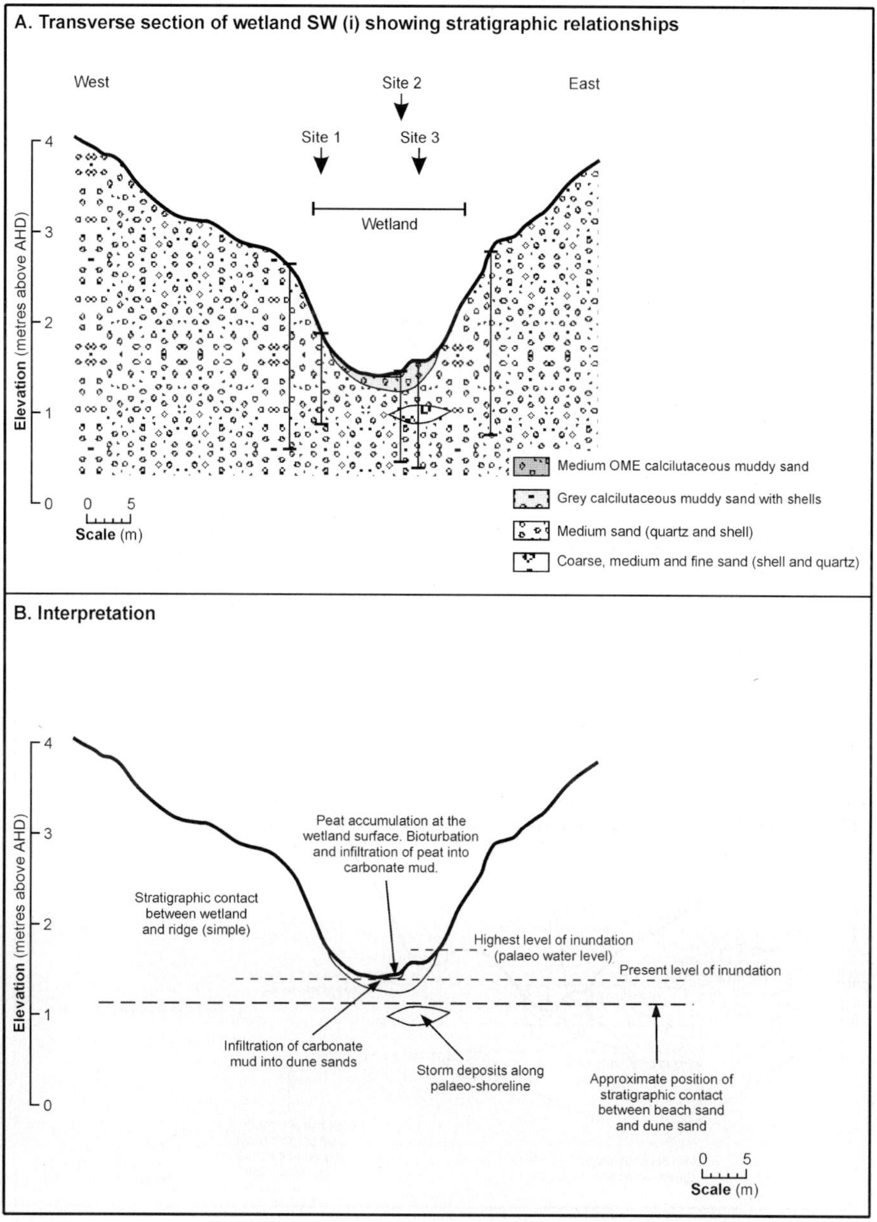

Figure 6-40. *Description of sedimentary wetland fills and interpretation of formative processes in wetland swi.*

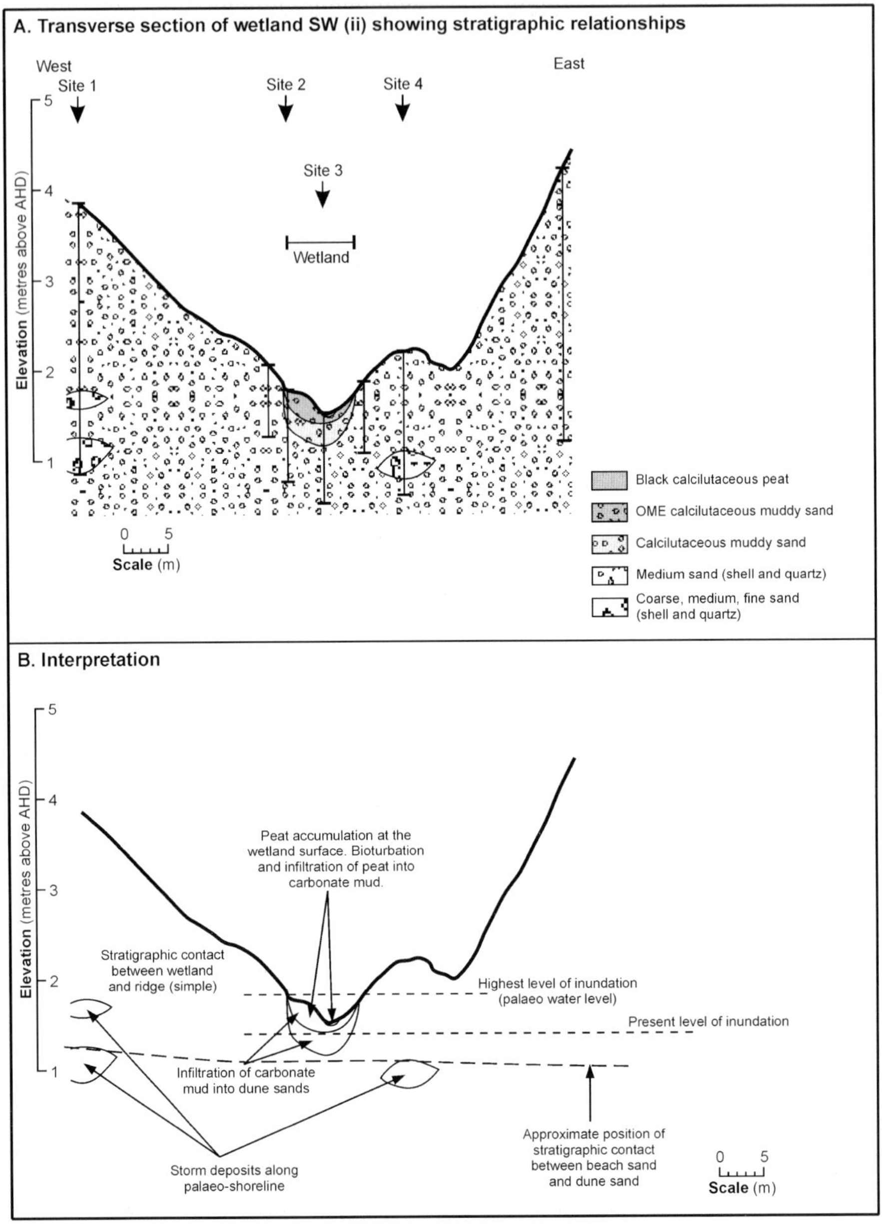

Figure 6-41. *Description of sedimentary wetland fills and interpretation of formative processes in wetland swii.*

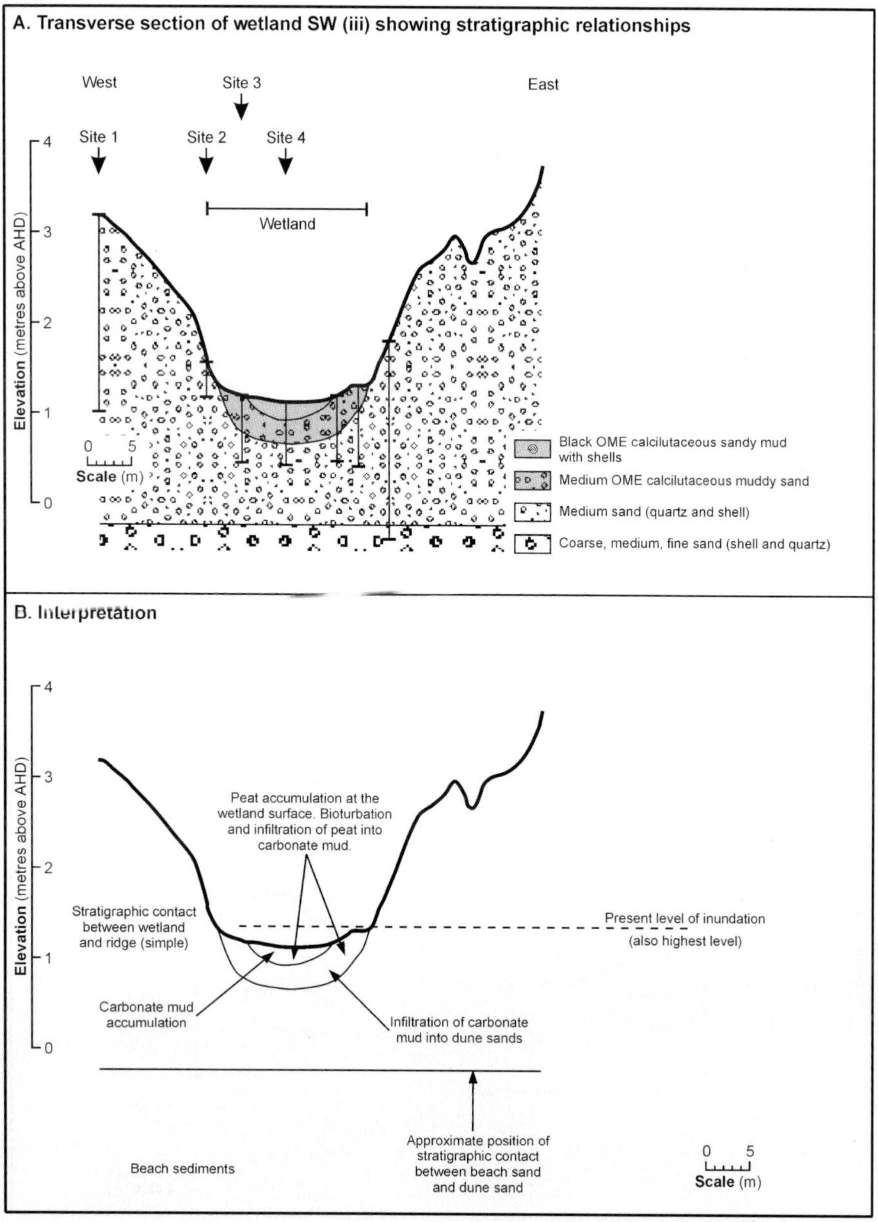

Figure 6-42. *Description of sedimentary wetland fills and interpretation of formative processes in wetland swiii.*

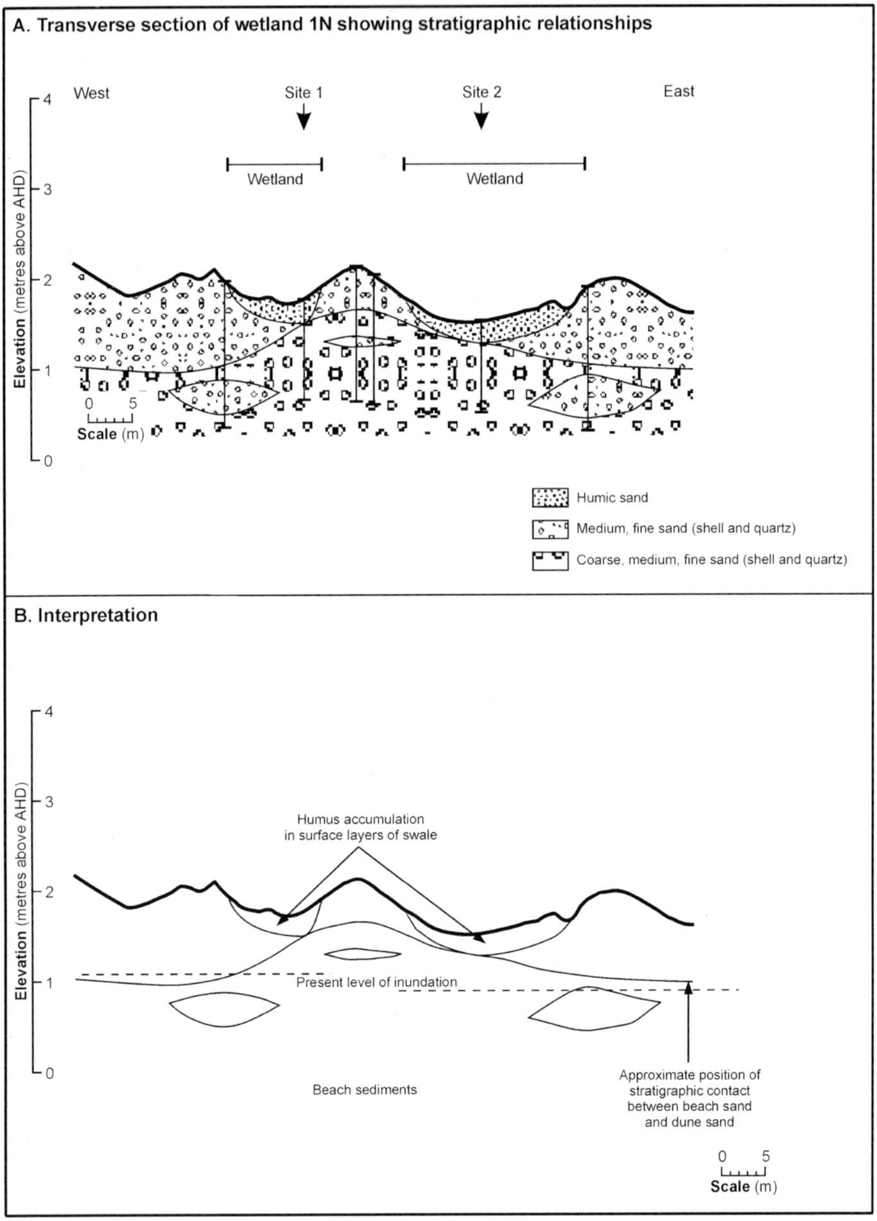

Figure 6-43. Description of sedimentary wetland fills and interpretation of formative processes in wetland 1N.

The breccioid structure is due to the formation of gravel size platy clasts from the calcilutite which are arranged in random orientations.

6.3.7 Granulometry

Granulometric separation of sediments down profile into major textural components of gravel, sand, and mud, was undertaken to characterise the sediment types (Figs. 6-44 to 6-56 A). The >2000 μm (gravel size) component was minor down profile across all wetlands, and did not occur at all in two of the wetlands (swii and swiii). It ranged from 0% to 8% by weight, with a mean of 2.4%. The 63-2000 μm (sand size) component ranged from 15% to 97% by weight, and was the dominant component in all but four of the wetlands. At a depth of approximately 60 cm (range 20-40 cm), there was a marked increase in the proportion of sand generally, signalling the beginning of the gradational contact with the basement sands. The <63 μm (mud size) component ranged from 6% to 88% by weight. It was dominant in wetlands 161, 162, 163, and WAWA and decreased in proportion in the younger wetlands. The mud component was greatest in the upper layers of each profile (0-50 cm), whether it was the dominant or sub-dominant component.

6.3.8 Composition of grain fractions

To further characterise the sediment types, and to examine trends down profile, determination of the percentage composition (by weight) of calcium carbonate, humus and plant material, and quartz in each of the major textural components (gravel, sand, and mud), was undertaken (Figs. 6-44 to 6-56 B). On the basis of this analysis, wetlands were categorised as mud, sand, or muddy sand dominated groups. This categorisation was then used in ordination of environmental features and hydrological studies.

Root material comprised the major part of the >2000 μm fraction, followed by shell. The shell component often comprised one or two complete fossil shells of *Glyptophysa* sp. and *Gyraulus* sp. Shells were distributed unevenly throughout the profiles, occurring predominantly in the surface horizons or in shell lag horizons.

In order of abundance, the sand fraction was composed of calcium carbonate (skeletal fragments), quartz grains, and plant material (seeds, stem and leaf detritus). The calcium carbonate content exceeded quartz in nearly all wetland sites to a depth of 60 cm and then the trend reversed (Table 6.5 and Figs. 6-48, 49, 52, 53). The exception was wetland WAWA in which the carbonate was negligible throughout.

Wetlands exhibited several patterns in calcium carbonate composition in the sand fraction down profile:

- a vertically consistent pattern *e.g.*, wetlands 161, WAWA, 63, 35, swi,ii,iii, 1N
- an expanding/contracting pattern *e.g.*, wetlands 163, 135, 72, 45, 9-6
- a fluctuating pattern *e.g.*, wetlands 162, 163, 142

These patterns may be attributed to several processes, *e.g.*, inundation frequency, carbonate dissolution, and carbonate buffering. The age of the wetland is also a factor, insofar as many of the processes are time dependent. In older wetlands (161, 162, 163, 142, 135, 72, 35, 9-6, WAWA), the percentage of calcium carbonate in the sands decreased with depth, while quartz increased. In younger wetlands (63, 9-11, swi, swii, swiii, 1N), the percentages of calcium carbonate and quartz remained relatively stable.

Below the top 10 cm, the carbon content of the sand was generally low with the exception of wetlands WAWA, 161, 45, 35, sites with a higher frequency of inundation.

The mud size component ranged from 6% to 88%. The mud fraction comprised calcium carbonate, peat and organic material, and quartz, with minor amounts of diatoms and traces of K-feldspar in the younger wetlands (Table 6.3 142, 9-5, swii). Calcium carbonate was dominant in all wetlands except WAWA, which was dominated by peat. Peat and organic material were minor overall, exhibiting maximum carbon content from 0-20 cm and thereafter remaining consistent down profile. Quartz content was variable: negligible in wetlands 161, 142, 72, 63, swii; consistent down profile in wetlands 162, 163, 35, 9-6, 9-11, swi, 1N; and increasing sharply at depth in wetlands WAWA, 135, 45, and swiii. The distribution of mud throughout the profiles was partly determined by the burrow mottled structure of the sediments, and the occurrence of specific structural or buried features.

Selected samples of carbonate mud and associated fine grained components from wetlands along three different east/west transects were analysed by XRD to determine mineralogy. These data show that the components of mud determined by X-ray diffraction are mainly calcitic, with minor aragonite and traces of quartz (Table 6.3). The non-carbonate component from selected samples also was investigated. After digestion of the carbon and carbonate in the mud, residues were analysed by XRD to determine their composition. Results are provided in Table 6.4.

The key gradational patterns for each wetland based on Figures 6-44 to 6-56, are described in Table 6.5. The data in Table 6.5 provide a link between the granulometry, the grain composition, and some sedimentary processes.

Table 6.3 *Constituents of carbonate mud determined by XRD*

Transect A site	*Constituents of mud*	*Transect B site*	*Constituents of mud*	*Transect C site*	*Constituents of mud*
Site 2 120 cm calcilutite	calcite; aragonite; trace quartz	Site 3 33 cm muddy sand	calcite; aragonite; trace quartz	Cud Swamp 42 cm calcilutite	calcite
Site 3 46 cm calcilutite	calcite	Site 4 38 cm calcilutite	calcite; trace quartz	Site 3 28 cm calcilutite	calcite; trace quartz
Site 4 88 cm calcilutite	calcite; magnesian calcite; aragonite; quartz	Site 6 43 cm calcilutite	calcite; aragonite; trace quartz	Site 4 23 cm muddy sand	calcite; aragonite; trace quartz
Site 6 20 cm muddy sand	calcite; magnesian calcite; aragonite; quartz	Site 7 31 cm calcilutite	calcite; trace aragonite; trace quartz	Site 5 60 cm calcilutite	calcite; trace quartz, iron sulphide, dolomite
Site 7 50 cm calcilutite	calcite; aragonite			Site 6 16 cm calcilutite	calcite
Site 7 90 cm calcilutite	calcite; magnesian calcite; aragonite; quartz				
Site 9 47 cm calcilutite	calcite; aragonite				

Table 6.4 *Composition of sediment remaining after carbon and carbonate removal*

	142 (10 cm)	*142 (20 cm)*	*9-5 (20 cm)*	*swii (20 cm)*
Amorphous silica (diatoms)	D		D	D
Quartz	A	Tr	A	A
K-feldspar	Tr-A		Tr-A	A
Plagioclase	Tr		Tr	Tr

D Used for the component apparently most abundant
A Components judged to be present at levels of 5-20%
Tr Components judged to be below 5%

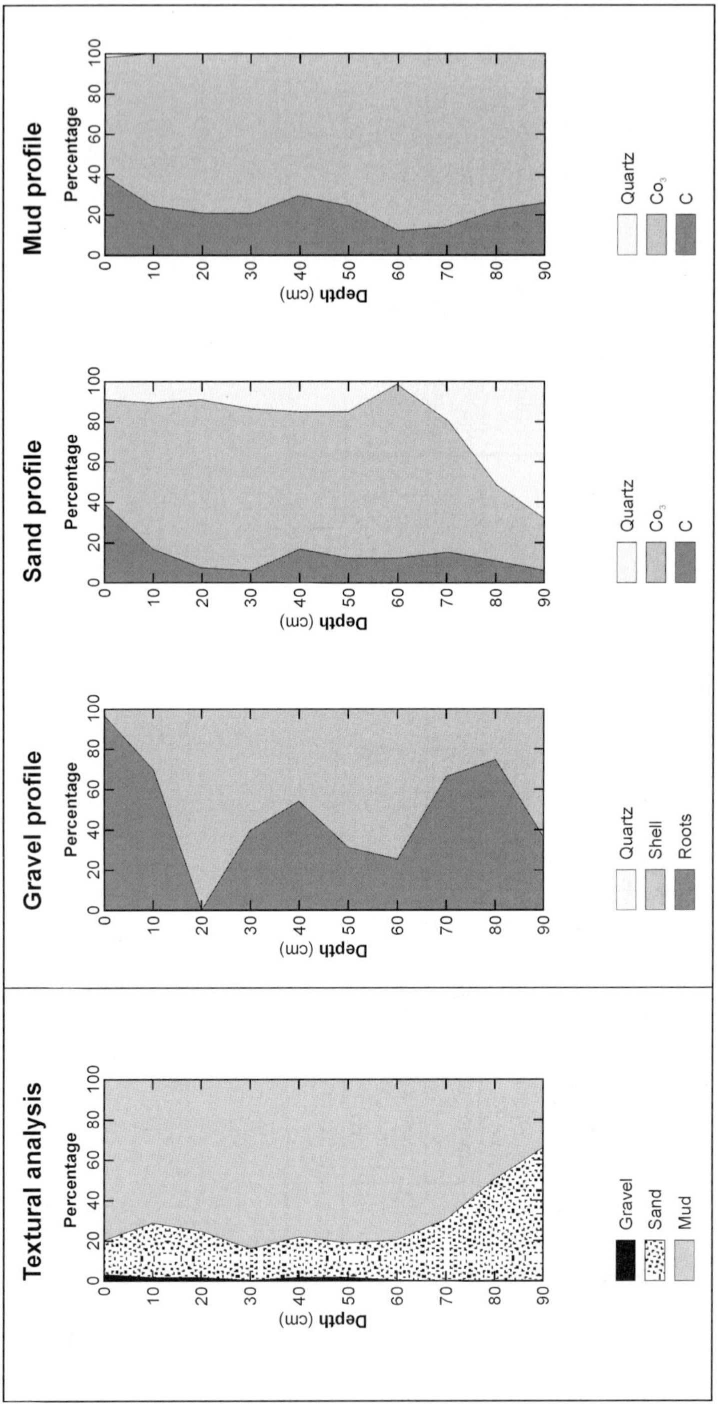

Figure 6-44. *Wetland 161: textural and compositional analyses.*

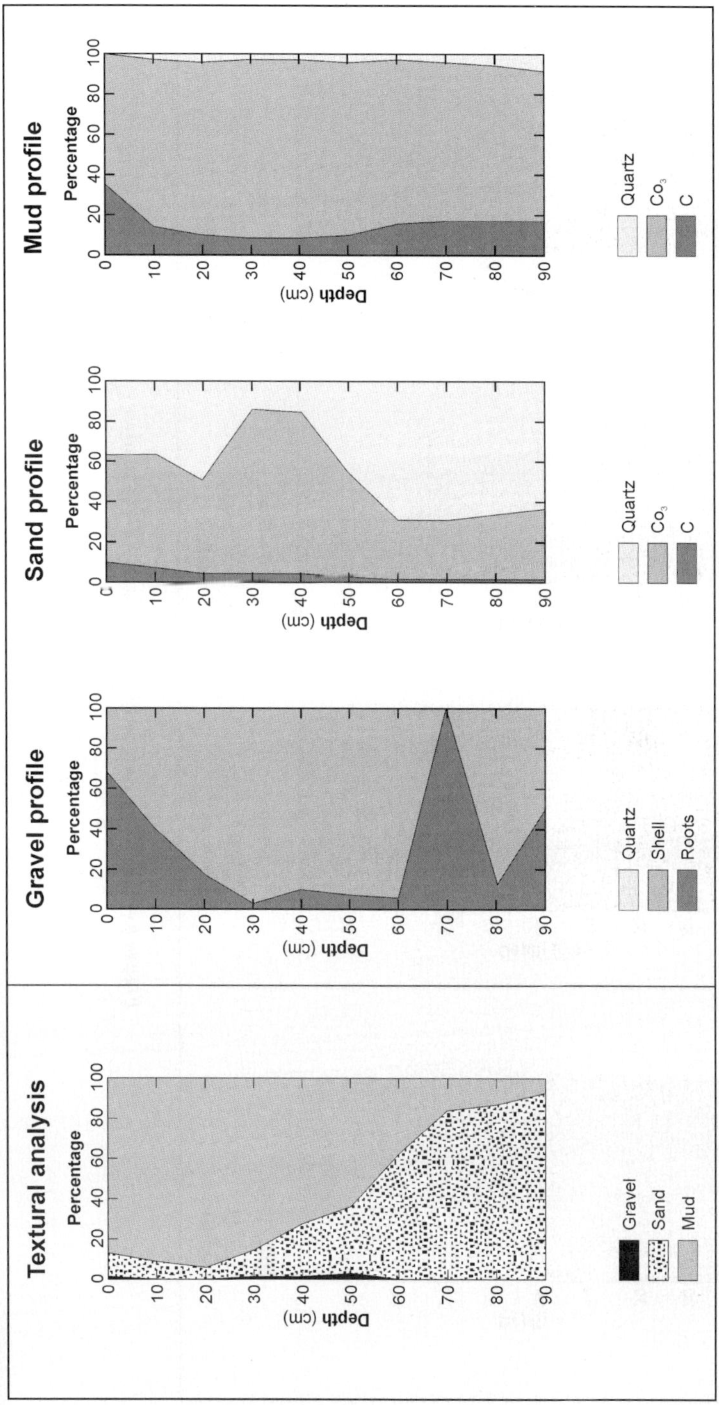

Figure 6-45. Wetland 162: textural and compositional analyses.

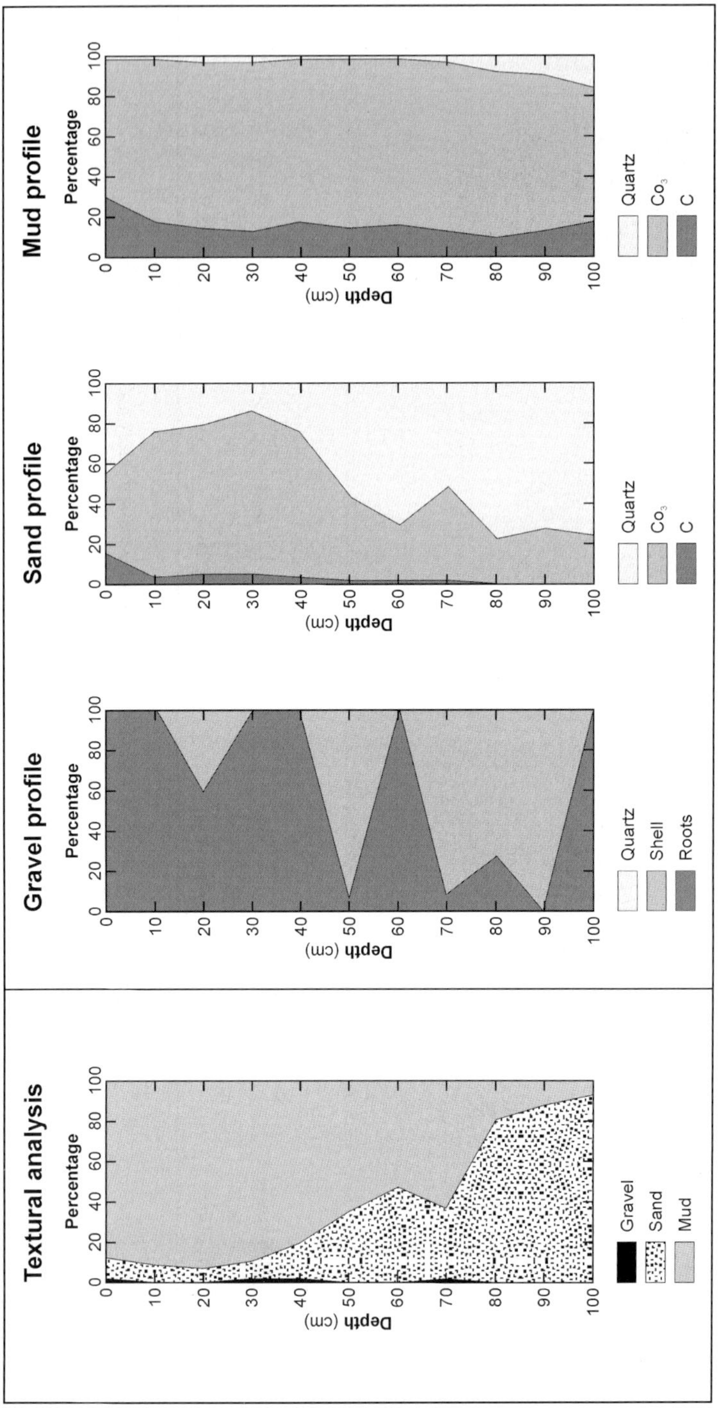

Figure 6-46. *Wetland 163: textural and compositional analyses.*

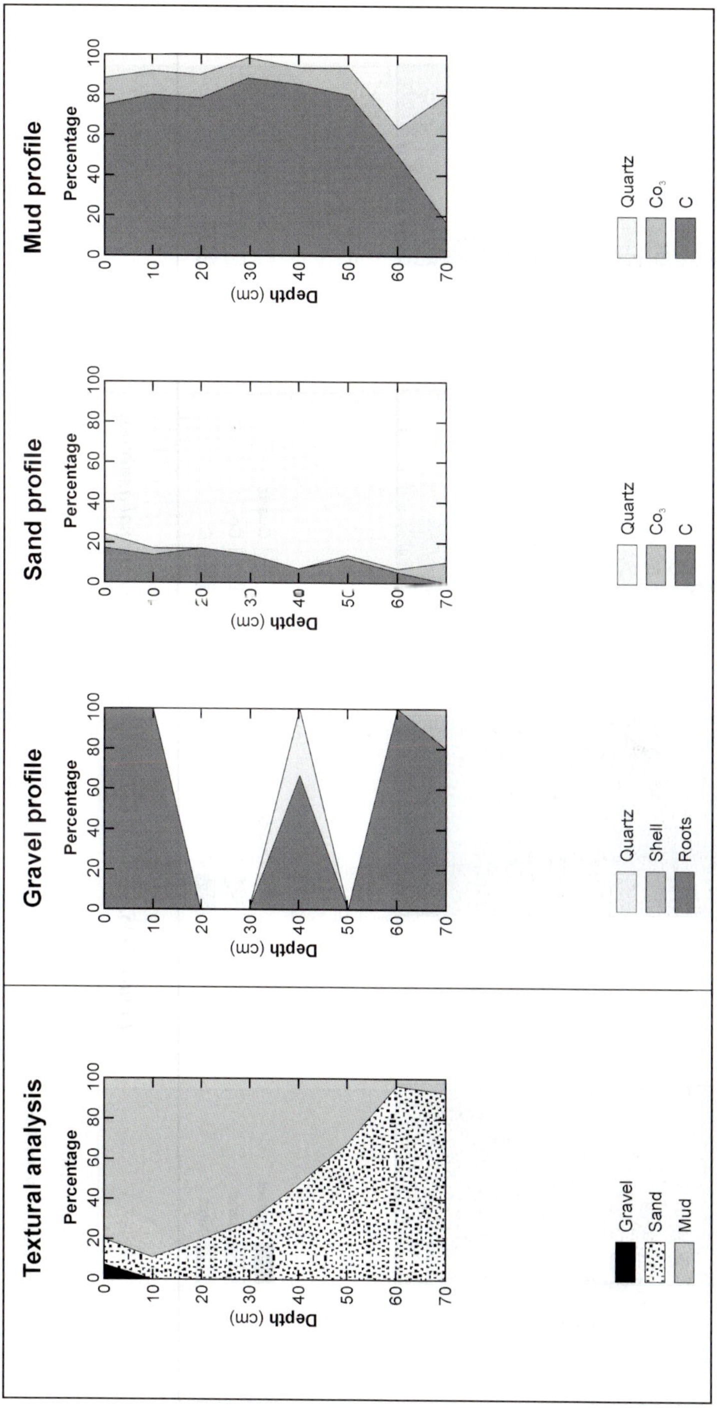

Figure 6-47. *Wetland WAWA: textural and compositional analyses.*

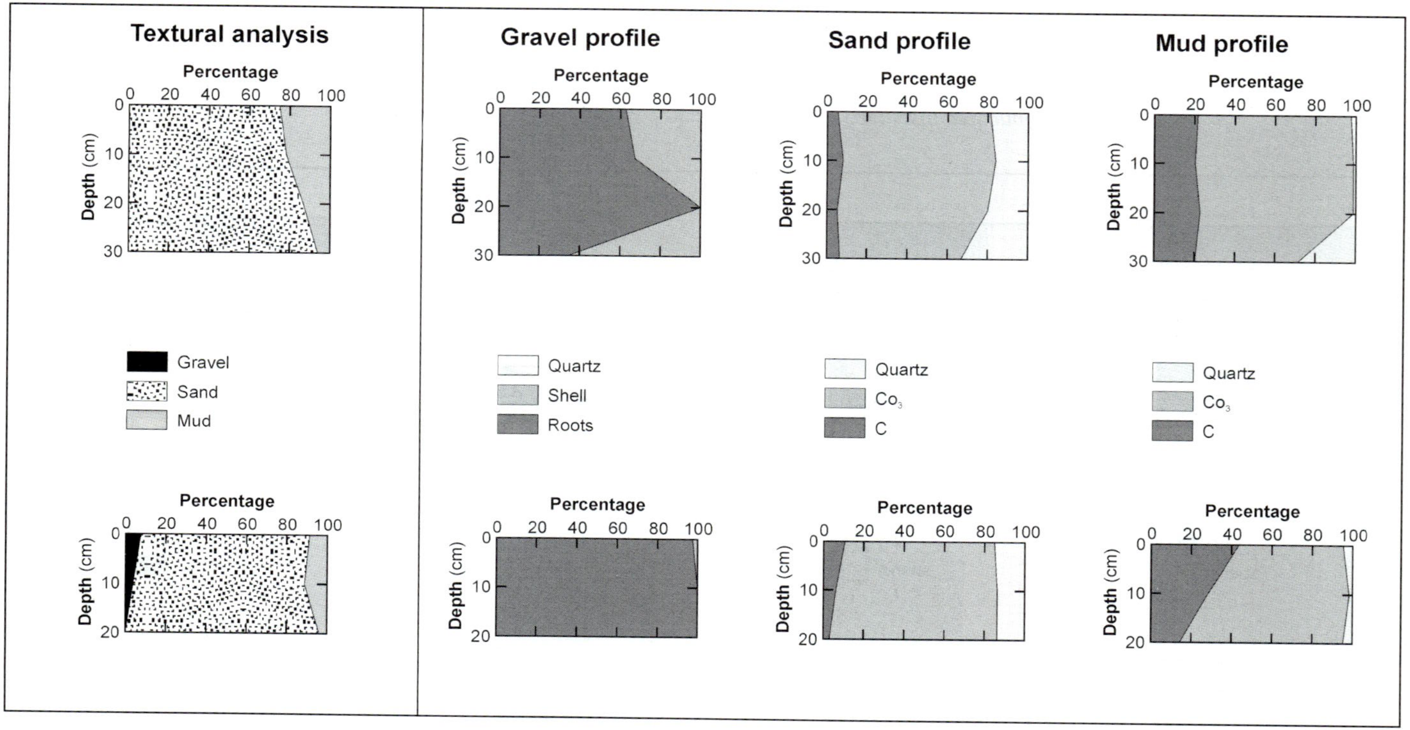

Figure 6-56. Wetlands swiii and 1N: textural and compositional analyses.

Table 6.5 *Granulometric patterns and compositional features of sediment profiles*

Wetland	*Patterns down profile*		*Comments*
161	Gravel	Gravel present throughout profile; freshwater shells to 60 cm; Live roots dominate 0-10 cm; non-living roots throughout profile	• Freshwater shells implies freshwater environment. There is a relationship between calcilutite and fossil shells. • Gravel inherited from beach sediments indicates basal sheet origin is beach. • Live *vs* dead root zone separates rhizosphere from buried rhizosphere. • Roots indicate root structured sediment • Mud-dominated profile. • Carbonate dominates all fractions • Carbon content of the mud fraction forms a low amplitude cyclic pattern.
	Sand	Sub-dominant component Sand content increases below 60 cm up to maximum > 60 % Carbonate dominant; carbon throughout profile; quartz dominates below 60 cm	
	Mud	Dominant component Mud content consistent to 60 cm, then decreases Carbonate dominant; carbon content fluctuates down profile; negligible quartz content	
162	Gravel	Gravel present throughout profile; freshwater shells to 60 cm; Live roots dominate 0-20 cm; layer of non-living roots at 70 cm	• Freshwater shells implies freshwater environment. There is a relationship between calcilutite and fossil shells. • Gravel inherited from beach sediments indicates basal sheet origin is beach. • Live *vs* dead root zone separates rhizosphere from buried rhizosphere. • Root structured sediment
	Sand	Co-dominant component Sand content increases below 50 cm up to maximum of > 90 % Variable ratio of carbonate to quartz down profile Carbonate and quartz dominant; minor carbon content down profile; quartz dominates below 50 cm	

Table 6.5 (cont.)

Table 6.5 (cont.)

Wetland	*Patterns down profile*	*Comments*
162	Mud Dominant component Mud is dominant to 50 cm, then decreases Carbonate dominant; all components consistent down profile	• Mud-dominated profile. • Carbonate dominates all fractions • Compositional proportions are related to small scale structural features
163	Gravel Gravel present in layers throughout profile; whole freshwater shells to 50 cm; fragments to 90 cm Live roots dominate 0-10 cm; non-living roots throughout Sand Sub-dominant component; variable down profile Sand increases below 70 cm up to maximum > 80 % Carbonate dominant to 50 cm, then quartz dominant; carbon throughout profile Mud Dominant component to 70 cm, then decreases Carbonate dominant; carbonate and carbon content consistent down profile; quartz content increases with depth	• Freshwater shells implies freshwater environment. There is a relationship between calcilutite and fossil shells. • Gravel inherited from beach sediments indicates basal sheet origin is beach. • Live *vs* dead root zone separates rhizosphere from buried soil. • Root structured sediment • Mud-dominated profile. • Carbonate dominates sand and mud fractions.
WAWA	Gravel Gravel present in layers throughout profile; no whole fossil shells, only thin facades. Live roots dominate 0-10 cm; non-living roots in layers throughout profile	• Live *vs* dead root zone separates rhizosphere from buried rhizosphere. • Root structured sediment. • Mud-dominated profile.

Table 6.5 (cont.)

Table 6.5 (cont.)

Wetland	*Patterns down profile*		*Comments*
WAWA	Sand	Co-dominant component Sand content increases down profile up to > 90 % Carbonate minor; carbon throughout profile; quartz dominant	• Quartz dominates the sand fraction • Peat dominates the mud fraction • Quartz peak at base of mud profile possibly due to aeolian input
	Mud	Co-dominant component Mud content decreases below 40 cm Carbon content dominant, > 80%; as mud content decreases, quartz and carbonate become dominant	
135	Gravel	Negligible component; whole freshwater shells to 40 cm Live roots dominate 0-10 cm; non-living roots throughout	• Freshwater shells implies freshwater environment. There is a relationship between calcilutite and fossil shells. • Gravel inherited from beach sediments indicates basal sheet origin is beach. • Live *vs* dead root zone separates rhizosphere from buried rhizosphere. • Root structured sediment • Sandy mud-muddy sand dominated profile. • Carbonate dominates sand and mud fractions. • Sand fill diluting surface mud layers. • Quartz peak at base of mud profile possibly due to aeolian input.
	Sand	Sub-dominant component Sand content fluctuating down profile; increases in surface layers and below 60 cm up to maximum 90 % Equal amounts of carbon, carbonate and quartz in surface layer; carbonate dominant to 50 cm; carbon throughout profile; quartz dominates below 60 cm; sands 50-70 cm have equal parts quartz and carbonate	
	Mud	Dominant component Dominant down profile to 60 cm, then decreases Carbonate dominant; carbon content minor but consistent down profile; negligible quartz to 40 cm then fluctuating	

Table 6.5 (cont.)

Table 6.5 (cont.)

Wetland	*Patterns down profile*	*Comment*
142	Gravel Gravel present throughout profile in variable amounts; whole freshwater shells to 50 cm Live roots dominate 0-10 cm; non-living roots throughout Sand Co-dominant component Sand content fluctuates down profile; increases in surface layers below 40 cm up to maximum 90 % Carbonate dominant in upper layers; carbon throughout profile; quartz dominates below 40 cm Variable carbonate to quartz ratio down profile Mud Co-dominant component Dominant down profile to 40 cm, then decreases Carbonate dominant; carbon content minor, variable down profile; negligible quartz	• Freshwater shells implies freshwater environment. There is a relationship between calcilutite and fossil shells. • Gravel inherited from beach sediments indicates basal sheet origin is beach. • Live *vs* dead root zone separates rhizosphere from buried rhizosphere. • Root structured sediment • Sandy mud-muddy sand dominated profile. • Carbonate dominates mud fraction. • Variable carbonate to quartz ratio down profile. • Sand fill diluting surface mud layers.
72	Gravel Gravel minor 0-10 cm, then negligible throughout profile Whole freshwater shells to 30 cm; Live roots dominate 0-10 cm Sand Dominant component Sand content increases below 20 cm up to maximum > 80 % Variable composition down profile: carbonate dominant in upper layers; carbon content peak at 30 cm; quartz increases with depth	• Freshwater shells implies freshwater environment. There is a relationship between calcilutite and fossil shells. • Root structured sediment • Sandy mud-muddy sand dominated profile. • Carbonate dominates mud fraction. • Variable carbon, carbonate, quartz ratios in sand down profile

Table 6.5 (cont.)

Table 6.5 (cont.)

Wetland	*Patterns down profile*	*Comments*
72	Mud Sub-dominant component Mud content consistent to 20 cm, then decreases Carbonate dominant; carbon content consistent down profile; negligible quartz content	• Sand fill diluting surface mud layers.
63	Gravel Gravel minor 0-10 cm, then negligible throughout profile Fragmented freshwater shells to 30-50 cm Live roots dominate 0-10 cm; non-living roots 30-40 cm Sand Dominant component Sand content increases down profile up to maximum > 80 % Carbonate dominant; carbon throughout profile; carbonate quartz ratio consistent down profile Mud Sub-dominant component Mud content decreases steadily down profile Carbonate dominant; carbon content decreases 0-20 cm, then consistent down profile; negligible quartz content	• Freshwater shells implies freshwater environment. There is a relationship between calcilutite and fossil shells. • Live *vs* dead root zone separates rhizosphere from buried rhizosphere. • Roots indicate root structured sediment • Muddy sand dominated profile. • Carbonate dominates mud and sand fractions.
45	Gravel Gravel present throughout profile; whole freshwater shell laminae at 10, 30, and 60 cm Live roots dominate 0-20 cm; non-living roots throughout profile	• Freshwater shells implies freshwater environment. There is a relationship between calcilutite and fossil shells. • Root structured sediment • Sandy mud-muddy sand dominated profile..

Table 6.5 (cont.)

Table 6.5 (cont.)

Wetland	*Patterns down profile*	*Comment*
45	Sand Sub-dominant component Sand increases in surface layers and below 30 cm up to > 90 % Carbonate dominant; carbon content fluctuates down profile; quartz increases in surface layer, then again below 20 cm Mud Dominant component Mud content decreases sharply below 50 cm Carbonate dominant; carbon content fluctuates slightly down profile; negligible quartz content to 60 cm, then peak	• Carbonate dominates mud and sand fractions. • Significant carbon content in all fractions. • Quartz peak at base of mud profile possibly due to aeolian input.
35	Gravel Gravel present in layers throughout profile, minor 0-10 cm, then negligible; whole freshwater shell fragments to 50 cm Live roots dominate 0-10 cm and 30 cm; layer of non-living roots at 60 cm Sand Dominant component Sand increases in surface layers and below 30 cm up to > 90 % Carbonate dominant; carbon content minor throughout profile; quartz increases in surface layer, then again below 20 cm Mud Sub-dominant component Mud content decreases sharply down profile below 10 cm Carbonate dominant; carbon content peak at 0-10 cm, then consistent down profile; negligible quartz content	• Freshwater shells implies freshwater environment. There is a relationship between calcilutite and fossil shells. • Root structured sediment • Sandy mud-muddy sand dominated profile. • Carbonate dominates mud fraction. • Sand fill diluting surface mud layers.

Table 6.5 (cont.)

Table 6.5 (cont.)

Wetland		*Patterns down profile*	*Comments*
9-6	Mud	Dominant component Mud content decreases sharply below 30 cm Carbonate dominant; carbon content decreases 0-10 cm, then consistent down profile; negligible quartz to 50 cm, then sharp increase	
swi	Gravel	Gravel present throughout profile; Beach shell fragments to 30 cm Live roots dominate 0-10 cm; non-living roots at 30 cm	• Freshwater shells implies freshwater environment. There is a relationship between calcilutite and fossil shells. • Live *vs* dead root zone separates rhizosphere from buried rhizosphere. • Root structured sediment • Muddy sand dominated profile. • Carbonate dominates mud and sand fractions. • Significant carbon content in mud.
	Sand	Dominant component Sand content increases below 20 cm up to maximum > 70 % Carbonate dominant; carbon peak at 0-10 cm, then minor throughout profile; quartz increases slightly at 10 cm, then consistent down profile	
	Mud	Sub-dominant component Mud content decreases below 20 cm Carbonate dominant; carbon content fluctuates down profile; quartz content fluctuates down profile	
swii	Gravel	Gravel negligible. Live roots dominate 0-10 cm	• Root structured sediment • Muddy sand dominated profile. • Carbonate dominates mud and sand fractions.
	Sand	Dominant component	

Table 6.5 (cont.)

Table 6.5 (cont.)

Wetland	*Patterns down profile*	*Com*
swii	Sand Sand content increases up to maximum > 90 %Carbonate dominant; carbon peak at 0-10 cm; quartz consistent down profile Carbonate quartz ratio consistent down profile Mud Minor component Mud content decreases down profile Carbonate dominant; carbon content significant; negligible quartz content Carbon peak at 0-10 cm, then consistent down profile	• Significant carbon content in mud.
swiii	Gravel Gravel minor 0-10 cm, then negligible throughout profile Fragmented freshwater shells to 0-20 cm Live roots dominate 0-20 cm Sand Dominant component Sand content increases up to maximum > 90 % Carbonate dominant; carbon content consistent down profile; quartz increases down profile Mud Minor component Mud content decreases down profile Carbonate dominant; carbon content consistent down profile; quartz peak at 30 cm	• Roots indicate root structured sediment • Muddy sand dominated profile. • Carbonate dominates mud and sand fractions. • Significant carbon content in mud. • Quartz peak at base of mud profile possibly due to aeolian input.

Table 6.5 (cont.)

Table 6.5 (cont.)

Wetland	*Patterns down profile*	*Comments*
1N	Gravel Gravel present throughout profile 0-20 cm Live roots dominate 0-20 cm Sand Dominant component Sand content decreases 0-10 cm then increases up to maximum > 90 % Carbonate dominant; carbon content minor, decreases down profile; quartz dominates below 60 cm Carbonate quartz ratio consistent down profile Mud Minor component Carbonate dominant; carbon content 0-20 cm significant, but decreasing; variable quartz content	• Roots indicate root structured sediment • Muddy sand dominated profile. • Carbonate dominates mud and sand fractions. • Significant carbon content in mud.

Important textural and compositional patterns are related to the wetlands in which they occur (Table 6.6).

Table 6.6 *Summary of textural and compositional patterns in wetland sediments*

The main textural and compositional patterns	*Wetlands in which these patterns occur*
Accumulation of organic matter in surface layers	161, 162, 163, WAWA, 135, 142, 63, 45, 35, 9-11, 9-6, swi, swii, 1N
Significant quartz/carbonate ratio in sand	161, 162, 163, WAWA, 135, 142, 35, 9-11
Mud sized quartz peaks at base of wetland fill	WAWA, 163, 135, 45, swiii
Freshwater shells	161, 162, 163, 135, 142, 72, 63, 45, 35, 9-6, swiii
Fragmented marine shell material	135, 142, 9-11
Mud dominant profiles	161, 162, 163, WAWA, 135, 45, 9-6
Sand dominant profiles	63, 9-11, swi, swii, swiii, 1N
Buried humic horizon/rhizosphere	161, 162, WAWA, 45, 9-6

The main features of the Becher Suite wetlands deriving from the granulometric and compositional data are as follows:

- there are shallow rhizospheres
- humus production occurs in the surface layers
- organic mud size material has infiltrated into underlying layers
- carbonate is the dominant component of mud and sand
- there is a change in quartz/carbonate ratios down profile
- freshwater shells occur in the calcilutite
- marine and freshwater shell gravel layers occur in the calcilutite, and
- down profile there is a gradation from mud to muddy sand near the basal sheet.

6.3.9 Biota

There are two types of fossil shells in the wetland muds and muddy sands, which are distinct from beach shell material. They are the gastropods *Gyraula* sp. and *Glyptophysa* sp. The age of the gastropods, as determined by radiocarbon dating, ranges from *circa* 2,205-280 ^{14}C yrs BP. They occur scattered in the calcilutite, or form shell laminae beds. Both species are extant and endemic to the southwest of Western Australia, however, there has been little research about the habitat requirements of these snails. The species both appear to inhabit fresh to possibly brackish (?) shallow sumplands. They have been observed colonising macrophytes growing in fine loam or mud (*Pers. Comm.* S. Slack-Smith, WA Museum, 2000). They do not normally co-exist. A study of pulmonate snails in Nigeria revealed similar ecological requirements (Ndifon and Ukoli 1989). The species of *Gyraula* was found in seasonal freshwater wetlands underlain by sand and muddy sand, with water shallow enough to support macrophytes. Also, the species most commonly occurred in isolation or with one other species of gastropod. Diatoms also occur in small numbers in the calcilutite *e.g.*, wetlands 142 and swii.

6.3.10 Pedogenesis and synsedimentary diagenesis

There are several pedogenic and synsedimentary diagenetic processes operating in the wetland sediments: generation of humus and organic matter from the current colonising vegetation; bioturbation; colour mottling; cementation; disintegration of carbonate grains; and leaching of calcium carbonate.

Humus and organic matter

Humus is generated at the surface of damplands under aerobic conditions where organic matter decomposition increases with the degree of fragmentation. Sediment grains such as quartz and shell are coated with humus, and organic matter accumulates interstitial to the sand. Humus production is at its peak under the grass tree *Xanthorrhoea preissii* (Fig. 6-24), which tends to grow as monospecific clumps in incipient wetlands within the beachridge swales or in a ring on the outer edge of the wetlands. Soils under *X. preissii* exhibit various development of organic horizons ranging in depth from 20-100 cm. In wetland muds, organic matter is contributed predominantly by the sedges *Lepidosperma gladiatum, Baumea articulata/Typha* sp., and *Gahnia trifida*. The mean content of mud sized organic matter in the surface layers under selected wetland species is tabled below.

***Table 6.7** Organic content of surface sediment under various species*

Vegetation assemblage	*Mud size organic matter content*	*Number of sites*
Baumea articulata/ Typha sp.	49 ± 26	n = 2
Lepidosperma gladiatum	31	n = 1
Melaleuca teretifolia	19 ± 7	n = 5
Juncus kraussii	19 ± 4	n = 5
Centella asiatica	18 ± 6%	n = 5
Melaleuca rhaphiophylla	17 ± 6%	n = 4
Baumea juncea	12 ± 2	n = 4

Bioturbation

Bioturbation by plants and animals is evident both at the modern wetland surface and at depth. In the modern environment, in wetlands that are seasonally inundated and waterlogged, bioturbation commonly occurs in the dry part of the hydroperiod. Insects, such as ants and crickets, as well as some introduced vertebrate species, bioturbate the sandy sediments; other less common vertebrate burrowers such as the Southern Brown Bandicoot, are active only in some of the wetlands. Turn over of surface sediments by scratching or digging is to a depth of approximately 10 cm.

Within the stratigraphic profile, individual burrows are evident in the calcilutite layer. Bioturbation is expressed as texture mottling within a layer, or traversing a layer; as gradational contacts between some of the sediment types near the surface, and/or as homogeneous layers composed of mixed mud types. Layers of mixed composition are produced when infiltration of an overlying mud (type 1) into the lower horizon, composed of a second mud type (type 2), is followed by bioturbation.

Colour mottling

Within the calcilutite layers there may be variable grey and brown colour mottling. This is associated with humic mud infiltration, iron staining, oxidation-reduction processes in gley sediments (Wright and Platt 1995), decomposition of plant material, and burrowing.

Cementation

Within the wetland sediments there are two types of local cementation. The first is cementation within the muddy sand by fine grained calcite, associated with a buried "stromatolite" bed, and the second is cementation within muddy sand by calcrete. Both types of cementation occur within the zone of groundwater fluctuation (wetlands Cooloongup A and 9-3, 9-6).

In Cooloongup swale A, there is a buried cemented bed that resembles a "stromatolite", at a depth of 30-60 cm below the surface. In wetland 9, the cementation is a precipitate of calcrete (Read 1974), occuring as an indurated lens, (5-15 cm thick), or nodule layer, 30-50 cm deep within the muddy sand. The calcrete does not appear to be related to the current water level regime. The depth at which the calcrete layers occur (- 50 cm) corresponds roughly to the depth of the root system of *Melaleuca rhaphiophylla,* which is the only tree occurring in the wetland. Its occurrence, therefore, is best explained as a feature resulting from plant utilisation of vadose and phreatic waters causing precipitation of $CaCO_3$ (Semeniuk and Meagher 1981). This calcrete occurs at the vadose/phreatic zone interface within the carbonate sequence.

6.3.11 Age structure and rate of sedimentation

Radiocarbon dating was used to determine the Holocene age of the wetland deposits, the age structure of the sedimentary fills, the rates of accumulation of the various sediment types, and the variable ages of the compositional components in the mud-sized fraction of the sediment.

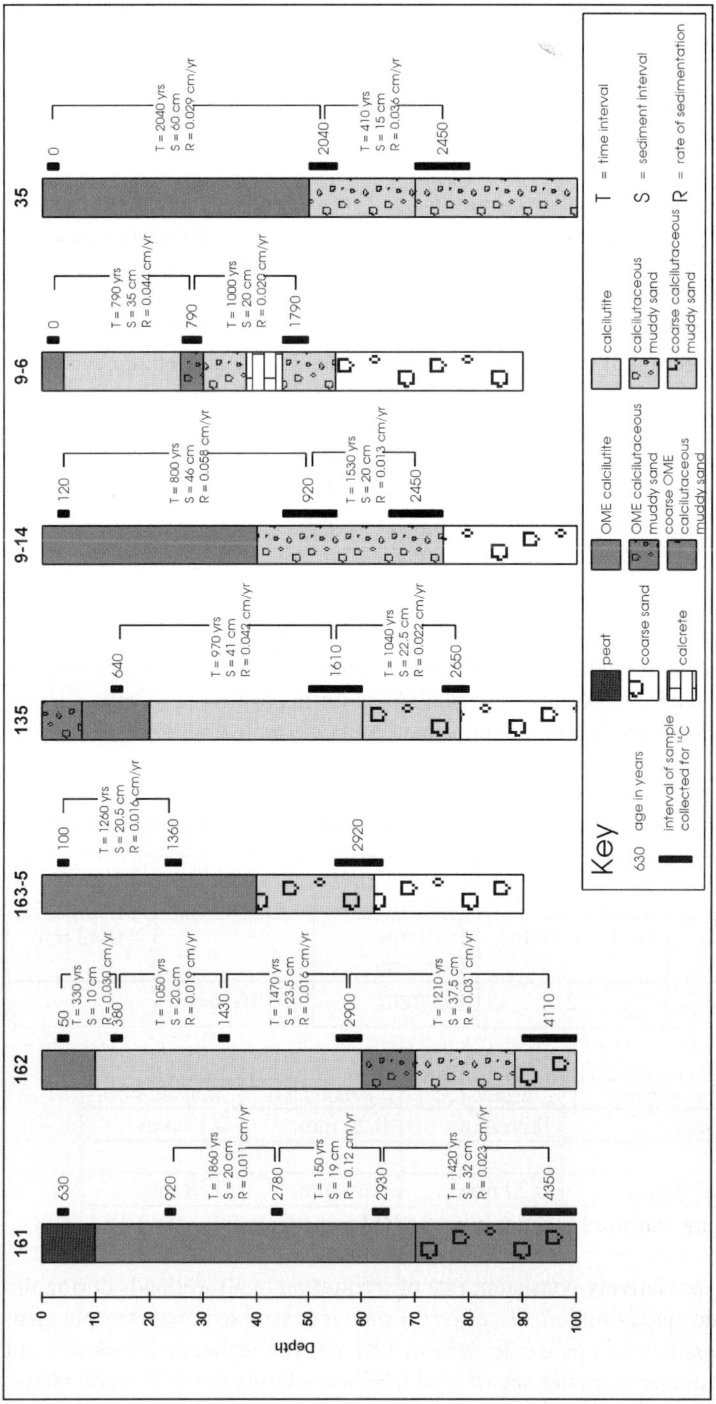

Figure 6-57. *Age structure of wetland fills, and interpretation of rates of sedimentation.*

Based on the oldest and youngest of the radiocarbon dates obtained for the calcilutite, it appears that deposition of calcilutite commenced east of Lake Cooloongup *circa* 5740 ^{14}C yrs BP, west of Lake Cooloongup between the arms of the spit barrier *circa* 4590 ^{14}C yrs BP, (Cooloongup A2), and *circa* 4350 ^{14}C yrs BP, under the oldest wetland in the Becher Suite, and continues up to the present (surface of wetlands 162, 163, 9-6, and 9-14) (Table 5.4).

While the Becher beachridge plain ranges in age from *circa* 7000 years BP to modern (Searle *et al.* 1988), all wetlands in the study area are middle to late Holocene in age, *i.e.*, generally younger than the surrounding local landscape. The oldest dates were derived from the base of wetlands located in topographic low swales in the eastern parts of the beachridge plain. Although the majority of dates were derived from the base of wetlands to determine their period of initiation in the context of subregional wetland evolution within the prograding beachridge plain, a number were specifically obtained to determine the age structure of the sedimentary fills. Dates, derived from wetlands 161, 162, 163, 135, 35, and 9, show a progessive younging upwards of the wetland sedimentary fills. The range of ^{14}C dates within several stratigraphic sequences provides a basis for determing the rates of sedimentation within the wetlands.

Rate of deposition of carbonate mud

Calculations for the rate of deposition of interstitial carbonate mud are based on radiocarbon dates for the top and base of muddy sand in central wetland sites (Fig. 6-57), and are tabled below. Additional calculations for deposition rates of carbonate mud are separated into near pure calcilutite and OME calcilutite (Table 6.8).

Table 6.8 *Rate of carbonate mud accumulation in mm/yr*

Sediment type	*wetland 161*	*wetland 162*	*wetland 163*	*wetland 135*
muddy sand	0.23 mm	0.31 mm	-	0.22 mm
calcilutite	0.11-1.23 mm*	0.19 mm 0.16 mm	-	~ 0.42 mm
OME calcilutite	-	0.3 mm	~ 0.16 mm	-

Sediment type	*wetland 35*	*wetland 9-6*	*wetland 9-14*
muddy sand	0.36 mm	0.20 mm	0.13 mm
calcilutite		-	-
OME calcilutite	0.29 mm	0.44 mm	0.58 mm

* rates of accumulation in interlayered calcilutite and peat

These data show a relatively consistent rate of infiltration in all wetlands during the phase of carbonate mud initiation, *i.e.*, *circa* 0.2 mm/yr of sand sediment was plugged. The rates of accumulation of pure calcilutite in wetlands varied, beginning slowly and being not too dissimilar from the rate of mud formation during the infiltration phase, *e.g.*, (*circa* 0.15 mm/yr) in wetlands 161 and 162. However, rates of accumulation increased in wetland 135 subsequent to 2000 years BP. The rate of mud accumulation in the sediment comprising mixed carbonate and organic material is two to four times that of pure calcilutite. Similar accumulation rates were extrapolated by Backhouse (1993) from cores of Holocene carbonate mud and peat in wetlands on Rottnest Island (offshore from Becher Point). Rates of accumulation were estimated to be 3 cm/100 years in the Holocene carbonate mud and 10 cm/100 years in the upper and lower peat (Backhouse 1993).

Where there was mixed carbonate mud and organic matter, there was opportunity to separate these components and derive dates from each. Radiocarbon dating of both the carbonate and peat components within small segments of the stratigraphy was undertaken for four Becher wetlands (Table 6.9, Fig. 4-23).

Table 6.9 *^{14}C dates for two mud fractions (carbonate mud and organic carbon) at various wetlands*

Sample type	*Wetland 161 (3-5 cm)*	*Wetland 161 (23-25 cm)*	*Wetland 162 (3-5 cm)*	*Wetland 162 (13-15 cm)*
carbonate mud	630 ± 110	920 ± 110	Modern 50 ± 110	380 ± 110
organic carbon	250 ± 110	990 ± 110	Modern	580 ± 110

Sample type	*Wetland 163 (3-5 cm)*	*Wetland 135 (13-15 cm)*
carbonate mud	Modern 100 ± 110	640 ± 110
organic carbon	Modern	480 ± 110

Since all these samples are from very young sedimentary accumulations, the measure of standard deviation is close to the determined age of the sample, recommending caution in interpretation. In addition, if the normal procedure of applying two standard deviations to radiometric dating is followed, then there is no significant difference between pairs. The discussion that follows assumes that the dates are valid based on one standard deviation. In wetlands 161 (3-5 cm), 162 (13-15 cm), and 135 (13-15 cm), the dates may be interpreted as indicating that the muds formed at different periods and that the current organic matter enriched carbonate mud is due to bioturbation in these layers. A period for each cycle of carbonate mud and organic matter accumulation of 70-380 years is suggested. In the other wetlands no significant difference in age could be determined.

Generally, climate controls the rate and nature of biogenic productivity, as well as influencing the rate and depth of penetration of illuviation. The variability in the rate of carbonate mud accumulation in wetland 161 may indicate that climate was gradually becoming wetter during the period 2500-1000 years BP. Radicarbon dates of 2600 and 1400 ^{14}C yrs BP from the base of carbonate mud horizons outside the current perimeter of wetlands WAWA and 163, respectively, indicate that carbonate mud deposition still prevailed in these basins during this period of wetland expansion. Thus, it can be seen from these data, that the micro-environment associated with conditions at a particular site may be as important as the macro-climate variation in determining rates and style of accumulation.

6.4 Reconstruction of palaeo-environmental and palaeo-sedimentological processes

From the foregoing descriptions of the sediments and stratigraphic sequences in the Becher Suite wetlands, several evolutionary changes to the wetlands may be deduced. In order of presentation, the aspects of wetland evolution discussed in this section are: infiltration of sediments; fossil calcilutite deposits; peat and humus deposits; and wetland deepening through grain dissolution.

6.4.1 Infiltration of sediments

Infiltration of mud, which has accumulated at the wetland surface under waterlogged or inundated conditions, into underlying sediments, is a common process in the Becher wetlands. Infiltration occurs in two ways: 1) illuviation and deposition, and 2) bioturbation. An early sedimentary process in wetland history was the alteration of the top of the beach or dune basement sand, *i.e.* the floor of the proto-wetland, through these wetland processes. Infiltration of carbonate mud into the basement sand formed a muddy sand basal sheet in all wetlands. In the intermediate layers, illuviation of surface mud by rainwater percolation and groundwater recession has resulted in compositional, textural and colour layering, *e.g.*, the infiltration of peat into the light cream calcilutite in cores 161, 162, 142, Cooloongup B4 (Fig. 6-3, 6-4, 6-7, 6-22 D).

Burrows also play a part in the infiltration of one sediment into another. Burrows often are filled with material texturally distinct from the sediment horizon in which they occur (Figs. 6-10, 11, 18, 21 C, D, E). The end result of burrowing and bioturbation is a homogeneous layer intermediate to the mud and sands. Within a wetland basin, a single horizon may exhibit the complete gradation, *i.e.*, burrows and thoroughly mixed compositional sediment.

Organic/carbonate horizons

Several textural and compositional gradations were found in the organic matter enriched carbonate muddy sand layers. These gradational types were linked with one of the following three processes:

1. infiltration of overlying carbonate mud into the humic sand horizon
2. infiltration of peat and organic matter into the calcilutaceous sand horizon at the surface
3. sheet wash of sand into the OME calcilutite horizon

The first process occurred in sediment horizons at the base of the wetland fill. These humic sands represent former soil surfaces of swales which have been buried by wetland fill and infiltrated by the overlying carbonate mud. The second process occurred in the modern surface at the centre and margins of wetlands. A change in sediment style, from carbonate mud accumulation to humus and peat production, is the underlying reason why this type of infiltration occurred. The third process occurred in horizons at the margins of the wetlands, following sheet wash of aeolian sediment into the wetland, or instances of disturbance to the beachridge slope (*e.g.*, fire, erosion, trampling). In each case the result is a layer of humic/carbonate muddy sand. These layers, although lithologically similar, are not related stratigraphically, being diachronous sedimentary layers signifying independent stages of development from wetland to wetland.

6.4.2 Calcilutite

The calcilutites in the wetlands of the Becher area are composed dominantly of calcite with subsidiary Mg-calcite and traces of aragonite. The particles of carbonate mud are silt- and clay-sized, 1-20 μm, with a range from 0.2 μm to 63 μm. Calcilutites from the Coolongup Suite are dominantly clay-sized, 0.4-2 μm in size, also with a range from 0.2 μm to 63 μm (Fig. 6-58). The muds have a porosity range of 0.49-0.6. SEM photographs of the calcilutite show that the particles are skeletal in origin, composed of charophyte, ostracod, molluscan, and crustacean fragments (Fig. 6-59 A-L). Since deposition, the carbonate grains comprising carbonate mud have undergone disintegration and chemical corrosion. Disintegration is evident in the gradation of grain sizes from those of diameter 40 μm to 2 μm (Fig. 6-60). Chemical corrosion is evident in the pitted surface of the majority of grains (Fig. 6-61), such pitting being controlled by the skeletal grain architecture and calcite crystal cleavage.

The calcilutite sequence contains sedimentary structural features consistent with ephemeral wetlands (Platt and Wright 1992). Features consistent with periodic exposure include intercalations of allochthonous material, brecciation, pulmonate snails, layers of reworked shells, and thin zones of cementation. Features consistent with periodic inundation include burrows, fossil pulmonate snails reworked into layers, root structures, and colour mottling.

The calcilutite filling wetland basins is an intra-basinal accumulation, *i.e.*, it does not occur under the beachridge/dunes, nor outside the margin of a given wetland. Its mudstone fabric suggests low energy sedimentation. Absence of lamination and the abundance of root and burrow structures grading to homogeneous sediments, implies bioturbation of sediments, and low salinity, oxygenated, bottom waters (Platt 1989).

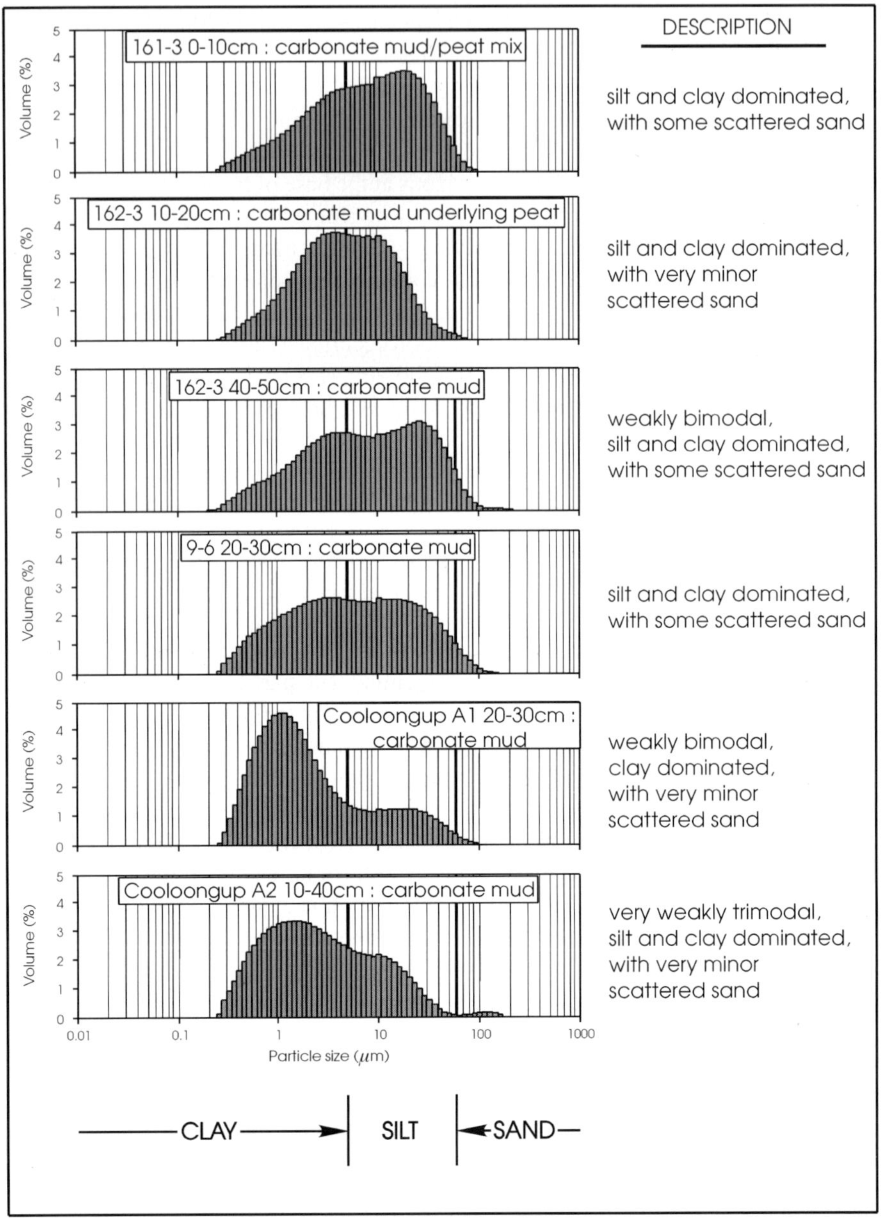

Figure 6-58. *Particle size distribution of carbonate mud in the Becher area. Size classes after Wentworth-Udden in Folk (1974).*

Lake carbonates, formed through biogenic or bio-induced precipitation in shallow water elsewhere, suggest that carbonate production is the result of one of the following processes (Flugel 1982; Wetzel 1983; Scholle *et al.* 1983, Anadon *et al.* 1991):

1. bioherms built by blue/green algae and cyanobacteria
2. aggregates of molluscan or ostracod shell material
3. carbonate encrustation of reeds or other macrophytes through photosynthetic uptake of carbon dioxide.

In the Becher Suite wetlands the mud is formed from the *in situ* disintegration of carbonate materials within the wetland basins. In scientific literature, the carbonate muds most closely approximating those at Becher were freshwater marls of Holocene age in an alluvial landscape in Maryland (Shaw and Rabenhorst 1997). The marl had formed in ponds through inorganic and biogenic processes associated with the green algae *Chara* sp. and contained gastropods, bivalves, and algae, resulting in extremely high calcium carbonate content. *Chara* sp. accumulates carbonate internally through metabolic processes and externally by photosynthetic removal of CO_2 (Scholle *et al.* 1983). Much of the calcite formed from *Chara* may not be recognisable as biological remains, although the calcareous cortication tubules, reproductive organs and stems (Fig. 6-59) are preserved in low energy environments (Bathurst 1981, Scholle *et al.* 1983). In these settings, the stems can act as nucleation sites for precipitation of calcium carbonate, which overprints their original geometry. In Maryland, the marl development was intermittent and interspersed with buried soil horizons with higher organic content (Shaw and Rabenhorst 1997). This sedimentary sequence is similar to those in the Becher wetlands. The intermittent nature of the marl development is a plausible explanation for the occurrence of the calcilutaceous muddy sand layer underneath the buried soil in wetland 162. It appears that early in the development of this particular wetland, conditions conducive to calcilutite production temporarily ceased, and were replaced by organic accumulation as a result of macrophyte colonisation.

There were three phases of carbonate mud accumulation. The first phase, termed the inundation phase, denotes the period of regular seasonal wetland inundation as opposed to seasonal waterlogging. During this period, carbonate mud accumulated above the basement sands with the initial inundation of the proto-wetland. The second phase, termed the clogging phase, refers to the period in which the carbonate mud was subsequently washed into the underlying sand by rainfall and the fall of the water table, such that the pores of the basement sands became progressively filled with mud until illuviation was impeded. The third phase, termed the true fill phase, refers to the period in which the basement sands were relatively clogged and impermeable to illuviation, carbonate mud then began to accrete upwards.

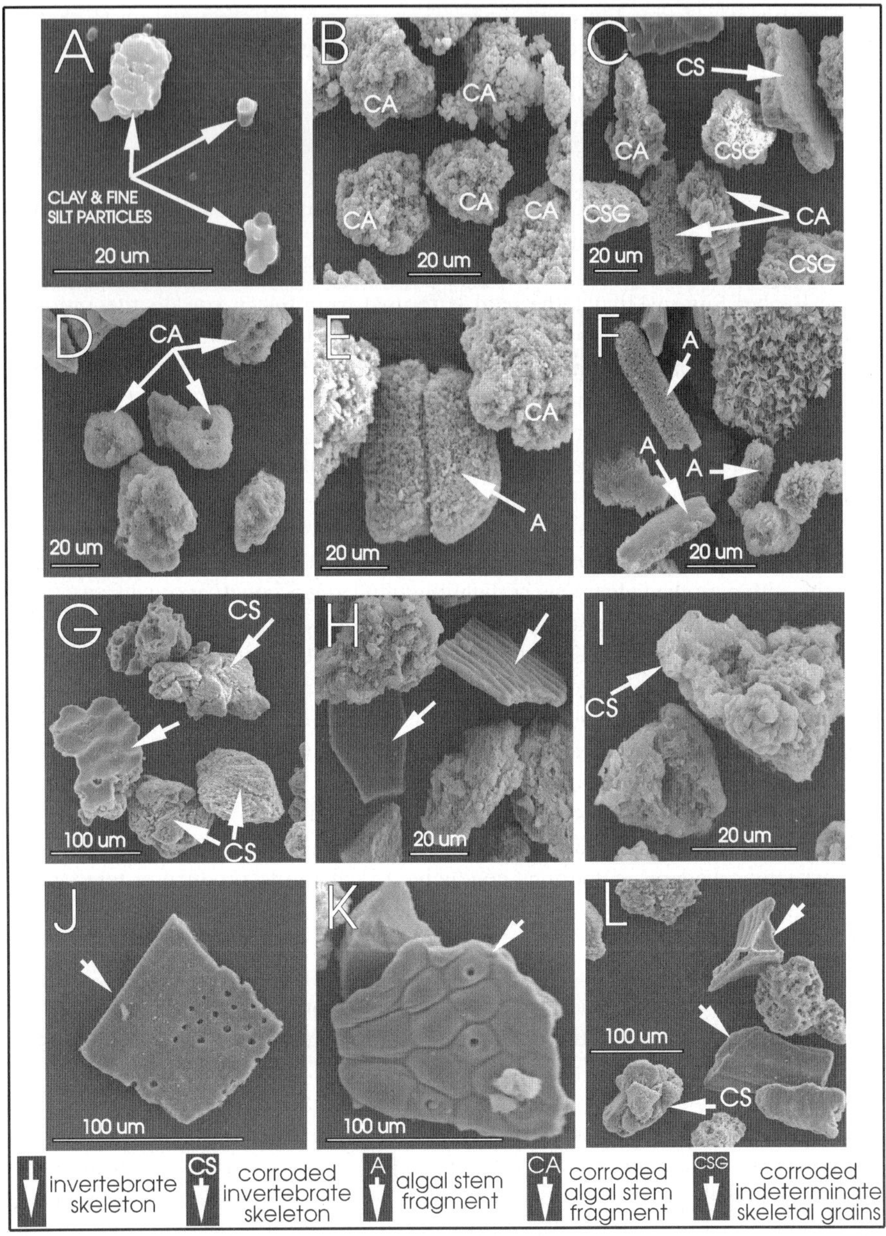

Figure 6-59. SEM photomicrographs showing various types of particles that comprise carbonate mud (silt and clay).

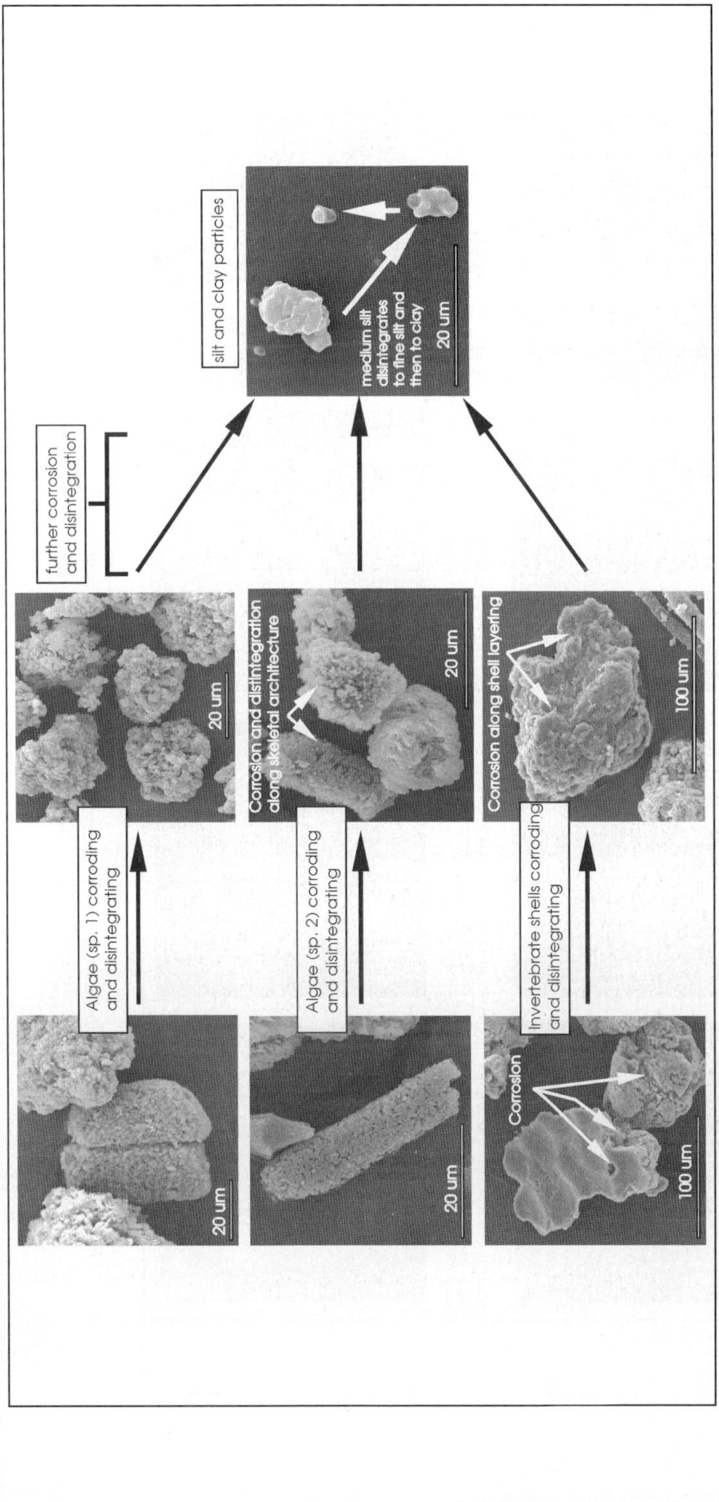

Figure 6-60. *Sequence of SEM photomicrographs showing gradation of algal and invertebrate skeletons corroding and disintegrating to carbonate silt and clay.*

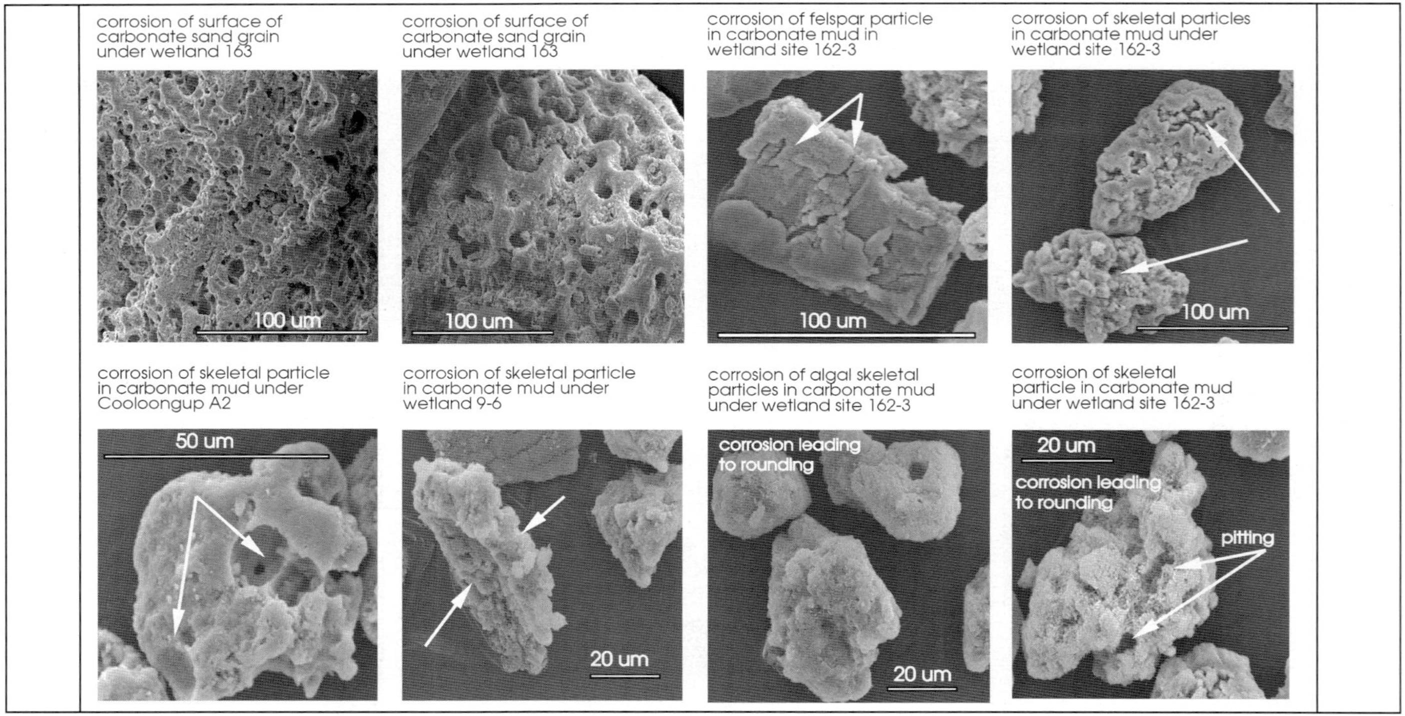

Figure 6-61. *Evidence for dissolution/corrosion of carbonate grains and felspar in mud and in sand under wetlands (corrosion sites are arrowed).*

6.4.3 Peat and humus deposits

Black mud sized decayed plant material (peat and humus) is currently being generated in all wetlands, covered by a 2-3 cm surface layer of decaying and undecayed stems, leaves and seeds. This decayed plant material forms peat in the sumplands and humus in the damplands. Minimum inundation required to maintain production of peat appears to be 6-7 months per year. As these conditions are not consistently fulfilled in all wetlands, the rate of accumulation is variable. In most wetlands, peat deposits are only 10 cm thick, however, in wetland WAWA, the peat fill is 50 cm. The rate of accumulation in wetland WAWA was 0.23 mm/year which is lower than the rate calculated for the surface sediments. The base of the peat in wetland WAWA was dated *circa* 2,200 ^{14}C yrs BP, however, this date does not represent the commencement of peat accumulation in all Becher Suite wetlands. Other wetlands at that time were environments in which calcium carbonate was accumulating, *e.g.*, wetlands 135, 142, 35, 9, swi, swii, swiii.

The development and accumulation of peat is important in the history of a wetland in that its occurrence will alter the water chemistry within wetlands. Groundwater residing in the calcareous beach or dune sand, or the overlying wetland fill of carbonate mud deposits has pH 8-8.5. As the groundwater rises to the surface, there is a change from calcium carbonate saturated waters to waters containing humic acids resulting in pH 7-7.8.

6.4.4 Subsidence of wetland through dissolution of carbonate

In the majority of wetlands, calcilutite accumulation has been replaced in near surface layers by varying thicknesses of peat and humus. This effect is attributed to a change in saturation with respect to the carbonate ion (indicated by a lower pH value) of the groundwaters. A coincident change in water penetrating the surface layers has also occurred. In the older wetlands, the oxidation of organic matter in the sediments has increased the hydrogen activity of the surface water, so that it is now approximately pH 7. Dissolution of carbonate is indicated by a decrease in pH to 7 and an increase in Ca ions (Scholle *et al.* 1983). When the pore fluids in the surface layers are undersaturated with respect to carbonate mineralogy, the carbonate sediments in and underlying these horizons undergo dissolution (Tucker and Wright 1990). Abundant CO_2 in solution and organic acids formed by decay dissolves carbonates. As a result, there has been, for the Becher wetlands, an on-going loss of wetland sediment in the older wetlands through dissolution of carbonate grains. The geometry of the area or zone of dissolution suggests that this fluid moves as a plume through the wetland sediments and then westwards.

Evidence for this, involving relative heights of the contact between beach and dune sediments under beachridge/dunes and wetlands, comparison of carbonate to quartz grain ratios under beachridge/dunes and wetlands, wetland WAWA stratigraphy, and SEM photographs of carbonate mud particles, is described below.

The two hydrographs exhibit very different patterns considering that these variations occur within a depth of the fill of 1 metre. The major reason for the differences is related to the texture of each wetland fill.

7.3.2 Some case studies on the effect of composition and texture on subregional groundwater table patterns

As intimated above, textural differences in sediment layers can affect rates of infiltration and can be used to explain small scale variability of rising and falling water tables, of groundwater response to rainfall, and of water chemistry, in different sedimentary fills. Four wetland basins containing differing sediment types were selected for more detailed analysis of water table configuration (*i.e.*, geometry and slope) and water table response under wetlands and adjoining dunes throughout the seasons. The selected wetland basins and their sediment types were:

- peat (wetland WAWA);
- OME calcilutite (wetlands 163, 45);
- calcilutite (wetlands 161, 162);
- calcilutaceous muddy sand (wetland 63).

Three small scale features in the rising and falling water tables were selected to demonstrate variable hydrological pathways in different sedimentary fills: mounds, depressions, and gradients. The analysis of groundwater level data was undertaken for the years 1992 and 1994 for specific conditions:

- low rain, low water table which corresponded with autumn;
- high rain, low water table which corresponded with early winter;
- high rain, high water table which corresponded with late winter;
- low rain, high water table which corresponded with spring;
- low rain, medium water table which corresponded with summer.

The results are presented in Figure 7-3 and Table 7.2. In all situations, the groundwater table sub-regionally is sloping down from east to west, and the various basins with their sedimentary fill perturbate this pattern in different ways.

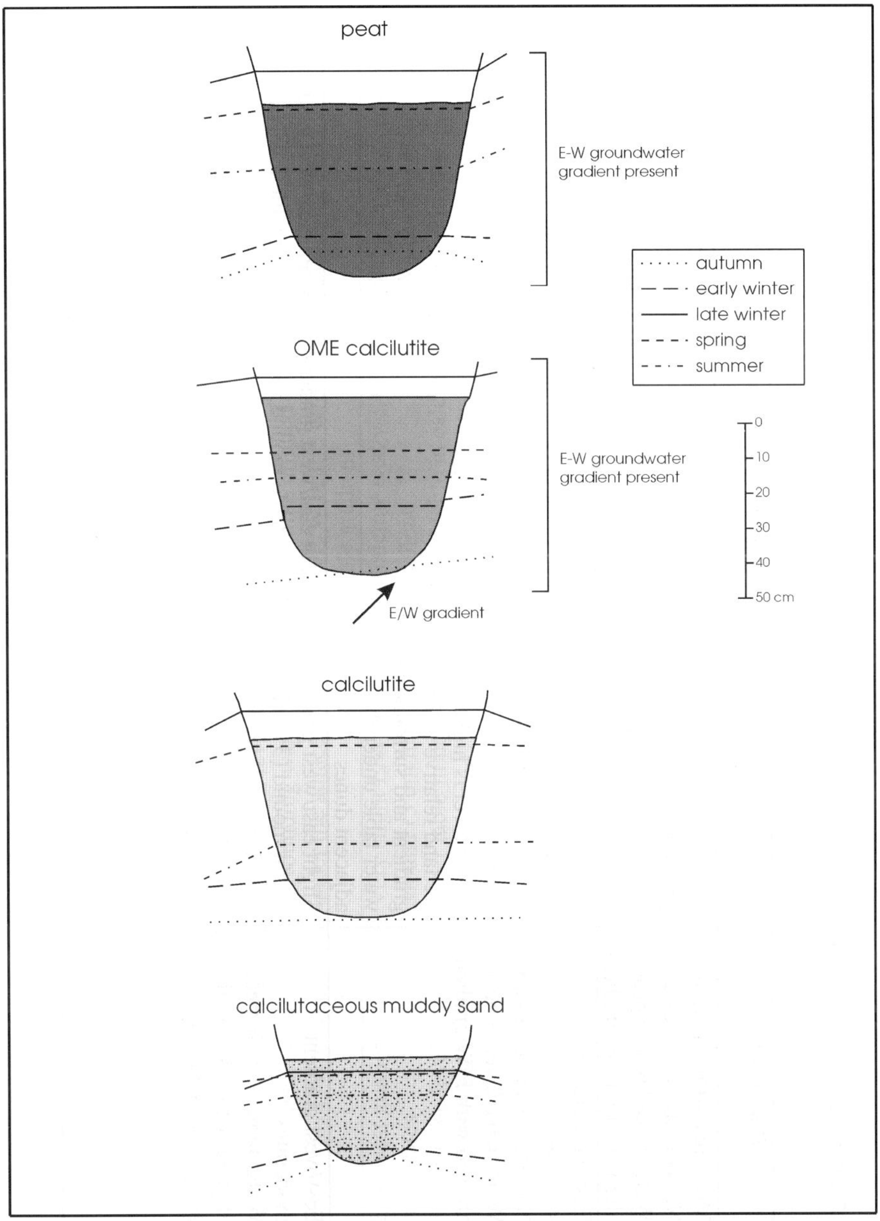

Figure 7-3. *Summary of the effects of various lithologies on groundwater movement, and on the water table morphology, i.e., development of mounds, troughs, gradients.*

Table 7.2 (cont.). *Patterns in groundwater levels in 1994 with respect to types of wetland fills for various times of the hydrocycle*

	Peat (WAWA)	*OME calcilutite (163, 45)*	*Calcilutite (161, 162)*	*Calcilutaceous muddy sand (63)*
Autumn minimum (April)	Medium east/west gradient Water table below peat horizon	Medium east/west gradient Mound (3-10 cm high) in water table in OME calcilutite	No east/west gradient Mound (5-8 cm high) in water table in calcilutite Water level is higher in wetland relative to E/W gradient and surrounding water table under the adjacent dunes	Moderate east/west gradient Water table below muddy sand horizon
Early winter (June)	Medium east/west gradient Water table below peat horizon	Slight east/west gradient Slight mound (2-5 cm high) in water table in OME calcilutite Water level is higher in wetland relative to E/W gradient and surrounding water table under the adjacent dunes	Slight east/west gradient Mound (2-9 cm high) in water table in calcilutite Water level is higher in wetland relative to E/W gradient and surrounding water table under the adjacent dunes	Moderate east/west gradient Water table below muddy sand horizon

Table 7.2 (cont.)

Table 7.2 (cont.)

	Peat (WAWA)	*OME calcilutite (163, 45)*	*Calcilutite (161, 162)*	*Calcilutaceous muddy sand (63)*
Late winter-early spring maximum (August)	Steep east/west gradient Water level in wetland is lower relative to E/W gradient	Steep east/west gradient Mound (3-20 cm high) in water table in OME calcilutite Water level is higher in wetland relative to E/W gradient and surrounding water table under the adjacent dunes	West/east gradient Water level is higher in wetland relative to E/W gradient	No east/west gradient Water table level in muddy sand and adjacent dunes
Spring (October)	Medium east/west gradient Slight mound (3-9 cm high) in water table in peat Water level is higher in wetland relative to E/W gradient and surrounding water table under the adjacent dunes	No east/west gradient Mound (5 cm high) in water table in peaty carbonate mud Water level is higher in wetland relative to E/W gradient and surrounding water table under the adjacent dunes	West/east gradient Water level is higher in wetland relative to E/W gradient	Moderate east/west gradient Water table below muddy sand horizon
Summer (December)	East/west gradient moderates Water table level in peat is aligned with E/W water table gradient	Very slight E/W gradient Water level is higher in wetland relative to E/W water table gradient	Moderate east/west gradient Water table in calcilutite below E/W water table gradient	Moderate east/west gradient Water table below muddy sand horizon

The effects of the various sedimentary fills within wetlands in perturbating the groundwater patterns are summarised below.

The main effects of the peat fill on groundwater were 1) the lack of development of mounds and depressions in the water table under the wetland with respect to the regional groundwater gradient throughout the seasons, and 2) extension of the period of waterlogging through the mechanisms of rapid recharge and slower discharge.

The main effects of the OME calcilutite fill on groundwater were 1) mounding of approximately 10 cm in the wetland above the water table levels under the adjoining dunes, and 2) the quickest initial response to rainfall.

The main effects of the calcilutite fill on groundwater movement were 1) delay in initial water table rise and 2) greatest overall rise in water table and 3) most rapid discharge from September to December.

The main effects of the calcilutaceous muddy sand fill on groundwater movement were 1) a slightly higher water table level in the muddy sands than in the calcareous sands under the adjoining dunes, 2) minimal development of mounds or depressions in the water table under the wetland with respect to the regional groundwater gradient throughout the seasons, 3) lowest overall rise in water table, 4) slowest response to rainfall and 5) medium rate of discharge from September to December.

Response of variable basin fills to winter rainfall and summer discharge
Winter rainfall and summer discharge cause water table rise and fall respectively. Rapidity of response, rate of fall, and magnitude of water table rise in each of the wetland fills are tabled below. Rates of water table fall in the various sediments varied between wet and dry years and therefore the order from highest to lowest also changed, but although the magnitude of water table rise in the various sediments also varied between wet and dry years, the order from largest to smallest did not change.

Table 7.3 *Water table responses in various wetland fills to the presence and absence of rainfall*

Response of water table to recharge from rainfall, from most rapid to least rapid	*Magnitude of water table rise in the various sediments, from largest to smallest*	*Rates of water table fall in the various sediments in order of magnitude, from highest to lowest*
OME calcilutite	calcilutite	calcilutite or OME calcilutite
peat	OME calcilutite	
calcilutite	peat	peat
carbonate muddy sand	carbonate muddy sand	carbonate muddy sand

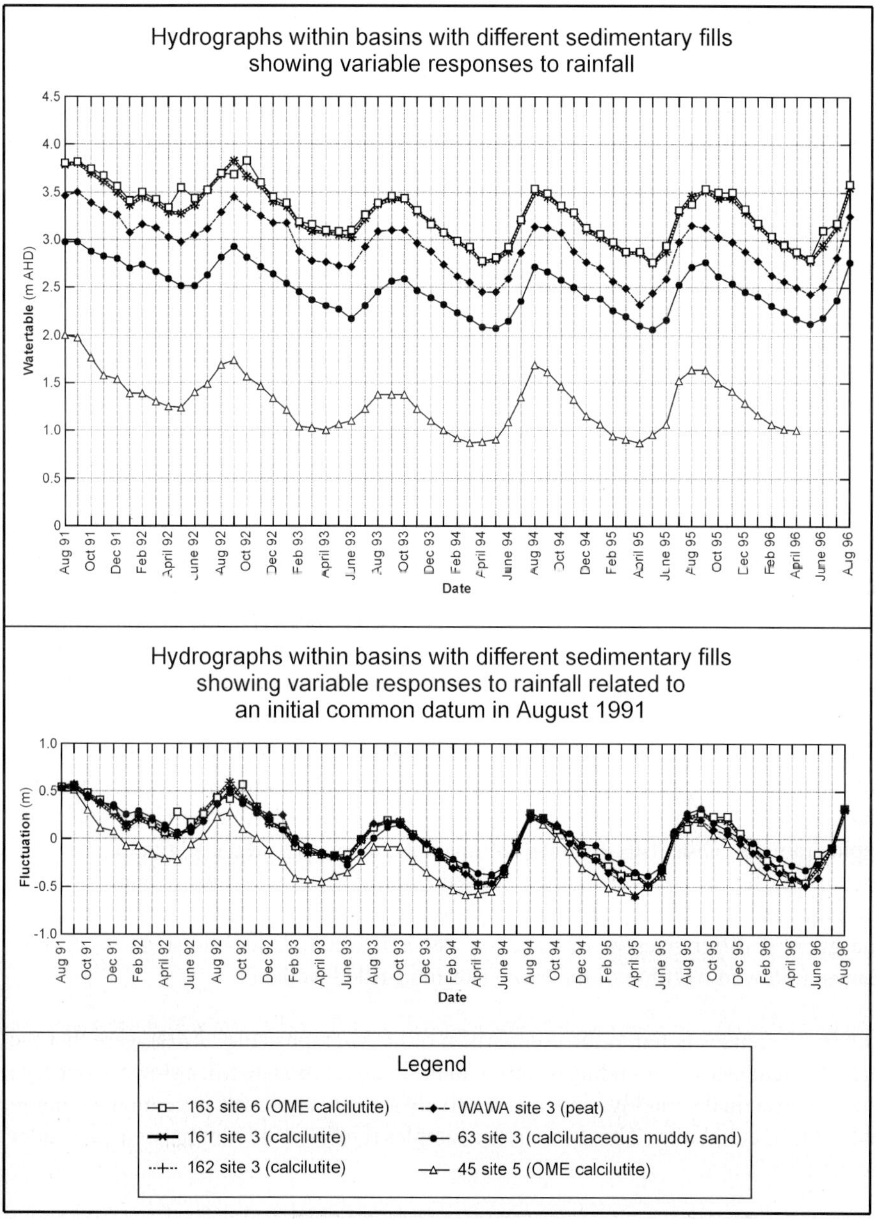

Figure 7-4. *A. Comparison of hydrograph trends graphed against AHD, and B. from a common initial datum.*

Again at the basin scale, small scale variability in the groundwater response to rainfall was analysed, this time using hydrographs (Fig. 7-4) spanning the period August 1991 to August 1996. In Figure 7-4A the hydrographs are presented separately in relation to ground surface, in Figure 7-4B, they have been presented without reference to either ground level or AHD, in order to compare patterns. The patterns are presented in Table 7.4.

Under the rainfall regime of 1991-1996, inundation patterns were compared for basins in which the water table levels were similar, using 2 categories:

1. minimum water level at 90 cm: examples include the peat (wetland WAWA), the OME calcilutite (wetland 45), the calcilutite (wetland 161);
2. minimum water level at 125 cm: examples include the OME calcilutite (wetland 163), and the carbonate muddy sand (wetland 63).

From the hydrographs it can be seen that the peat basin fill was inundated regularly 4 out of 6 years, whereas the calcilutite and OME calcilutite basins (wetland 161 and 45) were inundated slightly less, *i.e.*, 3:6 years and 2:5 years respectively. Even in the driest winter 1993, the peat basin was waterlogged to the surface in contrast to the other basins. Neither the OME calcilutite basin (wetland 163) or the carbonate muddy sand basin (wetland 63) were inundated, but the extent of waterlogging in the OME calcilutite basin was greater.

Comparison between annual groundwater fluctuation patterns also reveals some differences. In both years (1991-92 and 1994-95), the minimum fluctuation occurred in the carbonate muddy sand basin, which is subject to negligible perching and more rapid drainage. Other fills exhibited comparable ranges in water level fluctuation. The greatest difference in fluctuation between the wetter and drier year occurred in the peat basin fill (Table 7.4). Associations of peat-filled wetlands similarly showed the greatest differences in water levels beneath wetlands and adjacent beachridge dunes, followed by those of the carbonate-mud-filled wetlands.

Other differences noted in the comparison of hydrographs (Fig. 7-4B) were that the highest incidence of mounding occurred in the peaty/carbonate fill, a slower discharge rate occurred in the muddy sand, and relatively static water levels were more common in the OME calcilutite. These concrete examples reinforce the summarised data under Section 7.3.2.

7.3.3 Effect on groundwater of lateral contacts between beachridge/dune and wetland

Lateral contacts between wetland sediments and adjacent calcareous sands underlying the beachridge/dunes also perturb the hydrologic processes at the margins of the basins. The nature of the lateral contact can influence pathways (*e.g.*, deflection) and

***Table 7.4** Comparison of hydrographs in different wetland fills*

Wetland sedimentary fill type	*Inundation patterns*	*Annual fluctuations (cm) for 1991-92 (wet year) and 1994-95 (dry year)*	*Comparison between beachridge/dune and wetland water levels*
Peat: WAWA	Inundated or totally saturated every year	1991-92 - 53 1994-95 - 87	*w > d (west) 1-21 cm except spring, late autumn w < d (east) all year
OME calcilutite: 163	Not inundated Waterlogged. Max. water table-10 cm	1991-92 - 49 1994-95 - 73	w > d (west) 3 cm except in low rainfall months
OME calcilutite: 45	Inundated only in wet years	1991-92 - 75 1994-95 - 82	w > d (west) 4 cm except spring and late autumn
Calcilutite: 161	Inundated 3 out of 6 years	1991-92 - 51 1994-95 - 82	
Calcilutite: 162	Inundated 2 out of 6 years	1991-92 - 53 1994-95 - 75	w > d (west) 1-11 cm except at odd times in late spring and early summer
Carbonate muddy sand 63	Not inundated Waterlogged. Max. water table-30 cm	1991-92 - 45 1994-95 - 65	w > d (west) 3 cm 0-13 cm except spring and early summer

*w > d, *w < d water level under wetland is higher than or lower than water level under the beachridge/dune.

rates of water infiltration. In settings with different lateral stratigraphic relationships, small scale variability in the rising and falling water tables between the wetland margin and wetland centre was identified. While some aspects of this variability could be linked to 1) perching in the wetland, 2) east/west gradients, 3) evapo-transpiration and 4) depth to water table, other aspects of variability related to the nature of the lateral contact. The response of the water table was compared in settings with four different types of lateral stratigraphic relationships (Fig. 7-5):

1. simple juxtaposition of wetland and ridge sediments termed "simple" (*e.g.*, wetlands 9-7, swii);
2. simple interfingering along dune and wetland contacts termed "interfingering I" (*e.g.*, wetlands 162, 9-5);
3. complex interfingering along dune and wetland contacts termed "interfingering II" (*e.g.*, wetlands 161, 163, 142); and
4. benching of wetland sediments along the wetland margin (*e.g.*, wetland 35).

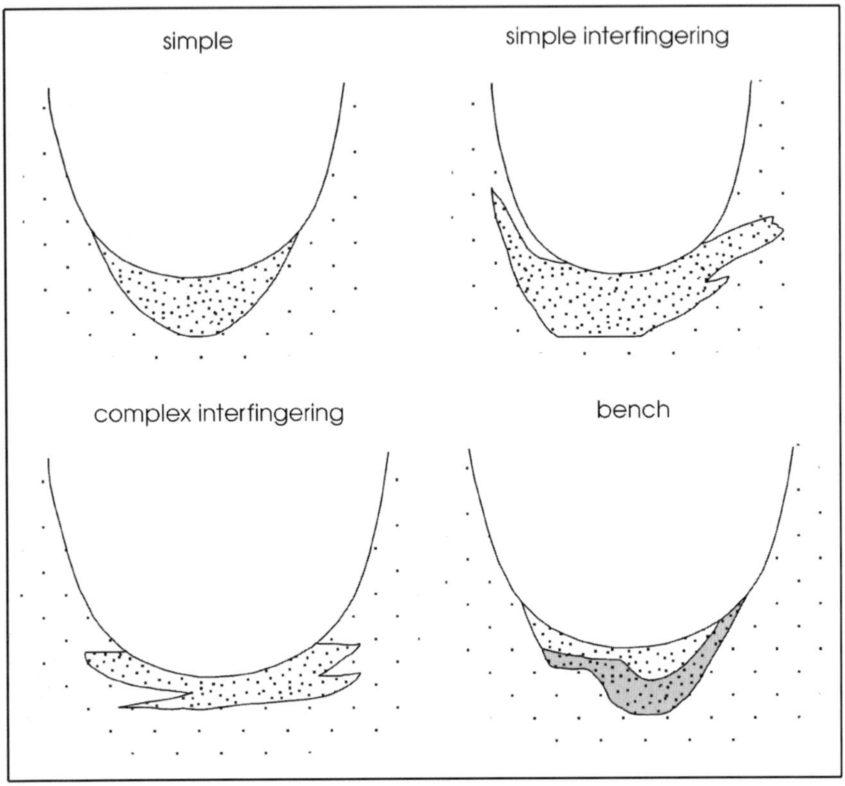

Figure 7-5. *Types of lateral stratigraphic contacts to wetland fill.*

Seasonal (early winter, late winter, spring, mid-summer, and autumn) water levels for marginal sites representing each category were compared to those of the central wetland sites during 1992 (Table 7.5). Seasonal divisions mirror changes in rainfall, and groundwater recharge and discharge patterns.

In this subsample, the water levels under the margins of wetlands exhibiting interfingering were consistently higher than those in the wetland during the full range of seasons, showing that these types of margins channel large and small flows toward the wetland. In contrast, the water levels under the margins of wetlands which were simple or bench type, showed either temporary inequalities, or no differences to those in the central wetland, indicating that these margins intermittently promote flow into the wetlands during early and late winter or have no observable effect.

Table 7.5 *Water level in the marginal site relative to the wetland site*

	Simple		*Interfingering I*		*Interfingering II*			*Bench*
Date	*9-7*	*swii*	*162*	*9-5*	*161*	*163*	*142*	*35*
May-92 autumn	~	<	≥	≥	~	>	>	<
Jun-92 early winter	≥	<	>	≥	~	>	>	~
Sep-92 late winter	≥	>	~	>	~	>	>	<
Oct-92 spring	~	~	>	>	~	>	>	<
Dec-92 summer	~	=	>	≥	>	>	>	<

< the water table under the margin is lower than under the wetland site
> the water table under the margin is higher than under the wetland site
= the water table under the margin is the same as under the wetland site
~ the water table under the margin is sometimes lower, sometimes higher, and sometimes the same as under the wetland site

The data base used in a fuller analysis comprised the monthly water level measurements for August 1991 to July 1996.

Patterns of groundwater response for the four stratigraphic settings
A summary of groundwater response in three of the lateral stratigraphic settings is tabled. The fourth setting, the bench margin is discussed separately below.

Summary of water level response at the margin where there is simple juxtaposition
In autumn, the water levels at the simple margin cannot be separated from the east/west gradient, *i.e.*, eastern margins have slightly higher water levels and western margins slightly lower. At the beginning of winter, water levels are similar to autumn but the gradient is slightly steeper (Fig. 7-6). At end of winter, water levels under the wetland margin approximate or are slightly higher than the wetland. When levels begin to fall, the response again is variable (slightly above to considerably less) relative to the wetland. In summer, water levels under the wetland margin are generally the same as in the wetland.

Summary of water level response at the margin where there is simple interfingering (interfingering I)
In autumn, water levels at the margin are higher than the wetland. At beginning of winter, water levels at the margin are slightly to considerably higher than the wetland (Fig. 7-6). At end of winter, water levels are variable relative to the wetland. When levels begin to fall, water levels at the margin are higher than the wetland. In summer, water levels at the margin are also higher than the wetland.
Summary of water level response at the margin where there is complex interfingering (interfingering II)
In autumn, water levels at the margin are higher than the wetland. At beginning of winter, water levels at the margin are still higher than the wetland but less so (Fig. 7-6). At end of winter, water levels at the margin are higher than the wetland. When levels begin to fall, water levels at the margin are higher than the wetland In summer, water levels at the margin are higher than the wetland.

Water level response at the margin where there is a bench of wetland sediments shows that in autumn, water levels at the margin are lower than the wetland. At the beginning of winter, water levels are variable, sometimes higher due to short term deflection or ponding, sometimes lower, than the wetland (Fig. 7-6); at end of winter, water levels are lower than the wetland. When levels begin to fall, water levels are lower than the wetland, and in summer, water levels are lower than the wetland. The benched margin, at all times it was investigated, showed water levels less than in the the centre of the wetland. As there is only one representative of this type of setting in the study wetlands, and this wetland exhibits a persistent east/west gradient, it appears that over this short time scale, any effects on the hydrological patterns created by the existence of the bench cannot be separated from variation produced by the local gradient.

The simple margin for most of the year exhibited a water level contiguous with the local water level under the wetland centre and adjacent beachridges. Where differences were noted, they were inconsistent between settings and between years, and therefore cannot be attributed to the nature of the marginal contact.

For settings with both simple and complex types of interfingering, the water levels were consistently higher at these margins than in the wetland. In autumn, early and late winter, and as water levels begin to recede, infiltrating water was preferentially directed to the margin of some of the wetlands, forming a local mound. This process is linked to the occurrence of surface sedimentary layers with contrasting permeability and transmissivity, such that the upper layer becomes saturated, and vadose percolation through coarser layers, encountering a less transmissive layer at shallow depth, results in lateral flow. These conditions occur at margins where wetland muds interfinger with dune sands or where calcrete is interlayered with muddy sands.

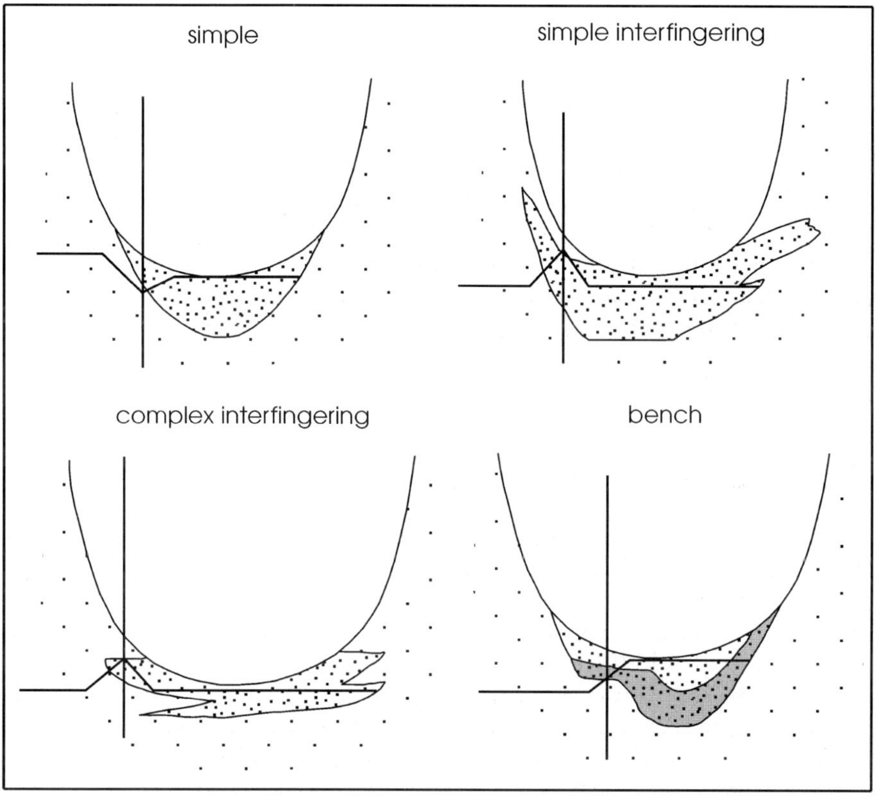

Figure 7-6. Groundwater response at the beginning of winter rain.

Preferential recharge to the groundwater via the complex interfingering margin, particularly since these margins occasionally have coarser sand layers, creates an earlier rise in water levels in settings where beachridge/dunes are relatively high, and wetland sediments relatively impervious. In spring, water from the adjoining beachridge/ dunes is preferentially directed to the wetland margin in the lateral contact where muds interfinger with sands, and in the summer, intermittent rainfall is able to penetrate to the shallow water table, via the contact zone rather than remain stored in the vadose zone or be evaporated from the wetland surface. This lateral recharge maintains higher water levels to the early part of the summer.

Summary of patterns of groundwater response to variable rainfall conditions for the four lateral stratigraphic settings

To investigate small scale variability in the groundwater response, under variable rainfall conditions in settings with different lateral stratigraphic relationships, a second analysis of the marginal sites was undertaken. Water level data were analysed under high and low monthly rainfall, and for undersaturated and saturated conditions (Figs. 7-7, 8). The results are drawn from the following periods:

- high rainfall conditions: June 1992, June 1994, July 1995, July 1996;
- low rainfall conditions: August 1993, September 1994, 1995, 1996;
- undersaturated conditions: February 1992, May 1995, December 1995, June 1996;
- saturated conditions: July 1992, August 1993, July 1994, July 1995, August 1996.

Patterns of water table response under these conditions are summarised below.

High rainfall conditions (Fig. 7-7A)

Simple margin	Water table rise at the western margin approximates or is greater than in the wetlands
Simple interfingering margin	Water table rise at the western margin approximates that in wetlands
Complex interfingering margin	Water table rise at the western margin is greater than in wetlands
Benched margin	Water table rise at the western margin is sometimes less than and sometimes greater than in the wetland (no net effect)

Low rainfall conditions (Fig. 7-7B)

Simple margin	Water table rise is variable and inconsistent, *i.e.*, at different times it approximates, is greater than or less than in wetlands
Simple interfingering margin	Water table rise at the western margin is inconsistent, *i.e.*, sometimes greater and sometimes less than in wetlands
Complex interfingering margin	Water table rise at the western margin is greater than in wetlands
Benched margin	Water table rise at the western margin is less than in the wetland

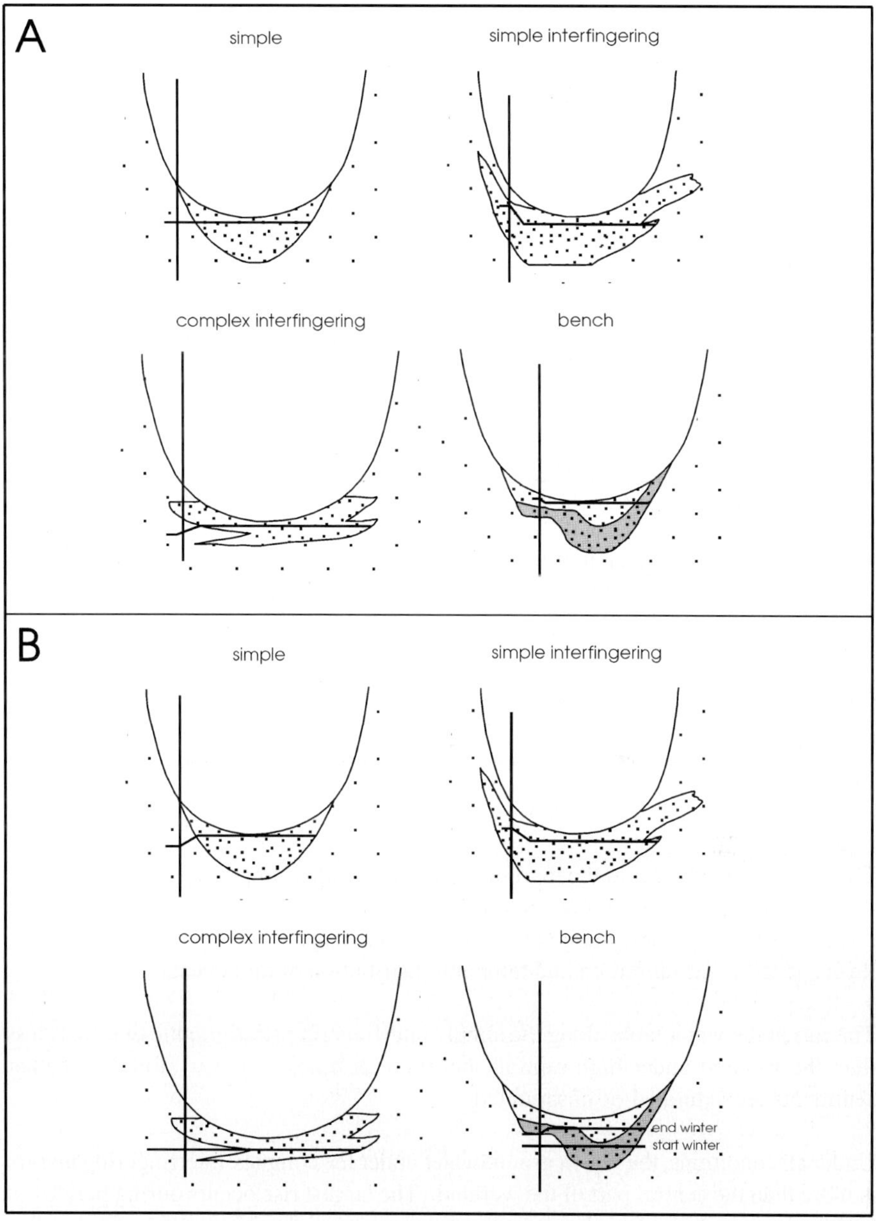

Figure 7-7. *Response of groundwater to A. high rainfall B. low rainfall.*

Undersaturated conditions (Fig. 7-8A)

Simple margin	Water table rise at the eastern and western margin is slightly more than in wetlands that are underlain by impervious sediments
Simple interfingering margin	Water table rise at the western margin is greater than in wetlands
Complex interfingering margin	Water table rise at the western margin is greater than in wetlands
Benched margin	Water table rise at the western margin is less than in the wetland. In Feb. 1992 recharge was significantly less

Saturated conditions (Fig. 7-8B)

Simple margin	Water table rise at the margin is variable and inconsistent, *i.e.*, at different times it approximates, is greater than or less than in wetlands
Simple interfingering margin	Water table rise at the western margin is greater than in wetlands
Complex interfingering margin	Water table rise at the western margin is equal to or greater than in wetlands
Benched margin	Water table rise at the western margin is less than or equal to the wetland

The rise in groundwater at the margin of wetlands with a simple stratigraphic margin displays most similarity to the central parts of the wetland under conditions of high rainfall. At other times, there are a variety of recharge rates depending on relative levels of sediment saturation and temporal distribution of rain events.

The rise in the water table along the simple interfingering stratigraphic margin is less than the wetland under high rainfall, but more at times of lower rainfall, whether sediments are saturated or unsaturated.

Under all conditions, the rise in groundwater under the complex interfingering margin is more than the central part of the wetland. The largest rise occurs during periods of high or low rainfall when the sediments are unsaturated prior to the event, but this margin type also acts as a conduit under saturated conditions.

Water table rise under the benched margin is less than in the central part of the wetland, under all conditions. Only with very high rainfall, when the water table is relatively high, is the situation reversed.

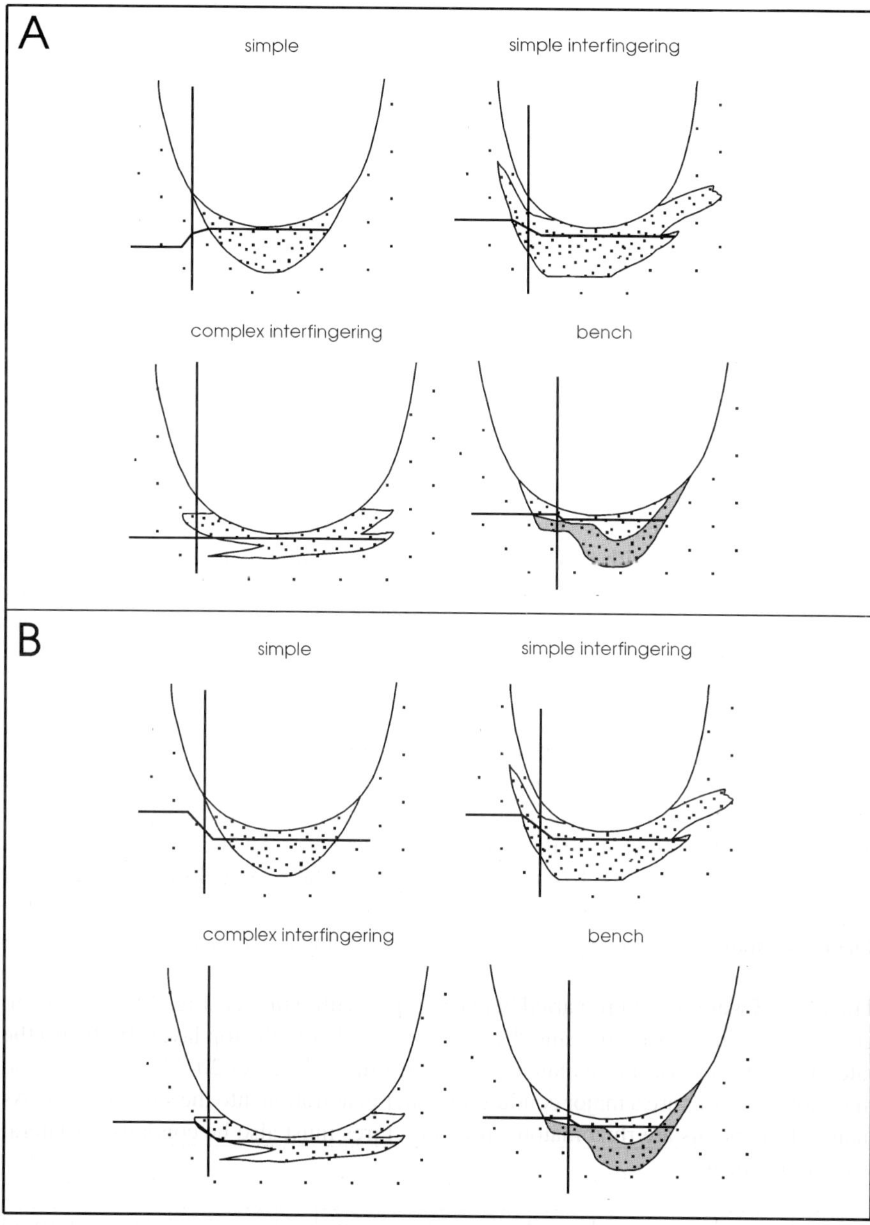

Figure 7-8. *Response of groundwater to rainfall under A. undersaturated and B. saturated conditions.*

Macropore efficiency is a function of their geometry, size, number, and the pore volume of the sediment. Depending on the rate of precipitation, macropores are likely to fill with water before the matrix is saturated. At this stage, lateral transfer from macropore to matrix commences and is the dominant flow control (Chen and Wagenet 1992). Harvey and Nuttle (1995) found that infiltration and evaporation water fluxes were segregated between macropores and matrix pores. In the Becher wetland sediments, macropores include root and rhizome channels, burrows, shell gravel deposits, sediment layering, and texture mottling due to bioturbation.

7.5 Summary and discussion

Various attributes of sedimentary stratigraphic sequences affect hydrology at the regional, basin, and bed scale. At the regional scale, the mud fill (textural attribute) is the major determinant influencing water movement within an unconfined sand aquifer. At the basin scale, the structural, textural and compositional variations down profile affect the rate of movement of water, the direction of water flow, and the wetland hydroperiod. The effect of a varied stratigraphy on groundwater properties and flows has been demonstrated in a number of other studies (Wilcox *et al.* 1986, Phillips and Shedlock 1993, Shedlock *et al.* 1993, Steinwand and Fenton 1993; Zeeb and Hemond 1998, Eser and Rosen 1999, Anderson *et al.* 2000; Mann and Wetzel 2000b).

The main principles that other studies share in common with this study are:

- there can be, and often are, separate flow systems and velocities in separate, relatively shallow aquifers (Steinwand and Fenton 1993; Mann and Wetzel 2000b);
- wetlands can be influenced by regional, intermediate and local flow systems at the same time (Anderson and Munter 1981, Cherkauer and Zager 1989, Kenoyer and Anderson 1989, Shedlock *et al.* 1993, Phillips and Shedlock 1993);
- wetlands may be situated in a landscape where there is more than one flow system occurring, but may only be influenced by the superficial aquifer (D'Amore *et al.* 2000);
- even in the situation where recharge is from meteoric water only, a varied stratigraphy will cause a heterogeneous response in soil moisture, rate of recharge to water table and thickness of the zone of capillary rise within the one wetland basin (Gerla 1992; Anderson *et al.* 2000);
- the interaction between groundwater and each sediment layer in the stratigraphic profile will change under wet and dry conditions, and this change will be further modified by the conditions in the period preceding the current one (Winter and Rosenberry 1998, Zeeb and Hemond 1998).

The underlying explanation for many of these phenomena lies in the nature and position of the zone of tension saturation (the capillary fringe). Gillham (1984) demonstrated that a disproportionate rise in the water table can occur in response to even meagre rainfall, depending on the degree of saturation in this zone. This localised disproportionate rise in the water table can initiate transient radial or reverse flows in settings which otherwise would be subject to very slow unidirectional lateral flow. Not only do these flows differ in direction but they can also differ in intensity, moving water much more quickly than would be assumed. The dimensions of the capillary fringe will depend upon the porosity resulting from textural, structural (aggregational) and compressional characteristics of the different sediments as well as the position of the water table within the stratigraphic profile (Gillham 1984). In sands this may be relatively small, but in finer sediments the capillary fringe is able to intersect the ground surface when the water table lies at the base of any carbonate or peat muds, *i.e.*, up to 1.2 m. During the rainfall season itself, re-wetting of sediments after a dry spell will have different effects on water table response to that of an equal volume of rain during consistent falls, *i.e.*, the zone of capillary rise in a particular sediment is to some extent dependent on the recent rainfall history. The implications for the Becher wetlands and their heterogeneous stratigraphy are clear. The response to any rain event will be variable throughout the wetland suite. The availability of water for plants will also vary. With the loss of the same amount of water, either by drainage or evapotranspiration, the water table will return to its original position, and transient flows will cease.

At the bed scale, root and burrow structures significantly increase the rate and direction of water flow. Many factors affect the rate of water movement or hydraulic conductivity within a sediment. The main factors are sediment granulometry, hydrostatic head, preferred conduits, degree of saturation of the aquifer, all of which have been described above. Other minor factors include temperature, ionic concentration of the water, and entrapped air (Bouwer 1978). Of these three minor factors, temperature has not been considered in this discussion because it is assumed that the variability in temperature of the groundwater at the same depth below the surface will be negligible. The ionic composition of the water has not been addressed either, although it could be important during the first flush at sites where the sodium content of the infiltrating water is significant. The presence of entrapped air is considered to be a function of sediment structure, texture, and composition, and in this sense has been included in the preceding discussion.

The direction of water flow is controlled by conduits, layering, and hydraulic gradients. In a typical multi-layered stratigraphic profile, such as 1) OME calcilutite overlying calcilutite, or 2) carbonate muddy sand or sand overlying calcilutite, which is common at the wetland margins, the direction of water flow can be changed from dominantly vertical to lateral. This can occur at the boundary between two layers with different permeability characterstics, or in response to a hydraulic head associated with mounding. In calcilutite and peat, a hydraulic head is created by the difference between the saturated surface layer and the water table creating predominantly vertical movement in sediments of low permeability. A heterogeneous sedimentary sequence will exacerbate the variability between wetlands given similar hydrological mechanisms.

For wetland plants with roots in the top 10 cm of the sediment, small scale changes in flow direction or flow rate result in a change to the hydroperiod to which they are adapted. Localised hydroperiod changes for plants experiencing a period of water deficit can be beneficial or crucial to survival. The result of preferential flow to various parts of the wetland basin, and/or of impounding or retardation of infiltration in the surface layers, can provide real and timely adjustments to the normal cycle of water availability in a wetland basin.

8. WETLAND HYDROLOGY

8.1 Introduction

The aims of this chapter are to identify and describe the hydrological processes which maintain the wetlands on the Becher cuspate foreland. A part of this objective is the resolution of the question of whether the Becher wetlands are simply surface expressions of groundwater in a surficial homogeneous aquifer, recharged by direct infiltration from rainfall and discharged through evapo-transpiration, or whether they are isolated closed systems with their own internal balance of water input and output. Although there has been a general assumption by various workers that groundwater recharged by meteoric infiltration is the underlying hydrological mechanism sustaining these wetlands, no connection has thus far been demonstrated between rainfall and groundwater rise and fall.

This simple hypothesis of rainfall recharge, evapo-transpiration discharge does not explain observed variability in wetland water levels and annual hydroperiod. These latter phenomena testify to the influence of smaller basin scale processes occurring which have so far not been identified. Using data obtained from the study wetlands, the aim was to identify wetland hydrological processes at all scales, and to document their contribution to the overall functioning of the wetlands. Vertical pathways, by which rainfall recharges, and evapo-transpiration discharges water within the wetland basin, were identified, and the role and significance of lateral flow were examined.

Longer term water level behaviour was examined in order to identify trends, and to assess the significance of short term water level variability in wetlands. Water levels under beachridge/dunes and wetlands were compared, in order to investigate differences in recharge/discharge mechanisms, and to identify interactions at wetland margins. Measurements of hydrological parameters were made with respect to wetland vegetation formations, and small scale hydrological processes which affect water availability were targetted, *e.g.*, seasonal soil water content and diurnal water table response to the onset of precipitation. These results assisted in determining the role of hydrological processes in the evolution of the wetlands, and in delineating some of the small scale, site-specific, hydrological processes which affect wetland plant selection and sustainability.

The chapter begins with an overview of regional hydrological features, *i.e.*, rainfall volume and frequency, evaporation and its seasonality, and regional gradients of throughflow. This is followed by studies of hydrological mechanisms at the basin and bedding scale, preliminary discussion of findings with respect to the long term climatic cycles, and lastly, discussion of the effects of small scale hydrological variability on plant distribution within wetland basins.

8.2 Regional hydrological features

Rainfall is the lifeblood of the Becher Suite wetlands. Rainfall recharges the groundwater within the Safety Bay Sands. Groundwater then rises and intersects the ground surface in scattered, discrete, low topographical areas within the Becher Cuspate foreland of beachridges and swales, creating wetlands with hydric soils, inundation or waterlogging conditions and habitats for wetland plant species. To understand the nature and functioning of these wetlands, a description of the regional and local precipitation and evaporation patterns is required.

Wetland hydrological monitoring took place from August 1991-August 2001, with the most intense monitoring occurring between 1991-1996, therefore the rainfall and evaporation data for this period are directly related, but this period in the context of longer term data is also relevant.

8.2.1 Long term rainfall

Long term rainfall data used in this study were from the Perth, Perth Airport, and Mt Lawley meteorological stations. Figure 8-1 shows the annual rainfall for Perth from 1876 to 2001, the 120 year annual Perth rainfall average (869 mm), and the line representing the series of 10-year backward averages. The pattern of the 10-year backward moving average, viewed over approximately 120 years, is weakly sinusoidal with a period of 18-20 years, and begins and ends with below average rainfall. From 1991-1996, the period of intense fieldwork for this study, the Perth region was continuing to experience rainfall conditions below the long term average, but the moving average had passed the trough and was on the rise or upturn.

8.2.2 Regional rainfall

Regional rainfall figures were taken from the two closest meteorological stations to Rockingham, Medina (north east) and Mandurah (south). As the location and establishment of the rain gauge at Rockingham itself had been inappropriate (*pers. comm.* Meteorological Bureau), it was dismantled halfway through 1993. Rainfall figures from Medina and Mandurah were compared to the limited set from Rockingham in order to determine the extent of correlation between all three stations and to determine which data set would be appropriate for the study area (Fig. 8-2). The mean of the volumes at Medina and Mandurah was selected to approximate the Becher locality. Rainfall totals for the periods between monthly sampling from 1991-2001 were graphed (Fig. 8-3). Rainfalls for one week, and three days, prior to the date of sampling, were plotted to further clarify the hydrological response to rainfall of the groundwater.

The annual pattern of rainfall is seasonal, concentrated between the months of May and November, with events outside this period being sporadic, unreliable, and usually insignificant. During 1991-1996, the main period of field sampling, the rainfall deviated from this annual pattern, decreasing in amount, and exhibiting a

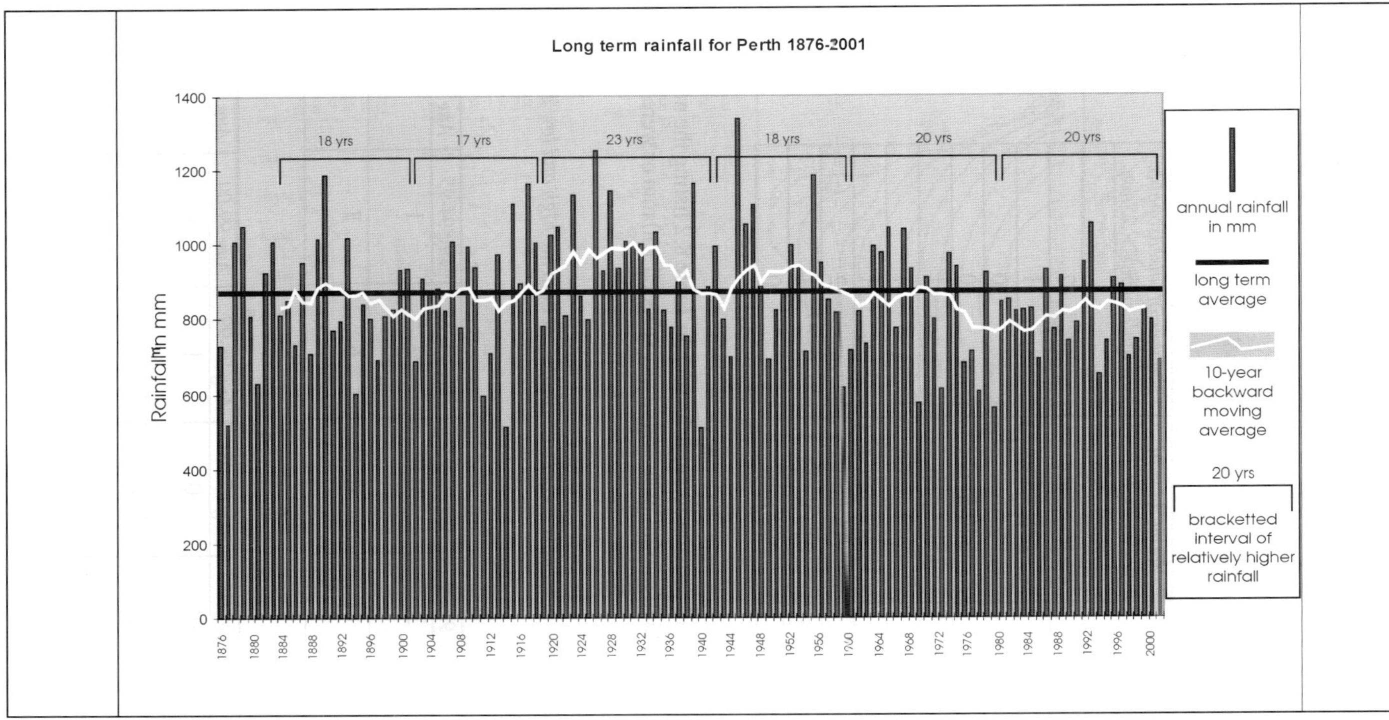

Figure 8-1. *Annual rainfall for Perth 1876-2001, showing the long term average, the 10-year backward moving average, and bracketed intervals of relatively higher rainfall periods recurring on a ~ 20-year turn-around (for further discussion, see Semeniuk & Semeniuk 2005).*

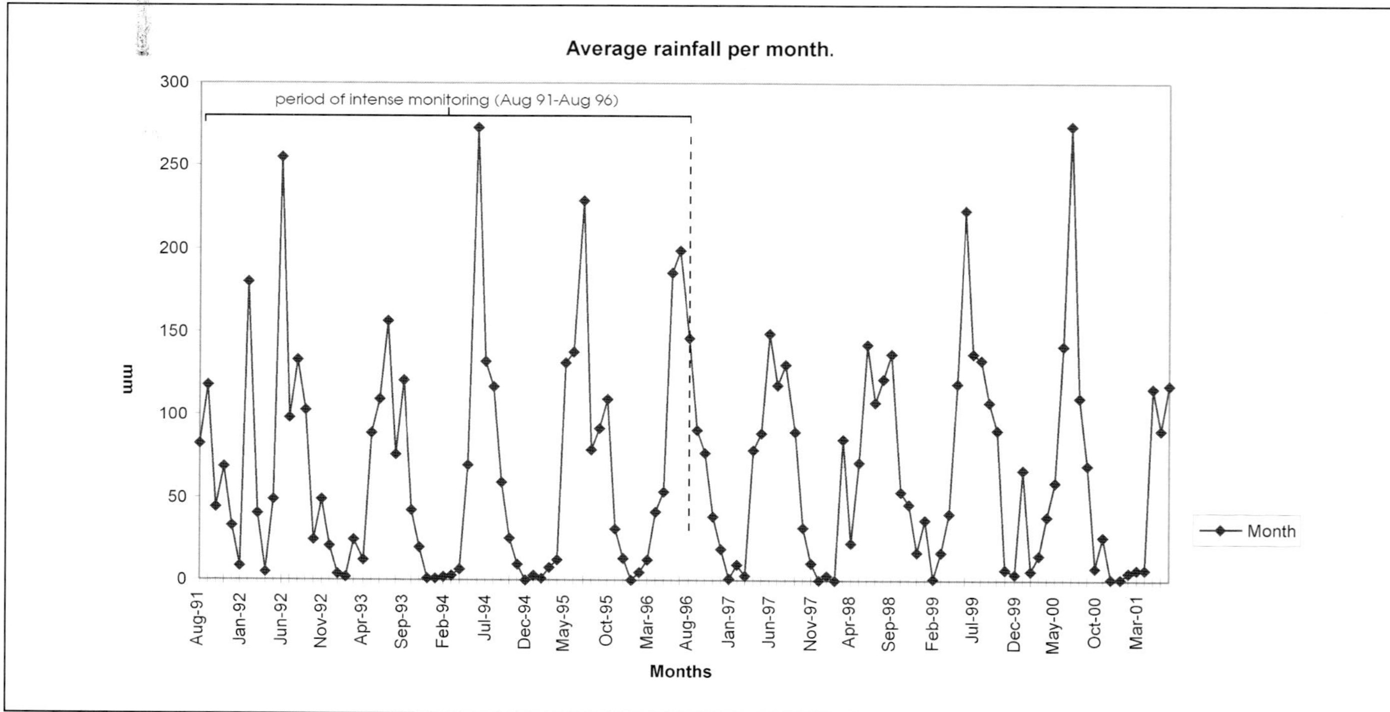

Figure 8-3. *Rainfall recorded between monthly monitoring events, based on the mean of rainfall volumes recorded at Medina and Mandurah Stations.*

8.2.3 Local rainfall

Rainfall gauges were established at three sites over the Becher cuspate foreland at varying distance from the coast (Fig. 8-4A) to compare local variability with the approximation of regional rainfall used here. Rainfall at the three sites was found to be spatially variable (Fig. 8-4B), with an increase towards the relatively high ground of the inland Spearwood Ridge (Walyungup site), but no consistent trend at the other two sites. Rainfall at the site nearest the coast (Becher track) was the most capricious (Fig. 8-4B), probably due to the wind factor. Statistical analysis of these monthly measurements showed variation of $F = 0.7092$ with $F_{.05}$ *crit.* = 2.7826 showing that variation between sites at Becher, and the Medina/Mandurah mean is not statistically significant.

8.2.4 Evaporation

From 1991-2001, the annual evaporation pattern was consistent (Fig. 8-5). The maximum monthly evaporation was approximately 300 mm and the minimum monthly evaporation close to 50 mm. There was a stochastic rise in evaporation through spring and early summer, and a primary fall between February and March (summer to autumn). The maximum evaporation occurred in the months of January or February, associated with the period of highest monthly mean temperatures and greatest activity of onshore/offshore breezes. From December to February, SSW winds occurred approximately 20% of the time and wind speeds of 6-9 m/s were recorded.

8.2.5 Description of aquifer

This section describes the characteristics of the groundwater body in the Safety Bay Sand and Becher Sand under the Becher cuspate foreland. The morphology of the water body in the Safety Bay Sand aquifer on the northern cusp of the Rockingham twin cuspate system (Passmore 1970), as defined by water table contours, is an elongate mound approximately 4-5 m deep, sloping down north, west and east.

On the southern Becher cuspate foreland, the water body, as defined by water table contours, is roughly wedge shaped, approximately 25 m deep at maximum thickness, sloping west, northwest and southwest. The configuration of the surface varies with the volume of water in the aquifer and its height varies between 2.8 and 4.2 m AHD (Fig. 8-6 A, B, C). From midway to the apex of the cusp, there is a flattening of the water table. Generally, there is a steepening of the gradient closer to the shorelines, with the steepest slope to the northwest adjacent to Warnbro Sound. To the northeast there is a mound due to vegetation clearing, which has created a southwest perturbation in the general westward orientation of groundwater contours.

The base of the aquifer is the undulating ridge/swale topography of the Spearwood Dunes, modified by Pleistocene estuarine mud fill (Australind Formation) (Fig. 3-5). The eastern margin of the aquifer is a straight cliffed contact with the Tamala

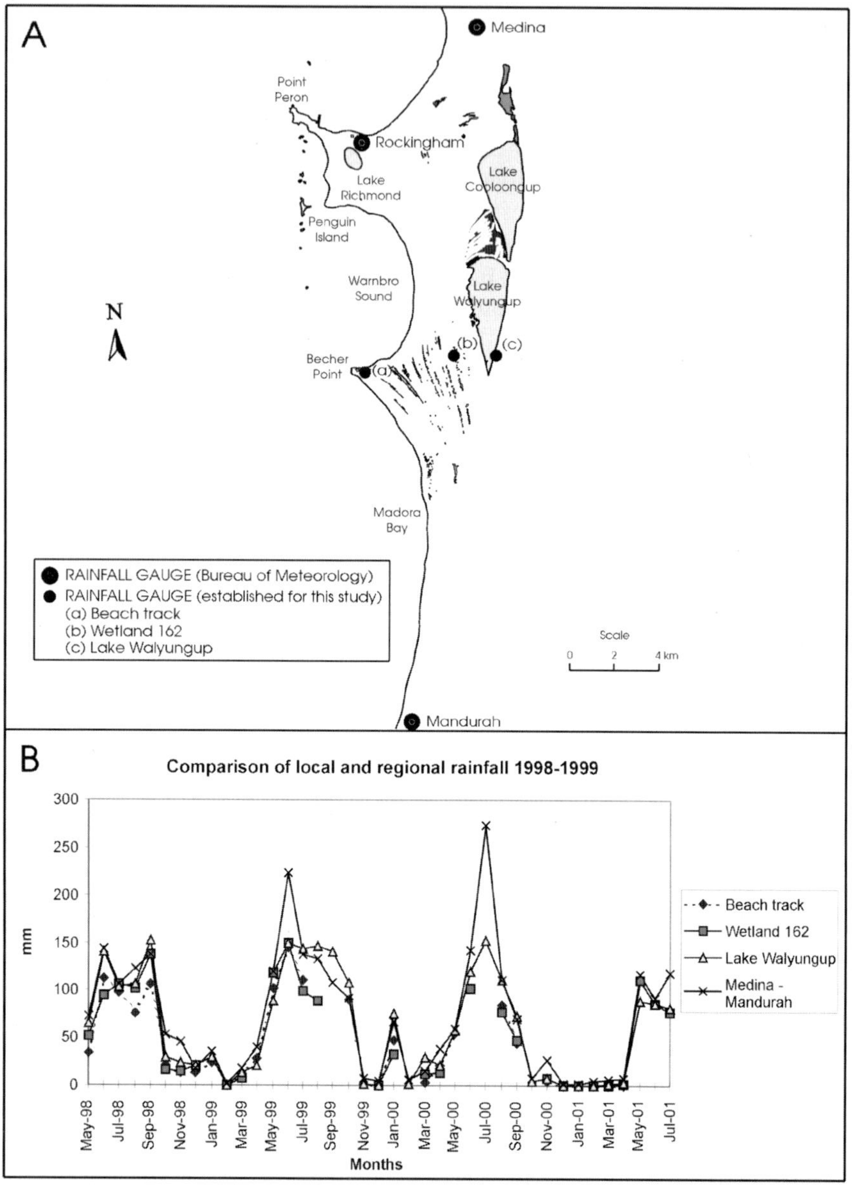

Figure 8-4. *A. Location of rainfall gauges used in this study. B. Comparison of rainfall volumes derived from the mean from Medina and Mandurah stations and local gauges on the Becher cuspate foreland.*

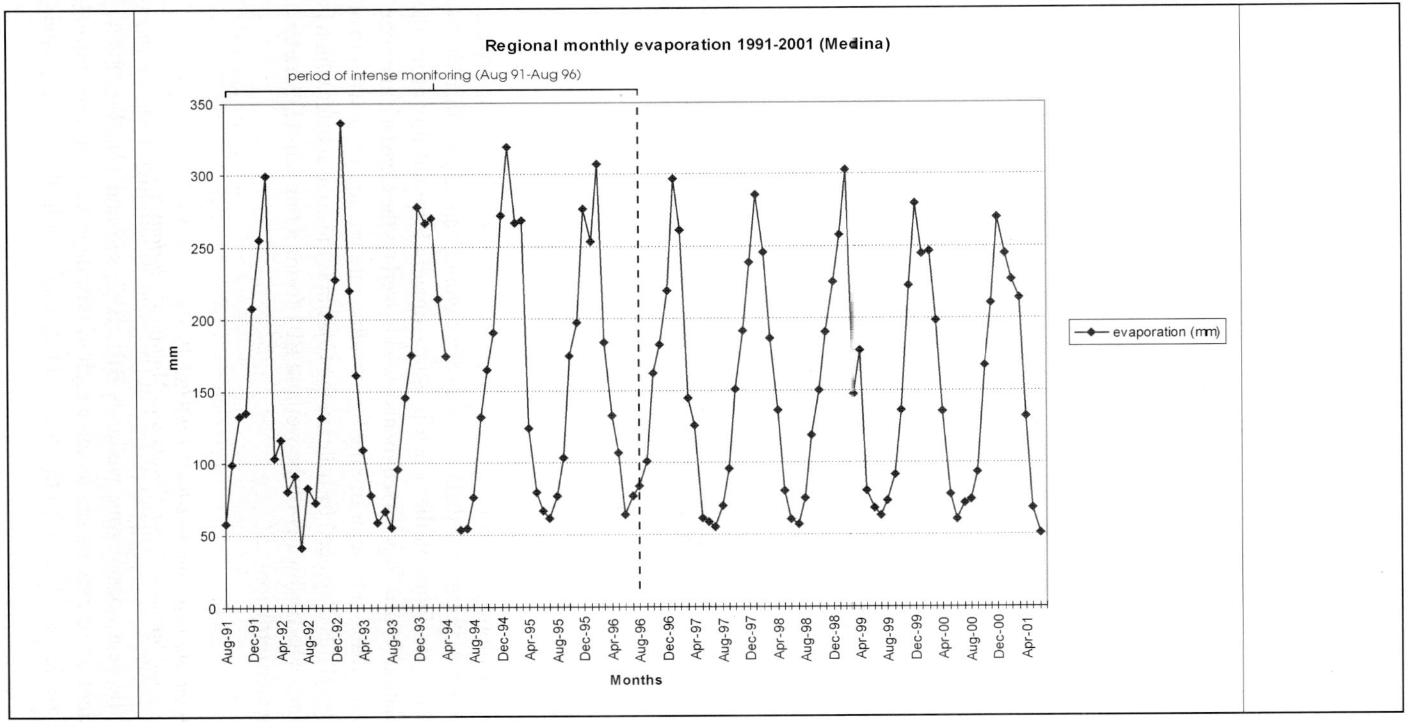

Figure 8-5. Monthly evaporation based on volumes recorded at Medina Station.

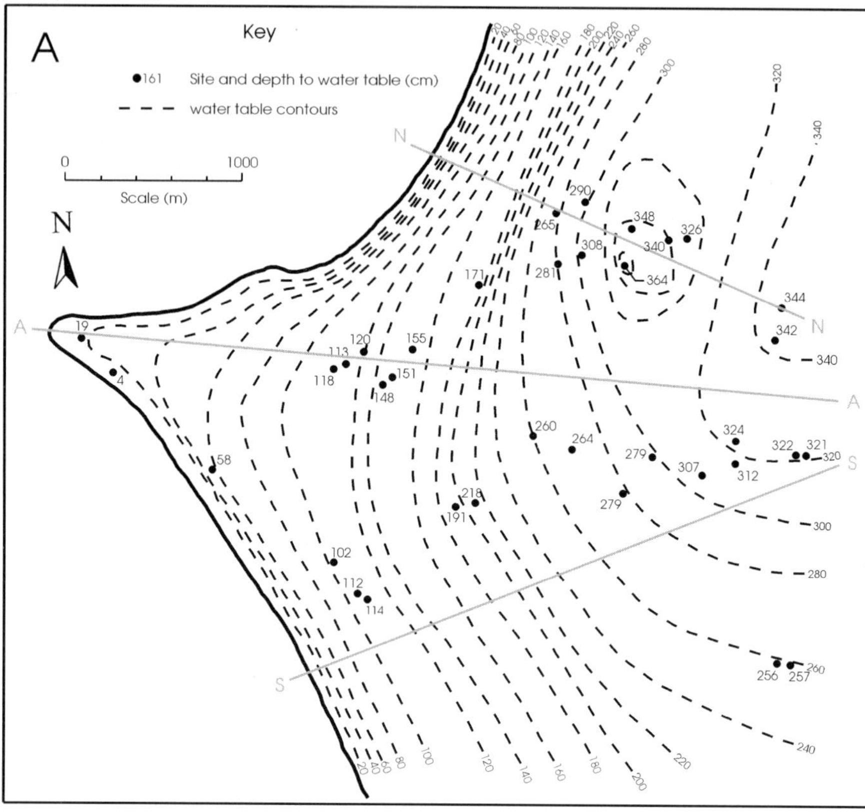

Figure 8-6A. September 1994 groundwater levels in Point Becher area.

Limestone or the Cooloongup Sand. At the western margin (site BP1), the Safety Bay Sand aquifer becomes continually shallower between wetland 9 and the northern beach near Becher Point. Its depth on the beach is approximately 9 m and it overlies the Australind Formation (coastal lagoon sediments), which overlays the Coastal Limestone. A salt water body creates a zone of mixing by intruding into the aquifer to a depth of 15 m, at which level it is displaced by water in the Coastal Limestone aquifer which is fresh to subhaline. The freshwater of the wetland BP1 forms a thin lens (1-2 metres) above the salt water intrusion.

Tidal influences on the western margin of the aquifer

Tidal influences extend as far inland as 600 m (wetland 9) although they are most pronounced in the near coastal zone wetlands BP1, BP2, swi and 1N, (Fig. 8-7A). Pressure cells depress or elevate sea levels which in turn depress or elevate diurnal water levels in wetlands adjacent to the coast. Over one month, these fluctuations

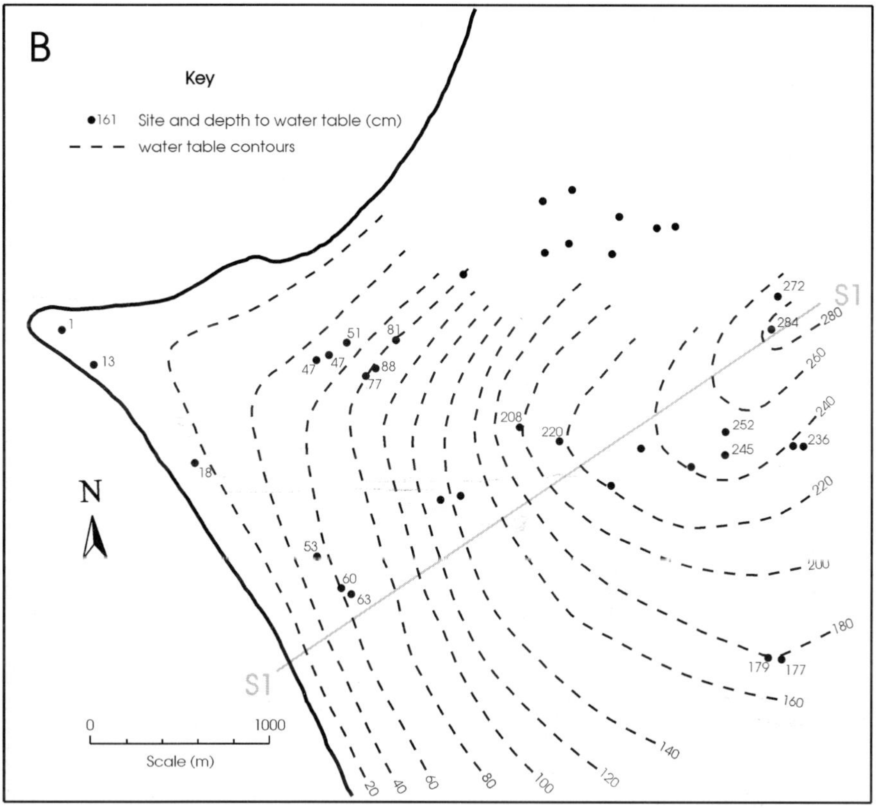

Figure 8-6 B. Groundwater levels during period of low water - April 1995.

are obscured, but a diurnal response can be in the order of 0.5 m. Measurements taken daily in May 1996, at wetlands BP1 and 135 (Fig. 8-7B, 7C), showed water levels in BP1 (1-4) responding to the oscillations in sea level, in contrast to the water level in wetland 135 which continued to fall to its pre-winter minimum. The water levels in wetland BP1showed a lag in response time of approximately 24 hours (Fig. 8-7C). A second series of measurements was taken during June/July 1996, the season of groundwater rise. Winter storms, sea level rise and concomitant near coastal groundwater rise were the prevailing conditions. Despite marked water level fluctuations at all sites in wetland BP1, oscillations were concordant with sea level. In wetland 1N diminished oscillations were evident in groundwater levels, whereas at wetland 135 water levels continued to rise steadily. On the 14th July 1996, the sand barrier protecting wetland BP1 was breached and several piezometers were lost, however, monitoring of the remaining sites showed that the patterns continued (Fig. 8-7D).

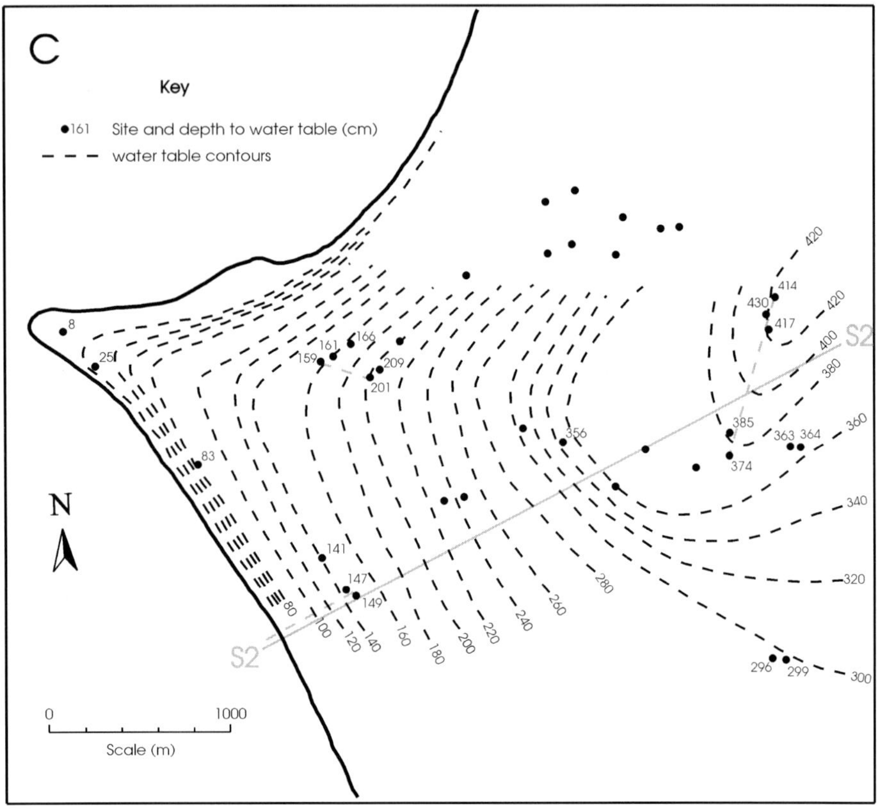

Figure 8-6 C. Groundwater levels during period of high water - October 1999.

8.2.6 Regional hydraulic gradients and flow paths

Groundwater contours under the Becher cuspate foreland were derived from the most comprehensive water level data base, September 1994, marking the end of winter. Supplementary contour maps were produced for April 1995 and October 1999 to investigate seasonal changes. Under all conditions there were varying spatial patterns. In September 1994, at the north and south coast, contours were tightly spaced, commensurate with coastal discharge. In the main body of the cuspate foreland, the contours formed three patterns: widely spaced furthest from the coast; closer together in the central part; and wider again landward of the protruding point. Wider spaced contours denote unimpeded flow, contours closing together in the central region denote some degree of impounding by down stream linear bodies of wetland sediments, and contours closing together at the coast denote more rapid discharge at the aquifer margin. At lowest water levels (April 1995), the contour spacing was even, suggesting unimpeded flow. At highest water levels (October 1999), a time when many of the wetlands were inundated, there were three patterns: a south-southwest trend inland separated from the main contours; evenly spaced contours in the central part; and closely spaced contours at the coast and landward of the protruding point. This picture also tends to suggest relatively unimpeded flow.

From a low groundwater mound located in the northeast of the area, the flow paths deduced from the contour patterns radiate seawards in an arc extending from northwest to southwest. From the groundwater contour maps, hydraulic gradients were calculated in three directions: northwest (to the northern shore); west (along the major axis of the Becher cusp); and southwest (to the southern shore) (Figs. 8-6, 8-8). The three transects each showed distinct changes in slope along their length. September 1994 contours indicate that the shortest and fastest flow path would have been to the northwest (gradient 1:622), but this has been impeded by an artificially created mound under the area cleared for urban development, and flow from wetlands 161, 162, 163, has been diverted from a northwest flow to southwest flow towards wetlands 135 and 136. The new gradient in the vicinity of these wetlands was between 1:1667-1:1838, and the resultant flow rates were low. Flows to the west and southwest were more moderate, with gradients in the order of 1:1000. At low water levels the contours were consistent with minimal lateral flow and minimal discharge at the coast (gradient 1:1173). At high water levels the contours were consistent with lateral flow and discharge to the line of wetlands approximating the 3000 year isochron, with a second zone of discharge at the coast (gradient 1:772).

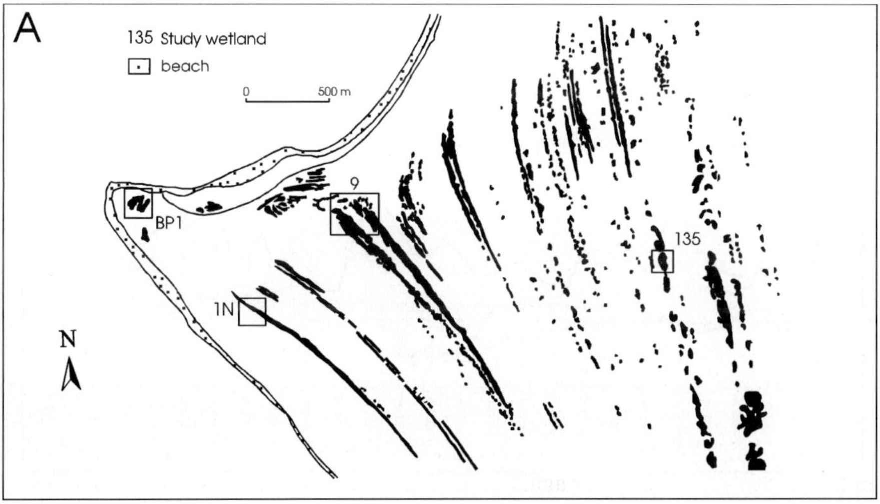

Figure 8-7. A. *Location of wetlands in which diurnal groundwater level changes were monitored.*

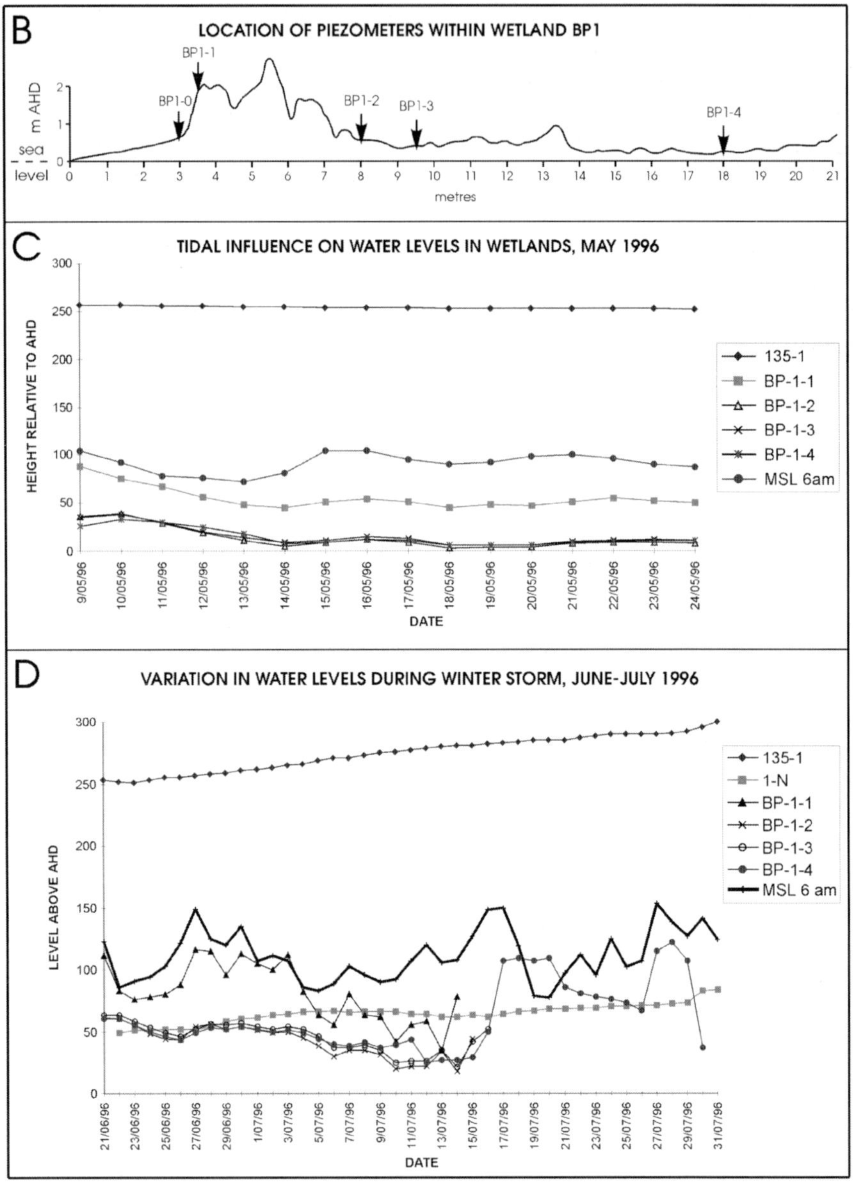

Figure 8-7 (cont.). *B. Location of sites in coastal wetland BP1. C. and D. Groundwater levels under coastal and inland wetlands showing diurnal response to tidal effects during periods of groundwater fall (C) and groundwater rise (D), respectively.*

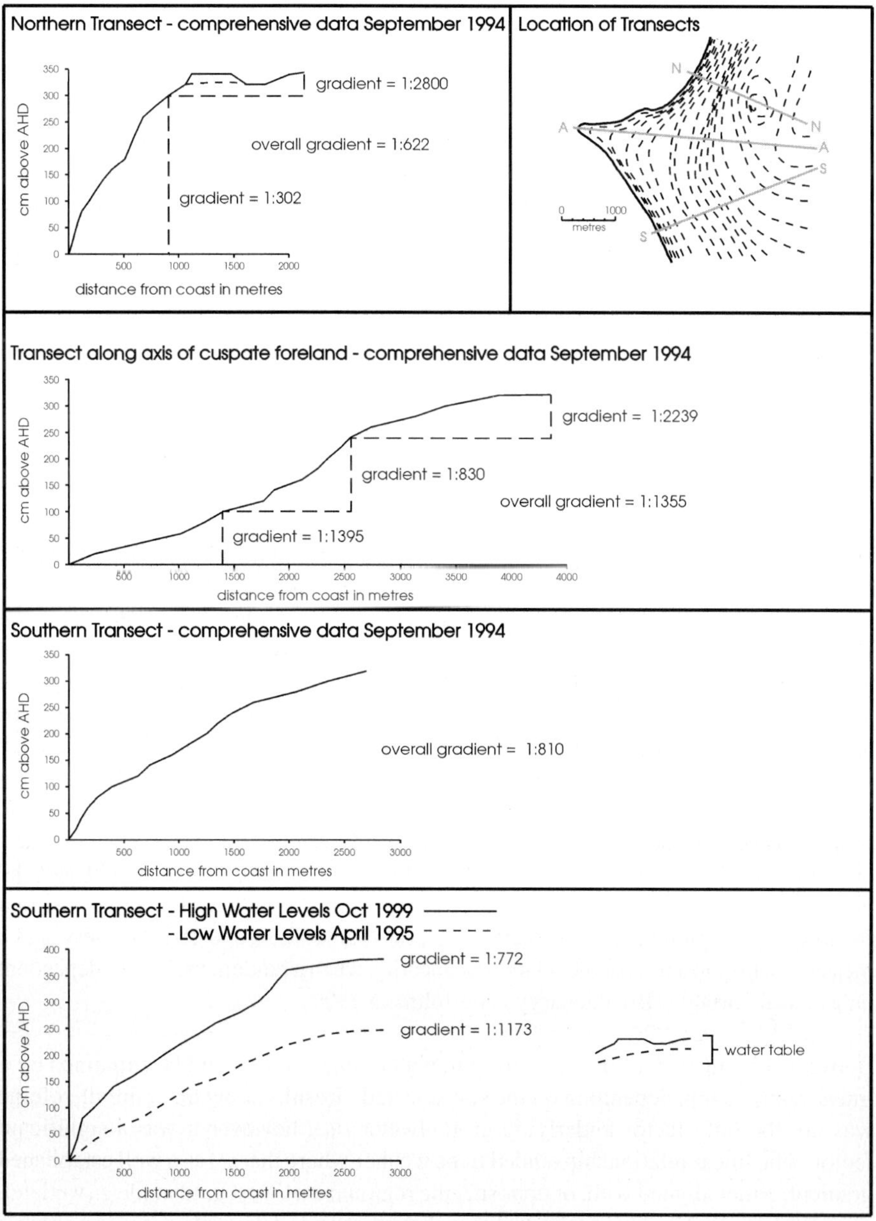

Figure 8-8. *Hydraulic gradients derived from groundwater contours in Figure 8-6 along Transects N, A, and S.*

8.3 Connection between rainfall and groundwater

Several attempts were made to identify the relationship between the rainfall and groundwater rise variables. There are several difficulties in undertaking statistical analysis of these variables because there are causal variables other than rainfall that influence groundwater response, which bring into question assumptions about linear relationships, constant error variance, and normal distributions. Additional causal variables include:

- permeability of sediment
- porosity of sediment
- depth to water table at beginning of winter
- effect of aseasonal events

Without discounting these caveats, the question of how the Becher Suite wetlands are maintained was central to this study and required some resolution. This necessitated testing the degree to which the groundwater response was related to the frequency and volume of winter rainfall. After several trials with Fourier analysis and linear regression, using both annual and winter rainfall volumes, and groundwater rise between minimum pre-winter position and maximum water level position, a decision was made to use linear regression to test the hypothesis that monthly surface and groundwater input into the wetland was the result of monthly rainfall, even though the assumption of "error independence" is compromised by using a time sequence. To test this hypothesis, eleven wetlands were selected, and the amount of water in each sediment layer, as well as surface water, was calculated using empirically determined porosity values. For example, the overall groundwater rise was calculated, the proportion of this rise in each sediment layer was determined, and the height of water contained in the sediment layer calculated by using the product of height and the porosity value for that sediment. The results were correlated with winter rainfall *i.e.*, rain from April to September, for that particular year (Tables 8.3A and 3B). "r^2" serves as a measure of the closeness of the relation to linearity, when a random variable is dependent on a causal variable (Bhattacharyya and Johnson 1977).

Between 17 and 86% of the variability in water table levels could be explained by a linear relationship, depending on the site selected. Results show that rainfall volume was not the only factor underlying groundwater rise, however it was a significant factor. The linear relationship tended to be weaker where there was a well established gradient, either aligned with, or opposite, the regional east/west gradient, *e.g.*, wetlands swii, WAWA, 135-2. This suggests that in these wetlands, lateral flow is also a process influencing groundwater levels.

Table 8.3A *Correlation between groundwater recharge and rainfall*

Site	r^2	*r*	*SE*	*Observations*
		Wetlands		
161	0.72	0.85	11.15	9
162	0.66	0.81	8.74	9
9-14	0.65	0.81	8.72	9
63-3	0.60	0.77	5.14	6
WAWA	0.60	0.77	10.73	9
135-2	0.56	0.75	11.05	9
35-4	0.51	0.71	10.44	9
72-3	0.47	0.69	11.53	7
swii	0.32	0.56	5.77	9
1N	0.17	0.41	5.70	9
		Beachridge/dunes		
135 ridge	0.52	0.71	10.27	9
9-9 ridge	0.37	0.61	8.28	9

When the results were ordered to correspond with spatial distribution of wetlands along the axis of the cuspate foreland, effectively mirroring distance from coast, and those wetlands with fewer than 9 observations were eliminated from the database, the value of r^2 decreased (Table 8.3B). This inverse correspondence with distance of wetland from the coast is interpreted to be the result of the increasing importance of throughflow nearer the coastal discharge zone. This aspect of groundwater recharge is further discussed in Section 8.4.2.

Table 8.3B *Decreasing linear regression values for wetlands ordered from inland to coast*

Site	r^2
161	0.72
162	0.66
WAWA	0.60
135-2	0.56
35-4	0.51
swii	0.32
1N	0.17

8.3.1 Recharge pertaining to specific rainfall events

The spatial and temporal variability in infiltration rates is outside the scope of this study, however, the response of the groundwater to a specific rainfall event was documented to illustrate the variable recharge from wetland to wetland across the

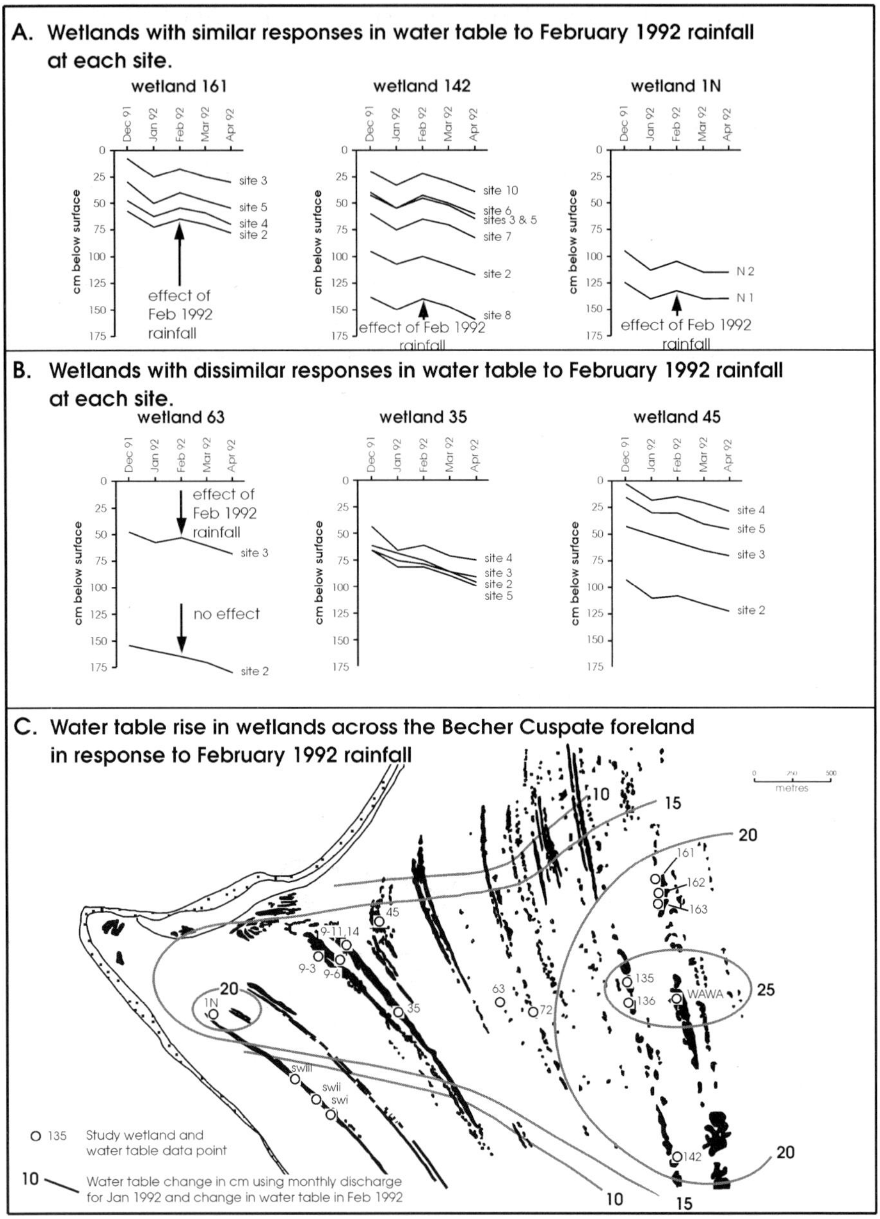

Figure 8-9. *Water table responses to aseasonal rainfall event in February 1992 - water level responses at various sites, and interpreted water level response sub-regionally.*

cuspate foreland. For this purpose, the single aperiodic rainfall event in the latter part of the dry season in February (1992), recording approximately 180 mm in 24 hours, was selected. February is at the end of summer and in the middle of the dry period. Examples of the monthly water table response, (March 1992), are presented in Figure 8-9.

Some wetlands registered similar recharge volumes at all piezometric sites, (161, 142, 1N, WAWA, Fig. 8-9A). Some wetlands registered dissimilar recharge volumes at all piezometric sites, (63, 35, 45, swiii, Fig. 8-9B). The overall response to the aseasonal rainfall was calculated by adding the monthly discharge for January 1992 to the change in water table height in February. These overall changes in the water table were plotted as isolines using data from the study wetlands. The resulting pattern showed that for this particular event, recharge was greater in the inland wetlands, increasing again in the vicinity of Becher Point. In the central part of the cuspate foreland, recharge was not only lower but more variable from site to site (Fig. 8-9C).

8.4 Groundwater under beachridges and wetlands

The hydrological components at the basin scale are rainfall, evaporation, and groundwater, but the processes of recharge and discharge are more complex in the wetland aquifers of mud and muddy sands. In a wetland basin, rainfall may or may not infiltrate the sediments to recharge the groundwater. If groundwater recharge through vertical flow does occur, it may be unimpeded and direct, impeded, or delayed. Groundwater in the wetland basin may also be recharged through lateral flow but only under certain conditions. Evapo-transpiration can be equally complex, and may occur from free standing surface water, from plants with different physiognomy, from sub-surface perched water, from interstitial water in the vadose zone, and from the water table via capillary rise. Groundwater hydrology, *i.e.*, its levels, fluctuations, rates of rise and fall and relationship with groundwater outside the basin, depends on the balance between rainfall recharge, lateral flow and evapo-transpiration discharge, and on the effects of the sedimentary stratigraphic sequence in which it resides.

The focus of this section is the detailed examination of groundwater response with a view to interpreting the causal processes underlying its variability. The framework for analysing groundwater behaviour at the basin scale is the division between beachridge and wetland habitat.

8.4.1 Seasonal changes to surface morphology of water table

The groundwater response to rainfall was measured as water levels and plotted with respect to AHD and then the local land surface (Figs. 8-10). The height of the water table above AHD, under any particular wetland, is determined by its distance from the aquifer's discharge zone, in this case, the shore, but, at equilibrium, this height will be the same under each discrete adjacent beachridge/dune and wetland.

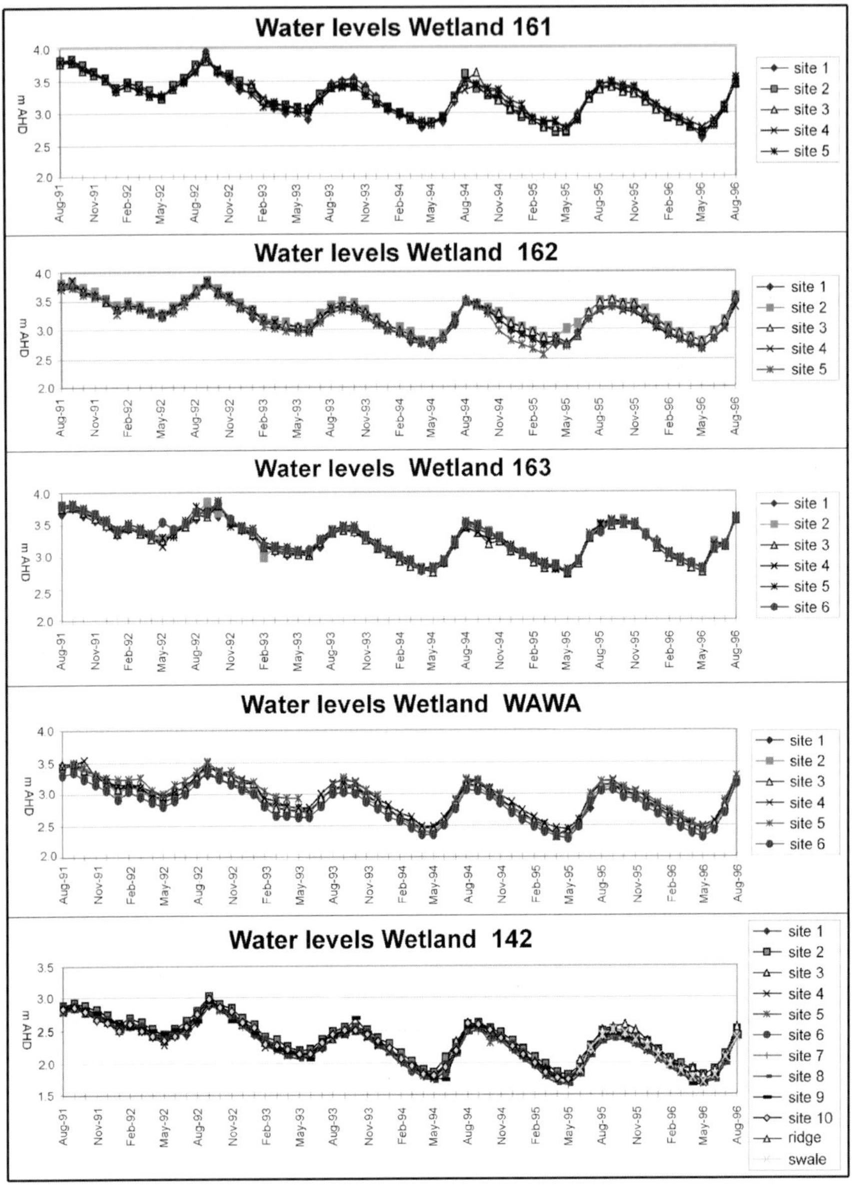

Figure 8-10. *Graphs showing relatively similar water levels and fluctuations under all wetland and beachridge sites (wetlands 161, 162, 163, WAWA, 142).*

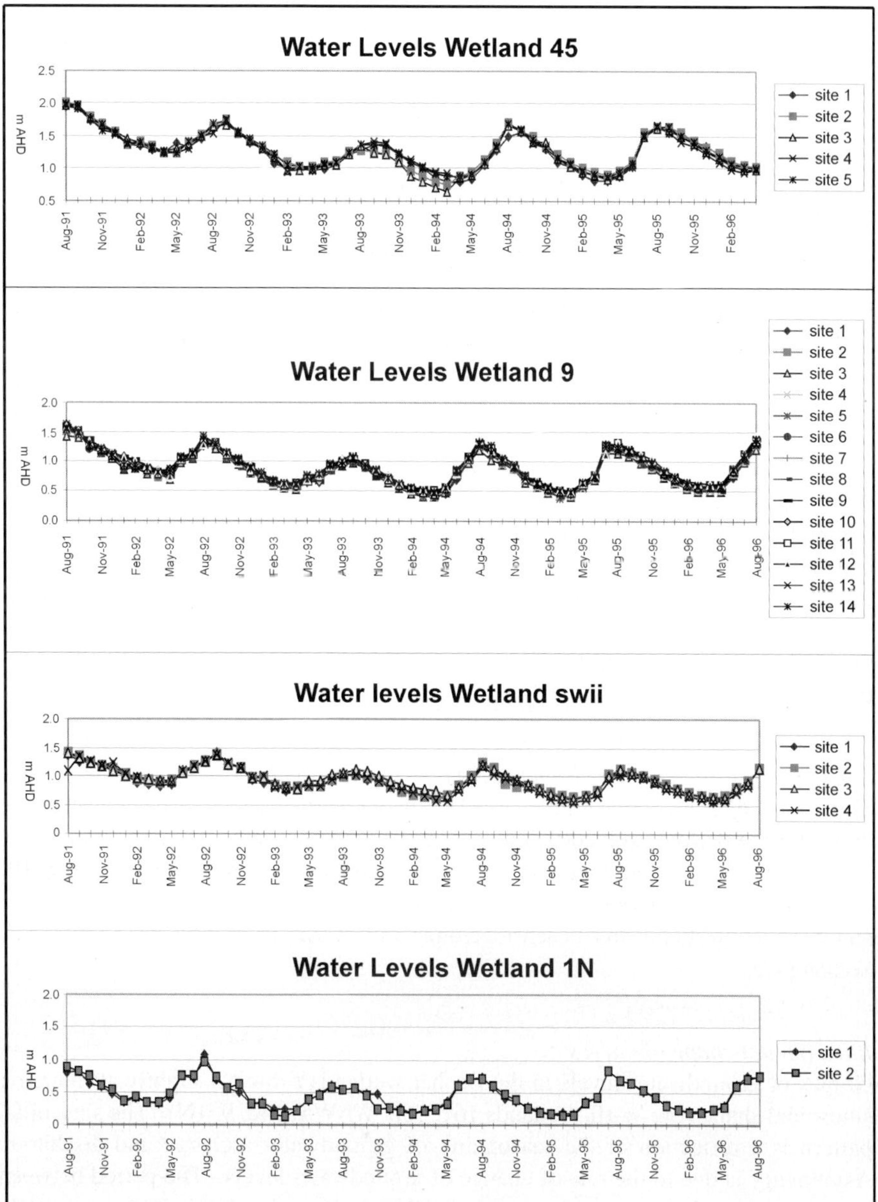

Figure 8-10 (cont.). *Graphs showing relatively similar water levels and fluctuations under all wetland and beachridge sites (wetlands 45, 9, swii, 1N).*

However, in hydrologically dynamic systems, the water table beneath the wetlands and adjacent ridges exhibits small scale mounds, troughs and gradients. Some of these features occur in response to a singular set of conditions, others occur seasonally.

In areas with distinct seasonal rainfall, there are features in the groundwater which relate to the change from summer drought to winter inundation. After the first three or four rain events, the water table is higher under the wetland than the adjacent ridges. This is due to at least four factors: 1) the contrasting topography and differences in depth to the water table, 2) the differences in the width of the zone of capillary rise and its depth below the ground surface, 3) the groundwater flows in the shallow subsurface at the wetland/ridge margins, and 4) the differences in volume of pellicular water in sands and muds. Water infiltrating through dunes needs to be accommodated through 3-6 m of sand as pellicular water before water table rise, whereas rain directly saturates wetland sediments often causing rapid water table rise. There is commonly a slight mound under one of the wetland sites indicating more rapid infiltration. At the end of the rainfall season, which approximates the latter part of spring, the water table may be higher under the beachridge/dune due to the time lag in ongoing groundwater recharge and the lower evapo-transpiration. There is a low east/west gradient evident in Autumn. In years with below average rainfall, there may be depressions in the water table at the wetland margins, (September/October), west/east gradients, and depressions under the wetland in the late spring or late summer growth period due to renewed evapo-transpiration.

8.4.2 Hydrographs under beachridge/dunes and wetlands

Water level measurements relative to AHD, under wetlands and adjacent beachridge/dunes for the period 1991-2001, showed that at the temporal scale of one month, the wetland and beachridge/dune water levels were generally synchronous (Figs. 8-10, 8-11). Small scale temporary disequilibria in water levels (15 cm) within the wetland and between wetland and beachridge/dune did occur; these will be discussed in Section 8.4.5.

Intra annual shape of curves

Graphs of groundwater levels in the Becher wetlands exhibit a slightly asymmetric sinusoidal shape (Fig. 8-10, wetlands 161, 163, WAWA, 142, 9, 1N). The sinusoidal pattern is a reflection of the seasonality of groundwater recharge and discharge. Asymmetry is due to the rate of change of groundwater levels. The period between initiation of water table rise and attainment of the maximum level is short, approximately 3-4 months. Incrementally there is an overall lowering of the water table taking between 7-10 months to reach minimum levels. Major aberrant rainfall occurrences, such as February 1992, result in a slight upward fluctuation in the incremental downward change in water level, a flattening of the slope, and a less extreme annual minimum level. Minor aberrant rainfall, such as March 1993, are not evident at this scale.

The mean annual water level fluctuations for wetland sites east to west are presented in Table 8.4 and centre around 0.70 m, however a difference of 1.0 m is not uncommon. Although the mean decrease in groundwater levels is 9 cm per month, rates two to four times this amount commonly occur between October and February.

Table 8.4 *Mean annual water level fluctuations*

Wetland Site	*Mean and standard deviation*	*Number of years of observation*
161	0.76 ± 0.16	n=10
162	0.75 ± 0.14	n=10
163	0.73 ± 0.12	n=10
WAWA	0.74 ± 0.11	n=10
135	0.74 ± 0.18	n=10
136	0.76 ± 0.16	n=10
142-3	0.76 ± 0.13	n=10
142-6	0.78 ± 0.14	n=10
72	0.55 ± 0.14	n=8
63	0.60 ± 0.10	n=7
45	0.72 ± 0.65	n=5
35	0.73 ± 0.85	n=10
9-14	0.76 ± 0.9	n=10
9-6	0.75 ± 0.9	n=10
9-3	0.74 ± 0.10	n=10
1N-1	0.57 ± 0.12	n=10
1N-2	0.60 ± 0.18	n=10
swi	0.52 ± 0.11	n=10
swii	0.51 ± 0.8	n=10
swiii	0.53 ± 0.8	n=10

Inter annual pattern - trends 1991-2001

Longer term (1991-2001) annual water levels (Fig. 8-11) all exhibited the same trend of a decline of water tables to their lowest position and thereafter a slight upturn.

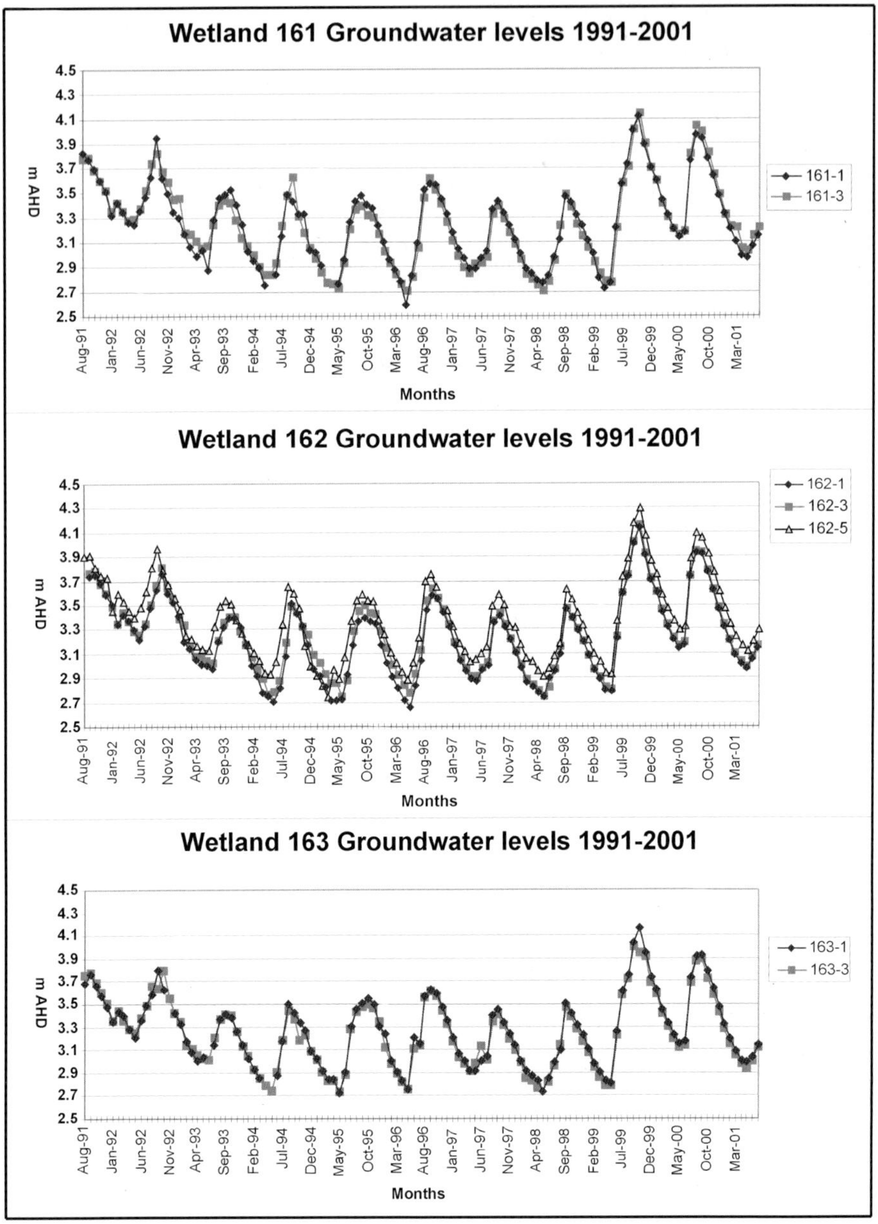

Figure 8-11. *Trends in water table maxima and minima under wetland and adjacent beachridge sites for ten years, 1991-2001 at wetlands 161, 162, 163.*

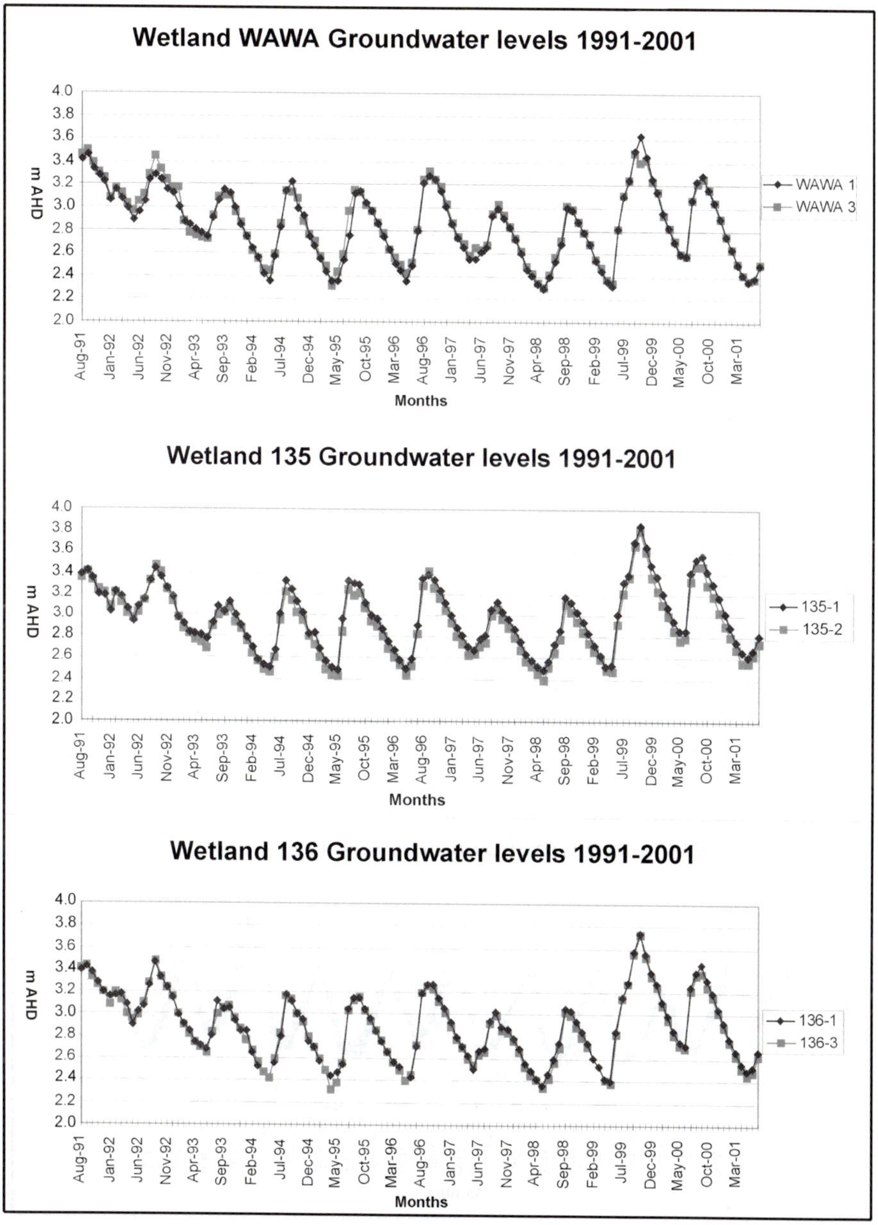

Figure 8-11 (cont.). *Trends in water table maxima and minima under wetland and adjacent beachridge sites for ten years, 1991-2001 at wetlands WAWA, 135, 136.*

Table 8.5 Comparison of water levels in shallow and deep nested piezometers in different seasons

Site	*Spring*		*Summer*		*Autumn*		*Winter*	
163 east	! s > d early spring s < d late spring (7 cm)	downward recharge leakage up	s < d (9 cm)	leakage up	s < d (6 cm)	leakage up	s = d early winter s > d late winter	downward recharge
163 west	s > d	downward recharge	s > d	downward recharge	s > d	downward recharge	s > d	downward recharge
135 east	s ≤ d (1 cm) s=d	leakage up	s ≤ d (1 cm)	leakage up	s > d	downward recharge	s > d	downward recharge
135 west	*s = d s > d	downward recharge	s ≤ d (1 cm)	leakage up	s = d		s = d	
35 east	s < d (4.5 cm)	leakage up	s < d (5 cm)	leakage up	s < d (4 cm)	leakage up	*s < d s > d (1.5 cm)	leakage up downward recharge
35 west	s < d (3 cm)	leakage up	s < d (4 cm)	leakage up	*s < d s > d (5 cm)	leakage up	s < d (3 cm)	leakage up
swii east	s < d (1 cm) s = d	leakage up	s = d	leakage up	s = d		s > d early winter s < d late winter	downward recharge leakage up

s = shallow piezometer, d = deep piezometer; ! s > d denotes the water level in the shallow pipe is higher than in the deeper pipe; * dominant condition

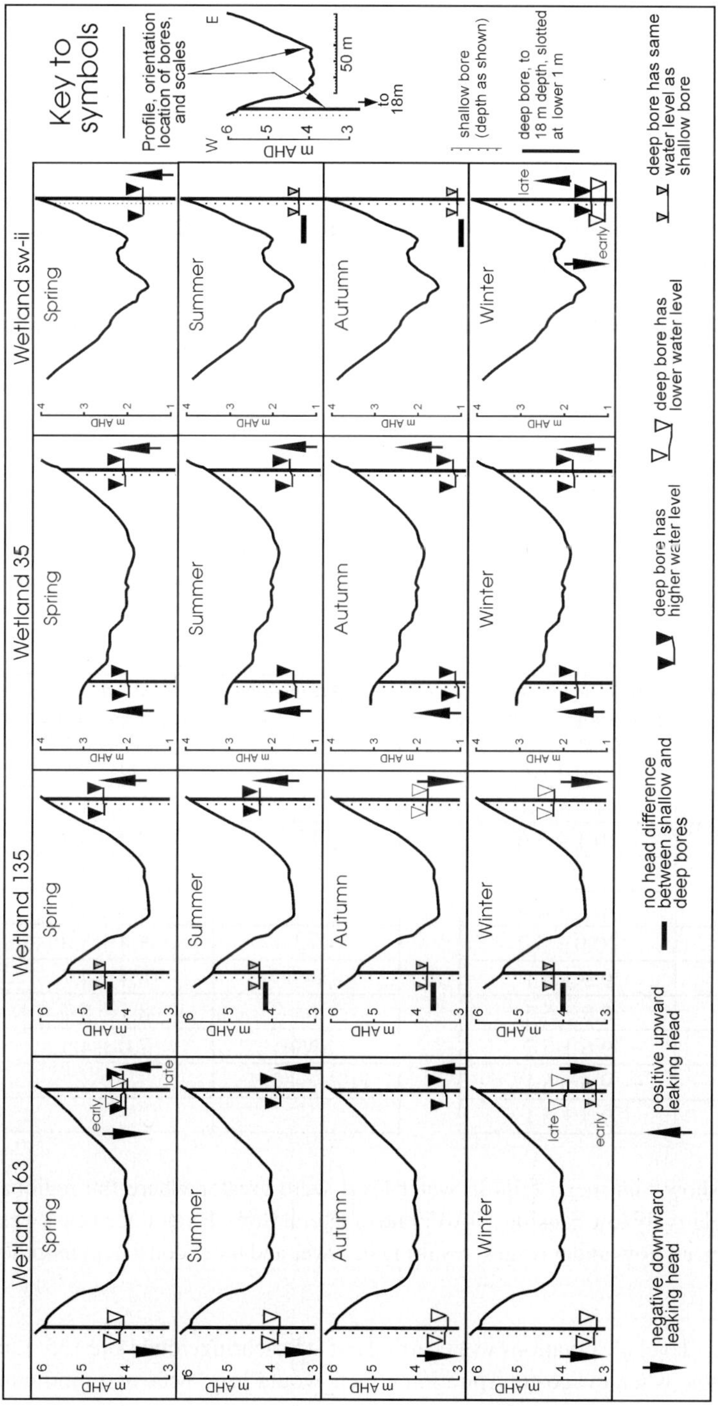

Figure 8-14. *Water levels in shallow and deep nested bores, east and west of selected wetlands 1999-2001.*

The conclusion drawn from the above data is that flow into and out of these wetland basins is initiated and directed by hydrological mechanisms controlled by seasonal conditions. The greatest effect of upward leakage occurred in spring and on the eastern side of any wetland in conjunction with a strong east/west gradient, *e.g.*, 163, 35 (Fig. 8-14). On the western side of the wetland throughflow continued as downward leakage. This process reached its maximum in summer. In autumn, all types of flow were reduced and in most cases, water levels could be viewed as approaching stasis. In winter, except during periods of infrequent rain, the dominant flow was vertically downward, driven by rainfall infiltration.

Water level falls under the beachridge/dunes were examined to discern discharge rates and patterns. The monthly discharge rate was such that mean falls in water levels were 9.4 cm/month between Aug 1991 and Aug 1996 (Table 8.6). Above average falls occurred in December and January when evaporation reached its maximum. In addition, coastal and central beachridge/dunes, (swii, swiii and 9, 35, 45 respectively), exhibited large falls in water level immediately following maximum groundwater levels, *i.e.* October and November, when local hydraulic gradients were steeper.

Table 8.6 *Mean monthly water level fall in groundwater under beachridge/dunes between October and April (in decreasing order).*

Beachridge/dune	*Mean water level falls above the mean (cm/month) 1991-1996*
161	11.6 ± 7.1
9-12	10.1 ± 5.6
9-4	10.1 ± 4.8
163	10.0 ± 4.4
35-6	10.0 ± 5.2
9-9	9.9 ± 5.1
142-9	9.9 ± 5.5
162	9.6 ± 5.2
9-1	9.4 ± 6.1
45	9.4 ± 6.1

Beachridge/dune	*Mean water level falls below the mean (cm/month) 1991-1996*
WAWA 1	9.3 ± 4.5
142-1	9.0 ± 4.2
136	9.0 ± 4.0
135	8.7 ± 4.9
72	8.4 ± 5.0
swii	8.4 ± 4.7
63	8.1 ± 5.7
swiii	7.0 ± 4.5

These results show that mean falls in water level were greater where the regional gradient was relatively low causing a lower rate of lateral flow. Pronounced east/west gradients enhanced flow under ridges, resulting in lower and more consistent monthly falls.

A series of water level observations were carried out at beachridge/dune site 135-1, in May 1996, a time when evapo-transpiration effects would have been at a minimum (water level was 3.25 m below the surface and diurnal temperature and wind were low to moderate). Water levels were monitored daily for 16 days, during which time they dropped from 2.57 m to 2.52 m AHD. At the time of monitoring, the local gradient was 1:5 and the rate of fall in water level was 0.5 cm/day to 0.2 cm/day. As evapo-transpiration is unlikely to have been the cause of the fall in water level, these figures are considered to indicate the effect of lateral flow.

In this study no attempt was made to separate effects of evaporation and transpiration on groundwater levels, but an area cleared for urbanisation in the northeast of the main study area provided opportunity to examine water level changes under unvegetated ridges and wetlands where evapo-transpiration would be negligible (Fig. 8-6). By comparing water level contours under cleared areas with those under nearby vegetated areas, the effect of transpiration could be quantified locally. The cleared area is located above the zone where the unaltered water table should have been 3.0-3.2 m AHD. As a result of clearing, the water table now resides between 3.26-3.60 m AHD, a rise of 6-60 cm. This is a measure of the transpiration effect of the heath and low shrub vegetation colonising the beachridges, and the sedge and shrub vegetation colonising the wetlands.

8.4.4 Intra-basin - groundwater under the wetlands

For the period of the study, the maximum position of the water table in the majority of wetland sites was 0-0.6 m below ground, and the minimum position was 0.6-1.2 m (Fig. 8-15). The zone of capillary rise in the calcilutite was 30-60 cm. Inundation occurred infrequently, most commonly in 1991, 1992, 1999, and 2000. The regularly inundated wetlands included 161, 162, WAWA, 135, 9-6,14 and swiii.

Groundwater levels under the wetlands rose and fell seasonally as a result of meteoric recharge, lateral flow, and evapo-transpiration, however, there were several paths of meteoric water movement in wetland environments, as described below:

1. Rainwater was perched by impermeable surface sediments until completely discharged by evaporation.
2. Rainwater infiltrated the surface sediments, then became perched or slowed by impermeable sediments in the shallow subsurface. A portion of this water was discharged by evapo-transpiration with a portion further infiltrating the sediments either to be stored there as interstitial water or to recharge the water table.
3. Rainwater infiltrated the sediments, then percolated to the water table causing groundwater to rise. Groundwater rose within the profile and above the surface, in both cases to be discharged by evapo-transpiration and lateral flow.

As well as occurring in different wetland basins, each of these processes occurred in different parts of the same wetland.

Rainwater, perched by impermeable surface calcilutite lasted 1-4 weeks before being completely discharged by evaporation *e.g.*, wetlands 9-3, 136 (Fig. 8-16). This occurred more commonly in the late autumn or early winter, but also after heavy rain following a number of rain free days.

Rainwater infiltrated the surface organic enriched sediment to 10 cm, and became perched or slowed by the underlying calcilutite *e.g.*, wetlands 161, 162, 142, 135, swiii, Cooloongup A, B, C. This was demonstrated using nested piezometers in the centres of wetlands 161, 162 and 135 to record winter water levels. Results are shown relative to AHD in (Table 8.7). Sub-surface perching and retardation were more common than surface water ponding.

Table 8.7 *Comparison of water levels in shallow and deeper nested piezometers in three wetlands over two months*

Site	*June water level (m AHD)*	*Difference in piezometric water level*	*July water level (m AHD)*	*Difference in piezometric water level*
161-deep	3.11 (surface 3.59)		3.22	
161-shallow	3.17	0.06 m	3.21	0.01 m
162-deep	3.08 (surface 3.83)		3.17	
162-shallow	3.50	0.42 m	dry	na
135 deep	2.59 (surface 3.44)		2.74	
135-shallow	dry	na	dry	na

When data from shallow and deep nested piezometers were compared for June, it was apparent that water levels in the shallow piezometers for sites 161 and 162 were higher, suggesting that vertical *in situ* infiltration was still in progress and had not yet reached the water table at the time of monitoring (to be registered in the deeper piezometers). Field trials to determine the vertical hydraulic conductivity of calcilutite showed that the mean rate of water penetration in the calcilutite devoid of root structures with a hydraulic head of 10 cm, was 1.26-2.7 cm/day. The lack of a piezometric level in the shallow bore in wetland 135 demonstrates the diversion of rain infiltration to sediment storage in the vadose zone (approximately 50-60 cm). By July, levels in deep and shallow bores in wetland 161 were approaching the piezometric level, while in the other two wetlands the situation remained unchanged.

Levels in 135 further suggest that lateral flow was the major form of recharge in the wetland. This interpretation is supported by the existence of a west/east gradient, discussed further in Section 8.4.5. In most wetlands underlain by calcilutite, the minimum water level resides in the regional Safety Bay Sand aquifer beneath the wetland fill. The preliminary recharge (May/June) to the groundwater is by infiltration via low beachridge/dunes, swales and wetland margins (Chapter 7), rather than infiltration through wetland muds. Thereafter, as the groundwater rises and upper sediments become saturated, a higher percentage of *in situ* infiltration reaches the increasingly

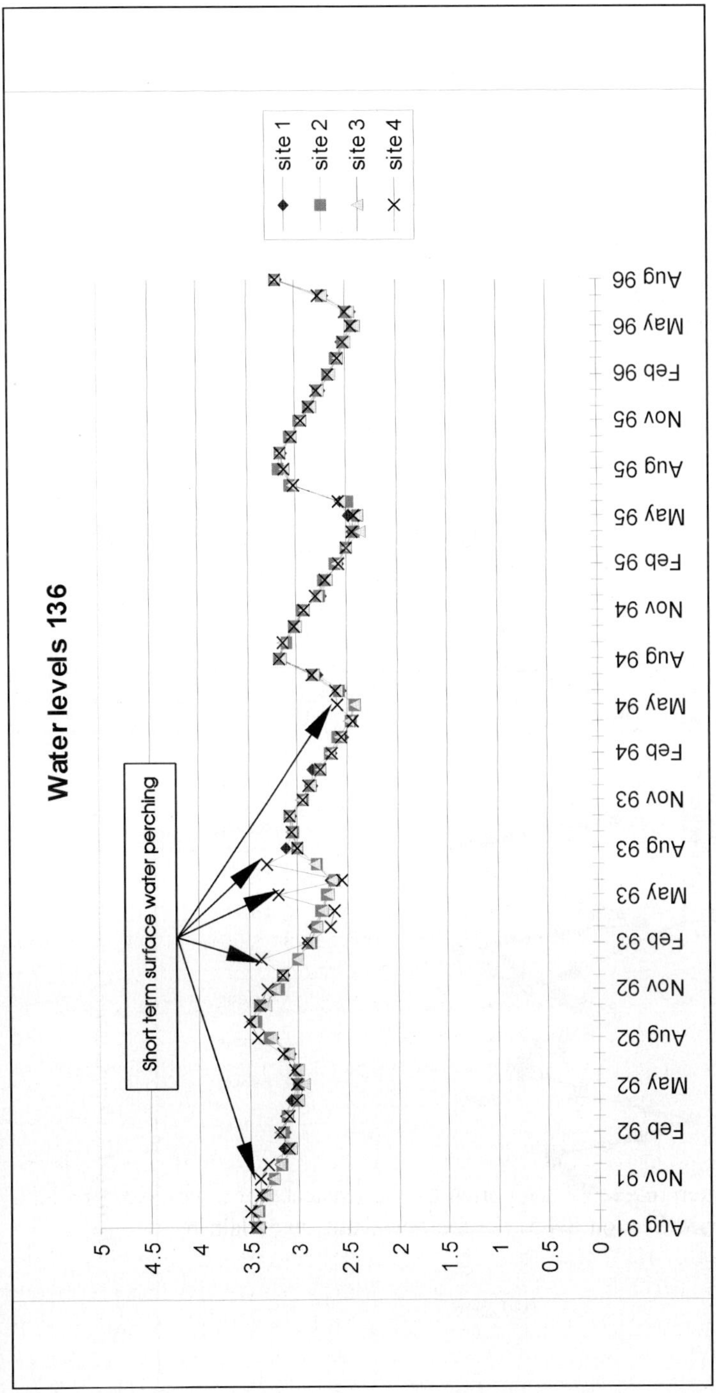

Figure 8-16. *Perched surface water in wetland 136.*

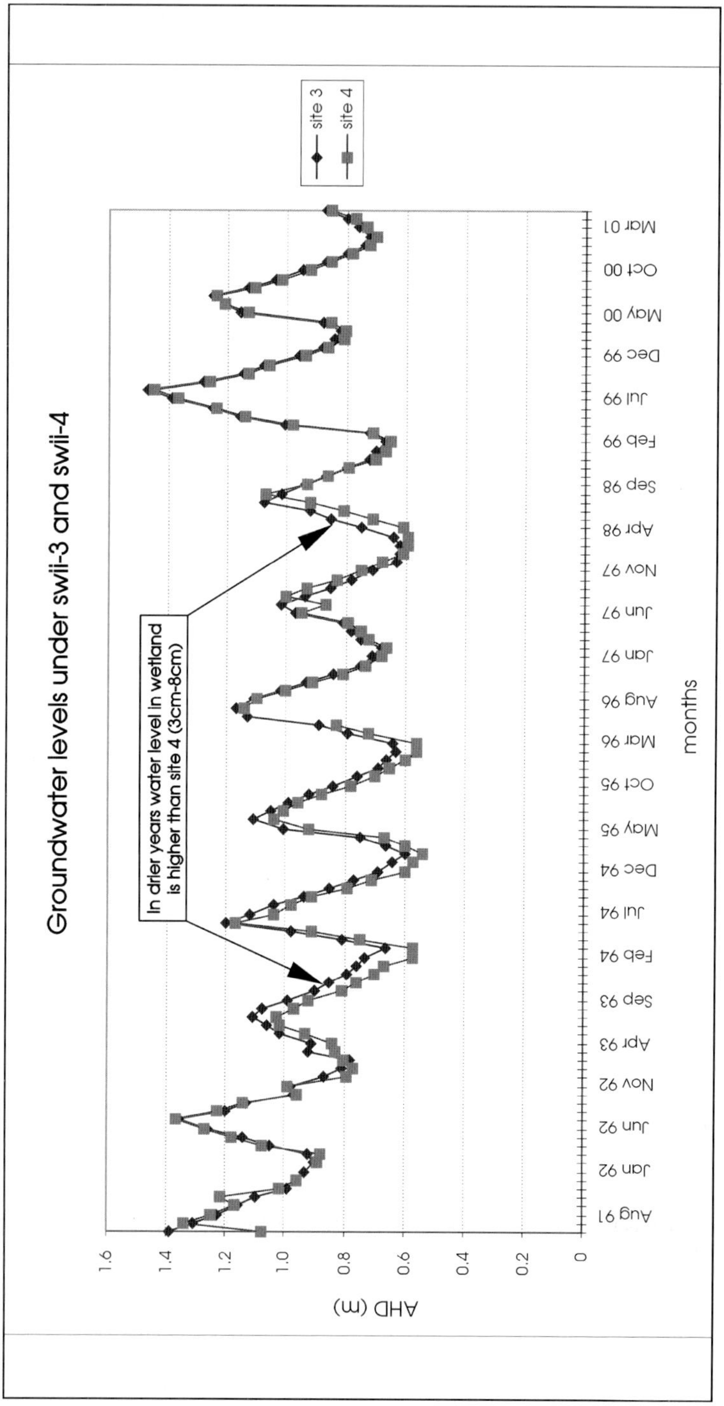

Figure 8-17. *Comparison of water levels under wetland (site 3) and adjacent low ridge (site 4) 9 metres to the east.*

shallow water table, while recharge under adjacent swales (WT = approximately 3 m) remains the same.

Sub-surface perching also occurred above the calcrete layer in wetlands 9-3, 9-6 and above the cemented muddy sand in Cooloongup A2. Retardation of vertical flow was more common at the beginning of winter and during months with low frequency or low volume rainfall. Wetlands with the greatest annual fluctuation were 9, 136, 142, 161, 162, all of which are underlain by calcilutite.

Retardation of vertical percolation also occurred in the peat filled basin (WAWA), because of the capacity of the sediment to absorb and store water. The soil moisture content measured as the ratio (by weight) of water to wet soil ranged between 0.5 in the dry season and 0.8 in the wet season.

In wetlands underlain by sandy mud, muddy sand or sand (163, 72, 63, 45, 9-11, swi, swii, 1N), rainwater infiltrated the sediments and percolated unimpeded to the water table, causing groundwater to rise. In these wetlands recharge was rapid, water table rise occurring regularly in April associated with spasmodic late autumn rainfall, the precursor to the winter rains. In contrast, groundwater recharge in the wetlands underlain by relatively impermeable sediments did not normally occur until May or June. Average annual water level fluctuations in wetlands underlain by sandy mud, muddy sand or sand were also less than for other wetlands, indicating faster discharge.

Comparison of water levels in wetland swii at site 3 (central wetland) and site 4 (9 m to the east) demonstrates the difference between unimpeded recharge to groundwater under wetland sediments (carbonate muddy sand) and under the adjacent low ridge (Fig. 8-17). In most months the water levels were the same under the two sites, but in 1993, 1994, 1995, and 1998, the years of below average rainfall, the water levels under the wetland site were 3-8 cm higher.

The importance of lateral flow differed between the wetlands underlain by permeable and impermeable sediments. In the former type of basin fill, inflow and outflow of water was unrestricted and the wetland was hydrologically "open", lateral flow occurring as long as the hydraulic head had sufficient potential energy. In the latter type of basin fill, inflows and outflows to the central basin were restricted and lower or higher basin water levels, out of equilibrium with regional water levels, occurred. Sometimes these differences were temporary; sometimes they persisted.

Evapo-transpiration was the major discharge mechanism in the wetlands. The greatest falls in water levels occurred when the cessation of winter rain (Oct/Nov) or the period of highest evaporation coincided with the water table being in the rhizosphere (Dec/Jan/Feb). At these times, water levels could fall more than 20 cm in a month (Fig. 8-18). As water levels reached a minimum level during March-May, the monthly incremental fall was reduced to several centimetres and then zero (Fig. 8-18). This is interpreted as the combined effects of reduced evapo-transpiration rates due to groundwater levels lying below the rhizosphere, reduced velocity of lateral groundwater flow due to flattening of local (dune/wetland) gradients, and equilibrium reached in the regional position of the water table relative to AHD.

8.4.5 Piezometric differences between ridges and wetland basins

Temporary to semi-permanent water level differences between ridges and wetlands ensued from the following states:

- differences in depth to water table
- increase in vadose air pressure during infiltration
- differences in the depth and thickness of the zone of capillary rise
- differences in infiltration rates and volumes brought about by variation in sediment characteristics
- preferential pathways for water flow, *e.g.*, rootlets or burrows that act as conduits, stratigraphic contacts at wetland margins
- different evapo-transpiration rates

Differences in depth to the water table affect water levels in two ways: 1) the thickness of the vadose zone, in some measure, determines the proportion of interstitial water stored therein, and 2) different depths vary the length of time for infiltrating meteoric water to reach the water table. These differences are temporary, but repetitive. A minor to substantial water table rise can occur in a monitoring bore extending below the water table when there exists a layer intermediate to the phreatic zone, and a surface saturated by recent rapid rainfall in which the pressure of entrapped air rises (Bianchi and Haskell 1966; Gerla 1992). This phenomenon usually only persists between 1-24 hours (Gerla 1992). Water level differences can also result from the impact of infiltration on sediments in which the depth and thickness of the zone of capillary rise varies (Gerla 1992). If there is very little aerated pore space remaining in the zone of capillary rise, only an incremental volume of water is necessary to attain saturation at atmospheric pressure. The resulting rise in water table can be equivalent to the thickness of the zone of capillary rise and disproportionate with infiltration (Gillham 1984; Gerla 1992). This effect is likely to be semi-permanent. The effect of variation in sediment type on recharge rates and water levels, and preferential pathways for water flow such as rootlets, burrows, and stratigraphic contacts at wetland margins that act as conduits, were discussed in Chapter 7. Different evapo-transpiration rates and volumes predominantly affect the rate of water discharge within the wetland stratigraphic sequence, and, at specific times of the annual cycle.

When any of the conditions, outlined above, prevail, the results are small scale changes in the morphology of the water table. Examples of morphological changes to the water table include 1) mounds, 2) troughs, and 3) reversal or sublimation of east/west gradient.

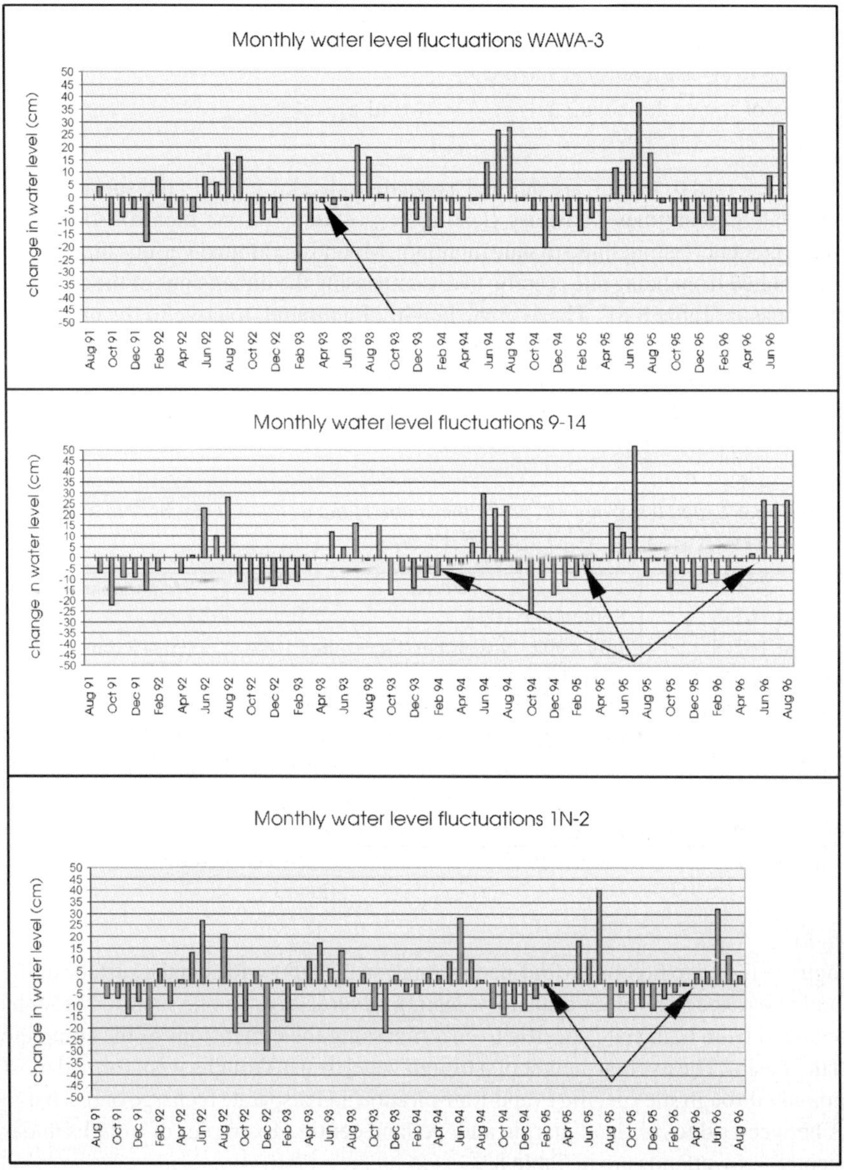

Figure 8-18. *Monthly discharge rates in groundwater in three wetlands. Arrows indicate declining discharge as water levels fall below the zone of evapo-transpiration and regional gradients flatten.*

Examples to illustrate the differences between wet and dry years, 1992 and 1994, with respective annual rainfalls of 964 mm and 697 mm, are presented for selected wetlands (Tables 8.8 to 8.12, Figures 8-19 to 8-25).

Mounds

Mounds, referred to herein, are defined as small scale elevations in the surface of the groundwater, sometimes transitory (lasting several days to one month), sometimes semi-permanent (lasting three to nine months). Mounds exhibited various dimensions as they ranged from being site specific to encompassing the dimensions of the complete wetland basin (Table 8.8). The average height of a mound, relative to the prevailing level of the subregional groundwater table, was 10 cm. Mounds were formed most commonly as a result of disparate recharge rates where juxtaposing sediments had different permeability characteristics. Mounds were evident under wetlands at all sites and in every year of monitoring.

Table 8.8 *Examples of mounds under the centre of the wetland (Figs. 8-19 to 8-25)*

Site	*Period*	*Height of mound*
wetland 161-3	September 1994	25 cm
wetland 162-3	July 1994, November-December 1994	5 cm, 12 cm
wetland WAWA-3	April-May 1992	7 cm
wetland 35-3, 4	May 1992, January to May 1994, September 1994	7 cm, 5-7 cm 5-10 cm
wetland swii-3	December 1992, February-April 1994, June-July 1994, October-November 1994	2 cm, 10-15 cm, 7-10 cm, 5-7 cm
wetland 9-3	August 1992	3-5 cm

Troughs

Troughs, referred to herein, are defined as small scale depressions in the surface of the groundwater; they are often transitory, lasting several days to one month. Troughs also ranged from being site specific to encompassing the dimensions of the complete wetland basin. The average height of a trough was 10 cm. Troughs were formed most commonly through site specific evapo-transpiration and disparate recharge or discharge rates between ridge, wetland margin and wetland centre. Examples of troughs under the centres of wetlands are in Table 8.9.

Reversal and reduction of regional gradient

The regional groundwater gradient slopes downward from east to west. However, there are instances when the gradient across a wetland from ridge to ridge is oriented the opposite way, west to east. This is termed herein a "reverse gradient". In the situation where the groundwater under both ridges and the intervening wetland is level, and the regional groundwater gradient is obscured, the term "reduced" gradient" is used. Reverse gradients persisted, lasting six to twelve months, whereas reduced gradients were most often transitory lasting up to one month. The height difference at either end of the gradient was 2-10 cm. Reverse gradients commonly formed where there was a significant height difference between the east and west ridges bordering the wetland. Examples of reversal and reduction of regional gradient are in Table 8.10.

Table 8.9 *Troughs in the water table under wetlands (Figs. 8-19 to 8-25)*

Site	*Period*	*Height of trough*
wetland 162-3	October 1992	3 cm
wetland WAWA-3	November 1994	3-10 cm
wetland 35-3, 4	January 1992, July 1992, October 1992 November 1994	10 cm, 3-5 cm, 3-7 cm 2 cm
wetland swii-3	September 1992 June 1992 August 1994, December 1994	3 cm, 3-5 cm 7 cm, 5 cm
wetland 9-3	February-March 1992, May-August 1992, January-March 1994, August-September 1994	5-8 cm, 5 cm, 3-5 cm, 10 cm
wetland 9-6	January-February 1992, July 1992, April 1994	3 cm, 3 cm 3 cm
wetland 9-11	January-February 1992, December 1992, March 1994, October 1994	5 cm, 8 cm 3 cm, 3 cm

Table 8.10 *Examples of reversal and reduction of regional gradient under wetlands and beachridge/dunes (Figs. 8-19 to 8-25)*

Site	*Reversal of gradient*	*Piezometric height differential*	*Reduction of gradient*
wetland 161-3			January 1994
wetland 162-3	July 1992 August-October 1994	2 cm 2-5 cm	
wetland swii-3	June 1992, January 1994 July-October 1994 December 1994	2 cm, 7 cm 2-7 cm 5 cm	
wetland 9-3 wetland 9-6 wetland 9-11	December 1992 November 1992	7 cm 10 cm	 August 1994 August 1992 July, December 1994

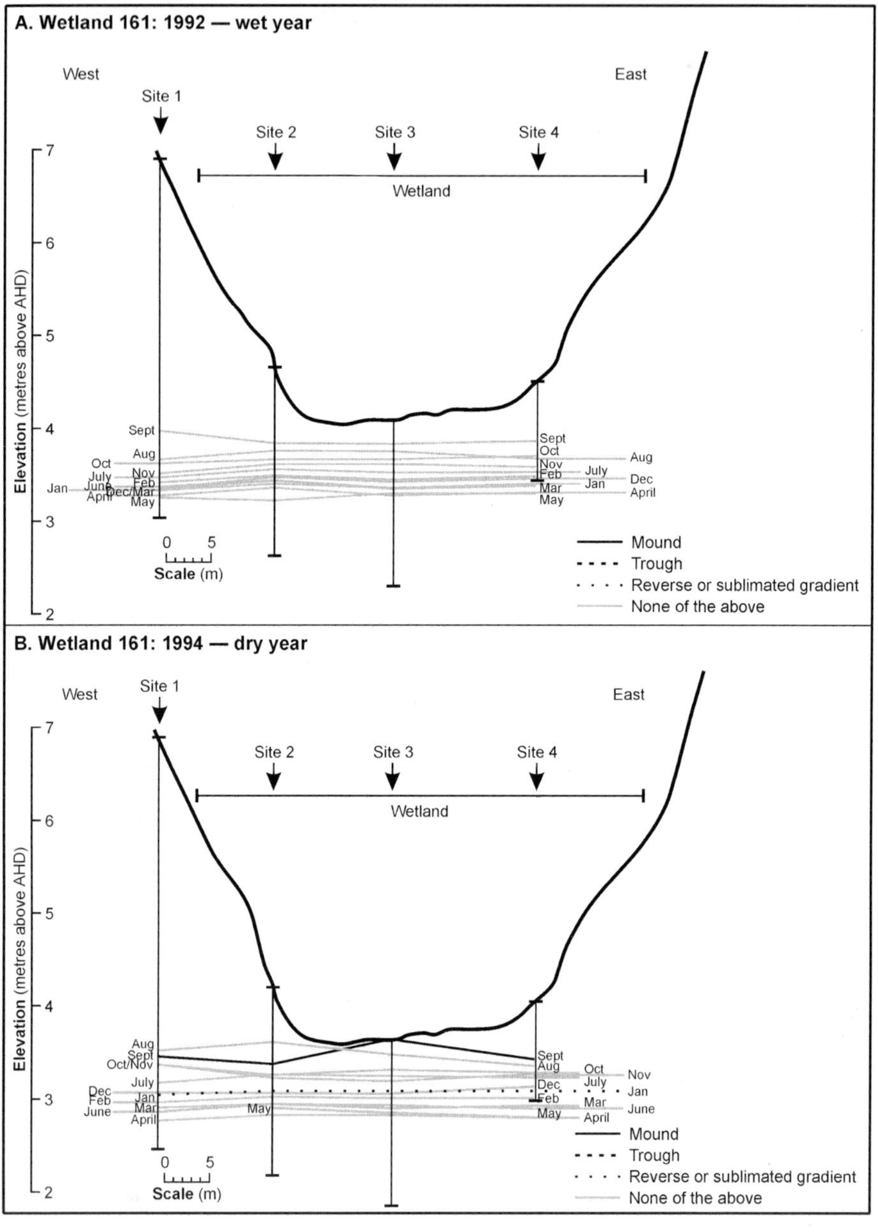

Figure 8-19. *Changing morphology of the water table under wetland 161 in a wet and dry year.*

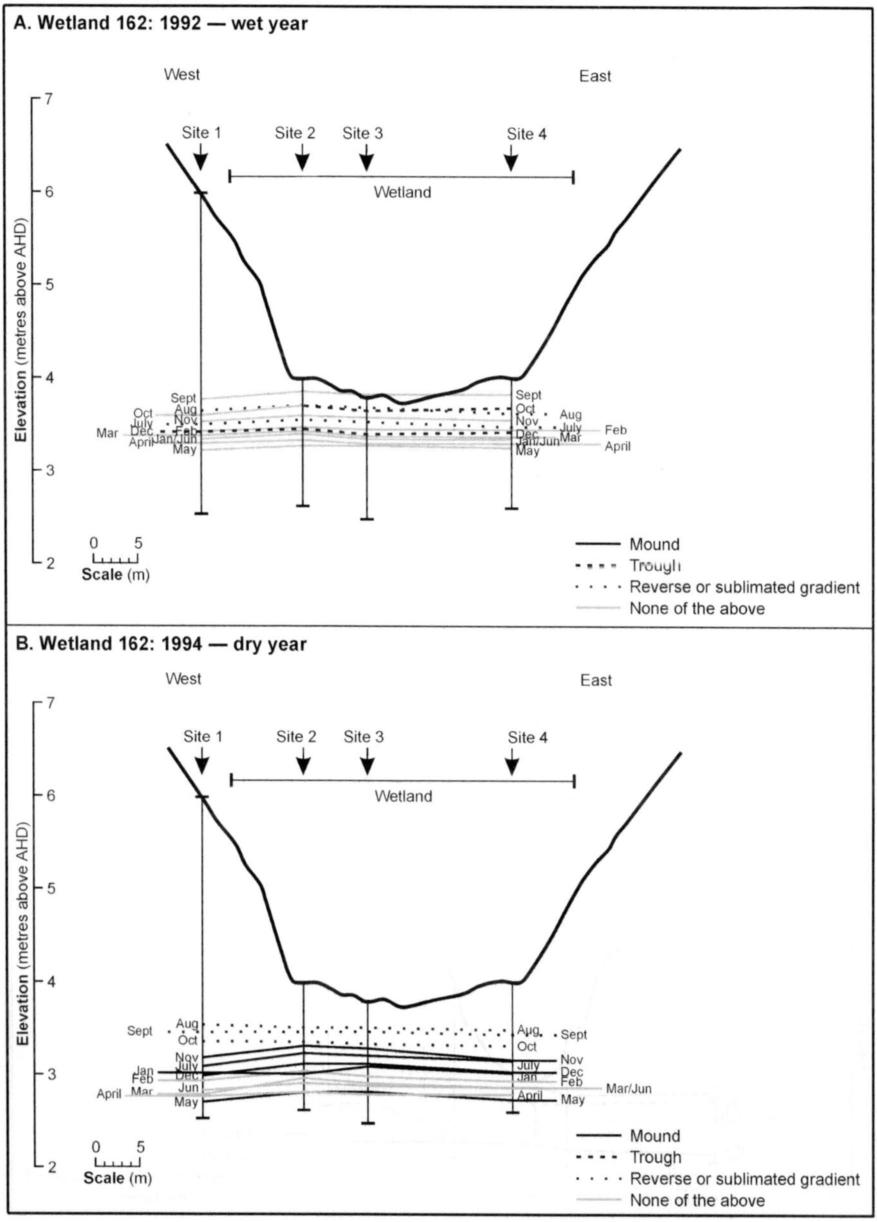

Figure 8-20. *Changing morphology of the water table under wetland 162 in a wet and dry year.*

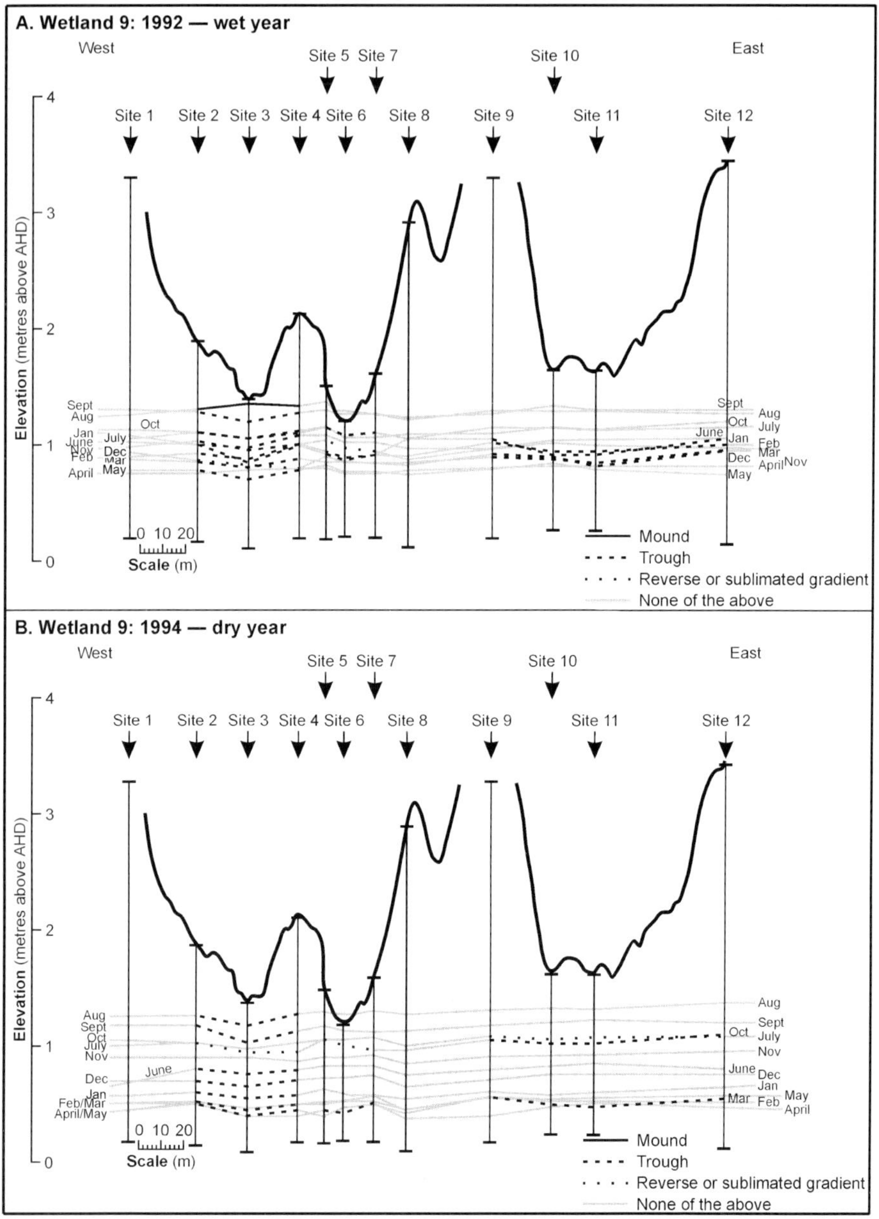

Figure 8-23. *Changing morphology of the water table under wetland 9 in a wet and dry year.*

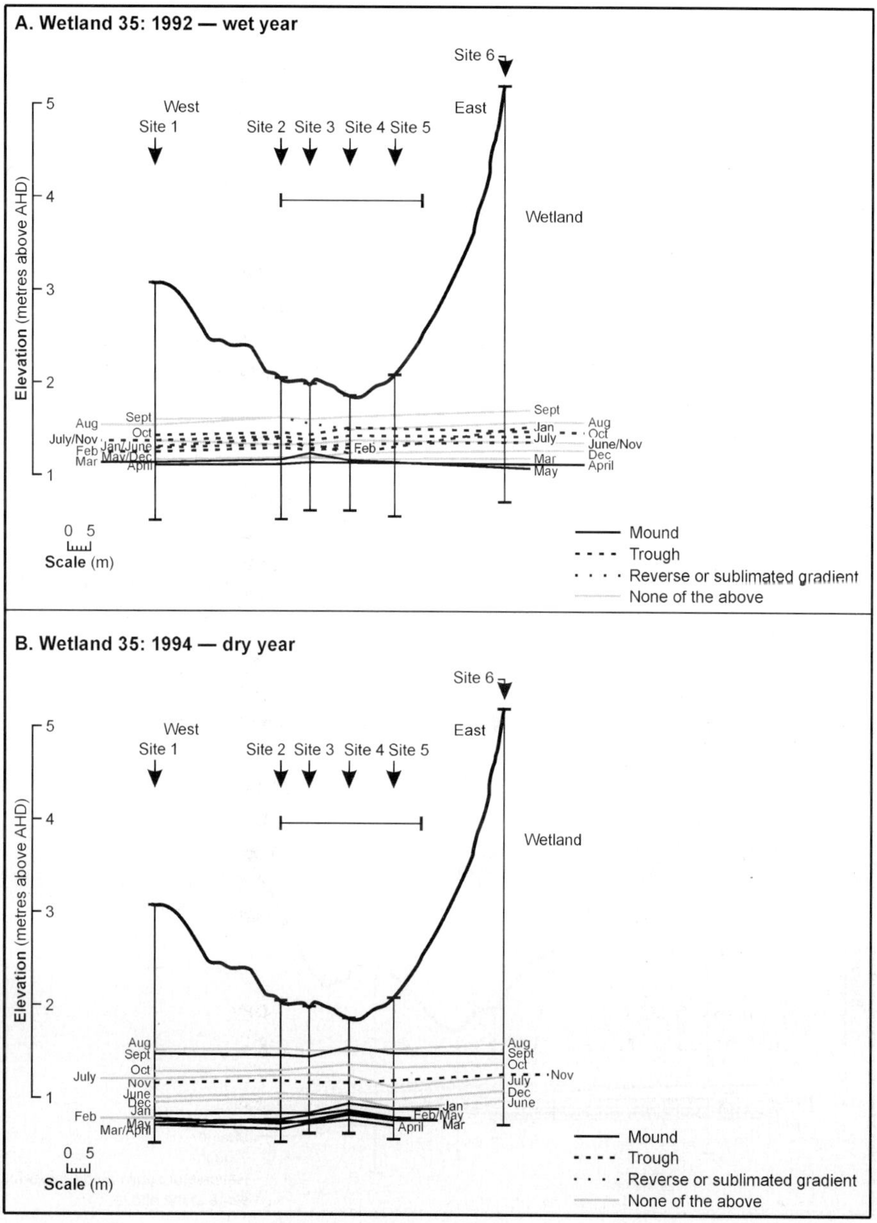

Figure 8-24. *Changing morphology of the water table under wetland 35 in a wet and dry year.*

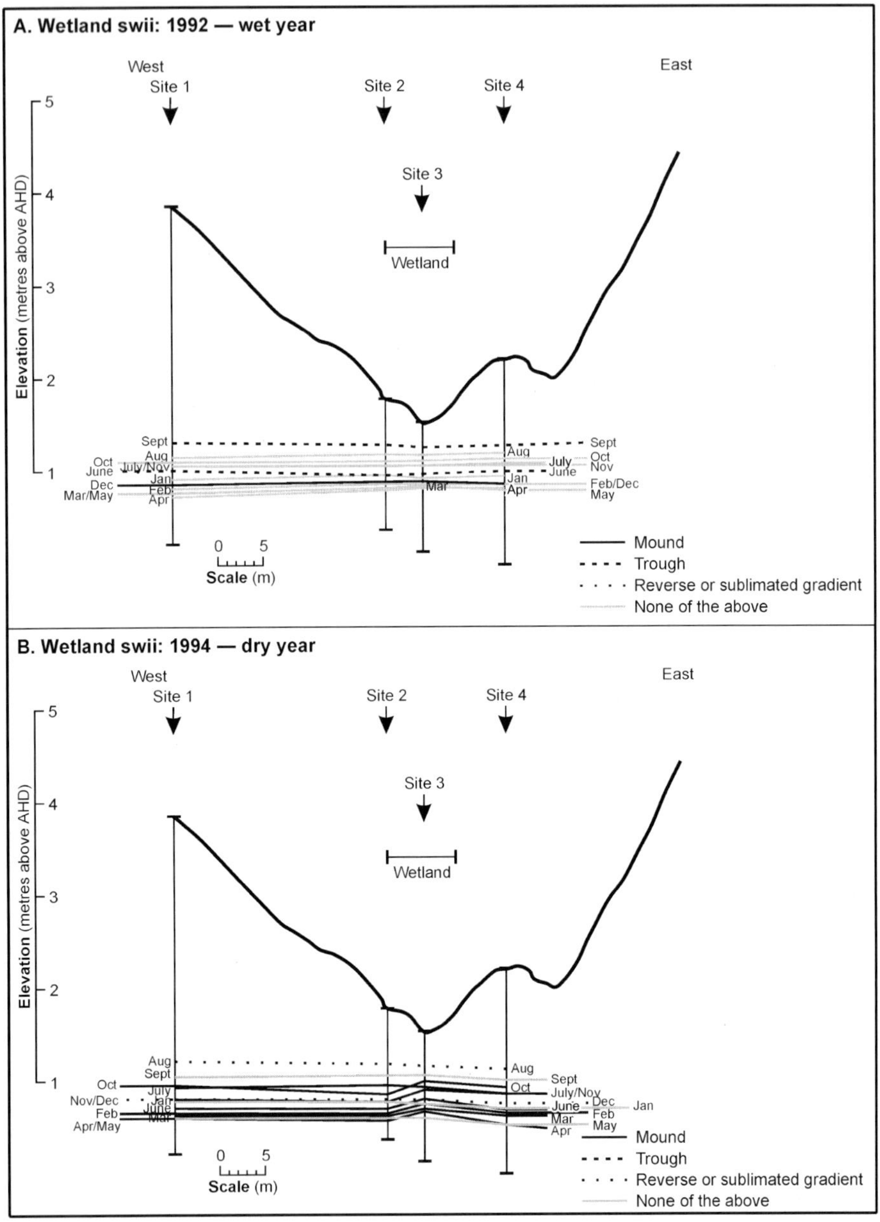

Figure 8-25. Changing morphology of the water table under wetland swii in a wet and dry year.

Table 8.11 *Description of water level responses and water table morphology under relatively wet conditions 1992 (Figs. 8-19 to 8-22)*

Month	*161*	*interpretation*	*WAWA*	*interpretation*	*135*	*interpretation*
January	mounds under wetland margins (5 cm) w > d	rapid recharge along w/d contact	E/W gradient (15 cm) water level under site 2 and d depressed w=d	western lateral flow from Ed	E/W gradient (4 cm)	
February	water level rises at all sites no change to water table surface water level depressed under w and d (5 cm) w=d	recharge to groundwater evapo-transpir-ation recharge lag under d	water level rises at all sites greater than average rise under d E/W gradient (7 cm) w=d	variable recharge to groundwater after rainfall recharge to all sites except Ed	water level rises at all sites level under both sites	high rainfall higher than average recharge
March	water level falls at all sites; no change to water table surface; mound under site 2 w=d		water level falls at all sites except Ed; E/W gradient (19 cm) w > d	Recharge from Feb rainfall to Ed	water level falls at all sites change to W/E gradient w < d (5 cm)	evapo-transpiration under *M. rhaphiophylla*
April	water level falls at all sites; no change to water table surface; mound under site 2 persists w=d	discharge by lateral flow	water level falls at all sites; E/W gradient (7 cm) mound under wetland increases (5-7 cm); w > d	recharge from sporadic pre-winter rainfall in central wetland; storage of infiltration in vadose zone under other sites	w < d (5 cm)	evapo-transpiration under *M. rhaphiophylla*

Table 8.11 (cont.)

Table 8.11 (cont.)

Month	*161*	*interpretation*	*WAWA*	*interpretation*	*135*	*interpretation*
May	water levels remain constant except under site 2 where it falls w > d (5 cm)	seepage into wetland	water level falls at all sites; no major change to water table surface; mound under wetland (5-7 cm) E/W gradient (10 cm)	recharge from sporadic pre-winter rainfall in central wetland; storage of infiltration in vadose zone under other sites	water level falls at all sites; change from W/E gradient to almost flat w > d (3 cm)	
June	water level rises at all sites; mound under site 2 w=d	recharge to groundwater; rapid recharge along w/d contact	water level rises at all sites; mound under wetland reduced; E/W gradient (18 cm)	recharge to all sites variable recharge rates	water level rises at all sites; w < d (6 cm)	recharge to all sites; more rapid recharge under d
July	water level rises at all sites; water level depressed under d; slight mound under site 2 persists; w > d (5 cm)	variable recharge rates	water level rises at all sites; water levels in wetland below gradient; E/W gradient (15 cm) w > d		level under both sites	

Table 8.11 (cont.)

Table 8.11 (cont.)

Month	*161*	*interpretation*	*WAWA*	*interpretation*	*135*	*interpretation*
August	water level rises at all sites; mound under site 2 and wetland; w > d (10 cm)	rapid recharge along w/d contact and into saturated sediments of wetland	water level rises at all sites; E/W gradient (10 cm); w > d (3 cm)	seepage from ridge to wetland	water level rises at all sites; level under both sites	
Septem-ber	water level rises at all sites; water level slightly depressed under wetland; mound under d w < d (15 cm)	*cf* Oct., Nov., 1994	water level rises at all sites; E/W gradient (20 cm); w > d (15 cm)	greater than average recharge; seepage from ridge to wetland; direct recharge to surface water table	water level rises at all sites w > d (3 cm)	higher than average recharge; possible short term perching
October	water level falls at all sites; fall under d greater than other sites; E/W gradient w > d (5 cm)	flow to site 2	water level falls at all sites; mound under E wetland margin; E/W gradient (19 cm) w > d (17 cm)	seepage from E dune to wetland margin	water level falls at all sites; E/W gradient w > d (5 cm)	

Table 8.11 (cont.)

Table 8.11 (cont.)

Month	*161*	*interpretation*	*WAWA*	*interpretation*	*135*	*interpretation*
November	water level falls at all sites; slight mound (3 cm) under site 2 and wetland; water level depressed under d $w > d$ (10 cm)	rapid recharge along w/d contact	water level falls at all sites; no change under ridges; E/W gradient (21 cm) $w > d$ (10 cm)	delayed recharge to water table under beachridges; higher than average evapo-transpiration in wetland	water level falls at all sites level under both sites	
December	greater than average water level falls at all sites mound under site 2 persists water level depressed under d $w > d$ (10 cm)	evapo-transpiration	water level falls at all sites fall under site 2 and d greater than other sites	western lateral flow from Ed	water level falls at all sites W/E gradient $w < d$ (4 cm)	evapo-transpiration under *M. rhaphiophylla*

d western beachridge/dune
Ed eastern beachridge/dune
$w = d$ the water levels under the centre of the wetland and the western beachridge/dune are at the same level (AHD)
$w > d$ the water levels under the centre of the wetland are higher than those under the western beachridge/dune (AHD)
w/d contact the contact between beachridge/dune sediments and wetland sediments
site 2 refers to the site at the western margin of the wetland

Table 8.11 *Description of water level responses and water table morphology under relatively wet conditions 1992 (Figs. 8-23 to 8-25)*

Month	*35*	*interpretation*	*swii*	*interpretation*
January	water level depressed under wetland; E/W gradient (25 cm)	evapo-transpir-ation in wetland	mound under site 2 (5 cm) w = d	
February	recharge to central wetland; water level falls at other sites	high rainfall; rapid but low recharge to wetland	E/W gradient (5 cm); sl depressed under d; w > d (5 cm)	discharge by lateral flow
March	water levels at all sites fall; E/W gradient maintained	evapo-transpiration	E/W gradient; w > d (7 cm)	discharge by lateral flow
April	water levels in wetland slightly above gradient; E/W gradient maintained		no change in water table surface; E/W gradient (7 cm); w > d (7 cm)	discharge by lateral flow
May	rise in water levels at all sites except Ed; mound under wetland (7-15 cm); continued fall under Ed; W/E gradient	variable recharge rates	recharge to 3 sites; depressed under E margin; mound under wetland (5 cm)	first rains; rapid recharge in central wetland where groundwater is shallow
June	water level rises at all sites; trough under wetland; E/W gradient restored	variable recharge rates	water level rises at all sites; greater rise under d; w < d (5 cm)	recharge to all sites; tidal and recharge effects
July	water level rises at all sites; E/W gradient maintained	variable recharge rates	water levels are the same for all sites	

Table 8.11 (cont.)

Table 8.11 (cont.)

Month	*35*	*interpretation*	*swii*	*interpretation*
August	water levels rise at all sites; water level depressed under central wetland; mound under marginal site (12 cm) E/W gradient maintained	variable recharge rates	water levels are the same for all sites	
September	water level rise at all except marginal site; E/W gradient maintained		water level depressed under wetland; w < d (5 cm)	tidal and recharge effects
October	fall in levels at all sites water level mound under central wetland; higher than average fall under Ed; E/W gradient maintained)	lower volume of rainfall recharge to wetland, but not Ed	E/W gradient (2 cm); w = d	discharge by lateral flow
November	all water levels fall; no E/W gradient; water level falls greater under marginal site	water levels under marginal sites reside in sand, *i.e.*, are below wetland fill	W/E gradient (2 cm); w < d (2 cm)	tidal effect or evapo-transpiration prefer former
December	all water levels fall; water level falls greater under marginal site; E/W gradient restored		slight mound under wetland; w > d (2 cm)	

Table 8.12 *Description of water level responses and water table morphology under relatively dry conditions 1994 (Figs. 8-19 to 8-22)*

Month	*161*	*interpretation*	*WAWA*	*interpretation*	*135*	*interpretation*
January	water levels are the same for all sites except under d; w > d (5 cm)		E/W gradient (10 cm) w < d	western lateral flow from Ed	W/E gradient w < d (3 cm)	
February	water level falls at all sites; no change to water table; w > d (5 cm)	discharge by evaporation and lateral flow	water level falls at all sites; no change to water table surface; w < d	discharge by evaporation and lateral flow	water level falls at all sites; W/E gradient; slight change to water table surface; w < d (6 cm)	evapo-transpiration under *M. rhaphiophylla*
March	water level falls at all sites; water level depressed under wetland (3-5 cm); w = d	evapo-transpiration greatest in wetland centre	water level falls at all sites; water level fall greater under ridges; water level depressed under site 2; E/W gradient (7 cm) w = d		water level falls at all sites; W/E gradient; w < d (10 cm)	evapo-transpiration under *M. rhaphiophylla*
April	water level falls at all sites; mounds under site 2 and wetland (2-7 cm) w > d	spasmodic pre-winter rainfall water held in vadose zone, resulting in small recharge to groundwater	water level falls at all sites water levels higher under ridges E/W gradient (5 cm) w > d (3 cm)	water held in vadose zone, resulting in small recharge to groundwater	water level falls at all sites W/E gradient w < d (8 cm)	evapo-transpiration under *M. rhaphiophylla*

Table 8.12 (cont.)

Table 8.12 (cont.)

Month	*161*	*interpretation*	*WAWA*	*interpretation*	*135*	*interpretation*
May	water level rises at all sites; no change to water table surface; w > d (5 cm)	small rainfall; water held in vadose zone, resulting in small recharge to groundwater	water table rise under east margin and wetland, fall under west margin and d; E/W gradient (11 cm); w > d (10 cm)	disparate recharge rates	water level falls at all sites; W/E gradient; w < d (9 cm)	
June	water level rises at all sites; no change to water table surface; slight mound under site 2 and wetland (2-7 cm)	recharge to groundwater	water level rises at all sites; higher recharge under d; E/W gradient (7-10 cm) w > d (3 cm)	storage of water in wetland vadose zone (peat aquifer)	water level rises at all sites; W/E gradient; recharge to all sites; w < d (11 cm)	more rapid recharge under d
July	water level rises at all sites; slight mound under site 2; w > d (2 cm)	preferential recharge at w/d contact	water level rises at all sites; no change to water table surface; E/W gradient (7 cm) w > d (3 cm)	higher than average recharge; direct recharge to water table; little storage in vadose zone	water level rises at all sites; W/E gradient; w < d (10 cm)	greater than average recharge
August	water level rises at all sites; mound under site 2 (10-12 cm); w < d (3 cm)	rapid recharge along w/d contact; greater recharge under ridges than in wetland	water level rises at all sites; no change to water table surface; E/W gradient (5-7 cm) w=d	higher than average recharge	water level rises at all sites; W/E gradient; w < d (14 cm)	greater than average recharge; more rapid recharge under d

Table 8.12 (cont.)

Table 8.12 (cont.)

Month	*161*	*interpretation*	*WAWA*	*interpretation*	*135*	*interpretation*
August	water level rises at all sites; mound under site 2 (10-12 cm); w < d (3 cm)	rapid recharge along w/d contact; greater recharge under ridges than in wetland	water level rises at all sites; no change to water table surface; E/W gradient (5-7 cm); w=d	higher than average recharge	water level rises at all sites; W/E gradient; w < d (14 cm)	greater than average recharge; more rapid recharge under d
Septem-ber	water level rises under eastern sites and falls under d and site 2; mound under wetland (25 cm); w > d (17 cm)	recent rainfall event, resulting in rapid recharge under central wetland and delayed recharge at deeper sites	water levels remain constant except under d; water level depressed under site 2 and level under ridges; w < d (10 cm)	low recharge	water level falls at all sites; no change in water table surface; W/E gradient; w < d (14 cm)	cessation of winter rain; low recharge
October	water level falls at all sites; reduction in mound under wetland; mound under d; w < d (5 cm)	disparate recharge rates	water level falls at all sites; larger than average fall under d; mound under wetland; w > d (7 cm)	delayed groundwater recharge under ridges	W/E gradient; small change in surface; w < d (11 cm)	evapo-transpiration
Novem-ber	water level falls at all sites; water level depressed under wetland; mound under d; w < d (17 cm)	evapo-transpiration in wetland	water level falls at all sites; larger than average fall under wetland; E/W gradient (5 cm) w < d (5 cm)	normal fall + evapo-transpiration	water level falls at all sites; W/E gradient; no change in surface; w < d (10 cm)	evapo-transpiration

Table 8.12 (cont.)

Table 8.12 (cont.)

Month	*161*	*interpretation*	*WAWA*	*interpretation*	*135*	*interpretatio*
Decem-ber	water level falls at all sites; water level fall greatest under d; eastern margin higher than other sites (8 cm); w=d	flow to site 2 E/W gradient induced seepage from Ed	water level falls at all sites; greater than average fall under d; E/W gradient (15 cm); w > d	western lateral flow from Ed	water level falls at all sites; fall under wetland less than fall under d; W/E gradient; w < d (7 cm)	

Table 8.12 *Description of water level responses and water table morphology under relatively dry conditions 1994 (Figs. 8-23 to 8-25)*

Month	*35*	*interpretation*	*swii*	*interpretation*
January	mound under wetland (5-10 cm); E/W gradient (8 cm)	evapo-transpiration under eastern margin	W/E gradient (7 cm); w < d (3-5 cm)	discharge by lateral flow
February	water level falls at all sites; mound under wetland persists; E/W gradient declines; w > d (5 cm)	lateral flow from Ed; evapo-transpiration under eastern margin	level under wetland unchanged; other levels fall; w > d (10 cm)	
March	water level falls at all sites; no change to water table surface	lateral flow from Ed; evapo-transpiration under eastern margin	fall under wetland greater than other sites; w > d (7 cm)	
April	water level falls at all sites except d; no E/W gradient; mound under wetland persists (5 cm)	spasmodic pre-winter rainfall recharges sites with shallow water table	fall under E margin greater; mound under wetland (10-16 cm)	rapid recharge under central wetland to shallow water table

Table 8.12 (cont.)

Table 8.12 (cont.)

Month	*35*	*interpretation*	*swii*	*interpretation*
May	water level rises at all sites; mound under wetland (8-10 cm); trough under eastern margin; E/W gradient re-established	lateral flow from Ed; recharge to groundwater at all sites	mound under wetland diminished (5 cm); W/E gradient (7 cm); w=d	variable recharge rates; lateral flow from wetland to adjacent western sites
June	water level rises at all sites; mound under sites 2 and 3 (8-10 cm); trough persists eastern margin; no E/W gradient	lateral flow from Ed; recharge to groundwater at all sites	water level rises at all sites; recharge to all sites; mound under wetland (10 cm); w > d	variable recharge rates
July	water level rises at all sites; no change to water table surface; E/W gradient re-established (5 cm)	similar recharge to all sites	W/E gradient (7 cm); w=d	discharge by lateral flow
August	water level rises at all sites; water levels slightly depressed under wetland; E/W gradient (10 cm); no trough under east margin	greater than average recharge	W/E gradient (7 cm); w < d (5 cm)	discharge by lateral flow
September	fall in levels at all sites except wetland; mound under wetland 5-10 cm); Slight E/W gradient	cessation of winter rains low recharge	fall in levels; w > d (3 cm)	rapid recharge to wetland

Table 8.12 (cont.)

Table 8.12 (cont.)

Month	*35*	*interpretation*	*swii*	*interpretation*
October	water level falls at all sites; water level depressed under site 2; E/W gradient (10 cm); slight mound under wetland	minor recharge to shallow water table	water level depressed under site 2, elevated under d; mound under wetland (7 cm)	
November	water level falls at all sites; trough under wetland; E/W gradient (12 cm)		greater than average fall under d; surface unchanged; w > d (5 cm)	
December	water level falls at all sites; no change in water table surface; water levels under wetland are level; E/W gradient (15 cm)		mound under wetland diminished; W/E gradient (5 cm); w=d	discharge by lateral flow

Drill bores under the zenith of four higher than normal ridges, WAWA, 142, 35, swii, were used to show that consistently elevated water level readings under eastern ridge sites were due to east/west gradients rather than semi-permanent or permanent groundwater mounds. Examples of water table features under ridges and swales were selected from maximum and minimum water levels for the period 1995 to 1996 in order to encompass a full data set of both wetland and ridge/swale sites (Figs. 8-26 to 8-29). Although groundwater levels under the adjacent eastern beachridge/dunes were consistently higher than under both the wetlands and the western ridge (2-20 cm), they were equivalent to those under the eastern swale and were within the parameters of local gradients (Figs. 8-26 to 8-29).

8.4.6 Water tables during prevailing wet vs dry conditions

Various hydrological effects and features, which are common to either the prevailing wetter or drier conditions, can be identified from the monthly water level data for 1992 and 1994. Water table characteristics for ridge and wetland sites repeated from site to site are presented in Tables 8.11, 8.12 and summarised in Table 8.13.

Table 8.13 *Morphological features of the water table and small scale hydrological processes common to high or low rainfall conditions*

Common characteristics
Position of maximum water level under wetlands and beachridge/dunes is 20-25 cm lower in the drier year
Position of minimum water level under wetlands and beachridge/dunes is 35-40 cm (up to 60 cm) lower in drier year
Groundwater fluctuation is greater in drier years
More perching of surface water occurs in wetter years
Mounding under wetland margins is consistent for both wet and drier years Mounding under wetlands is more common in dry years
Troughs occurred more frequently in the wetter years, relative to summer mounding under ridges
East/west gradients are not evident in some wetlands and are present for 8 out of 12 months in other wetlands, particularly in wet years East/west gradients are most obvious in months of March/April and Oct and Dec in wet years West/east gradients are most common in drier years particularly in months of Dec/Jan and July/Aug/Sept/Nov
Calcrete layers perch subsurface water in wet years and suppress the rising of water below calcrete in drier years

8.4.7 Flow between ridge and wetland

When there is an interface between a higher water table under a ridge and lower surface water in a wetland, there is potential for flow between ridge and wetland

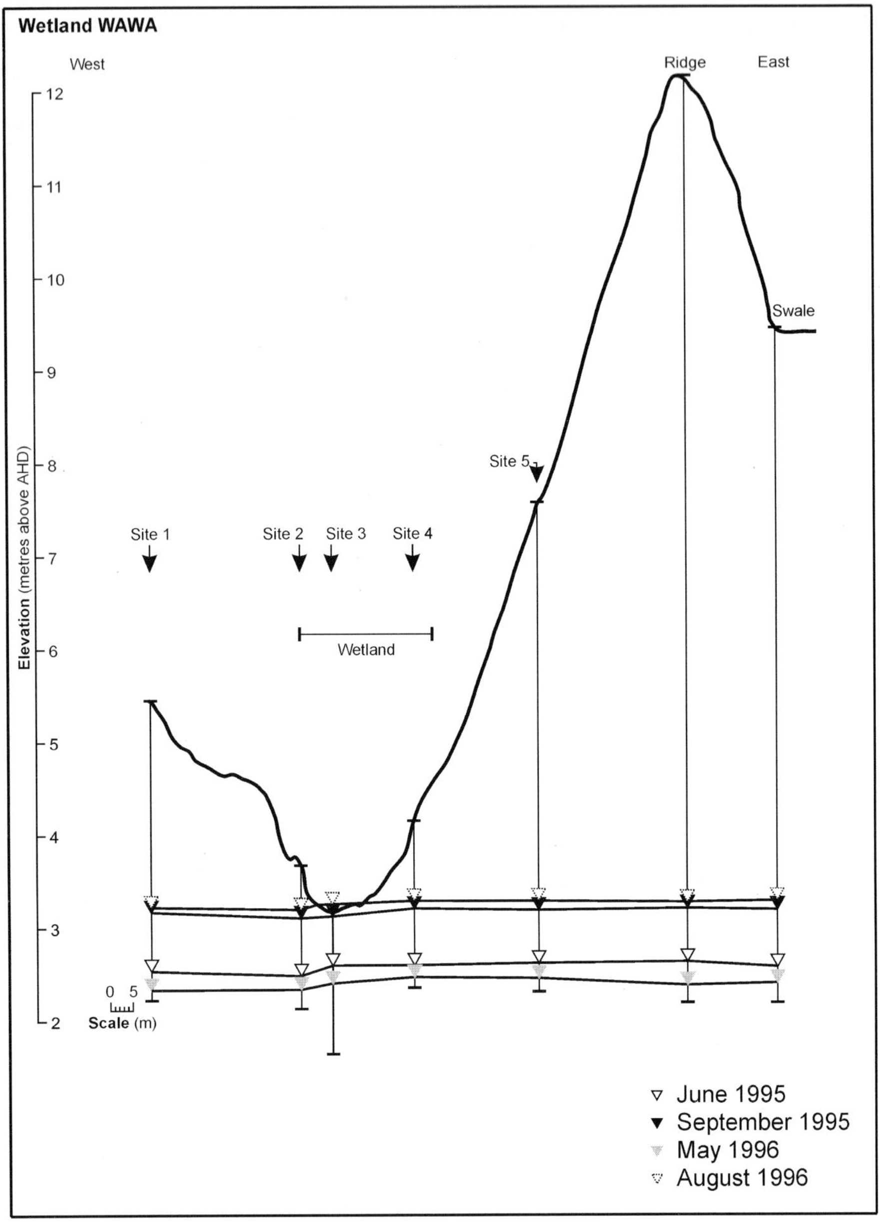

Figure 8-26. Maximum and minimum water levels under wetland WAWA and adjacent eastern ridge 1995, 1996.

Under the hydraulic gradient between ridge and wetland margin (Table 8.14), water velocity could reach 14 m/month in the coarse sands underlying the ridge at WAWA, and 4 m/month in the medium sands underlying the ridge at 35. Water level data showed that conditions suitable for movement between adjacent beachridge/wetland sites occurred frequently. Differences less than 10 cm produced a gradient similar to the local gradients and therefore lateral water movement was undetectable in a monthly time frame.

8.5 Wetland hydrology at bedding scale

Sampling of soil moisture content down profile at beachridge and wetland sites was undertaken to investigate hydrology at the bedding scale, *i.e.*, the processes that affect vegetation. Sampling took place in April and September, periods of water table minima and maxima. Results for three beachridge and fourteen central wetland sites are presented below.

8.5.1 Beachridge/dune soil moisture down profile

In all ridge sites the ratio of water to wet soil remained fairly constant down profile and between sites (Fig. 8-30), but varied slightly between the wet and dry seasons. In April, water movement in all the beachridge profiles was confined to the top 100 cm, with slow downward movement predominating below 25 cm. In September, slow downward movement predominated throughout the profile (Fig. 8-30). However there was a two to three fold increase in soil moisture between seasons in the top metre.

The moisture content under the beachridges showed that pore water at the end of winter (September) was 2-3 g in 50 g of sediment. In terms of storage this amounts to 48.3 kg water in the top cubic metre, which dropped to approximately 39 kg water prior to winter rains (April). These calculations were based on a bulk density of 1.15 g/cc derived from empirical measurements.

8.5.2 Wetland soil moisture down profile

Soil moisture content down profile in the wetland sediments ranged from 20-200 g in 50 g of sediment. This is up to two orders of magnitude higher than in the beachridges. There are some very obvious differences in soil water content down-profile between wetland sites (Fig. 8-31). The main patterns relate, firstly, to the effects of summer evaporation, aseasonal summer precipitation, and seasonal winter precipitation, and, secondly, to the stratigraphy and the effects of sediment composition or grainsize on the water retention capacity of various layers. The data in Figure 8-31 are presented against a backdrop of the stratigraphy so that the influence, where present, of the sediments on the down-profile content of soil moisture can be readily gauged.

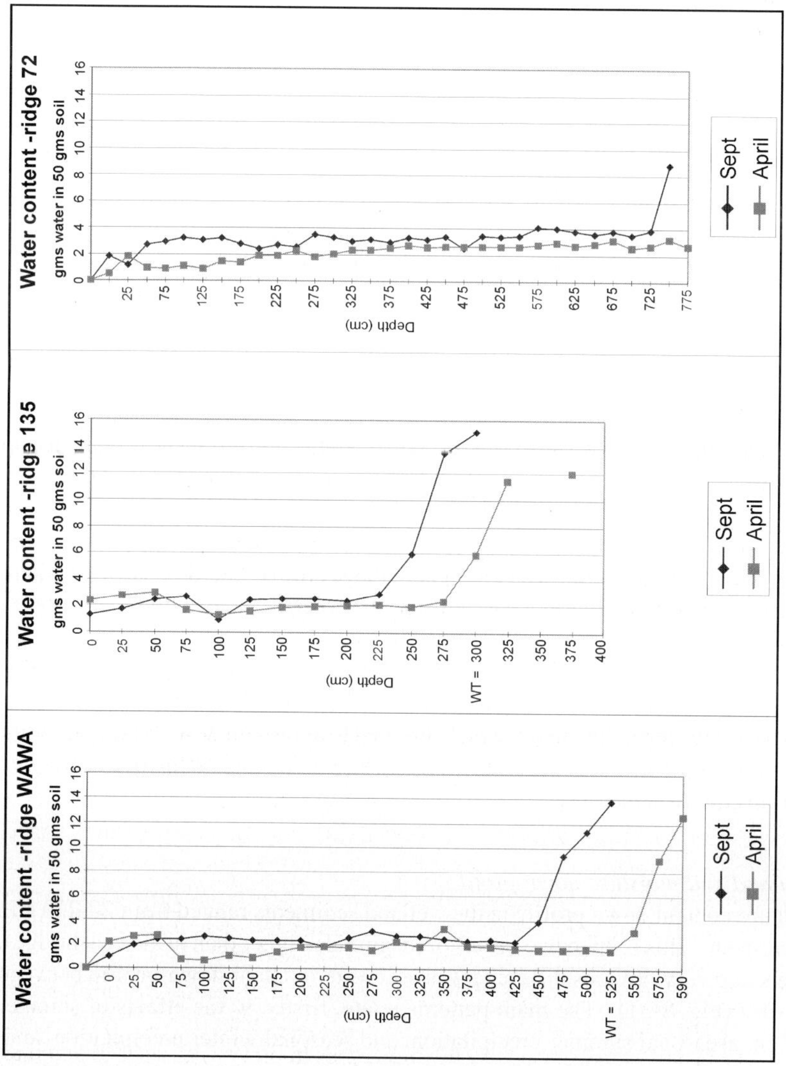

***Figure 8-30**. Soil moisture content down profile under beachridge/dunes. Weight of water per 50 g of sediment.*

The contrast between moisture depletion in summer and water retention in winter for the entire profile is best illustrated by wetlands 163, WAWA, 135, 142, 35, swii, and 1N. The contrast between moisture depletion in summer and water retention in winter for the surface layers is best illustrated by wetlands 162, 163, WAWA, 135, 42, 72, 63, 35, 9-6, 9-11, swii, swiii, and 1N. Soil water content is markedly affected by stratigraphy in wetlands 161, 162, 163, and WAWA. The change in soil moisture content near or at a stratigraphic boundary is best illustrated by wetlands 161, 163, WAWA, 135, and 35. The effect of organic rich upper layers in retaining water moisture, especially in the winter, is evident in most wetlands 161, 162, 163, WAWA, 135, 72, 63, 35, 9-6, 9-11, and swii. The effect of grainsize variability in the retention of soil moisture down profile is best illustrated in wetlands 9-11 and 1N.

Generally, at the end of summer, in all profiles the water tables were low, and the sediments were approaching a point of minimal water content (field capacity). Soil moisture content in the 0-10 cm interval followed one of three patterns: it decreased rapidly (wetlands 161, 35, 9-6), increased slightly as the beginning of a flux evident lower in the profile (wetlands 162, 63, swiii), or, if low already, remained constant (wetlands 142, 72, 1N). In the vadose zone below this level, the pore water content either fluctuated while decreasing overall, or remained constant. In the phreatic zone, the pore water content remained constant in all wetlands except wetland 161.

Generally, in winter, the water tables were high and the dominant hydrological process in the central part of the wetland was infiltration of rainwater. Soil moisture content in the 0-10 cm interval consistently increased. In the vadose zone below this level, the pore water content either fluctuated or remained constant. In the upper part of the phreatic zone, the pore water content continued to fluctuate or remained constant.

These patterns can be explained by the variable frequency of rain events in both summer and winter, by the dominant hydrological processes occurring in the wetland centre (*i.e.*, infiltration, evaporation, transpiration), and by the heterogeneous nature of the wetland sedimentary stratigraphic sequences. Rainfall events interspersed with dry periods create temporal fluctuations in pore water, which when viewed down profile appear as variation in moisture content. Changes in permeability of sediments over a relatively small sequence of wetland fill, influence the magnitude and location of these down profile variations. The dominant hydrological process in the vadose zone of the central parts of wetlands during periods of rainfall recharge is infiltration, causing downward movement of pore water. This drainage tends towards pore water constancy within any sediment layer. The differences in pore water content and rate of infiltration may both be explained by the heterogeneous nature of the wetland sediments. The dominant hydrologic processes in summer are surface evaporation, and near-surface soil water depletion by transpiration.

In terms of storage of water, the amount of interstitial water and/or pellicular water held in the upper parts of the sediments varied between sediments and between seasons. Water stored in the top half cubic metre of peat in winter was 269 kg falling to 99 kg in summer. Water stored in the top half cubic metre of calcilutite in winter varied from 119-266 kg, and in summer, varied from 83-170 kg, the amounts being similar to the water content in peat. Sediments from inundated wetlands had the highest water content, *e.g.* wetland WAWA followed by wetlands 161, 35, and 9-6, in that order. The sediments within the majority of other wetlands exhibited approximately half this water content. The lowest water content occurred in wetlands 1N and 142. Wetland 1N, composed of slightly humic sand, most closely approximated the texture of the beach ridges, but even here showed a five to tenfold increase in soil moisture compared to the ridges. The low soil moisture content in wetland 142 cannot be explained in terms of stratigraphic attributes, but may be due to an anthropogenically induced lower water table.

Within any wetland, soil moisture content was highest in the muds, then muddy sand, then sand (Fig. 8-31), thus generally decreasing down the profile following the stratigraphic sequence. In only two wetlands did this trend not occur, 72 and 9-6. In wetland 72, the sedimentary layers are very thin (20 cm) and differentiation between sandy mud and muddy sand over this interval may be insufficient to determine water content differences. Capillary rise processes between the various granulometrically differentiated layers ensures exchange of moisture. In wetland 9-6, the occurrence of calcrete in the profile at 50 cm formed a barrier to vertical water movement, and this was reflected in the increase in soil moisture in the upper layer in both winter and summer.

For the two sampling times, in different seasons, patterns of moisture content down profile and in the surface soils were similar in wetlands underlain by thin layers of calcilutite and thicker layers of muddy sand (*i.e.*, wetlands 63, 72, 9-6, 9-11, swii, and swiii). In wetlands which were underlain by calcilutite or peat, (*i.e.*, wetlands WAWA, 142, 135, and 35) patterns of moisture content down profile and in surface soils varied. Wetlands 142 and 135, underlain by calcilutite, showed different seasonal trends but similar moisture content, while wetlands WAWA and 163, underlain by peat, showed considerable moisture content differences. Wetland 35 showed variability in both characteristics. These differences can be related to the effects of wetting and drying in sediments of different composition.

The major differences in soil moisture content occurred in the top 20 cm of any profile, and proximal to the water table (Fig. 8-31). Soil water content in the top 10 cm in wetland centres and margins was sampled on a quarterly basis from 1991 - 1994 in order to document variation in this layer where the rhizosphere is best developed (Fig. 8-32). One interesting result was the consistency of the soil water content for most wetlands, in spite of variable rainfall amounts and distribution. The most consistent soil water content was found in the centres of the wetlands, *e.g.*, wetlands 63, 135, 142,

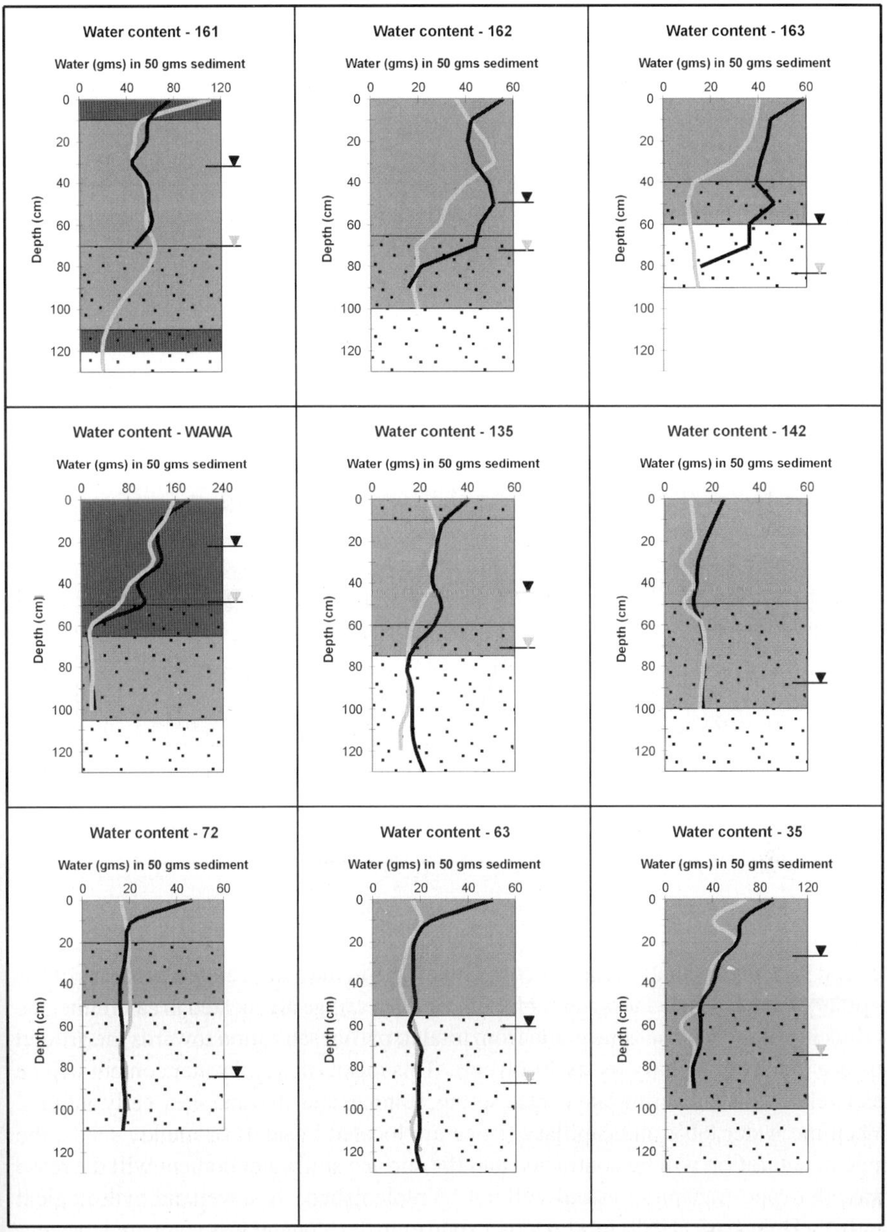

Figure 8-31. *Soil moisture content down profile under wetlands.*

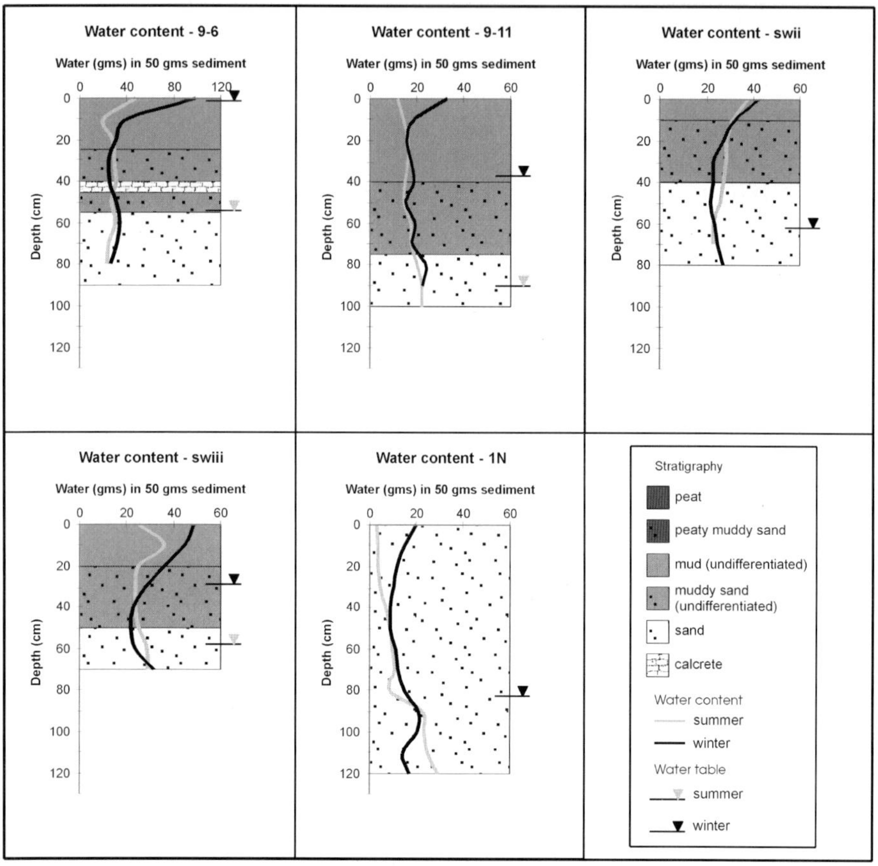

Figure 8-31 (cont.). *Soil moisture content down profile under wetlands.*

45, and 9. The explanation for this consistency lies in the nature and dimensions of the capillary fringe. When the water table and capillary fringe are located in calcilutite, the sediments of the vadose zone grade from total to partial saturation towards the ground surface (zone of capillary rise is 30-60 cm). This means that soil water content will be relatively consistent during late winter, spring, summer and, in some sites, early autumn. When the water table and capillary fringe are located in sands or muddy sands, the zone of saturation will be contracted and the surface soil water content will decrease through evapo-transpiration and will not be replenished. In a wetland hydrological study by Hunt *et al.* (1999) in which root zone moisture content was compared to water table position, both soil texture and the capillary fringe were found to be important determinants, a result comparable to the findings above. Where variation in surface soil moisture did occur, *e.g.*, sites 162-5, WAWA 3, swiii-4, 5, and at wetland margins, there were corresponding marked changes to vegetation in terms of its density, luxuriance, height, and composition (Chapter 10).

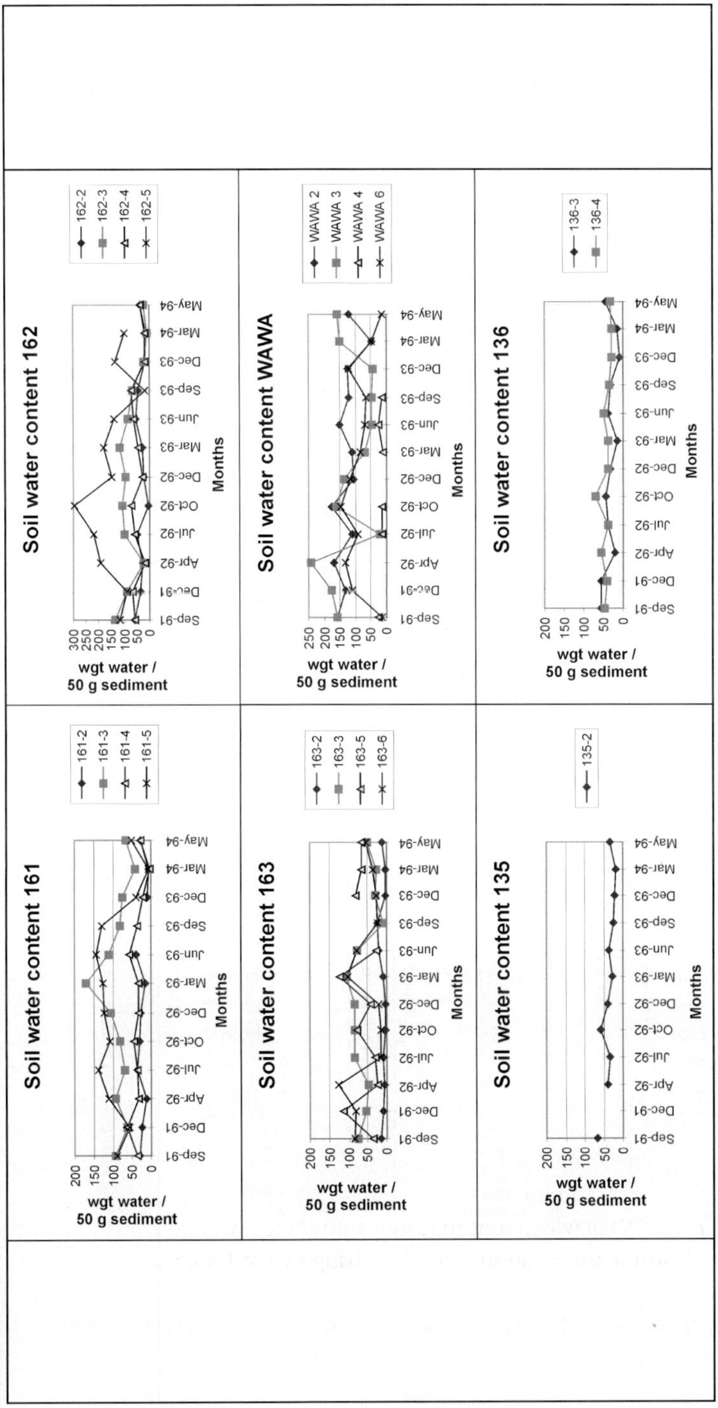

Figure 8-32. *Seasonal soil moisture (by weight/50 g sediment) in the surface layer of each wetland.*

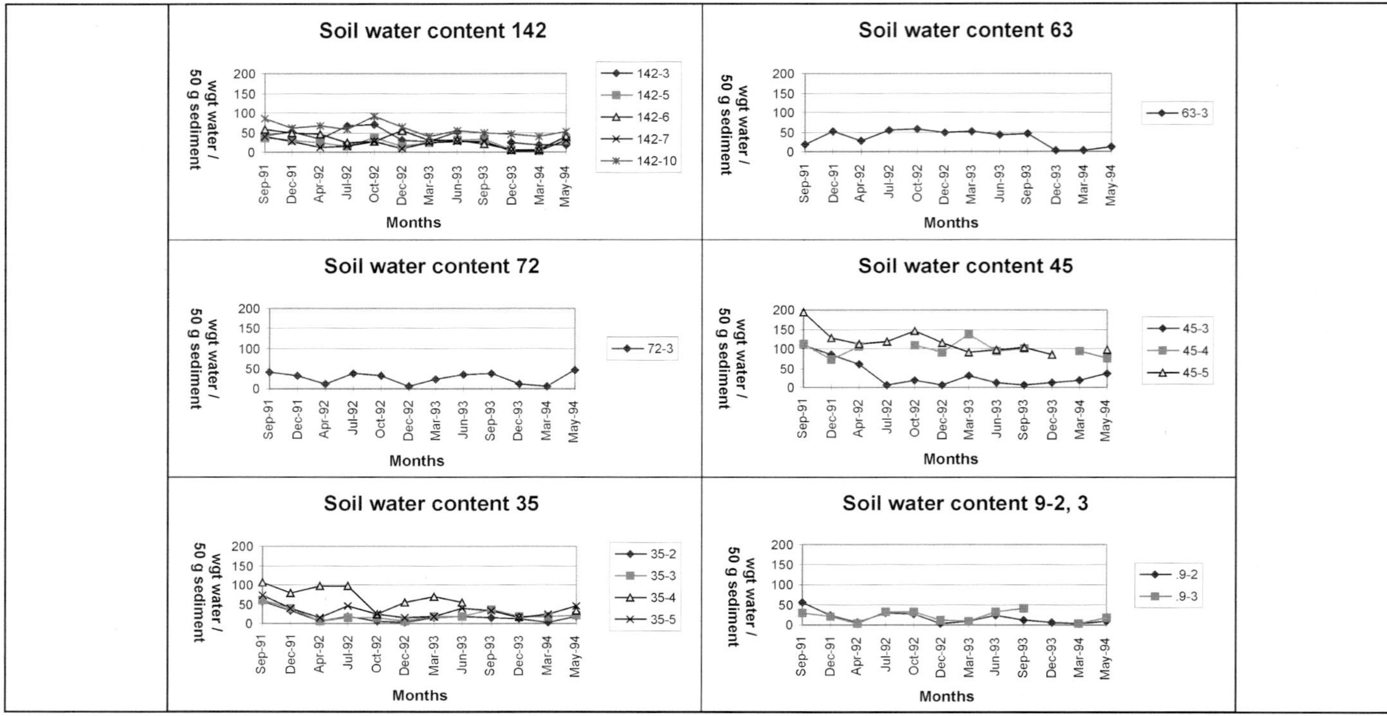

Figure 8-32 (cont.). *Seasonal soil moisture (by weight/50 g sediment) in the surface layer of each wetland.*

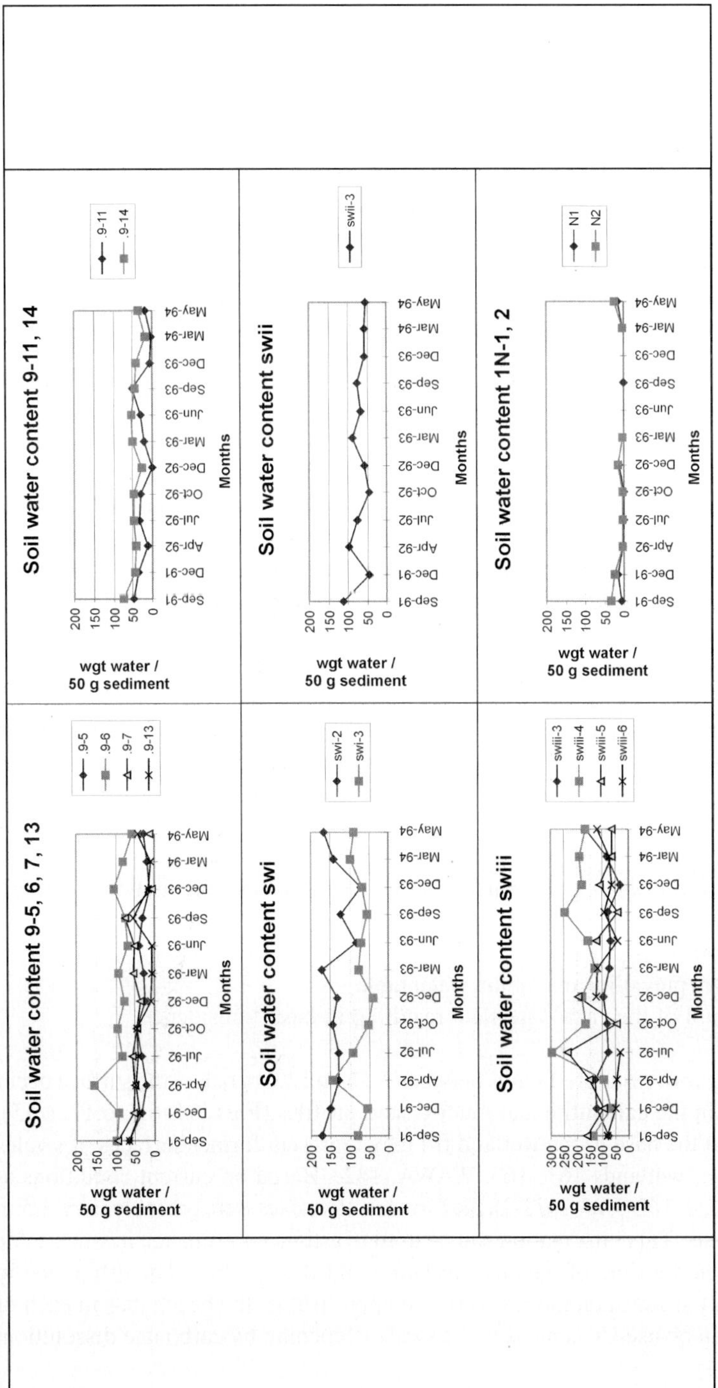

Figure 8-32 (cont.). *Seasonal soil moisture (by weight/50 g sediment) in the surface layer of each wetland.*

The reason for documenting these small scale patterns in soil water content was to identify the period in which groundwater is available for wetland plants. These conditions favour wetland plants with shallow roots (sedges, rushes and herbs), which can take advantage of small volumes of aperiodic precipitation ephemerally stored in the upper sediment layers before it is lost via downwards percolation or evaporation, and plants which can withstand alternating conditions of waterlogging and drought (species of *Melaleuca*). Although many plants exhibit broad tolerance to the variation in water availability, the differences in hydroperiod resulting from small scale permeability factors such as infiltration rates, and/or soil moisture content, rather than the groundwater level itself, may be the more important reason for species distribution.

8.6 Water level with respect to palaeo-surface

Regional groundwater rise resulting from seaward progradation of the coastal plain was the initial cause of wetland development on the Becher cuspate foreland. However, under variable climatic conditions involving annual rainfall varying some 200-300 mm on a *circa* 20 year turnaround, and on a turnaround period greater than 50 years, groundwater rise at the basin scale has been a fluctuating process. In any basin, there is evidence of former groundwater levels in the stratigraphic sequence. Sedimentation processes reflect the conditions concordant with changing groundwater levels. As groundwater rose, frequent waterlogging of the swale resulted in an accumulation of organic matter in the sediments. On regular inundation, a new sedimentation process began on the floor of a swale producing calcilutite. As calcilutite sedimentation requires inundation, it may be inferred that water levels in many wetlands were higher than at present. In many of the wetlands the current positions of maximum water levels lie 30-40 cm below the calcilutite surface.

Estimating the rise in water levels since wetland inception rests on three foundations:

- humic soil
- the current prevailing maximum water table
- dissolution of the carbonate sand resulting in basin subsidence.

Layers of humic root structured sands between 0.26 and 1.0 m below the surface occur under and within the calcilutite and muddy sand profiles (Figs. 6.3-6.7, 6.12, 6.15). When located at the base of the wetland fill they represent former surfaces of swales now buried, *e.g.*, wetlands 161, 163, WAWA, 142. Based on current conditions at similar sites (162-2, 142-8, 63-2, 72-2), the former water tables were probably *circa* 1.5 m below the surface. This information can be used to estimate the change in water table position between the time of wetland initiation and the present. The difference in height between the stratigraphic levels of the upper part of the beach unit in each of the wetlands can be used as a measure of swale deepening by carbonate dissolution.

This is considered to be a more accurate estimate than the difference in carbonate mud levels, as the latter are subject to alteration by bioturbation, sheet wash, and loss of interlayered peat through ignition. The water table rise since the inception of a particular wetland, was calculated using 1.5 m as the baseline water table below the swale, the current prevailing maximum water level in each wetland, and the estimate of wetland subsidence (Table 8.15). Even acknowledging that the baseline water table could in some instances have been nearer the surface than 1.5 m, the rise in water table is still considerable.

Table 8.15 *Height of buried swales in relation to present maximum prevailing water levels*

Site	*Depth of buried swale surface from current surface (m)*	*Height of buried swale surface AHD*	*Height of current prevailing maximum water level AHD*	*Difference in stratigraphic levels of beach/dune contact (m)*	*Calculated water table rise*
161	1.1-1.2 m	2.59 m	3.61 m	0.85 m	Minimum water level rise = 1.67 m
163	0.55-0.6 m	3.25 m	3.64 m	1.1 m	Minimum water level rise = 0.79 m
WAWA	0.9-1.05 m	2.27 m	3.25 m	0.85 m	Minimum water level rise = 1.63 m

In Cooloongup A, in the shallow subsurface 60-65 cm below the ground (1.9 m AHD), there is a brecciated layer of gravel sized grains which are carbonate mud intraclasts. The random orientation of these intraclasts indicates reworking of indurated or dried calcilutite at an erosional surface (Shinn 1983). As the modern surface is now higher than this layer and is composed of calcilutite consisting of organic debris, this indicates a subsequent minimum rise in water level of 60-65 cm. The current maximum water level relating to this period of below average rainfall lies approximately 32 cm below the calcilutite surface (2.18 m AHD).

8.7 Summary and discussion

There are several conclusions in relation to wetland hydrology, which derive from this study. At the largest scale, it is clear that rainfall infiltration to the water table is the major source of groundwater recharge and the cause of groundwater rise. Changes in frequency, intensity, and temporal distribution of winter rainfall govern the amount and period of groundwater recharge to the Safety Bay Sand aquifer (Fig. 8-10 and Tables 8.1, 8.2). Factors underlying long term rainfall cycles of above or below average rainfall determine the annual patterns of frequency and intensity of rainfall. Several long term cycles, which may relate to rainfall patterns, have been identified in the geomorphic and stratigraphic records in the coastal zone of Western Australia (Semeniuk 1995; Semeniuk and Semeniuk 2005), *e.g.*, the shifting of climatic regions over the past 7000 years related to Earth Axis Precession, the occurrence of higher than normal beachridges *circa* 250 years, and the 19-20 year beach erosion cycles related to the 18.6 year lunar nodal (Currie and Fairbridge 1985; Semeniuk 1995). In the 125 years of rainfall data for the Perth region, the 20 year cycles are evident in the patterns of above and below average rainfall (Fig. 8-1). 1991-1996, the period of intense field measurement for this study, occurred in the drier period 1980-2000. Local variability in rainfall, exemplified by the increase towards the relatively high ground of the inland Spearwood Dune Ridge (Walyungup site), also affects local *in situ* recharge.

Patterns at the smaller scales are:

- meteoric input to the groundwater system is altered by the lenses and ribbons of wetland fill through which rainfall must percolate to reach the water table; these lenses and ribbons influence the height and rate of groundwater rise and therefore the degree and length of period of waterlogging and inundation

- within the wetland sediments, small scale sedimentary structures facilitate domination of vertical flow over lateral flow, while interlayered sediments below the rhizosphere facilitate localised lateral flow

- lateral flow through a homogeneous impermeable layer of the wetland fill is negligible, and where this type of layer is well developed, lower or higher basin water levels, out of equilibrium with regional water levels, persist

- contacts between wetland and beachridge/dune determine preferential flow paths to the wetland margins

- under the central beachridge plain, evapo-transpiration of the groundwater from wetlands is the dominant mechanism of discharge

- discharge by gradient induced flow dominates near the coast and where local gradients are steep

- local gradients are related to the configuration of the water table which varies with the volume of water in the aquifer and its geographic position relative to AHD

At the scale of the individual layers in the wetland fill, the soil water processes identified in the Becher wetlands were: saturation of soils in the top 10 cm sediment; build up of infiltrating water at stratigraphic (sediment textural) boundaries; consistently low soil water content in the calcareous sand and muddy sand, which would indicate that field capacity is quickly achieved and imply little water movement; and little water movement in peat implied by the soil water content gradient.

The results show that there are a number of flow paths into and out of a wetland basin, and the dominance of any given pathway or flow rate is determined by a variety of factors and mechanisms, and their interaction. Factors include stratigraphy, precipitation, the water volume already residing in the wetland, and the amount of physical recharge and discharge. Mechanisms include infiltration, seasonal groundwater fluctuation, upwelling, throughflow, ponding and evapo-transpiration. These findings clearly refute the idea that the Becher wetlands are simply surface expressions of groundwater in a surficial homogeneous aquifer recharged by direct infiltration from rainfall and discharged through evapo-transpiration. However, the results also refute the idea that the Becher wetlands are isolated closed systems with their own internal balance of water input and output. In truth, the hydrological mechanisms maintaining the Becher wetlands are subject to seasonal variation and the nature of the basin fills. Firstly, this means that some mechanisms are short lived, such as reversal of flow and upwelling, and some mechanisms dominate in one season and become sub-dominant in another, *e.g.*, throughflow. Secondly, this means that the significant hydrological mechanisms differ in older and younger wetlands due to variation in thickness and composition of fill.

Few comparable studies exist in the extensive literature on wetland hydrology (Winter 1986; Mann and Wetzel 2000b). In studies of inter-dune wetlands (lakes, swamps, marshes) elsewhere, in which the groundwater has been the focus, and well installation has been of sufficient spatial density to obtain hydrological data at the small scale, findings are similar regarding the configuration of the water table and dynamic reversals of seepages at wetland margins (Erickson 1981 cited in Winter 1986; Winter 1986; Doss 1993; Phillips and Shedlock 1993). Changes to the seasonal configuration of the water table result from high and low levels of groundwater recharge in response to climatic conditions (Winter 1986; Doss 1993). Measured time lags in water table recharge (57 days) by meteoric infiltration between deep and shallow bores through a sand aquifer are comparable to the two month lag observed in this study. Rates of rise and magnitudes of water level increase in wells with variable depths to water table are also analogous. Finally, the range of measurements of hydraulic head, corroborated by groundwater chemistry presented by LaBaugh (1986), were directly comparable with the beachridge to wetland margin gradients recorded at Becher.

In other studies of wetland hydrology, it has been demonstrated that flow occurs between the adjacent dune or beachridge and the wetland margin (Grootjans *et al.* 1996; Richardson *et al.* 2001). It was apparent that even in small wetlands, this flow can be impeded by relatively impermeable wetland sediments resulting in little effect in the wetland centre. It was also recognised that inter-dune wetlands receive groundwater from very local recharge areas, *i.e.*, within 100-200 m (Grootjans *et al.*

1996). The results, expressed herein, emphasising local, and ridge-to-wetland over regional hydraulic gradients, are in agreement with these authors' results.

Other hydrological aspects of this study that have been described in the literature are:

- troughs, mounds, planar surfaces and gradients in the water table (Winter 1986; Phillips and Shedlock 1993; Eshleman *et al.* 1994)
- wetlands with both recharge and discharge functions (Doss 1993)
- varied hydrological responses in wetlands to wet and dry periods

Seasonal troughs, mounds, planar surfaces and changing local gradients were morphological features evident in the water table in a forested coastal plain drainage basin in Maryland and Delaware (USA), and in the sandhills of Nebraska (USA) (Phillips and Shedlock 1993; Winter 1986). In each study, these morphological features were observed because the investigation strategy was designed to probe a variable land surface and geology in a climatic regime characterised by temporal variability. In both studies, piezometers were installed within the terrain at sites selected to monitor wetland hydrology in the context of the landscape in which they were situated, *i.e.*, along transects which incorporated each wetland, its edge, wetland marginal sites, and ridge sites (Phillips and Shedlock 1993). This method resulted in 30 piezometers being installed in an area containing five wetlands (Winter 1983). Water levels were measured hourly, daily and monthly. In a separate study, an equally intense spatial design of piezometer installation was used to determine a 0.5 m groundwater mound below the marshland on an estuarine plain with extremely low topographic relief (Logan and Rudolph 1997). In the Becher wetland study, 119 permanent piezometers and 16 temporary piezometers were installed in wetland centres, margins, and ridges along transects and in supplementary vegetation quadrats. This provided a rich data set to document the seasonal troughs, mounds, planar surfaces and changing local gradients in this area.

The importance of these morphological features in the water table, documented at Becher and in the USA, is in determining groundwater flows, whether transient, seasonal, event based, or semi-permanent (Gillham 1984). Seepage across the wetland margins occurs when there are differences in groundwater recharge times between wetlands and ridges, or when there is a disproportionate recharge due to varying thickness in the capillary fringe (Novakowski and Gillham 1988). Seepage alters the soil moisture content in the surface layers and the rhizosphere and subtly modifies the hydroperiod. Differences in groundwater recharge times between high and low ridges create hydraulic gradients which may result in water flow to or from the wetland either enhancing or ameliorating discharge via the general throughflow. Hydrochemistry at the wetland margins can be strongly influenced by these types of alternating flows from wetland centre and ridge (Phillips and Shedlock 1993; Hayashi *et al.* 1998). Hydrologically induced sediment changes at the margins of wetlands, potentially leading to expansion or contraction of the wetland, can occur under the influence of these specific short distance flows. The flow paths at the margins of the Becher Suite wetlands are short, and hydraulic gradients are relatively small, but elsewhere, transient flows across wetland margins can result in large seasonal and event based changes in water table profiles (Phillips and Shedlock 1993).

That the hydrological function of wetlands can change from season to season and from one part of the basin to another is not widely documented elsewhere in the literature. In some wetlands, discharge and recharge zones did occur concurrently (Siegal and Glaser 1987; Gehrels and Mulamoottil 1990; Shedlock *et al.* 1993; Logan and Rudolph 1997; Mann and Wetzel 2000b), or recharge and discharge functions alternated at the same part of the basin (Cherkauer and Zager 1989; Doss 1993). In the first instance, separate recharge and discharge functions are likely to have been segregated between the centre and the margin of a wetland, or to be located in different parts of a wetland complex. In the second instance, the water exchange was governed by a water table mound down gradient but adjacent to the wetland (Winter and Pfannkuch 1985; Cherkauer and Zager 1989). During events when the groundwater was recharged, the size of the mound increased to sufficient height to create, at the boundary of this locally induced flow, a zone in which the hydraulic head was greater than that of the wetland, thus preventing seepage out of the wetland. After recharge ceased, the mound and outward flow dissipated, allowing seepage from the wetland to recommence. Factors documented elsewhere, which play a part in this phenomenon, are anisotropy of geologic materials, lake depth, and geometry of groundwater system (Winter and Pfannkuch 1985). The Becher Suite wetlands exemplify this process with the result that they change from dominantly throughflow (late winter) to discharge basins (early winter, spring) which capture marginal flow from upslope and downslope. That is, their hydrodynamics change from throughflow to upwelling or a combination of down turn flow to bypass the wetland and then upwelling.

Varied hydrological responses, which are the result of constantly changing areal and temporal distributions of recharge and discharge in wet and dry periods, have been documented in the Becher Suite wetlands and elsewhere (Winter and Rosenberry 1998; Zeeb and Hemond 1998; Mann and Wetzel 2000b). Water table gradients between wetlands and ridges often steepen as drought continues due to lower minimum water tables and higher evapo-transpiration in wetlands. Number and size of fluctuations within an annual cycle increase. Similar varied responses have been illustrated in a study of a riverine peatland (Zeeb and Hemond 1998): under average hydrologic conditions, the aquifer discharged to the wetland which discharged to the stream via a sand layer beneath the peat; under wet conditions the direct rainfall was conveyed to the stream as runoff and groundwater recharge flowed upwards through the peat; and under dry conditions, local infiltration of stream water occurred in the sand layer beneath the near stream section of wetland. Such examples serve to dispel simplistic ideas of hydrological interactions between wetlands and groundwater.

At least two palaeo water levels can be identified in the wetland fills, one below and one above the present prevailing water level. The position of the former is marked by a humic/root structured muddy sand layer at the base of the calcareous sand, the latter is marked by the upper levels of calcilutite. The lower water level is a baseline from which to measure water table rise concomitant with progradation. Since progradation of the cuspate foreland at Becher has largely ceased (Searle *et al.* 1988), the current mechanisms of groundwater rise and fall derive solely from climatic effects. The stratigraphic upper limit of the calcilutite is interpreted to be an indicator of higher former water levels and inundation. In order to predict the volume of rainfall required for inundation of each of the basins under present conditions, the rainfall data and maximum wetland water levels for the period of monitoring were used to aggregate the sites which were inundated or waterlogged under various rainfall volumes (Fig. 8-33). For regularly inundated basins, 600-700 mm of winter rainfall are required to maintain inundation. For intermittently inundated basins, more than 700 mm of winter rainfall are required to maintain inundation. For the majority of damplands to be inundated would require more than 900 mm of winter rainfall. Under present conditions, some of these basins are unlikely to be inundated even with an increase to 1100 mm annual rainfall, *e.g.*, 1N, because of its geographic position and the permeable nature of the underlying sediment. The interstitial calcilutite, present in the surface layers of this wetland, point to conditions that are out of phase with those of the last 120 years.

In addition to the volume required, the frequency of inundation would have to be matched to the recharge and discharge mechanisms operating in each wetland. For example, wetlands which are underlain by calcilutite require the highest frequency of winter rainfall between September and October. Wetlands with shallow depth to groundwater require the highest frequency of rainfall between July and August. Wetlands with rapid discharge, (*e.g.*, sw basins) would require a high frequency within a short period of time (less than one month) such that recharge exceeds discharge. In the period between 1876 and 2001, the regularly inundated basins are likely to have been annually inundated 87% of the time, the intermittently inundated basins less than 66% of the time and the damplands less than 36% of the time.

Expansion and contraction of wetlands is also a function of water level. When the water table gradually rises, wetland margins expand into proximal low lying areas which then develop wetland characteristics. Excluding water table rise consequent to coastal progradation, regional water table rise can be the result of shorter term changes to annual rainfall volumes and frequency, and sea level changes. Local changes also can affect wetland expansion and contraction. For instance, local diagenetic effects contribute to the expansion or contraction of wetland area by influencing the depth and period of inundation and the extent of the zone of capillary rise through 1) deepening of wetland through dissolution of sediment grains, 2) the compositional change from calcilutite to peat, and 3) bioturbation of sands from sheet wash shoals. The first process results in an increase in inundation and a concomitant expansion of marginal wetland area. The latter two processes, involving compositional changes which replace

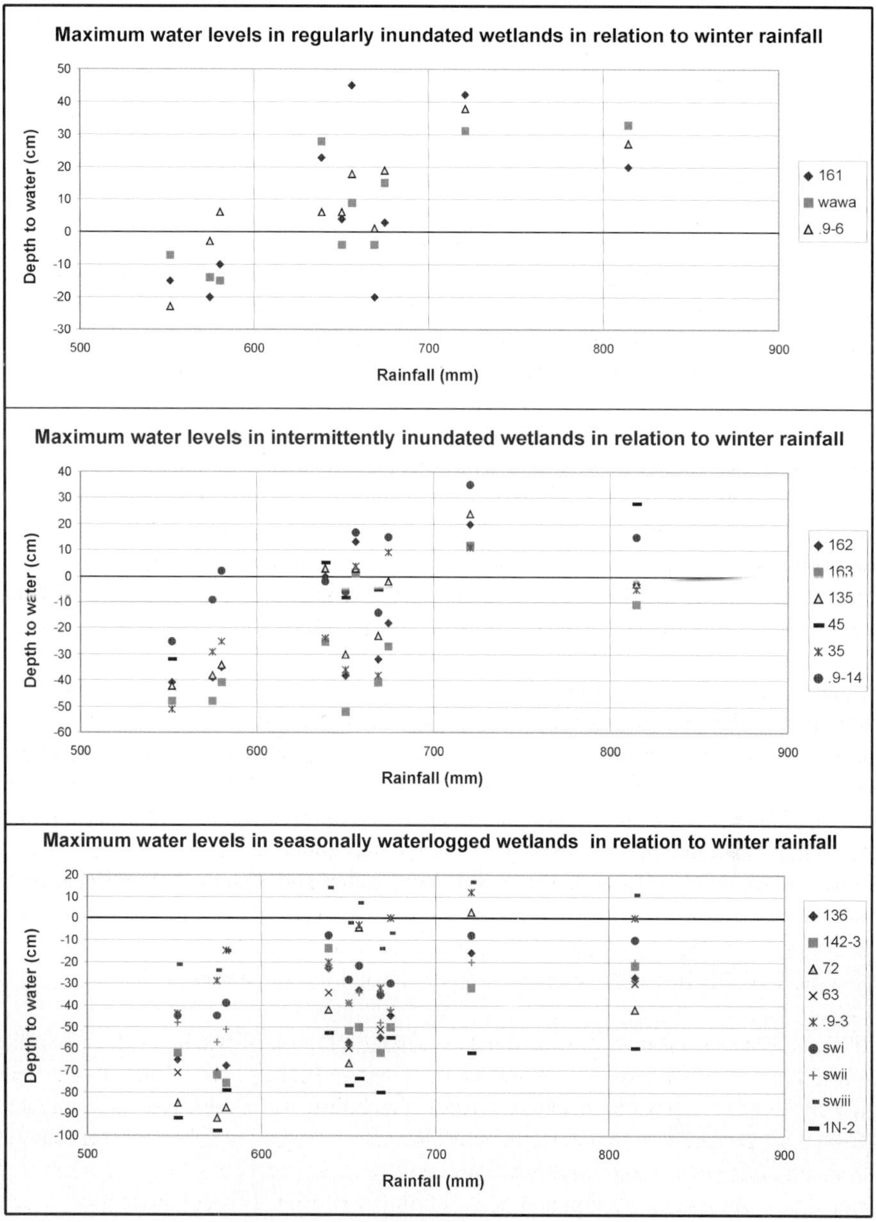

Figure 8-33. *Maximum water levels relative to wetland ground surface in relation to the annual winter rainfall 1991-2000.*

or dilute the carbonate muds, diminish the effect of surface and subsurface perching, which reduces the frequency and period of inundation and potentially contracts the size of the wetland. These three pedogenic processes are associated with the present humid climatic phase and could reverse and/or cease as a result of climatic change. In the event of a return to drier conditions, through the process of calcilutite accumulation, water levels could again increase within a basin, and potentially expand wetland boundaries.

The rise of the water table over time affects plant colonisation through pedogenic and diagenetic alteration of soil properties, particularly water retention capacity. In turn, plants alter the physical, chemical and biological nature of the sediments and water creating an evolving wetland habitat.

The consequences of lower wetland annual water table maxima and minima, which prevailed during much of this study, provided a framework to observe vegetation response. During drier years the maximum water level position was often below the rhizome layer. The normal periods of inundation and waterlogging of plant roots did not occur. Wetland species experienced similar soil conditions to non-wetland species, *i.e.*, they were dependent on the soil moisture content in the vadose zone. This resulted in competition and invasion into the wetlands of non-wetland species, *e.g.*, *Acacia saligna, A. cyclops*, and *Isolepis nodosa*. Woody remains of dead plants of these species in central zones, were noted at commencement of the study in 1991 showing that the encroachment and retreat have been a recurring event. At the wetland margin, where soil moisture content fluctuated most, the marginal wetland vegetation zones surrounding many of the wetlands disappeared, either through contraction of the wetland boundary, such that upland vegetation abutted central basin vegetation, or through plant demise and invasion by alien and endemic annual species. When higher water levels returned, some wetland species specifically belonging to the marginal zone returned.

9. WETLAND HYDROCHEMISTRY

9.1 Introduction

The scope of enquiry and the approaches used in this chapter to describe some aspects of the chemistry of the groundwater stem from the stratigraphic and hydrologic investigations described in the previous chapters, and are contained within that framework. This differs from the mainstream ecological approach to hydrochemistry of recent years, which has the general objectives of establishing mass balances, identifying chemical and biochemical processes and transformations, relating vegetation distribution patterns to a hydrochemical gradient, or quantifying anthropogenic nutrient enrichment in natural wetlands (Sjors 1952; Ponnamperuma 1972; Kemmers and Jansen 1988; Koerselman 1989; Reddy *et al.* 1999; and many others). In many of these studies, the stratigraphy and small scale hydrology were either not described or were briefly mentioned in the discussion as a variable affecting some aspect of the results.

It is important also to understand that the hydrochemical aspect of the present study is an adjunct to the main hydrological study, *i.e.*, it is a complementary study, not a detailed observation of any particular chemical element or hydrochemical process.

There were four objectives underlying the choice of chemical species, hydrochemical attributes, and methods of sampling and analyses in this study. They were:

1. to characterise the wetlands in terms of key chemical species;
2. to identify hydrological processes through the dynamics of chemical concentrations in the groundwater;
3. to explore any trends which might be related to wetland evolution; and
4. to investigate the relationship between vegetation distribution and aspects of sediment or water chemistry.

To achieve these objectives, a broad scale approach was instigated that would include all the wetland study sites rather than concentrate observations and sampling at one site. It entailed a comparison between groundwater hydrochemistry under beachridges and wetlands, and between wetlands. The sampling strategy itself was designed to emphasise seasonal patterns and patterns down the stratigraphic profile. It included an attempt to trace water movement between piezometers using cation concentrations in the groundwater at each site. Although water movement and cation exchange capacity proved to be too variable for this approach to succeed, the measurements recorded were a resource for investigating possible trends related to wetland evolution. Hydrochemical features of the groundwater and sediments were viewed as a reflection of the present evolutionary stage at which each wetland had arrived, or, as a result of ancestral processes in the wetland culminating in its present state. Details pertaining to the fourth objective are reported in Chapter 10.

The chemical components selected for monitoring in these coastal wetlands were: wetland salinity in groundwater and soil water; the concentrations of the cations Na^+, K^+, Ca^{++} and Mg^{++} in groundwater, interstitial water, plants and sediments; and the nutrient content in groundwater and sediments, with an emphasis on orthophosphate and total phosphorus. With respect to groundwater salinity, the sources identified included rainfall, chemical weathering of minerals, aerosol particulates, and fauna excreta (Wetzel 1983; Richardson *et al.* 1994). On the Becher cuspate foreland, the ionic composition of the wetland fresh waters is dominated by solutions of bicarbonate and carbonate compounds from the carbonate mud fills in the wetlands themselves, and from the basal calcareous sands, and by sodium from atmospheric precipitation derived from the ocean. The salinity of these waters expressed as *total dissolved solids* (TDS), is an estimation of inorganic materials dissolved in water (after Hutchinson 1957, 1975 Vol 2), measured in micro-siemens and converted to total dissolved solids using a calibration graph (Schlumberger 1985).

In detail, the objectives of the hydrochemical investigations and monitoring may be summarised as:

- To describe the salinity regimes and patterns which characterise the groundwater and soil water within the wetlands
- To identify processes (at the basin scale) which affect water salinity
- To relate processes and seasonal patterns to the developmental stage of a particular wetland
- To describe the cationic concentrations which characterise the groundwater and interstitial water within the wetlands
- To identify processes (at the basin scale) which affect cationic concentrations
- To relate processes and seasonal patterns to the developmental stage of a particular wetland
- To describe the orthophosphate concentrations which characterise the groundwater within the wetlands

These form the framework of sections presented in this chapter: salinity, cation concentrations and nutrients. The scope of these investigations is defined by the scale of the basin and the stratigraphic sequences of the wetland fill.

9.2 Water salinity

Various salinity regimes and patterns related to wetland hydrological processes characterise the groundwater and soil water within the wetlands.

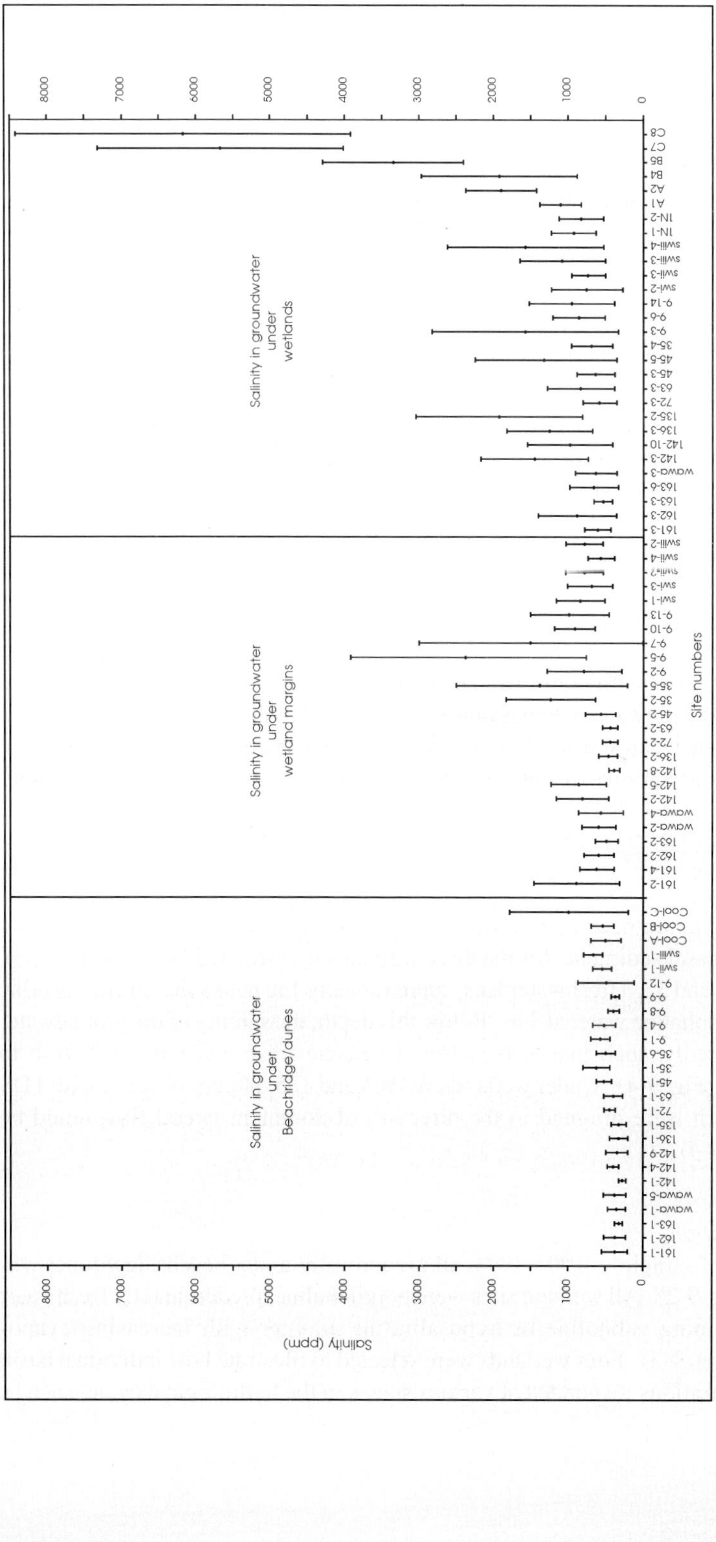

***Figure 9-1.** Graph of groundwater salinity under beachridge/dunes, wetland margins and wetlands [Mean salinities and standard deviation over 3 years (Becher sites) and 15 months (Cooloongup sites)].*

Rain collected on the coast, mid-way on the cuspate foreland and near Cooloongup, inland to the east, exhibited variable salinity. For winter rain, the mean TDS value was 99±116 ppm (n = 2) on the coast and 114 ±86 ppm (n = 10) mid-way along the axis of the cuspate foreland. For spring, the TDS value on the coast was 14 ppm (n = 1), mid-way was 43 ppm (n = 1), and at Cooloongup was 35 ppm (n = 1). For summer, the TDS value was 314 ppm (n = 1). The major cation contributing to salinity in winter and spring rain was sodium; in the single summer rain sample, sodium and calcium concentrations were similar.

9.2.1 Phreatic groundwater salinity

Spatial variation
For this study, two categories of freshwater were defined: low salinity freshwater encompassing values <500 ppm; and high salinity freshwater encompassing values 500 -1000 ppm. The subdivision separates the freshwater regime into that recently precipitated and that "aged" through various concentration processes over the year, or inter-annually.

Groundwater salinity, though largely in the freshwater field, varied spatially at the sub-regional and local scales. At the sub-regional scale, wetlands furthest from the coast were underlain by low salinity freshwater and wetlands nearest the coast were underlain by higher salinity freshwater to hyposaline water. Local variability was also exemplified by the contrast between low salinity freshwater within and under beachridge/dunes, and the low to high salinity freshwater ranging to hyposaline water, within and under the wetlands (Figs. 9-1, 2, 3).

Stratification
Sampling in late autumn, exhibited stratification between water interstitial to wetland sediments and basal sediments for the three wetlands investigated (WAWA, 135, 35). In wetlands 135 and 35, a freshwater lens, approximately 1 m below the surface, overlay a thin layer of subhaline water at 3 m. Below this depth, the salinity of the groundwater decreased markedly, such that at 6 metres it became fresh and remained fresh to 40 m at all sites (Fig. 9-4). Under wetlands WAWA and 135, plumes of water with TDS >400 ppm which were oriented in the direction of dominant lateral flow could be identified (Fig. 9-4).

Temporal variation
For the period of sampling, 1991-1994, all groundwater under beachridge/dunes was stasohaline (Fig. 9-2). All wetland sites were poikilohaline, predominantly freshwater in winter becoming subhaline or hyposaline in summer with increasing evapo-transpiration (Fig. 9-3). Four wetlands were selected to illustrate how individual basin salinity concentrations responded at various stages in the hydrological cycle, such as

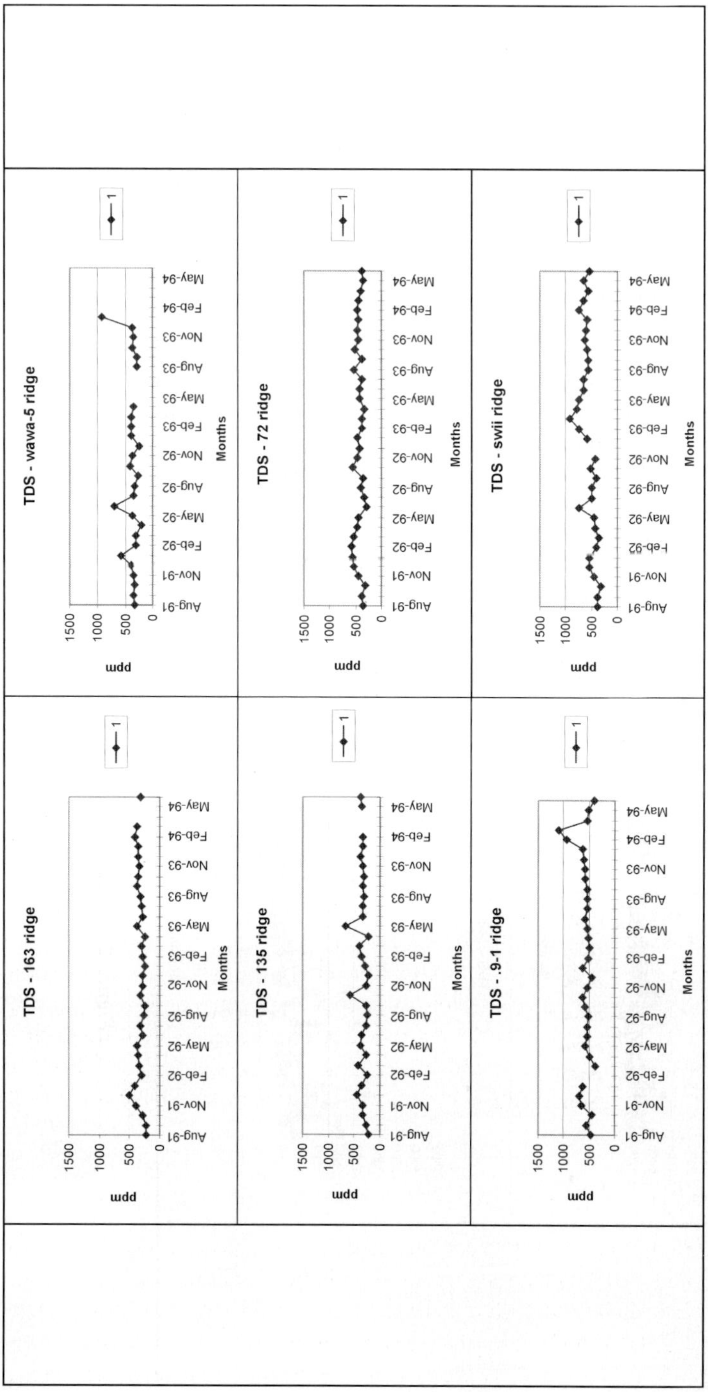

Figure 9-2. *Seasonal variation in TDS in groundwater under beachridge/dunes.*

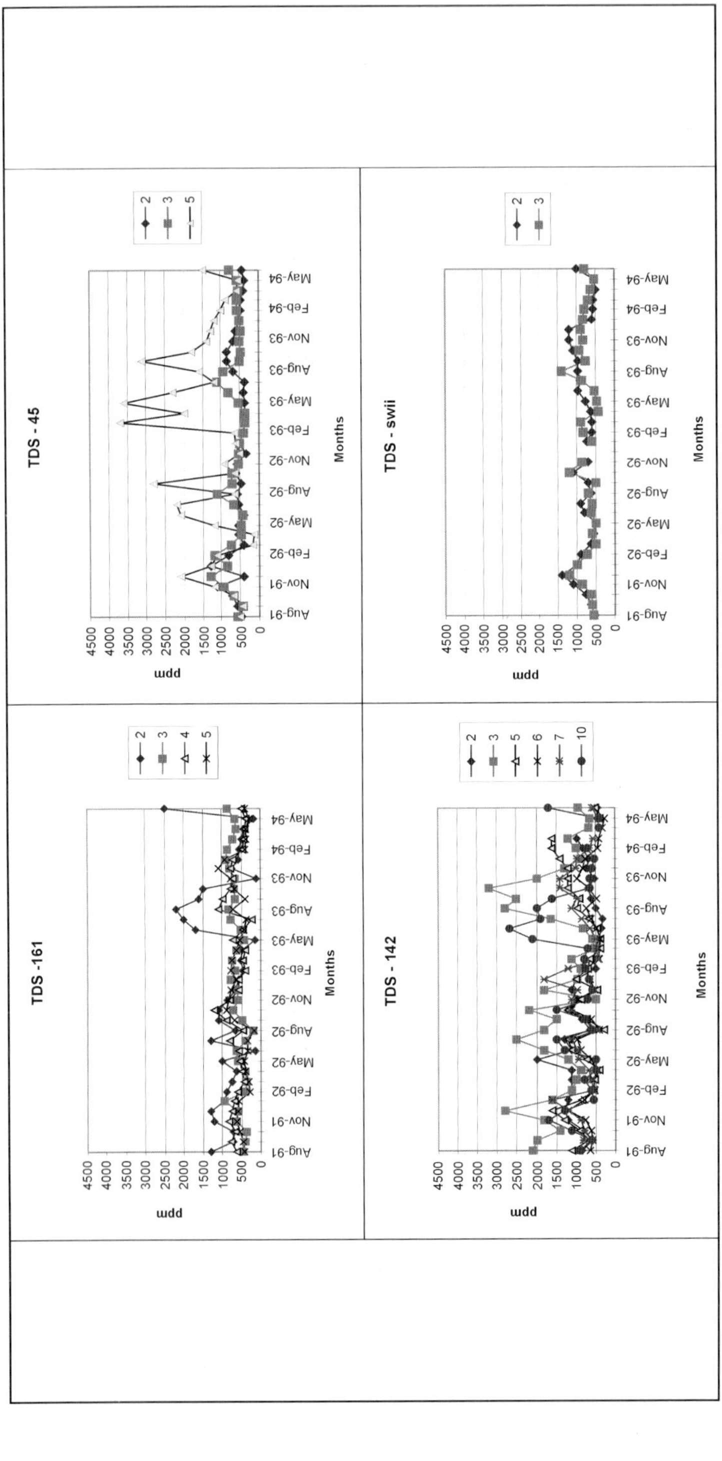

Figure 9-3. *Seasonal variation in TDS in groundwater under wetlands.*

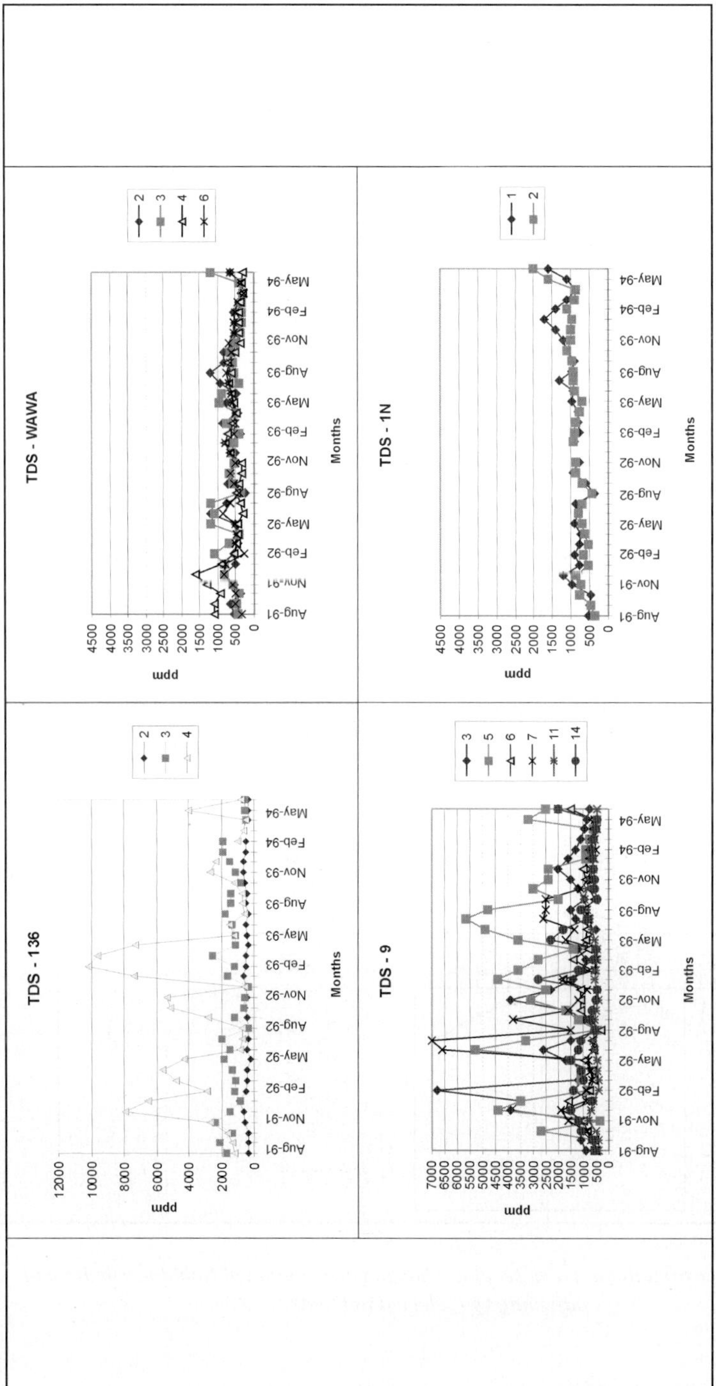

Figure 9-3(cont.). Seasonal variation in TDS in groundwater under wetlands.

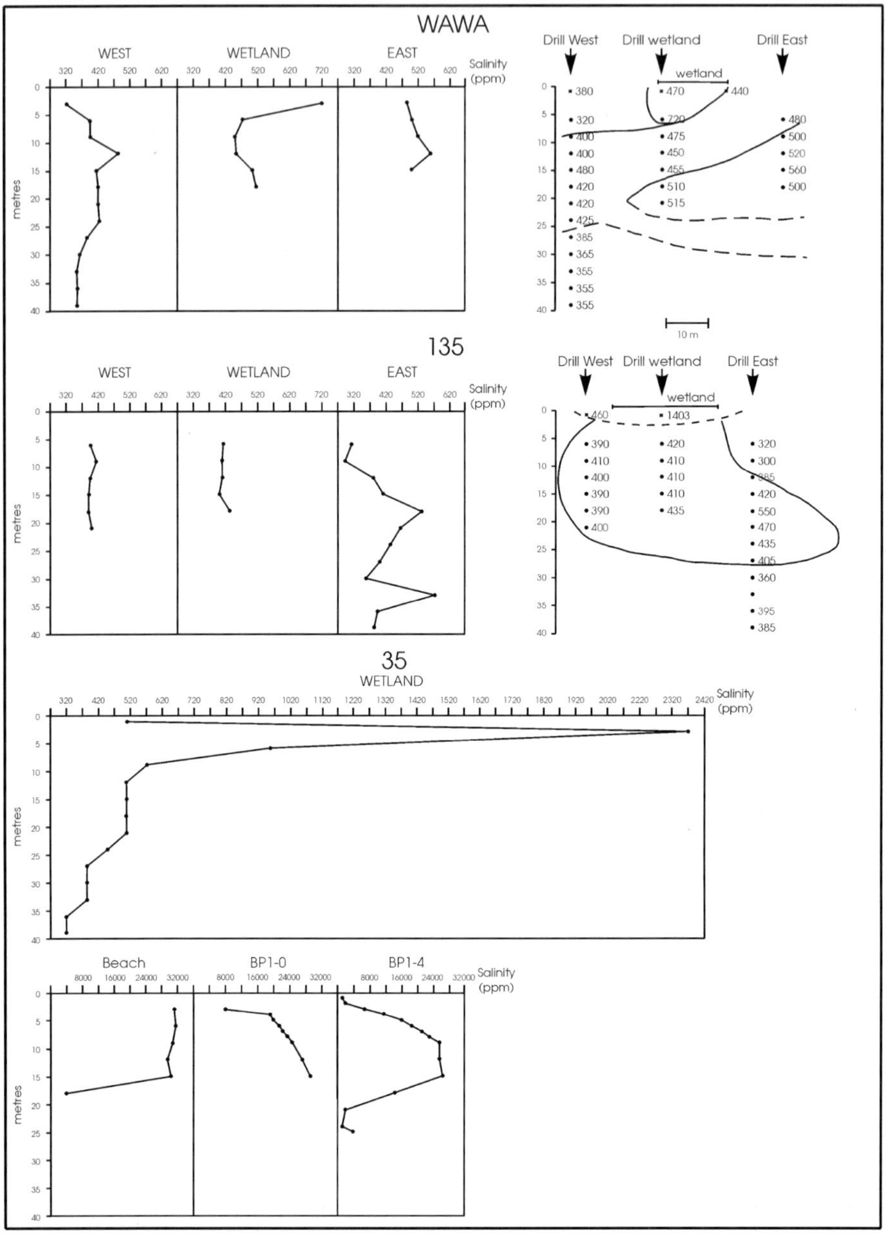

Figure 9-4. *Interpretational cross-sections showing distribution of isohalines under and adjoining two selected wetlands.*

wet or dry seasons, during surface water perching, groundwater level stillstands (April, May), and upward leakage or recharge (spring, summer), particularly at wetland margins: wetlands 161, 136, 9 (5, 6, 7), and swii.

- Wetland 161 is the oldest wetland and is representative of well established wetlands with a steep western slope and interfingering margin *viz.* 162, 163.
- Wetland 136 is of medium age and is representative of wetlands with a moderate western slope and simple western margin *viz.* 142, 135, 72, 63, and 45. Two contrasting wetland sites are cited for this study, one of which (site 4) exhibits surface water perching.
- Wetland 9 is a medium age wetland, with an interfingering margin on the western side and a simple eastern margin. The western margin (site7) is underlain by calcrete and exhibits perching.
- Wetland swii is a young wetland near the coast with simple margins and is representative of similar types *viz.* 1N, swi, swiii and 9-10, 11, 14.

Figure 9-3 illustrates several patterns. Salinity peaks occurring in winter (June/July) and summer (December/January) showed that the fluctuation in groundwater salinity was not simply related to seasonal evaporation changes. Sites of surface water perching exhibited four times the range in salinity of other sites. At times of groundwater level stillstands, salinities were lower and fluctuated less. Salinities at the eastern and western margins of wetlands differed (sites 2, 4 in wetland 161; 5, 7 in wetland 9). Salinity effects of upward leakage could not be differentiated.

Identification of hydrological processes in relation to salinity patterns

Various hydrological processes have an impact on, or affect, groundwater salinity. These include:

- rainfall frequency and volume
- evapo-transpiration
- groundwater fluctuation (rise May-September, fall October to March)
- groundwater level stillstands
- marginal effects, and
- upward leakage (spring, summer).

The pattern of monthly concentration and dilution, expressed as peaks and troughs, is evidence for the shaping of the pattern of salinity by evapo-transpiration and rainfall. Winter precipitation brings various salts in solution to wetlands, albeit in low concentrations and evapo-transpiration concentrates them.

Rainfall enters the wetland by two different pathways: vertically downward as direct infiltration and vertically upward as groundwater rise from meteoric recharge. An increase in the frequency and volume of rainfall changes the balance between meteoric and groundwater input in a given wetland, resulting in a decrease in salinity. Within the sediment profile of the wetland, frequent rainfall will have several affects: 1) flushing, 2) mobilisation of solutes within the profile, and, 3) in positive and negative combination with volume, an increase in the annual range of water salinity. Increased frequency of rain creates more permanent conditions of saturation with consequent increased solubility and hydration of previously adsorbed ions. As saturated downward flow is initiated, mobilisation of solutes occurs, resulting in sediment leaching (Figure 9-3: WAWA, 45, 9, swii, in Feb-March 1992, 135-2, in March-April 1992). Decreased frequency of rain events creates conditions of alternate pore water saturation and undersaturation. These successive swings can set up conditions alternately favouring dissolution and precipitation of saline compounds. Interrupted infiltration, apart from retarding downward flow, can also allow evaporation to reverse soil water movement, such that salts are transported to other levels within the profile. Variable volume of rainfall can have several affects: 1) changes in amounts of salts in groundwater through dilution or concentration (dilution affects in wetlands WAWA, 142, 136, 9, swiii, 1N August 1992), 2) greater or lesser water table rise, and 3) accumulation or depletion of salts in sediments.

Frequency and volume of rainfall are functions of climatic variability. During the period of monitoring, the climatic cycle was in a stage of relative aridity and many years prior to, and during 1991-1994, had below average rainfall and aseasonal rain events which had a cumulative effect on the salinity. This was expressed in higher salinity values towards the middle and latter part of 1993 and in a change from isolated salinity peaks and spikes to prolonged or extended high values. Garcia *et al.* (1997), in their study of temporary saline lakes, reported similar results in response to different annual hydrological budgets.

Elevated salt concentrations in the groundwater occur during December to March due to evapo-transpiration. The agents of evapo-transpiration discharge and its magnitude, under similar atmospheric conditions, vary with changes in the position of the water table relative to the ground surface. Free surface water and a shallow water table, at or near the ground surface, are subject to direct solar radiation and wind induced evaporation, as well as transpiration from plants in vegetated areas. A deeper water table, if subject to any evaporation, is likely to be affected through capillary rise induced by evaporation at the ground surface (Eghbal *et al.* 1989).

In almost all wetlands there is a decrease in groundwater salinity concomitant with groundwater rise. This occurs whether the water table rises above the ground surface or remains in the subsurface. However, as the groundwater fluctuates in response to varying rainfall and recharge, the salinity both increases and decreases. The change in salinity depends on the path to the water table taken by infiltrating water, the degree of sediment saturation, and the volume of infiltrating water. Initially, there are likely to be preferred conduits through continuous macropores, but these have been shown to

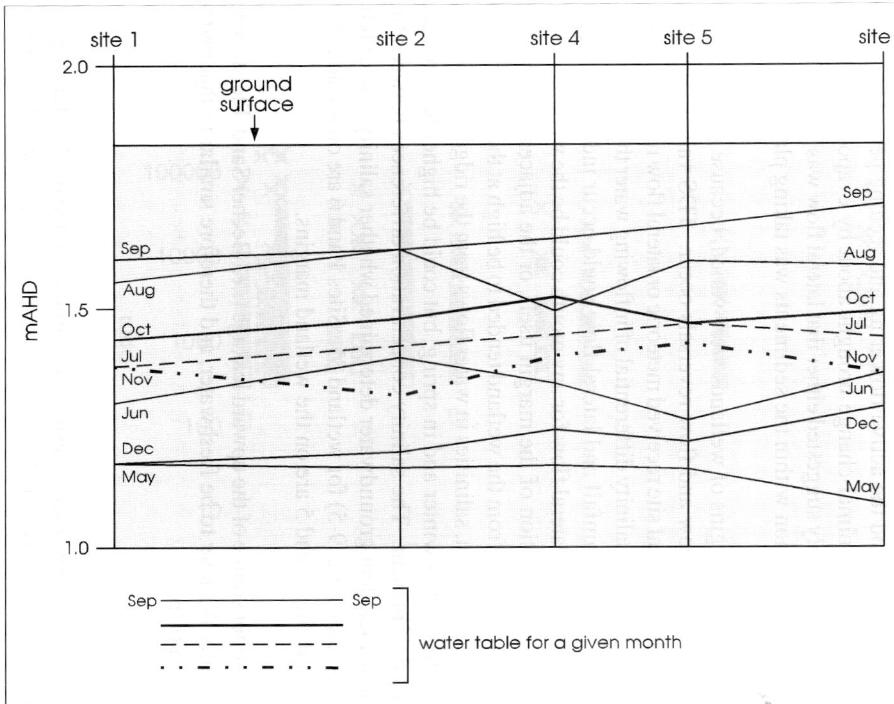

	Salinity (ppm)	
	Site 2	Site 5
May	1300	575
June	2000	2400
July	2000	2000
Aug	2700	2800
Sept	2300	1200
Oct	2600	1100
Nov	1200	1200
Dec	840	540

TDS in reponse to water level flow		
	site 2	site 5
May/June	large recharge	less recharge
June/July	water level the same TDS unchanged	recharge and seepage from ridge TDS increases
July/Aug	large recharge no lateral flow TDS increases	medium recharge no lateral flow TDS increases
Aug/Sept	no net recharge (recharge balances discharge) slight decrease in TDS	small recharge seepage from 6 of lower salinity water (500 ppm) large decrease in TDS
Sept/Oct	medium groundwater fall flow from Wetland not registered TDS rises	large groundwater fall flow from Wetland not registered TDS falls
Oct/Nov	large groundwater fall flow from ridge site 1 of lower salinity water (650 ppm) TDS falls	small groundwater fall no inward flow TDS unchanged
Nov/Dec	medium groundwater fall seepage from Wetland of lower salinity water (330 ppm) TDS falls	large groundwater fall seepage from ridge site 6 of lower salinity water (480 ppm) TDS falls

Figure 9-5. *Water table configuration, flows and salinities in wetland 35 (May-December 1992).*

times the corresponding groundwater salinity (Fig. 9-6). Summer subsampling of the top 10 cm layer into 0-1 cm, 5 cm and 10 cm layers, showed that a salinity gradient existed from the surface to 10 cm, and that the highest salinity occurred in the very surface layer (Fig. 9-7). Locally, the salinity of the surface layer exceeded 300,000 ppm, with halite, gypsum, and carbonate precipitates. Although high salinity concentrations were present for up to three months of any year, they were removed by substantial rain events, so that no build up of salt in the sediments occurred over time.

Spatial variation
The majority of measurements of total dissolved solids in interstices or pore waters of soils for wetland sites were between 1,000 and 20,000 ppm (hyposaline), although the sample range spanned freshwater to hypersaline (600 ppm to 418,000 ppm) (Figs. 9-6, 8). All marginal sites recorded marked fluctuations and peaks but the highest salinities were recorded at sites where perching occurred, *i.e.*, where calcilutite or the shallow calcrete layer was present. Other comparatively minor peaks in soil water salinity occurred at sites supporting woodland and low forest vegetation.

Temporal variation
Soil water salinity was measured over the 32 month period 1991-1994 (Fig. 9-8). In summary, the graphs of temporal variation exhibited consistent peaks annually, although the duration and intensity of these peaks varied between sites. There was an increase in the number and duration of soil water peak salinities during 1994, which was the second year of below average rainfall (Fig. 9-8). At this time differentiation between wetland and marginal sites became apparent. Peaks corresponded to periods of elevated evaporation and low water table positions, which lowered the soil moisture content of the surface horizons. Soil moisture content in the surface layers approached zero in a number of wetlands at this time inducing salt precipitation (sites 161-2, 4, 142-5, 6, 7, 63-3, 35-2, 45-3, 9-2, 3, 11, 1N).

In the four wetlands described above *viz.* 161, 136, 9 (5, 6, 7) swii, soil water salinity values were low during the winter and spring, but increased during the summer reaching their maximum in autumn. During the months of 1992-93, when salinity values peaked, the site of surface water perching (136-4) exhibited salinity values almost three times that at the other central wetland site (136-3), and wetland marginal sites also exhibited much higher concentrations (Fig. 9-8).

Identification of hydrological processes in relation to salinity patterns
After the initial period of consistent rainfall, the salts reached the water table, *i.e.*, there was a transference from soil water to groundwater, resulting in a peak annual record for groundwater salinity.

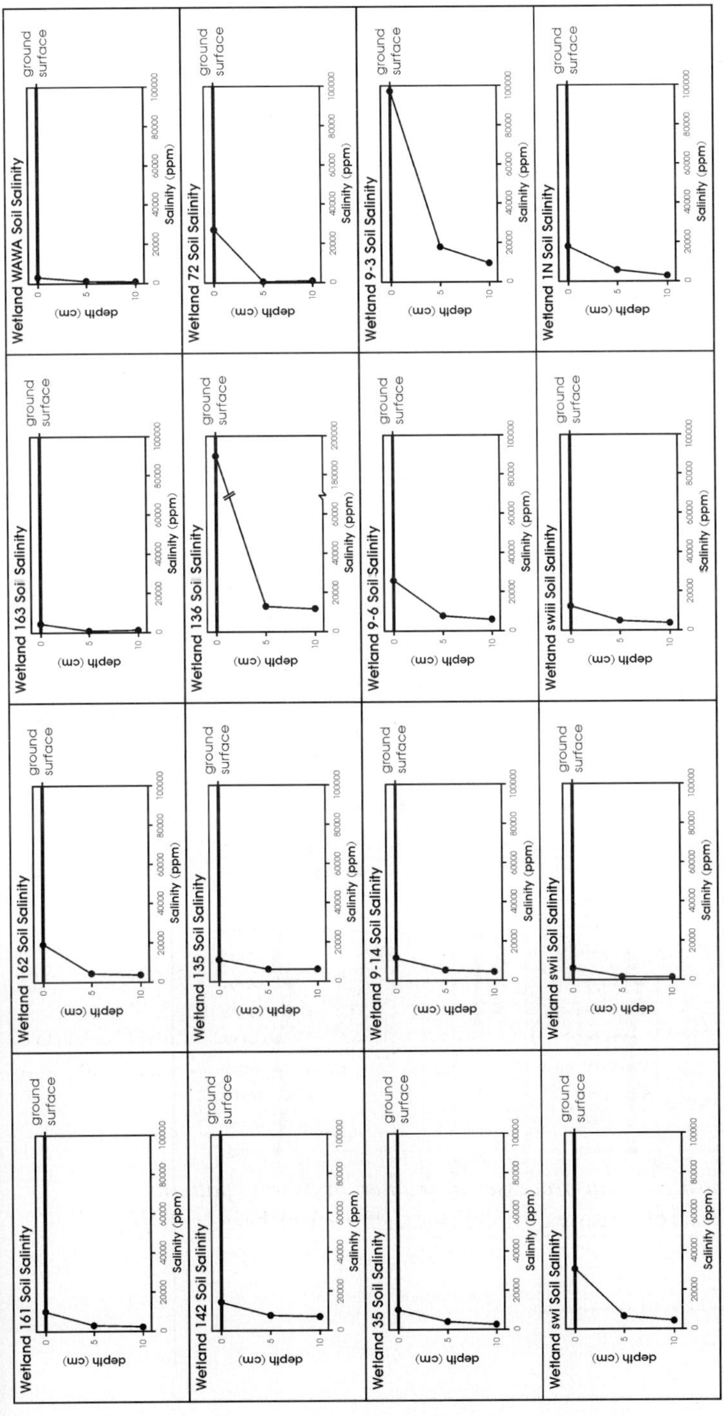

Figure 9-7. *Profiles of soil water salinity, 0-10 cm, for the various wetlands.*

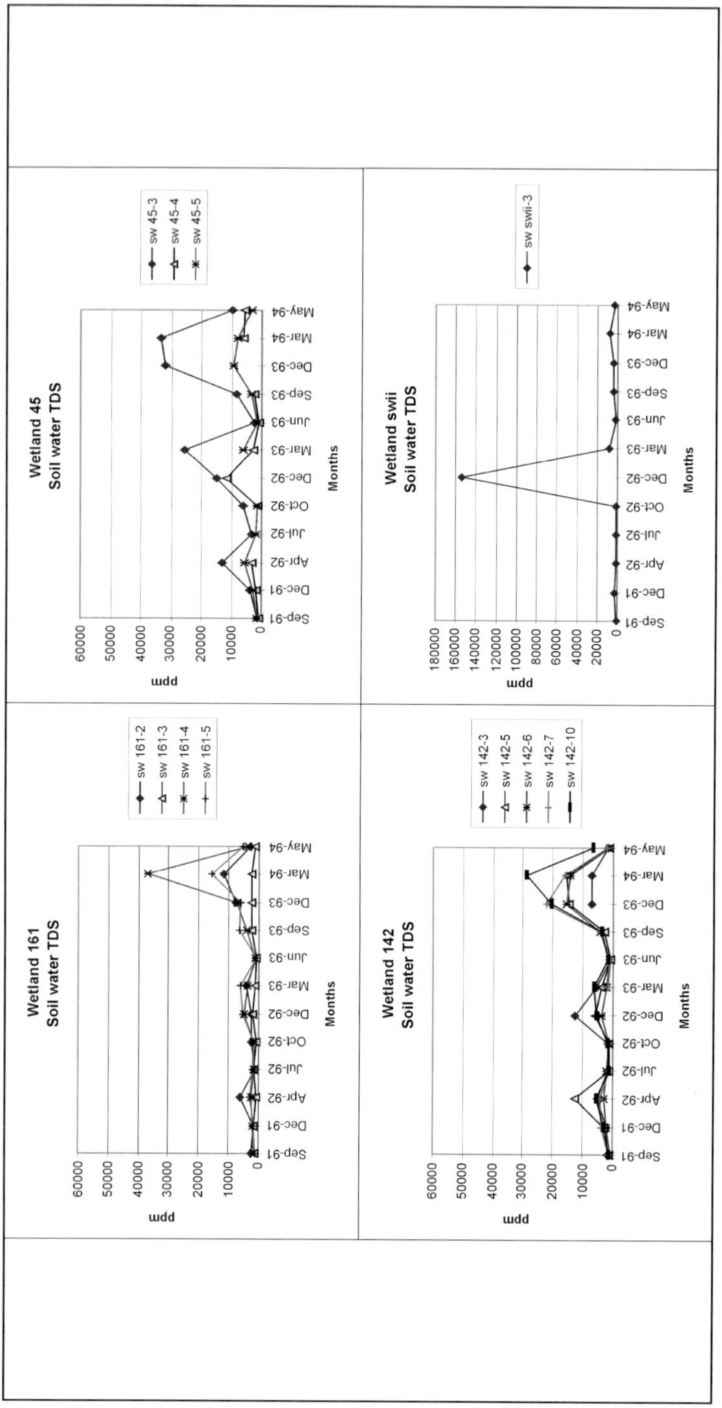

***Figure 9-8.** TDS of soil water in wetlands and wetland margins.*

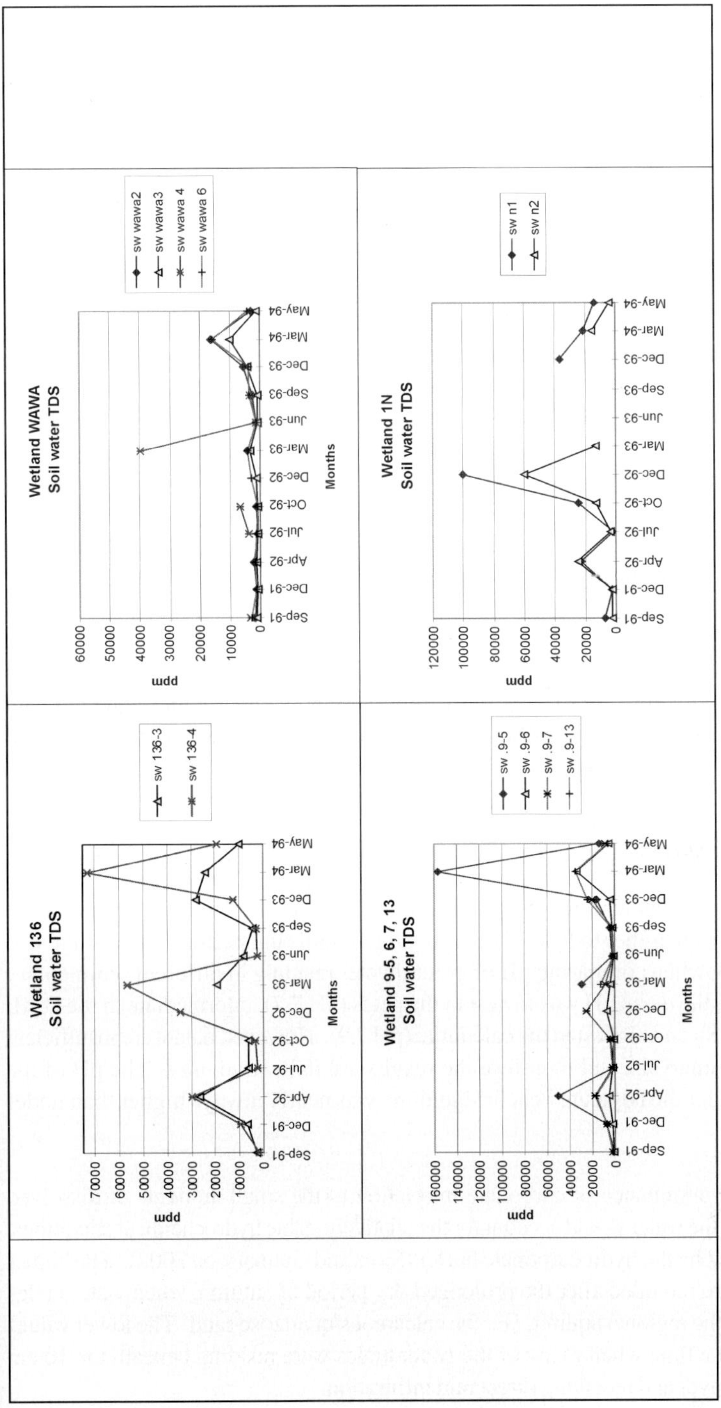

Figure 9-8 (cont.). TDS of soil water in wetlands and wetland margins.

Eghbal *et al.* (1989), in their study of endorheic lakes underlain by fine loamy calcareous and montmorillonite soils, found the highest salt concentrations in the subsoil, indicating that leaching of salts was dominant over evaporative rise from a water table. In a study by Arndt and Richardson (1989), soil water salinities in shallow inundated and waterlogged wetlands underlain by dolomite and shales, and dominated by freshwater recharge in a subhumid setting in North Dakota, were found to be non-saline, with the chemistry of the interstitial waters resulting from evapo-transpiration, recharge hydrology (downward saturated flow, leaching), ionic mobility, and exchange relationships. The Becher pattern accords with these two examples.

9.2.3 Salinity and developmental stage of wetland
Groundwater salinity does not appear to be related to wetland developmental stage because salts are generally exported annually from the system via water transport. This transport is initially vertical and then lateral. Groundwater at depth exhibited stratification, indicating very slow lateral movement away from the wetland. However, salinity is affected by the products of diagenesis, which form during periods of exposure, or through vegetation induced precipitation throughout the history of a particular wetland, especially those which induce perching.

The wetlands, which tended towards the lower end of the range of soil water salinities, fell into two groups, those which were regularly inundated, and those which were the youngest of the wetlands. In each case, the relatively high water tables and abundant leaf litter, which inhibits evaporation, may be the underlying reason for the stable soil water salinities.

9.3 Groundwater pH

The pH of the groundwater under wetlands ranged between 7.1 and 8.3. There was an overall change from higher to lower pH during the monitoring period. From August to November 1993, when comparing pH of groundwater residing in different sedimentary layers (Fig. 9-9), the mean pH was lowest in the peats (pH 7.7), intermediate in the OME calcilutite (pH 7.8), and highest in the calcilutite (pH 7.9). However, data were insufficient to replicate this analysis, and therefore the results are indicative only. The pH of the groundwater under the adjacent beachridge/dune was nearly always higher than under the wetland.

The quantitative dominance of carbonate in relation to the small amounts of dissolved organic matter in the water would account for the relatively stable hydrochemical conditions, with pH regulated by the hydrocarbonate buffer (Sjors and Gunnarsson 2002). The higher values of pH were recorded after the prolonged dry period of autumn, when water tables were residing in the regional aquifer, *i.e.*, the calcareous/quartzose sand. The lower values were recorded at a time when many of the water tables were residing beneath the 10 cm surface organic layer and receiving direct rain infiltration.

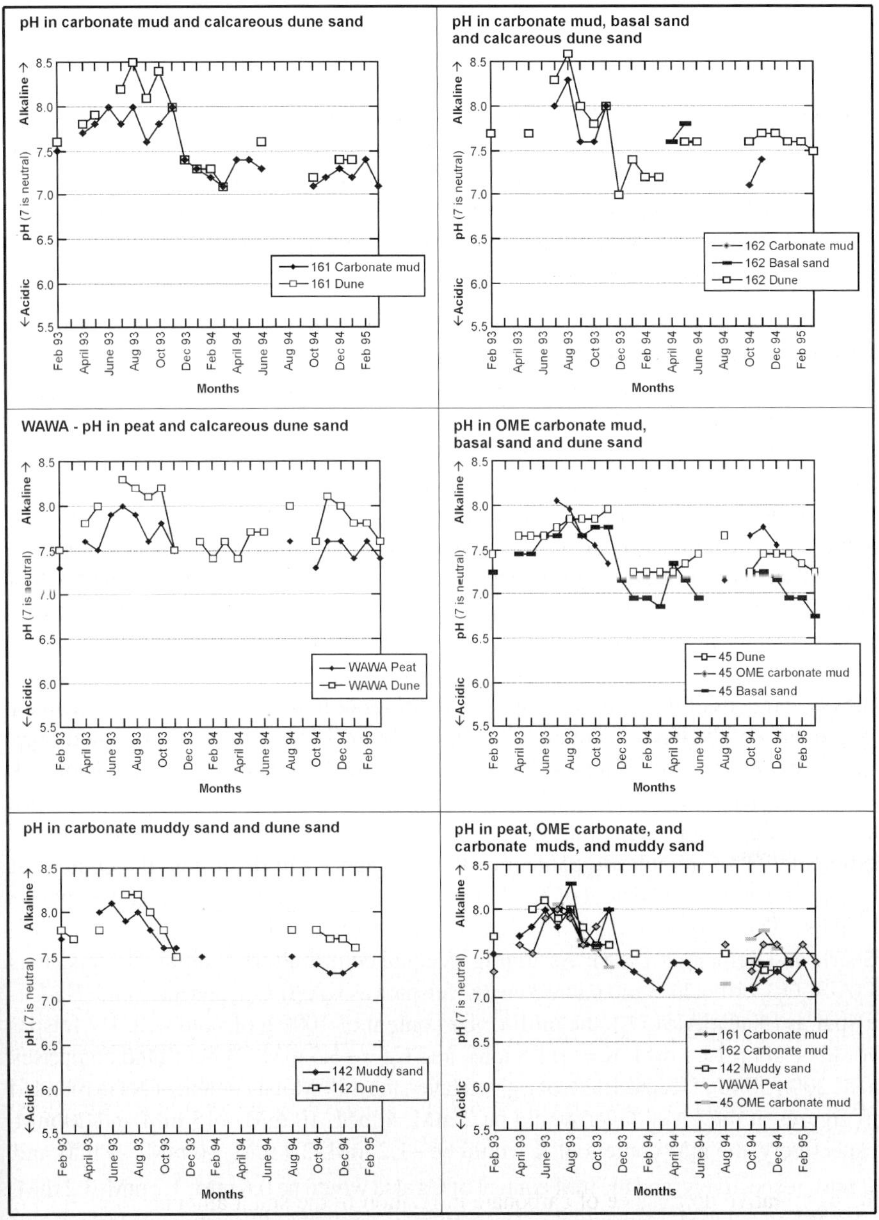

Figure 9-9. Comparison of pH of groundwater in different wetland stratigraphic fills.

9.4 Cation content

This part of the study was undertaken as an adjunct to the main hydrological study, and incorporates the first two objectives. Although the study did not fully achieve the second objective at the basin scale, the results of the investigation provided data about ionic water chemistry resulting from the interaction between vertical groundwater movement and the range and chemistry of sediment types intercepted. Throughout this section, cation concentrations are expressed either as parts per million (ppm) or as millimoles per litre (abbreviated as mM/L).

9.4.1 Sources of metal ions

Some of the background patterns of metal ion occurrence in wetland sediments, and metal lability and up-take, either determined from the literature or determined *a priori*, are presented so that the results of this study can be viewed within an established framework of the occurrence of metals in sediments. The main sediment types of the various stratigraphic fills and the underlying parent material include calcareous quartzose sand, humic sand, OME calcilutaceous muddy sand and mud, calcilutite and peat. There are minerals, grains and organic components in these sediments that naturally contain Ca, Mg, Na, and K, or contain some of the locally precipitated salts of these metals in residual pockets such as foraminiferal chambers.

The occurrences of the four metal species within sedimentary grains and sedimentary components of the Becher wetland sequences (Table 9.1) are drawn from mineralogic texts (Deer *et al.* 1966; Bathurst 1975), plant chemistry texts (Boyd 1978; Klopatek 1978; Wheeler *et al.* 1992), beach sand petrology of coastal beaches of southwestern Australia (Searle and Semeniuk 1988), XRD results, and microscopic examination of sediments. The siliciclastic content of the sediments will determine the content of felspar, and hence the content of Na, K and Ca, residing in the alkaline and plagioclase felspars. The felspar content of beach sand in this region is < 2% and mostly ~ 1% (Searle and Semeniuk 1988). Assuming an equal contribution of albite (Na-felspar as $NaAlSi_3O_8$), orthoclase and microcline (K-felspars as $KAlSi_3O_8$), and andesine (Na-Ca-felspar as $[NaCa]AlSi_3O_8$), the milliMolar content of 1000 g of sand with 1% felspar would be Na = 19.1 mM, K = ~11.8 mM, and Ca = ~6.5 mM. Where sand comprises 10%, 20%, 50%, 75% and 100% of a given layer, the milliMolar content of Na in rounded off figures at the given layer would be 2 mM, 4 mM, 10 mM, ~15 mM and 20 mM, respectively, the mM content of K would be ~1.2 mM, 2.4 mM, ~6 mM, ~9 mM and 12 mM, respectively, and the mM content of Ca also would be 0.6 mM, 1.2 mM, 3.2 mM, 4.5 mM and 6.4 mM, respectively. The carbonate content of the sediments largely will determine the amount of Ca and Mg in the profile.

Table 9.1 *Occurrence of the four metal species within various sedimentary materials*

Metal	*Source/occurrence of the metal species in the sediment in the Becher wetlands*
Ca	1. mainly occurring in carbonate mud (*i.e.*, calcite, aragonite, and Mg-calcite); in natural situations, and potentially labile, especially under acidic conditions 2. occurring in carbonate grains in the sand (*i.e.*, skeletal grains of calcite, aragonite, and Mg-calcite); in natural situations, and potentially labile, especially under acidic conditions 3. very minor occurrence in calcic felspars, *e.g.*, andesine, in sand fraction; largely locked into sediment, but will be detected in chemical analyses as < 1% Ca 4. occurrence in plant material, and hence in decaying/decayed plant material in soils, in peat, and in peaty sand
Mg	1. mainly occurring in carbonate mud (*i.e.*, Mg-calcite); in natural situations, and potentially labile, especially under acidic conditions 2. occurring in carbonate grains in the sand (*i.e.*, skeletal grains of Mg-calcite); in natural situations, and potentially labile, especially under acidic conditions 3. occurrence in plant material, and hence in decaying/decayed plant material in soils, in peat, and in peaty sand
Na	1. occurring in sodic felspar in sand, *e.g.*, albite; largely locked into sediment, but will be detected in chemical analyses as <1% Na 2. occurrence in plant material, and hence in decaying/decayed plant material and detritus in peat, peaty sand, sand and mud 3. possible minor local pockets occurring as NaCl within intra-skeletal cavities such as foraminiferal chambers precipitated from interstitial waters but not fully removed by the de-ionised water flushing (this source, however, is unlikely)
K	1. as a constituent in plants (comprising c. 1%), in plant detritus in sand and mud, and in decaying/decayed plant material in soils, peat, and peaty sand 2. occurring in potassic felspar in sand, *e.g.*, orthoclase and microcline; largely locked into sediment, but will be detected in chemical analyses generally as <1% K 3. possible very minor local pockets occurring as KCl within intra-skeletal cavities such as foraminiferal chambers precipitated from interstitial waters but not fully removed by the de-ionised water flushing; an artifact of the processing (this source, however, is unlikely)

For the sediments, a strong relationship can be expected between some metal content and the type of sediment (*viz*., sand with felspar will have concentrations of Na and K elevated above that occurring in carbonate muds, sediments with plant detritus

Table 9.2 Occurrence of metal species in components of various sediment types

Sediment	*Components*	*Anticipated metals species*
sand	felspar carbonate grains minor plant detritus	Ca in carbonate grains, plant detritus and to some extent in feldspars Mg in carbonate grains, plant detritus Na in plant detritus and felspars K in plant detritus and felspars
muddy sand	felspar carbonate grains carbonate mud minor plant detritus	Ca in carbonate grains, calcilutite, plant detritus, and, to some extent, in feldspars Mg in carbonate grains, plant detritus Na in plant detritus and felspars; also potentially as a very minor residue as NaCl precipitated in cells of plant detritus in very surface layer, but not removed by the de-ionised water flush K in plant detritus and felspars; also potentially as a very minor residue of KCl precipitated in cells of plant detritus in the very surface layers, but not removed by the de-ionised water flush
calcilutite	calcite > Mg-calcite > aragonite; minor sand grains; minor plant detritus; scattered sand grains	Ca mainly in carbonate mud Mg mainly in carbonate mud Na in plant detritus, or a minor residue as NaCl in skeletal cavities and plant cells, in the very surface layer, but not flushed out during rinsing K in plant detritus, or as minor KCl residue in skeletal cavities and plant cells in the very surface layer, not flushed out during rinsing; some Na, K, and Ca in felspars, and some Ca in carbonate grains in sand
peat	organic matter and plant detritus; scattered sand grains	Ca, Mg, K and Na reside in plant detritus, or are residual as calcite, Mg-calcite, KCl and NaCl, minor precipitates from original interstitial waters residing locally in cavities of plant cells, in the very surface layer, but not flushed out during rinsing; some Na, K, and Ca in felspars, and some Ca in carbonate grains in sand

and organic matter may have elevated K, and carbonate muds will have elevated Ca and Mg). However, while Ca, K and Na can be contributed from felspars in sand, and Ca and Mg can be contributed from carbonate skeletons in sand, their contribution is only of note where sand dominates, and where they constitute sand grains in peaty sequences. If sand comprises <10% of the sediment, the contributions of these

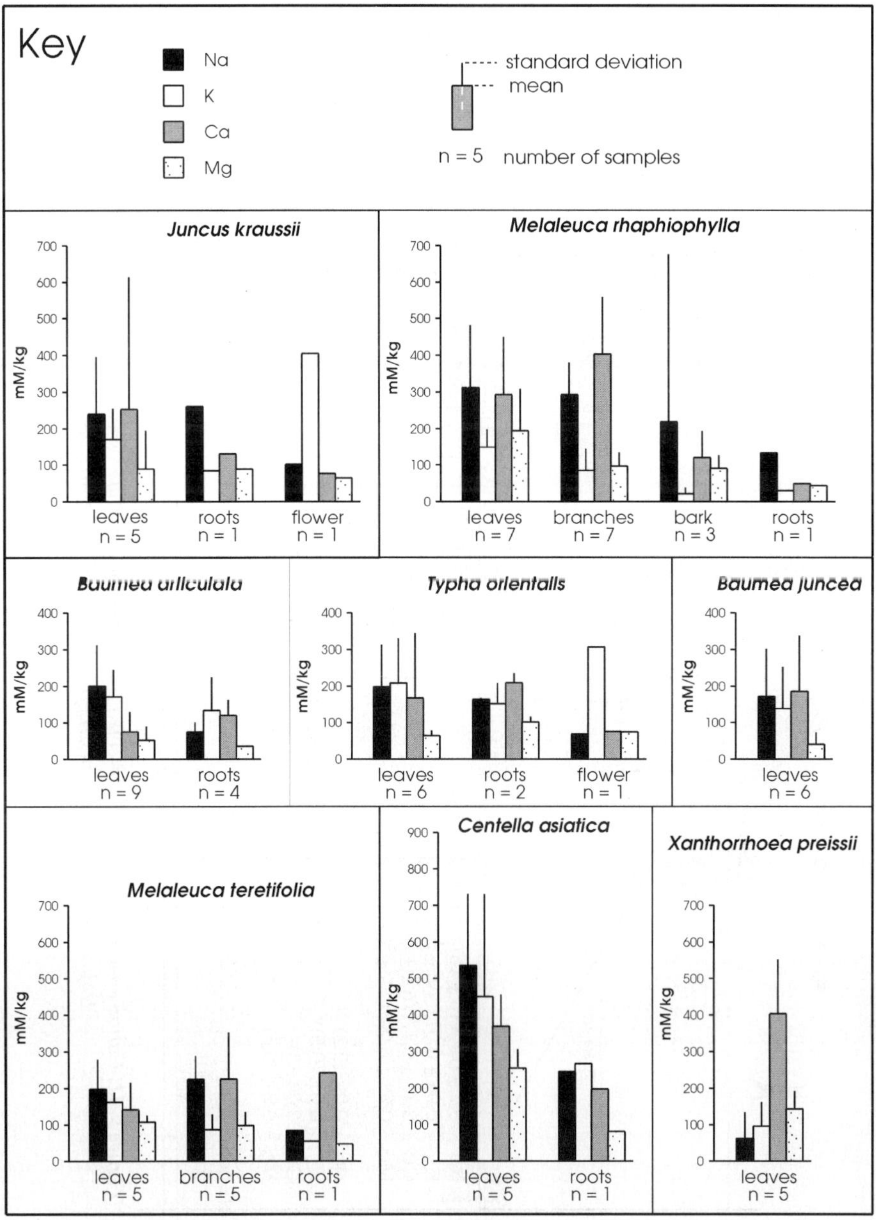

Figure 9-10. *Na, K, Ca and Mg content of various wetland plants.*

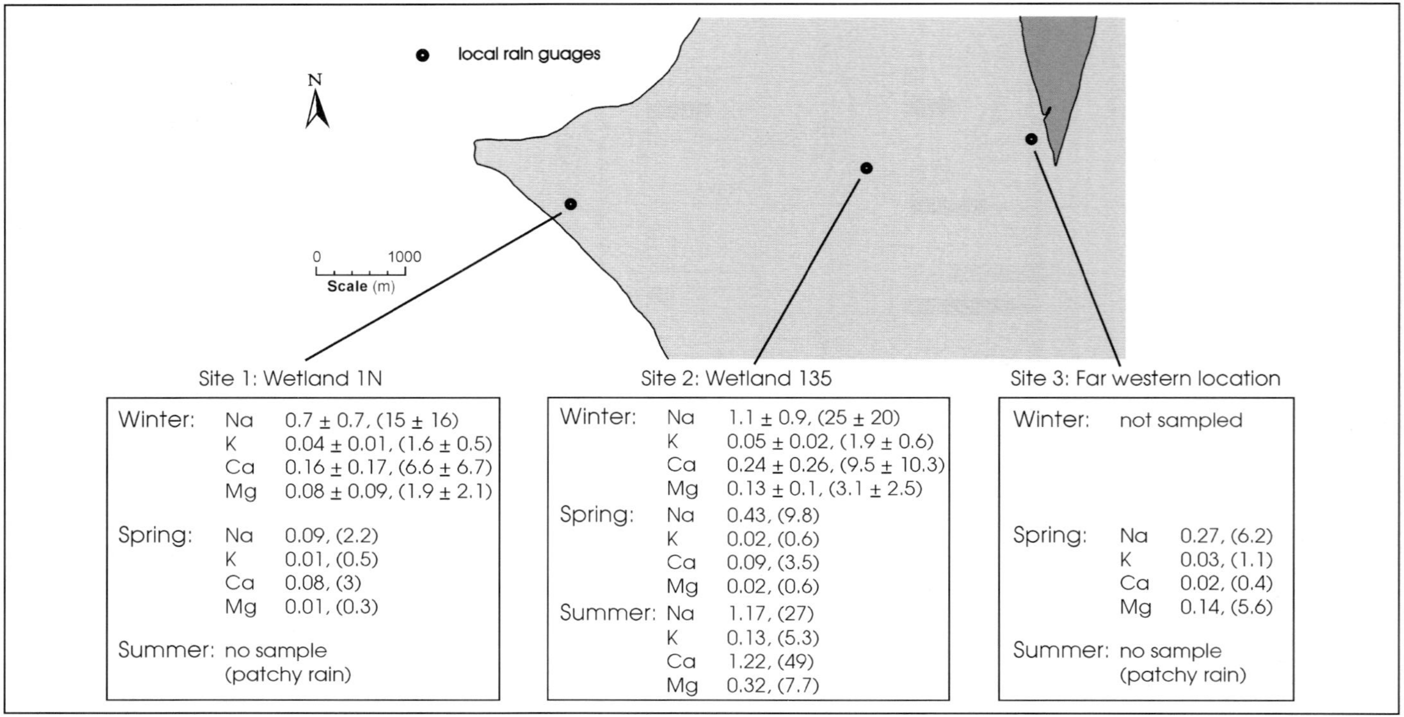

Figure 9-11. *Cation content of rainwater collected from three sites in winter, spring and summer 1996 Unbracketed values are mM/L, and bracketed values are ppm.*

metals to the chemistry of the sediments is overshadowed by carbonate mud or by the metal content in organic accumulations. Note also that the mud-sized fractions of the wetland sediments are carbonate mud and quartz and not phyllosilicate clays, and so are not likely to bond cations.

The components of the various sediment types and their metal species are described in Table 9.2. The content of Na, K, Ca and Mg in key wetland plants is graphed in Figure 9-10.

9.4.2 Cation concentrations in rainfall

Sodium dominated the cationic concentration in winter and spring rain, with a weak trend towards higher concentrations of all cations inland (Fig. 9-11). For summer, all cationic concentrations increased and calcium became the dominant cation in solution. This is attributed to the dust content of the atmosphere, which increases in summer and is composed of calcium carbonate in this region.

9.4.3 Cation concentrations in groundwater

In the majority of sample sites under beachridges and wetlands, the major cations are present in the following abundance: $Na^{+} > Ca^{2+} > Mg^{2+} > K^{+}$. In terms of millimole content, the ratios of the cation concentrations are approximately 10:5:2:1. Generally, in groundwater under the wetlands, the peak TDS values corresponded with peaks in a range of cation concentrations, indicating contribution by a variety of salts in solution, but in wetlands 161, 162, WAWA, 63, (Fig. 9-12) 1N, 9, 142, 163, TDS peaks corresponded to sodium and calcium peaks. The sodium ion concentration in groundwater is sodium chloride in solution, and the calcium ion concentration in groundwater is derived from dissolution of calcium carbonate.

Spatial variation of cation concentrations in groundwater under wetlands

Mean sodium concentrations in groundwater under the wetlands for 1992 - 1993 ranged between 1.0 and 46.0 mM/L (23-1350 ppm). The higher mean concentration occurred in 1992 but the greatest variability occurred in 1993 . All wetlands exhibited high variability. With the exception of the grazing sites, potassium concentration ranged between 0.01 and 0.77 mM/L (0.5-30 ppm), which was equal to, or less than that under the ridges . Variability at all sites was high. Calcium concentration ranged between 0.4 and 10.8 mM/L (16-434 ppm). Magnesium concentrations ranged between 0.33 and 8.4 mM/L (8-204 ppm). Mean concentrations are shown in Figures 9-13, 9-14, 9-15 and 9-16. The greatest difference between cation concentrations in groundwater under beachridges and wetlands was apparent in magnesium values (Fig. 9-16).

Overall, the greatest concentrations of cations in the groundwater occurred at the wetland margins (Fig. 9-17).

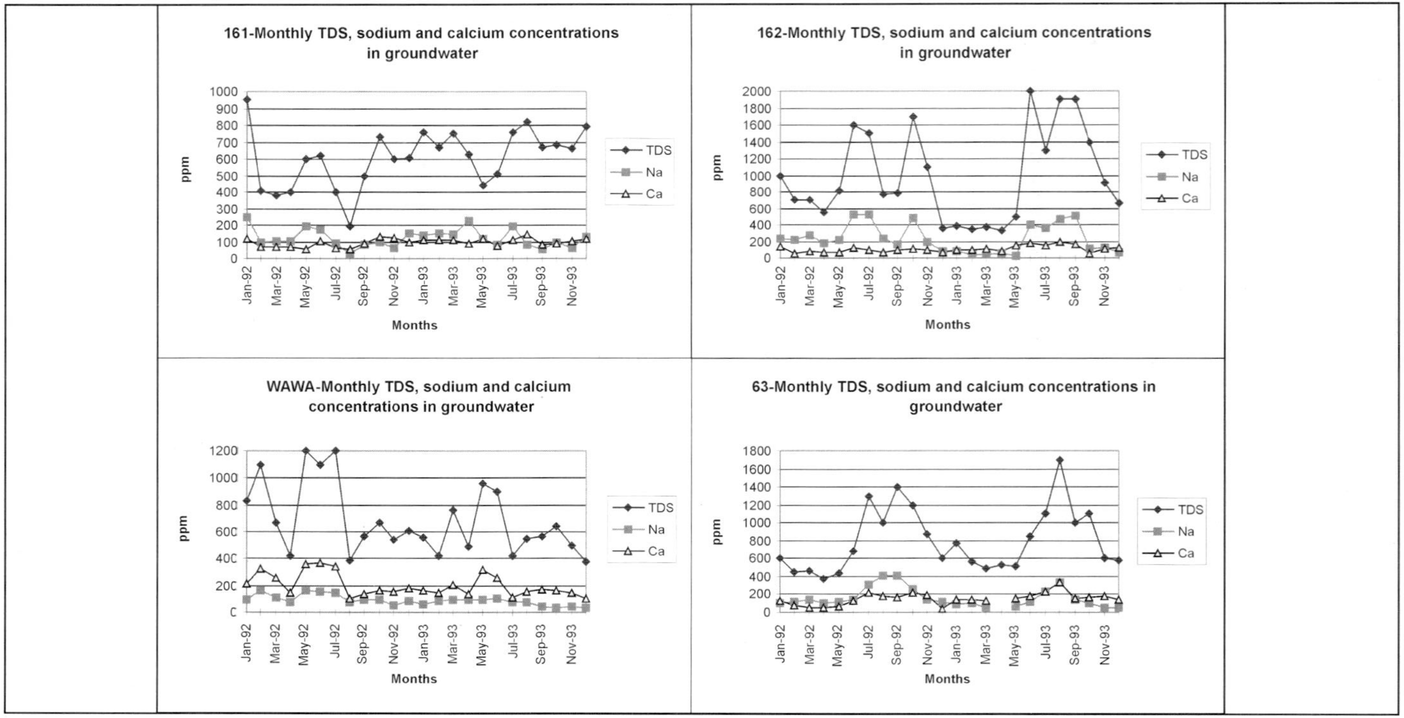

Figure 9-12. *Monthly TDS, sodium and calcium concentrations in groundwater.*

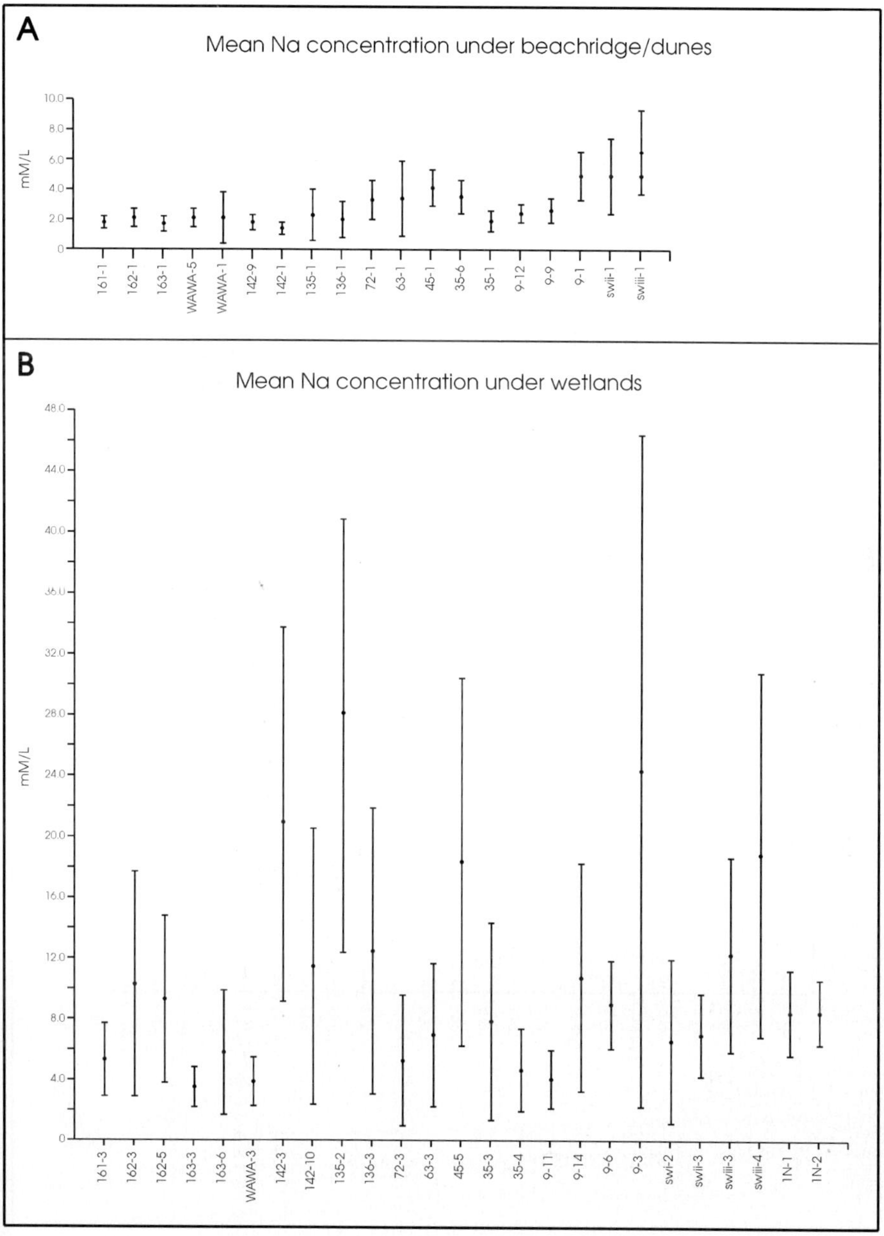

Figure 9-13. *Mean concentration of sodium in groundwater under beachridge/dunes and wetlands.*

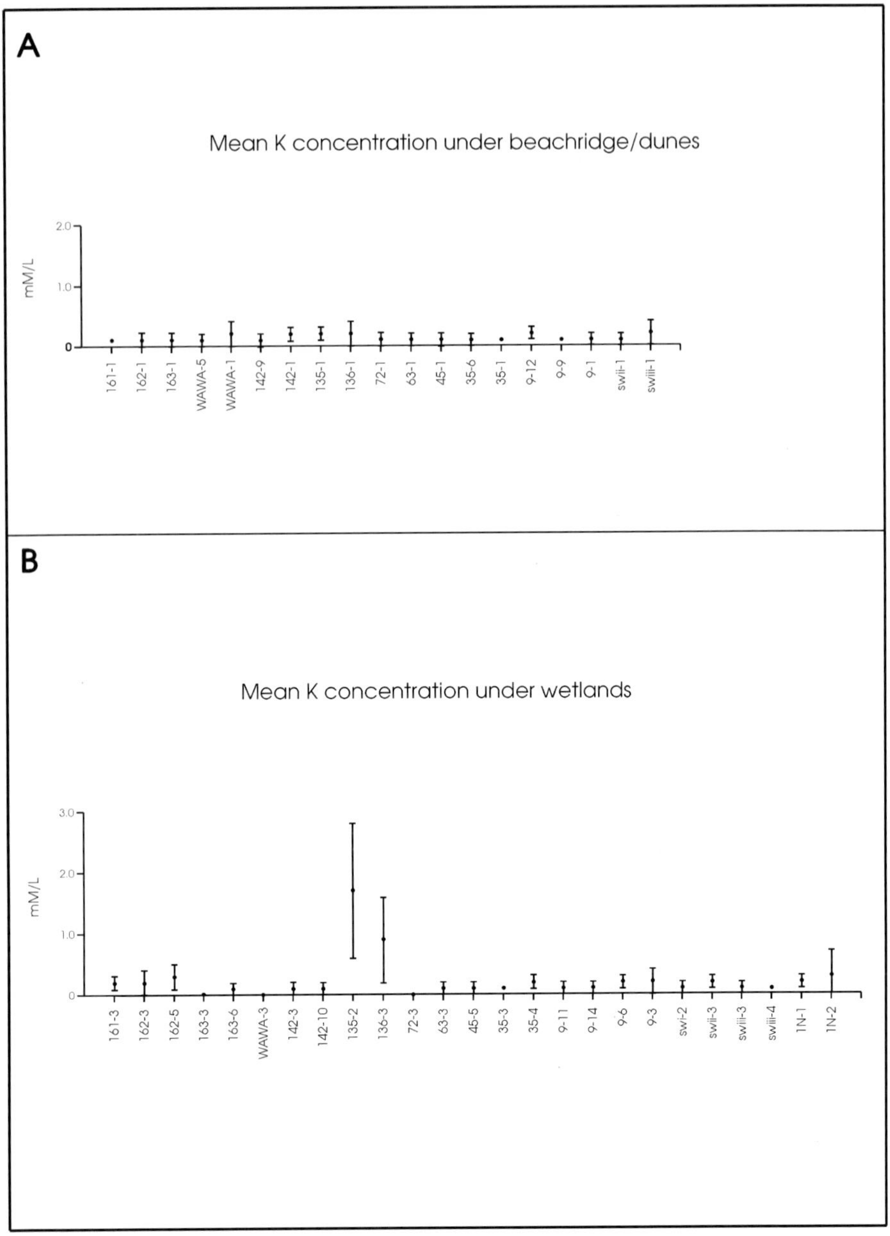

Figure 9-14. *Mean concentration of potassium in groundwater under beachridge/dunes and wetlands.*

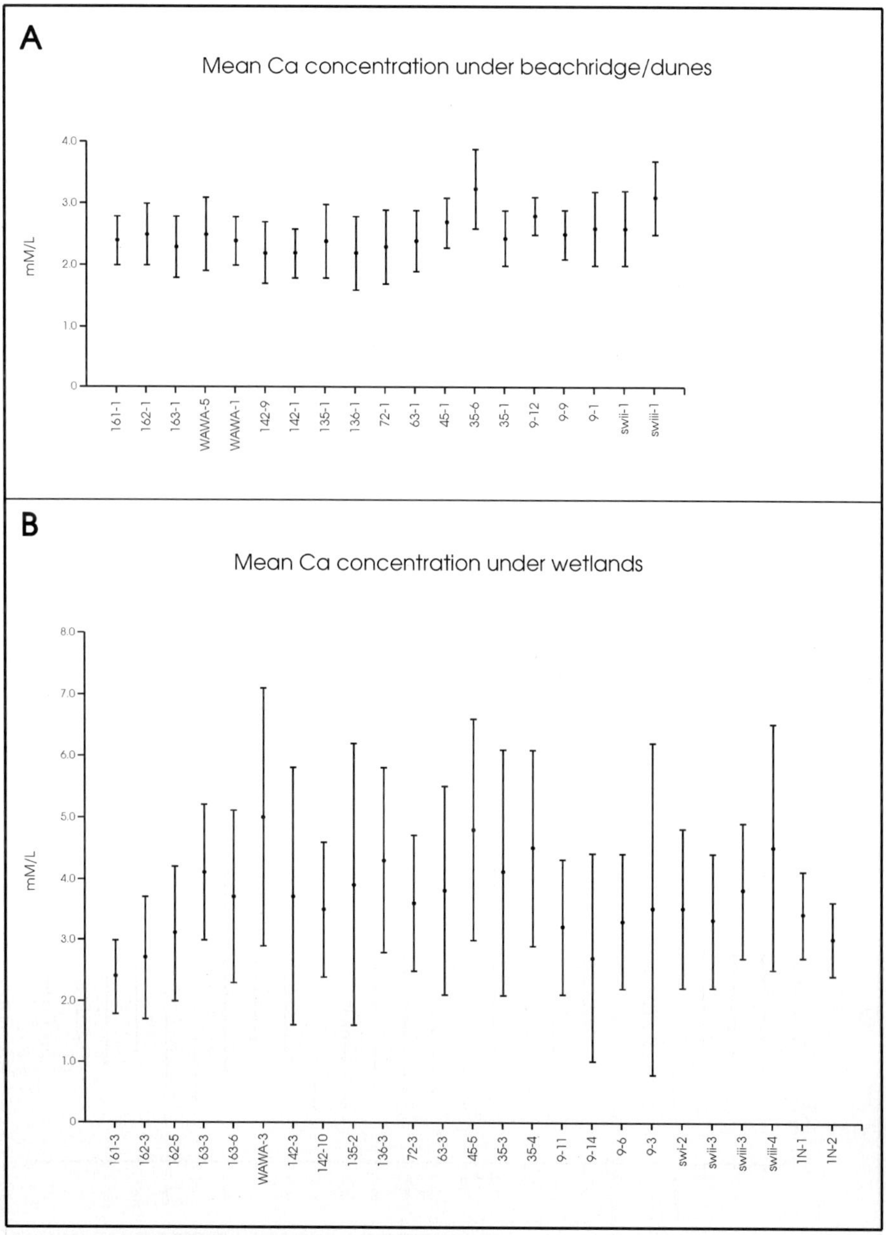

Figure 9-15. *Mean concentration of calcium in groundwater under beachridge/dunes and wetlands.*

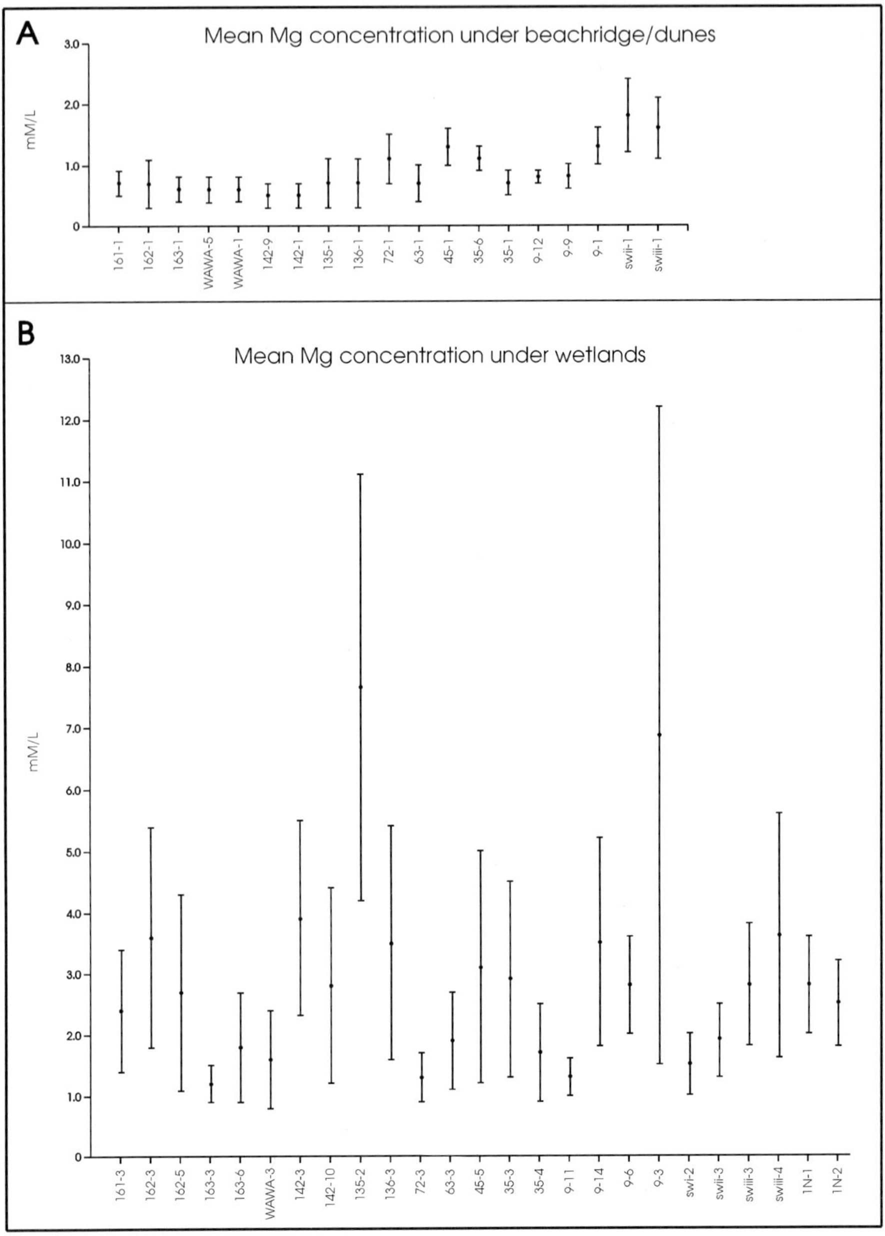

Figure 9-16. *Mean concentration of magnesium in groundwater under beachridge/dunes and wetlands.*

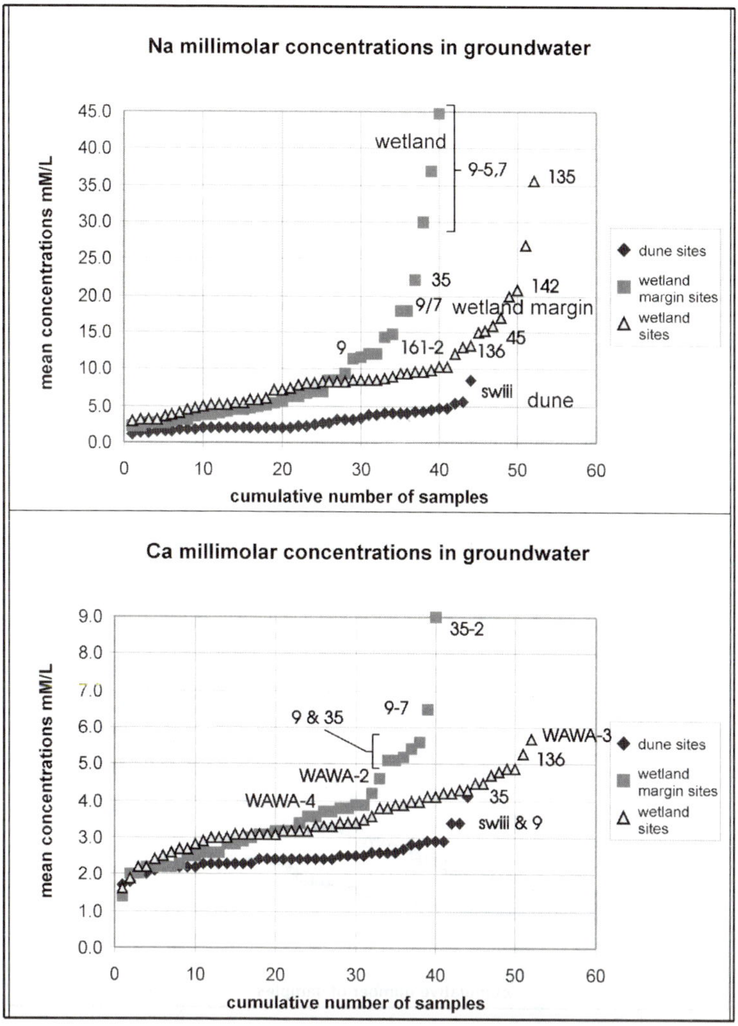

Figure 9-17. *Mean millimolar concentrations of sodium and calcium in groundwater under wetlands, wetland margins and adjacent beachridge/dunes, showing general trends and outliers.*

Temporal variation - Seasonal patterns comparing beachridge/dunes and wetlands

Seasonal patterns of the ionic concentrations in the groundwater (1991-1993) under beachridge/dunes (site 1) and wetlands (site 3) are presented in Figures 9-18 to 9-35. Figures are presented at a common scale for comparison, followed by a more appropriate scale where necessary.

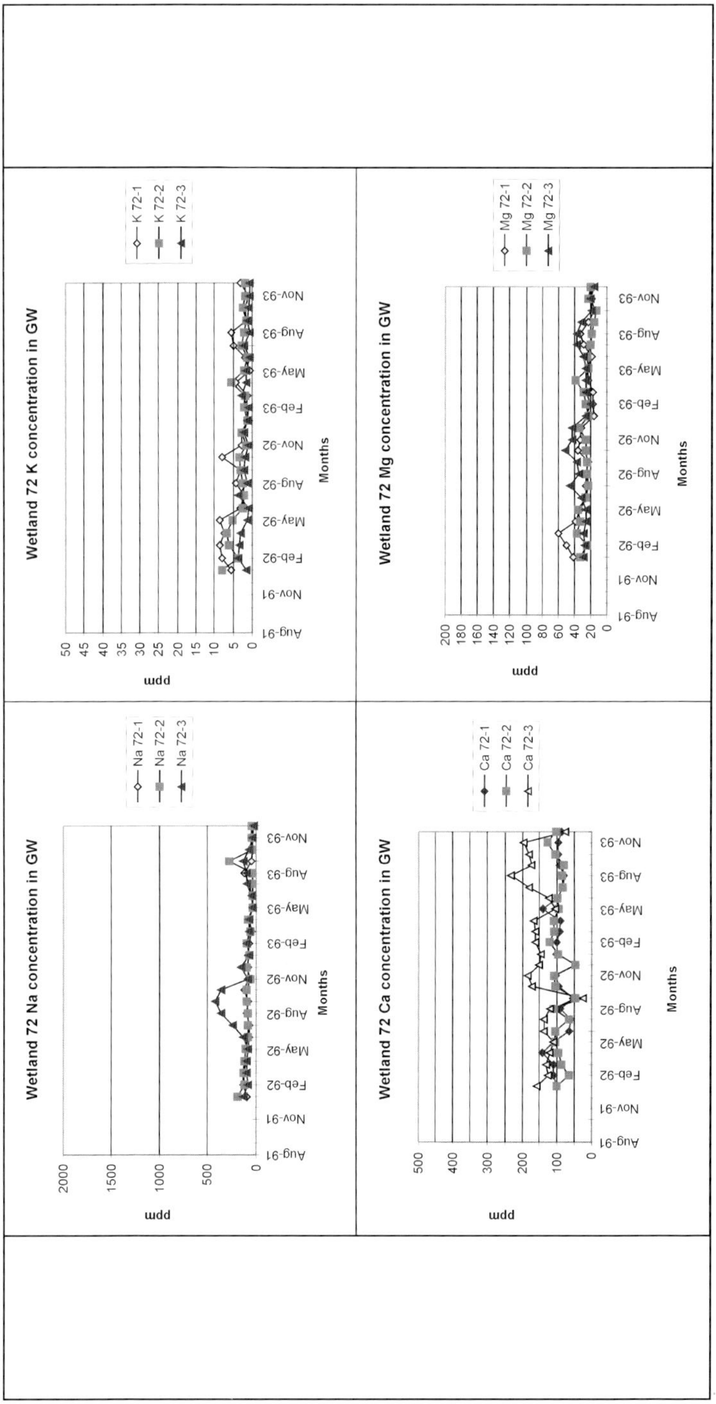

Figure 9-25*. Cation concentrations in groundwater under beachridge/dunes, wetland margins and wetlands. (Aug 1991 - Dec 1993).*

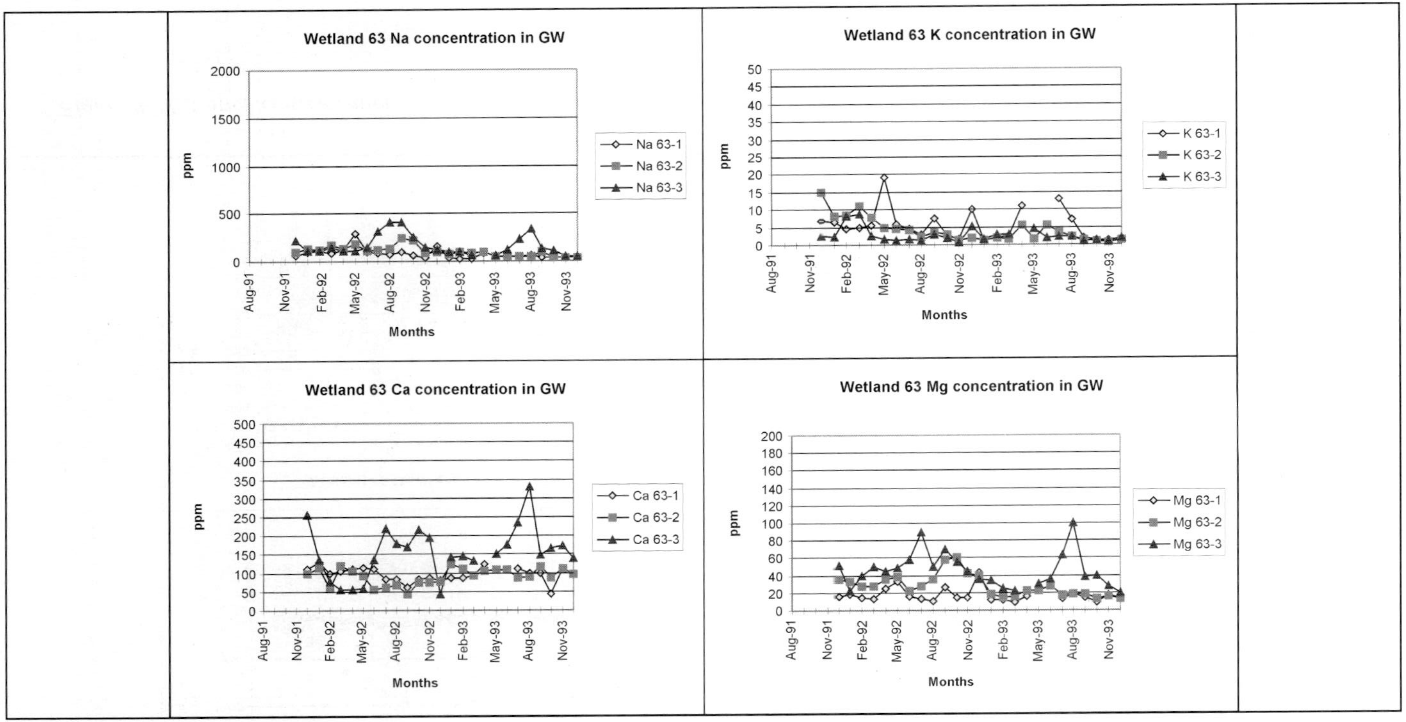

Figure 9-26. *Cation concentrations in groundwater under beachridge/dunes, wetland margins and wetlands. (Aug 1991 - Dec 1993).*

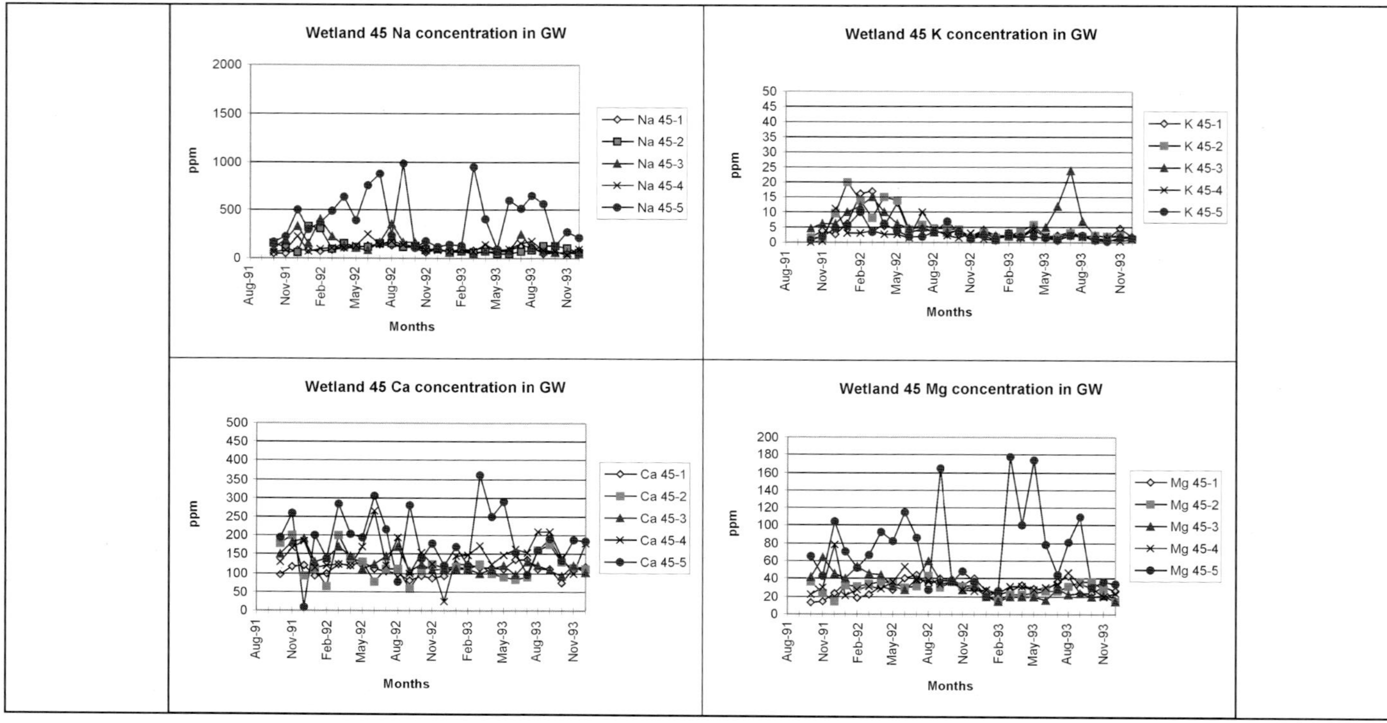

Figure 9-27. *Cation concentrations in groundwater under beachridge/dunes, wetland margins and wetlands. (Aug 1991 - Dec 1993).*

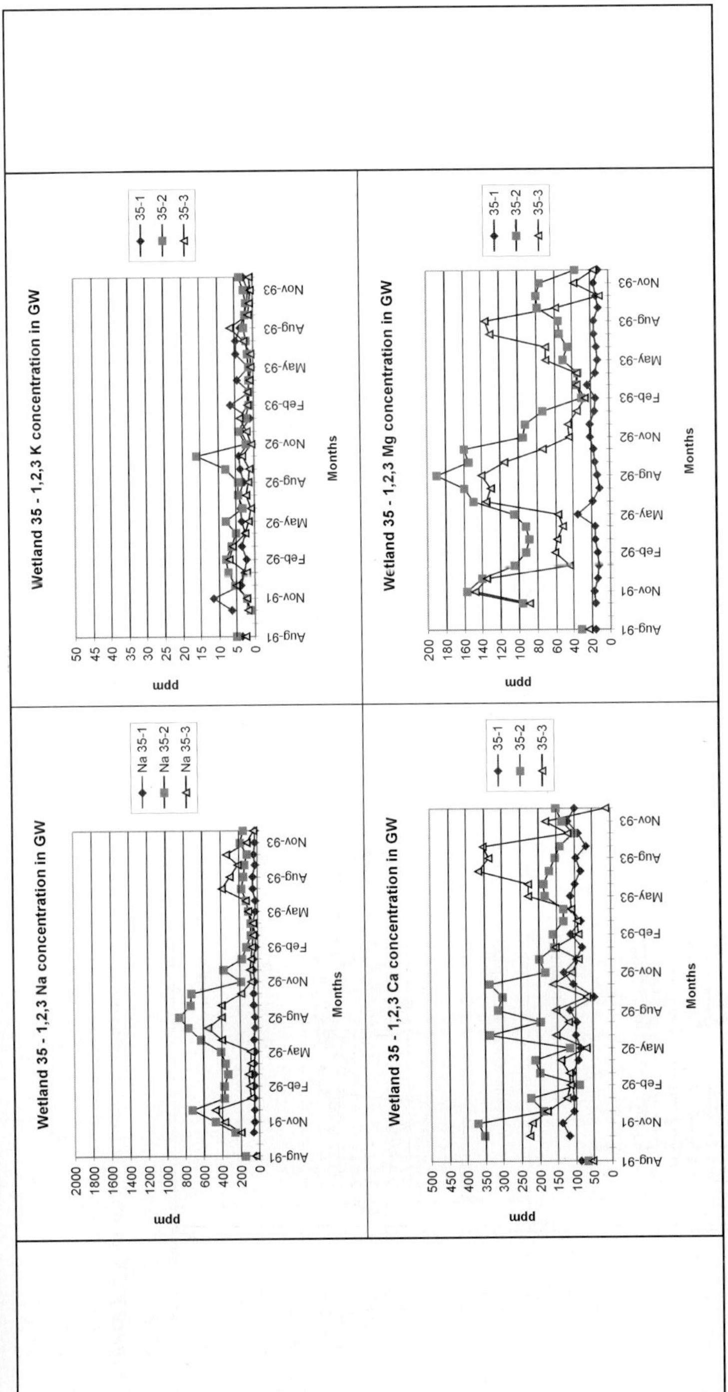

Figure 9-28. *Cation concentrations in groundwater under beachridge/dunes, wetland margins and wetlands. (Aug 1991 - Dec 1993).*

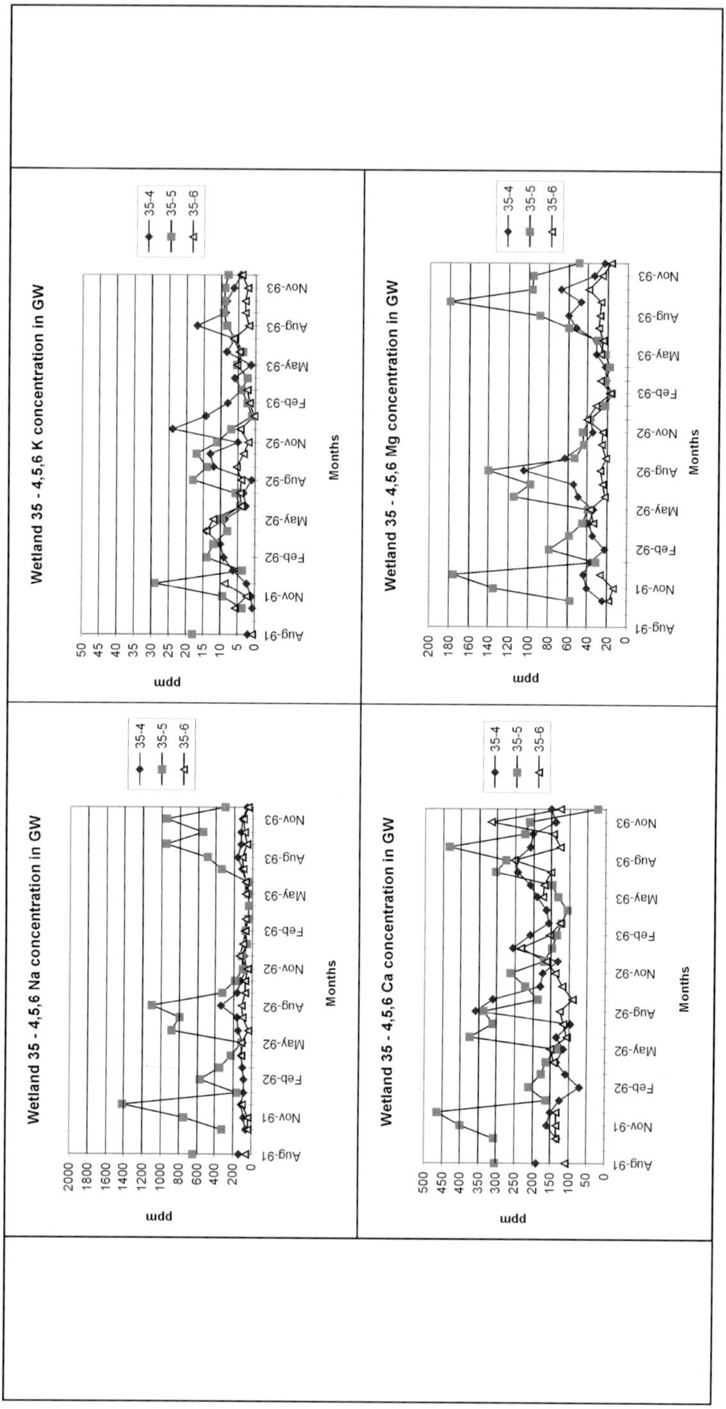

Figure 9-28 (cont.). *Cation concentrations in groundwater under beachridge/dunes, wetland margins and wetlands. (Aug 1991 - Dec 1993).*

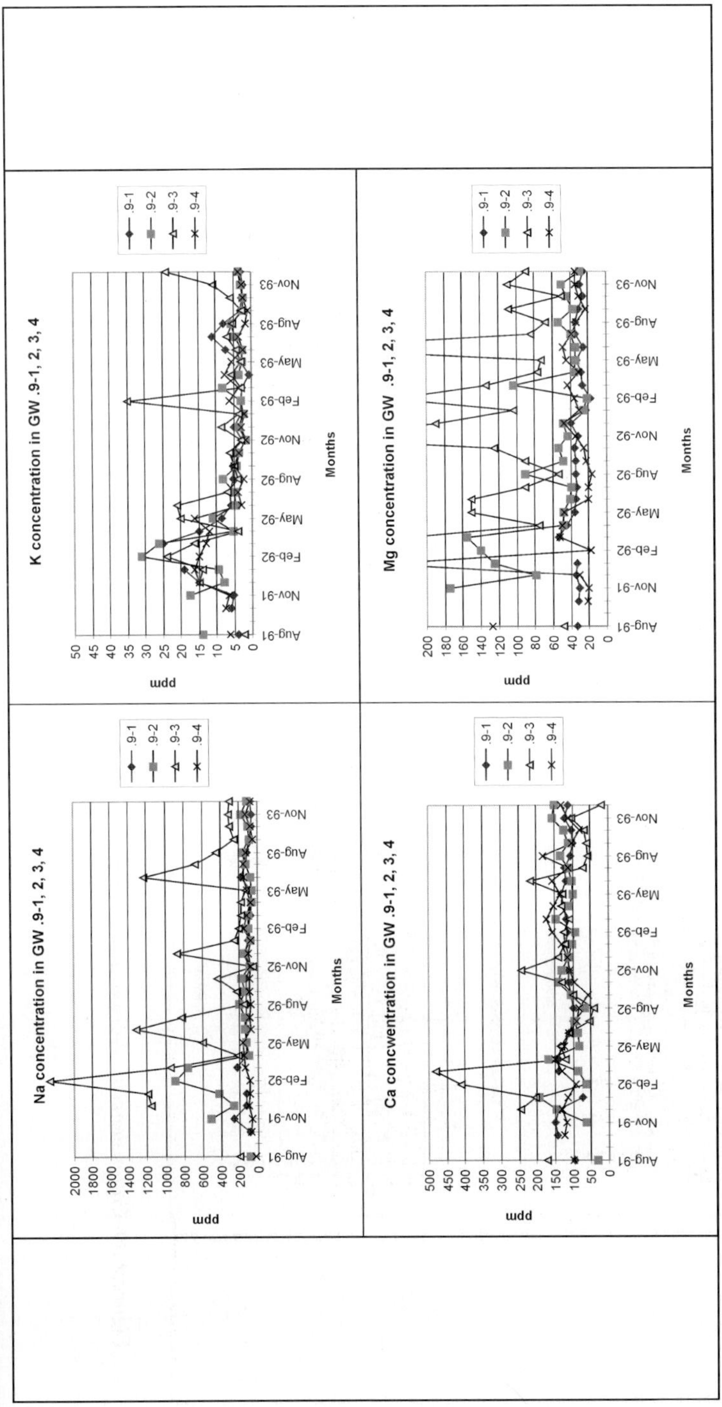

Figure 9-29. *Cation concentrations in groundwater under beachridge/dunes, wetland margins and wetlands. (Aug 1991 - Dec 1993).*

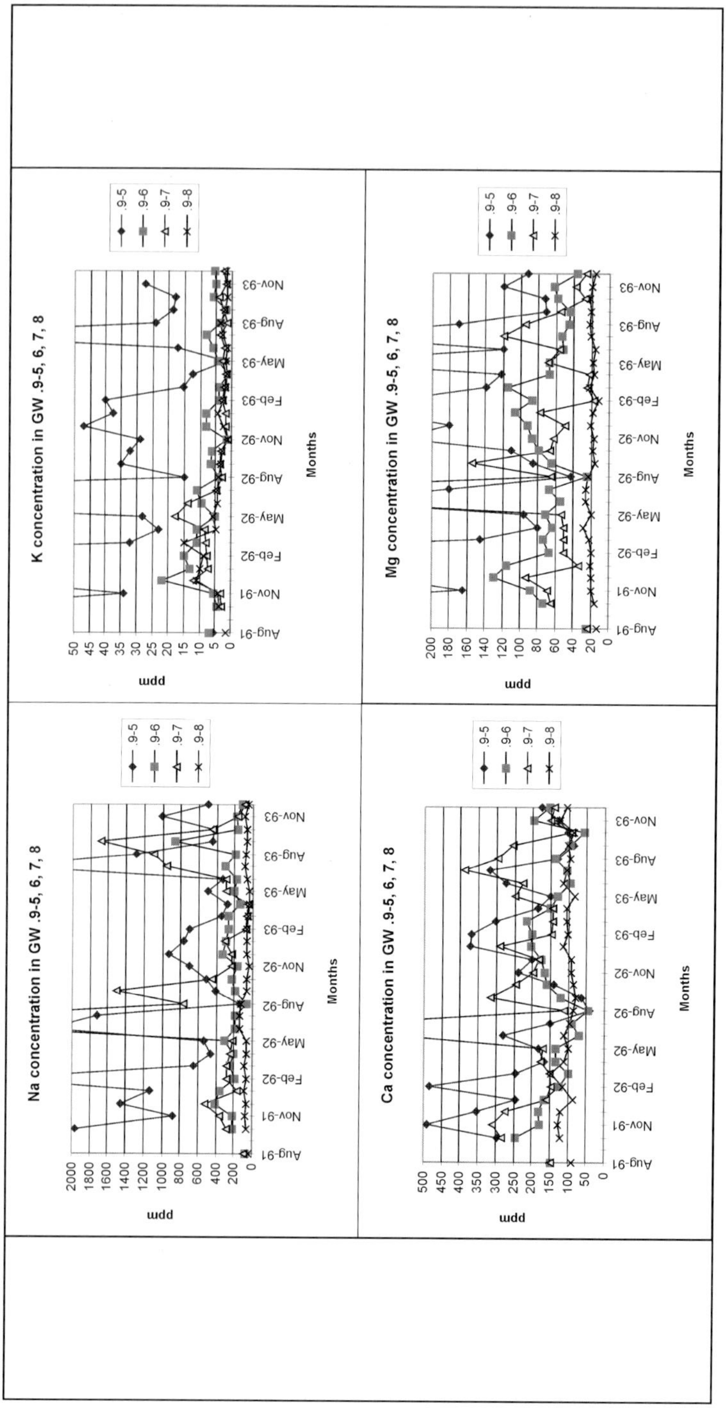

Figure 9-30. *Cation concentrations in groundwater under beachridge/dunes, wetland margins and wetlands. (Aug 1991 - Dec 1993).*

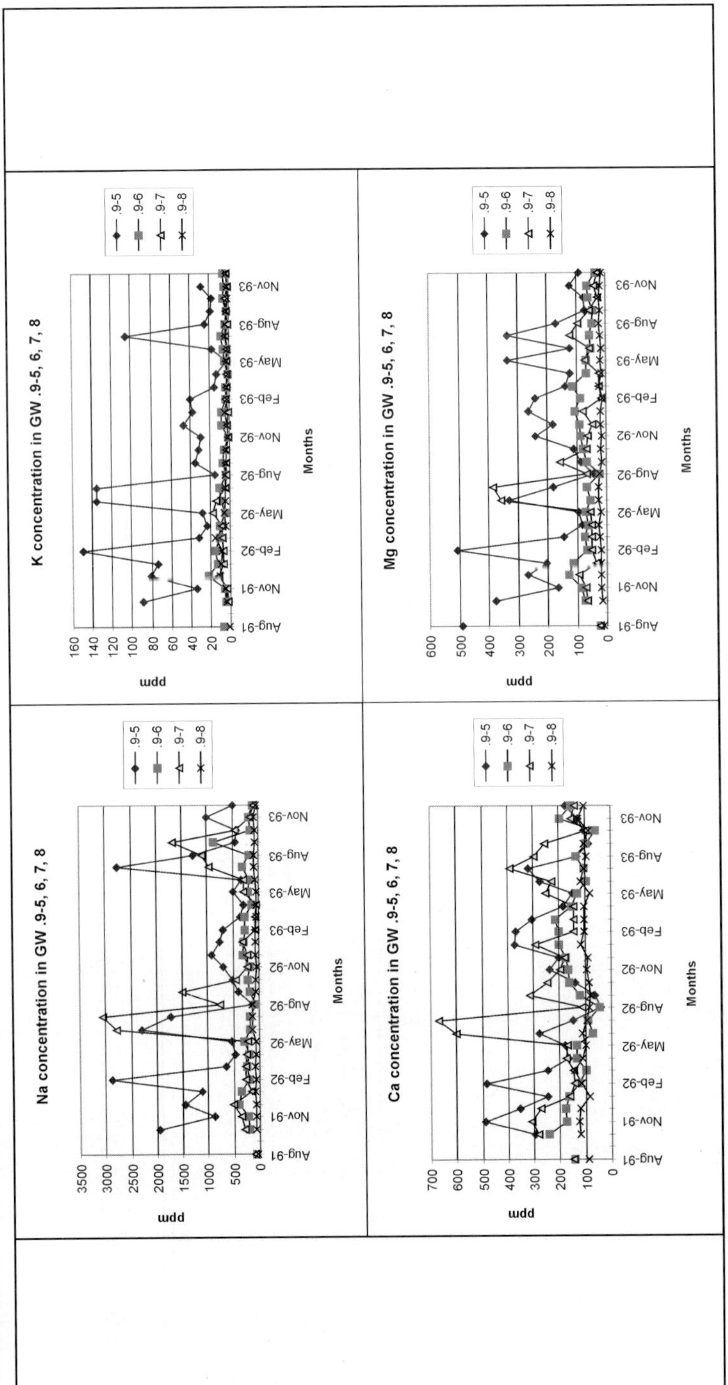

Figure 9-30 (cont.). *Cation concentrations in groundwater under beachridge/dunes, wetland margins and wetlands. (Aug 1991 - Dec 1993).*

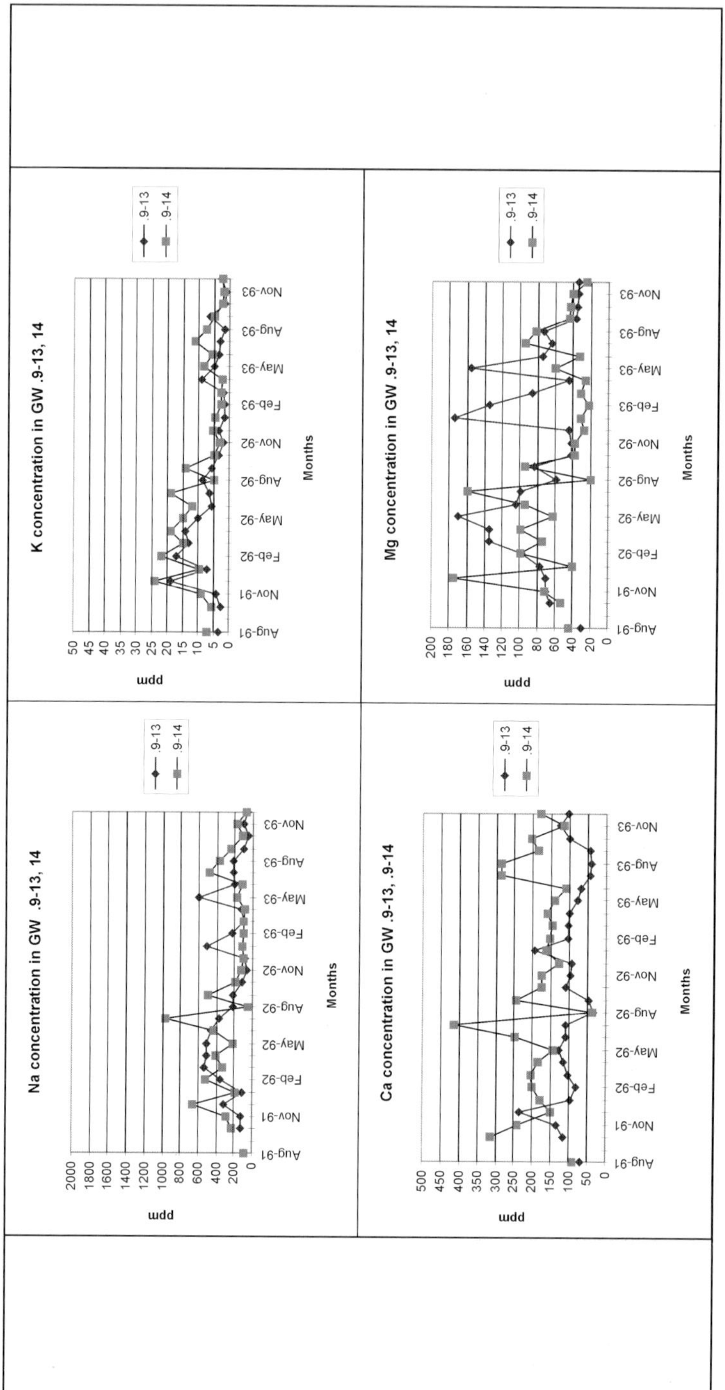

Figure 9-31. *Cation concentrations in groundwater under beachridge/dunes, wetland margins and wetlands. (Aug 1991 - Dec 1993).*

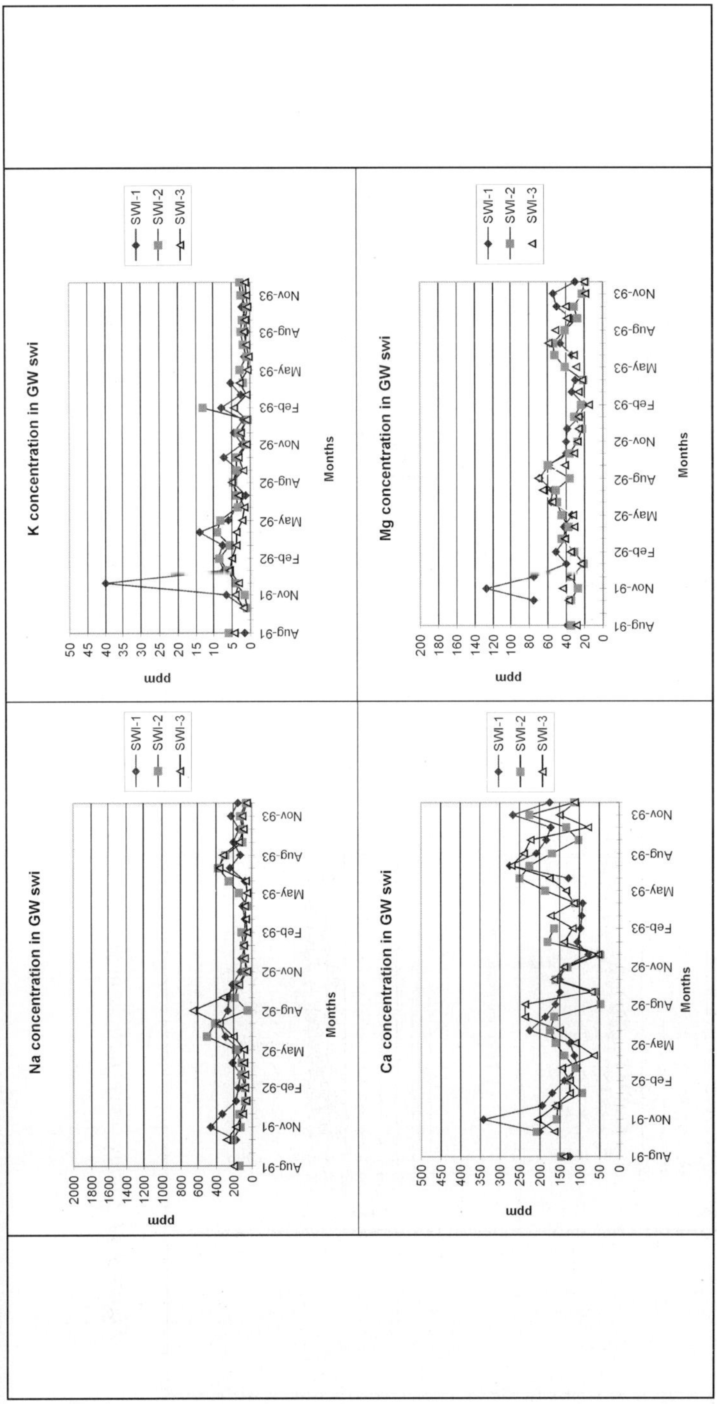

Figure 9-32. *Cation concentrations in groundwater under beachridge/dunes, wetland margins and wetlands. (Aug 1991 - Dec 1993).*

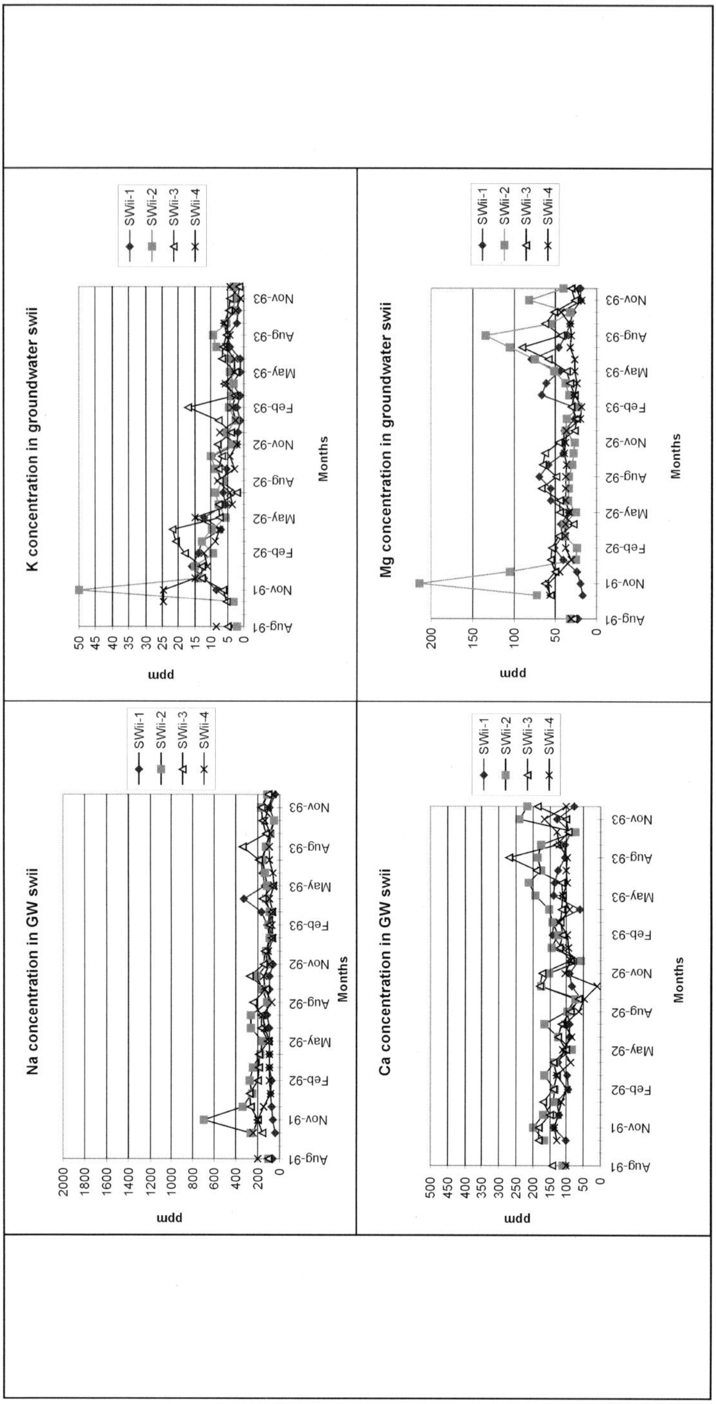

Figure 9-33. *Cation concentrations in groundwater under beachridge/dunes, wetland margins and wetlands. (Aug 1991 - Dec 1993).*

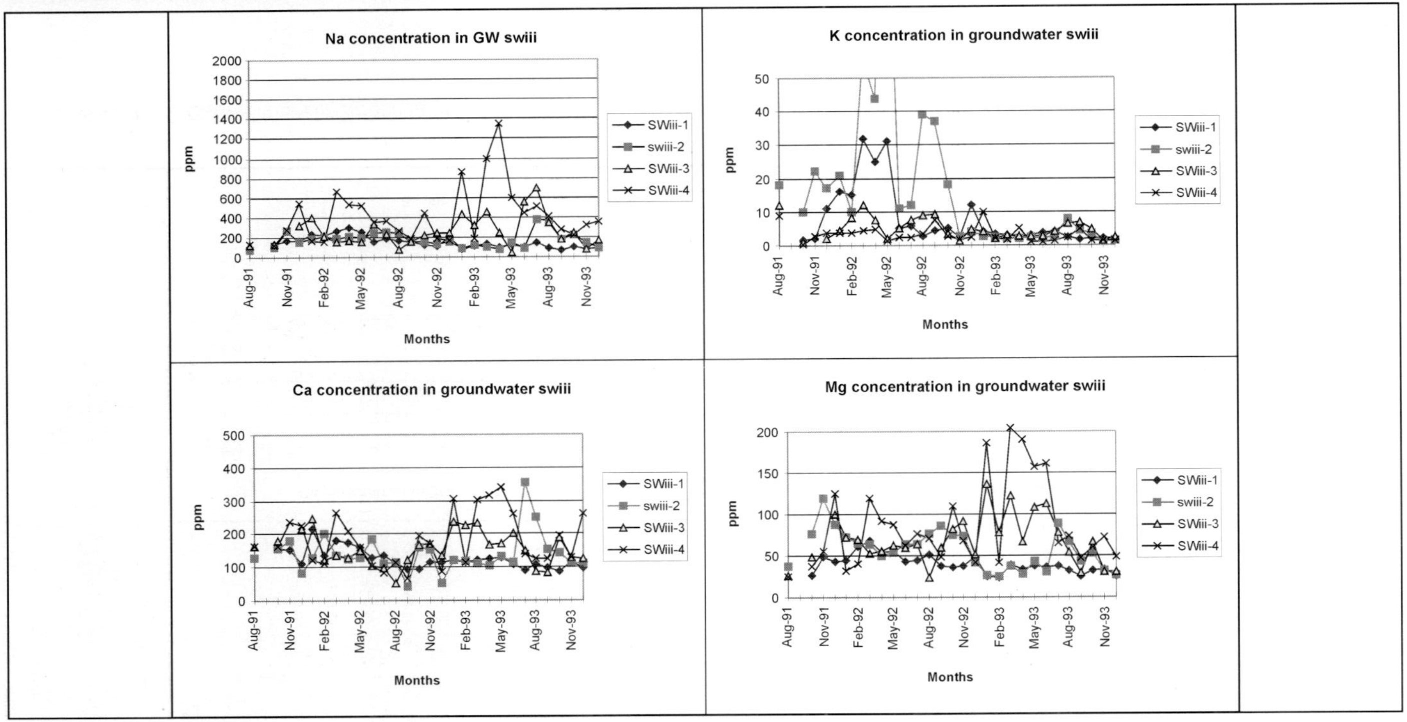

Figure 9-34. *Cation concentrations in groundwater under beachridge/dunes, wetland margins and wetlands. (Aug 1991 - Dec 1993).*

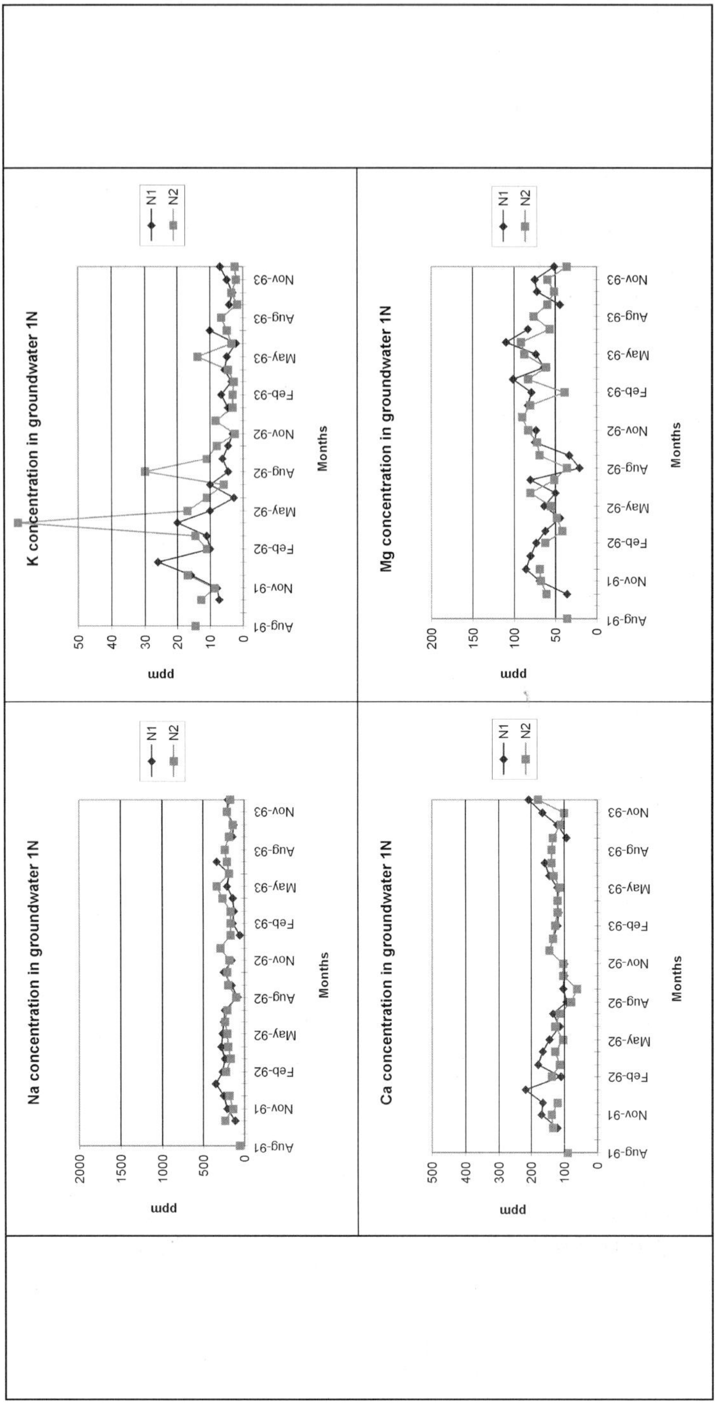

Figure 9-35. *Cation concentrations in groundwater under beachridge/dunes, wetland margins and wetlands. (Aug 1991 - Dec 1993).*

consistent identifiable peaks, May and October, and one consistent trough, November (Figs. 9-18 to 9-35). Calcium concentrations under the wetlands were variable from month to month but exhibited two identifiable troughs, February and September and two minor peaks, May and November (Figs. 9-18 to 9-35).

Magnesium concentrations under wetlands showed gentle modulations rather than peaks (Figs. 9-18 to 9-35). In several of the wetlands there were similar seasonal patterns in cationic concentration in the groundwater between sites 2-3 metres apart, *e.g.*, wetland 142 sites 5, 6, 7, wetland 35 sites 2, 3, wetland swi sites 1, 2, 3, wetland swiii sites 5, 6, and wetland 1N sites 1, 2 (Figs. 9-24, 28, 32, 34, 35).

9.4.4 Cation concentrations in wetland sediments and interstitial waters

To ascertain the patterns of cation concentrations in sedimentary profiles, an analysis of Ca, Mg, Na, and K content of wetland fills was undertaken. The sediments included were peat, OME calcilutite, calcilutite, calcilutaceous muddy sand, and calcareous sand, located in wetlands WAWA, 162, 163, 142, 1N (Figs. 9-36A to 9-40A). Metal content was determined for each sediment type in the sedimentary profiles and these results provided the basis for interpretation of the hydrochemistry of corresponding interstitial waters. While there may be a strong relationship between metal species content in the sediment and the type of sedimentary particle, interstitial waters do not necessarily reflect the chemistry of the sediment. Where interstitial water chemistry is related to the sediment particle type, solubility is a key factor. The main materials, therefore, that will yield cations into the groundwater or interstitial water are plant materials, carbonate grains in sand, and carbonate mud particles in calcilutite. To confirm that the hydrochemistry of interstitial waters is, in fact, related to sediment type through the processes of cation leaching by groundwaters and interstitial waters, a series of leaching experiments was undertaken. The results are shown in Figures 9-41, 9-42, 9-43.

The first experiment approximated the leaching that would occur if the sediments were inundated by rainwater to a depth of 2 cm. The results then were transformed to cation content as mM/L that would be present in interstitial water in a water-saturated sediment. A second leaching experiment was undertaken in which dune sands, thoroughly rinsed with de-ionised water to free the sand of labile cations that might be present in pellicular water, or as precipitated salts, were left to drain to allow leaching of cations from the sand grains into the film of pellicular water remaining. Experiments using de-ionised water were also carried out on some dried, comminuted wetland plant material to determine how rapidly and to what extent cations could be leached from such matter. The water samples from the experiment were analysed for Na^+, K^+, Ca^{++}, and Mg^{++}.

The results of simulating leaching under inundation by rainwater, show that cations can be readily leached by water from the surface soils and organic matter (Fig. 9-41). Leaching after one day showed weak mobilisation of cations from the sediment, and

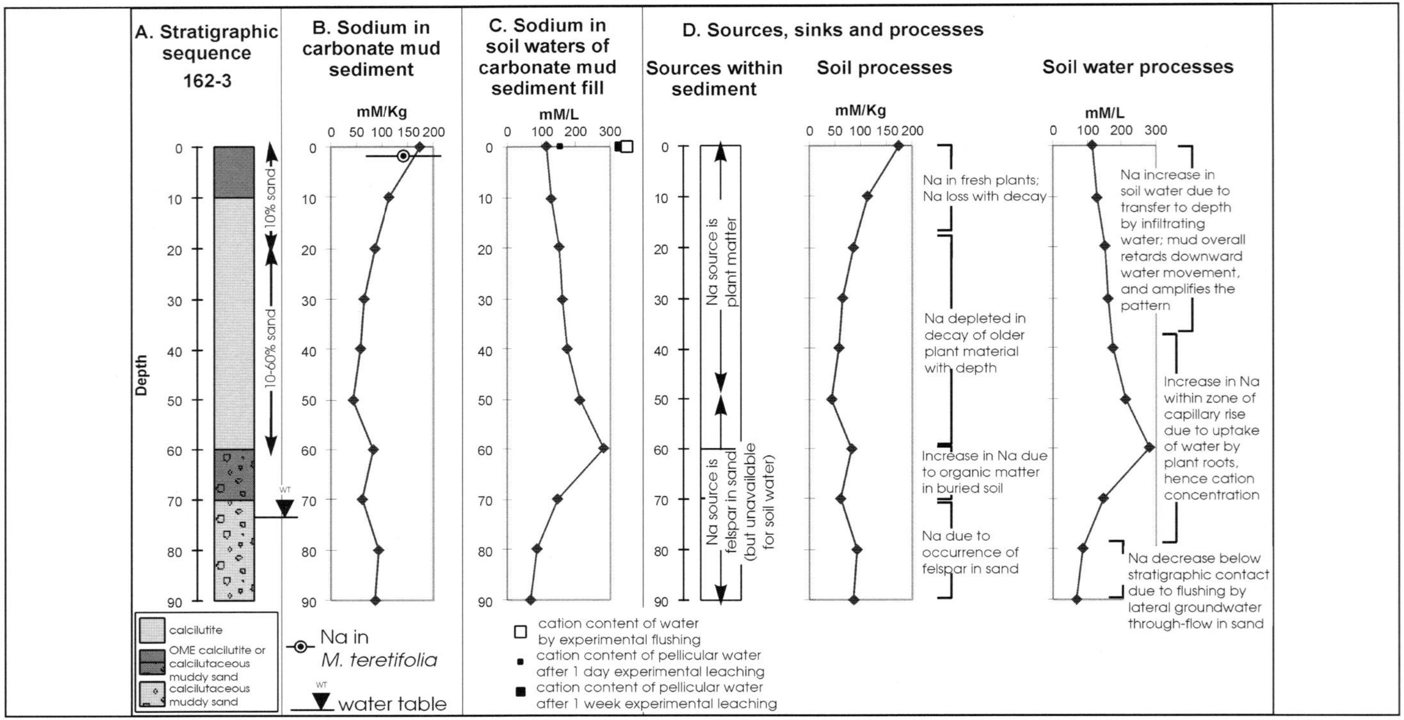

Figure 9-36A. *Concentrations of sodium in carbonate mud and its interstitial waters, with a stratigraphic context and interpretation of processes leading to their concentration distributions, at wetland 162-3 site.*

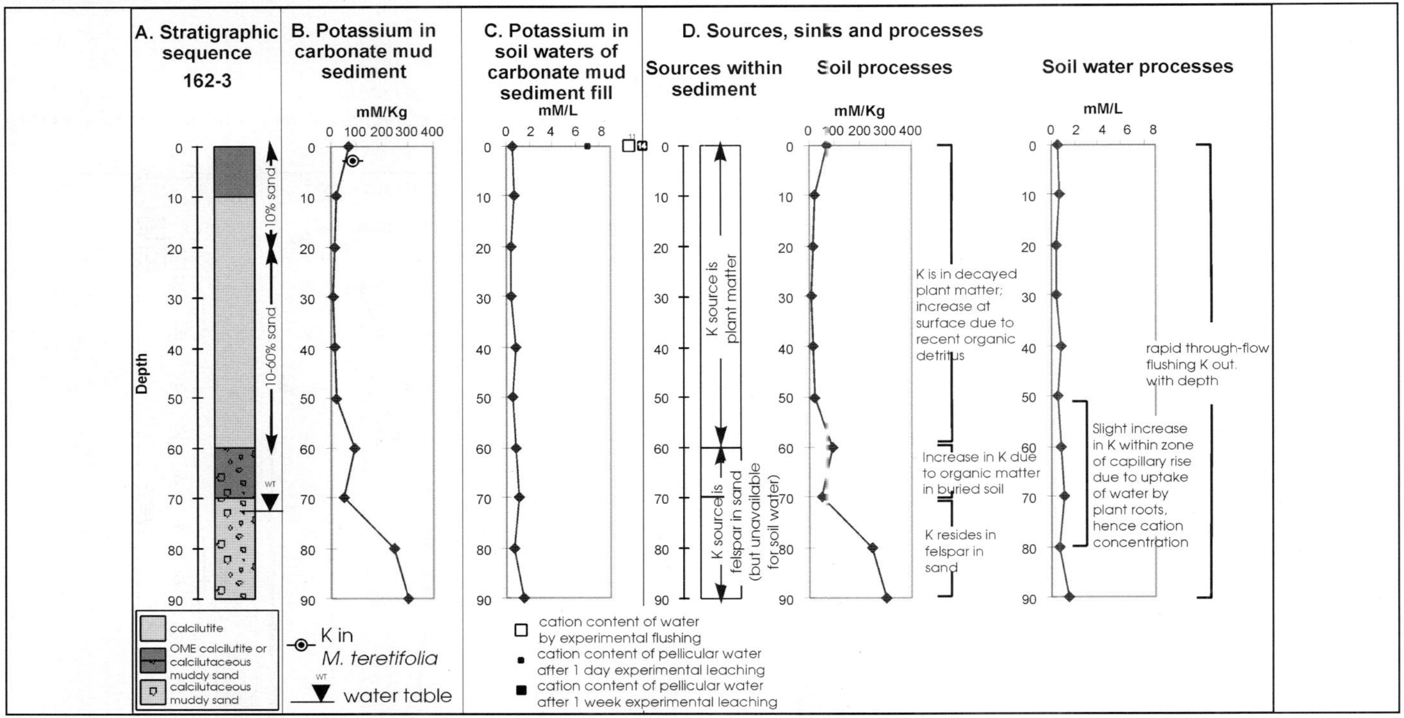

Figure 9-36B. *Concentrations of potassium in carbonate mud and its interstitial waters, with a stratigraphic context and interpretation of processes leading to their concentration distributions, at wetland 162-3 site.*

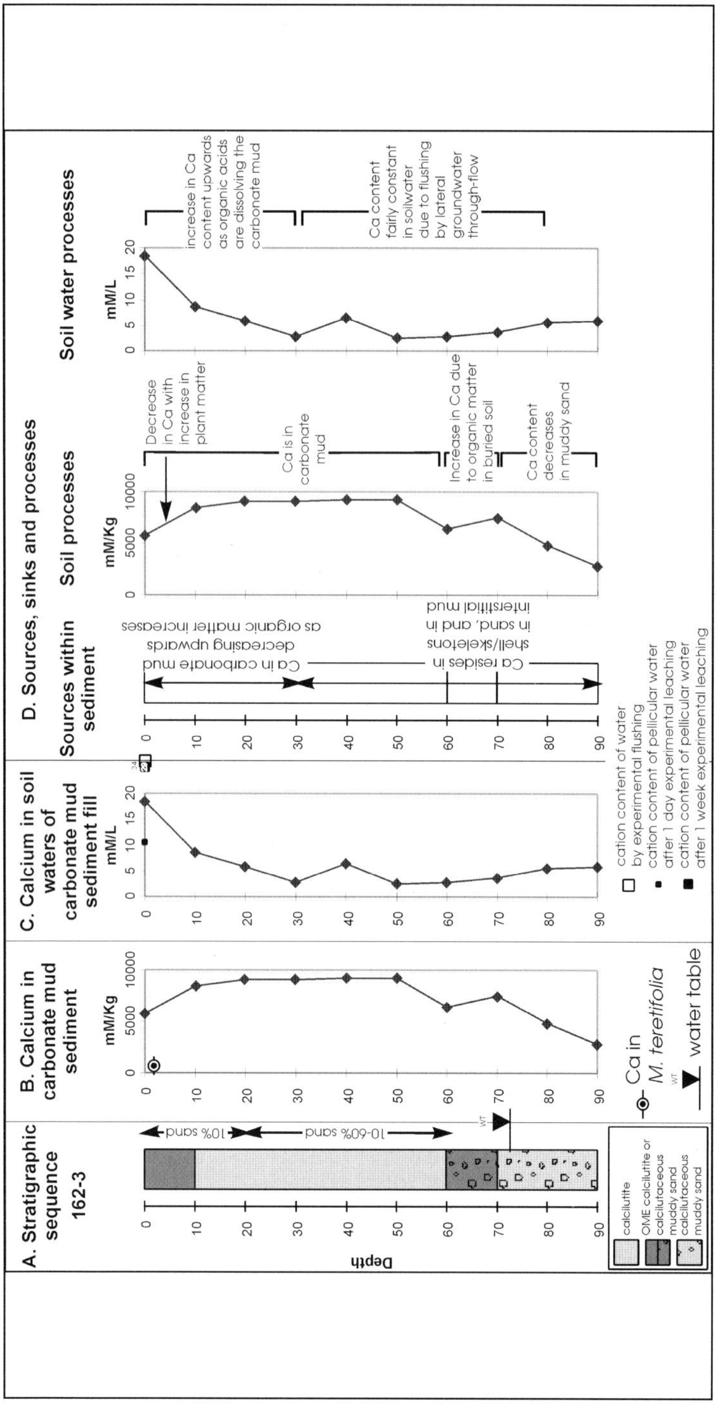

Figure 9-36C. *Concentrations of calcium in carbonate mud sediment and its interstitial waters, with a stratigraphic context and interpretation of processes leading to their concentration distributions, at wetland 162-3 site.*

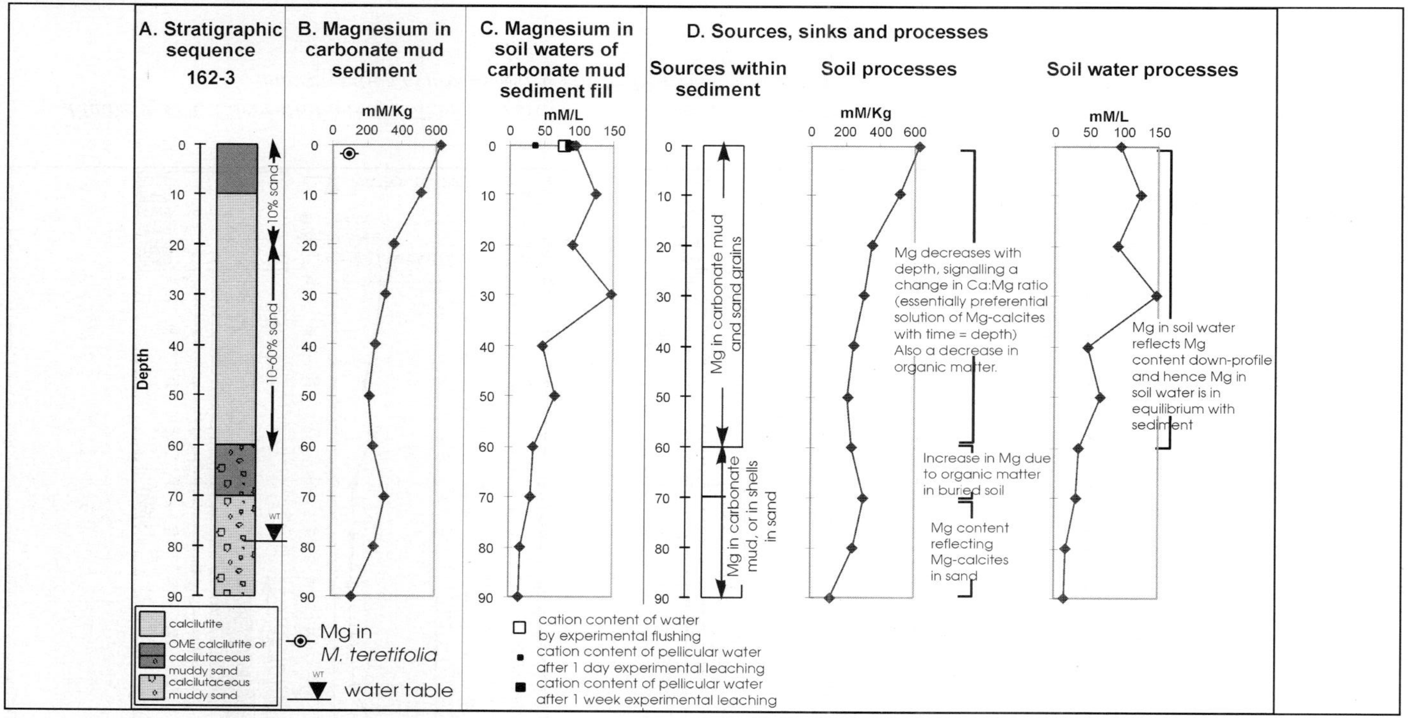

Figure 9-36D. *Concentrations of magnesium in carbonate mud sediment and its interstitial waters, with a stratigraphic context and interpretation of processes leading to their concentration distributions, at wetland 162-3 site.*

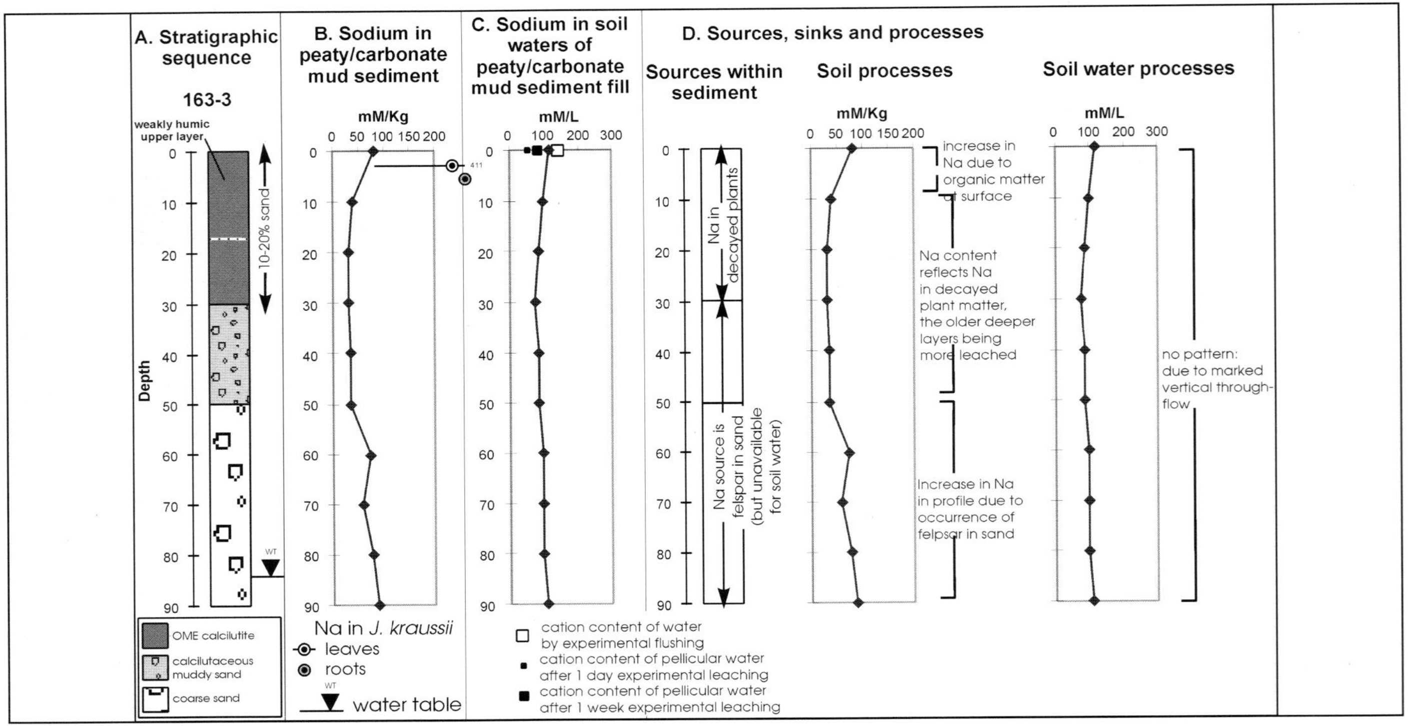

Figure 9-37A. *Concentrations of sodium in peaty/carbonate mud sediment and its interstitial waters, with a stratigraphic context and interpretation of processes leading to their concentration distributions, at wetland 163-3 site.*

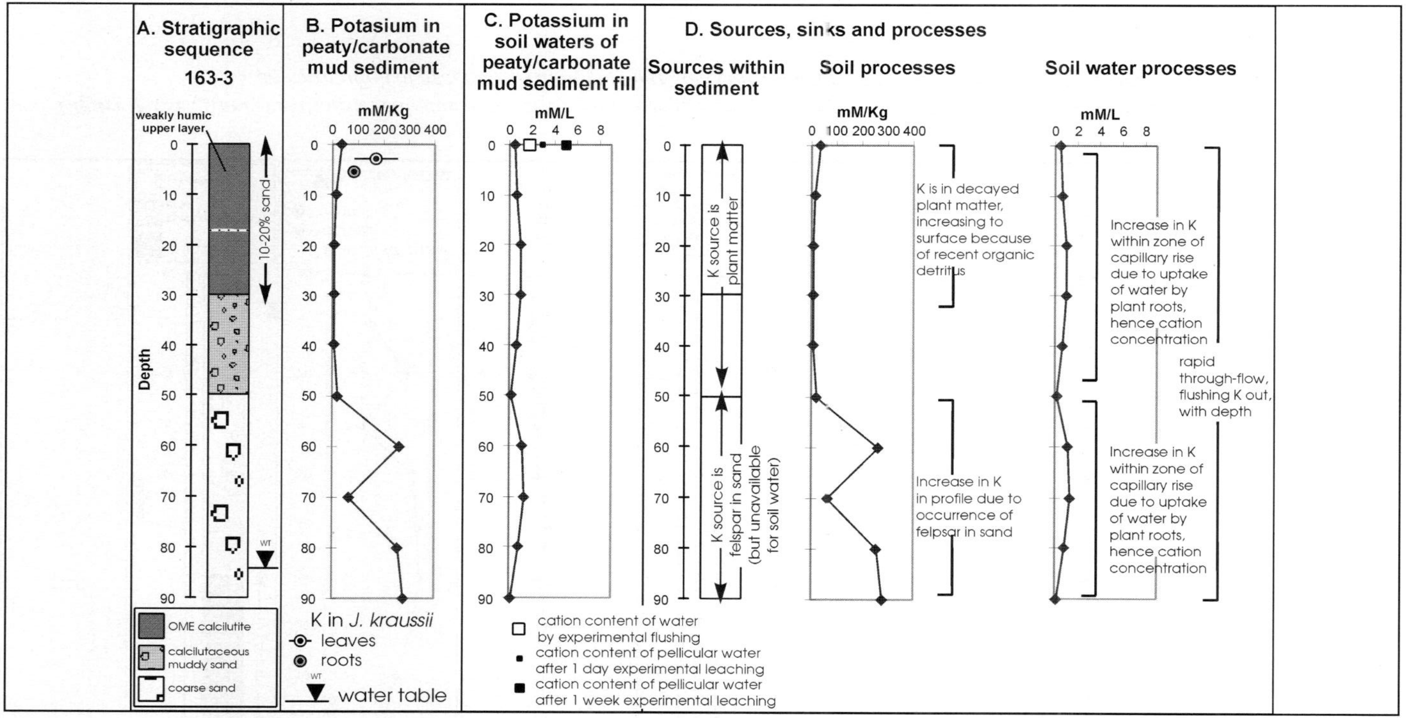

Figure 9-37B. *Concentrations of potassium in peaty/carbonate mud and its interstitial waters, with a stratigraphic context and interpretation of processes leading to their concentration distributions, at wetland 163-3 site.*

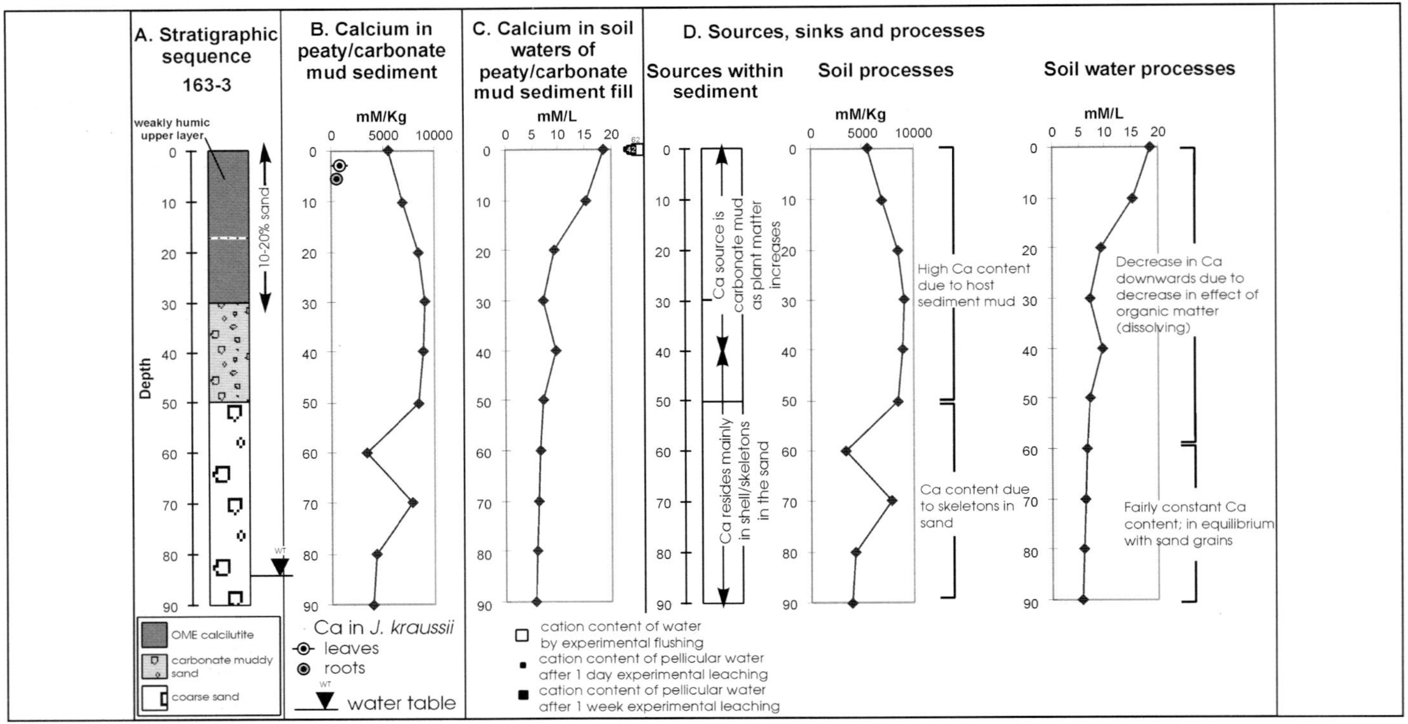

Figure 9-37C. *Concentrations of calcium in peaty/carbonate mud sediment and its interstitial waters, with a stratigraphic context and interpretation of processes leading to their concentration distributions, at wetland 163-3 site.*

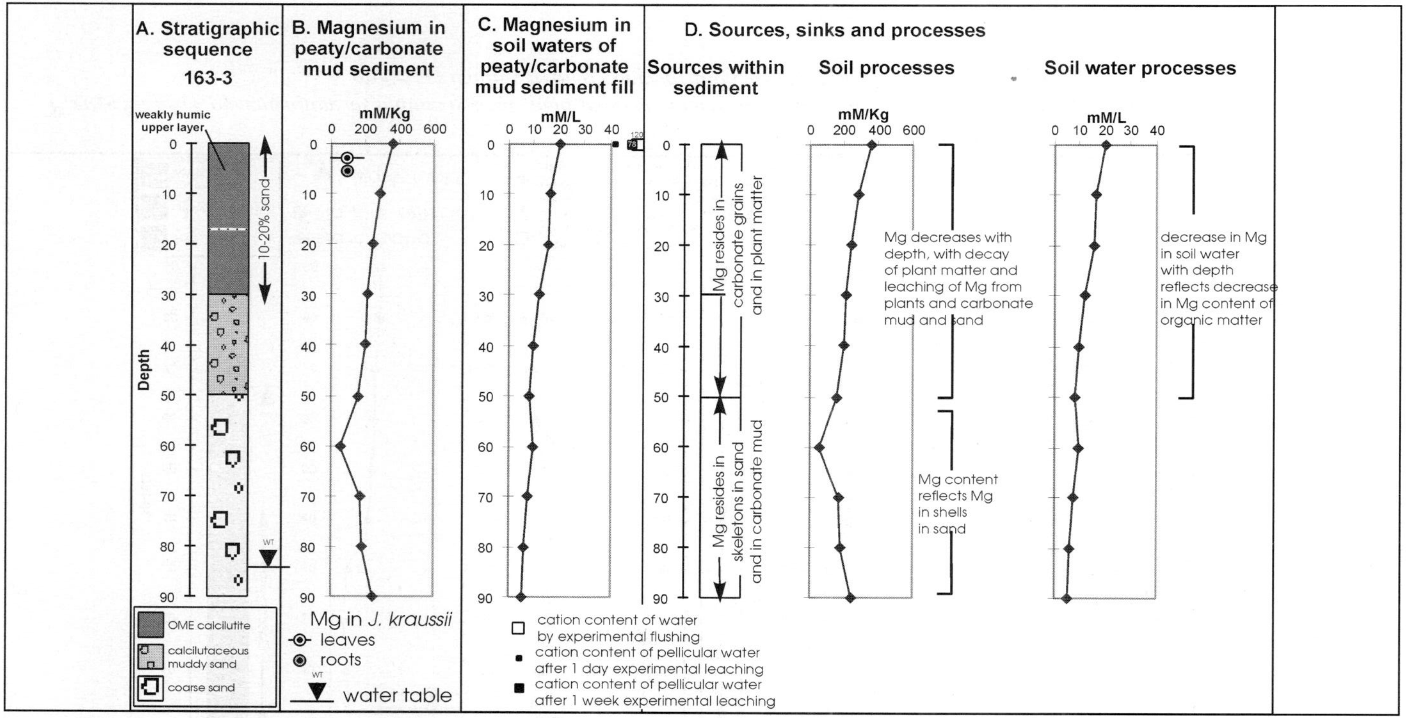

Figure 9-37D. *Concentrations of magnesium in peaty/carbonate mud sediment and its interstitial waters, with a stratigraphic context and interpretation of processes leading to their concentration distributions, at wetland 163-3 site.*

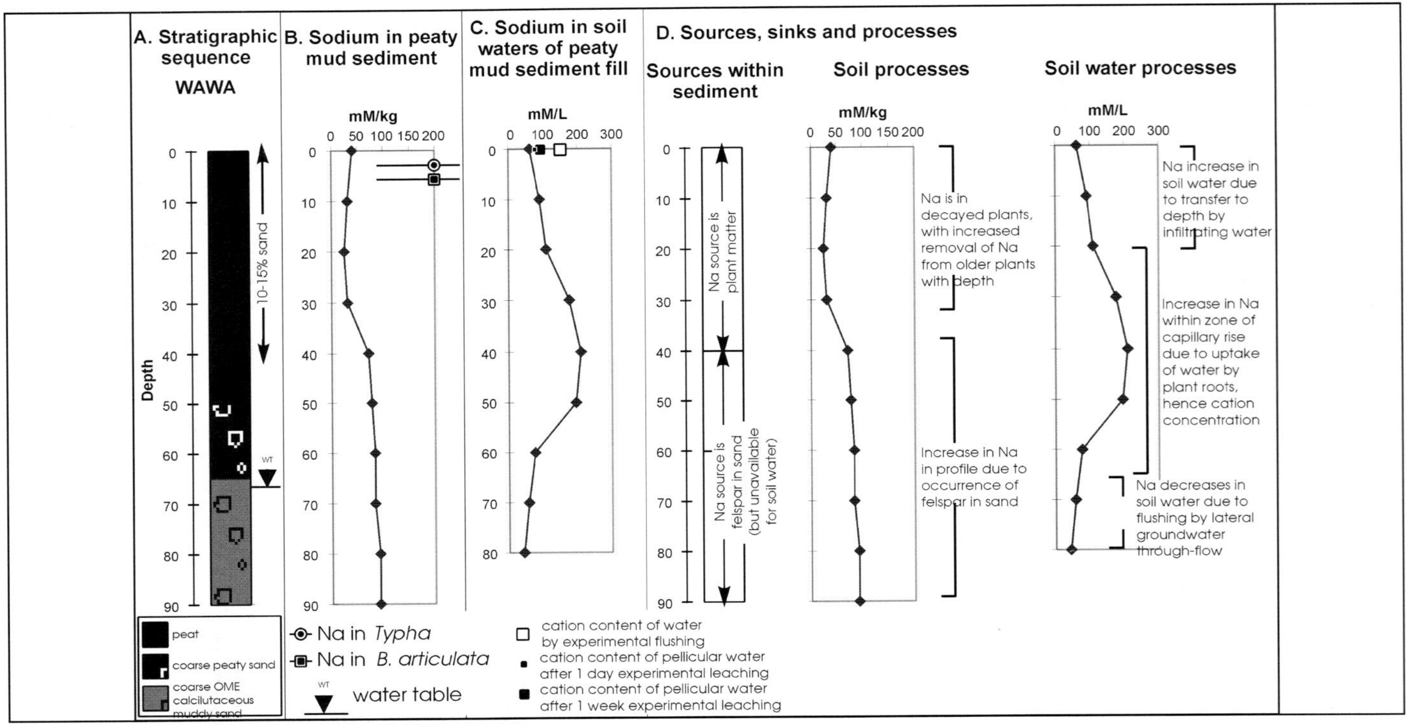

Figure 9-38A. *Concentrations of sodium in peaty mud sediment and its interstitial waters, with a stratigraphic context and interpretation of processes leading to their concentration distributions, at wetland WAWA site.*

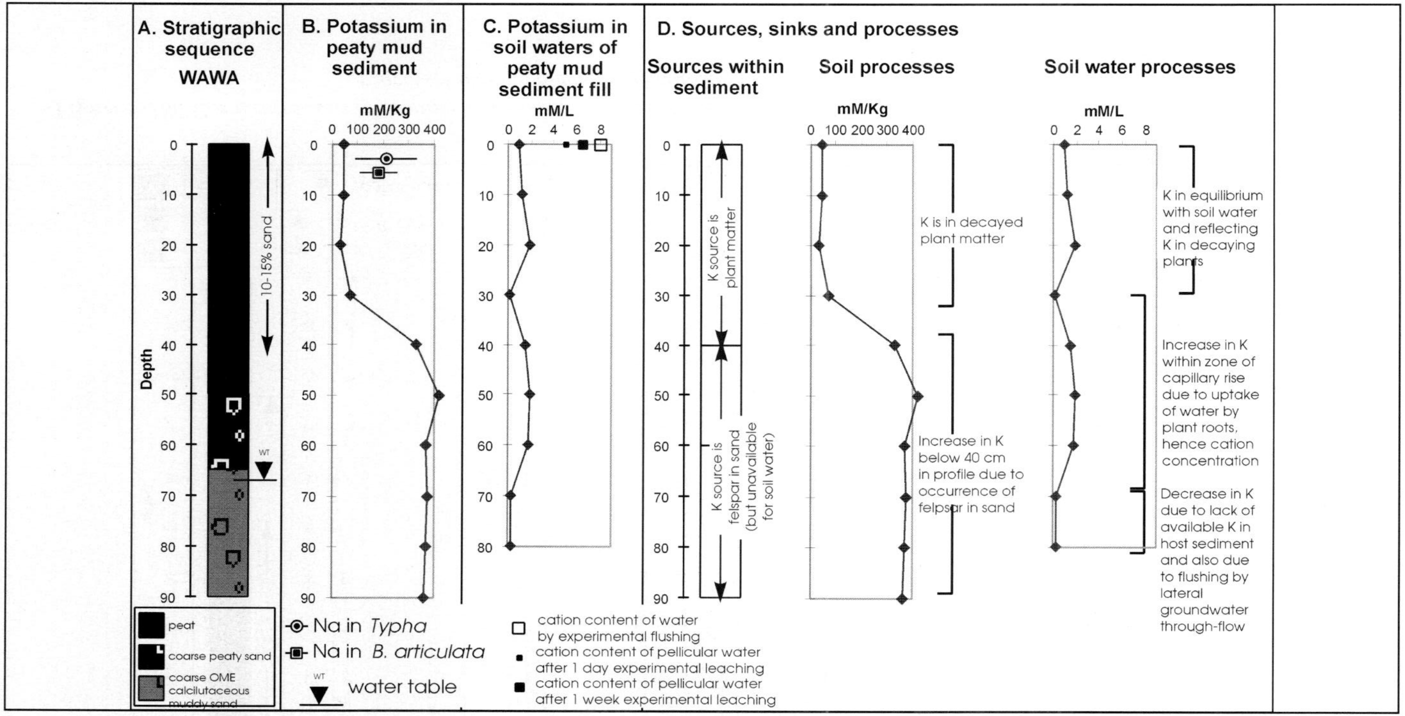

Figure 9-38B. *Concentrations of potassium in peaty mud sediment and its interstitial waters, with a stratigraphic context and interpretation of processes leading to their concentration distributions, at wetland WAWA site.*

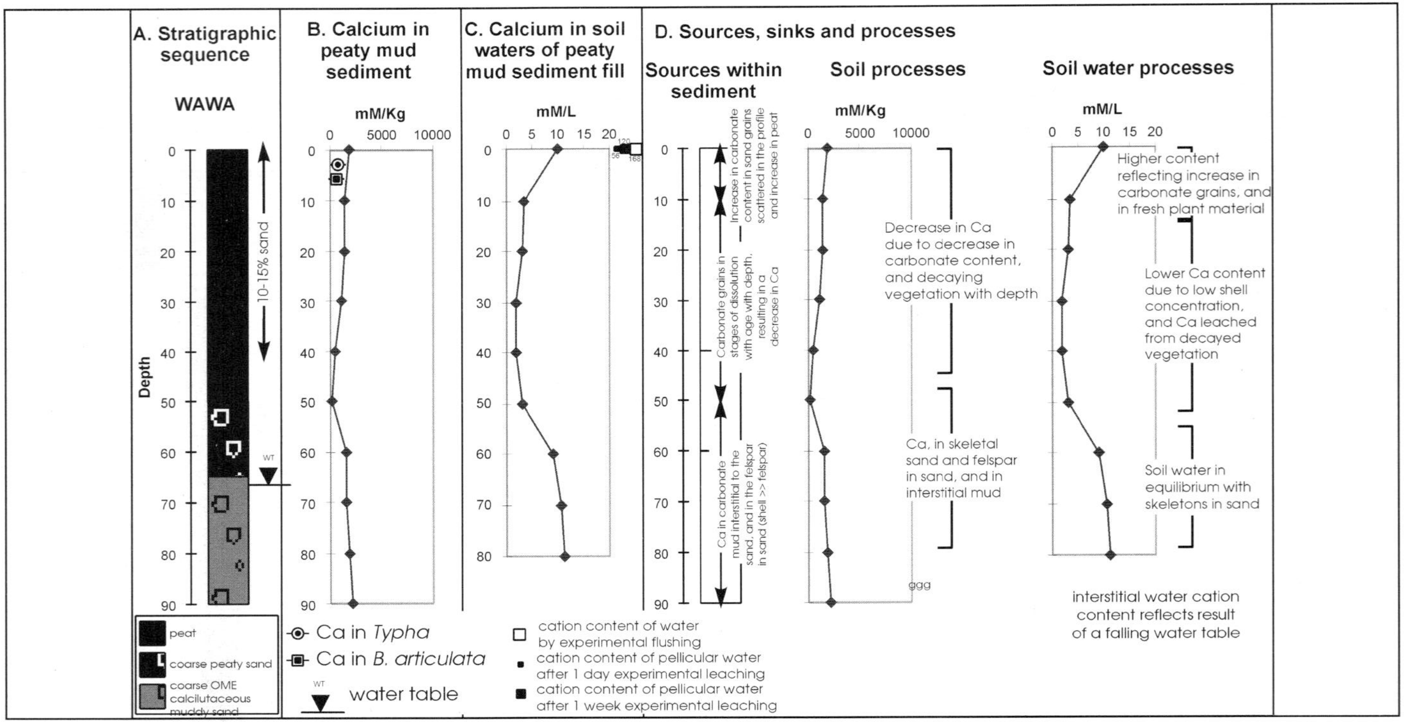

Figure 9-38C. Concentrations of calcium in peaty mud sediment and its interstitial waters, with a stratigraphic context and interpretation of processes leading to their concentration distributions, at wetland WAWA site.

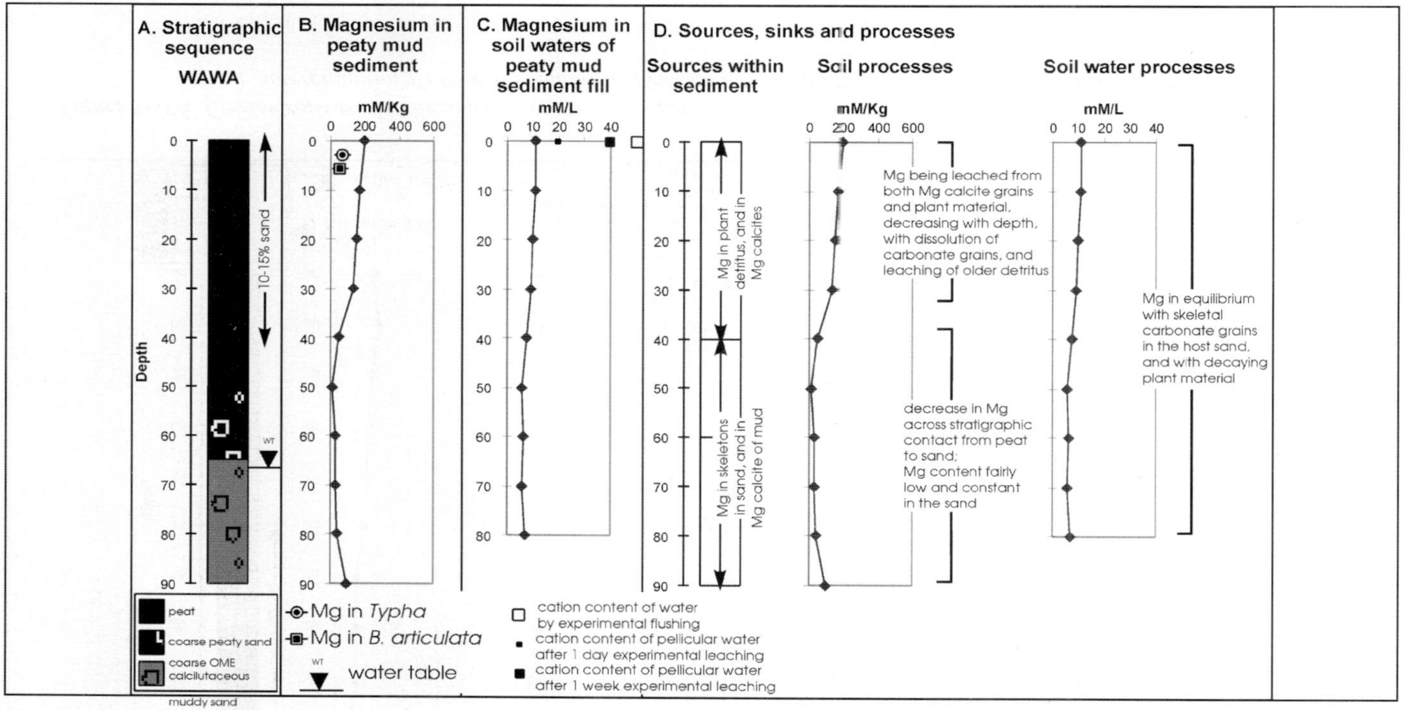

Figure 9-38D. *Concentrations of magnesium in peaty mud sediment and its interstitial waters, with a stratigraphic context and interpretation of processes leading to their concentration distributions, at wetland WAWA site.*

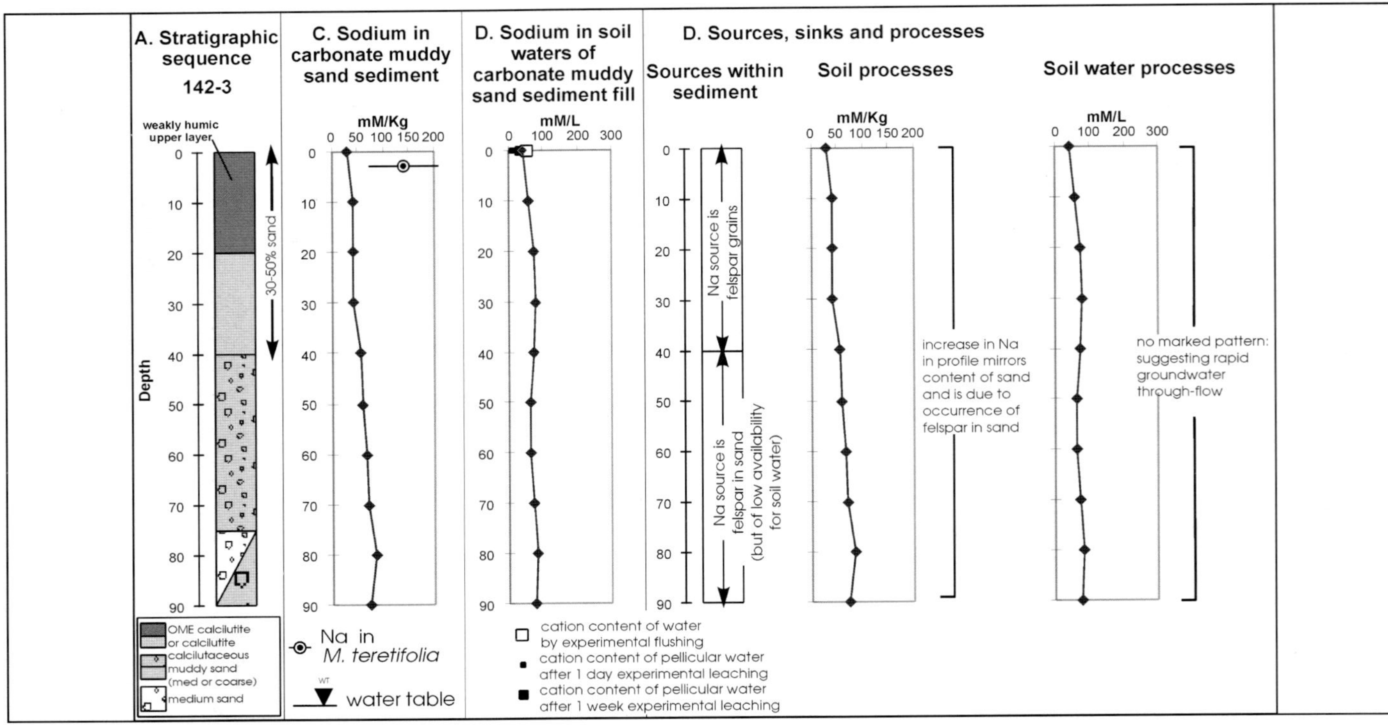

Figure 9-39A. *Concentrations of sodium in carbonate muddy sand sediment and its interstitial waters, with a stratigraphic context and interpretation of processes leading to their concentration distributions, at wetland 142-3 site.*

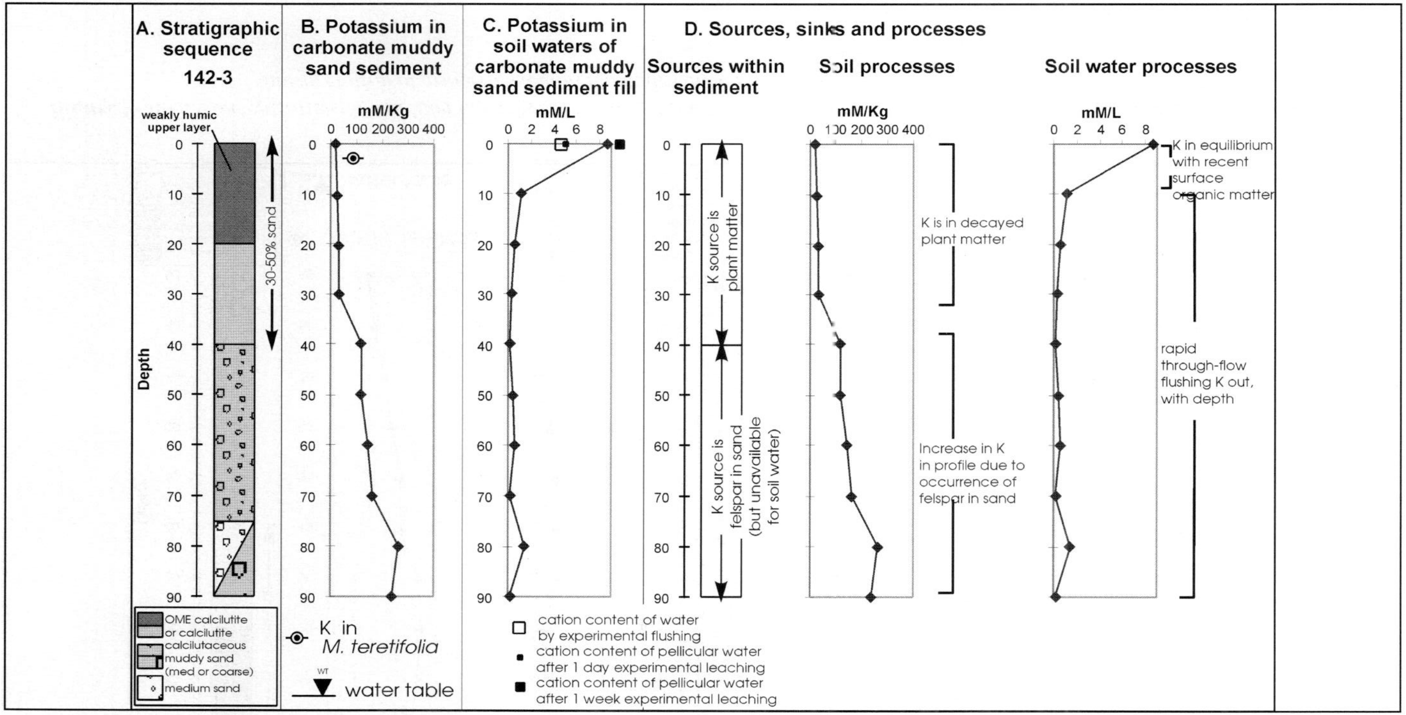

Figure 9-39B. *Concentrations of potassium in carbonate muddy sand sediment and its interstitial waters, with a stratigraphic context and interpretation of processes leading to their concentration distribution, at wetland 142-3 site.*

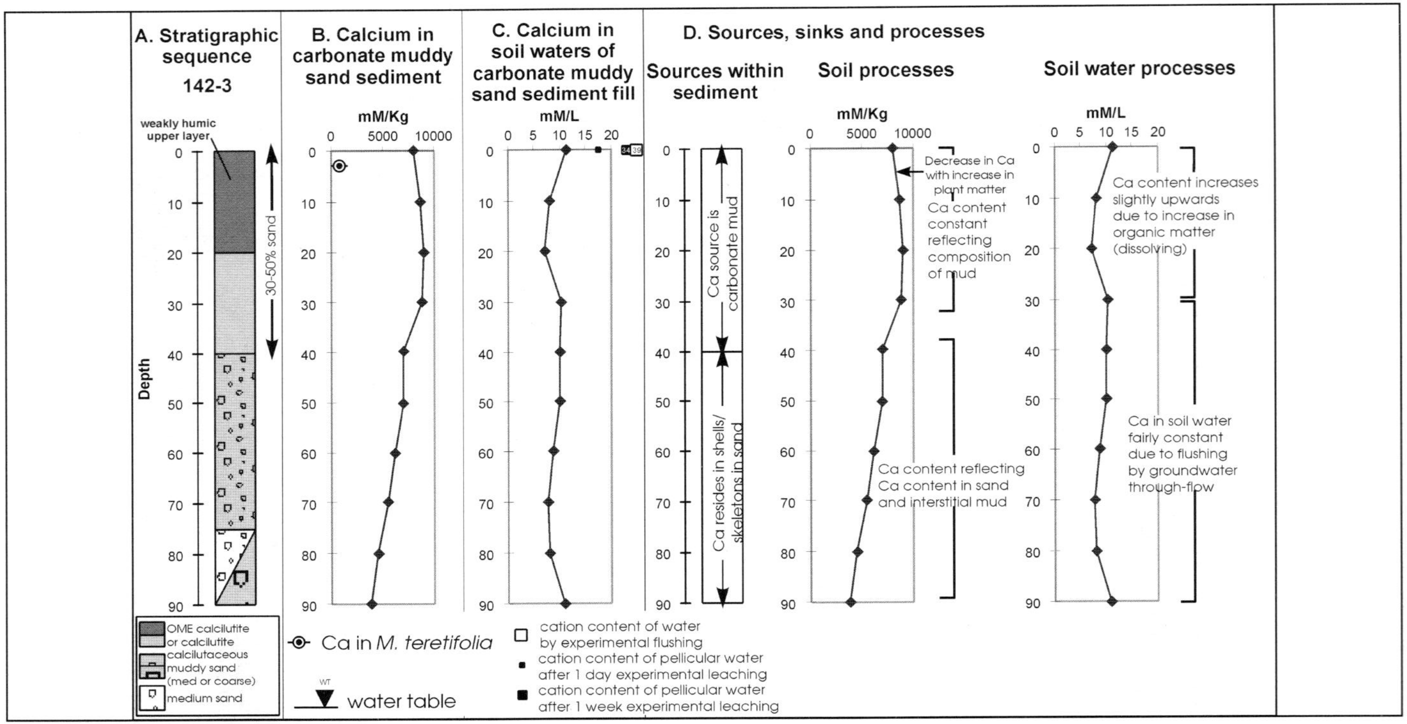

Figure 9-39C. *Concentrations of calcium in carbonate muddy sand sediment and its interstitial waters, with a stratigraphic context and interpretation of processes leading to their concentration distributions, at wetland 142-3 site.*

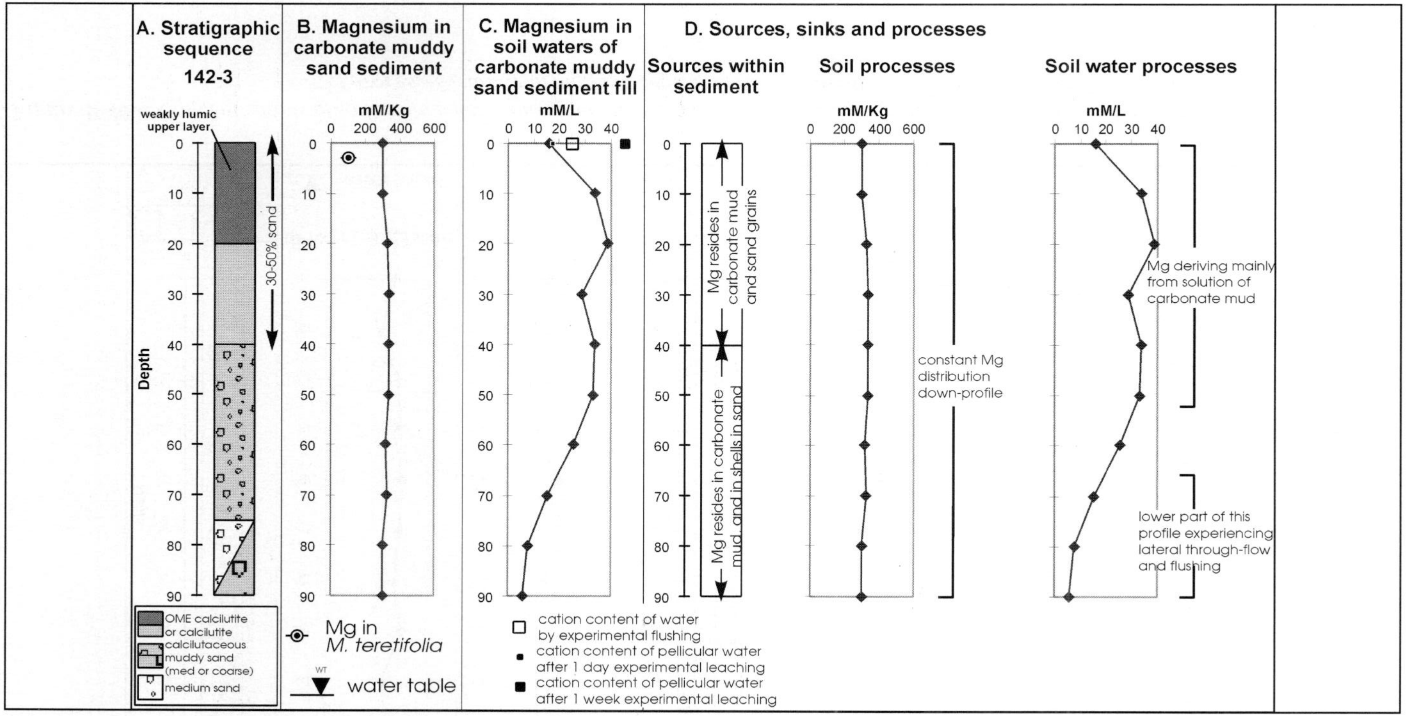

Figure 9-39D. *Concentrations of magnesium in carbonate muddy sand sediment and its interstitial waters, with a stratigraphic context and interpretation of processes leading to their concentration distributions, at wetland 142-3 site.*

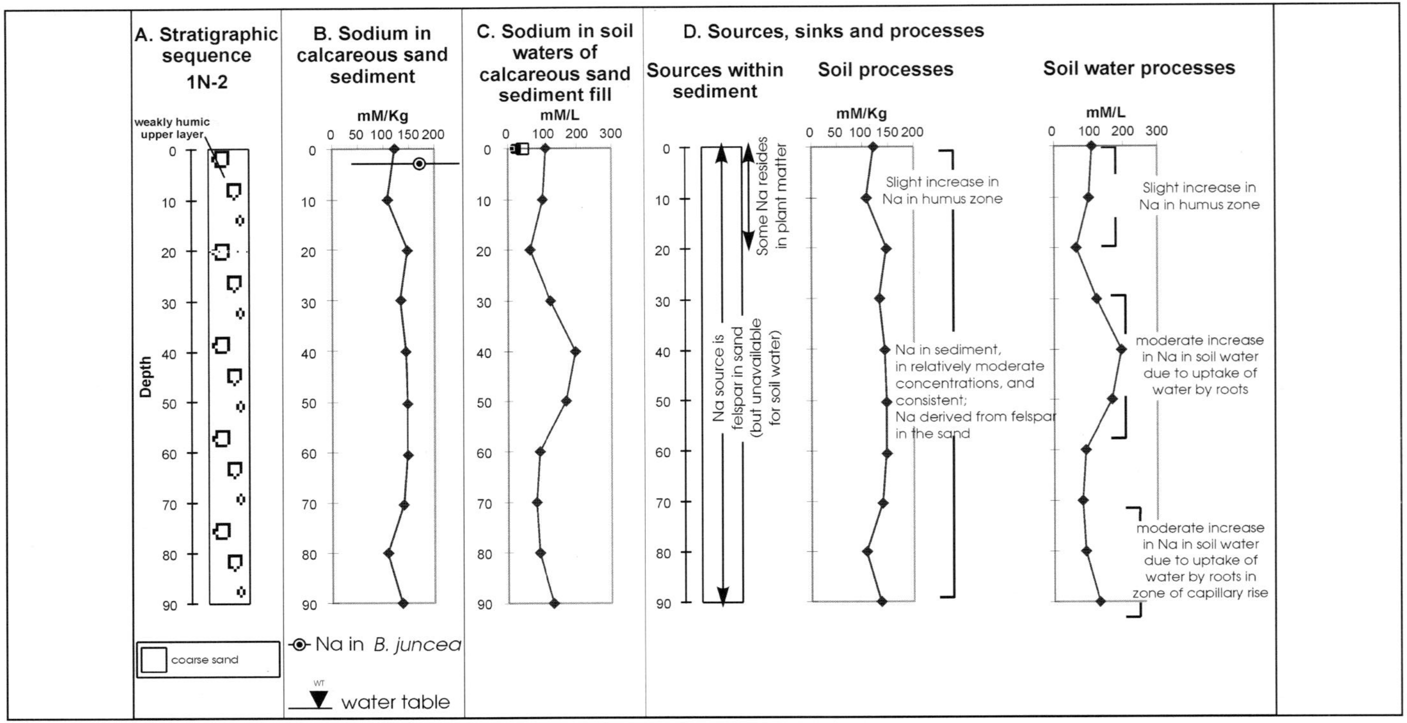

Figure 9-40A. *Concentrations of sodium in calcareous sand sediment and its interstitial waters, with a stratigraphic context and interpretation of processes leading to their concentration distributions, at wetland 1N-2 site.*

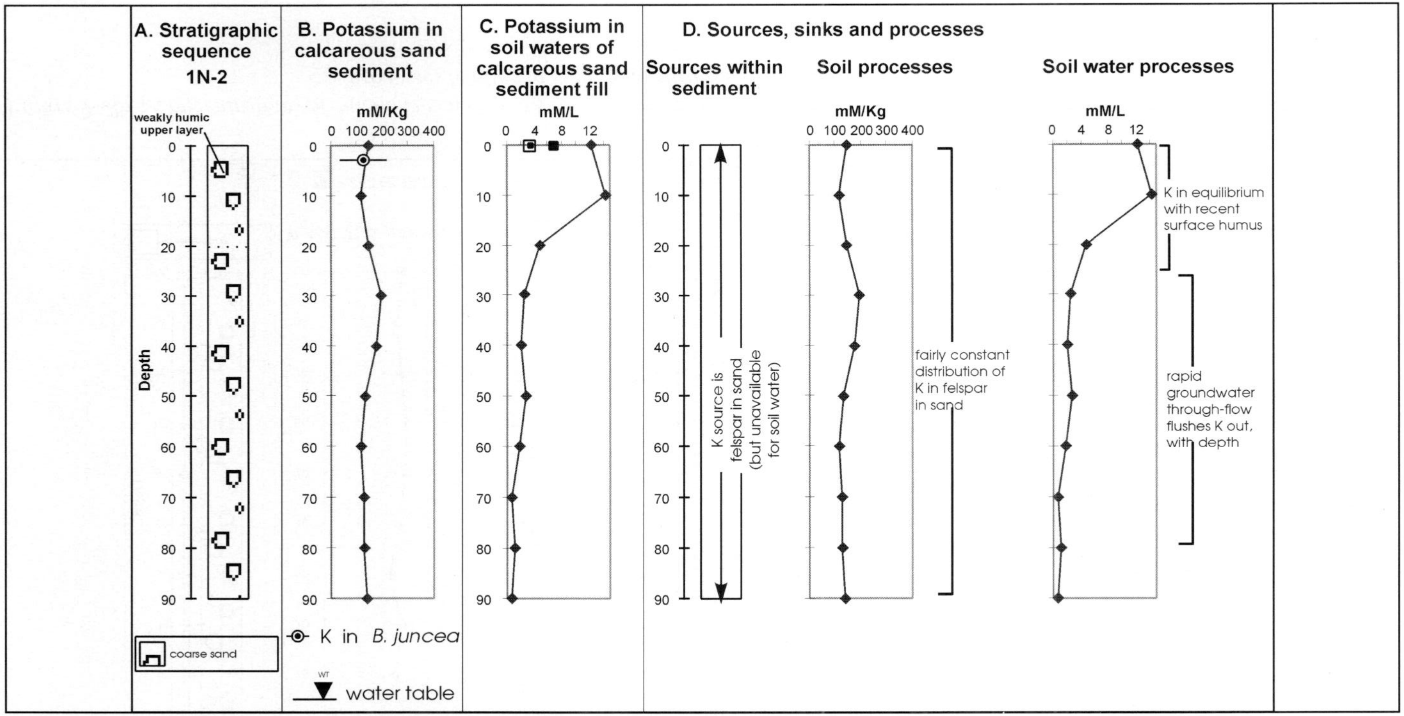

Figure 9-40B. *Concentrations of potassium in calcareous sand sediment and its interstitial waters, with a stratigraphic context and interpretation of processes leading to their concentration distributions, at wetland 1N-2 site.*

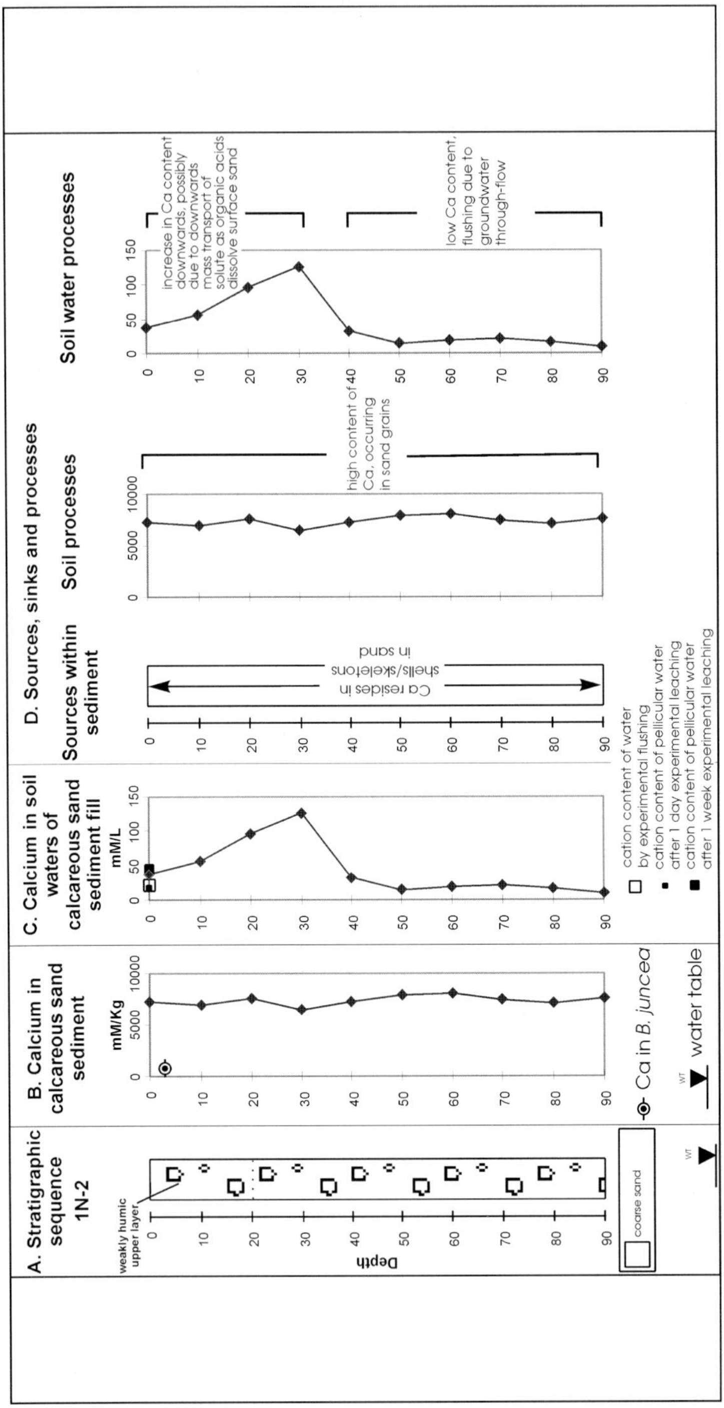

Figure 9-40C. *Concentrations of calcium in calcareous sand sediment and its interstitial waters, with a stratigraphic context and interpretation of processes leading to their concentration distributions, at wetland 1N-2 site.*

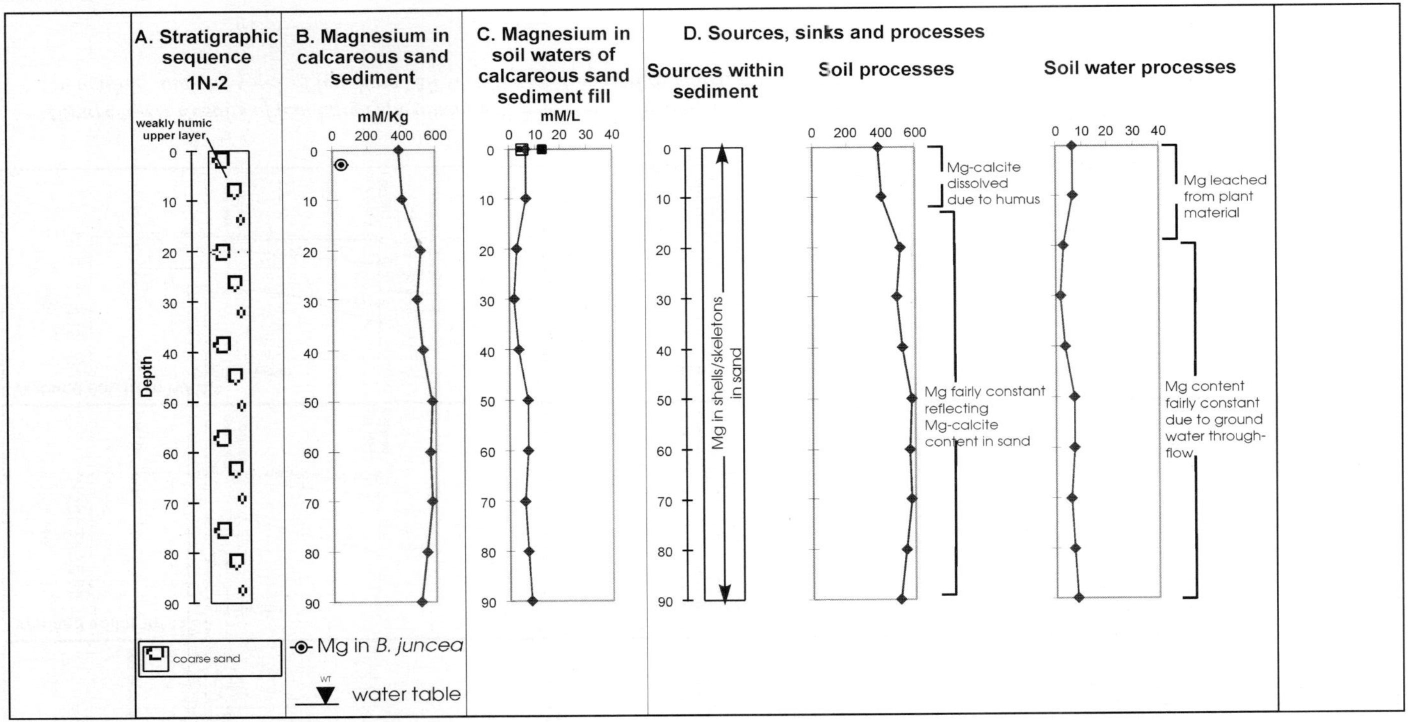

Figure 9-40D. *Concentrations of magnesium in calcareous sand sediment and its interstitial waters, with a stratigraphic context and interpretation of processes leading to their concentration distributions, at wetland 1N-2 site.*

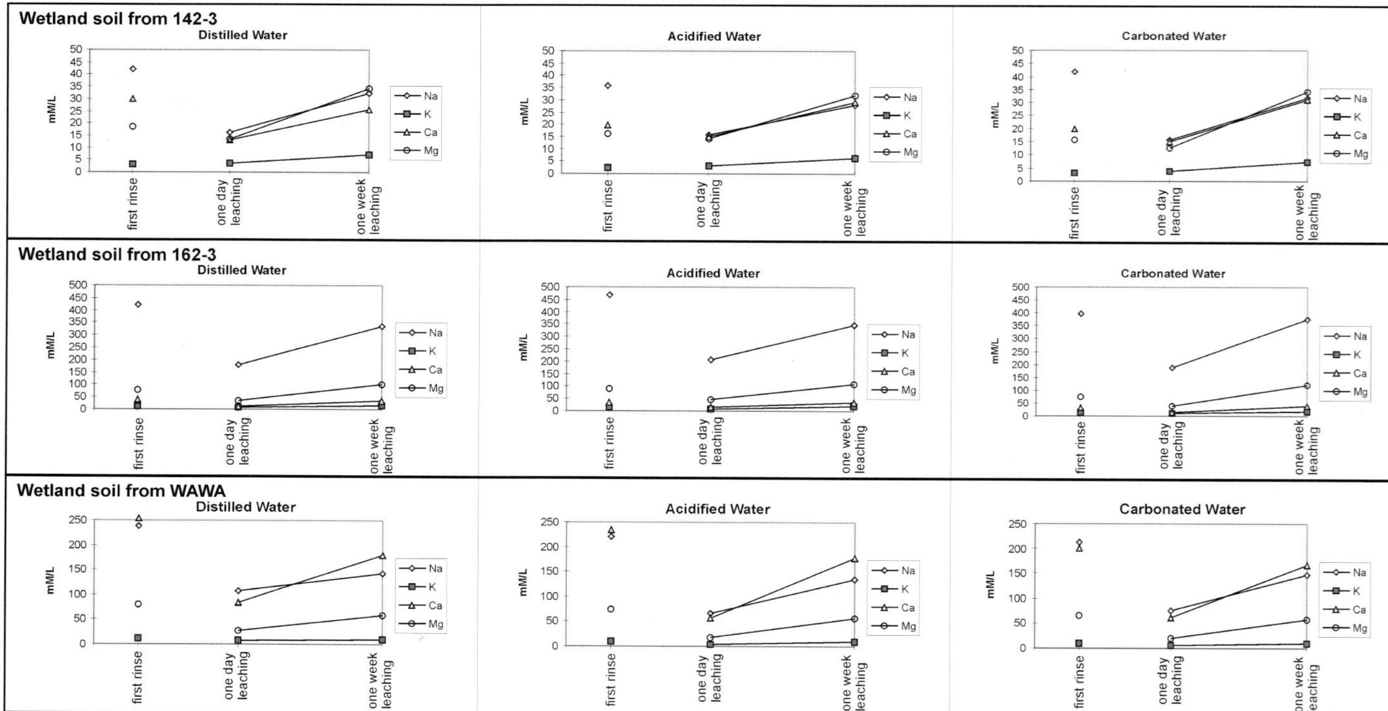

Figure 9-41. *Results of leaching experiment using de-ionised water, acidified water and carbonated water, following first rinse, one day of leaching, and one week of leaching. All results standardised to mM/L concentration derived from 100 ml of water overlying 1kg of soil.*

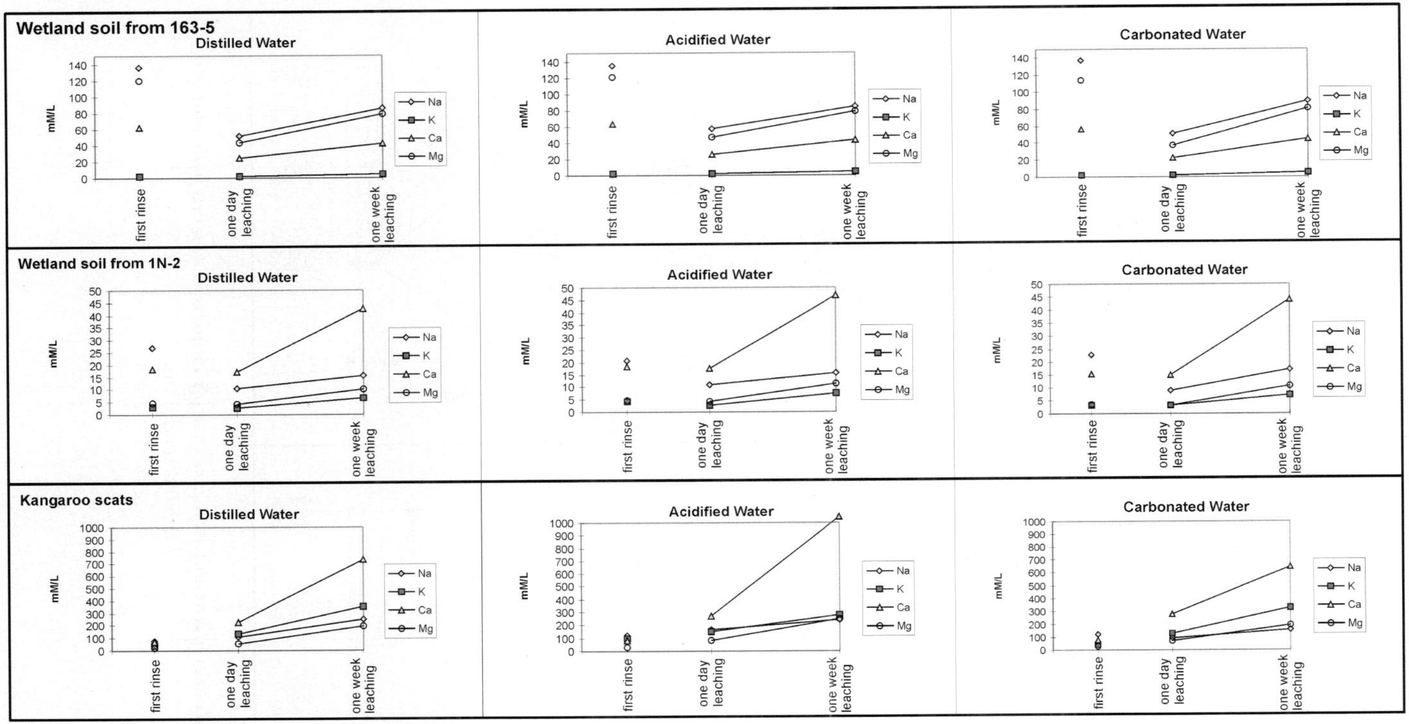

Figure 9-41 (cont.). *Results of leaching experiment using de-ionised water, acidified water and carbonated water, following first rinse, one day of leaching, and one week of leaching. All results standardised to mM/L concentration derived from 100 ml of water overlying 1kg of soil.*

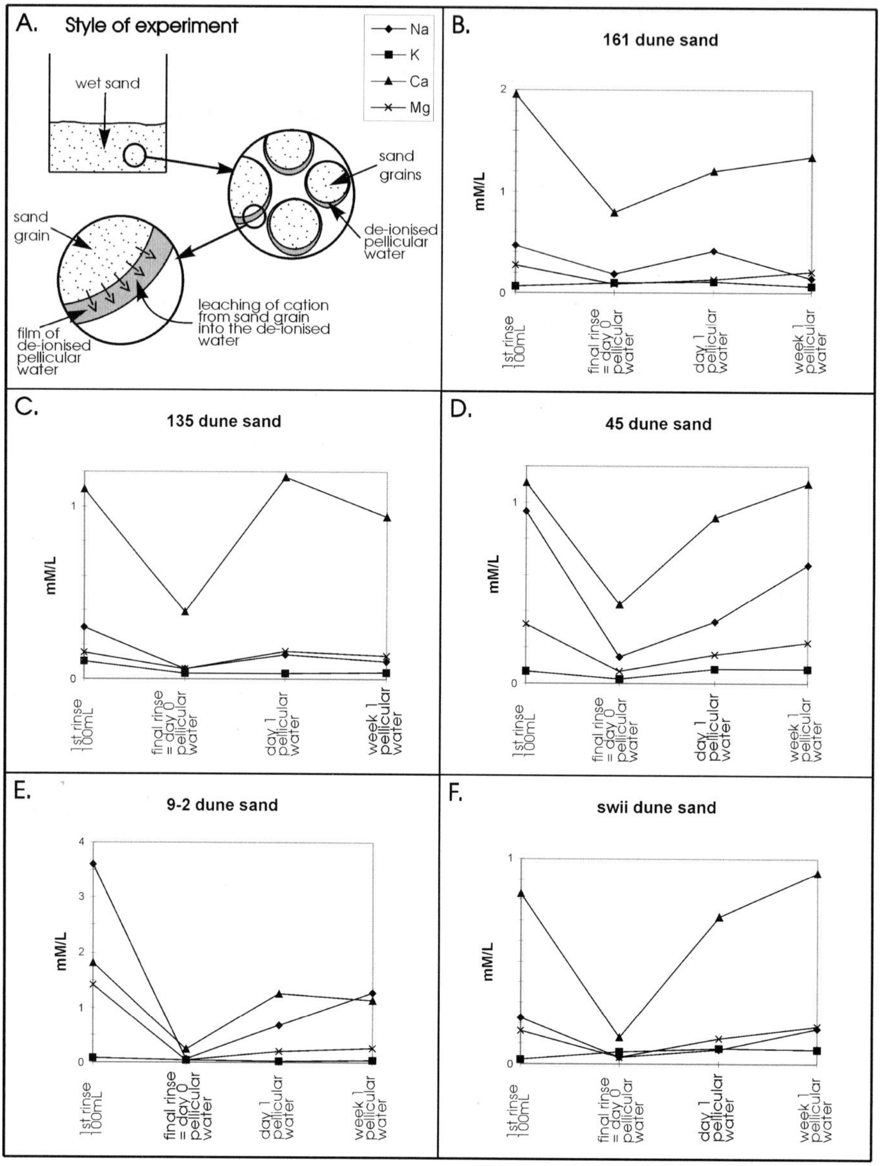

__Figure 9-42.__ Results of leaching experiment, using de-ionised water as pellicular water around sand grains.

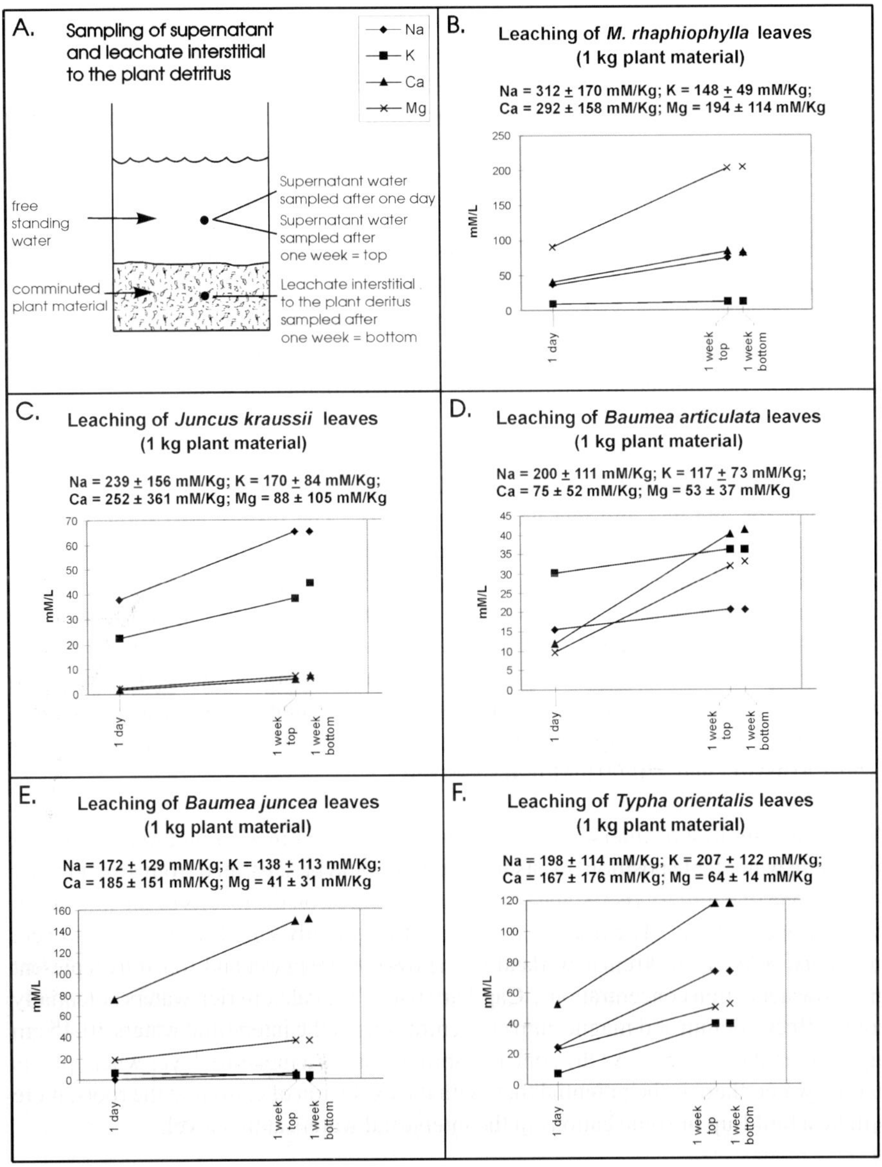

Figure 9-43. *Results of leaching experiment of comminuted plant material, using de-ionised water following one day leaching and one week leaching.*

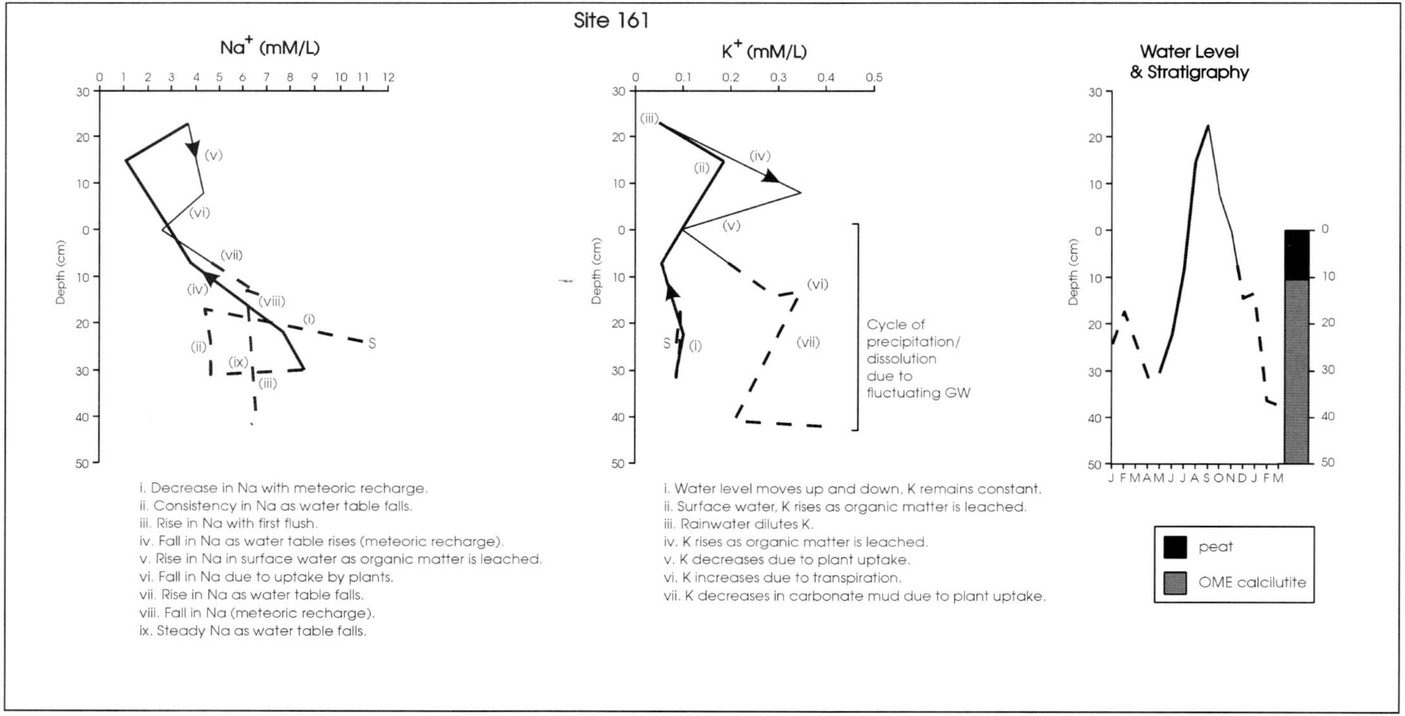

Figure 9-46. *Groundwater cation concentrations in relation to recharge/discharge processes and vertical movement of water table through a heterogeneous stratigraphy. A. Sodium B. Potassium.*

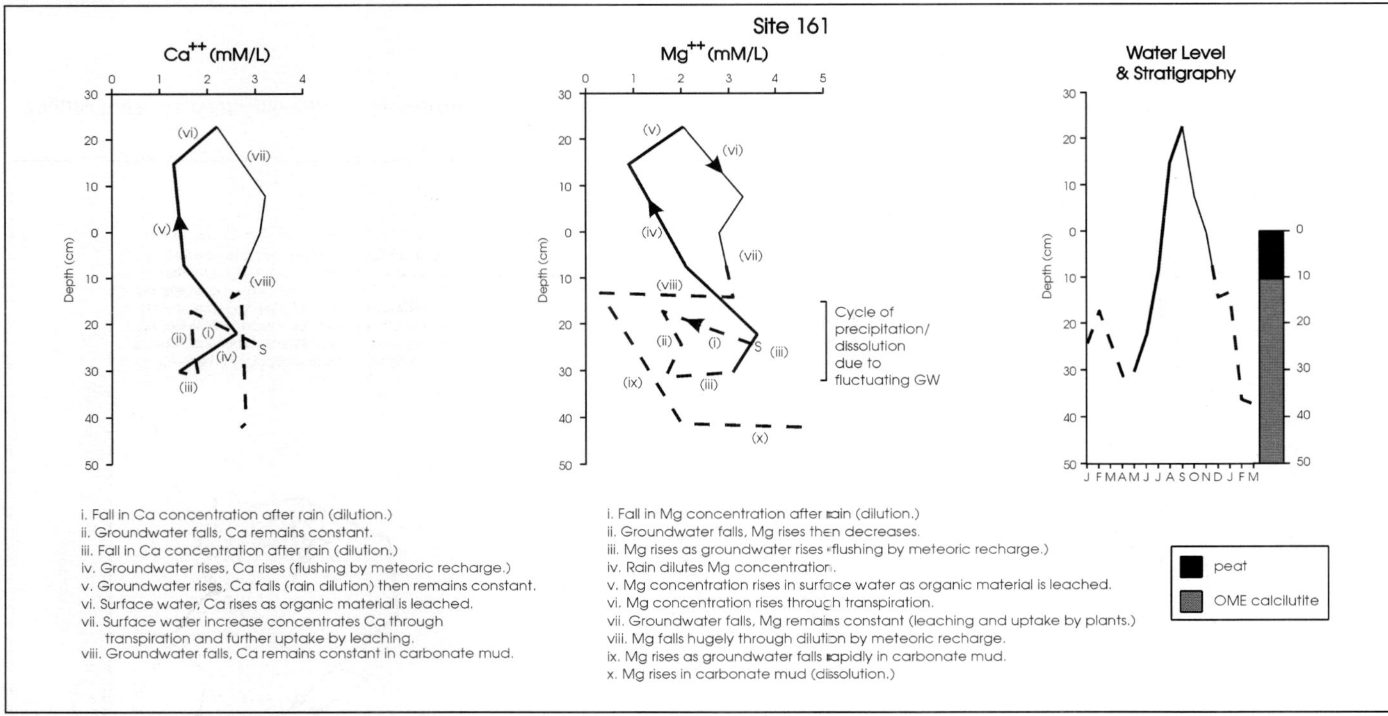

Figure 9-46 (cont.). *Groundwater cation concentrations in relation to recharge/discharge processes and vertical movement of water table through a heterogeneous stratigraphy. A. Calcium B. Magnesium.*

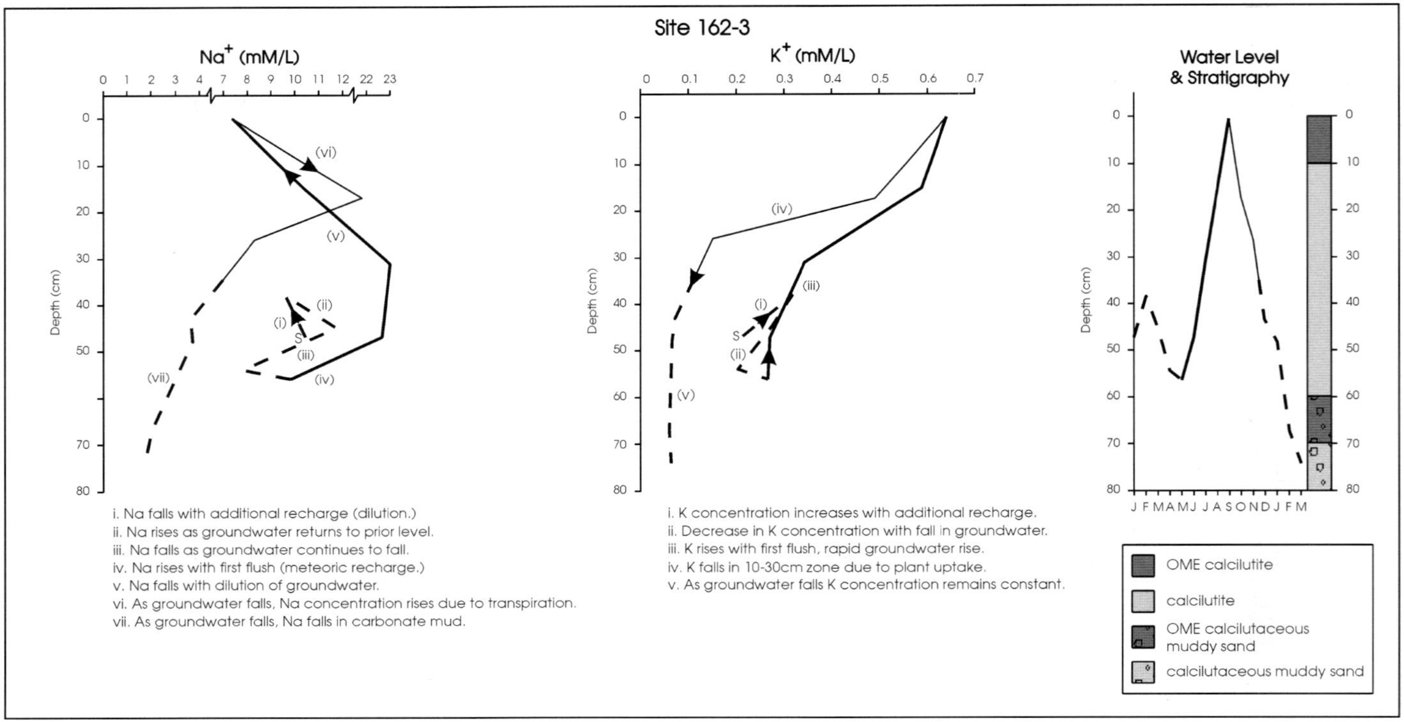

Figure 9-47. *Groundwater cation concentrations in relation to recharge/discharge processes and vertical movement of water table through a heterogeneous stratigraphy. A. Sodium B. Potassium.*

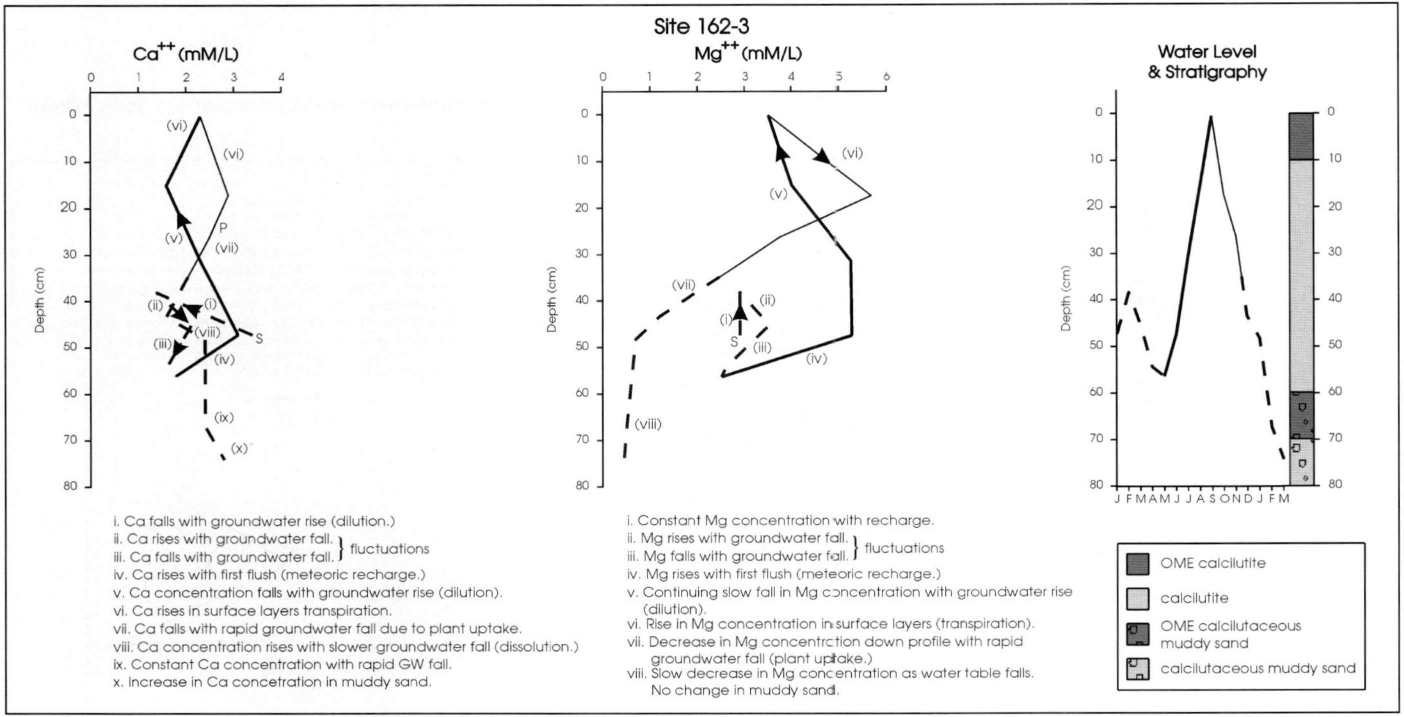

Figure 9-47 (cont.). *Groundwater cation concentrations in relation to recharge/discharge processes and vertical movement of water table through a heterogeneous stratigraphy. A. Calcium B. Magnesium.*

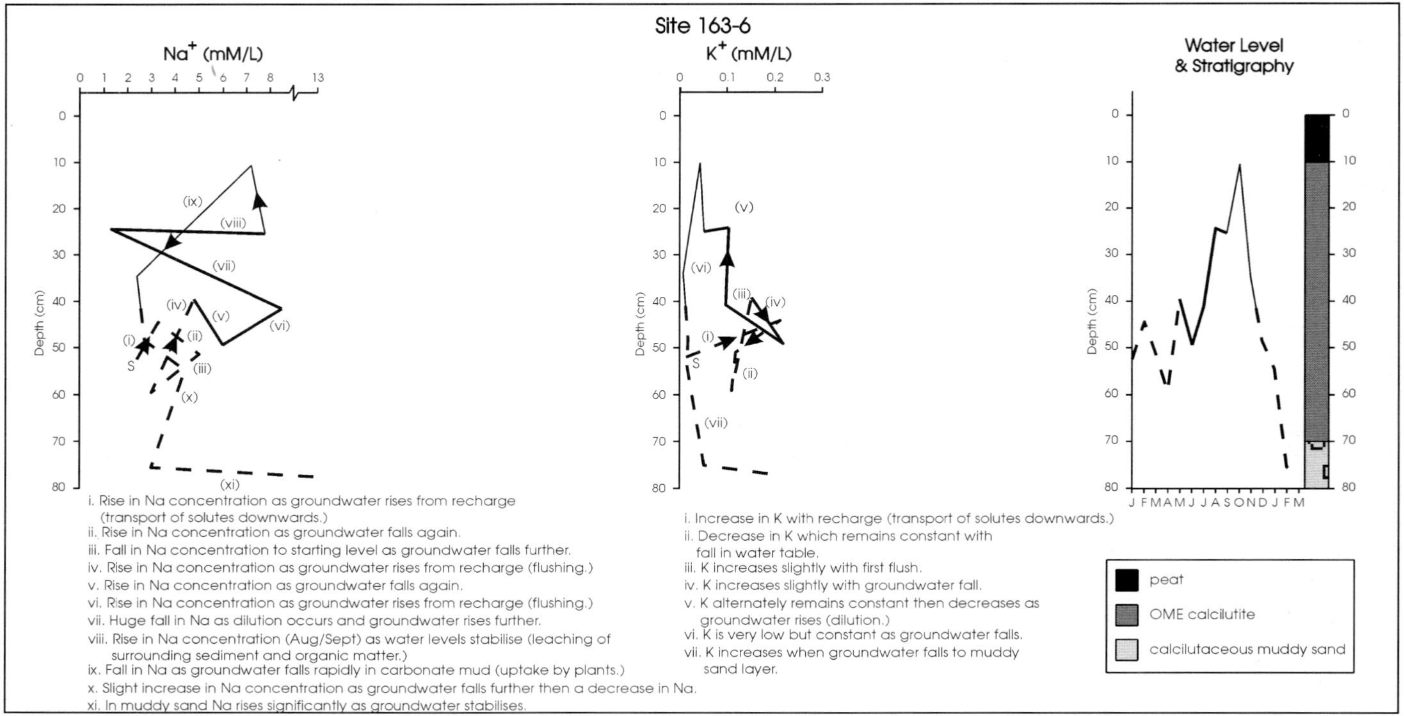

Figure 9-48. *Groundwater cation concentrations in relation to recharge/discharge processes and vertical movement of water table through a heterogeneous stratigraphy. A. Sodium B. Potassium.*

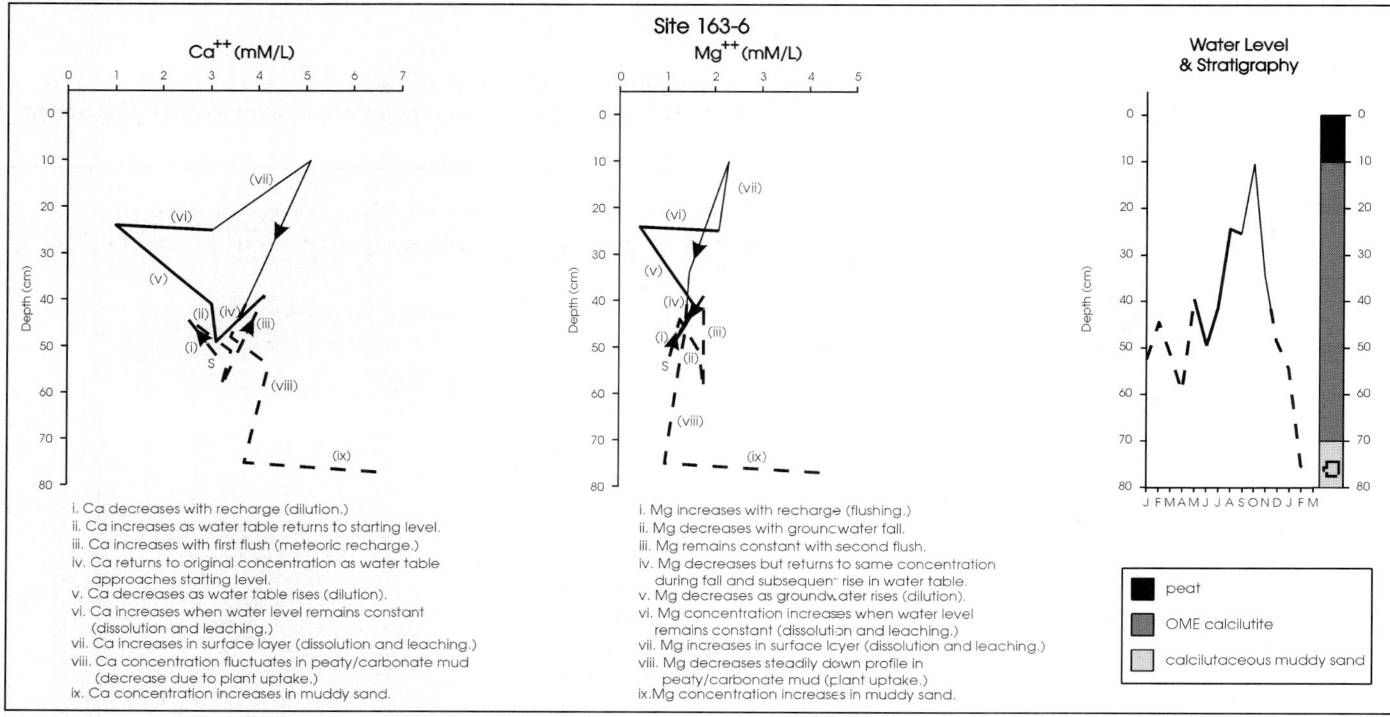

Figure 9-48 (cont.). *Groundwater cation concentrations in relation to recharge/discharge processes and vertical movement of water table through a heterogeneous stratigraphy. A. Calcium B. Magnesium.*

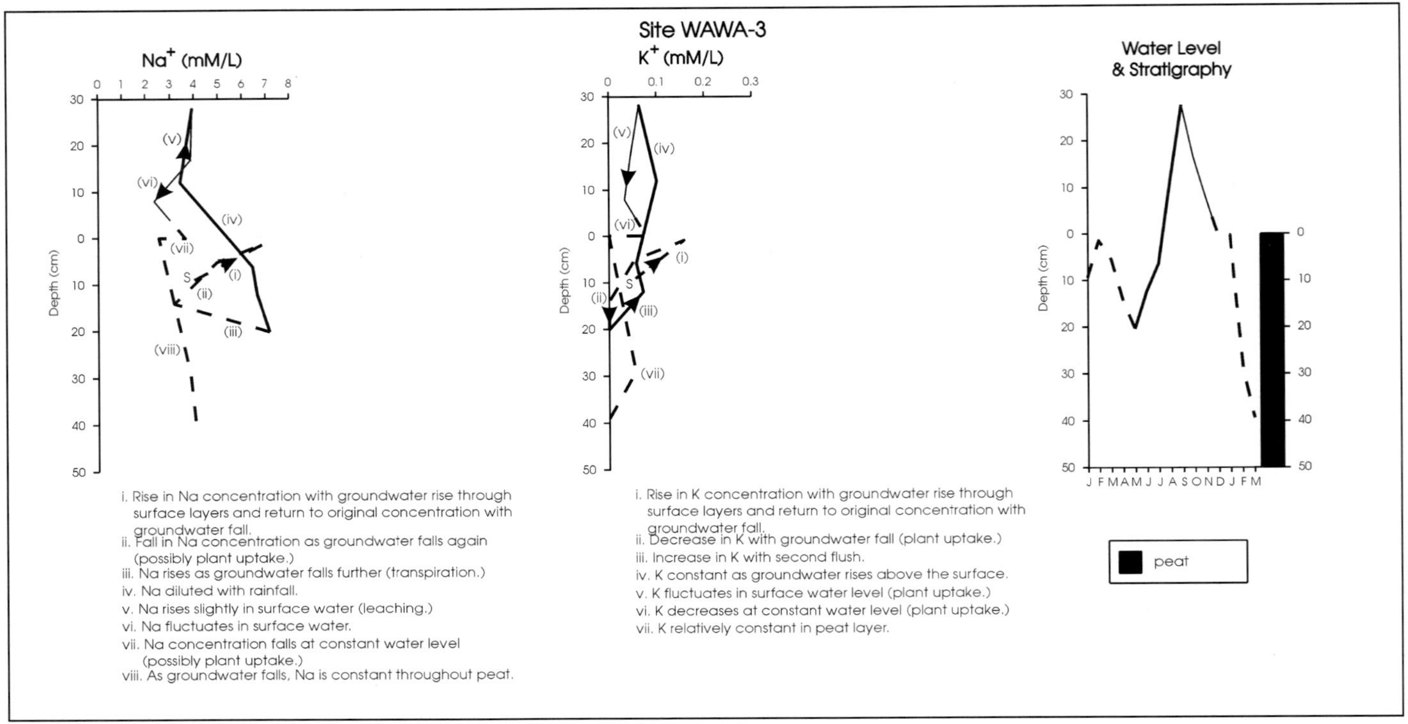

Figure 9-49. *Groundwater cation concentrations in relation to recharge/discharge processes and vertical movement of water table through a heterogeneous stratigraphy. A. Sodium B. Potassium.*

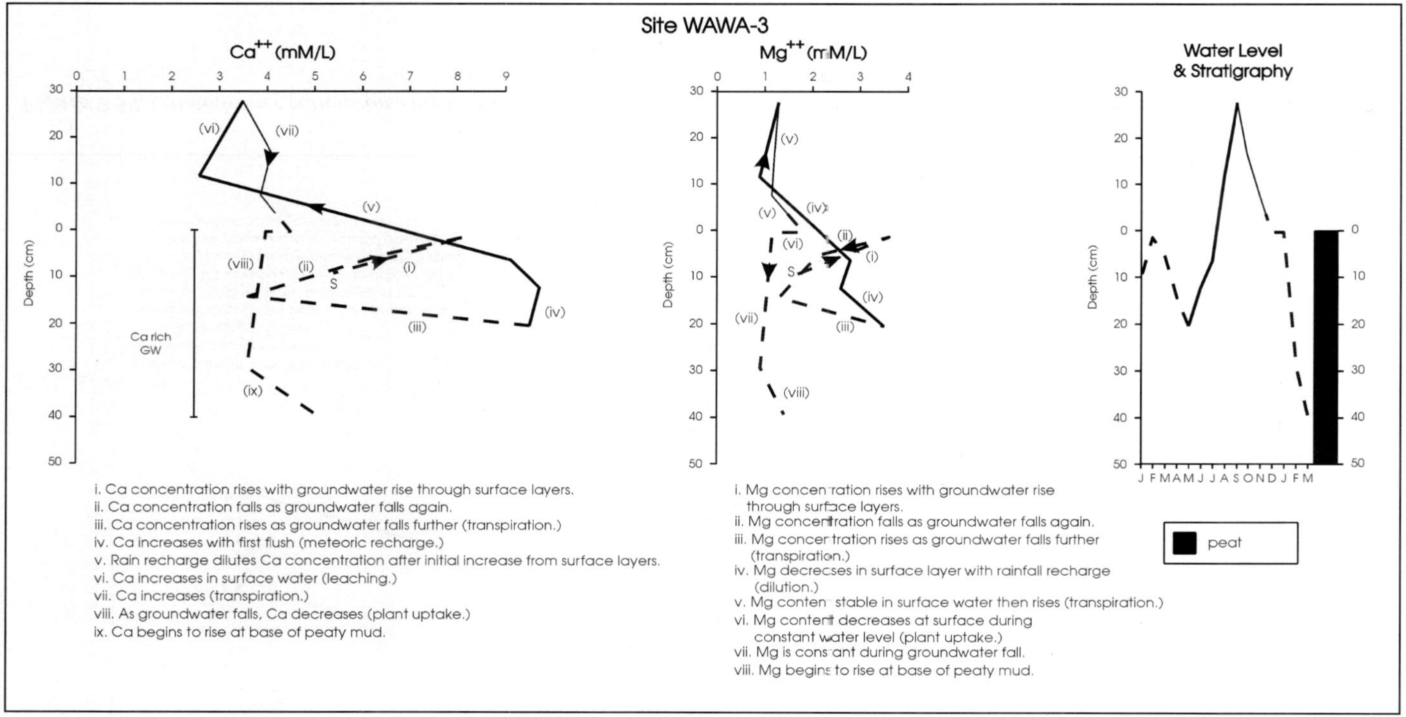

Figure 9-49 (cont.). *Groundwater cation concentrations in relation to recharge/discharge processes and vertical movement of water table through a heterogeneous stratigraphy. A. Calcium B. Magnesium.*

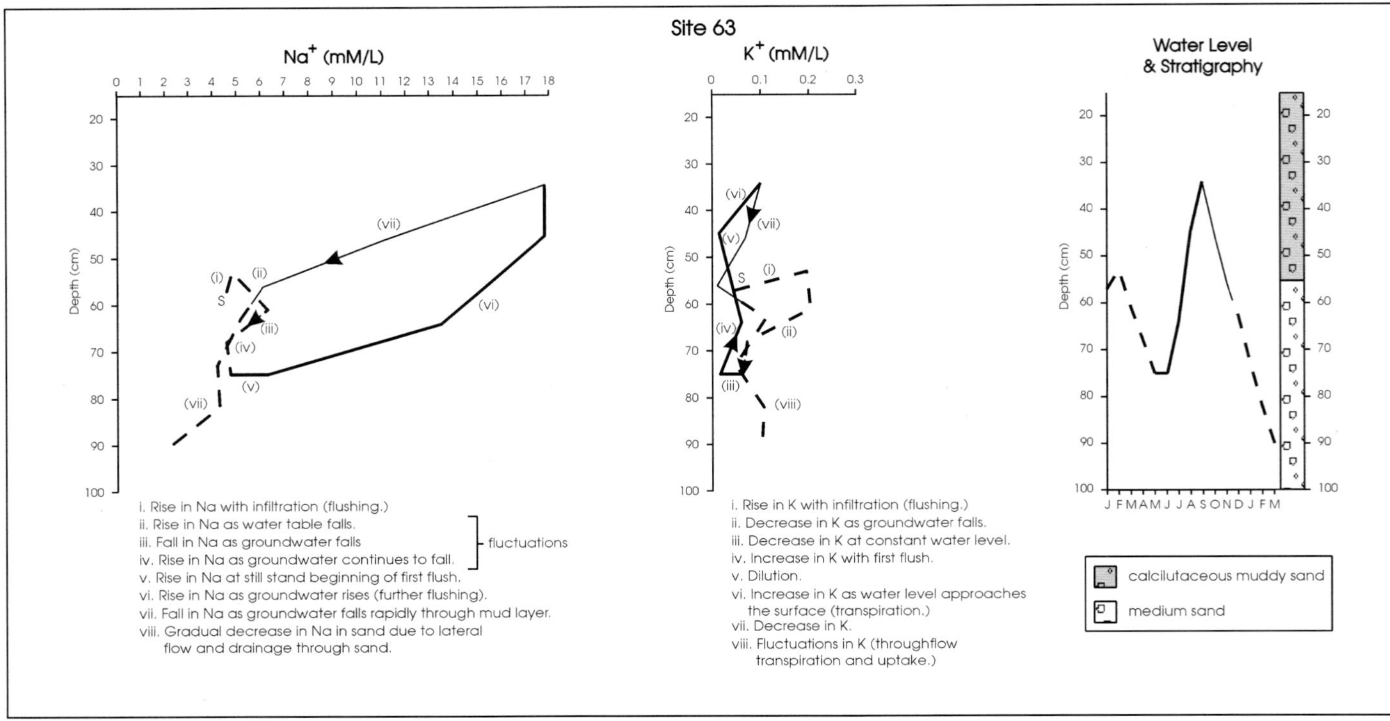

Figure 9-50. *Groundwater cation concentrations in relation to recharge/discharge processes and vertical movement of water table through a heterogeneous stratigraphy. A. Sodium B. Potassium.*

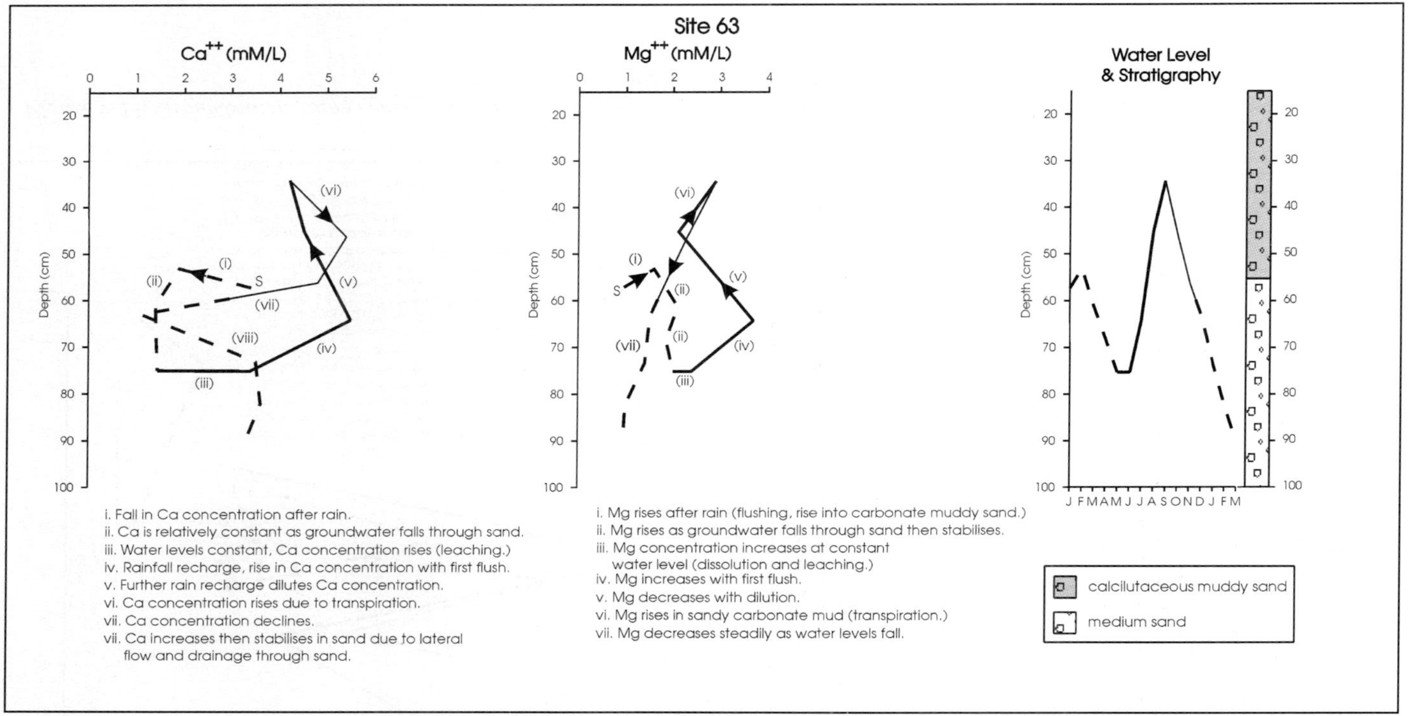

Figure 9-50 (cont.). *Groundwater cation concentrations in relation to recharge/discharge processes and vertical movement of water table through a heterogeneous stratigraphy. A. Calcium B. Magnesium.*

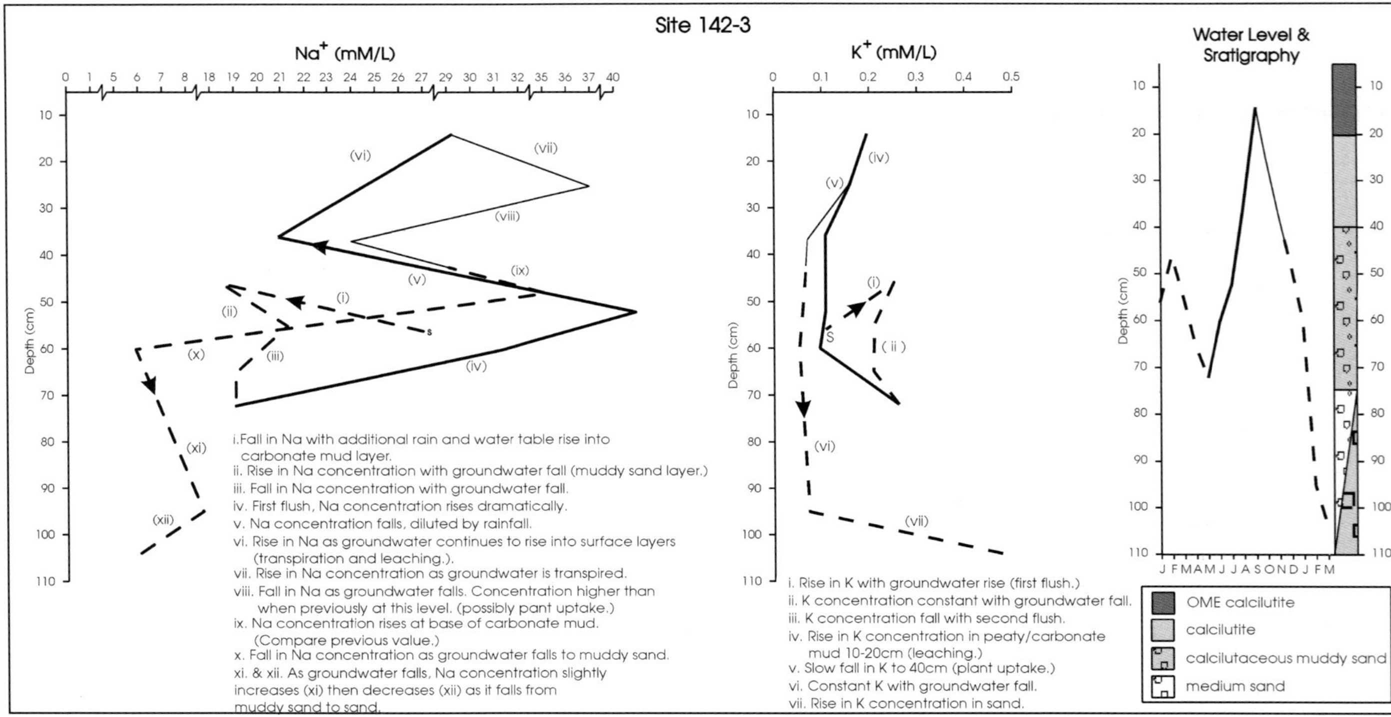

Figure 9-51. *Groundwater cation concentrations in relation to recharge/discharge processes and vertical movement of water table through a heterogeneous stratigraphy. A. Sodium B. Potassium.*

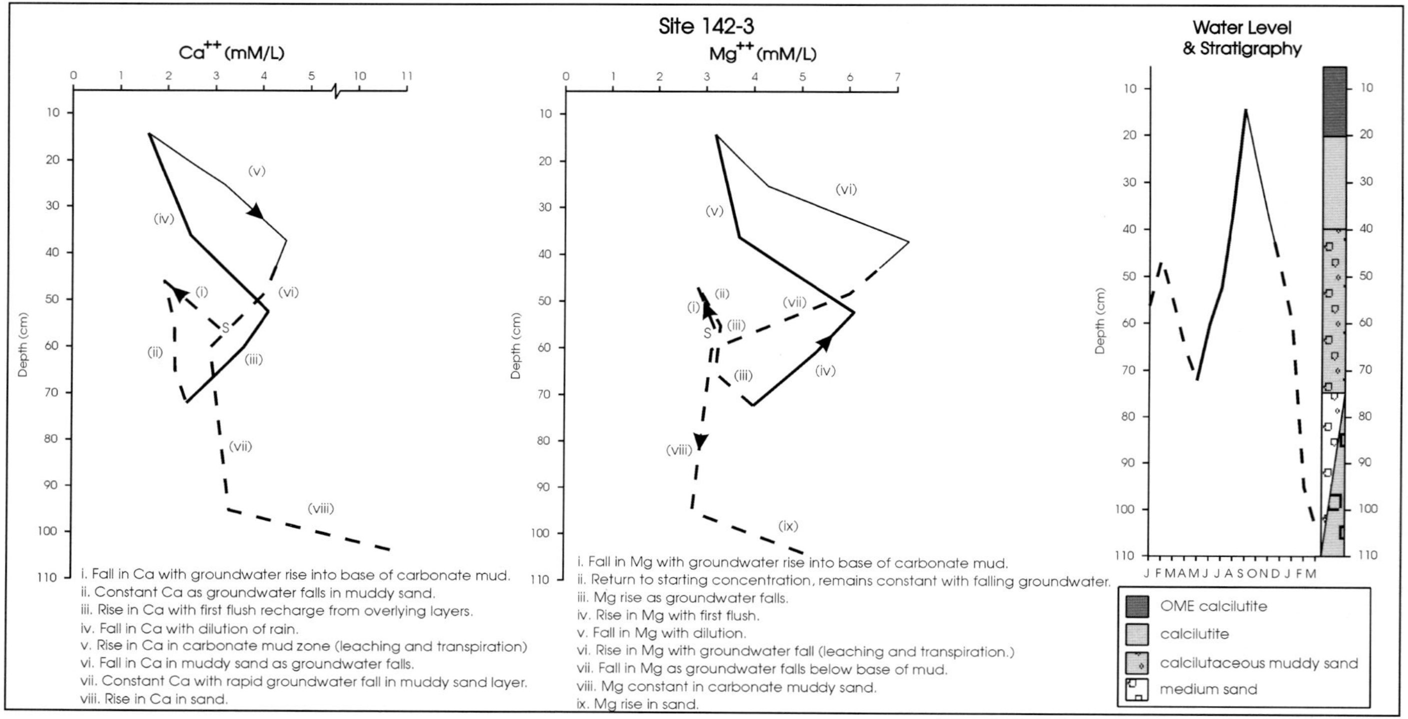

Figure 9-51 (cont.). *Groundwater cation concentrations in relation to recharge/discharge processes and vertical movement of water table through a heterogeneous stratigraphy. A. Calcium B. Magnesium.*

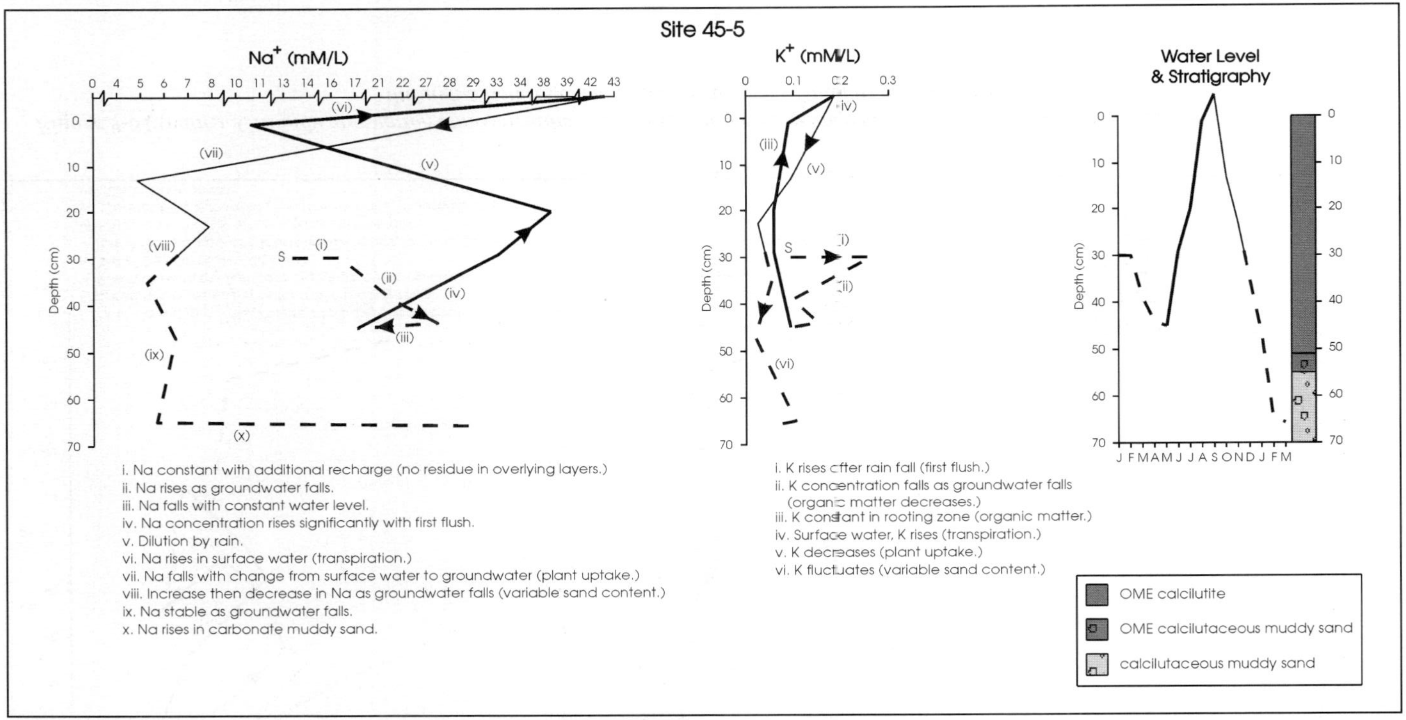

Figure 9-52. *Groundwater cation concentrations in relation to recharge/discharge processes and vertical movement of water table through a heterogeneous stratigraphy. A. Sodium B. Potassium.*

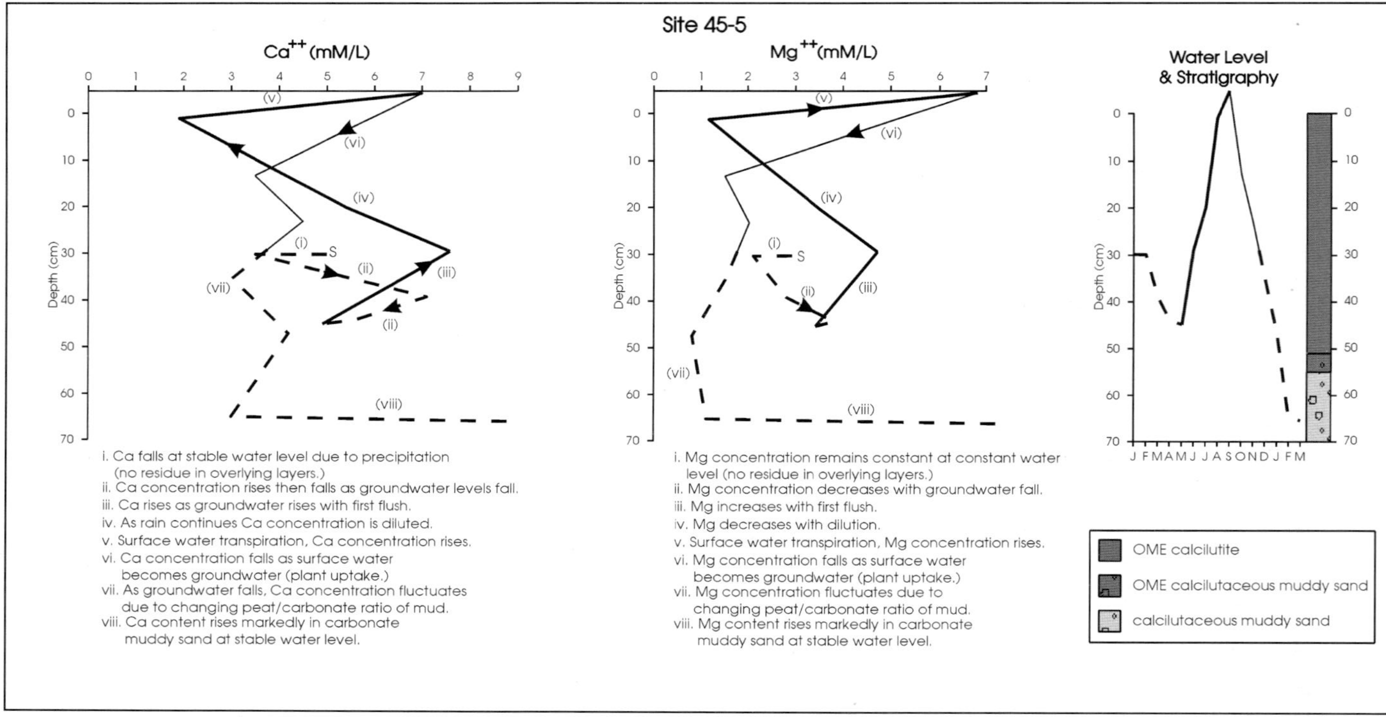

Figure 9-52 (cont.). *Groundwater cation concentrations in relation to recharge/discharge processes and vertical movement of water table through a heterogeneous stratigraphy. A. Calcium B. Magnesium.*

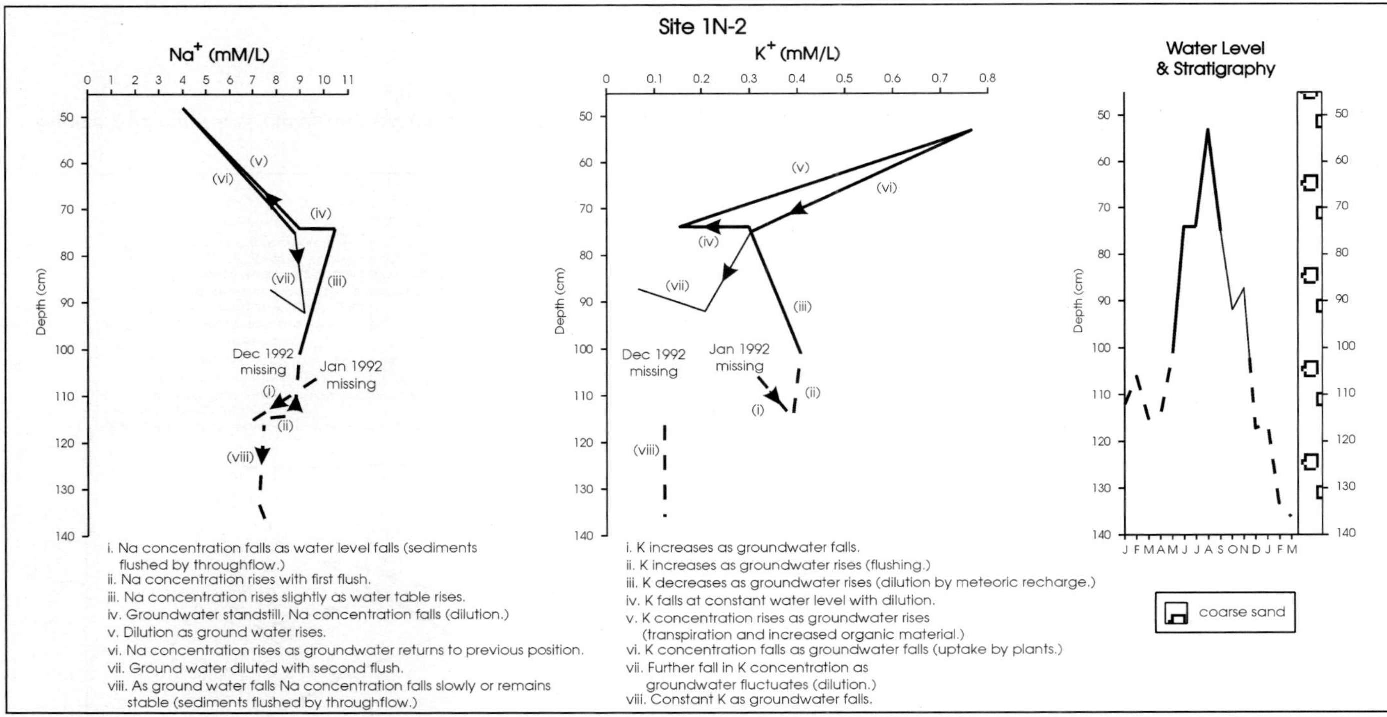

Figure 9-53. *Groundwater cation concentrations in relation to recharge/discharge processes and vertical movement of water table through a heterogeneous stratigraphy. A. Sodium B. Potassium.*

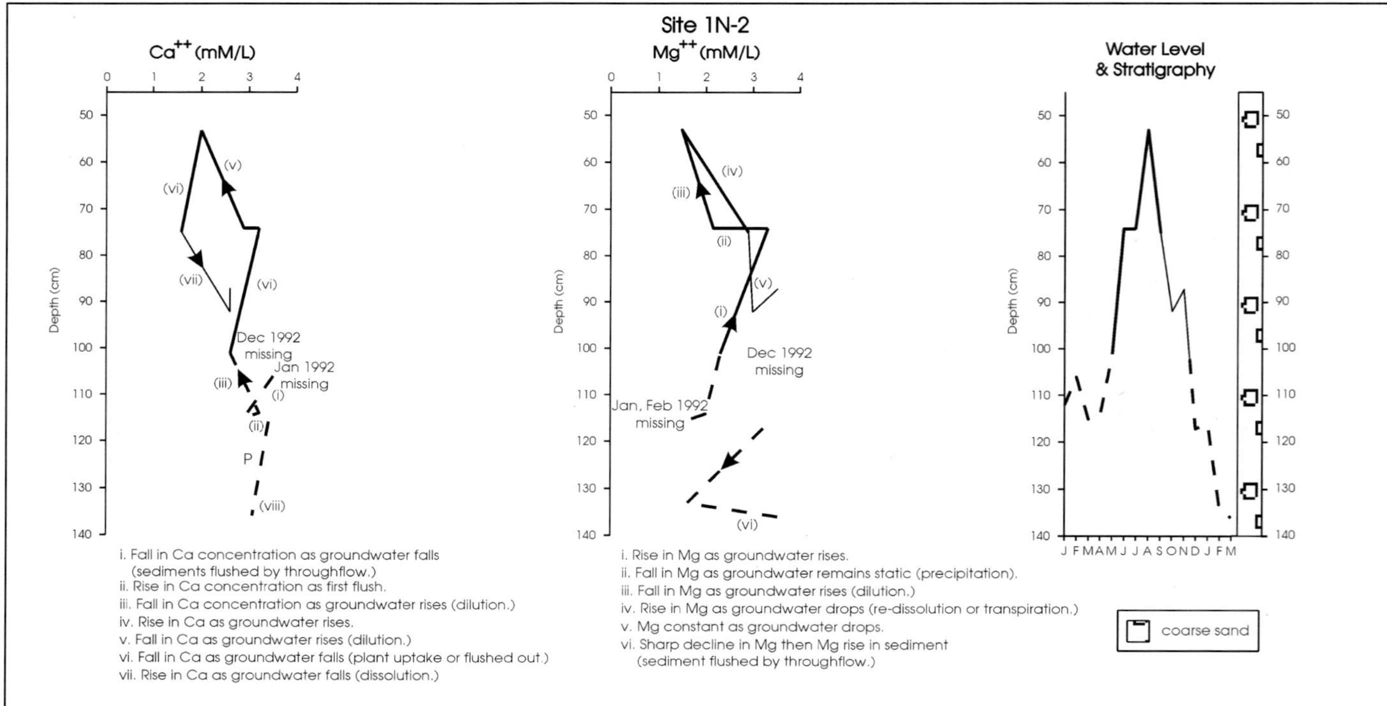

Figure 9-53 (cont.). *Groundwater cation concentrations in relation to recharge/discharge processes and vertical movement of water table through a heterogeneous stratigraphy. A. Calcium B. Magnesium.*

9.5 Nutrients

9.5.1 Background

Historically, much of the attention and consequent work on nutrients generally has been within a single component of the wetland ecosystem, *i.e.* either the water, the sediment, or the biosphere (Anderson 1976; Brinson 1977; Day 1982; Attiwill and Leeper 1987; Moore and Reddy 1994; Nixdorf and Deneke 1997; Rhue and Harris 1999). In those studies, where facets of more than a single component have been investigated, *e.g.*, the surface water and phytoplankton, or the surface water and macrophytes, or the soil/water interface (Wetzel 1983, 1999; Martinova 1993), the major part of work on seasonal nutrient exchanges between sediments and water has been carried out in either estuarine, lacustrine, or permanently waterlogged settings, where the water sediment interface is relatively stable *i.e.*, it is either permanently under water or varies less than 5-10 cm (Mortimer 1941, 42; Duursma 1967; Gore 1983; Wetzel 1983; Koerselman and Verhoeven 1992; Verhoeven *et al.* 1994; Smith *et al.* 1995; Ramm and Scheps 1997; and others). This situation is substantially different from the sediment/water interface in this study, where, seasonally there is interaction with plant detritus, variable oxygen availability, and, when the water level descends below the sediment surface, with stratified sediment. The seasonal hydrological dynamics in some peatlands are comparable to those of the Becher Suite wetlands (McKnight *et al.* 1985; Shotyk 1988; Wheeler *et al.* 1992; Verhoeven *et al.* 1994; Ross 1995; and others), particularly where peatlands are recharged by meteoric infiltration and groundwater discharge, however, the chemical environment of the peat substrate and the predominance of biochemical processes contrast with the hydrochemical processes in the calcareous mineral substrates.

As described in the methods, phosphorus concentrations in sediments in the centre of each wetland were derived from three samplings, and groundwater sampling frequency was based on the timing of specific hydrological events 1992-1994: water table maxima and minima; the end of the first month of winter rainfalls (referred to herein as first flush); and the late spring which marks the end of winter rainfall, the fall of groundwater levels, and the vegetative growth period of many wetland plants. The rationale underlying this was that studies have shown that water movement can influence phosphorus concentrations (Klopatek 1978; Carter *et al.* 1979; Howard-Williams 1985; Carter 1986; LaBaugh 1986; Mitsch and Gosselink 1986).

9.5.2 Phosphorus input and export

From a long term perspective, most of the phosphorus in wetlands can be regarded as allochthonous. Allochthonous sources are burrowing and grazing macrofauna and invertebrate fauna, roosting avifauna, sediments from sheet wash, distal ash from fires, pollen, and both meteoric water and groundwater. Recycled allochthonous sources (termed by some authors as autochthonous; *cf.*, Martinova 1993) include phytoplankton, zooplankton, macrophytes, ash from fires and wetland fill. The

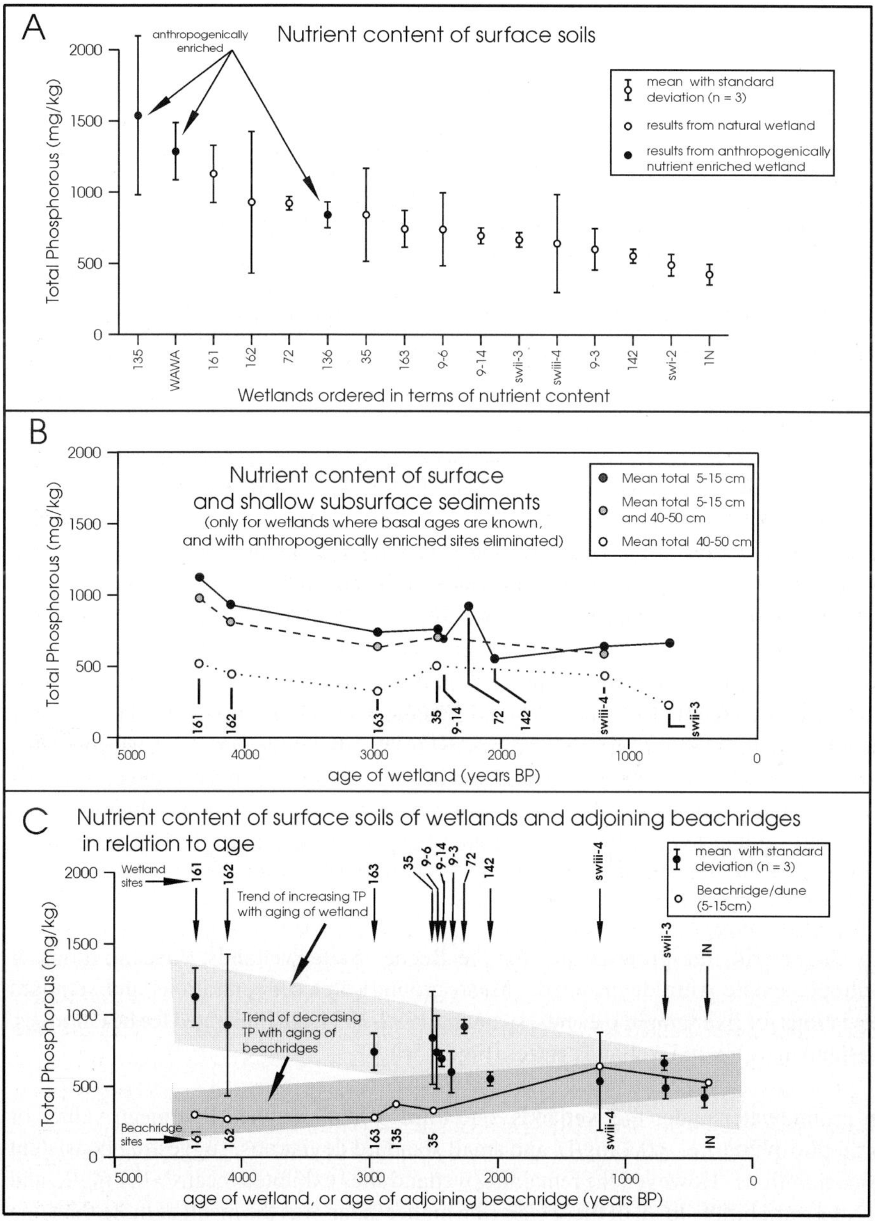

Figure 9-55. *Total phosphorous (TP) content of surface soils in the various wetlands, TP in surface and shallow subsurface soils in relation to wetland age, and comparison between TP in surface wetland sediments and adjoining beachridges with respect to isochrons.*

magnitude and form of allochthonous phosphorus in the wetland varies annually and seasonally (*e.g.*, kangaroo scats contained approximately 6,230 mg/kg phosphorus, of which 1.9% was orthophosphate). Values of phosphorus in various parts of selected wetland macrophytes, as collected in summer, ranged from 130-1,370 mg/kg. In the monocotyledons, most of the phosphorus was stored in the living roots; for dicotyledons, most of the phosphorus was stored in the leaves and fruit. As the biomass of fruits in the dicotyledons is low, the most important source of phosphorus for recycling was in the leaves. The export of phosphorus occurs predominantly through leaching by meteoric percolation, followed by groundwater throughflow and downward leakage, and by fauna which harvest wetland plants. The magnitude of these processes varies annually and seasonally.

9.5.3 Total phosphorus in sediments

In order to characterise the nutrient status of wetlands, concentrations of total phosphorus were measured in the sediments at 5-15 cm and 40-50 cm, and at 5-15 cm for selected beachridge/dunes (Fig. 9-55A). In the wetland sediments, phosphorus ranged from 0.15% of sediment (dry weight) to 0.02%. In all wetland sites, the total phosphorus content in the sediments decreased with depth, which is attributed to decreasing organic content down profile (Fig. 9-55B). This conclusion is supported by two observations: 1) the greatest difference occurred in wetlands underlain by calcilutite in which there was minimal penetration of organic material below the top 15 cm, *e.g.*, wetlands 161, 162, 135, and 2) there was little change in wetland WAWA which is underlain by peat and muddy sand to 50 cm. However, decreasing organic content alone does not explain the decrease in total phosphorus in the younger wetlands as suggested in Figure 9-55C. The thicker deposits of calcilutite in the older wetlands 161, 162, appear to contribute to the total phosphorus content of the sediment.

9.5.4 Orthophosphate in groundwater

To characterise the groundwater for the Becher Suite wetlands, concentrations of orthophosphate were determined. Mean groundwater concentrations and standard deviations for five sampling events, spanning 1992-1994, are illustrated for beachridges, wetland margins and wetland centres (Fig. 9-56).

In groundwater under the wetlands, one third of the sites had low mean values of orthophosphate, *i.e.*, <0.1 mg/L. and small standard deviations, suggesting consistent concentrations. However, the remaining wetland sites exhibited means > 0.1 mg/L, and greater variability. In all of these sites, the higher mean was the result of more than one high concentration. Therefore, orthophosphate in groundwater under the wetlands is considered to be characteristically variable from low to high, especially during the first flush of rain and the growing season.

In groundwater under the wetland margins, half the sites had low mean values of orthophosphate, *i.e.*, <0.1 mg/L and small standard deviations, suggesting relatively

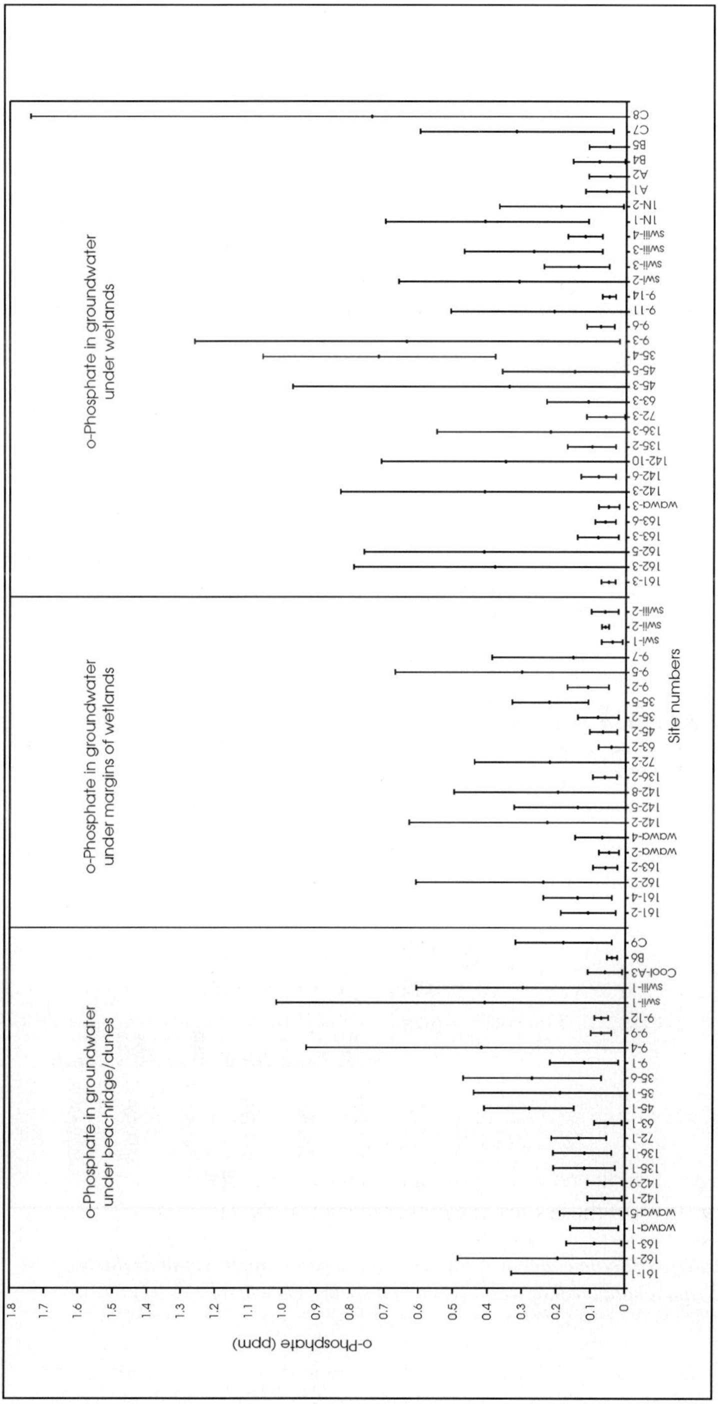

Figure 9-56. *Mean orthophosphate concentrations in groundwater under beachridge/dunes, wetland margins and wetlands. (1992-94).*

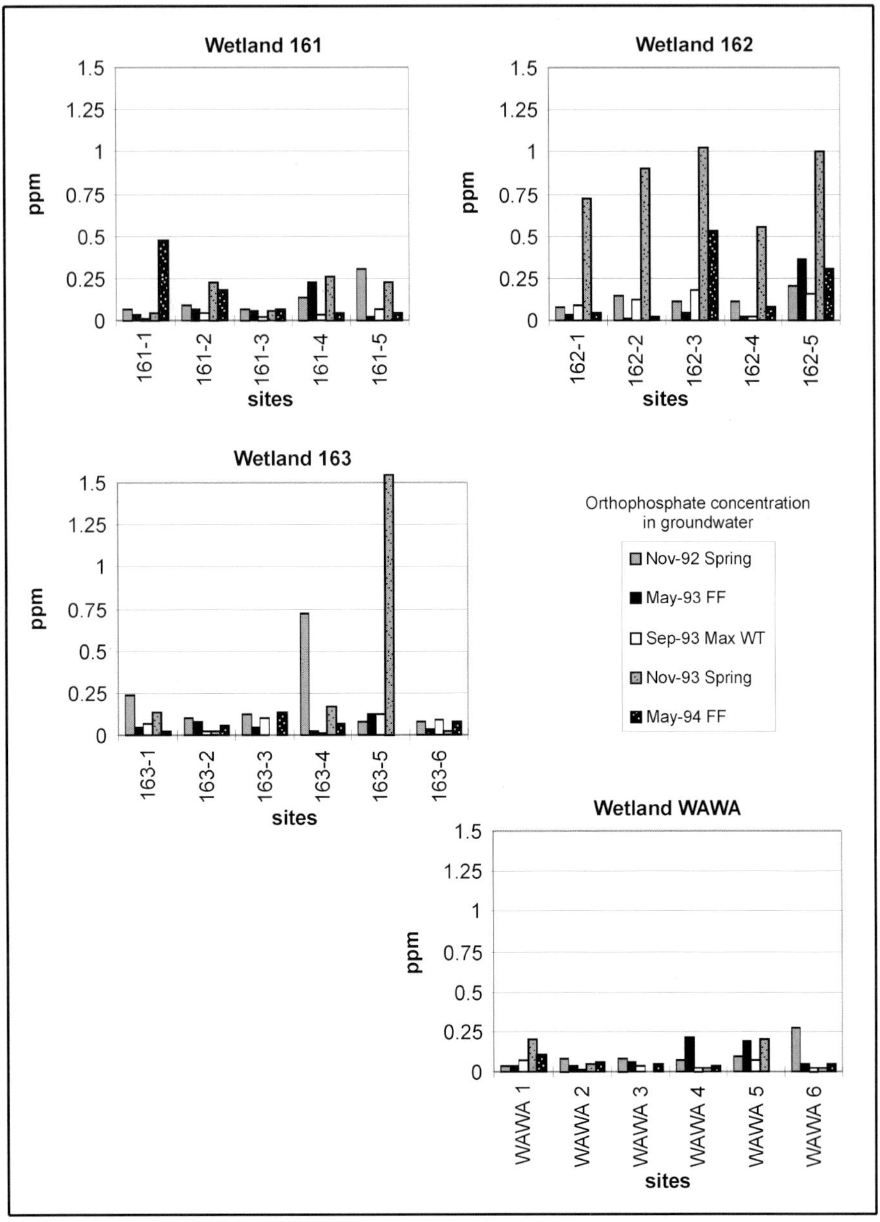

Figure 9-57. *Orthophosphate concentrations in groundwater under wetlands during first flush, maximum level of water table and spring growth. (1992-94).*

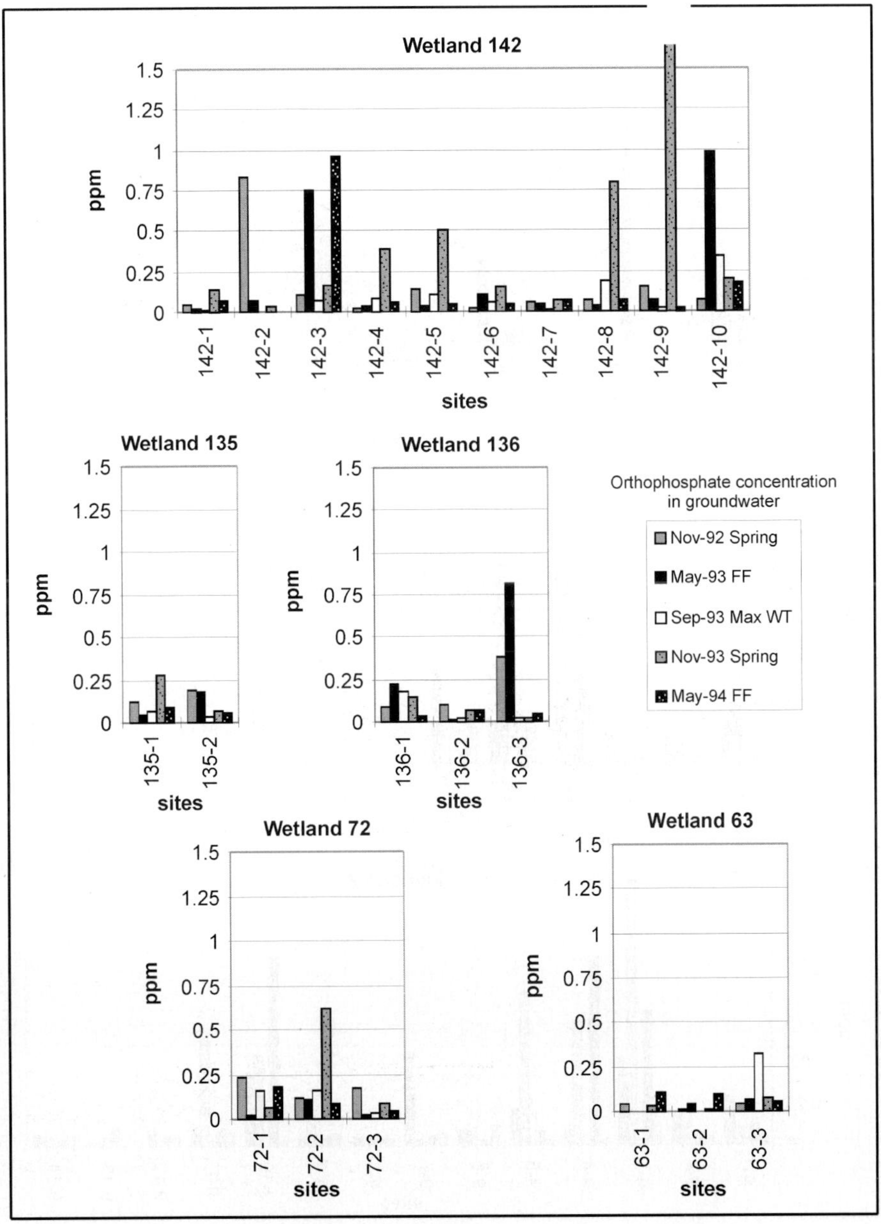

Figure 9-57 (cont.). *Orthophosphate concentrations in groundwater under wetlands during first flush, maximum level of water table and spring growth. (1992-94).*

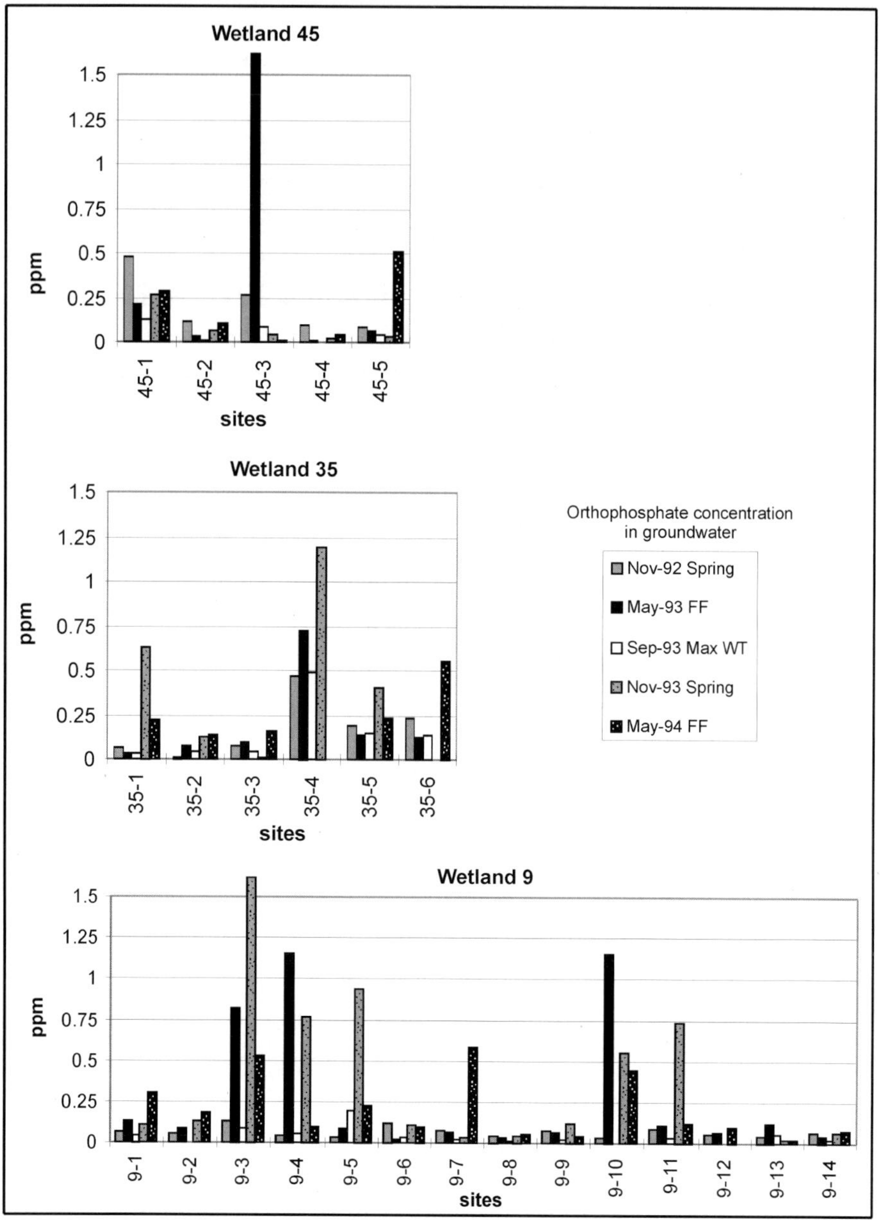

Figure 9-57 (cont.). *Orthophosphate concentrations in groundwater under wetlands during first flush, maximum level of water table and spring growth. (1992-94).*

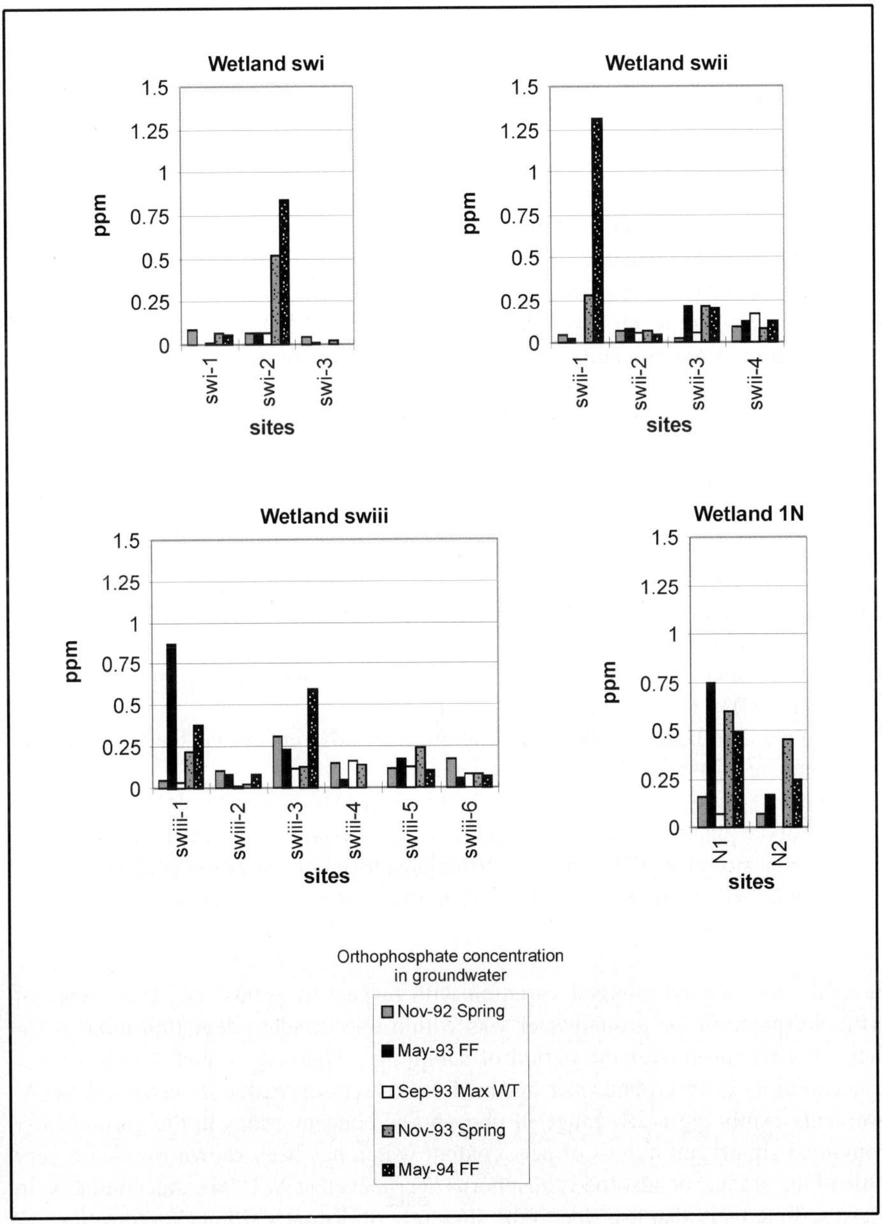

Figure 9-57 (cont.). *Orthophosphate concentrations in groundwater under wetlands during first flush, maximum level of water table and spring growth. (1992-94).*

consistent concentrations. However, other wetland margin sites exhibited means, >0.1 mg/L, and greater variability. In all of these sites, the higher mean was the result of one high concentration, predominantly in November. Therefore, orthophosphate in groundwater under the wetland margins is considered to be characteristically low, except during the growing season.

9.5.5 Patterns in groundwater orthophosphate concentrations relating to specific hydrological and ecological events

Generally, concentrations of orthophosphate <0.01 ppm occurred at both maximum and minimum water levels (September and May) (Fig. 9-57). High concentrations of orthophosphate predominantly occurred in the growing season (GS) (November) with minor occurrences associated with the first winter flush (FF) (April/May/June) (Fig. 9-57). Results for selected wetlands are summarised below:

1. some wetlands exhibited a low range in orthophosphate levels (161, 163, WAWA, 72, 9, swii, Cool-A1, B4, B5);
2. some wetlands exhibited a high range in orthophosphate levels (162, 142, 136, 45, 35, swiii, Cool-C7, C8);
3. some wetlands exhibited a similar range over two years (161, 163, WAWA, 142, 72, 35, 9, swii Cool-A1, B4, B5);
4. some wetlands exhibited a similar pattern over two years (161, 162, 142, 35, 9, Cool-A1, B4, B5);
5. in general, orthophosphate concentrations were different even in the same wetland from site to site and from year to year;
6. some wetland sites (163-6, 142-3, 45, swii, 1N, Cool-A1, B4, B5) had an orthophosphate concentration peak coinciding with the first flush of rain;
7. the majority of wetlands had an orthophosphate concentration peak coinciding with the end of winter rain, falling water levels, and the growing season (162, 136, 72, 45, 35, 9, swiii, Cool-C7, C8).

Several sites showed minimal variation with respect to events, *i.e.*, their range of orthophosphate in the groundwater was within one standard deviation interval for their specific mean over the period of sampling. These sites tended to have low concentrations in the groundwater, and medium to high concentrations in the sediments. Wetlands exhibiting a low range in phosphorus concentrations in the groundwater contained significant humus or peat content which has been shown to release very little of the organic or adsorbed phosphorus (Verhoeven *et al.* 1994). Accumulation in these sediments is also low due to the slow rate of decomposition (Verhoeven *et al.* 1994; Ross 1995).

The high range in groundwater orthophosphate concentration suggests active release and adsorption, and input and export of phosphorus. Four wetlands out of seven with a high range in groundwater orthophosphate concentration, had high levels of phosphorus in the surface sediments due to reasonable levels of litter production and root density, and all had calcilutite fills ≥ 50 cm. Under these same conditions, the potential for loss of soluble phosphorus from the groundwater, either by lateral water flow (*e.g.*, wetlands 162, 142, 35), or by adsorption or cation exchange in the calcilutite fill was also relatively high. Which process was dominant at any particular time was dependent on groundwater rise and fall and concomitant oxygen availability.

Wetlands exhibiting a similar range in groundwater orthophosphate concentration over the two years were commonly those with a small range and/or low content in sediments, indicating inherent low rates and levels of accumulation or tightly bound forms in the organic matter.

Spatial and temporal variability within a single wetland basin was partly related to the interaction between the water table and the sedimentary sequence. In comparisons between the relatively wet and dry years (1992 and 1993 respectively), not only were there quantitative differences in orthophosphate concentrations but even the patterns differed. The lower annual rainfall of 1993, combined with the lower minimum water levels at the beginning of the wet season, resulted in a drop of 30-40 cm in the maximum groundwater levels in all study wetlands (Figs. 9-58 to 9-64). Therefore, at any given sampling time, the orthophosphate concentration in the groundwater could be related to a different sedimentary layer; at the very least, a different depth within the stratigraphic sequence. Lower rainfall also reduced the eluviation effect.

The initial recharge to the groundwater from winter rain was signalled by higher than normal concentrations of orthophosphate, *e.g.*, 45, swiii, 1N (Figs. 9-57, 9-62, 63, 64), although in some wetlands (161, 162, 163), this did not occur until June or July because of below average rainfall.

The major increase of orthophosphate in the groundwater occurred in spring 1993 (Fig. 9-57). This is attributed to a number of conditions: 1) the release of carbonate bound orthophosphate in an environment with decreased redox potential (Holtan *et al.* 1988), 2) widespread plant deaths as a result of dry conditions and low ground water levels during autumn and winter 1993, leading to greater accumulation of leaf litter on surface layers and dead roots in the rhizosphere, and lower net uptake of orthophosphate by plants due to their lower vegetative biomass, and 3) rapid water level fall in the upper sediments resulting in flushing.

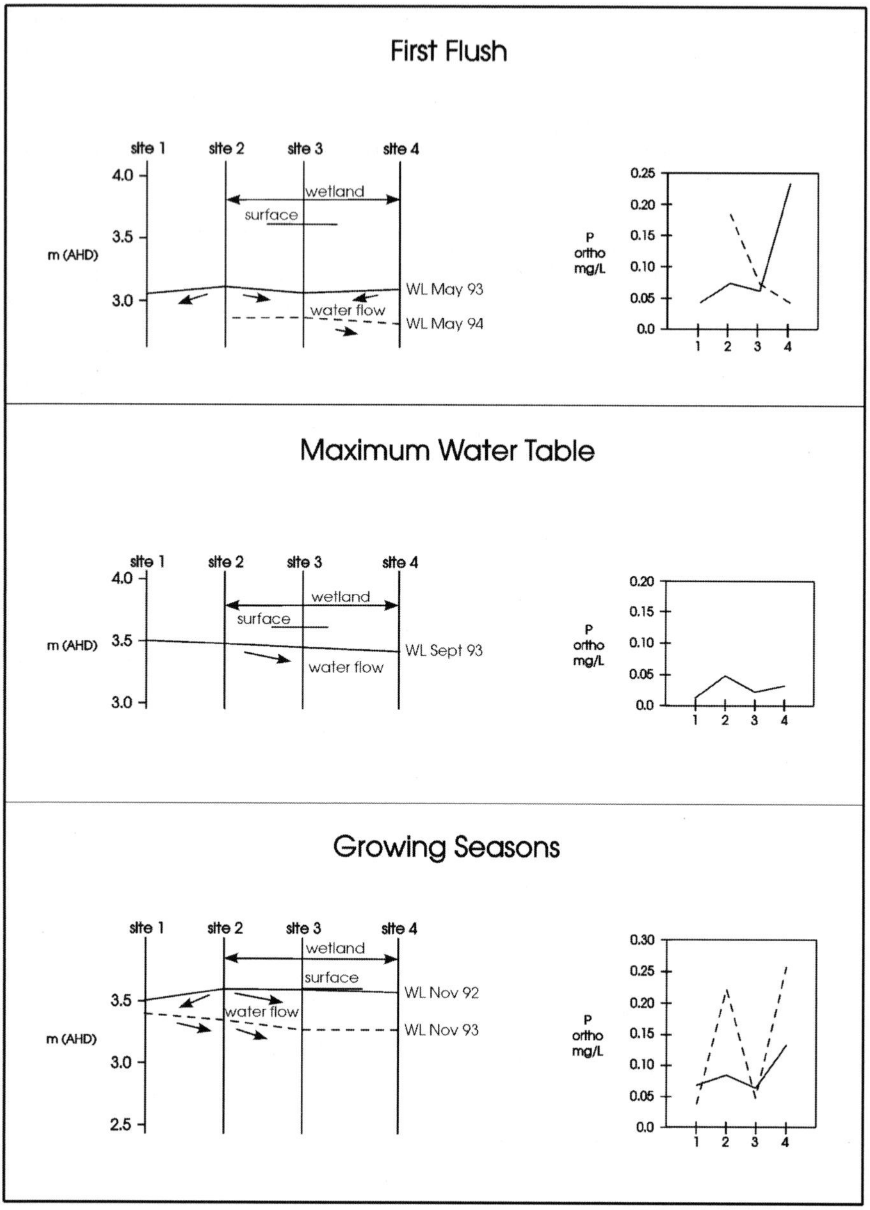

Figure 9-58. *Groundwater paths and orthophosphate concentrations in groundwater in selected wetlands during first flush (May), maximum level of water table (Sept) and spring growing season (Nov). Site 161.*

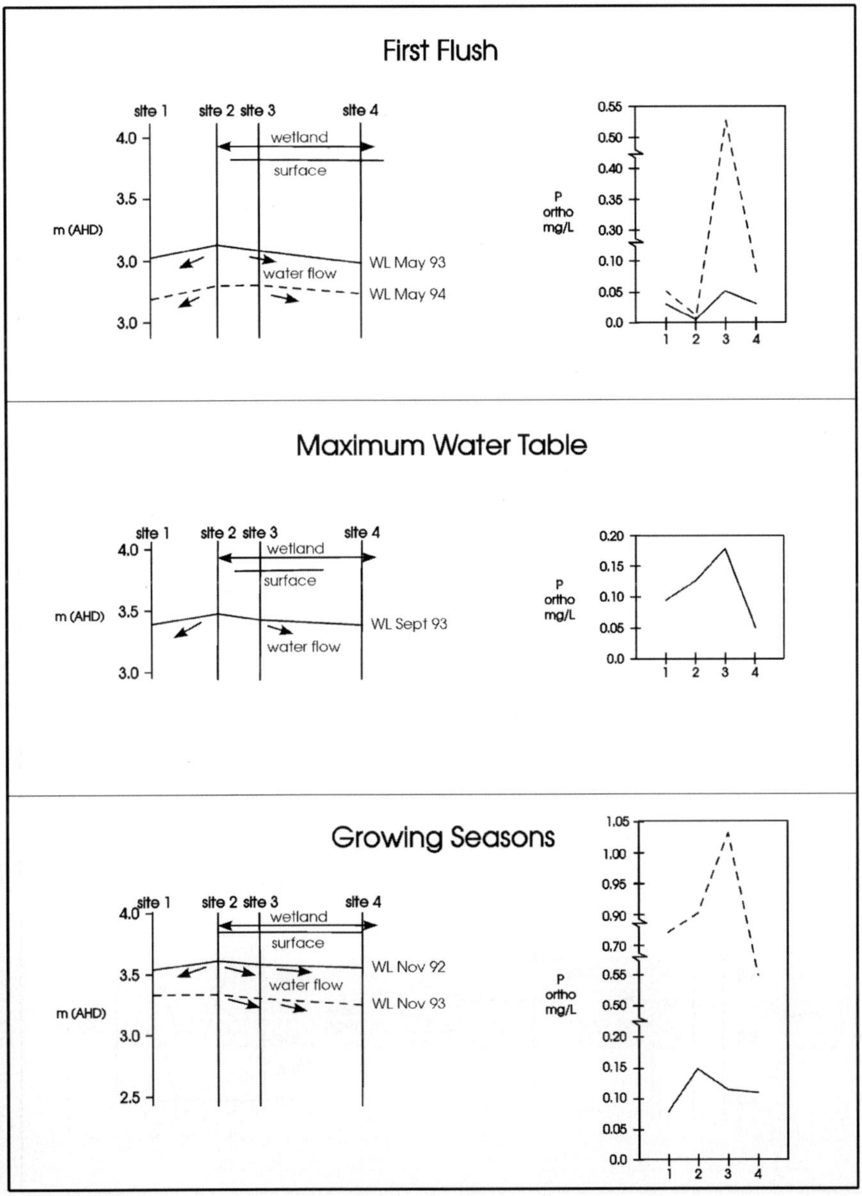

Figure 9-59. Groundwater paths and orthophosphate concentrations in groundwater in selected wetlands during first flush (May), maximum level of water table (Sept) and spring growing season (Nov). Site 162.

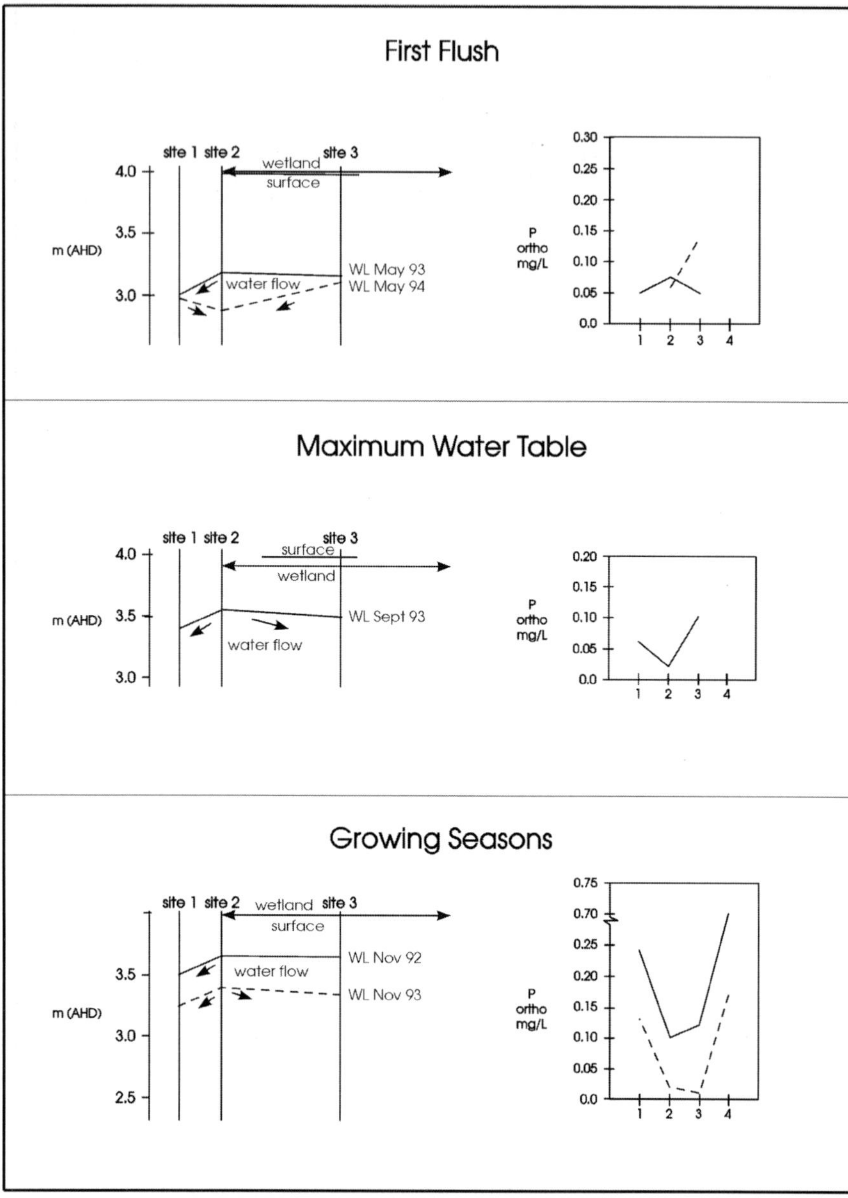

Figure 9-60. *Groundwater paths and orthophosphate concentrations in groundwater in selected wetlands during first flush (May), maximum level of water table (Sept) and spring growing season (Nov). Site 163.*

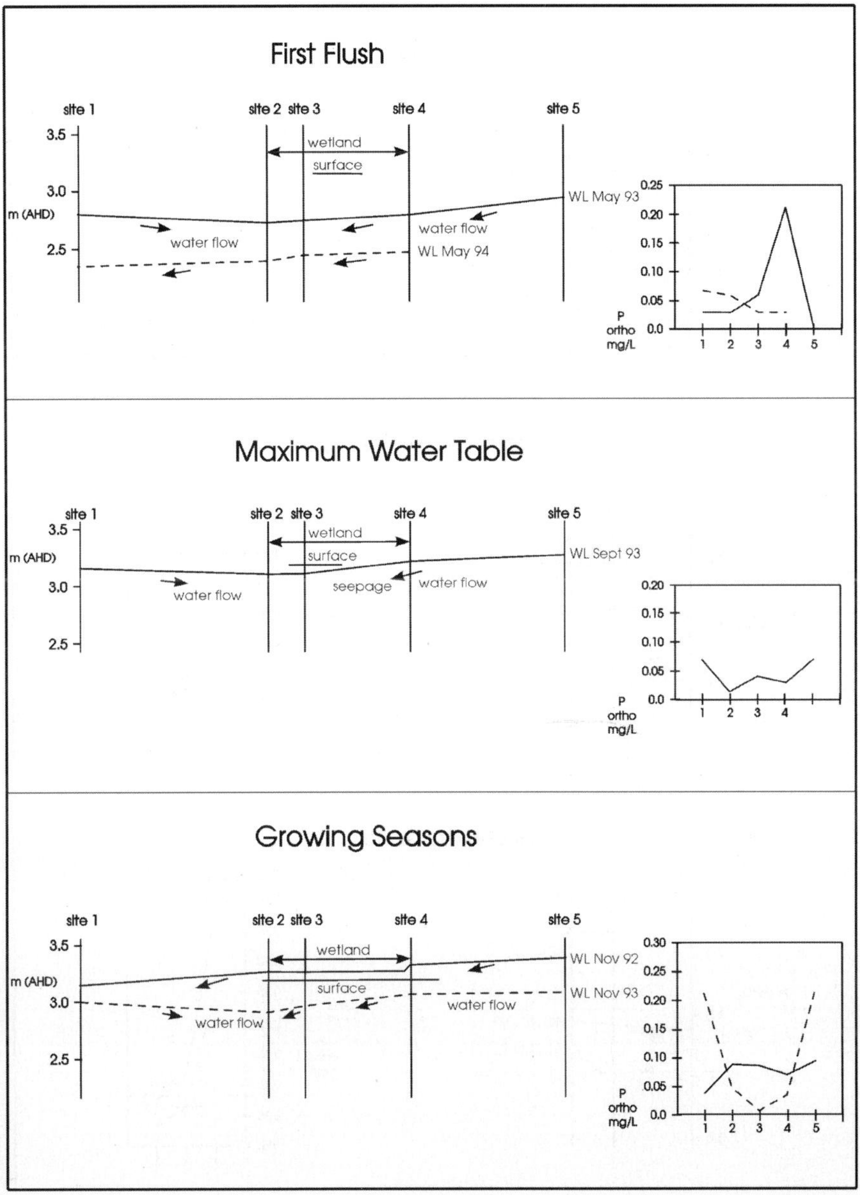

Figure 9-61. *Groundwater paths and orthophosphate concentrations in groundwater in selected wetlands during first flush (May), maximum level of water table (Sept) and spring growing season (Nov). Site WAWA.*

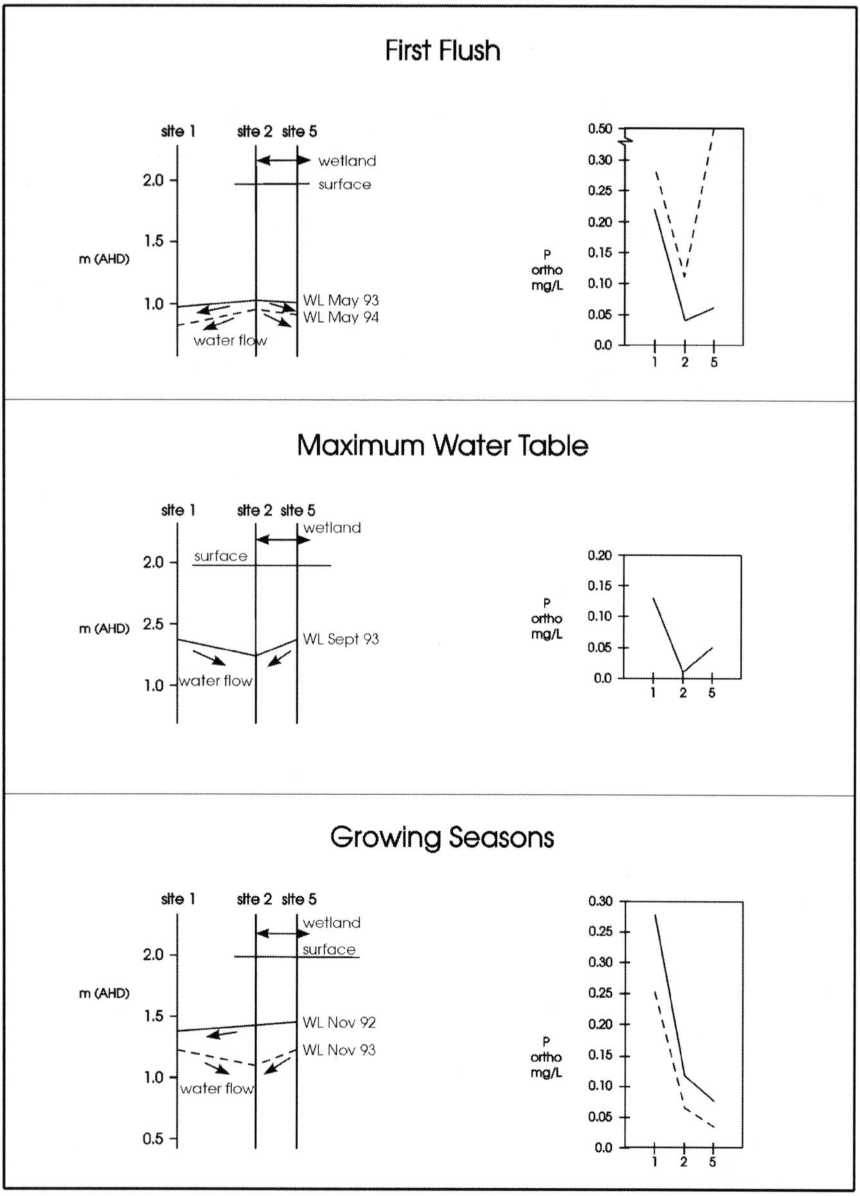

Figure 9-62. *Groundwater paths and orthophosphate concentrations in groundwater in selected wetlands during first flush (May), maximum level of water table (Sept) and spring growing season (Nov). Site 45.*

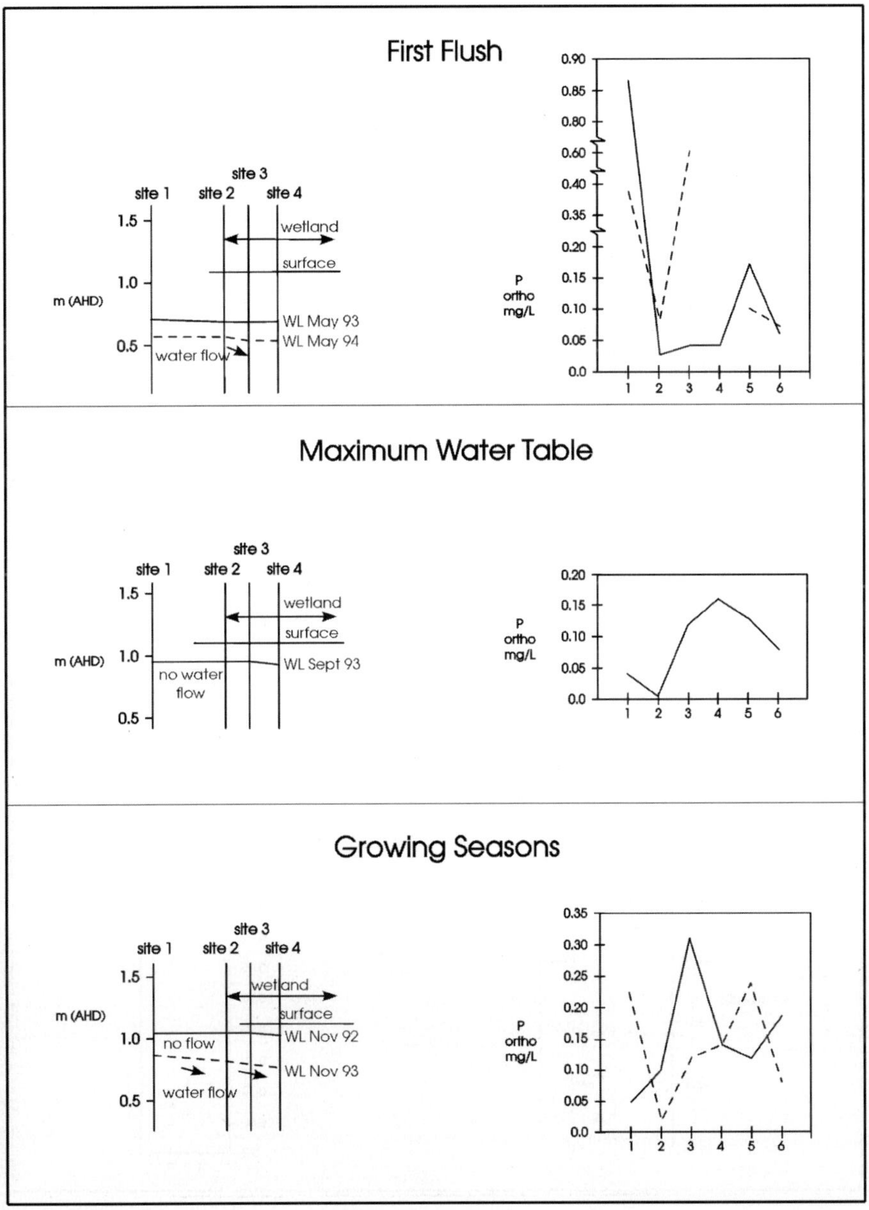

Figure 9-63. *Groundwater paths and orthophosphate concentrations in groundwater in selected wetlands during first flush (May), maximum level of water table (Sept) and spring growing season (Nov). Site swiii.*

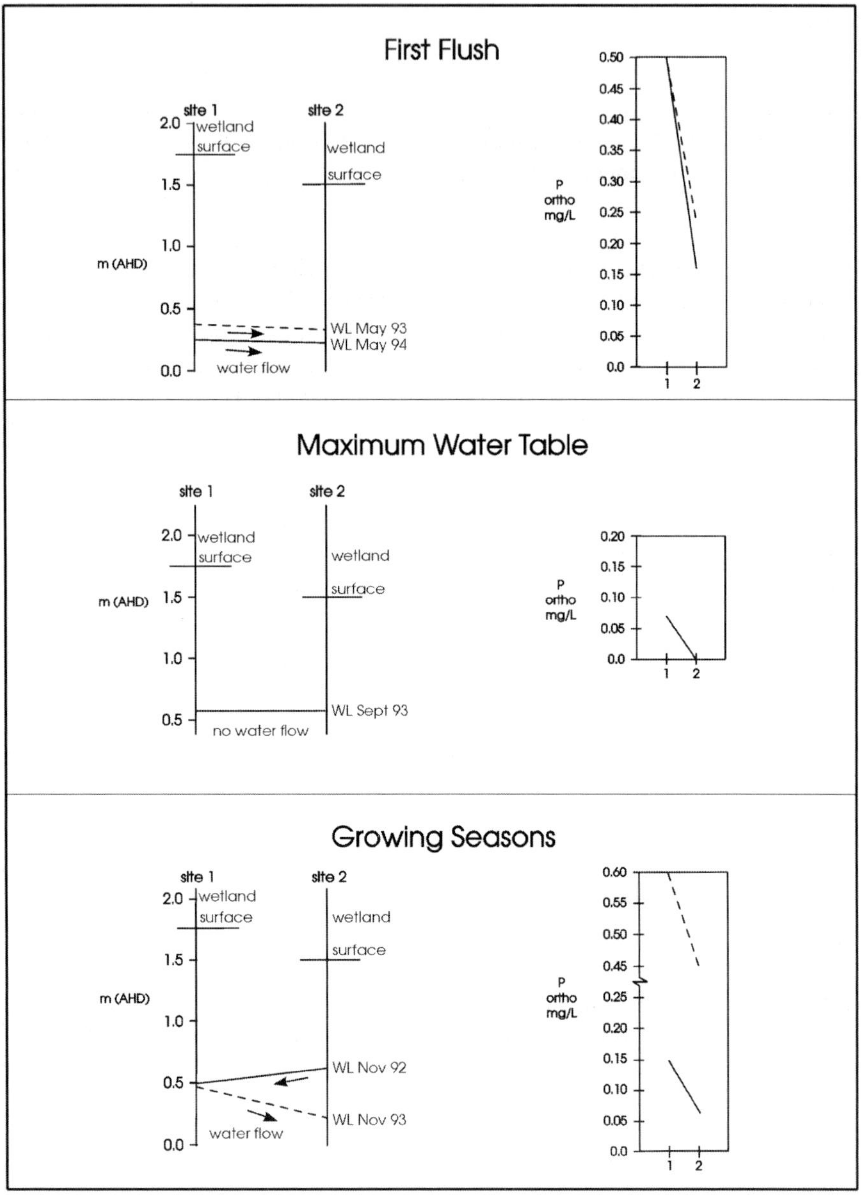

Figure 9-64. *Groundwater paths and orthophosphate concentrations in groundwater in selected wetlands during first flush (May), maximum level of water table (Sept) and spring growing season (Nov). Site 1N.*

9.6 Summary

The investigation of the salinity of wetland waters was undertaken to characterise the hydrochemical environment during the period of monitoring. This was followed by more detailed investigations into the contributions to water salinity of the major cations and one of two important nutrients. The results of these studies pertained to a period of below average rainfall 1992-1994, and characterise the wetland hydrochemistry under these conditions. Results show that groundwater, under both wetlands and beachridge/dunes, was freshwater, dominated by sodium and calcium ions, low in potassium ions and low in orthophosphate. The soil waters in the wetland surface sediments ranged from hyposaline to hypersaline, and interstitial waters down profile generally exhibited the following concentrations (autumn): sodium 50-200 mM/L, potassium 0-2 mM/L, calcium 5-20 mM/L and magnesium 5-40 mM/L. The majority of the wetlands had soil waters dominated by sodium but within this category there was a split based on whether the second most abundant cation was calcium or magnesium. Wetlands characterised by high Mg concentration in soil waters were younger wetlands with abundant leaf litter or a lower ratio of carbonate mud to organic material. Wetland WAWA, with the peat fill, was the only wetland with soil water dominated by calcium. Sodium also was the dominant cation in the interstitial waters, and its concentration was similar for each wetland. Similarly, calcium and potassium showed similar ranges in concentration in all wetland fills, contrasting with magnesium concentrations which varied between wetlands, depending on the proportion of carbonate mud in the fill.

Hydrological processes were identified through shifts in salinity, cation or orthophosphate concentrations. In all cases, at the basin scale, the hydrological mechanisms identified were similar and their effects were observed to occur in the same style and direction. For example, under the wetlands there was a salinity shift to subhaline or hyposaline waters for short periods in response to variable rainfall. This increase in salinity was mirrored by increases in concentration in each cation and in orthophosphate. The occurrences were linked to leaching of litter and upper layers of sediment in early winter with subsequent eluviation by infiltration of meteoric waters during sediment saturation, and to concentration as a result of evapo-transpiration. However, at the bedding scale, the timing and magnitude varied according to the nature of the sedimentary fill, the position of the water table below the surface, the position of the site within the wetland basin, the depth of the rhizome, and the interplay between plants and waters.

In terms of evolution, very few patterns could be recognised. The chemical components selected for study were labile and were observed to move between organic material, sediment, and water. Sodium, magnesium and calcium were transferred from surface layers to the water table and were then transported from the wetland by throughflow. However, there were corresponding trends in nutrient content and age of wetland, suggesting that sediment storage was growing over time. Explanations for these corresponding trends can be found in the works of Smith *et al.* (1985), who concluded that exchangeable potassium released from the parent material was incorporated into the organic cycle, and Ernst *et al.* (1996) in the study of dune slack pedogenesis which demonstrated that the increase in the pool size of phosphorus was related to the increase in the humus layer. In regard to the latter result, the indication in the Becher wetlands is that phosphorus may also be related to the thickness of the wetland fill. Ernst *et al.* (1996) found phosphorus to be present in inorganic compounds as orthophosphate, and as calcium phosphate in shell fragments.

A second pattern was noted in relation to vegetation. The cation and phosphorus contents in *Centella asiatica*, *Typha orientalis*, and the species of *Melaleuca* were higher than in the sedges and rush, suggesting that these species have the propensity to change the wetland hydrochemistry through their uptake and recycling, not only through changes to absolute amounts and balances, but also through their effect on microfauna and cation exchange.

9.7 Discussion

Temporal chemical variability in surface and groundwaters is widely documented in the literature (White 1971; Congdon and McComb 1976; Cole and Fisher 1979; Briggs *et al.* 1985; Arndt and Richardson 1989, 1993; Gehrels and Mulamoottil 1990; Proctor 1995; LaBaugh *et al.* 1996; Garcia *et al.* 1997; Malcolm and Soulsby 2001) and is almost universally related to seasonal hydrological dynamics. In their study of inter-annual variability in temporary endorheic lakes, Garcia *et al.* (1997) found that precipitation (rainfall) events and general hydrological budget were determining factors in salinity and nutrient dynamics. They found that, not only volume, but sequence, frequency, and duration of rainfall events were important. The role of other processes such as dilution by throughflow of snow melt in the spring, groundwater rise, recharge hydrology (downward saturated flow, leaching), ionic mobility, or precipitation, and concentration by evapo-transpiration and exchange relationships, has also been recognised. Ionic chemistry cycles within wetlands, alternately dominated by calcium and bicarbonate, and sulphate respectively, have been linked to wet and dry climatic cycles (Kemmers and Jansen 1988; LaBaugh *et al.* 1996), with an emphasis on the importance of alternating conditions which may be conducive to reduction and oxidation of common compounds.

Temporal chemical variability in surface waters in wetlands, as a result of seasonal vertical water table fluctuation and seasonal lateral water flow, has also been addressed in the context of a water balance or chemical mass budget (Crisp 1966; Hemond 1980; Verry and Timmons 1982; Kenoyer and Anderson 1989; Gehrels and Mulamoottil 1990). However, this study stands alone in its attempt to relate temporal variability in groundwater to the interaction of a dynamic water table with a variable stratigraphy.

With the exception of phosphorus and nitrogen, the source of solutes in the groundwater and surface water has only been partially addressed in the literature (Cowgill 1973).

Frequently, the focus has been on one component of the wetland, either vegetation (White 1971; Boyd 1978; Klopatek 1978; Price and Watters 1989; Warwick and Bailey 1997), or geology (Gasse *et al.* 1987; Malcolm and Soulsby 2001), and in many studies, the starting point has been the chemical composition of the groundwater itself. In this study, as a result of examination of cation concentrations in sediments, interstitial waters, groundwater, and plants, it was demonstrated that cation movement in wetlands is influenced by hydrological, geochemical, and biotic processes.

The results of leaching of Na and K from the sand samples are particularly interesting indicating that felspars contribute small amounts of Na and K to wetland groundwaters and interstitial waters. Weathering and cation mobility from and into felspars have been subjects of study in recent years (White & Brantley 1995), and the results obtained on the Becher cuspate foreland and in the experiments described above, largely conform to the patterns described in the literature in terms of concentrations and inferred dissolution processes. Such ideas support observations in the Becher wetlands, *viz.*, that Na and K are released from felspars, that the first flush can release a relatively large amount of these cations from the exterior of the crystals (Fig. 9-42), and that some K and Na can be resorbed back into the crystal exterior (Figs. 9-42B, D, F). The presence of organic material accelerates felspar dissolution, and in the case of the Becher wetlands, the presence of peats, humic soils, and organic-acid charged waters would contribute to a more ready dissolution of felspars and release of Na, K and Ca.

As Kadlec (1999) found for phosphorus, different mechanisms dominate in a specific situation, but both macro and micro cycles interact with wetland water flows to provide long term availability. In various media, metals exist in dissolved or solid form, some portion of which is freely interchanged between the water, sediment, and biosphere, and some portion of which is unavailable either temporarily or in the longer term.

Similar patterns, to the ones identified in the Becher Suite, were observed in a study by Eghbal *et al.* (1989). In this earlier study, at the margin of the lake basin, concentrations of Na, Ca, Mg increased with depth as a function of more intense leaching, while, in profiles close to the centre of the basin, higher concentrations were found towards the surface due to capillary movement from groundwater. The more soluble Na salts were readily redistributed seasonally. In relation to Ca, Egbahl *et al.* (1989) showed that concentrations in the upper profile decreased after the dry season, reflecting dilution and concentration processes following the seasons. Overall, Ca and Mg patterns were similar.

Although many studies exploring volumes and concentrations of calcium, bicarbonate, and carbonate ions in various water bodies, and their transfer to and from the sedimentary profile have been undertaken (*e.g.*, Kemmers and Jansen 1988; Almendinger and Leete 1998), the exploration of the role of calcareous sand in wetland hydrochemical processes has been less common (Kling 1986; Ernst *et al.* 1996; Grootjans *et al.* 1996; Sival 1996). Calcareous sand has been most important in situations which are groundwater dominated, and in which the lateral groundwater movement is relatively slow. Studies on the effects of dissolution of calcareous grains by ground and surface waters in the wetland environment have shown that in sediments with even minor components of carbonate such as occur in dune slacks, groundwater flow paths can be distinguished by the zone of carbonate dissolution (Grootjans *et al.* 1996; Sival and Grootjans 1996). In this respect, the Becher wetlands provide a further example of this type of interaction between basal sediment, groundwater hydrochemistry, and changes to chemical conditions within the rhizome layer as a result of groundwater rise, even though the mechanics of calcium carbonate dissolution are not presently fully resolved. It has been demonstrated that calcium is removed from the sediment profile by plants but the calcium amount in the biomass is small compared to the store in the calcareous sand. Calcium is also leached from the sediment, leaf litter, and organic material by percolating rainwater, and this cumulative effect is important. Calcium carbonate grains in the sediment exhibit surface morphology (pitting) consistent with acid etching, affirming the conclusion that dissolution is the result of acidification. This process begins with exudation of calcium by roots, which is then translocated from living and decomposing organic material in the surface of the sedimentary profile by percolating meteoric water to the calcareous sand underlying the wetland fill, a view supported by several other authors (Stuyfzand 1993; Ernst *et al.* 1996).

In relation to potassium, Egbahl *et al.* (1989), showed that this element was removed by plants from different depths and concentrated at the surface where plants died and recycled it, and that in lower horizons, the source of potassium was probably the groundwater. The patterns from this study at Becher suggest that potassium is continually being removed from the soil water, and the cation history plots show that it is removed from the groundwater, and that its dynamics affect the normal seasonal fluctuation in cation content. Ernst *et al.* (1996) investigated concentrations of a range of elements in sediments, interstitial waters, and plants in a study of the development of vegetation and soil in dune slacks. They showed that the nutrients per unit area increased as organic matter accumulated, the greatest effects occurring in potassium and magnesium, because, although calcium initially increased with new material, it declined thereafter. As a result, the interstitial water over time exhibited an increase in the concentration of potassium, a constancy in magnesium and sodium, and a decrease in calcium. The potassium content in leaves of sedge and grass species was nearly constant, while that of sodium fluctuated, and that of magnesium increased over time. These responses of vegetation to the accumulation of organic material with its consequent changes in nutrients per unit surface area and in interstitial waters is similar to the type of influence pedogenesis exerts on coastal wetland vegetation in the Becher setting.

The pH of groundwaters in the Becher wetlands lies within the range of neutral to alkaline. The upper limit defined by calcium carbonate in equilibrium with pure water and atmospheric conditions of CO_2 (Shotyk 1988) was measured in a sample of groundwater residing in carbonate mud 40 cm below the surface. The lower limit of pH 6.5, in this region, is defined by rain. In spite of buffering due to calcareous substrates under wetlands and beachridge/dunes, pH variations are the result of the corresponding variations in the chemical composition of the fluctuating groundwaters as they interact with a heterogeneous stratigraphy (Fig. 9-9), are recharged by meteoric infiltration, and move in and out of the rhizosphere. This concurs with findings elsewhere, in wetlands where peat overlies a calcareous substrate (Ingram 1983; Shotyk 1988; Grootjans *et al.* 1991; Proctor 1995; Ross 1995; Sjors and Gunnarsson 2002).

It is well known that nutrient dynamics in wetlands is a function of both water movement and water chemistry (Carter *et al.* 1979; Howard-Williams 1985; Carter 1986; LaBaugh 1986). As the ionic composition of water is essentially different from both possible sources, *i.e.*, rain and groundwater, it is clear that the water regime not only affects nutrient dynamics of wetlands in a direct way, but that it also may have important indirect effects on the soil nutrient availability.

In the first flush of rain, the dominant hydrological process is infiltration of rain water, and in regard to phosphorus, several processes then operate: 1) the leaf litter at the surface is leached of phosphorus by meteoric water (Nichols and Keeney 1973; Howard-Williams *et al.* 1978; Klopatek 1978), 2) calcium phosphate compounds are potentially dissolved by acidic infiltrating waters (Boyer and Wheeler 1989), 3) carbonate bound orthophosphate is potentially released as pore space is filled by infiltrating water, and 4) mineralisation of organic matter occurs while the soil oxygen concentration remains sufficient to sustain aerobic microbial populations (Kieft *et al.* 1987). All of these processes increase the concentration of phosphorus, firstly, in the interstitial waters and, finally, in the groundwater as a result of recharge by infiltrating water with relatively higher concentrations. This effect is dependent on one or more of the following factors: significantly high rainfalls at the beginning of the wet season; relatively shallow depth to groundwater; fresh leaf litter accumulation; and the presence of calcium phosphate compounds.

In the growing season there is less frequent rain, a fall in the water table, and concomitant aeration of previously waterlogged sediments. The major process potentially operating is the flux from decomposing organic matter into interstitial water (Klopatek 1978; Richardson *et al.* 1978).

The two years of study of phosphorus in the Becher wetlands were dominated by variability in frequency and volume of rainfall, which resulted in different rates and fluctuations of groundwater due to variable recharge. Wetlands experienced different periods of inundation and waterlogging, and resultant changes to oxygen availability in any sediment layer. As a result, the timing of the hydrological processes determining phosphate content in the groundwater was out of phase between years. Only where phosphate concentrations were in the low range were these effects ameliorated and seasonal patterns over the two years similar.

From the comparison of groundwater and sediment analyses, it appears that most of the phosphorus is locked into the organic material and the carbonate sediments which accords with previous findings (Richardson 1985; Gehrels and Mulamoottil 1989). Overall variability of phosphorous in the waters is small, the major variation being the result of changes in standing biomass. For example, plant demise reduces uptake and increases fresh leaf litter accumulation which is subsequently leached. Since metabolic activity is highest at the sediment surface, due to high degradability and favourable oxygen conditions, dissolved phosphate has its peak at or near the sediment surface.

Wetland sites with mean orthophosphate concentrations <0.1 mg/l and low standard deviations were colonised by similar species, sedges or rushes dominated by the genera *Baumea* and *Juncus* (Table 9.3). In contrast, sites invaded by *Typha orientalis* had elevated levels of orthophosphate in the groundwater. These results reflect the store of phosphorus found in the different species of plants, as well as the relative contribution to cover abundance of each species. As native sedge species do not store significant phosphorus in their structural components, their rate of decomposition is slow and phosphorus is not released into the groundwater in large quantity.

Table 9.3 *Vegetation cover at sites with low mean values of orthophosphate in the groundwater*

Site	*Mean concentration of orthophosphate in groundwater mg/L*	*Vegetation cover*
9-14	0.05 ± 0.02	rush - *Juncus kraussii*
163-3	0.08 ± 0.06	rush - *Juncus kraussii*
163-6	0.06 ± 0.03	rush - *Juncus kraussii*
142-6	0.06 ± 0.05	sedge - *Baumea juncea*
72-3	0.07 ± 0.06	sedge - *Baumea juncea*, and herb, *Centella asiatica*
9-6	0.08 ± 0.04	sedge - *Baumea juncea*, and herb, *Centella asiatica*
WAWA	0.05 ± 0.03	sedge - *Baumea articulata, Typha orientalis, Schoenoplectus validus*
161-3	0.13 ± 0.2	sedge - *Baumea articulata, Typha orientalis*
swiii-4	0.12 ± 0.05	sedge - *Schoenoplectus validus, Typha orientalis*

10. VEGETATION

10.1 Introduction

As the variability in spatial and temporal distribution of wetland vegetation was one of the major themes of this study, there were a number of objectives inherent to this arm of the research programme. They were:

- to define the extant wetland vegetation associations, and describe their pattern within the wetland basins (Semeniuk *et al.* 1990);
- to identify and order in a sequence the determining factors (environmental attributes) underlying the species distribution;
- to identify the determining factors (environmental attributes) underlying the spatial and temporal variability in species distribution;
- to identify structural and physiognomic plant adaptations to the seasonal wetland conditions in the Becher Suite; and
- to examine the inter-relationships between vegetation and stratigraphy, and vegetation and hydrology.

Methods used for selection of wetlands, collection of data on species abundance, and determination of vegetation associations are described in detail in Chapter 2 and briefly summarised here.

As there were over 200 wetland basins in the Becher Suite, a sub-sample of vegetated wetlands was selected for detailed study. The vegetation of these wetlands was classified in terms of both floristics (abundance, specifically plant frequency and constancy) and structure (Specht 1981). Generally, quadrats of size 1 m^2 were used for quantifying the vegetation, but because of tree structures in some wetlands, nested quadrats of 1 m^2 and 5 m^2 were used in wetlands 135, 136, 142.

10.1.1 Scale of vegetation study

It is generally recognised that vegetation, as part of the biosphere, forms a continuum with no permanent or easily recognised boundary, therefore divisions into communities and assemblages are related to scale, which, in respect to vegetation, is usually defined within the framework of given objectives. In this study, an objective similar to that underlying the global divisions has been applied, *i.e.*, habitat differentiation. Wetland vegetation stands in the Becher area are isolated from their surroundings in mosaic patterns in basins, isolated by hummocks, hollows and dune ridges. Although the distances between patches of the same community type may be as little as a few metres, clear discontinuities, as defined by van der Maarel (1988), occur in the distribution of wetland plant species. This demarcation between wetland and upland vegetation is underscored by habitat differentiation such as three-dimensional landform geometry (geomorphology), hydrological conditions (prevailing waterlogging and inundation), and the location and extent of hydric soils (carbonate muds, muddy sands, peat). Some wetlands are discrete *e.g.*, wetland 63, and some are a series of coalescing, undulating, interdune depressions whose boundaries are determined by criteria pertaining to height of intervening ridge (Semeniuk 1987), *e.g.*, the linear series of wetlands swi, swii, swiii. Within the study area, wetland vegetation is leptoscale to microscale, *i.e.*, contained within a 20 x 20 m frame ranging up to a 40 x 200 m frame (Semeniuk 1987).

10.1.2 Hierarchical classification

Wetland vegetation in the Becher area displays considerable diversity of physiognomy and plant species, some wetlands exhibiting sharp margins and some gradational margins. To highlight this diversity, the wetland vegetation has been classified hierarchically. The first level of classification denotes the variability of vegetation cover pattern and form using categories which incorporate percent of wetland vegetation cover, arrangement of vegetation within the wetland, and degree of structural and compositional homogeneity (Semeniuk *et al.* 1990). The second level of classification categorises the types of vegetation structures which are present, based on height of plant, density and life-form attributes (Specht 1981). The third level of classification categorises the dominant species which characterise any structural group. Many vegetation associations typically comprise a single structure which corresponds with the occurrence of a single species. Wetland plant species, together with beachridge/dune species, which commonly occur around the wetland margins and invade during drier periods, are in Tables 10.1, 10.2.

Wetland plant species are essentially obligate in function, but from time to time wetland basins are invaded by facultative species such as *Acacia cyclops, A. pulchella, Isolepis nodosa,* and *Pelargonium capitatum*, particularly during prolonged periods of below average rainfall (Table 10.2). These species do not truly reflect wetland vegetation as they are not truly adapted to wetland conditions, and their adventive nature means they occur spasmodically. As their inclusion in the quantitative analyses of wetland vegetation in this study would disrupt the results, these species have been excluded. However, for completeness in the general qualitative description of wetland vegetation associations that follows, they have been included.

10.2 Classifying wetland vegetation associations

Classification of vegetation was undertaken using multivariate analysis to differentiate vegetation associations on the dual attributes of species composition and species cover abundance. The data for analyses were collected in 1 m^2 quadrats along east west transects across wetlands in 1991. Cluster analysis was performed using the computer programme PATN (Belbin 1995) to identify similarity between quadrats (see Methods chapter). This analysis resulted in the recognition of 23 plant communities (Fig. 10-1). Species more closely associated with upland vegetation together with

Table 10.1 *Wetland plant species*

Family	*Species*
Apiaceae	*Centella asiatica*
Brassicaceae	**Brassica tournefortii*
Chenopodiaceae	*Halosarcia halocnemoides.*
Compositae	**Sonchus asper*
Crassulaceae	**Crassula* spp.
Cyperaceae	*Baumea articulata*
	B. juncea
	Isolepis nodosa
	Lepidosperma gladiatum
	Schoenoplectus validus
	Isolepis cernua
	Typha domingensis
	**Typha orientalis*
Geraniaceae	**Pelargonium capitatum*
Iridaceae	**Romulea* spp.
Juncaceae	*Juncus kraussii*
Juncaginaceae	*Triglochin* spp. *striata or mucronata*
Lobeliaceae	*Lobelia alata*
Myrtaceae	*Melaleuca cuticularis*
	M. viminea
	M. rhaphiophylla
	M. teretifolia
Papilionaceae	**Trifolium* spp *(clover)*
Poaceae	*Sporobolus virginicus*
Primulaceae	*Samolus repens*
Xanthorrhoeaceae	*Xanthorrhoea preissii*

* invasive and alien species

Table 10.2 *Species which invade wetlands during drier periods*

Family	*Perennial species*
Euphorbiaceae	*Adriana quadripartita*
Geraniaceae	**Pelargonium capitatum*
Mimosaceae	*Acacia cyclops*
	A. lasiocarpa
	A. pulchella
	A. rostellifera
	A. saligna

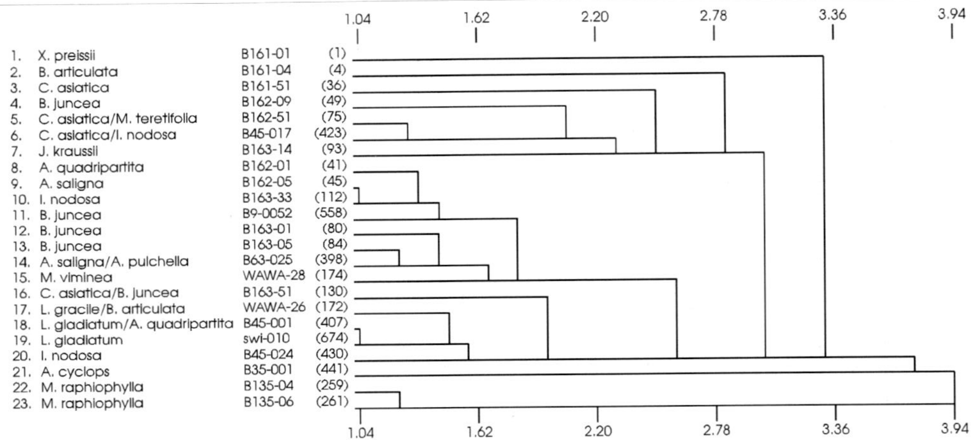

	Assemblage name.	Dominant species.	Other elements.
1.	X. preissii	X. preissii	B. juncea/S. virginicus/S. flavescens
2.	B. articulata	B. articulata	A. quadripartita/S. virginicus
3.	C. asiatica	C. asiatica	I. nodosa/L. gracile/B. articulata
4.	B. juncea	B. juncea	M. teretifolia/C. asiatica/S. virginicus
5.	C. asiatica/M. teretifolia	C. asiatica/M. teretifolia	
6.	C. asiatica/I. nodosa	C. asiatica/I. nodosa	
7.	J. kraussii	J. kraussii	C. asiatica
8.	A. quadripartita	A. quadripartita	A. saligna/I.nodosa
9.	A. saligna	A. saligna	X. preissii/B. juncea/I. nodosa
10.	I. nodosa	I. nodosa	C. asiatica/B. juncea/J. kraussii
11.	B. juncea	B. juncea	P. capitatum
12.	B. juncea	B. juncea	S. flavescens/A. quadripartita
13.	B. juncea	B. juncea	C. asiatica/A. quadripartita

	Assemblage name.	Dominant species.	Other elements.
14.	A. saligna/A. pulchella	A. saligna/A. pulchella	
15.	M. viminea	M. viminea	C. asiatica/L. gracile
16.	C. asiatica/B. juncea	C. asiatica/B. juncea	J. kraussii
17.	L. gracile/B. articulata	L. gracile/B. articulata	
18.	L. gladiatum/A. quadripartita	L. gladiatum/A. quadripartita	
19.	L. gladiatum	L. gladiatum	
20.	I. nodosa	I. nodosa	P. capitatum/A. cyclops
21.	A. cyclops	A. cyclops	
22.	M. raphiophylla	M. raphiophylla	I. nodosa
23.	M. raphiophylla	M. raphiophylla	

Fig 10-1. *Dendrogram of Becher wetland assemblages based on percentage cover (omitting total cover).*

transitional assemblages of upland and wetland species, *i.e.*, occurring at the margins of wetlands, were identified and separated from those which were truly representative of wetland habitats. Where two associations were composed of a single dominant species with variable understorey, they were amalgamated.

The resulting 11 wetland plant associations identified were:

1. Low forest - *Melaleuca rhaphiophylla,* (understorey - *I. nodosa)*
   Low forest - *Melaleuca rhaphiophylla,* (no understorey)
2. Sedgeland - *Baumea juncea,* overstorey *M. teretifolia,* understorey - *C. asiatica*
   Heath - *Melaleuca teretifolia,* understorey - *C. asiatica*
3. Heath - *Melaleuca viminea,* understorey - *C. asiatica, L. gracile*
4. Heath - *Xanthorrhoea preissii,* understorey - *B. juncea, S. virginicus*
5. Sedgeland - *Baumea articulata,* understorey - *C. asiatica, L. gracile*
6. Sedgeland - *Baumea juncea,* understorey - *C. asiatica, S. virginicus*
7. Sedgeland - *Schoenoplectus validus*
8. Sedgeland - *Lepidosperma gladiatum*
9. Rushland - *Juncus kraussii,* understorey - *C. asiatica*
10. Herbland - *Centella asiatica*
    Herbland - *Centella asiatica,* overstorey - *B. juncea, J. kraussii*
11. Sedgeland/Herbland - *B. juncea/C. asiatica*

The wetland vegetation in 20 wetland basins exhibited variability in terms of mosaics, zonation, extent and composition, even though all wetlands had 100% cover. The main pattern is maculiform but gradiform is also common. Dominant structure varies from sedgeland to herbland to low forest with minor occurrences of heath. The distribution of these vegetation associations, as well as gradational zones, where present, were mapped in each wetland basin (Figs. 10-2 to 10-22) and described in Chapter 4, Wetland Descriptions, using terminology from Semeniuk *et al.* (1990).

From east to west, *i.e.*, generally from oldest wetland to youngest, the pattern of vegetation changes from low forest, scrub, heath, sedge, rush, and herbs in eastern areas, to low forest, single trees, fewer areas of scrub, heath, sedge, rush, and herbs in central areas, to sedge, rush and herbs in the western area. The youngest wetlands support sedge of *B. juncea*, *S. validus*, and *L. gladiatum*, the rush, *J. kraussii,* and the herb, *C. asiatica*. Definitely not present are assemblages with *M. rhaphiophylla, M. teretifolia*, *M. viminea*, and the sedge, *B. articulata*. The oldest wetlands have all the species, *i.e.*, there are no species that occur exclusively in the younger wetlands. Table 10.3 lists all wetland species highlighting the spatial distribution of wetland associations and species across the Becher cuspate foreland.

Table 10.3 *Distribution of associations according to location within eastern, central, and western parts of the Becher Cuspate foreland (using all species)*

Eastern areas: oldest wetlands	*Central areas: middle-aged wetlands*	*Western areas: youngest wetlands*
Low forest - *Melaleuca rhaphiophylla,* understorey *Isolepis nodosa* Low forest - *M. rhaphiophylla,* no understorey	Low forest - *M. rhaphiophylla,* understorey *I. nodosa* Low forest - *M. rhaphiophylla,* no understorey	
Sedgeland - *Baumea juncea* overstorey *Melaleuca teretifolia,* understorey *Centella asiatica, Sporobolis virginicus* Heath - *M. teretifolia* understorey *C. asiatica*	 Heath - *M. teretifolia* understorey *C. asiatica*	
Heath - *Melaleuca viminea,* understorey *C. asiatica, L. gracile*	Heath - *M. viminea,* understorey *C. asiatica, L. gracile*	
Heath - *Xanthorrhoea preissii,* understorey *B. juncea, S. virginicus*	Heath - *X. preissii,* understorey *B. juncea, S. virginicus*	
Sedgeland - *Baumea articulata,* understorey *C. asiatica, Lepidosperma gracile*		
Sedgeland - *B. juncea,* understorey *C. asiatica, S. virginicus*	Sedgeland - *B. juncea,* understorey *C. asiatica, S. virginicus*	Sedgeland - *B. juncea,* understorey *C. asiatica*
Sedgeland - *Schoenoplectus validus*		Sedgeland - *S. validus*
Sedgeland - *Lepidosperma gladiatum*	Sedgeland - *L. gladiatum*	Sedgeland - *L. gladiatum*
Rushland - *Juncus kraussii,* understorey *C. asiatica*	Rushland - *J. kraussii,* understorey *C. asiatica*	Rushland - *J. kraussii,* understorey *C. asiatica*
Herbland - *C. asiatica* Herbland - *C. asiatica,* overstorey *B. juncea, J. kraussii*	Herbland - *C. asiatica* Herbland - *C. asiatica,* overstorey *B. juncea, J. kraussii*	Herbland - *C. asiatica* Herbland - *C. asiatica* overstorey *B. juncea, J. kraussii*
Sedgeland/Herbland - *B. juncea/C. asiatica*	Sedgeland/Herbland - *B. juncea/C. asiatica*	

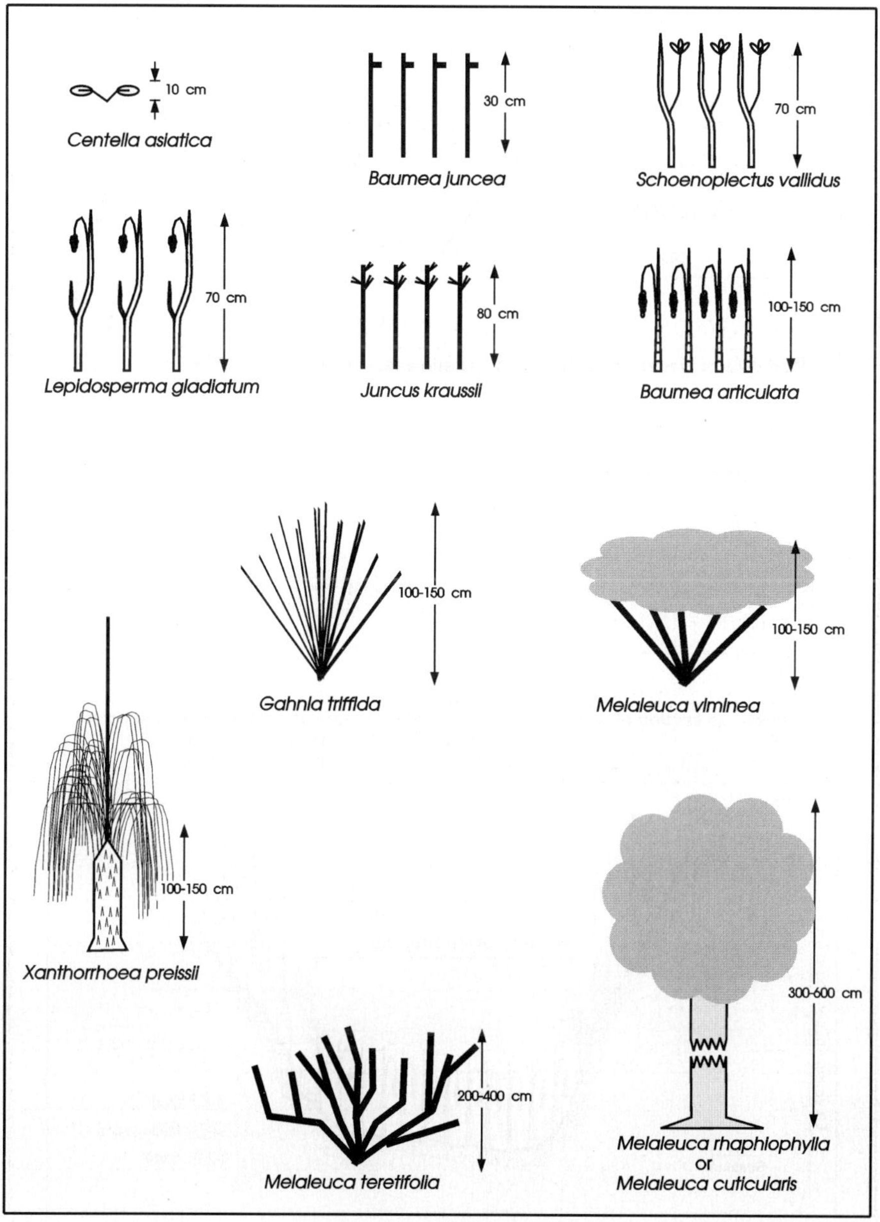

Figure 10-2. *Legend to the key species of wetland plants within and marginal to the wetlands.*

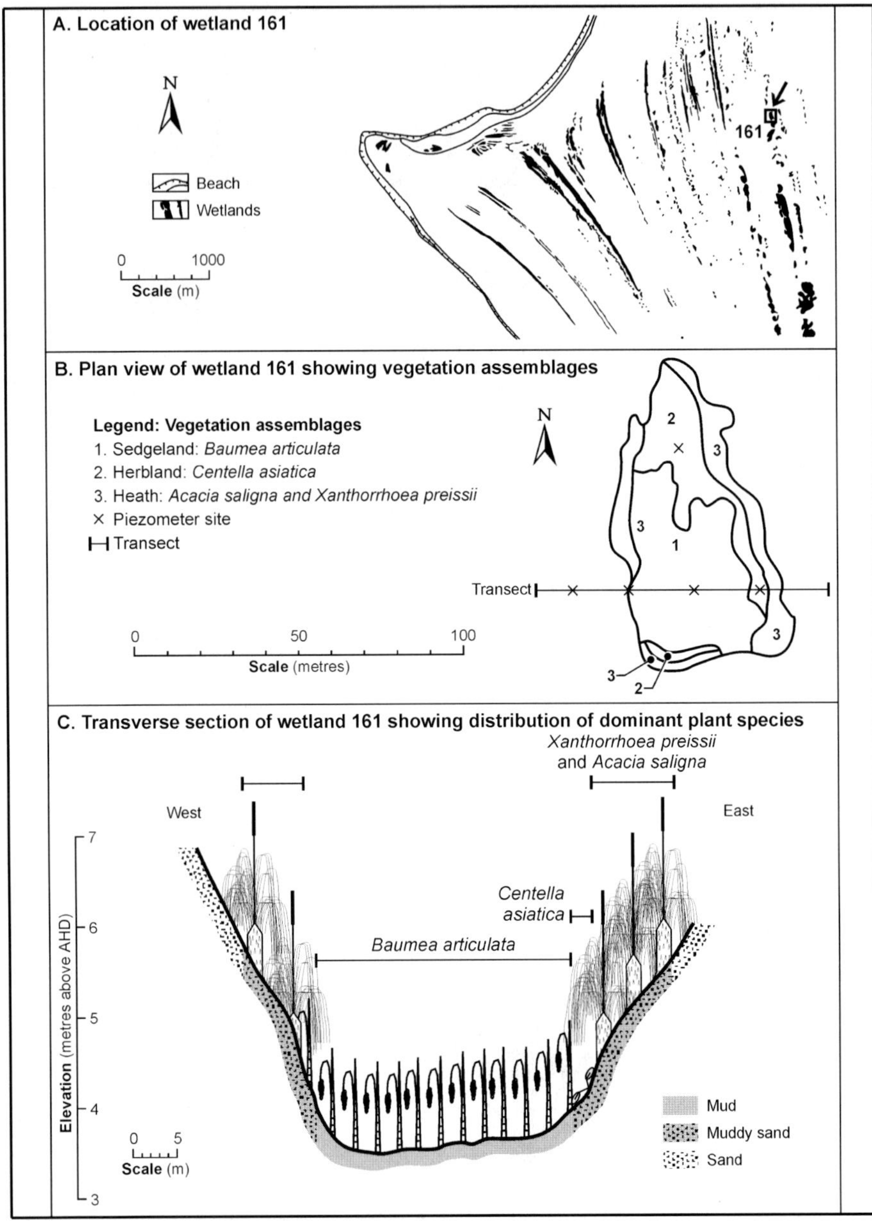

Figure 10-3. *Plan view and cross section showing distribution of wetland vegetation in wetland 161.*

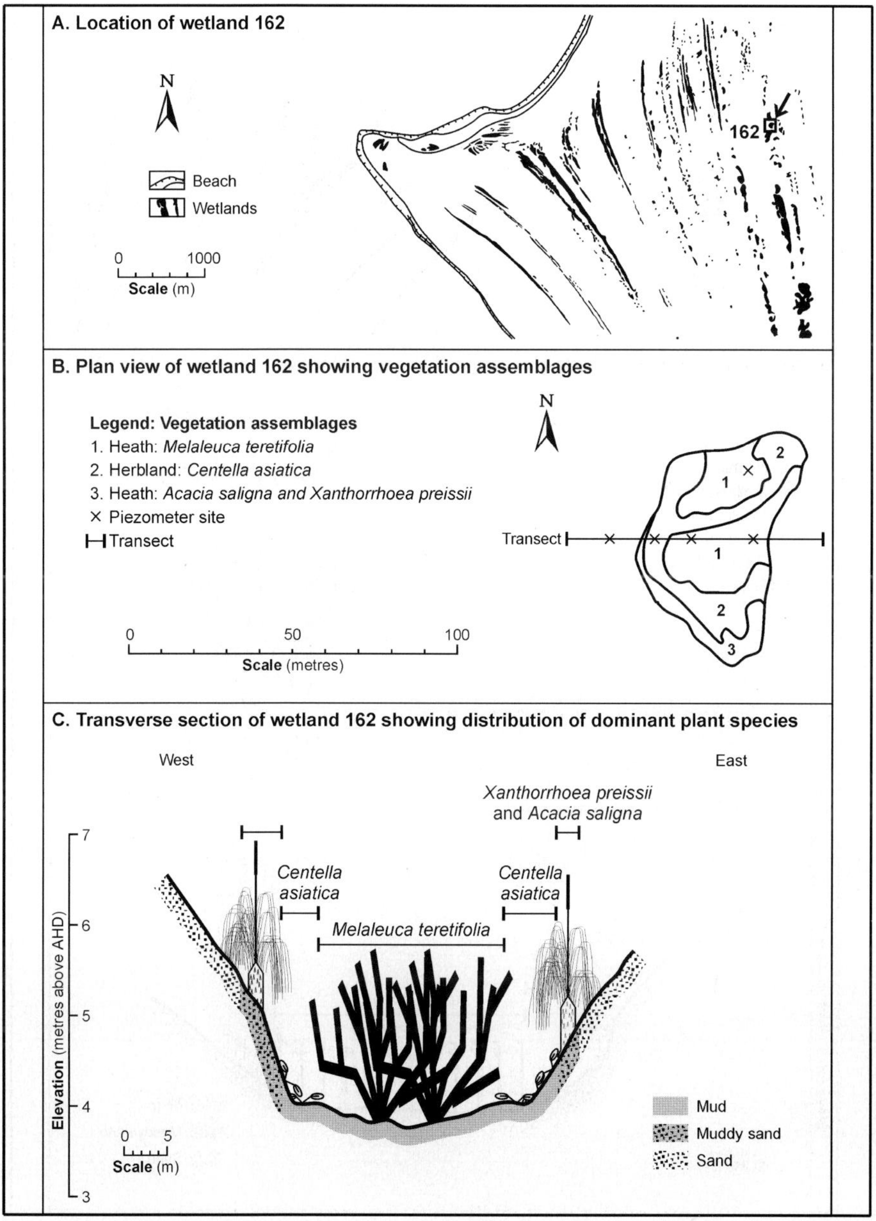

Figure 10-4. *Plan view and cross section showing distribution of wetland vegetation in wetland 162.*

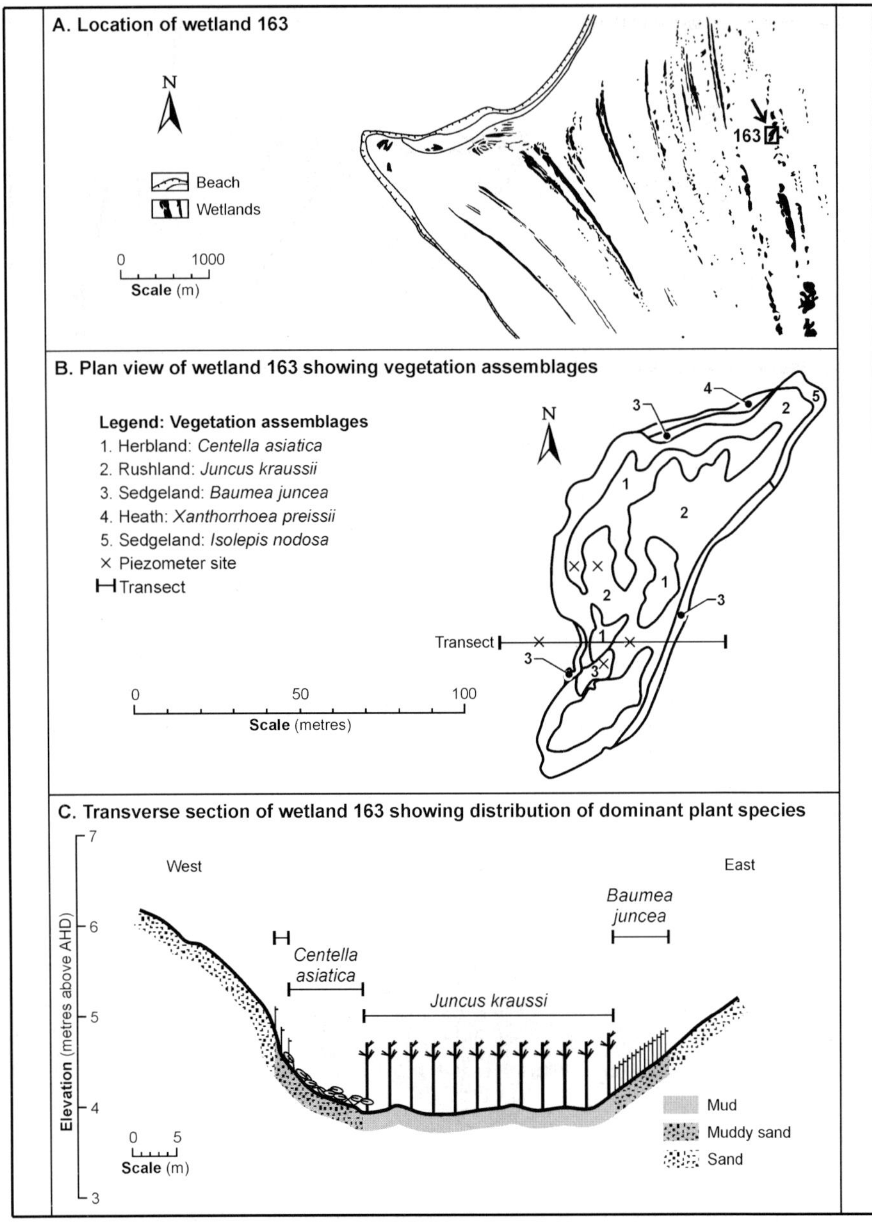

Figure 10-5. *Plan view and cross section showing distribution of wetland vegetation in wetland 163.*

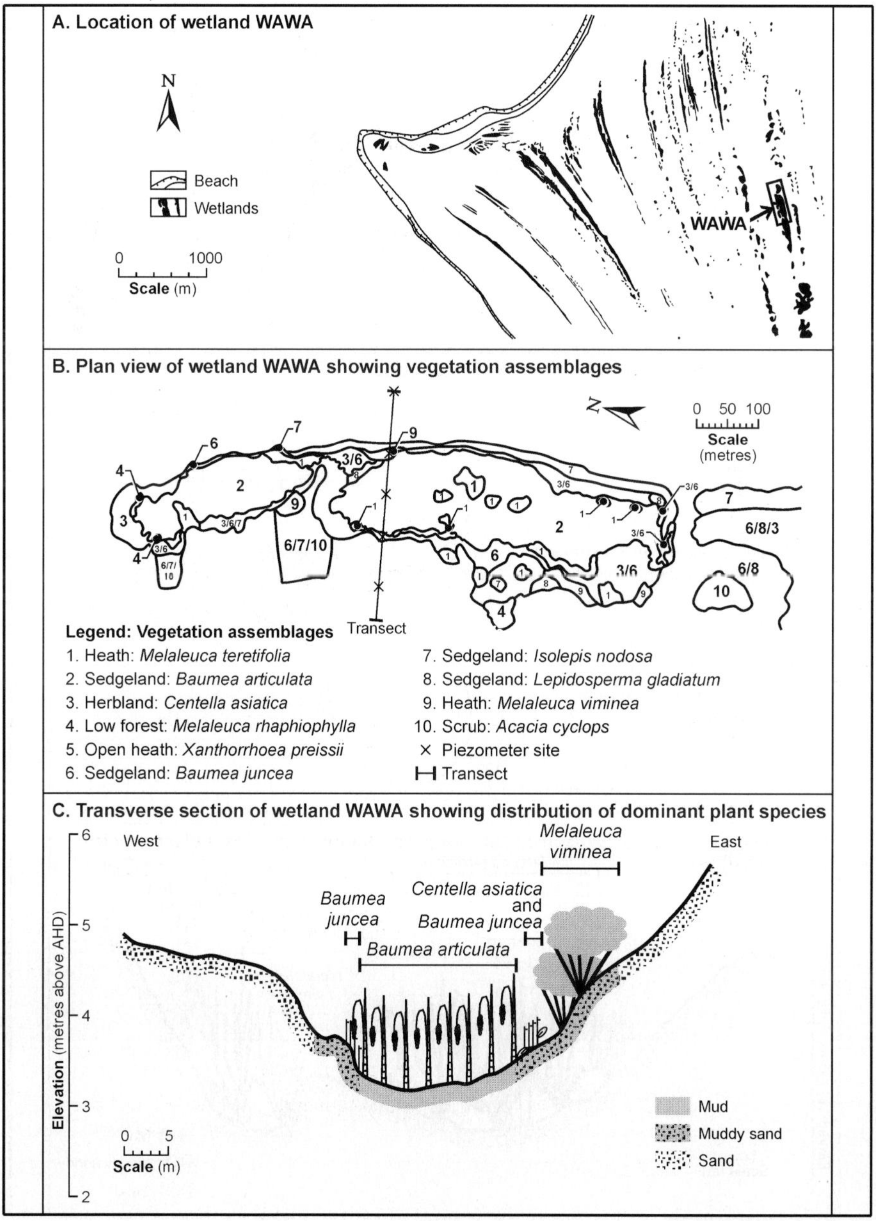

Figure 10-6. *Plan view and cross section showing distribution of wetland vegetation in wetland WAWA.*

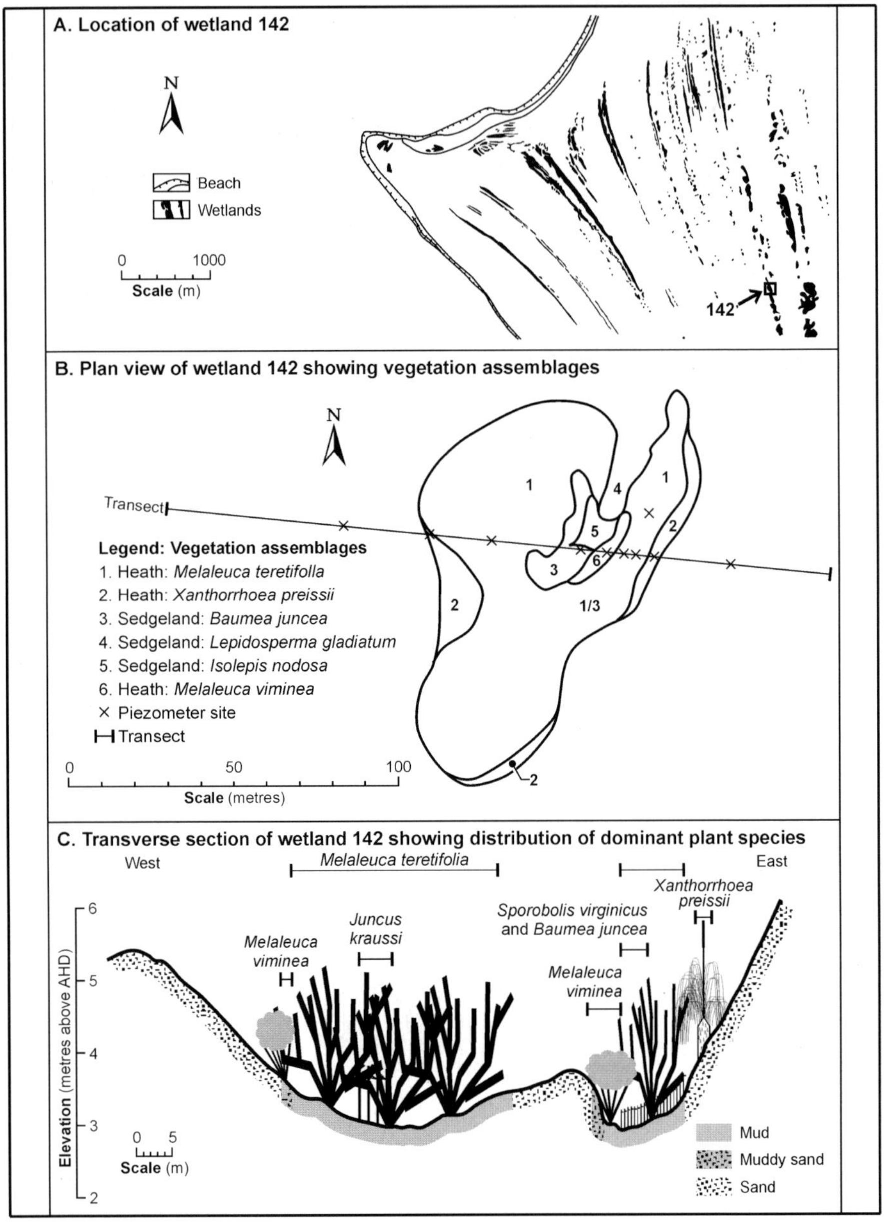

Figure 10-7. Plan view and cross section showing distribution of wetland vegetation in wetland 142.

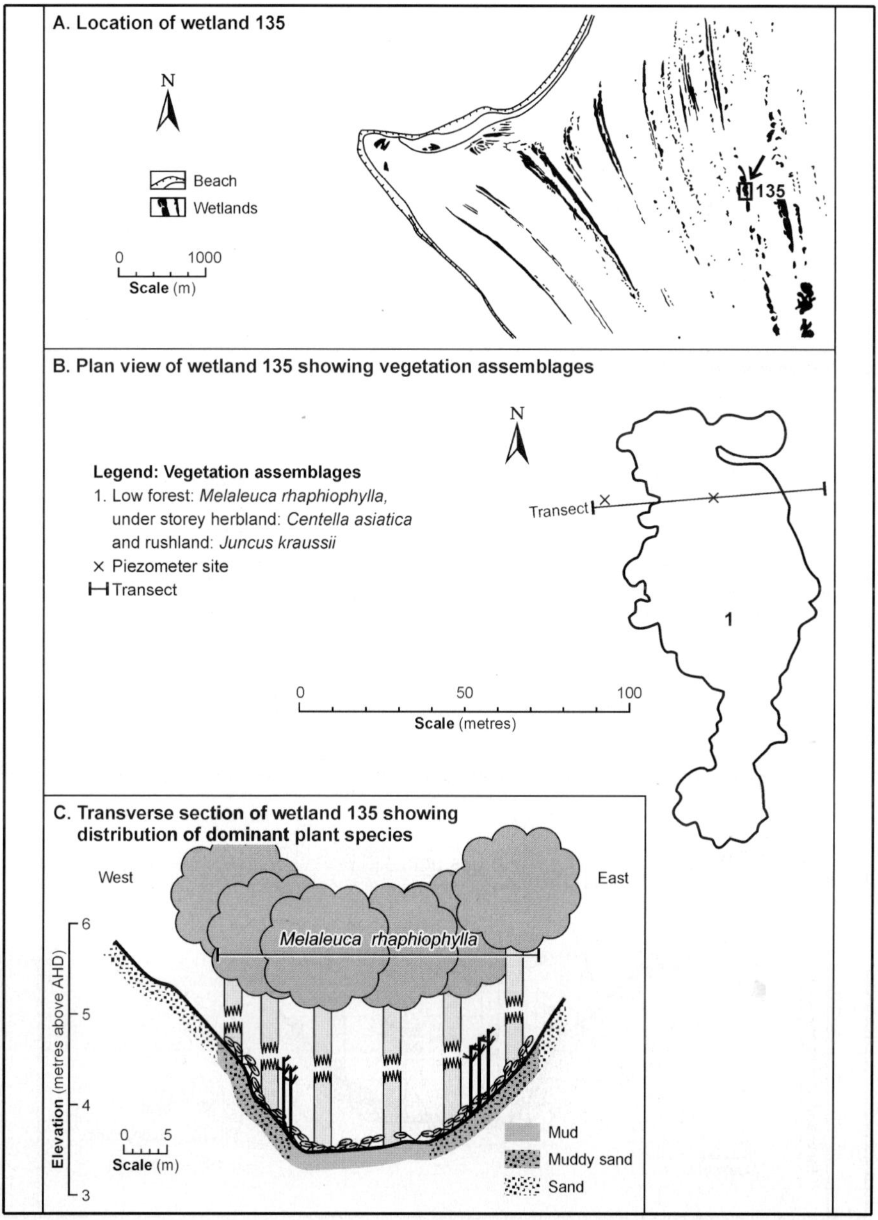

Figure 10-8. *Plan view and cross section showing distribution of wetland vegetation in wetland 135.*

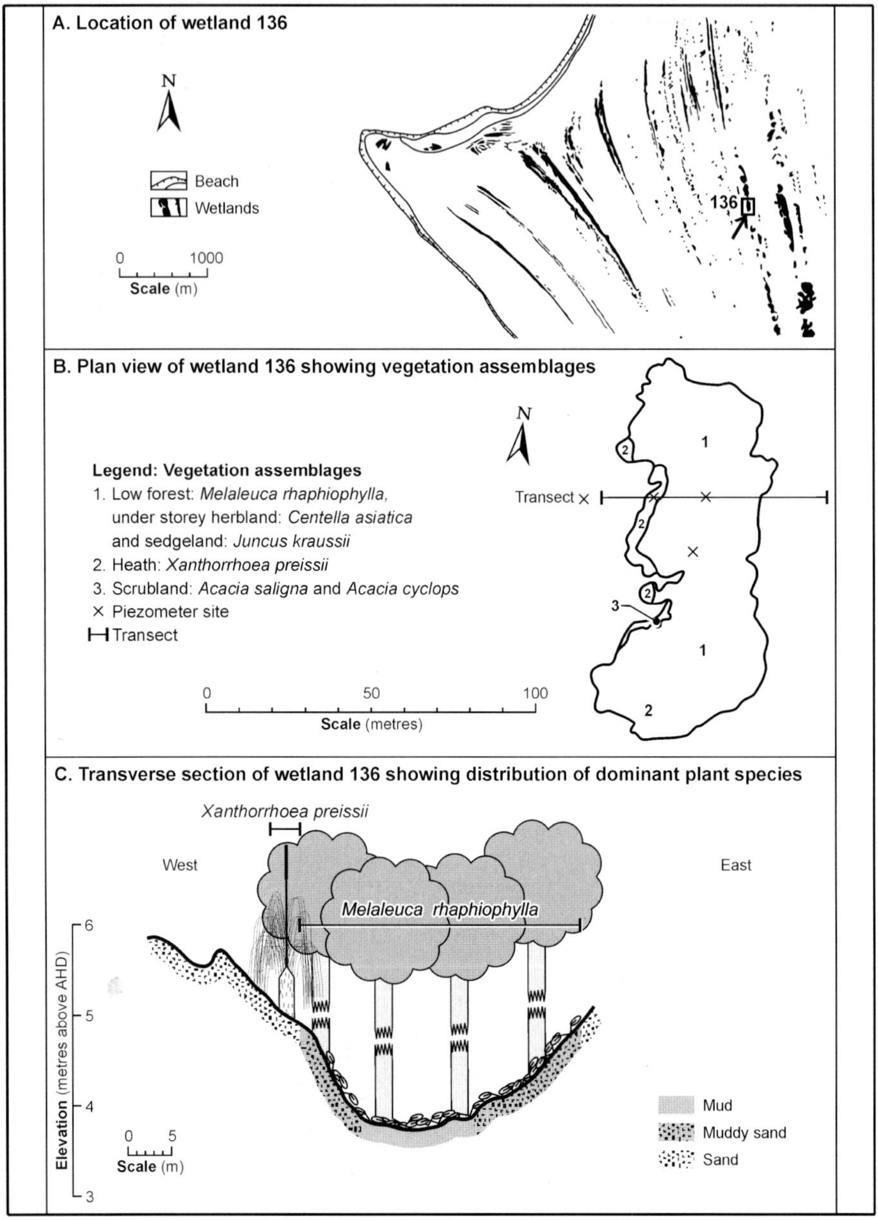

Figure 10-9. *Plan view and cross section showing distribution of wetland vegetation in wetland 136.*

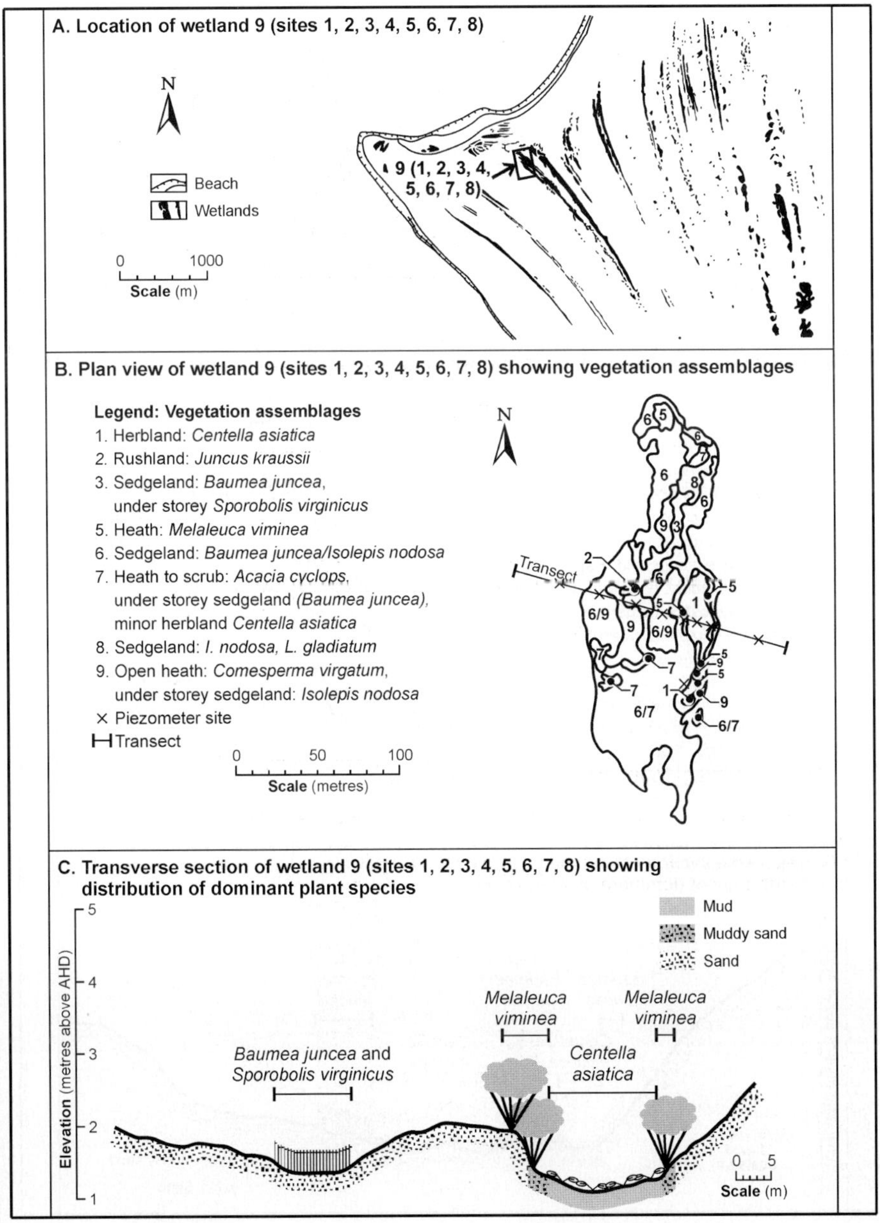

Figure 10-14. *Plan view and cross section showing distribution of wetland vegetation in wetland 9 (1, 2, 3, 4, 5, 6, 7, 8).*

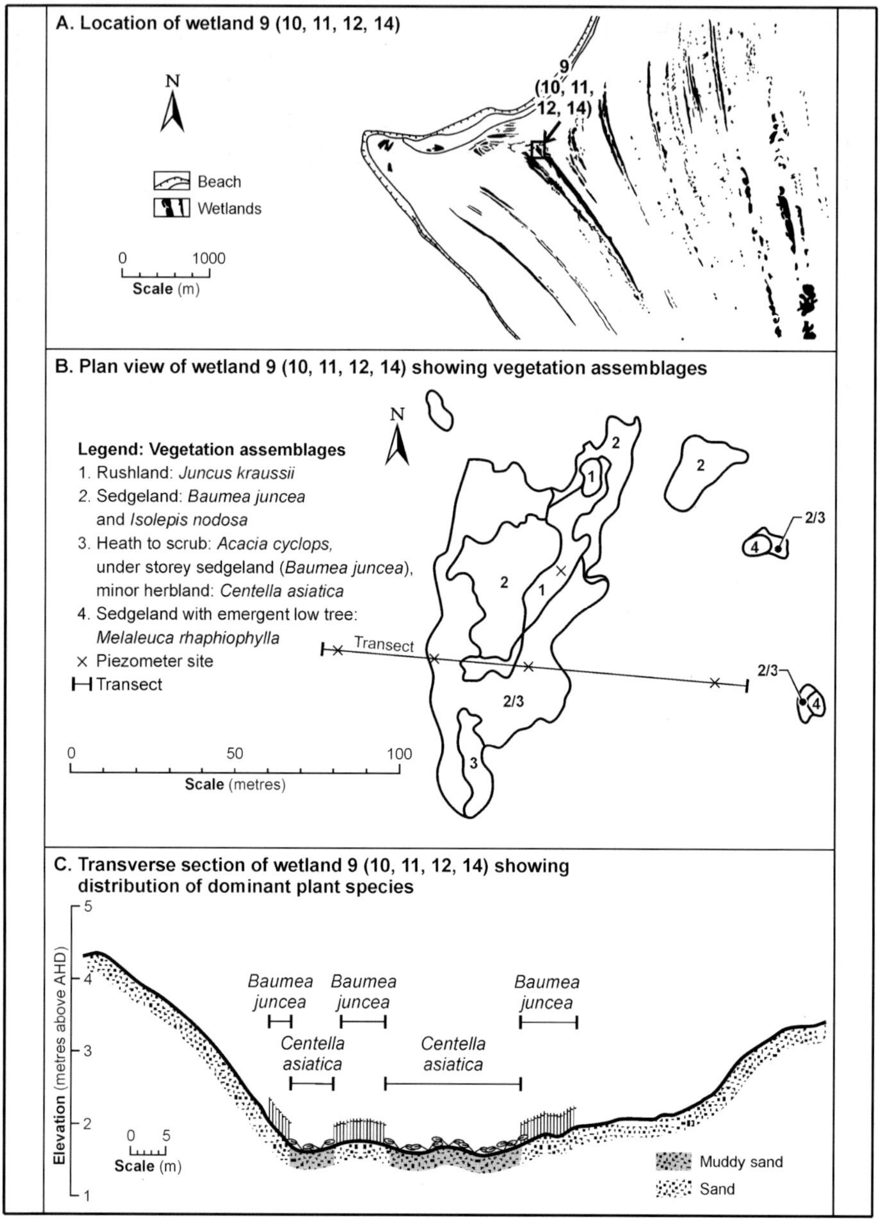

Figure 10-15. *Plan view and cross section showing distribution of wetland vegetation in wetland 9 (sites 10, 11, 12, 14).*

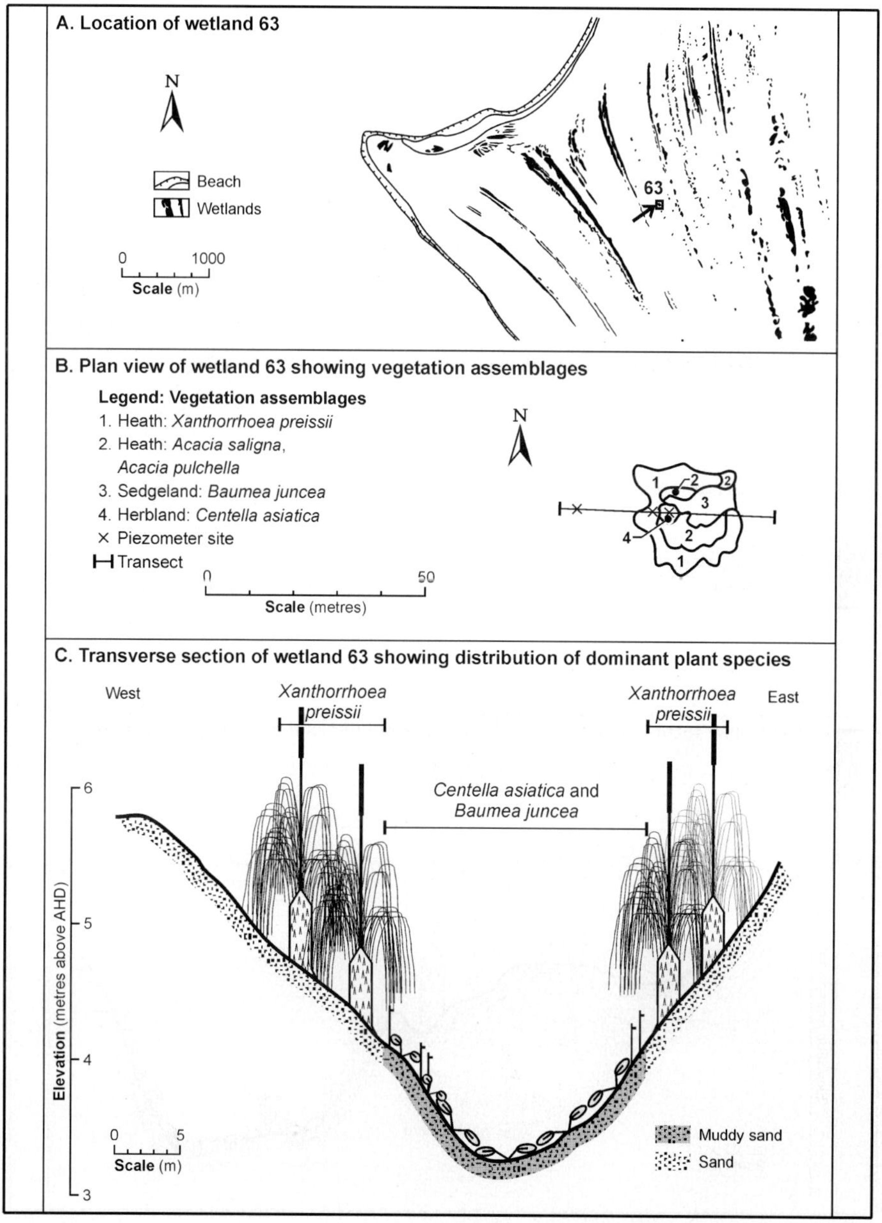

Figure 10-11. *Plan view and cross section showing distribution of wetland vegetation in wetland 63.*

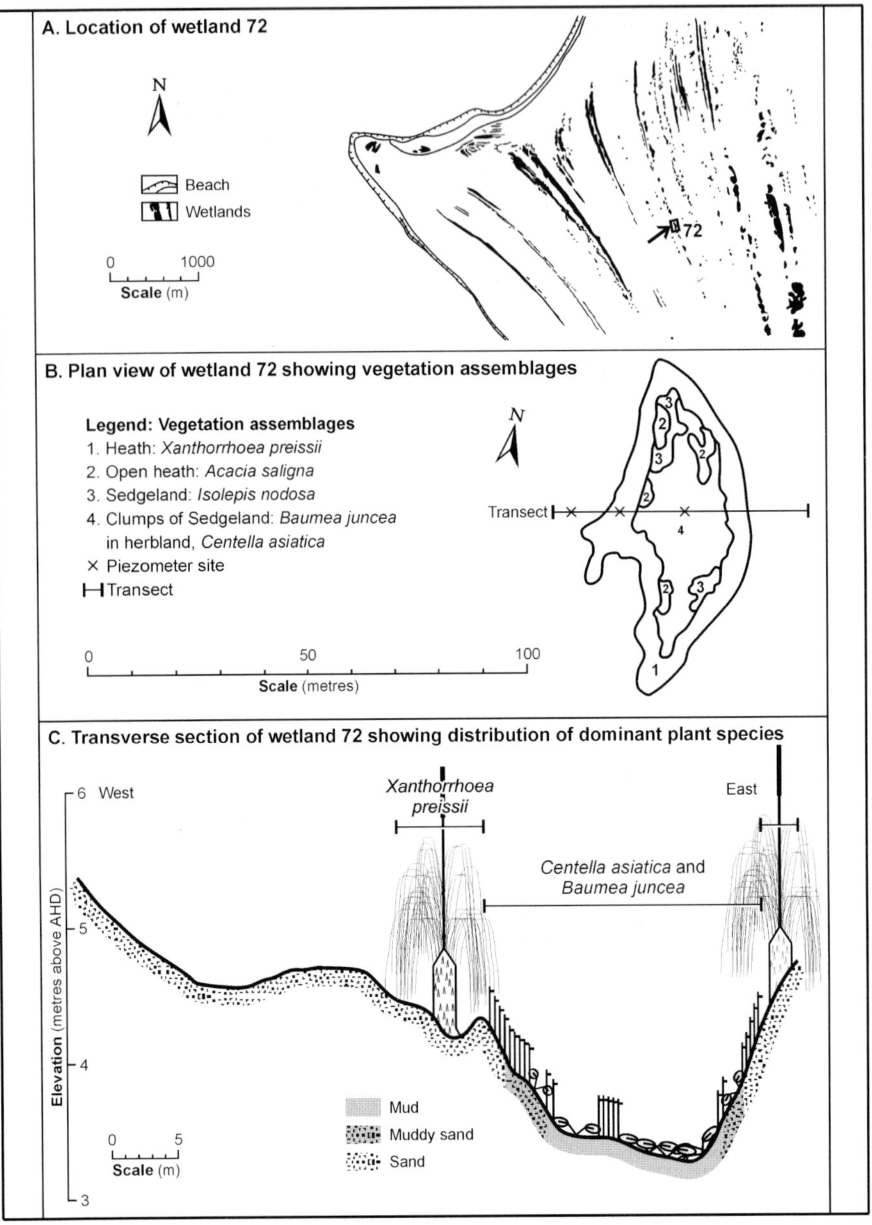

Figure 10-10. *Plan view and cross section showing distribution of wetland vegetation in wetland 72.*

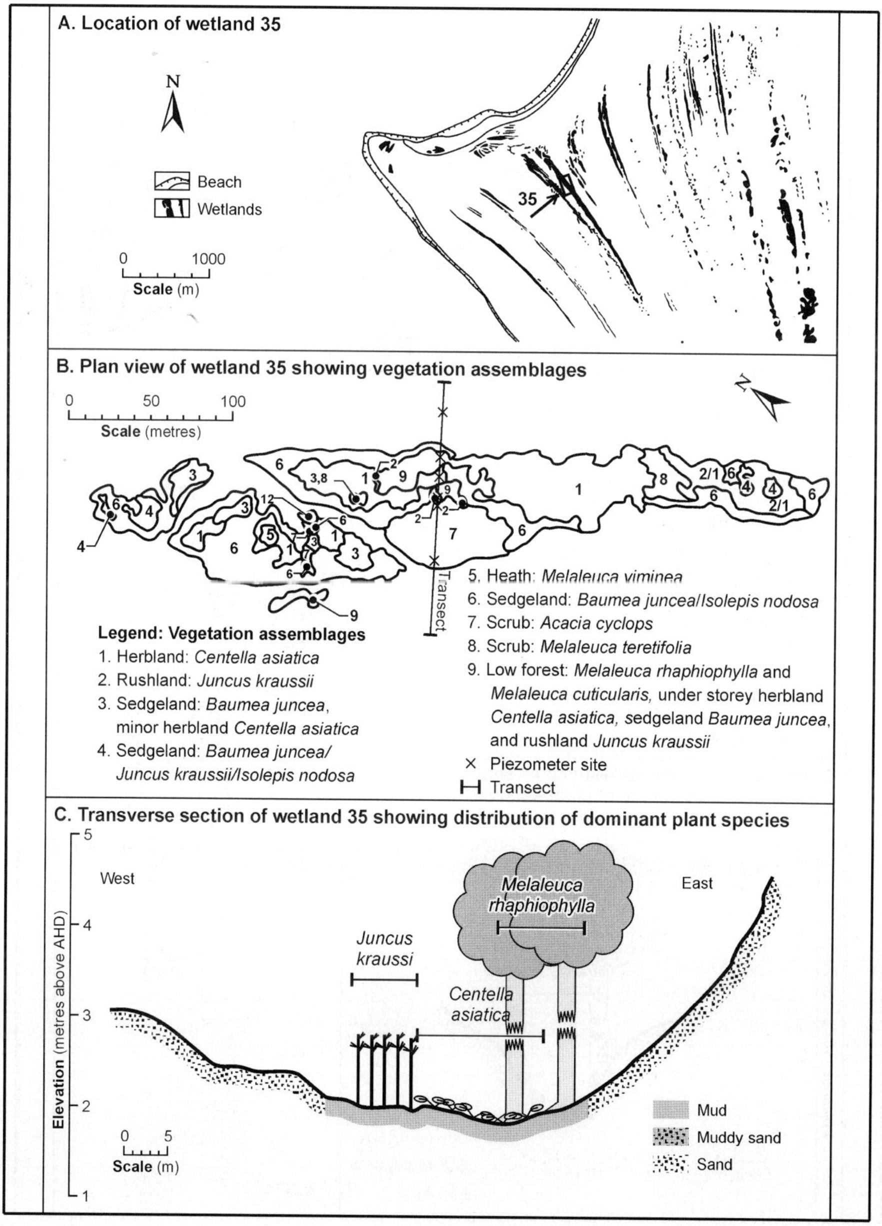

Figure 10-13. Plan view and cross section showing distribution of wetland vegetation in wetland 35.

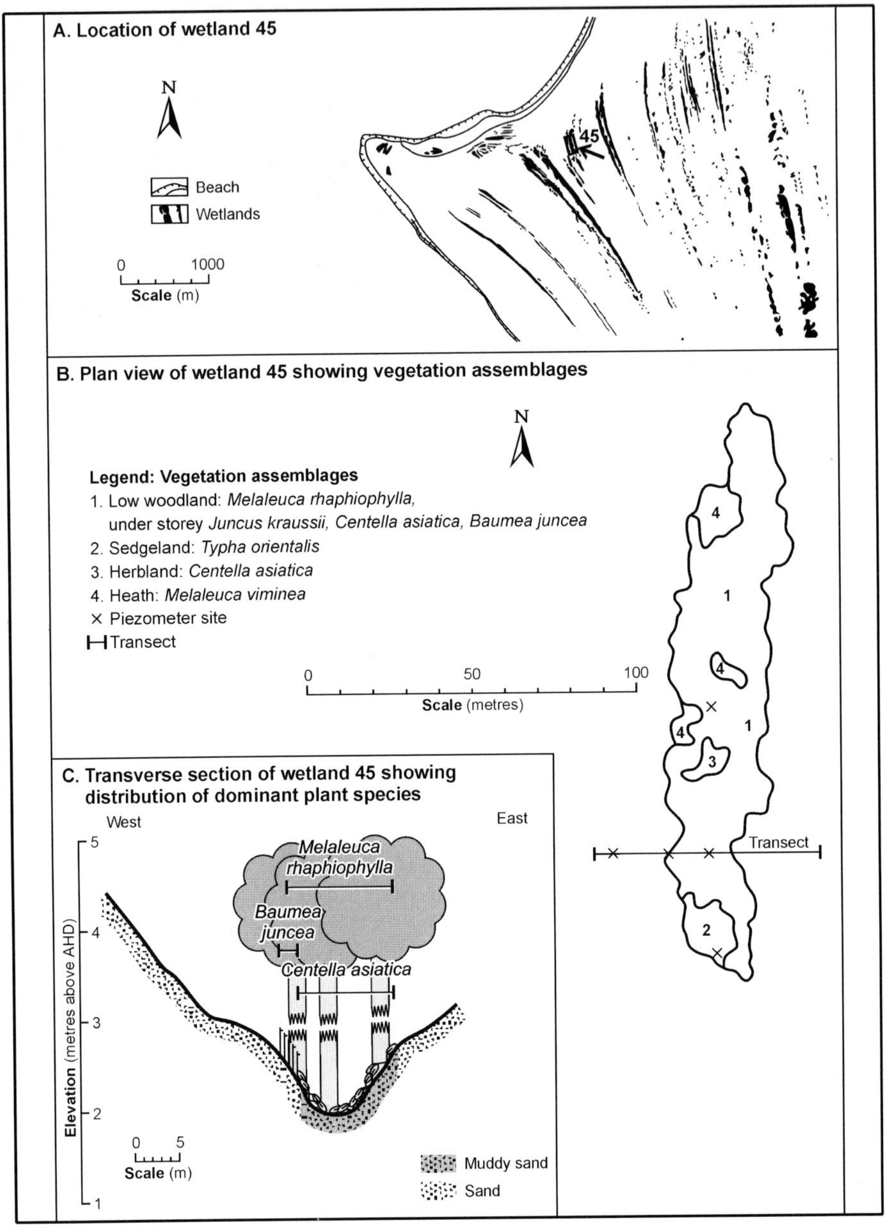

Figure 10-12. Plan view and cross section showing distribution of wetland vegetation in wetland 45.

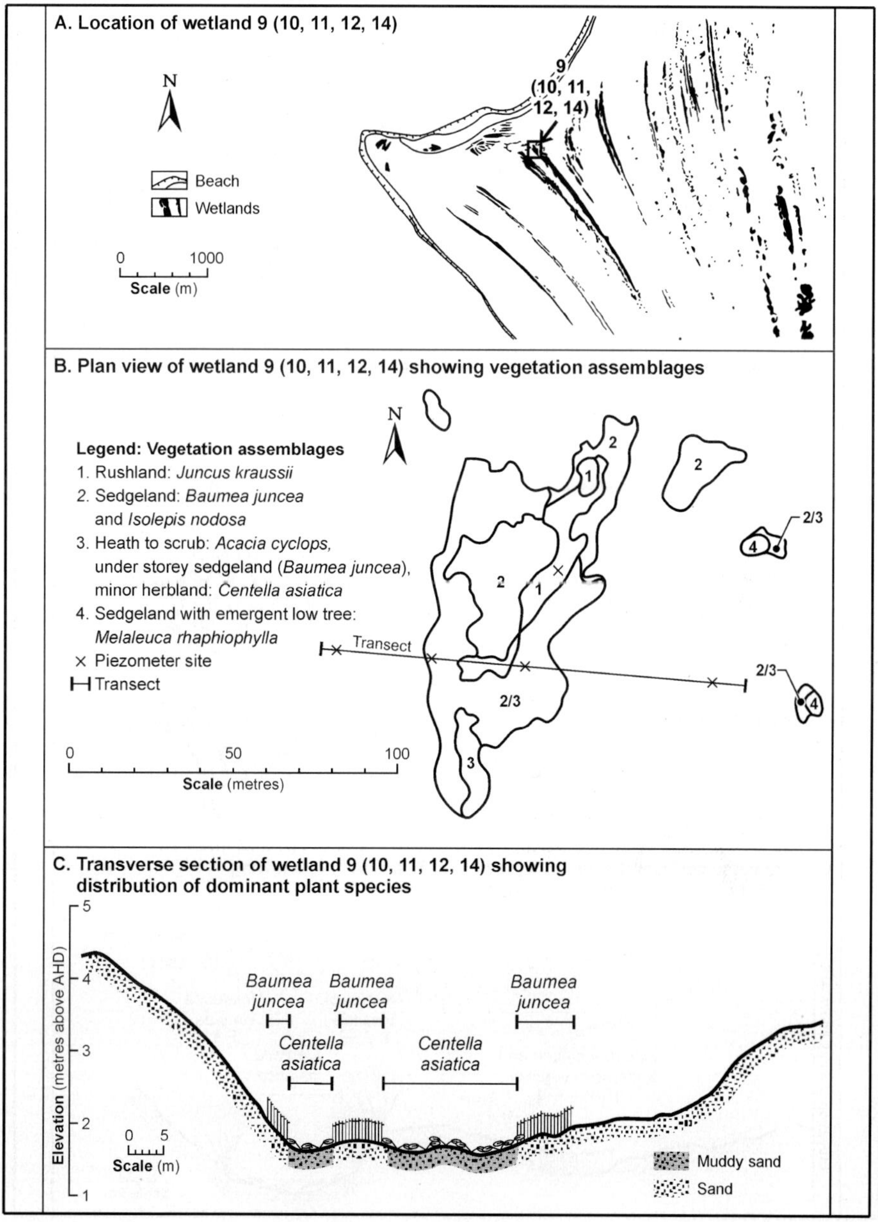

Figure 10-15. *Plan view and cross section showing distribution of wetland vegetation in wetland 9 (sites 10, 11, 12, 14).*

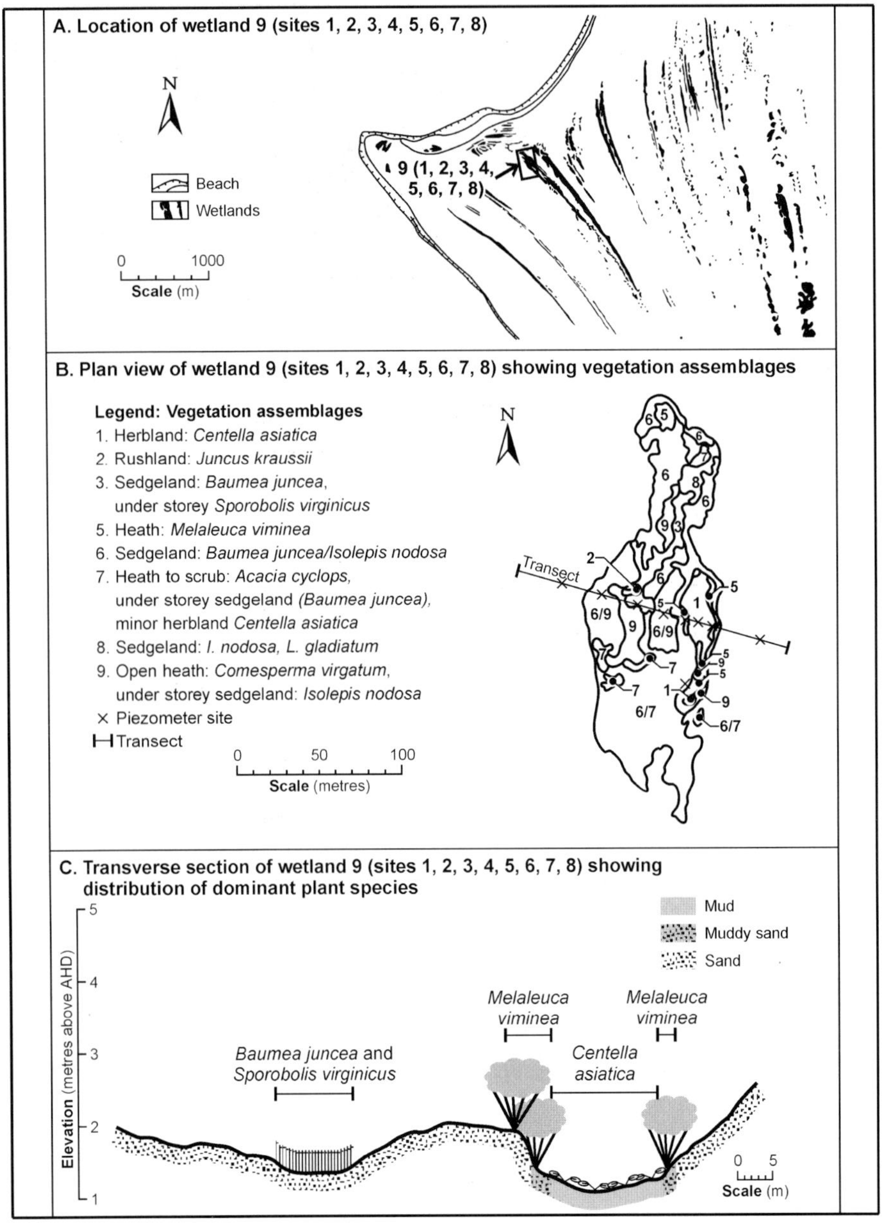

Figure 10-14. *Plan view and cross section showing distribution of wetland vegetation in wetland 9 (1, 2, 3, 4, 5, 6, 7, 8).*

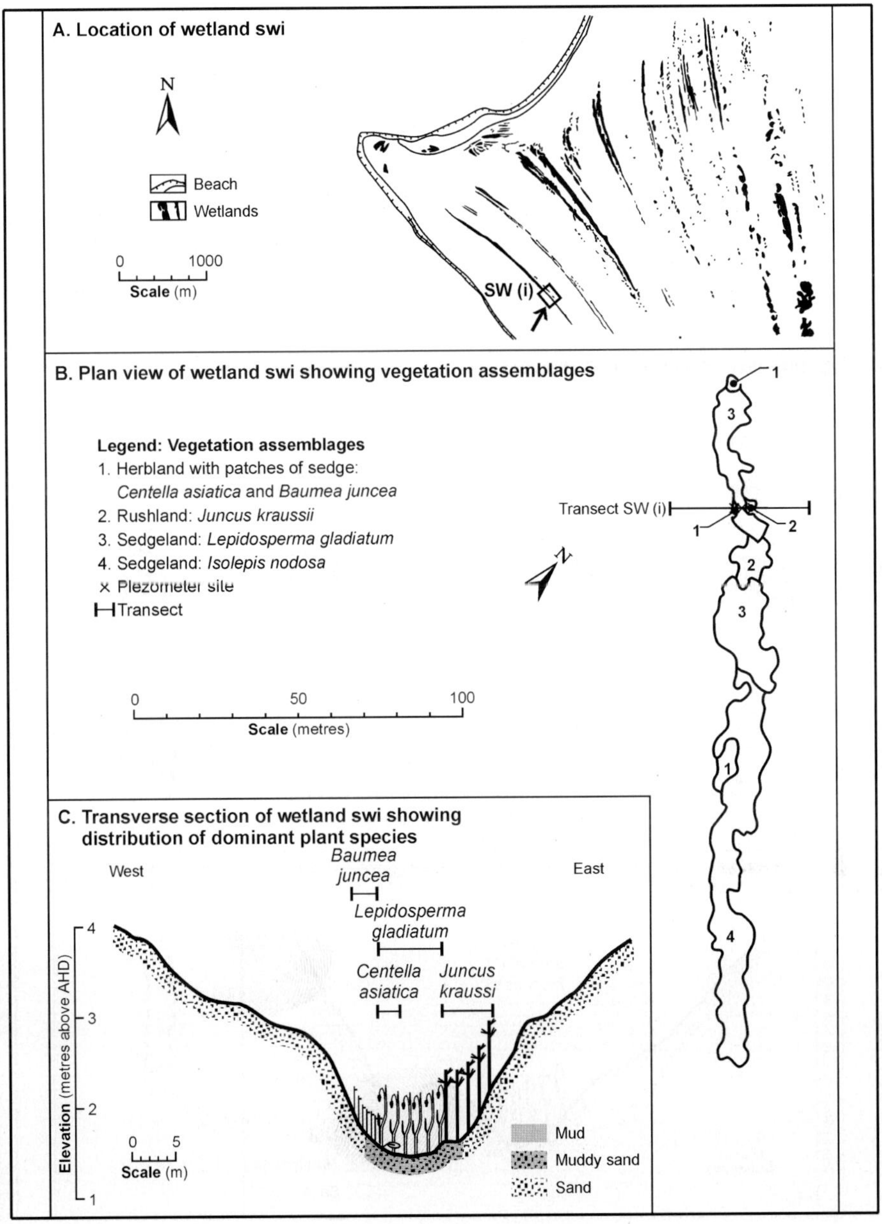

Figure 10-16. *Plan view and cross section showing distribution of wetland vegetation in wetland swi.*

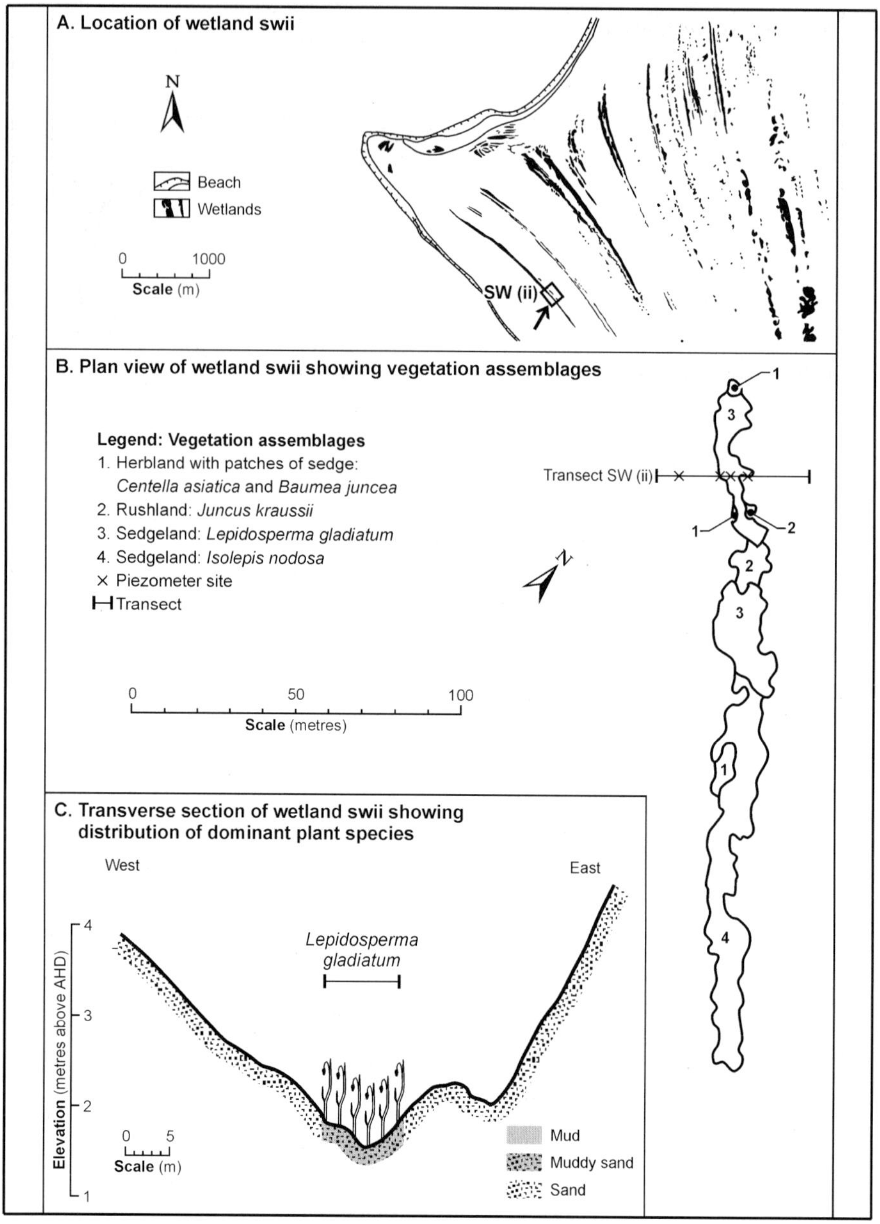

Figure 10-17. *Plan view and cross section showing distribution of wetland vegetation in wetland swii.*

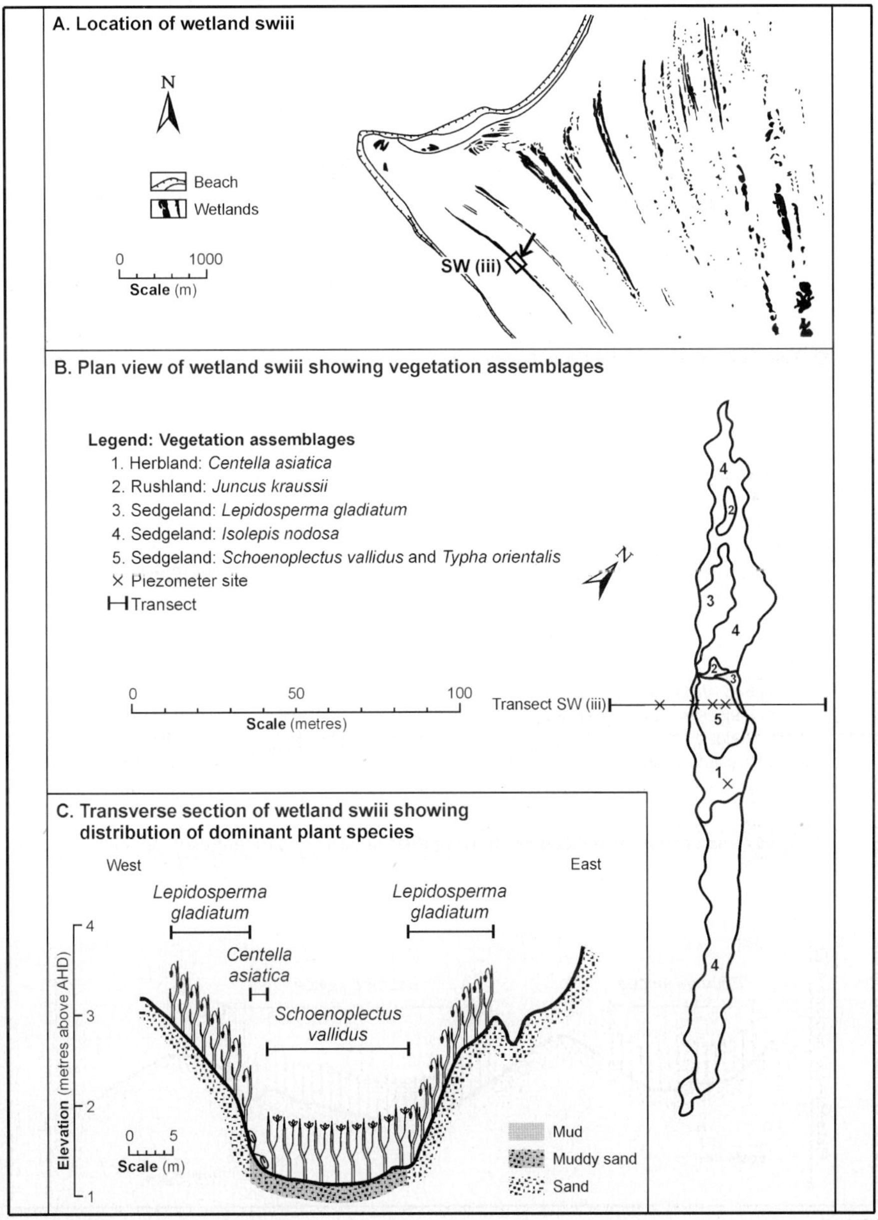

Figure 10-18. *Plan view and cross section showing distribution of wetland vegetation in wetland swiii.*

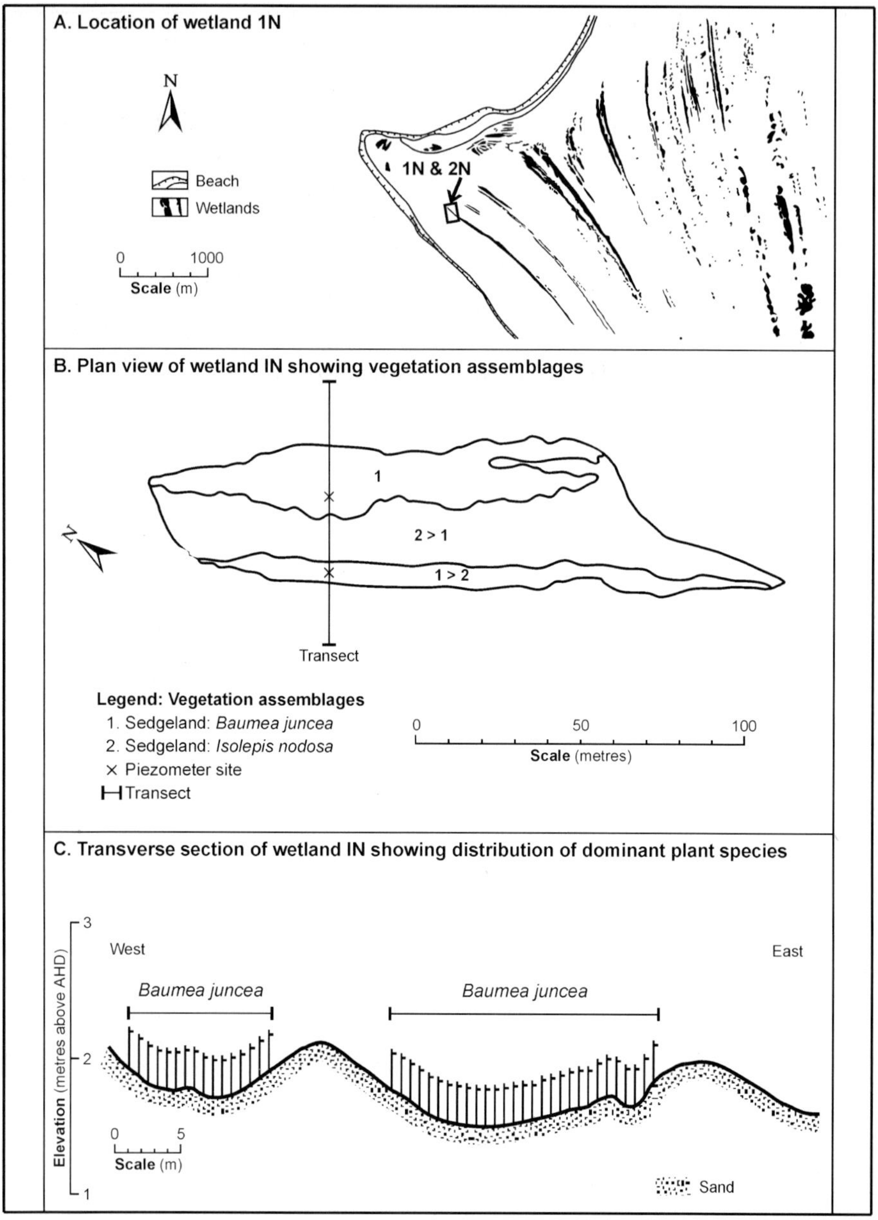

Figure 10-19. *Plan view and cross section showing distribution of wetland vegetation in wetland 1N.*

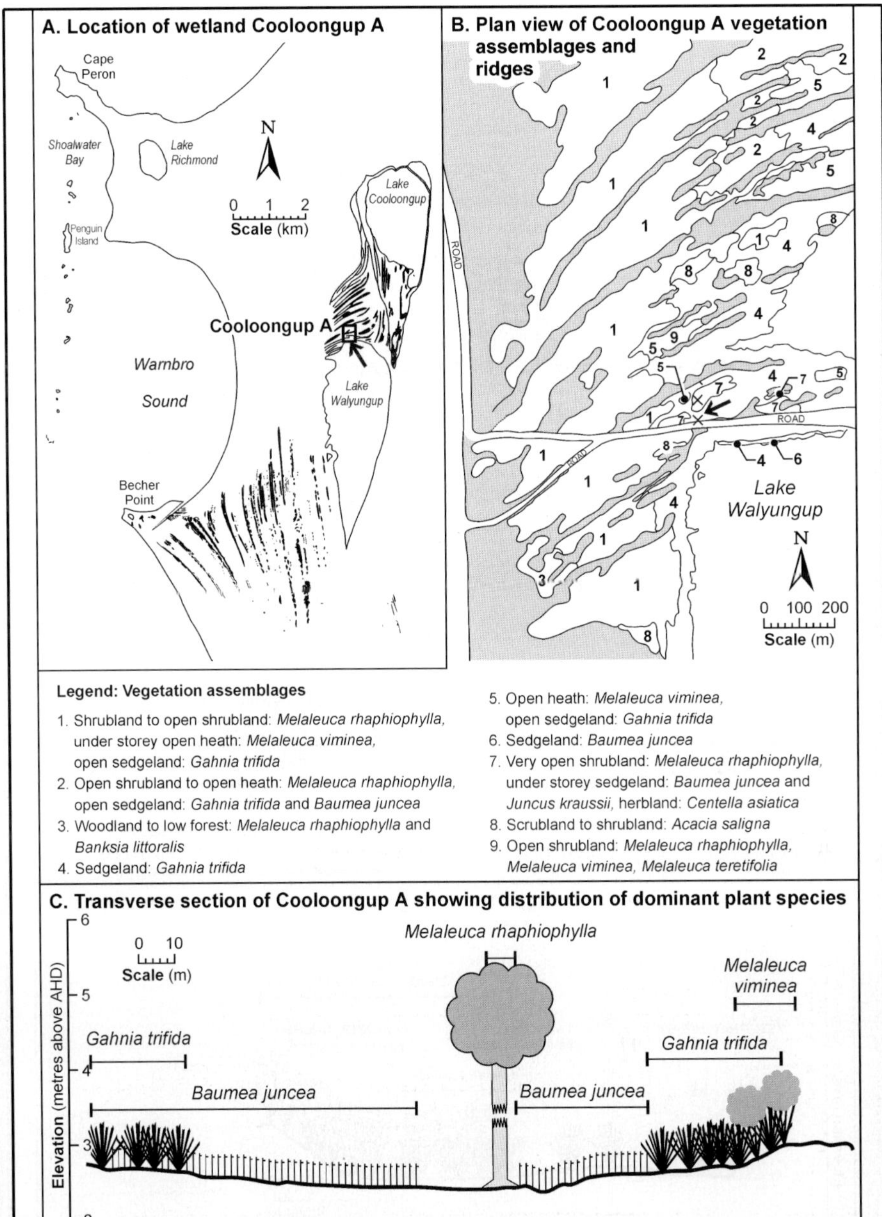

Figure 10-20. *Plan view and cross section showing distribution of wetland vegetation in wetland Cooloongup A.*

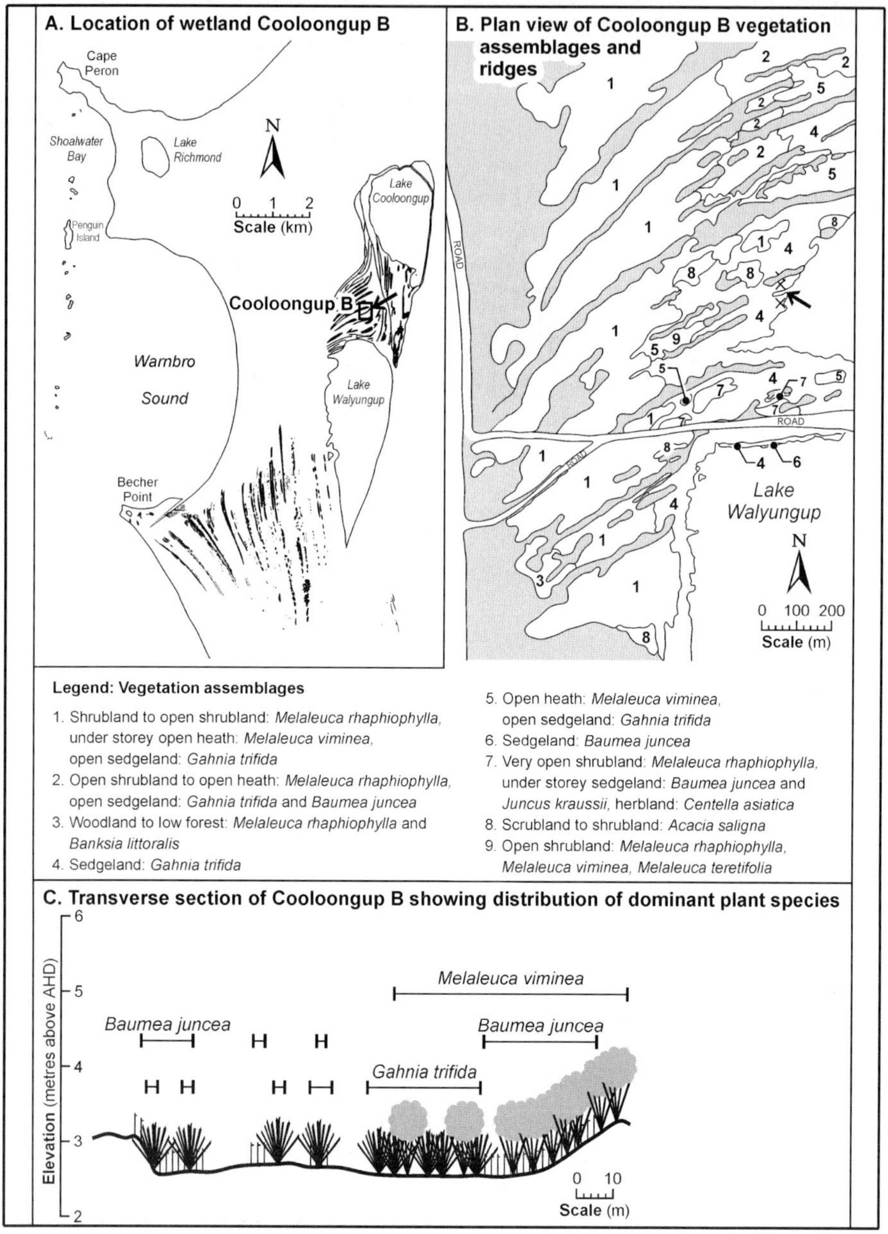

Figure 10-21. *Plan view and cross section showing distribution of wetland vegetation in wetland Cooloongup B.*

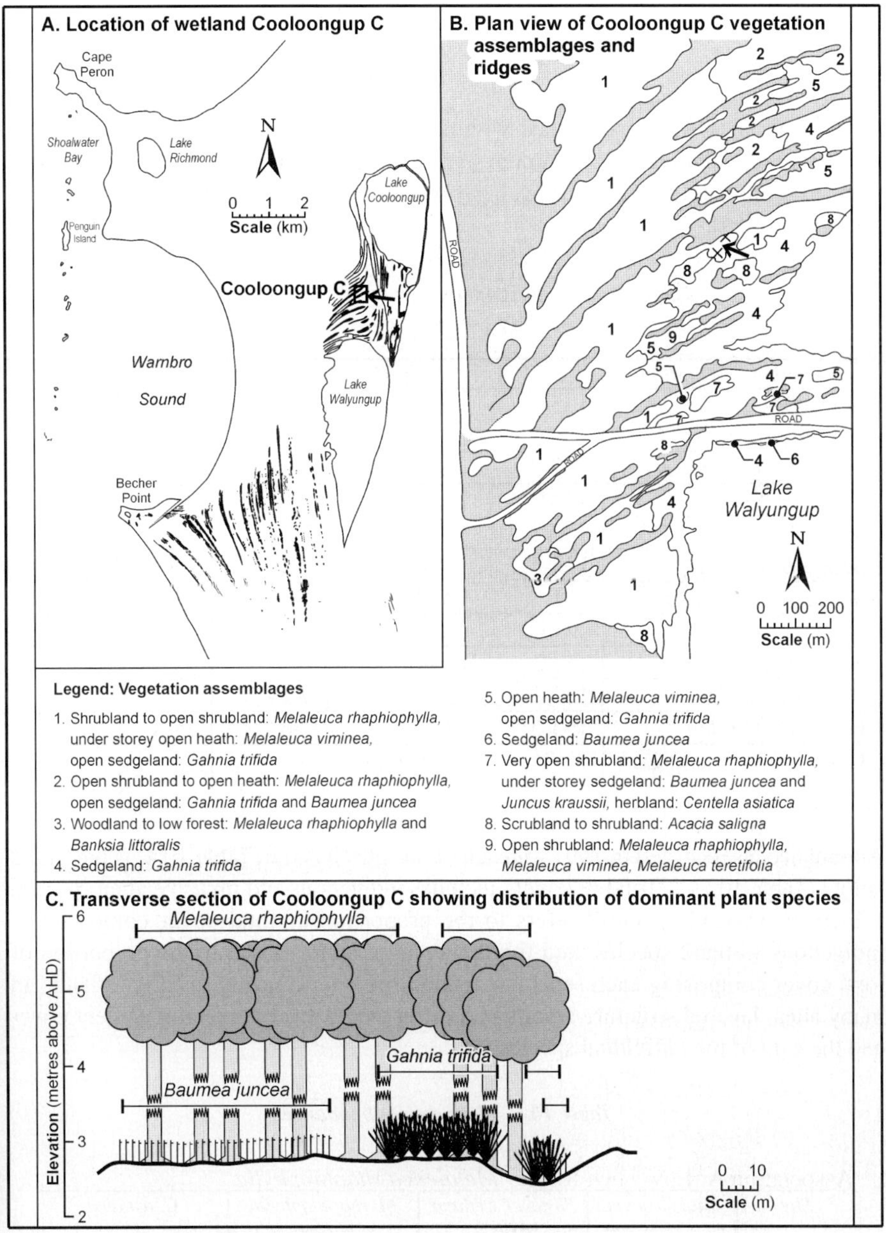

Figure 10-22. Plan view and cross section showing distribution of wetland vegetation in wetland Cooloongup C.

10.3 Multivariate analysis

10.3.1 Vegetation quadrats

Monitoring sites were established within the selected wetlands in each of the 11 vegetation associations. A maximum of five sites containing each assemblage was selected for longer term monitoring, however, some associations only occurred at two sites (Table 10.4).

Table 10.4 *Site designations for specific associations*

Vegetation associations	*Sites where association occurs*				
Forest-*M. rhaphiophylla*	135-2	136-3	45-3	45-5	35-5
Heath-*M. teretifolia*	162-3	162-5	142-3	142-6	142-10
Heath-*M. viminea*	WAWA 4	142-5	9-5	9-7	9-13
Heath-*X. preissii*	142-8	136-2	72-2	63-2	
Sedge-*B. juncea*	163-4	142-6	9-3	1N-1	1N-2
Sedge-*B. articulata*	161-3	WAWA 3			
Sedge-*S. validus*	swiii-3	swiii-4			
Sedge-*L. gladiatum*	swi-2	swii-3			
Rush-*J. kraussii*	163-3	163-6	9-14	35-3	swi-3
Herbs-*C. asiatica*	163-5	63-3	9-6	35-4	swiii-5
Herb/sedge-*C. asiatica/ B. juncea*	72-3	9-11			

Abundance, as recorded in 1991 for each of the sites listed in Table 10.4, is presented below (Table 10.5). "Total % cover" includes indigenous and invasive alien species, "% cover wetland species" refers to the proportion of total cover composed of indigenous wetland species, and the remaining columns indicate the proportion of total cover comprising each separate wetland species occurring in that wetland. In many sites, layered structure results in a difference between the total percent cover and the sum of the individual species.

Table 10.5 *Plant cover abundance*

Association A: Low open forest - *Melaleuca rhaphiophylla*

Site	*total % cover*	*% cover wetland species*	*M. rhaphiophylla %*	*C. asiatica %*
135-2	100	12	12	3
136-3	80	80	55	30
45-3	70	70	70	2
45-5	100	100	50	50
35-5	60	60	60	0

Table 10.5 (cont.)

Association B: Heath - *Melaleuca teretifolia*

Site	*total % cover*	*% cover wetland species*	*M. teretifolia %*	*C. asiatica %*	*J. kraussii %*	*B. juncea %*
162-3	100	60	20	50		10
162-5	80	80	26	55		
142-3	53	53	33	22	10	
142-7	60	30	10			30
142-10	70	70	70			

Association C: Heath - *Melaleuca viminea*

Site	*total % cover*	*% cover wetland species*	*M. viminea %*	*C. asiatica %*	*I. nodosa %*	*B. juncea %*
WAWA 4	88	66	66			20
142-5	64	20	15			5
9-5	76	24	12		2	12
9-7	71	56	6	17	25	8
9-13	56	51	16	5		30

Association D: Heath - *Xanthorrhoea preissii*

Site	*total % cover*	*% cover wetland species*	*X. preissii %*	*C. asiatica %*	*A. saligna %*	*B. juncea %*
136-2	50	50	50			
142-8	40	40	40			
72-2	75	75	75			
63-3	85	85	75		10	

Association E: Sedge - *Baumea juncea*

Site	*total % cover*	*% cover wetland species*	*I. nodosa %*	*A. saligna %*	*B. juncea %*
163-4	100	26			26
142-6	100	30			30
9-3	80	33			33
N1	50	28			28
N2	100	60	20	5	35

Association F: Sedge - *Baumea articulata*

Site	*total % cover*	*% cover wetland species*	*B. articulata %*	*C. asiatica %*
161-3	100	100	99	1
WAWA 3	100	100	98	2

Table 10.5 (cont.)

Association G: Sedge - *Schoenoplectus validus*

Site	*total % cover*	*% cover wetland species*	*S. validus %*	*C. asiatica %*
swiii-3	100	40	26	22
swiii-4	90	33	36	24

Association H: Sedge - *Lepidosperma gladiatum*

Site	*total % cover*	*% cover wetland species*	*L. gladiatum %*	*J. kraussii %*	*C. asiatica %*
swi-2	60	60	25	10	25
swii-3	40	40	40		

Association I: Sedge - *Juncus kraussii*

Site	*total % cover*	*% cover wetland species*	*J. kraussii %*	*C. asiatica %*	*B. juncea %*
163-3	100	100	92	2	6
163-6	100	100	99	1	
35-3	55	55	55	< 1	
9-14	100	100	100		
swi-3	100	100	100		

Association J: Herb - *Centella asiatica*

Site	*total % cover*	*% cover wetland species*	*C. asiatica %*	*M. cuticularis %*	*B. juncea %*
163-5	100	100	100		
63-3	65	65	60		5
35-4	100	100	100	10	
.9-6	100	100	100		
swiii-5	100	100	100		

Association K: Herb/Sedge - *Centella asiatica/Baumea juncea*

Site	*total % cover*	*% cover wetland species*	*C. asiatica %*	*B. juncea %*
72-3	85	85	55	30
9-11	95	82	70	12

10.3.2 Ordination

The reason that particular species grow together in a particular environment is usually related to the fact that they have similar requirements for co-existence in terms of environmental factors such as light, temperature, water, drainage and soil nutrients (Kent and Coker 1992). This assumption forms the basis of the null hypotheses generated for both the multivariate analysis technique of ordination, and the ANOVA multi-factor variance analyses used herein.

Ordination of environmental attributes

On the principle that wetland plants are fundamentally dependent on water, it was assumed that variability in the spatial distribution of wetland plants in the Becher wetlands would reflect variation in both the availability and hydrochemistry of the groundwater, therefore emphasis was placed on these attributes. Annual mean water levels and annual mean groundwater salinity were treated as separate factors to distinguish between wet and dry years. Salinity of groundwater during the plant growing season was separated out as a measure in order to weight the concentrations of groundwater salinity during the period of direct plant access. Annual variability is an integral aspect of seasonal wetlands, therefore measurements of several factors were included in the form of coefficient of variation. As the sediments generally contained organic enriched calcilutites and muddy sands, the organic content was selected as the factor to characterise differences in wetland sedimentary fills. Finally, the radiocarbon ^{14}C age of each wetland was selected to investigate the effect of time on wetland vegetation distribution. Ordination was undertaken using a total of 23 environmental attributes:

1-5. mean depth to water table during each year from 1991 to 1996 to include wetter and drier years;
6. number of years during the study that inundation occurred (1991-2001);
7. mean annual period of inundation (1991-2001);
8. annual period of access to capillary water in the vadose zone represented by mean annual period of waterlogging (<30 cm below ground surface) (1991-2001);
9. mean soil moisture content (1992-94);
10. soil moisture variability expressed as coefficient of variation (1992-94);
11. range in soil moisture (1992-94);
12-13. mean concentration of total dissolved solids in groundwater (1992-94);
14. mode concentration of total dissolved solids in groundwater (1992-94);
15. mean concentration of total dissolved solids in groundwater during the growing seasons August to February (1992-94), the period of groundwater availability to plants;
16. groundwater salinity variability expressed as coefficient of variation (1992-94);
17. orthophosphate concentration in soil water;

18. mean sodium, potassium, magnesium and calcium concentrations in soil water (1992-94);
19. variability in potassium concentrations in soil water expressed as coefficient of variation (1992-94);
20. variability in calcium concentrations in soil water expressed as coefficient of variation (1992-94);
21. macro organic composition of 0-10 cm sediment layer;
22. depth of the wetland fill; and
23. radiocarbon ^{14}C age of base of wetland fill;

Semi-strong hybrid multidimensional scaling (SSH) (Belbin 1992) was the ordination programme selected (see Methods Chapter) using the Bray and Curtis association measure. SSH fits output distances to input distances rather than the squared counterpart to avoid overly weighting the larger input distances (Belbin 1995). Three dimensions were selected and a ratio/ordinal cut requested at 0.9.

Results

Several ordinations were examined, the first using all 23 attributes and the latter four using sub-sets of related attributes in order to ascertain which aspects of water availability, groundwater salinity, soil water chemistry and soils were most strongly affecting the ordination of quadrats. The results are presented in Figures 10-23 to 10-27. In order to facilitate interpretation of the figures, geometric symbols are used to designate quadrants in which sites and attributes are located within the 3-dimensional space.

In the ordination using all 23 attributes, sites were clearly divided along a salinity gradient between sites experiencing fluctuations in groundwater salinity from freshwater to hyposaline, and sites experiencing more moderate fluctuations within the category of freshwater (Fig. 10-23). Aligned with groundwater salinity were the concentrations of individual cations which make up the solutes. There were secondary clusters associated with soil moisture and low water tables. The distribution of sites with respect to environmental attributes was reasonably widespread, which suggests that the diversity of the wetlands and vegetation assemblages in the Becher Suite is due to the combination of many factors. There were few sites outside the main group, and these were associated with inundation and waterlogging frequency.

The results further showed that sites with similar vegetation in disparate wetlands are more likely to exhibit an overall similarity in environmental attributes than sites which support different vegetation assemblages within the same wetland. The attributes which contribute the greatest variability to the variation between sites are intra-basinal.

In the sub-set of attributes relating to water permanence and depth, the sites which showed the greatest and most consistent relationship with water availability were

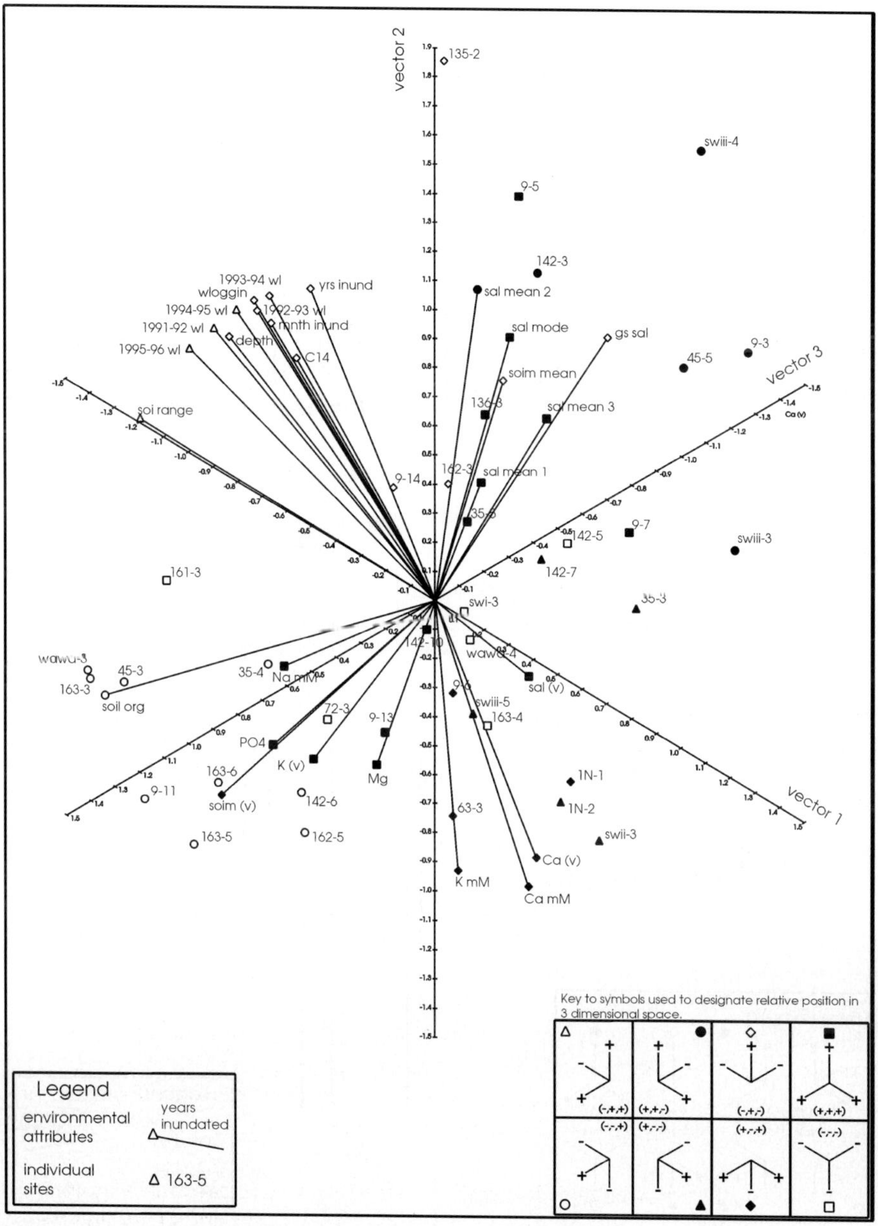

Figure 10-23. *Three dimensional ordination showing relation between wetland vegetation and 23 environmental attributes.*

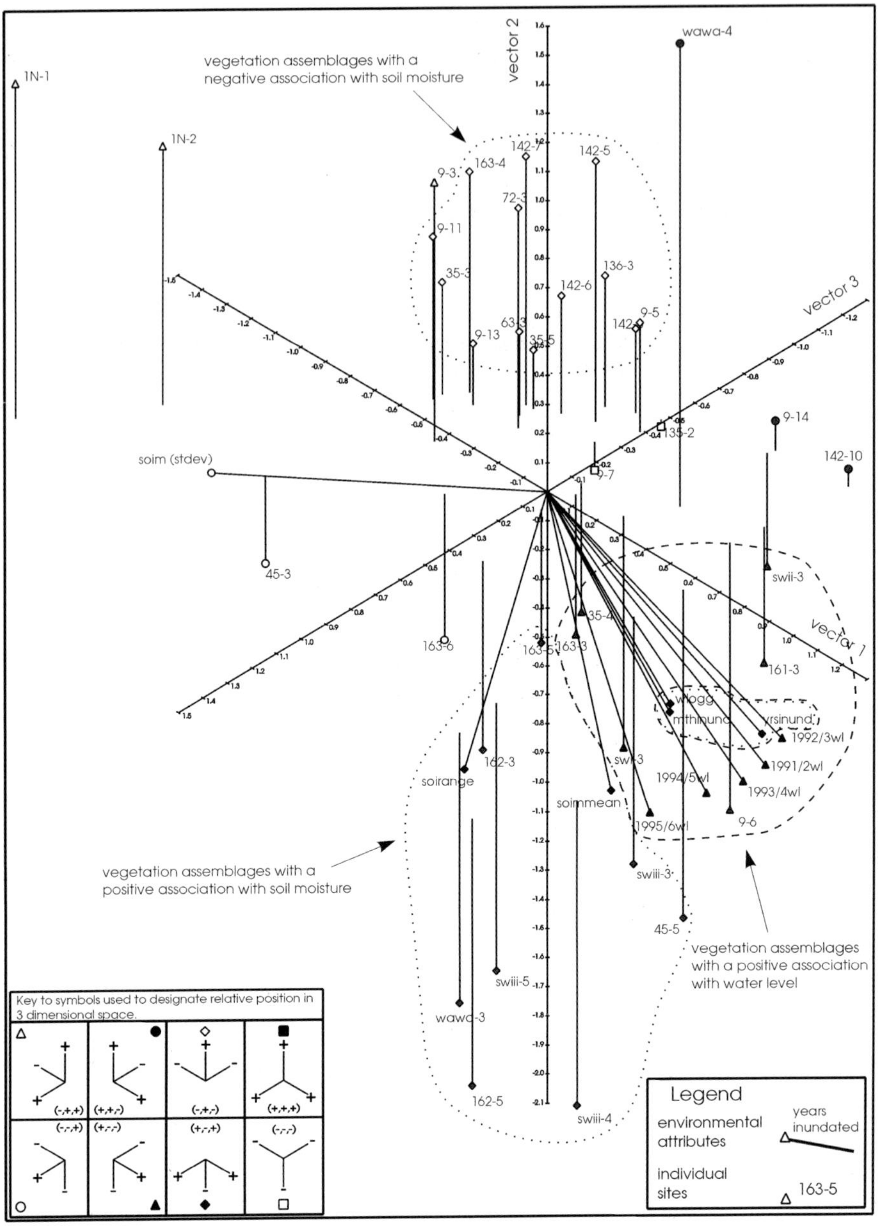

Figure 10-24. Three dimensional ordination showing relation between wetland vegetation and hydrologic attributes.

those under the herb *Centella asiatica, e.g.,* site 163-5, 162-5, 45-5, 35-4, 9-6, swiii-5. Of the 11 measures used in the ordination, the three measures of soil moisture were particularly relevant (Fig. 10-24). Sites which were more closely related to water levels were those which experienced seasonal inundation, *i.e.*, site 161, 163-3, 9-6, 35-4. These sites support a range of plant assemblages, *e.g.*, *B. articulata, J. kraussii, L. gladiatum* and *C. asiatica*. Sites under all species of *Melaleuca* appeared to be tolerant of variable water levels and soil moisture as did those under the sedge *B juncea, e.g.,* sites 142-5,6,7, 136-3, 35-5, 9-5.

In the sub-set of attributes pertaining to groundwater salinity, the sites separated into two groups, those tolerant of elevated groundwater salinity, and those intolerant of elevated groundwater salinity (Fig. 10-25). In the former group were sites under all species of *Melaleuca,* and the sedge *Schoenoplectus validus*. In the latter group were sites under both species of *Baumea, L. gladiatum,* and *C. asiatica*. The rush *Juncus kraussii* was represented in both groups, suggesting that groundwater salinity is not a factor which strongly influences its distribution.

In the sub-set of attributes pertaining to sediment characteristics (Fig. 10-26), none of the sites appeared to be strongly associated with the selected environmental variables.

In the sub-set of attributes relating to soil water chemistry attributes (Fig. 10-27), there were several recognisable groups of sites: 1) sites associated with variability of cation concentration in the interstitial water of the surface sediments, and 2) four groups of sites associated with the mean cationic concentration in the interstitial water of each of the major cations. None of these groups corresponded to vegetation distribution. The separation of sites more closely aligned with wetland age, *i.e.*, younger sites, were clustered around the environmental attributes pertaining to consistent calcium and potassium concentrations, and older sites were clustered around the environmental attributes pertaining to variable calcium and potassium concentrations. *B. juncea, L. gladiatum* and *J. kraussii* were the assemblages most commonly associated with stable calcium and potassium concentrations, and *C. asiatica* was the assemblage most commonly associated with cation variability.

Sites related to sodium concentration in interstitial water were similar to those pertaining to groundwater salinity and included the sites dominated by *M. rhaphiophylla*. There were no clusters related to magnesium or orthophosphate.

Interpretation of results

In the ordination of all 23 attributes, the sites which were outliers were not aberrant in any way from the rest of the group (Fig. 10-23), however, sites did exemplify strong associations with a particular attribute when specific stratigraphically determined hydrological processes were either present or absent at a particular site, and/or if the site was undergoing a change in the relative abundance of the dominant vegetation.

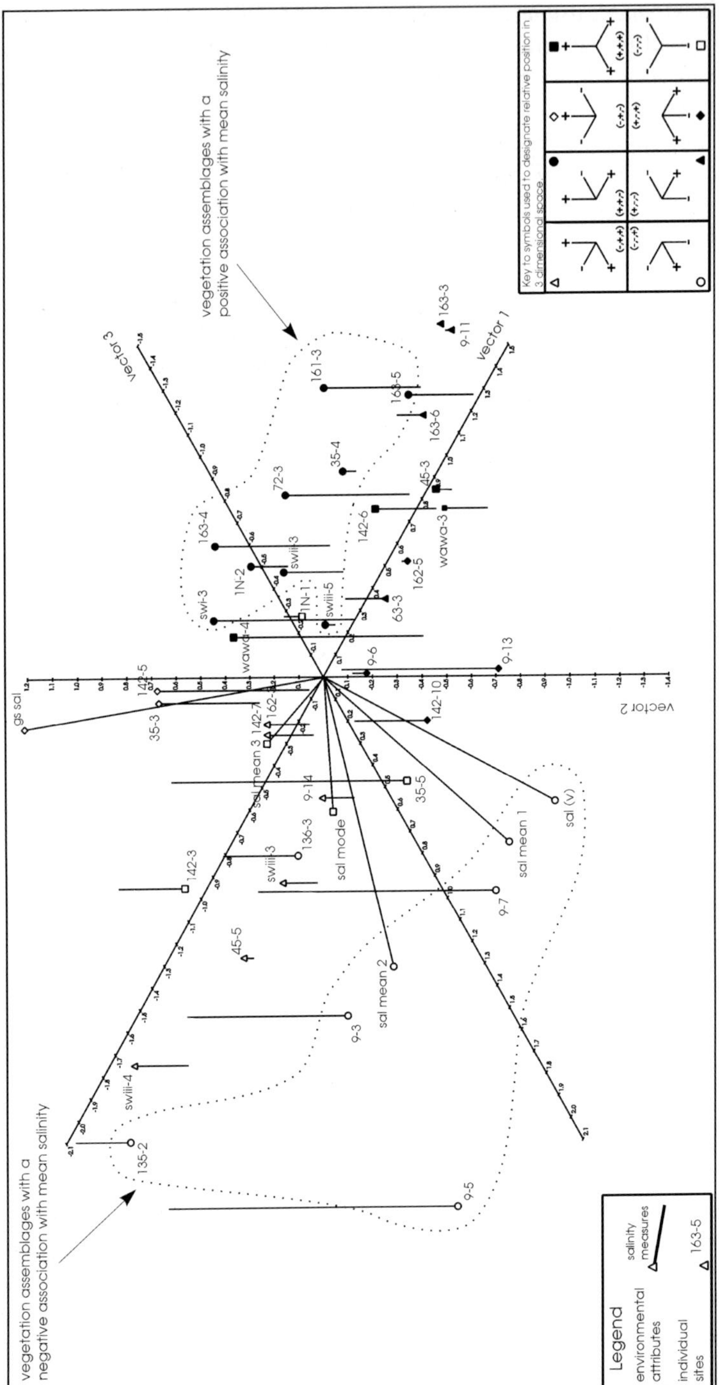

***Figure 10-25**. Three dimensional ordination showing relation between wetland vegetation and salinity attributes.*

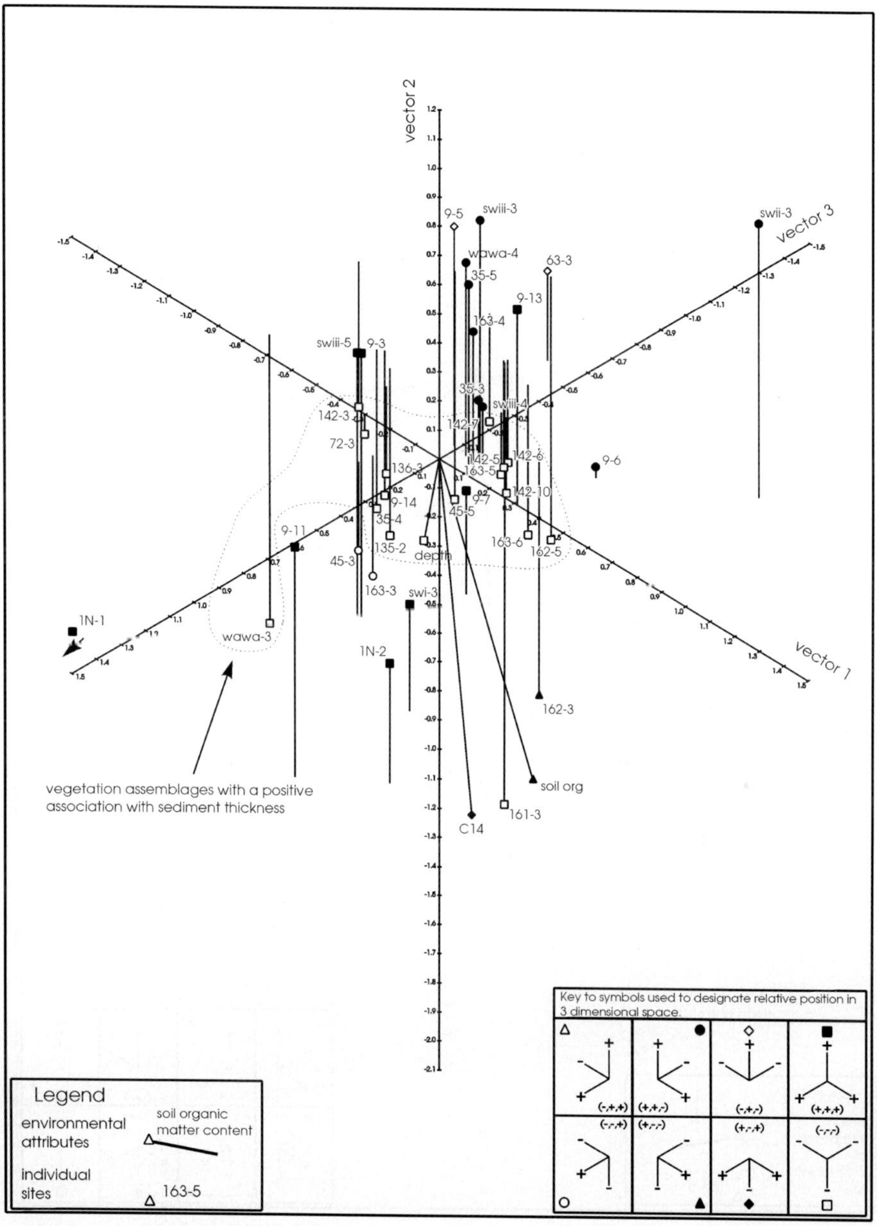

Figure 10-26. *Three dimensional ordination showing relation between wetland vegetation and soil data, age of wetland, organic matter, and depth of wetland fill.*

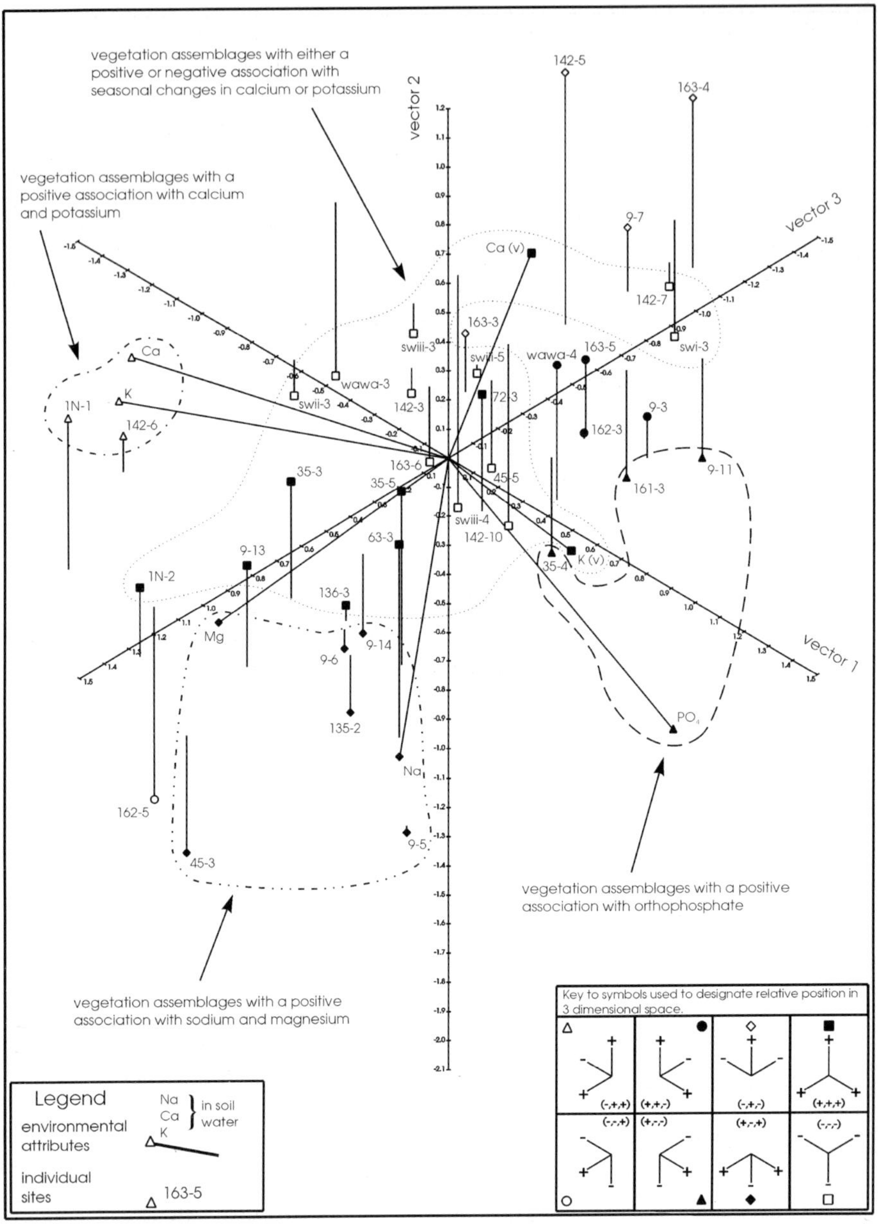

Figure 10-27. *Three dimensional ordination showing relation between wetland vegetation and soil water attributes.*

For instance: sites 9-3 and 9-5 correlated strongly with mean groundwater salinity due in part to the underlying calcrete; WAWA 3 (peat wetland) was strongly associated with the organic content of the surface layer; sites 1N-1, 2, which experience dynamic lateral water flow due to a strong, local hydraulic gradient were associated with water level; site 45-3 showed signs of significant stress due to a water deficit during the period of monitoring, and considerable loss of vegetation cover; site WAWA 4, during the period of monitoring, ceased to have any cover of wetland species as a result of the contraction of the wetland during this period of below average rainfall; and site swii-3 was the only example included in the sites which supported a sedgeland cover of *L. gladiatum*. These anomalies and perturbations manifested themselves in considerable variation in many of the hydrological and hydrochemical parameters, which then reflected in the resulting ordination.

For some wetlands there is considerable intra-wetland variability due to the specific combination of stratigraphic, hydrologic and hydrochemical processes which operate in discrete sections of any given wetland basin. The distribution of species within any basin is determined by the interplay between the water table and the geomorphology which determines the depth to water and the depth of wetland fill at any point in the basin. This interplay creates microhabitats that different species favour, hence there are central basin species and marginal basin species. There are species which preferentially colonise deeper central basins and those which colonise shallower central basins, as well as those species which colonise the slope at the margin and species which colonise the flats at the margin. This distribution is modified by inter-species competition during periods of duress. During such periods, encroachment from margin to centre and centre to margin, as well as expansion from patches, occurs.

One interesting aspect of the ordination was the relation between plant distribution at several sites (1N-2, swii, swiii, 35-3, and 142-7) and the lack of water, signified by low water levels. Sensitivity to water stress may set the limit to distribution for some species.

For some species (*Melaleuca* spp., *C. asiatica, B. articulata*, and *S. validus*), one aspect of the hydrological/hydrochemical regime, *i.e.*, water salinity or soil water chemistry, may be more important than all the others. For other species *(J. kraussii, B. juncea*, and *L. gladiatum*), a combination of factors appears to be more important. The ordination of the attributes relating to water permanence and availability suggests that, within this system of dominant freshwater, the most important attribute is soil moisture. The ordination of groundwater and soil water chemistry attributes suggests that groundwater salinity, and calcium and potassium concentrations in the soil water are the dominant attributes affecting vegetation response. The lack of correlation between any one plant species and orthophosphate suggest that it is a limiting factor for plant growth, a finding not uncommon in wetlands (Schot and Wassen 1993; Russell and Maltby 1995).

One of the characteristics which influenced intra-group similarity among sites with the same plant assemblage, was the variation in species abundance between sites. This was particularly important in the *M. teretifolia*, *M. viminea* and *B. juncea* groups. In sites with more than one species present, the sub-dominant species may have had more stringent environmental requirements than the dominant species. In the example of *M. teretifolia* heath, the understorey is the herb *C. asiatica*. In many of the ordinations, this herb was related strongly to one aspect of the environment, while *M. teretifolia*, although having similar requirements, had a broader relationship which included that attribute but others also. Therefore, the sites grouped under *M. teretifolia* were not always positioned in close proximity because the ratio of cover abundance of the two species present at those sites varied from site to site.

10.3.3 Refining of hypotheses - ANOVA

Generation of hypotheses

From the ordinations undertaken, six hypotheses were generated with regard to species distribution and abundance.

Hypothesis 1: Species presence is related to depth to groundwater
Hypothesis 2: Species presence is related to soil moisture in the 0-10 cm layer
Hypothesis 3: Species presence is related to groundwater salinity
Hypothesis 4: Species presence is related to soil water cation concentration
Hypothesis 5: Species presence is related to orthophosphate content
Hypothesis 6: Species abundance is related to position in wetland basin.

The objective was to determine to what extent the selected environmental variable, under conditions of randomly drawn measures, contributed to the differences in species composition from site to site. Single factor analysis of variance of the data (Crow *et al.* 1961), obtained from measures of several environmental attributes, was undertaken with respect to species. Analysis of the resulting ordination of attributes and species composition described above, suggested that depth to water, soil moisture content, groundwater salinity, and calcium and potassium concentrations in soil water would be appropriate factors. To calculate analysis of variance for sites where there were unequal sample sizes, the formula used for calculation was modified to the following (Crow *et al.* 1961):

Sum of the squares among treatments:

$$\sum_i \frac{\Sigma_t (x_{it})^2}{m_i} - \frac{(\Sigma_{i,t}\, x_{it})^2}{n}$$

where n is the total number of observations, and
m is the number of replicates for each subject.

In undertaking ANOVA, it is important to clearly identify the factors of randomness and independence which are contained within the assumptions underpinning the use of analysis of variance. It was recognised that the mean value of depth to groundwater represented an average of measures over time, and that these measures are not independent, each consecutive level being a factor of the previous level. The monthly annual water level measurements would generally prescribe a sinusoidal curve as a result of their seasonal nature. It is argued that the vegetation response is one which relates to the overall seasonal cycle rather than the diurnal or monthly incremental adjustments that the vegetation has adapted to this annual pattern of seasonality, and its dynamics are in response to fluctuations in the annual cycle. From this perspective, the mean value of the depth to water annual cycle, determined from monthly measures, is considered to be an independent measure for each site.

It is important to also understand the inter-relationship between environmental factors. In the Becher wetlands, water levels, in combination with the properties of their aquifer, determine to varying extent, most other environmental attributes. For example, water level position affects soil moisture content in the rhizosphere, which, in turn, changes soil water salinity, ionic concentrations, and nutrient concentrations, which affect vegetation response. In a second example, the chemistry of the groundwater is a result of the chemistry of the vertical and lateral recharge which is determined to a large extent by the organic matter content in the sediments, which, in turn, depends on the productivity and decay of plants and the abundance of oxygen, and both of these variables are functions of the groundwater level. Patently, the environmental factors are not independent and this condition of the analysis is not satisfied. In the light of this, results and the following discussion are regarded as guidelines for experimental design.

The first null hypothesis stated that there was no significant difference in mean depth to groundwater (as measured on a monthly basis over 12 months) between sites with different vegetation cover. Results are given in Tables 10.6 and 10.7.

Table 10.6 *Analysis of variance of depth to groundwater between August 1991 and July 1992*

Source of variation	*Sum of squares*	*df*	*Mean square*	*F*	*P-value*	*F-crit*
Between groups	54123.2	10	5412.3	11.542	0.05	2.177
Within group	13598.8	29	468.9			
Total variance	67722.0	39				

For both the wet and dry year, the null hypothesis was rejected, the differences between the mean depths to groundwater under each vegetation assemblage being significant at the 5% level. It was interesting to note that the variability between the means increased in the wetter year 1991-1992. This is explained by the fact that wetland sediments and stratigraphy affect water levels in the wetter years, whereas the calcareous sands of the parent materials affect water levels in the drier years.

Table 10.7 *Analysis of variance of depth to groundwater during August 1995 and July 1996*

Source of variation	*Sum of squares*	*df*	*Mean square*	*F*	*P-value*	*F-crit*
Between groups	46860.7	9	5206.7	8.875	0.05	2.25
Within group	15842.4	27	586.8			
Total variance	62703.0	36				

Wetland sediments cause different hydrological processes to occur and in wetter years these occur more frequently *e.g.*, perching of both surface water and infiltrating water, and retention of a greater volume of interstitial water in the muds. These phenomena create different water level patterns to those created by uninhibited flow, *e.g.*, in calcareous sands. Therefore, the different depth of wetland fill in various positions in the basin will affect water level, which in turn, may affect vegetation composition and abundance.

As the F test showed the difference in mean water levels between the vegetation assemblages to be significant at $\alpha = 0.05$, a comparison of pairs of means was undertaken using Multiple-*t* confidence intervals (Bhattacharyya and Johnson 1977). Analysis of the contribution to the overall variability of water levels made by each vegetation type, where either a spatial or temporal field relationship was observed, was undertaken (Table 10.8). The formula used to establish Multiple-*t* confidence intervals for *m* number of pairwise differences ($\mu_1 - \mu_2$) at $100(1 - \alpha)\%$ is given by

$$(\text{mean j'th treatment - mean j'th treatment}) \pm t_{\alpha/2m}\, s\sqrt{1/n_j + 1/n_{j'}}$$

where s = MSE, m = number of confidence statements and $t_{\alpha/2\mu}$ = the upper $\alpha/2$m point of t with d.f. = n-k. Using this procedure, the probability of all the m statements being correct is at least $(1 - \alpha)$ (Bhattacharyya and Johnson 1977).

These confidence intervals indicate that water levels under the species of *Melaleuca* do not differ appreciably, the difference between *M. teretifolia* and *M. viminea* being the greatest. Similar confidence limits occur between *C. asiatica* and the mixed assemblage of *B. juncea* and *C. asiatica*. This means that in terms of water level requirements, these assemblages are interchangeable. The confidence interval between *C. asiatica* and *B. juncea* is the smallest of the pairs selected. Field observations testify to these species often occurring together. The species have been observed to inter-relate in two ways: 1) in a dynamic co-existence with moderate expansion and contraction of clumps within the assemblage and 2) in a dynamic relationship with expansion of one species at the expense of the other. The confidence intervals for water levels under species *B. articulata* and *S. validus* are relatively wide, showing

Table 10.8 *Confidence intervals for pairs of vegetation types based on mean water levels*

Vegetation Type	*Confidence Intervals Mean water levels 1991-92*		*Confidence Intervals Mean water levels 1995-96*	
M. rhaphiophylla-M. teretifolia	7.5 ± 42.7	-35, 50	4 ± 47.7	-44, 52
M. viminea - M. rhaphiophylla	9 ± 40.8	-32, 50	6 ± 45.6	-40, 52
M. viminea - M. teretifolia	16.5 ± 37.5	-21, 54	2 ± 41.9	-40, 44
B. juncea - C. asiatica	40 ± 35.3	5, 75	39 ± 39.49	0, 78
B. juncea - B. juncea/C. asiatica mix	12 ± 46.7	-35, 59		*
B. juncea/C. asiatica mix - C. asiatica	28 ± 46.7	-19, 75		*
B. articulata - S. validus	9 ± 55.8	-47, 65	6 ± 62.4	-56, 68
M. rhaphiophylla - C. asiatica	16 ± 40.8	-25, 57	23 ± 45.6	-23, 69

*Sites destroyed.

that the mean water level differences between these groups may preclude them from co-existing in the long term, other environmental factors being equal. In the drier year, the patterns for all vegetation assemblages were similar, with the exception of the pair *C. asiatica* and *M. rhaphiophylla* which showed an increase in the confidence interval. This is interpreted to mean that the different tolerance of each species to water requirements made them incompatible during the drier period.

The second null hypothesis stated that there was no significant difference in mean annual soil moisture (as measured on a quarterly basis over 12 months) between sites with different vegetation cover.

Table 10.9 *Analysis of variance of soil moisture (wt water/50 g sediment) sampled quarterly during 1991-1992*

Source of variation	*Sum of squares*	*df*	*Mean square*	*F*	*P-value*	*F-crit*
Between groups	36674.1	9	4074.9	4.67	0.05	2.28
Within group	21802.9	25	872.1			
Total variance	58477.0	34				

This null hypothesis was rejected, finding the differences between the mean soil moisture under each vegetation assemblage to be significant at the 5% level (Table 10.9). The variability in the annual range in soil moisture under different vegetation types was also analysed. The differences were found to be within the variation about the population mean at the 5% level, and the null hypothesis was accepted.

Table 10.10 *Confidence intervals for pairs of vegetation types based on mean soil moisture*

Vegetation Type	*Confidence Intervals for mean soil moisture levels*	
M. rhaphiophylla-M. teretifolia	12 ± 58.1	-46, 70
M. viminea-M. rhaphiophylla	4 ± 55.6.1	-52, 60
M. viminea-M. teretifolia	16 ± 51.1	-35, 67
B. juncea-C. asiatica	68 ± 48.1	20, 116
B. juncea-B. juncea/C. asiatica mix	9.5 ± 63.7	-54, 73
B. juncea/C. asiatica mix-C. asiatica	58.5 ± 55.6	3, 114
J. kraussii-C. asiatica	27 ± 48.1	-21, 75
J. kraussii-B. juncea	41 ± 48.1	-7, 89
L. gladiatum-C. asiatica	24 ± 63.7	-40, 88
L. gladiatum-B. articulata	0.5 ± 76.1	-76, 77
B. articulata-S. validus	12 ± 76.1	-64, 88
M. rhaphiophylla-C. asiatica	46 ± 63.7	-18, 110

These confidence intervals indicate that soil moisture levels under the species of *Melaleuca* do not differ appreciably, the mean of *M. teretifolia* exhibiting the most difference (Table 10.10). The confidence limits for the pair *C. asiatica* and *M. rhaphiophylla* are wide and suggest that these species, although occurring together, have very different requirements of soil moisture. The mean soil moisture levels under the pair *C. asiatica* and *B. juncea* are very different and could be the underlying cause for the dynamic changes in cover abundance between the two species and the mixed assemblage. The means for soil moistures under sedge species *B. articulata* and *S. validus* and for *B. articulata* and *L. gladiatum* are similar but the confidence intervals are relatively wide. This is interpreted to mean that although all the sedge species require similar soil moistures, the tolerance level of each species is different, *L. gladiatum* having the widest tolerance and *S. validus* the smallest. The means for soil moisture levels under assemblage pairs *C. asiatica* and *J. kraussii,* and *C. asiatica* and *L. gladiatum,* are similar, and *C. asiatica* does occur within each of these assemblages, but in minor numbers.

The third null hypothesis stated that there was no significant difference in mean annual groundwater salinity (as measured on a monthly basis over 12 months), between sites with different vegetation cover.

Table 10.11 *Analysis of variance of groundwater salinity during August 1992-July 1993*

Source of variation	*Sum of squares*	*df*	*Mean square*	*F*	*P-value*	*F-crit*
Between groups	3733093	10	373309	2.11	0.05	2.18
Within group	5122844	29	176650			
Total variance	8855938	39				

At the 5% level, the null hypothesis is accepted, but at the 10% level, the null hypothesis is rejected (Table 10.11). In this case, the reality is probably that at some sites the groundwater salinity is a factor in the vegetation response and at others it is not necessarily the most critical factor. As the F test showed the difference in mean water levels between the vegetation assemblages to be significant only at $\alpha = 0.1$, a comparison of pairs of means could not be undertaken using Multiple-*t* confidence intervals (Bhattacharyya and Johnson 1977).

Analysis of variance of mean orthophosphate concentration in groundwater (sampled quarterly) during August 1992-July 1994 and mean calcium, potassium, and magnesium content in soil waters 1992-93 showed that no differences existed among the population means. The null hypothesis in each case was accepted, the probability of F being greater than, or equal to, the tabled value (F critical) if the null hypothesis was true, being less than 0.05. It cannot be concluded from these results that neither of these factors affect vegetation response. The result may be a consequence of either the sampling time interval (perhaps too large), or the particular parameter that was chosen for analysis, *i.e.*, mean value rather than range or variability.

10.3.4 Monthly observations of hydrology, hydrochemistry and vegetation cover

If vegetation were to be viewed as a continuum throughout the Becher Suite wetlands, even though not linked spatially (Avis and Lubke 1996), then the results of the multivariate analyses identify general patterns in plant distributions, however, the finding that the attributes which corresponded to the greatest spatial variability were intra-basinal meant that more precise investigations could be undertaken. Detailed monitoring and sampling were undertaken to try to determine the quantitative limits of tolerance for several plant species in the habitat setting of the Becher cuspate foreland. The species selected were the sedge *B. articulata*, the shrubs *M. teretifolia* and *M. viminea,* and the rush *J. kraussii.* For each species, five quadrats were selected of 1 m^2 in which there were varying densities of cover from 0% to 100%. In each case, the quadrat with 0% cover of the selected species had a ground cover of the herb *C. asiatica.* From August 1997 to September 1998, the following attributes were measured monthly: soil water content, depth to water table, groundwater pH, and groundwater TDS (Figs. 10-28 to 10-32). The following hydrochemical attributes were measured once: orthophosphate, nitrate, and ammonium concentration in groundwater (Figs. 10-28 to 10-32). Analysis of the variance in monthly measurements for each quadrat was compared to the variance between quadrats to ascertain which factors might be significant in determining plant density.

Results

Soil water content and depth to water table were determined to be critical factors affecting vegetation density for all selected trial species (Tables 10.12, 10.13). Groundwater pH was related to density of *M. teretifolia*, and *J. kraussii*, while groundwater TDS was significant in relation to density of *M. viminea* (Tables 10.14, 10.15).

Table 10.12 *Analysis of variance of soil water content over fourteen months*

Vegetation assemblage	*df*	*F*	*F-crit*	*P*
B. articulata	69	45.79	2.51	3.08E-18
M. teretifolia	69	22.35	2.51	1.22E-11
J. kraussii	69	7.27	2.51	6.69E-05
M. viminea	67	19.46	2.52	1.81E-10

Table 10.13 *Analysis of variance of depth to water table over fourteen months*

Vegetation assemblage	*df*	*F*	*F-crit*	*P*
B. articulata	74	10.25	2.50	1.37E-06
M. teretifolia	74	7.64	2.50	3.61E-05
J. kraussii	74	4.05	2.50	0.005
M. viminea	72	18.46	2.51	2.62E-10

Table 10.14 *Analysis of variance of groundwater pH over fourteen months*

Vegetation assemblage	*df*	*F*	*F-crit*	*P*
B. articulata	69	2.06	2.51	0.096
M. teretifolia	69	5.105	2.51	0.001
J. kraussii	69	3.04	2.51	0.023
M. viminea	63	2.44	2.52	0.056

Table 10.15 *Analysis of variance of groundwater TDS over fourteen months*

Vegetation assemblage	*df*	*F*	*F-crit*	*P*
B. articulata	62	1.69	2.53	0.16
M. teretifolia	64	1.98	2.53	0.108
J. kraussii	64	1.89	2.53	0.12
M. viminea	62	2.66	2.53	0.042

For species which exhibited significant variation between quadrats, the range of tolerance to environmental factors is discussed, based on the 14 monthly measures. For *B. articulata*, 100% cover required both consistent and high soil water content (50% by weight) (Fig. 10-28). Soil water content below 30% (by weight) caused changes to vegetation density, and where this occurred frequently enough, changes to species composition.

For *M. teretifolia*, 100% cover required both consistent and high soil water content (52% by weight). Soil water content fluctuations below 50% (by weight), resulted in lower vegetation density (Fig. 10-29), and soil water content consistently below this level resulted in altered species composition.

For *J. kraussii*, 100% cover occurred in sediments with slightly lower soil water content (45-49% by weight), and seasonal fluctuation. Soil water content was not clearly differentiated between sites and did not appear to be a major factor underpinning density (Fig. 10-30).

For *M. viminea*, 100% cover occurred in sediments with slightly lower soil water content (24-33% by weight), and seasonal fluctuation. Soil water content fluctuations below this amount resulted in lower vegetation density (Fig. 10-31), and soil water content above this amount resulted in a change in species composition.

Depth to water table varies considerably between seasons (*circa* 120 cm) and this intra-site variability overshadows the variance between sites. Depth to water table is an important factor also in determining vegetation density. For 100% cover of *B. articulata* sedge, water levels were within 50 cm of the surface for most of the year. Persistent levels between 50 and 100 cm resulted in loss of plants and decreased vegetation density (Fig. 10-28). Persistent levels below 100 cm resulted in changes to species composition. For 100% cover of *M. teretifolia* shrub, water levels were between 15-80 cm from the surface for most of the year. Persistent levels below a depth of 80 cm resulted in loss of stems, branches and leaves and decreased vegetation density (Fig. 10-29). Persistent levels less than 15 cm result in direct and successful competition from understorey species. For 100% cover of *J. kraussii* rush, water levels were between 55-105 cm of the surface for most of the year. Persistent levels above or below this depth had little impact on vegetation density (Fig. 10-30). For 100% cover of *M. viminea* shrub, water levels were between 60-125 cm from the surface for most of the year. Persistent levels below a depth of 125 cm resulted in reduced stature and luxuriance of plants (density of foliage) and decreased vegetation density (Fig. 10-31). Persistent levels above a depth of 60 cm resulted in direct and successful competition from understorey species.

There were no significant differences between the mean groundwater pH values for any series of quadrats under any selected vegetation assemblage.

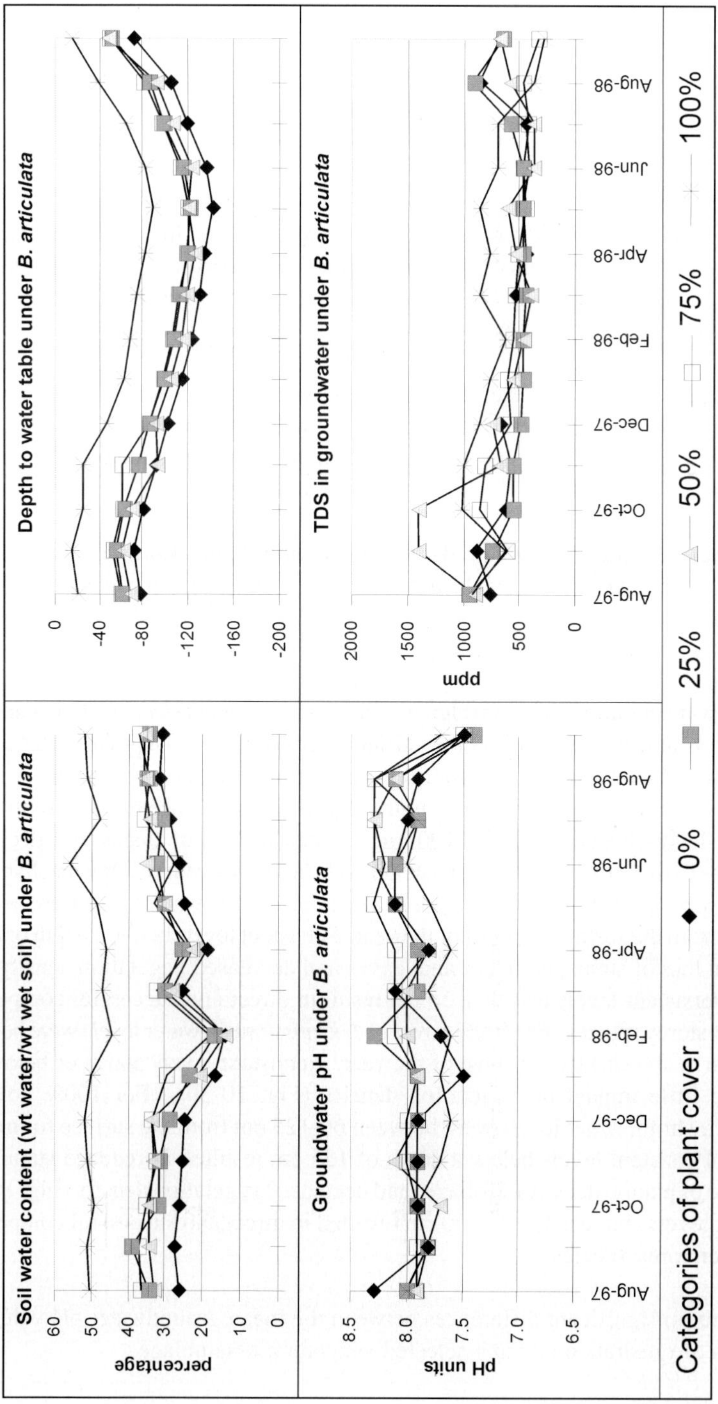

Figure 10-28. *Results of field experiments under Baumea articulata.*

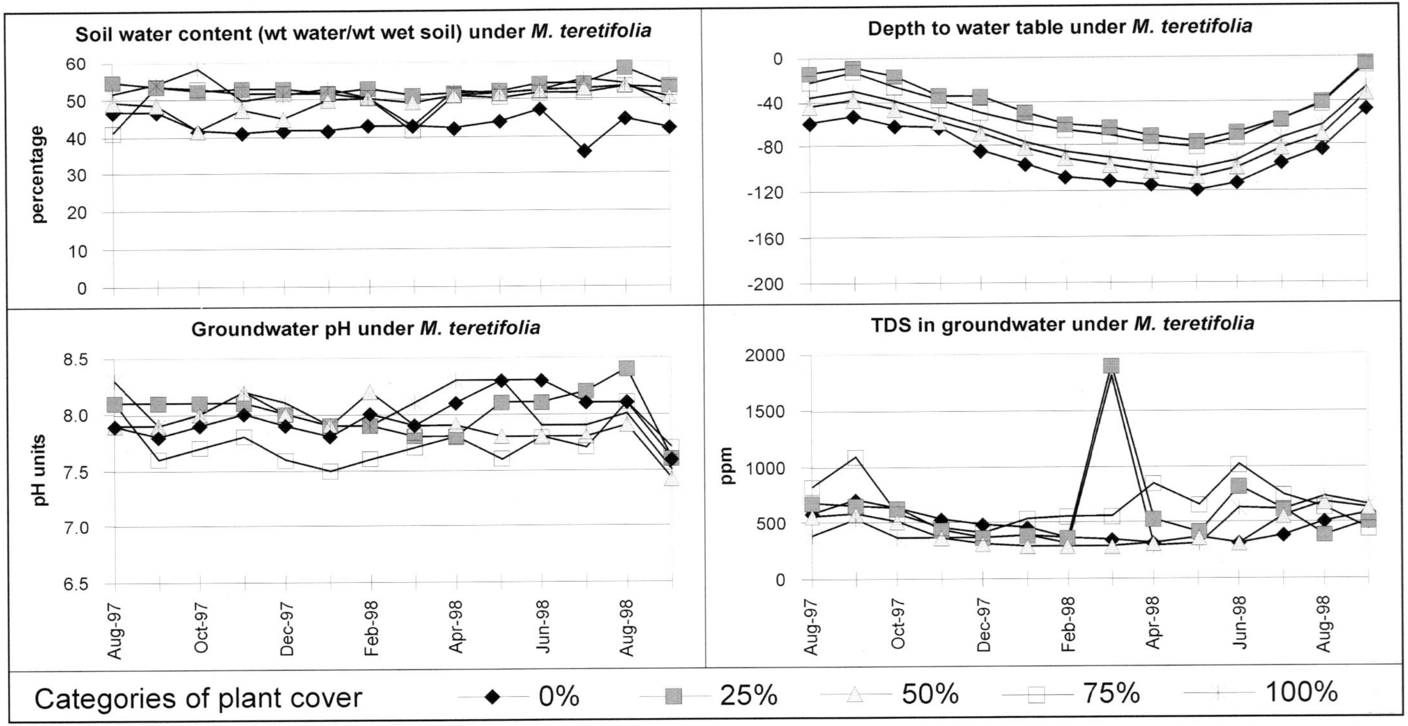

Figure 10-29. Results of field experiments under Melaleuca teretifolia.

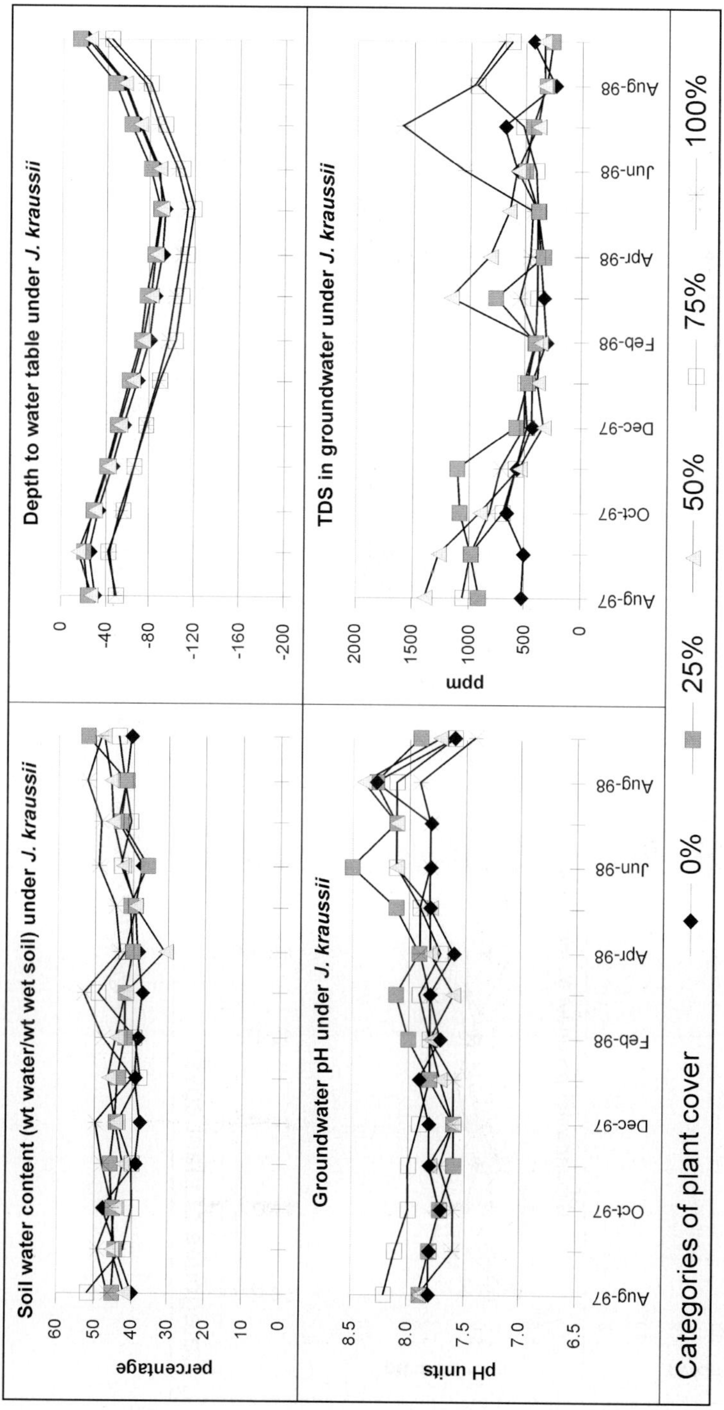

Figure 10-30. *Results of field experiments under Juncus kraussii.*

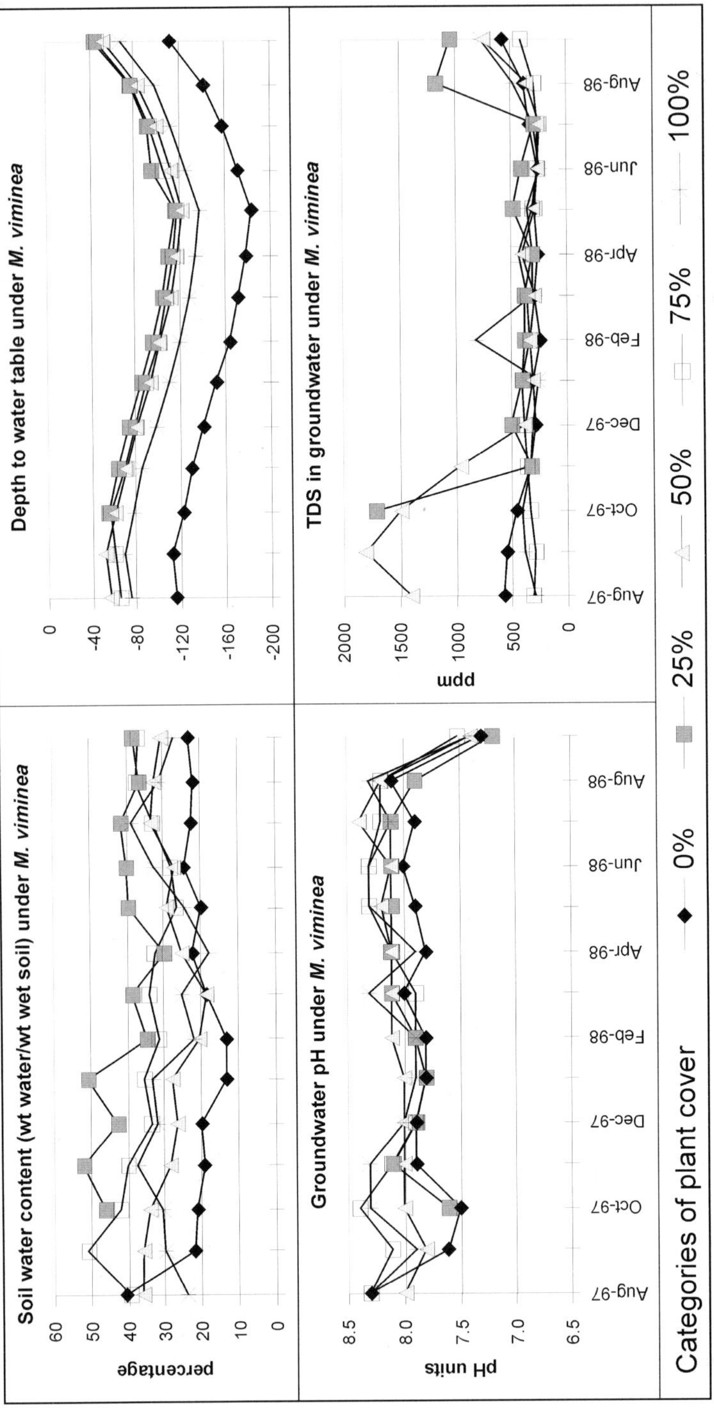

***Figure 10-31**. Results of field experiments under Melaleuca viminea.*

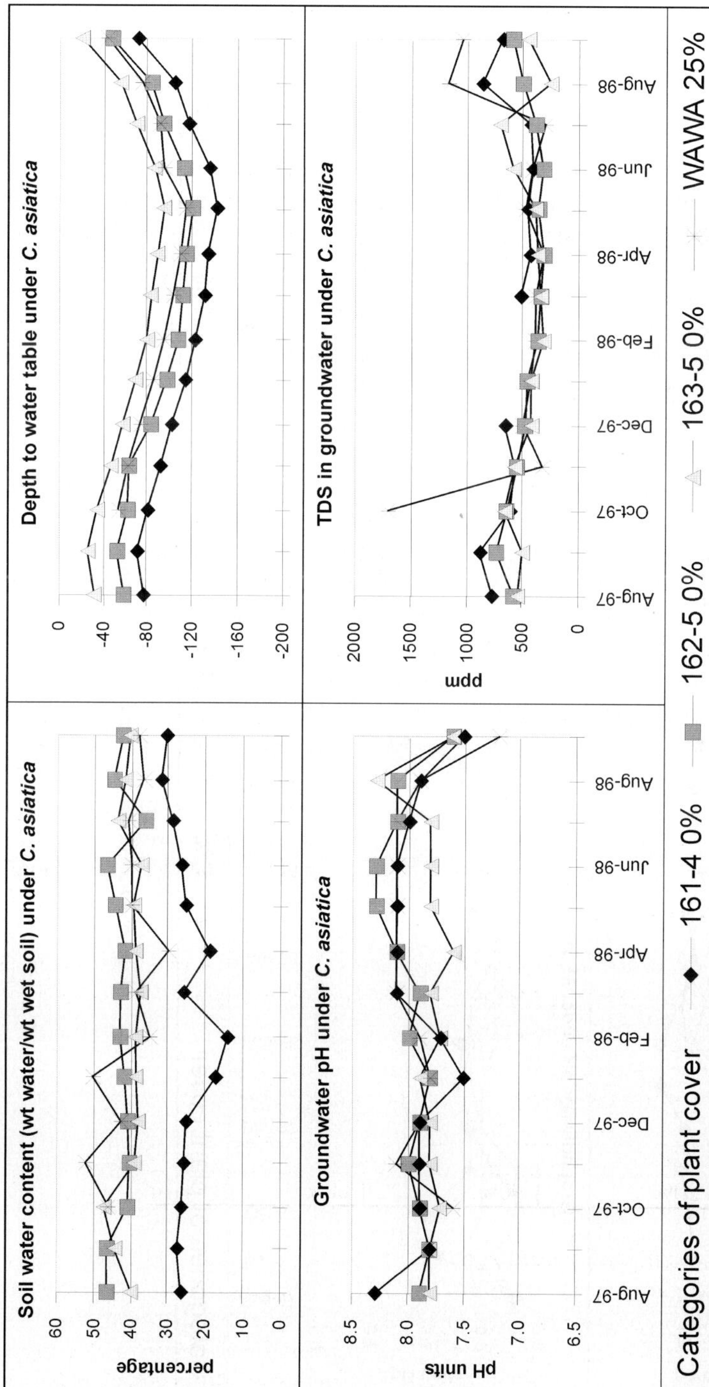

Figure 10-32. *Results of field experiments under Centella asiatica.*

The groundwater for all species was freshwater (<1000 ppm salt). All species showed a tolerance to short term (1 month) salinity increases in groundwater (*e.g.*, elevated from freshwater to subhaline), expressed as lack of change to vegetation density. Variation in groundwater TDS between sites showed more affinity with geographic location within a wetland basin than with vegetation density. For example, groundwater TDS consistently was highest under the 100% quadrat under *B. articulata* sedge, highest under the 75% quadrat under *M. teretifolia* shrub, highest under the 50% and 100% quadrats under *J. kraussii* rush, and highest under the 25% quadrat under *M. viminea* shrub (Figs. 10-28 to 10-31). In each of these examples, the site is in the centre of the wetland basin, (or nearer the centre than other sites in the given vegetation series); this position is inundated annually and subject to direct evaporation. The 4 remaining sites in all of the series exhibited groundwater TDS values very similar to each other (Figs. 10-28 to 10-31).

It should be noted that only a single episode of nutrient sampling in groundwater was undertaken, and therefore all comments and interpretations pertaining to vegetation response to nutrients should be treated as tentative. Orthophosphate and nitrate concentration in groundwater under *B. articulata* sedge were highest under the 100% quadrat (Fig. 10-33). No pattern was recognised for ammonium. Under *M. teretifolia* shrub, no patterns relating nutrient concentrations in groundwater to vegetation density were observed (Fig. 10-33). Under *J. kraussii* rush, orthophosphate and ammonium concentrations could be separated into two groups, one of low concentration under all quadrats where *J. kraussii* rush was present, and the other of higher concentration where *J. kraussii* rush was absent, but no interpretable pattern was recognised for nitrate (Fig. 10-33). Under *M. viminea* shrub, orthophosphate and nitrate in groundwater decreased in concentration as vegetation density increased, inferring a positive relationship between this species and these environmental factors (Fig. 10-33). This is not to suggest, however, that these are causal factors of the spatial distribution of this species.

Patterns were also observed for the quadrats that registered 0% of the target species under study, although these were located in different wetland basins. These quadrats exhibited varying vegetation cover (Table 10.16).

Table 10.16 *Percent of quadrat covered by C. asiatica at four wetland sites*

Site	*161*	*162*	*163*	*WAWA*
% cover	20%	90%	60%	70%

The quadrat with the highest percentage of cover occurred in sediments with only a slightly fluctuating soil water content of 37-46% (by weight). Soil water content fluctuations below this amount resulted in a change in species composition (Fig. 10-32). The quadrat with the highest cover of *C. asiatica* herb required water levels to be between 60-105 cm from the surface for most of the year. However, *C. asiatica*

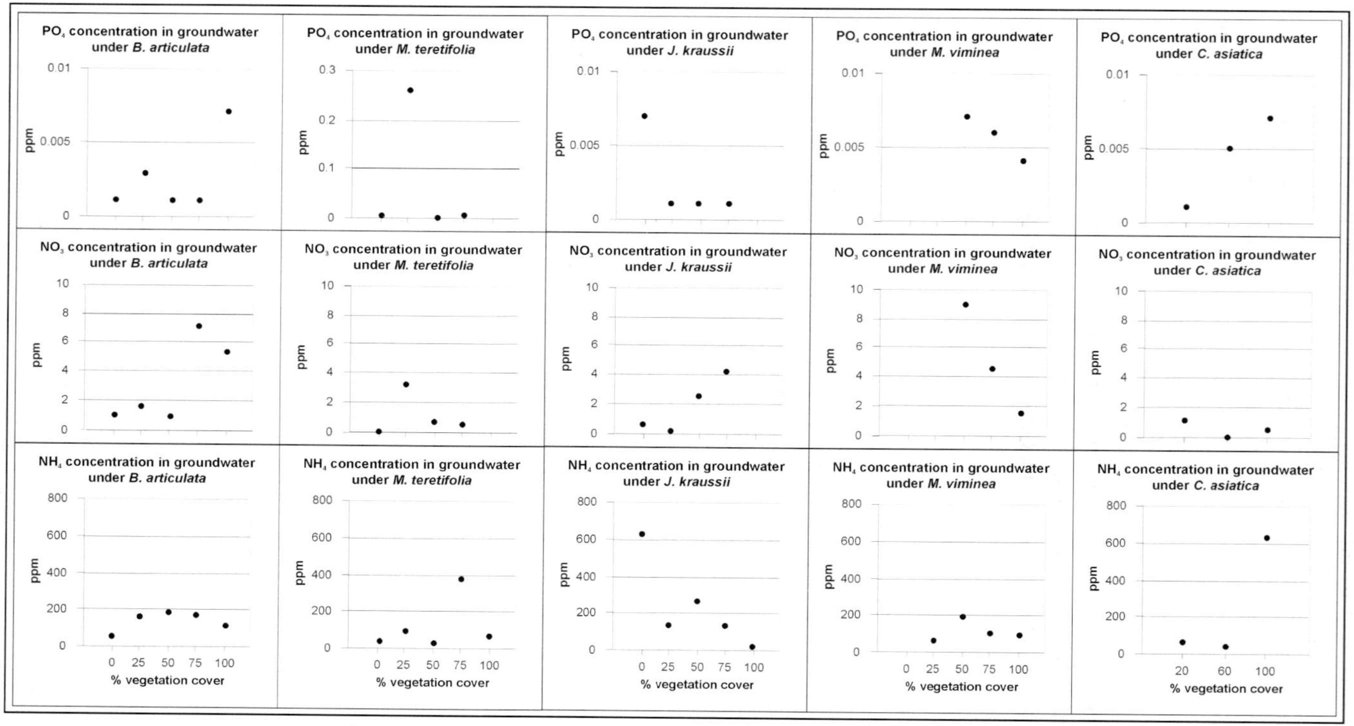

Figure 10-33. *Concentrations of orthophosphate, nitrate and ammonia in groundwater under vegetation quadrats.*

is adaptable to changing hydrological conditions and rapidly changes density, height of plant, and size of leaf to accommodate such changes. Persistent water levels below a depth of 110 cm resulted in reduced stature and luxuriance of this species, an increase in density of small plants, or a decrease in density of large plants. Persistent levels less than 60 cm resulted in increased stature and luxuriance (increased leaf area) of plants or increased density for the species. *Centella asiatica* was shown to be intolerant to groundwater salinity increases, except over short periods. Optimum groundwater salinity ranges between 400-500 ppm (Fig. 10-32).

Under *C. asiatica* herbs, only nitrate concentration in groundwater showed any pattern relating to vegetation cover, *viz.* decreasing concentration corresponding to increasing cover (Fig. 10-33). Again, a positive relationship between this species and these environmental factors is inferred, but it is not suggested that these are causal factors of the spatial distribution of this species.

10.3.5 Importance of environmental attributes in determining species distribution

Results from the ordination, the analysis of variance and the detailed monthly observations are consistent. From each subsequent method of analysis the information on the influence of the selected environmental attributes on particular species became more apparent. From the ordination of environmental attributes, the configuration of sites was largely determined by the attributes pertaining to groundwater salinity, with a distinction between vegetation sites which were tolerant, intolerant, or indifferent to excessive and/or variable groundwater salinities. Other factors, which were subdominant in the ordination of environmental attributes, were mean annual depth to water table, mean period of inundation, and calcium and potassium concentrations in the interstitial waters.

Ordination of subsets of environmental factors, analysis of variance, and the field trials showed several species were strongly affected by one or several of the selected environmental attributes.

- The herb *C. asiatica* was intolerant of changes in groundwater salinity, and was responsive to changes in depth to water and soil moisture content, preferring the zone of capillary rise to waterlogged conditions. Waterlogging and shallow inundation were both tolerated for periods up to three months. These conditions are typical of shallow central basins and flats between the centre and slope of deeper basins.

- The sedge *B. articulata* was more tolerant of changes in groundwater salinity than *C. asiatica*, but still preferred constant freshwater, whether surface or groundwater. It was responsive to changes in depth to water and soil moisture content, preferring conditions which were waterlogged. Shallow inundation was tolerated for periods up to seven months. These conditions are typical of deeper central basins.

- The sedge *B. juncea,* exhibited similar tolerance to changes in groundwater salinity fluctuations to *B. articulata*, but was more tolerant to changes in depth to water and soil moisture content, preferring the zone of capillary rise to waterlogged conditions. Shallow inundation was not tolerated. These conditions were often found at wetland margins.

- The sedge *S. validus* was tolerant of variable groundwater salinity, but responsive to changes in depth to water and soil moisture content, preferring conditions which were constantly waterlogged. Shallow inundation was tolerated for periods up to four months. These conditions are typical of central basins.

- The sedge *L. gladiatum* was intolerant of variable groundwater salinity, but responsive to changes in depth to water and soil moisture content, preferring conditions which were constantly waterlogged. Shallow inundation was not tolerated. These conditions are typical of central basins where throughflow predominates.

- The rush *J. kraussii* exhibited indifference to slight groundwater salinity fluctuations, and changes in depth to water, but was responsive to varying soil moisture content, preferring the zone of capillary rise to waterlogged conditions. Brief shallow inundation was tolerated. These conditions were often found in wetland margins where the water table was maintained by upwelling or throughflow.

- The shrub *M. teretifolia* exhibited indifference to slight groundwater salinity fluctuations, and marked changes in depth to water, but was responsive to varying soil moisture content, preferring the zone of capillary rise to waterlogged conditions. Shallow inundation was tolerated for periods up to four months. These conditions were often found in deeper wetland basins.

- The shrub *M. viminea* was tolerant of excessive groundwater salinity fluctuations, greater depth to water, and lower soil moisture content. These conditions were often found on the slopes of wetland basins. This species appeared to be dependent on nutrient availability.

- The tree *M. rhaphiophylla* was tolerant of large groundwater salinity fluctuations, greater depth to water, as well as marked changes in depth to water but relatively constant soil moisture content. Shallow inundation was tolerated. These conditions were often found on the slopes of wetland basins.

Although none of the assemblages showed strong association with cation content in the soil water, many exhibited loose positive and negative associations with two or more cations, particularly calcium and potassium, and particularly with their variability over time. The results of the field trial show that the density of some species, *i.e.*, *C. asiatica* and *M. viminea* was related to nutrient content in the groundwater.

This is probably true of all the dicotyledons present in the wetlands, but given the complexity of the hydrology in these seasonal wetlands, the exact relationship would have to be clarified in a further study.

The general conclusions drawn about vegetation distribution in this study are considered to be a preliminary step in the investigation. As in the majority of wetland studies, an association was found to exist between water availability and vegetation distribution. The definition of this association proved to be as elusive as it has in many other wetland studies. Seasonal swings from water surplus to deficit appeared to be a factor, as did soil moisture content. Some species were shown to be more tolerant of salinity (*M. cuticularis, M. viminea, S. virginicus*) than others (*C. asiatica, B. articulata*). Some species were shown to be more tolerant of waterlogging (*S. validus, B. articulata, C. asiatica*) than others (*B. juncea*). Some vegetation types were shown to be more reactive to changes in available nutrients (sedges) than others (trees and shrubs). However, it was not within the scope of this study to show how the effects of fluctuating water level regimes were translated into physical or chemical benefits, or hardships for any particular species.

10.4 Short term changes in vegetation associations

During the time of field sampling, as conditions became drier, there were several notable changes in species composition and cover. The first change was a decrease in the total cover at many sedgeland and herbland sites, *e.g.*, 161-3, WAWA 3 (*B. articulata*), swiii-5, 9-6, (*C. asiatica*), WAWA 4 (*M. viminea*). Invasive species were quick to take advantage of the reduced native plant cover. The second change in response to conditions becoming drier were changes in native species composition. Composition favoured more drought tolerant species, and hence overall composition of the assemblages changed, *e.g.*, 72-3, 9-11 (*B. juncea* replaced *C. asiatica*), swiii-6, WAWA 6 (*L. gladiatum* replaced *C. asiatica*), 9-3 (*S. virginicus* replaced *B. juncea*), WAWA 3 (*S. validus* replaced *B. articulata*). The third change was an increase in total cover. Some species became more luxuriant at a particular site, and an increase in cover was noted, *e.g.*, 162-5 (*M. teretifolia*) and incursion into the wetland by *M. viminea* downslope of WAWA 4. There also was a general contraction of wetland species in all wetlands between 1991 and 1996, and incursion by marginal species such as the sedge *Isolepis nodosa*, the shrub *Pelargonium capitatum*, shrubs *Acacia saligna, A. cyclops* and *A. pulchella*.

The results of total cover, and percent cover for selected species in the various wetlands for the years 1991, 1995, and 2000 are presented for comparison in Table 10.17.

Table 10.17 *Vegetation changes 1991-2000 (Domin cover scale)*

Site	*Total % cover 1991*	*Total % cover 1995*	*Total % cover 2000*	*1991 % cover*	*1995 % cover*	*2000 % cover*	*1991 % cover*	*1995 % cover*	*2000 % cover*	*1991 % cover*	*1995 % cover*	*2000 % cover*
135-2	10 (100%)	10 (100%)	10 (100%)	*M.rhap* 5	*M.rhap* 5	*M.rhap* 0	*C. asiat* 3	*C. asiat* 0	*C. asiat* 0			
136-3	9 (80%)	7 (50%)	9 (90%)	*M.rhap* 8	*M.rhap* 7	*M.rhap* 7	*C. asiat* 6	*C. asiat* 0	*C. asiat* 9			
45-3	8 (70%)	8 (70%)	-	*M.rhap* 8	*M.rhap* 8	-	*C. asiat* 3	*C. asiat* 0	-			
45-5	10 (100%)	9 (80%)	-	*M.rhap* 7	*M.rhap* 7	-	*C. asiat* 7	*C. asiat* 0	-			
35-5	8 (60%)	8 (60%)	6 (30%)	*M.rhap* 8	*M.rhap* 8	*M.rhap* 5	*C. asiat* 0	*C. asiat* 0	*C. asiat* 0			*I. nodo* 4
162-3	8 (60%)	6 (30%)	6 (32%)	*M. teret* 5	*M. teret* 5	*M. teret* 6	*B. junc* 4	*B. junc* 4	*B. junc* 0	*C. asiat* 7	*C. asiat* 4	*C. asiat* 5
162-5	9 (80%)	9 (80%)	10 (100%)	*M. teret* 6	*M. teret* 5	*M. teret* 5	*B. junc* 0	*B. junc* 0	*B. junc* 0	*C. asiat* 8	*C. asiat* 0	*C. asiat* 0
142-3	8 (53%)	8 (53%)	6 (30%)	*M. teret* 6	*M. teret* 6	*M. teret* 5	*B. junc* 0	*B. junc* 0	*B. junc* 0	*C. asiat* 5	*C. asiat* 5	*C. asiat* 0
142-7	8 (60%)	6 (30%)	8 (55%)	*M. teret* 4	*M. teret* 4	*M. teret* 4	*B. junc* 6	*B. junc* 5	*B. junc* 7	*C. asiat* 0	*C. asiat* 0	*C. asiat* 2
142-10	8 (70%)	8 (70%)	9 (90%)	*M. teret* 8	*M. teret* 8	*M. teret* 9	*B. junc* 0	*B. junc* 0	*B. junc* 0	*C. asiat* 0	*C. asiat* 0	*C. asiat* 0

Table 10.17 (cont.)

Table 10.17 (cont.)

Site	*Total % cover 1991*	*Total % cover 1995*	*Total % cover 2000*	*1991 % cover*	*1995 % cover*	*2000 % cover*	*1991 % cover*	*1995 % cover*	*2000 % cover*	*1991 % cover*	*1995 % cover*	*2000 % cover*
WAWA 4	8 (66%)	9 (88%)	7 (40%)	*M. vim* 8	*M. vim* 5	*M. vim* 0	*B. junc* 5	*B. junc* 5	*B. junc* 4	*C. asiat* 0	*C. asiat* 0	*C. asiat* 0
142-5	8 (64%)	5 (25%)	7 (50%)	*M. vim* 5	*M. vim* 5	*M. vim* 7	*B. junc* 4	*B. junc* 4	*B. junc* 4	*C. asiat* 0	*C. asiat* 0	*C. asiat* 3
9-5	9 (76%)	8 (51%)	8 (65%)	*M. vim* 5	*M. vim* 5	*M. vim* 6	*B. junc* 5	*B. junc* 5	*B. junc* 4	*C. asiat* 0	*C. asiat* 5	*C. asiat* 4
9-7	8 (71%)	8 (56%)	7 (48%)	*M. vim* 4	*M. vim* 4	*M. vim* 5	*B. junc* 4	*B. junc* 4	*B. junc* 5	*C. asiat* 5	*C. asiat* 5	*C. asiat* 4
9-13	8 (56%)	51		*M. vim* 5	*M. vim* 5	*M. vim*	*B. junc* 6	*B. junc* 6	*B. junc* 0	*C. asiat* 4	*C. asiat* 4	*C. asiat* 0
136-2	7 (50%)	7 (50%)	4 (10%)	*X. preis* 7	*X. preis* 7	*X. preis* 0			*I. nodo* 4			
142-8	7 (40%)	7 (50%)	8 (75%)	*X. preis* 7	*X. preis* 7	*X. preis* 8						

Table 10.17 (cont.)

Table 10.17 (cont.)

Site	*Total % cover 1991*	*Total % cover 1995*	*Total % cover 2000*	*1991 % cover*	*1995 % cover*	*2000 % cover*	*1991 % cover*	*1995 % cover*	*2000 % cover*	*1991 % cover*	*1995 % cover*	*2000 % cover*
swi-2	8 (60%)	7 (35%)	7 (50%)	*L. glad* 5	*L. glad* 5	*L. glad* 5	*J. krau*	*J. krau* 4	*J. krau* 6	*C. asiat* 5	*C. asiat* 1	*C. asiat* 0
swii-3	7 (40%)	7 (40%)	5 (25%)	*L. glad* 7	*L. glad* 7	*L. glad* 5	*J. krau* 0	*J. krau* 0	*J. krau* 0	*C. asiat* 0	*C. asiat* 0	*C. asiat* 3
163-4	10 (100%)	7 (40%)	7 (45%)	*B. junc* 6	*B. junc* 5	*B. junc* 7	*I. nodo*	*I. nodo* 0	*I. nodo*	*S. vir*	*S. vir* 7	*S. vir* 0
142-6	10 (100%)	7 (35%)	7 (48%)	*B. junc* 6	*B. junc* 5	*B. junc* 6	*I. nodo* 0	*I. nodo* 0	*I. nodo* 0	*S. vir* 0	*S. vir* 4	*S. vir* 5
9-3	9 (80%)	6 (30%)	7 (50%)	*B. junc* 6	*B. junc* 4	*B. junc* 7	*I. nodo* 0	*I. nodo* 0	*I. nodo* 0	*S. vir* 0	*S. vir* 5	*S. vir* 4
N1	7 (50%)	6 (28%)	5 (22%)	*B. junc* 6	*B. junc* 6	*B. junc* 5	*I. nodo*	*I. nodo* 0	*I. nodo* 0	*S. vir*	*S. vir* 0	*S. vir* 0
N2	10 (100%)	8 (60%)	6 (30%)	*B. junc* 7	*B. junc* 7	*B. junc* 5	*I. nodo*	*I. nodo* 5	*I. nodo* 5	*S. vir* 0	*S. vir* 0	*S. vir* 0
161-3	10 (90%)	7(45%)	5 (15%)	*B. art* 10	*B. art* 7	*B. art* 5	*T. or* 1	*T. or* 1	*T. or* 0	*S. val* 0	*S. val* 0	*S. val* 0
WAWA 3	10 (100%)	7 (50%)	7 (35%)	*B. art* 9	*B. art* 6	*B. art* 5	*T. or* 4	*T. or* 0	*T. or* 4	*S. val* 3	*S. val* 4	*S. val* 5

Table 10.17 (cont.)

Table 10.17 (cont.)

Site	*Total % cover 1991*	*Total % cover 1995*	*Total % cover 2000*	*1991 % cover*	*1995 % cover*	*2000 % cover*	*1991 % cover*	*1995 % cover*	*2000 % cover*	*1991 % cover*	*1995 % cover*	*2000 % cover*
swiii-3	10 (100%)	10 (100%)	10 (100%)	*S. val* 6	*S. val* 4	*S. val* 2	*C. asiat* 5	*C. asiat* 4	*C. asiat* 2			*Sch* sp. 5
swiii-4	9 (90%)	9 (88%)	10 (100%)	*S. val* 7	*S. val* 6	*S. val* 4	*C. asiat* 5	*C. asiat* 3	*C. asiat* 0			*Sch* sp. 0
163-3	10 (100%)	10 (100%)	10 (100%)	*J. krau* 10	*J. krau* 10	*J. krau* 10	*B. junc* 4	*B. junc* 4	*B. junc* 3	*C. asiat* 2	*C. asiat* 3	*C. asiat* 0
163-6	10 (100%)	10 (100%)	10 (100%)	*J. krau* 10	*J. krau* 10	*J. krau* 10	*B. junc* 0	*B. junc* 0	*B. junc* 0	*C. asiat* 1	*C. asiat* 1	*C. asiat* 0
35-3	8 (55%)	9 (82%)	5 (25%)	*J. krau* 8	*J. krau* 9	*J. krau* 5	*B. junc* 0	*B. junc* 0	*B. junc* 0	*C. asiat* 0	*C. asiat* 2	*C. asiat* 0
9-14	10 (100%)	8 (75%)	7 (50%)	*J. krau* 10	*J. krau* 8	*J. krau* 7	*B. junc* 0	*B. junc* 0	*B. junc* 0	*C. asiat* 0	*C. asiat* 0	*C. asiat* 0
swi-3	10 (100%)	10 (100%)	8 (60%)	*J. krau* 10	*J. krau* 10	*J. krau* 8	*B. junc* 4	*B. junc* 0	*B. junc* 0	*C. asiat* 0	*C. asiat* 0	*C. asiat* 0
163-5	10 (100%)	10 (100%)	10 (100%)	*C. asiat* 10	*C. asiat* 8	*C. asiat* 0	*B. junc* 0	*B. junc* 5	*B. junc* 10			

Table 10.17 (cont.)

Table 10.17 (cont.)

Site	*Total % cover 1991*	*Total % cover 1995*	*Total % cover 2000*	*1991 % cover*	*1995 % cover*	*2000 % cover*	*1991 % cover*	*1995 % cover*	*2000 % cover*	*1991 % cover*	*1995 % cover*	*2000 % cover*
35-4	10 (100%)	10 (100%)	4 (5%)	*C. asiat* 10	*C. asiat* 10	*C. asiat* 0	*B. junc* 0	*B. junc* 0	*B. junc* 0			
9-6	10 (100%)	10 (100%)	3 (3%)	*C. asiat* 10	*C. asiat* 8	*C. asiat* 3	*B. junc* 0	*B. junc* 0	*B. junc* 0			
swiii-5	10 (100%)	10 (100%)	0 (0%)	*C. asiat* 10	*C. asiat* 0	*C. asiat* 0	*B. junc* 0	*B. junc* 0	*B. junc* 0			
72-3	9 (85%)	9 (80%)	9 (90%)	*C. asiat* 8	*C. asiat* 7	*C. asiat* 1	*B. junc* 6	*B. junc* 6	*B. junc* 9			

The changes in overall vegetation cover and the cover for selected species, as related to changes in the environmental factor such as groundwater table level, salinity, and soil water salinity, are described below within a framework of the climate changes during the period.

From 1991-95, the annual rainfall decreased and depth to water table increased. In conjunction with rainfall decrease, groundwater salinity at some sites increased, soil water content in the surface layers decreased, and soil water salinity increased (Table 10.18). Vegetation cover was observed to respond to these changes in one of four ways:

1. by maintaining individual numbers but decreasing foliage cover (*M. rhaphiophylla*);
2. by maintaining individual numbers but decreasing number of branches (*M. teretifolia*);
3. by maintaining individual numbers but reducing plant height and spread (*C. asiatica*, *L. gladiatum* and *M. viminea*); and
4. by reducing the population numbers (*C. asiatica, B. articulata, B. juncea* and *S. validus*).

Melaleuca rhaphiophylla exhibited no net change in cover from 1991-1995, but the shrubs, herbs and sedges exhibited reductions in net cover ranging from 10% to over 50 % of that recorded in 1991 (Table 10.17). The greatest decreases in vegetation cover occurred in the sedges *B. articulata* and *S. validus* through direct plant demise.

From 1995-2000, the annual rainfall generally increased. While 1998 had below average rainfall, this was counteracted by above average rainfall in 1999 and 2000. As a result, during this period, plants had increased access to groundwater as depth to water table decreased. In conjunction with a decrease in depth to water, soil water content in the surface layers increased, and soil water salinity decreased. Species were observed to have different rates of response to these changes, and to respond in a variety of ways:

1. by maintaining individual numbers of established plants while increasing foliage cover and at the same time increasing the number of juvenile plants (*M. rhaphiophylla*);
2. by maintaining individual numbers of established plants while increasing foliage cover (*M. teretifolia*);
3. by further reduction in numbers due to competition from alien species (*C. asiatica, S. validus* and *L. gladiatum*);
4. by further reduction in numbers (*B. articulata*); and
5. by expansion of population (*M. viminea* and *B. juncea*).

Table 10.18 *Corresponding changes in hydrological factors and vegetation cover 1991-2001*

Site	*Change in vegetation cover* 1991-1995	*Change in vegetation cover* 1995-2000	*Change in mean depth to water table (cm)* 1991-1995	*Change in mean depth to water table* 1995-2000	*Change in mean ground water salinity (ppm)* 1991-94	*Change in mean ground water salinity (ppm)* 1997-98	*Change in mean soil water content* 1991-94	*Change in mean soil water content* 1997-98	*Change in mean soil water salinity (ppm)* 1991-94
M. rhaphiophylla									
135-2	12-12% 0%	12-0 -12%	-36	+14	1496; 2513; 1863	-	46; 40; 24	-	13,967; 9,626; 21,473
136-3	55-50% -5%	50-80 +30%	-41	+11	1562-1162	-	42; 32; 25	-	11,982; 10,568; 17,080
35-5	60-60% 0%	60-25 -35%	-30	nd	1723; 907; 4569	-	43; 24; 30	-	16,764; 26,086; 20,584
M. teretifolia									
162-3	20-20% 0%	20-30 +10 %	-34	+24	956-923	665	86; 100; 31	52	1,874; 2,446; 7,063
162-5	26-15% -11%	15-15 0%	-38	+27	798; 802; 653		154; 190; 84		1,859; 1,850; 4,656

Table 10.18 (cont.)

Table 10.18 (cont.)

Site	Change in vegetation cover 1991-1995	Change in vegetation cover 1995-2000	Change in mean depth to water table (cm) 1991-1995	Change in mean depth to water table 1995-2000	Change in mean ground water salinity (ppm) 1991-94	Change in mean ground water salinity (ppm) 1997-98	Change in mean soil water content 1991-94	Change in mean soil water content 1997-98	Change in mean soil water salinity (ppm) 1991-94
M. teretifolia									
142-3	33-33% 0%	33-13 -20%	!-52	0	1681; 1193; 1577	-	48; 44; 20	-	2,603; 5,128; 5,399
142-7	10-10% 0%	10-10 0%	!-54	nd	921; 860; 920	-	23; 23; 16	-	2,676; 1,465; 11,110
142-10	70-55% -15%	55-90 +35%	!-50	nd	950; 1147; 895	-	68; 62; 47	-	2,627; 3,817; 15,007
M. viminea									
142-5	15-15% 0%	15-50 +35 %	!-59	+4	913; 648; 1101	-	27; 27; 18	-	4,546; 2,988; 8,661
WAWA 4	46-66 +20%	66-0 -66%	-37	nd	795; 554; 435		20; 17; 14		3,654; 15,763; 3,256
9-5	12-12% 0%	12-30 +18%	-23	nd	2090; 2940; 1999	-	38; 34; 21	-	15,181; 11,737; 47,712

Table 10.18 (cont.)

Table 10.18 (cont.)

Site	*Change in vegetation cover* *1991-1995*	*Change in vegetation cover* *1995-2000*	*Change in mean depth to water table (cm)* *1991-1995*	*Change in mean depth to water table* *1995-2000*	*Change in mean ground water salinity (ppm)* *1991-94*	*Change in mean ground water salinity (ppm)* *1997-98*	*Change in mean soil water content* *1991-94*	*Change in mean soil water content* *1997-98*	*Change in mean soil water salinity (ppm)* *1991-94*
M. viminea									
9-7	6-6% 0%	6-25 +19%	-23	nd	1936; 1493; 1058	-	65; 45; 32	-	7,297; 9,888; 10,016
9-13		16-	-26	nd	1234; 868; 875	-	48; 17; 30	-	6,881; 5,262; 17,047
B. articulata									
WAWA-3	60-30% -30%	30-15% -15%	-42	-1	769; 608; 508		149; 104; 100		1,115; 1,654; 4,096
161-3	70-48% -22%	48-12 % -36%	-44	+37	446-587	729	80; 117; 66	47	1,252; 1,211; 2,080
S. validus									
swiii-4	33-30% -3%	30-8% -22%	-27	nd	1025; 2423; 1434	-	150; 159; 196	-	1,852; 19,583; 4,507
swiii-3	40-10% -30%	10-3 % -7%	-29	+17	849-920	-	103; 81; 90	-	1,754; 17,560; 3,036

Table 10.18 (cont.)

Table 10.18 (cont.)

Site	*Change in vegetation cover 1991-1995*	*Change in vegetation cover 1995-2000*	*Change in mean depth to water table (cm) 1991-1995*	*Change in mean depth to water table 1995-2000*	*Change in mean ground water salinity (ppm) 1991-94*	*Change in mean ground water salinity (ppm) 1997-98*	*Change in mean soil water content 1991-94*	*Change in mean soil water content 1997-98*	*Change in mean soil water salinity (ppm) 1991-94*
B. juncea									
163-4	26-20% -6%	20-15% -5%	-29	nd	584; 663; 852	-	25; 19; 7		3,694; 6,174; 8,058
142-6	30-25% -5%	25-28% +3%	-55	+4	719; 606; 583	-	44; 36; 18		2,060; 2,010, 8,903
1N-1	28-28% 0%	28-22% -6%	-12	-2	650; 809; 1108	-	5; 6; 6		8,835; 62,765; 24,098
1N-2	35-35% 0%	35-18% -17%	-20	+2	783; 837; 1178	-	16; 7; 14		8,432; 28,970; 10,091
9-3	33-5% -28%	5-45% +40%	-24	+10	2260; 1357; 1244	-	21; 22; 20		106,064; 53,950; 23,830
J. kraussii									
163-3	92-92% 0%	92-25% -67%	-32	+27	540; 492; 667	-	65; 87; 30		2,156; 2,616; 7,783

Table 10.18 (cont.)

Table 10.18 (cont.)

Site	*Change in vegetation cover* 1991-1995	*Change in vegetation cover* 1995-2000	*Change in mean depth to water table (cm)* 1991-1995	*Change in mean depth to water table* 1995-2000	*Change in mean ground water salinity (ppm)* 1991-94	*Change in mean ground water salinity (ppm)* 1997-98	*Change in mean soil water content* 1991-94	*Change in mean soil water content* 1997-98	*Change in mean soil water salinity (ppm)* 1991-94
J. kraussii									
163-6	99-99% 0%	99-99% 0%	-37	nd	573; 753; 734		76; 54; 35		2,295; 4,713; 6,303
35-3	80-80% 0%	80-25% -55%	-33	nd	844; 818; 1003	-	30; 14; 24		4,752; 7,326; 7,693
9-14	100-100% 0%	100-50% -50%	-21	+10	1078; 1084; 768	-	52; 44; 35		3,919; 4,480; 7,332
swi-3	100-100% 0%	100-60% -40%	-24	+10	692; 742; 740	-	93; 60; 78		1,867; 46,237; 7,446
L. gladiatum									
swi-2	25-25% 0%	25-20% -5%			728; 746; 887	-	145; 133; 125	-	2,116; 16,312; 5,101
swii-3	40-40% 0%	40-22% -18%	-26	+10	705-833	-	83; 65; 61	-	2,028; 41,329; 4,210

! difference between mean water levels in 1991 and 1995 although fluctuations occurred in between
nd not determined

Table 10.18 (cont.)

Table 10.18 (cont.)

Site	*Change in vegetation cover* 1991-1995	*Change in vegetation cover* 1995-2000	*Change in mean depth to water table (cm)* 1991-1995	*Change in mean depth to water table* 1995-2000	*Change in mean ground water salinity (ppm)* 1991-94	*Change in mean ground water salinity (ppm)* 1997-98	*Change in mean soil water content* 1991-94	*Change in mean soil water content* 1997-98	*Change in mean soil water salinity (ppm)* 1991-94
C. asiatica									
163-5	95-78 -17%	78-0 -78%	-33	nd	484; 675; 482		51; 67; 70		2,049; 1,549; 7,080
swiii-5	100-45 -55%	45-0 -45%	-30	nd	718; 881; 919	-	162; 121; 70	-	2,660; 24,804; 3,100
63	75-60 -15%	*nd	-32	*nd	787; 827; 970	-	37; 50; 16	-	3,216; 4,965; 20,737
9-6	100-50 -50%	50-5 -45%	-23	+5	864; 963; 818	-	107; 84; 80	-	2,741; 3,413; 11,233
35-4	100- 100% 0%	100-0 % -100%	-28	+17	530-721	-	95; 51; 26	-	3,600; 5,305; 6,868

* Site destroyed

Changes to net cover ranged from an 82 % decrease due to competition to a 24% increase due to greater water availability, and reduced soil water salinity compared to percent cover in 1995 (Tables 10.17, 10.18). *Baumea juncea* increased relative to *S. virginicus* with increased access to water. Increases also occurred in all species of *Melaleuca* but *S. validus* and *J. kraussii* continued to decline. The greatest changes occurred in *M. viminea* and *C. asiatica* assemblages, both of which occupy wetland marginal zones.

Information on changes in hydrological factors and vegetation cover for selected species in the range of wetlands for the period 1991-2001, were separated into the dry period 1991-1995, and the relatively wetter period 1995-2000 (Table 10.18 and Figs. 10-34 to 10-38). In Figure A, the annual data (from August to July) for 1991 and 1995 are presented, and in B, the trends over the four years are illustrated. For some species of wetland plants, the delay between cause and effect *i.e.*, between change in environmental condition and response of vegetation, exceeded the four year span, so these species were omitted from the following analysis, *e.g.*, *M. rhaphiophylla* and *Juncus kraussii*. Both of these species exhibited large changes in cover abundance in the period 1995-2000.

For *M. teretifolia* (Fig. 10-34), sites 162-5 and 142-10 showed changes (decrease) in vegetation cover and sites 162-3, 142-3, and 142-7 remained constant. For the sites exhibiting decreased cover, the hydrological attributes which underwent the greatest changes were mean depth to water, mean soil moisture content and mean soil water salinity. However, although the depth to the water table influences the hydrological conditions in the surface layers, it is not a simple linear relationship. Site 162-5 experienced the greatest decrease in soil water content, but not the greatest increase in soil water salinity, or the greatest change in depth to water table. The position of the water table was also an important factor. At both sites experiencing vegetation cover change, the position of the water table at commencement of monitoring was close to the surface (≤20 cm). The final position of the water table was > 40 cm, well below the rhizosphere. At the other three sites, the position of the water table at commencement of monitoring was already below the rhizosphere.

For *M. viminea* (Fig. 10-35), site WAWA 4 showed changes in vegetation cover (increase and decrease) and sites 142-5, 9-5, and 9-7remained constant. For the site exhibiting changes in cover, the hydrological attributes which underwent the greatest changes were mean depth to water and mean groundwater salinity. The position of the water table, again, was an important factor. Vegetation cover decreased around the monitoring bore but increased further down slope, encroaching onto the basin floor of the wetland. The position of the water table in the monitoring bore at commencement of monitoring was 120 cm; the final position was 140 cm below the surface. This decrease coincided with the movement of the groundwater from wetland fill to underlying parent material.

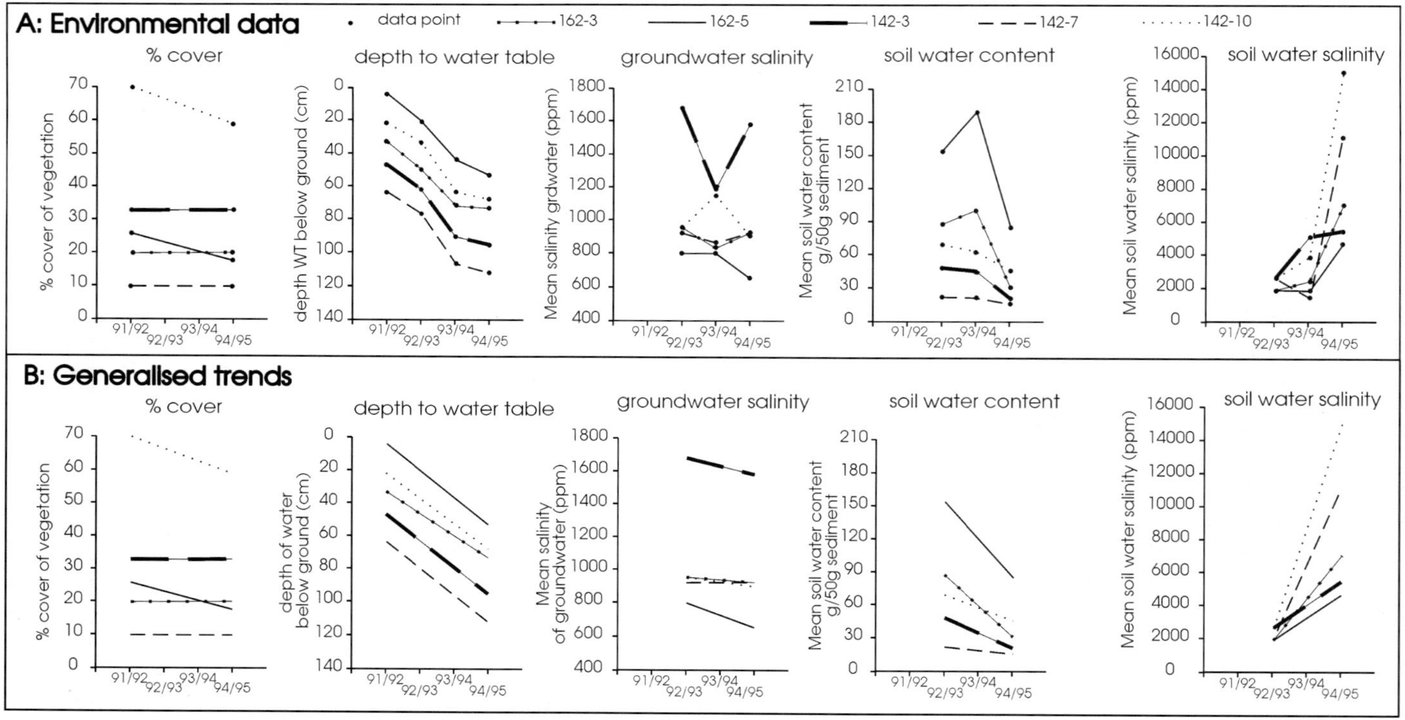

Figure 10-34. *Changes in % cover of M. teretifolia between 1991-1995 in relation to changes in water level, salinity, soil water content, and soil water salinity.*

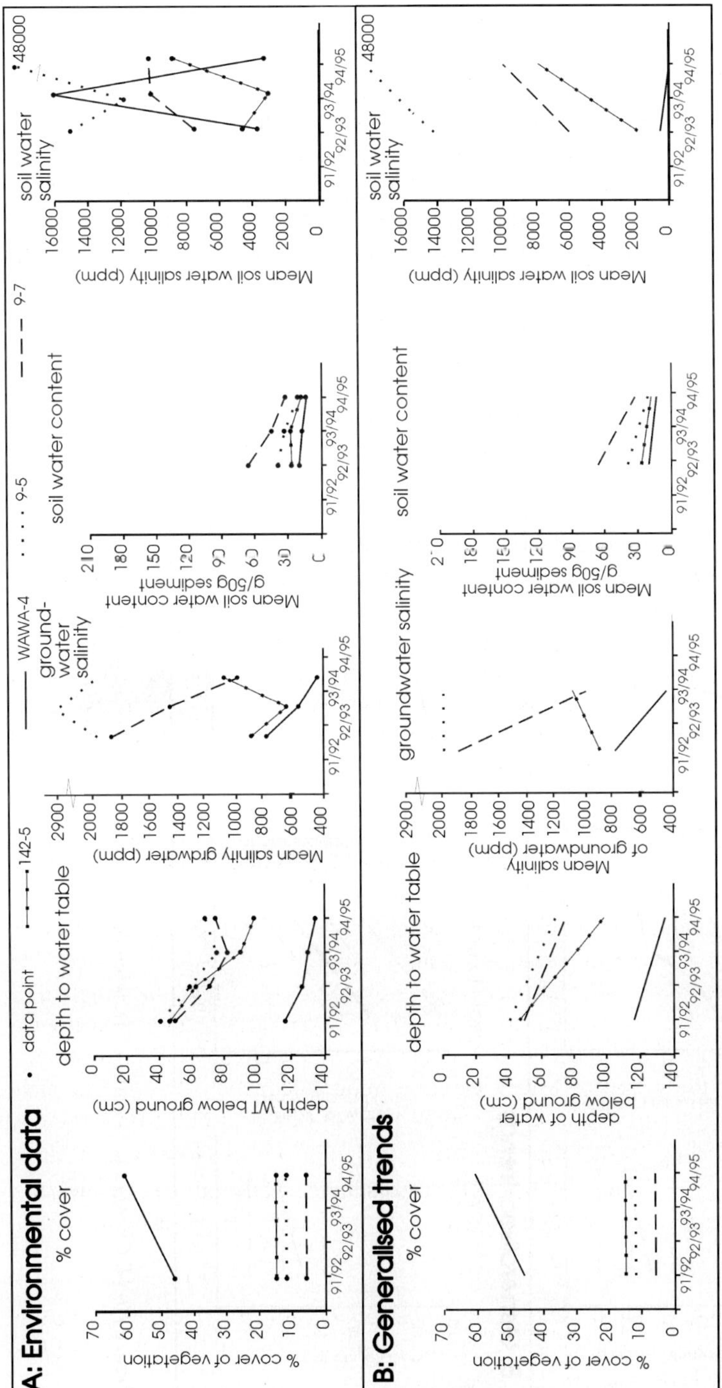

Figure 10-35. *Changes in % cover of M. viminea between 1991-1995 in relation to changes in water level, salinity, soil water content, and soil water salinity.*

For *B. articulata* (Fig. 10-36), sites 161-3 and WAWA 3 showed changes (decrease) in vegetation cover. For these sites, the hydrological attributes which underwent the greatest changes were mean depth to water, mean soil moisture content and mean soil water salinity. Site WAWA 3 experienced the greatest decrease in soil water content and vegetation cover, and the greatest increase in soil water salinity. Depth to the water table influences the hydrological conditions in the surface layers, and the position of the water table was an important factor. At both sites experiencing vegetation cover change, the position of the water table at commencement of monitoring was close to the surface ($\leq$10 cm). The final position of the water table was > 40 cm, well below the rhizosphere.

For *B. juncea* (Fig. 10-37), sites 142-6, 163-4, and 9-3 showed changes (decrease) in vegetation cover and sites 1N-1,2 remained constant. For sites 142-6, 163-4, which exhibited changes in cover, the hydrological attributes which underwent the greatest changes were mean depth to water, mean soil moisture content and mean soil water salinity. Sites 142-6 and 163-4 experienced similar decreases in vegetation cover and soil water content, and similar increases in the depth to water and soil water salinity. The position of the water table was a factor in that the site, where the falling water table was nearest the surface, exhibited the greatest increase in soil water salinity. At the third site, 9-3, the greatest changes in vegetation cover occurred. These changes coincided with lower water tables, low soil water content and extremely high soil water salinities. At this site, a calcrete layer prohibits vertical movement of water, both of the water table and in the zone of capillary rise. There is a separation of hydrological processes above and below this layer, the groundwater below being part of the regional aquifer and the vadose zone above sitting in isolation. At all three sites, *Baumea juncea* was replaced by the more salt tolerant species, *Sporobolus virginicus*.

For *C. asiatica* (Fig. 10-38), sites 163-5, 63, 9-6, and swiii-5 showed changes in vegetation cover (decrease) and site 35-4 remained constant. For the sites exhibiting changes in cover, the hydrological attributes which underwent the greatest consistent changes were mean depth to water and mean soil water salinity. The depth to the water table influences the hydrological conditions in the surface layers. Although sites swiii-5 and 9-6 experienced the greatest decrease in vegetation cover, they did not experience the greatest increase in soil water content, or the greatest increase in soil water salinity. The position of the water table was again an important factor. At both the sites experiencing the greatest change in vegetation cover, the position of the water table at commencement of monitoring was close to the surface ($\leq$20 cm). The final position of the water table was >40 cm, well below the rhizosphere. At the other three sites, the position of the water table at commencement of monitoring was already below the rhizosphere.

It is important to note that the change in abundance cover of *C. asiatica* was due to vigorous competition by alien species during the change in hydrological conditions, rather than a simple decrease in the population itself.

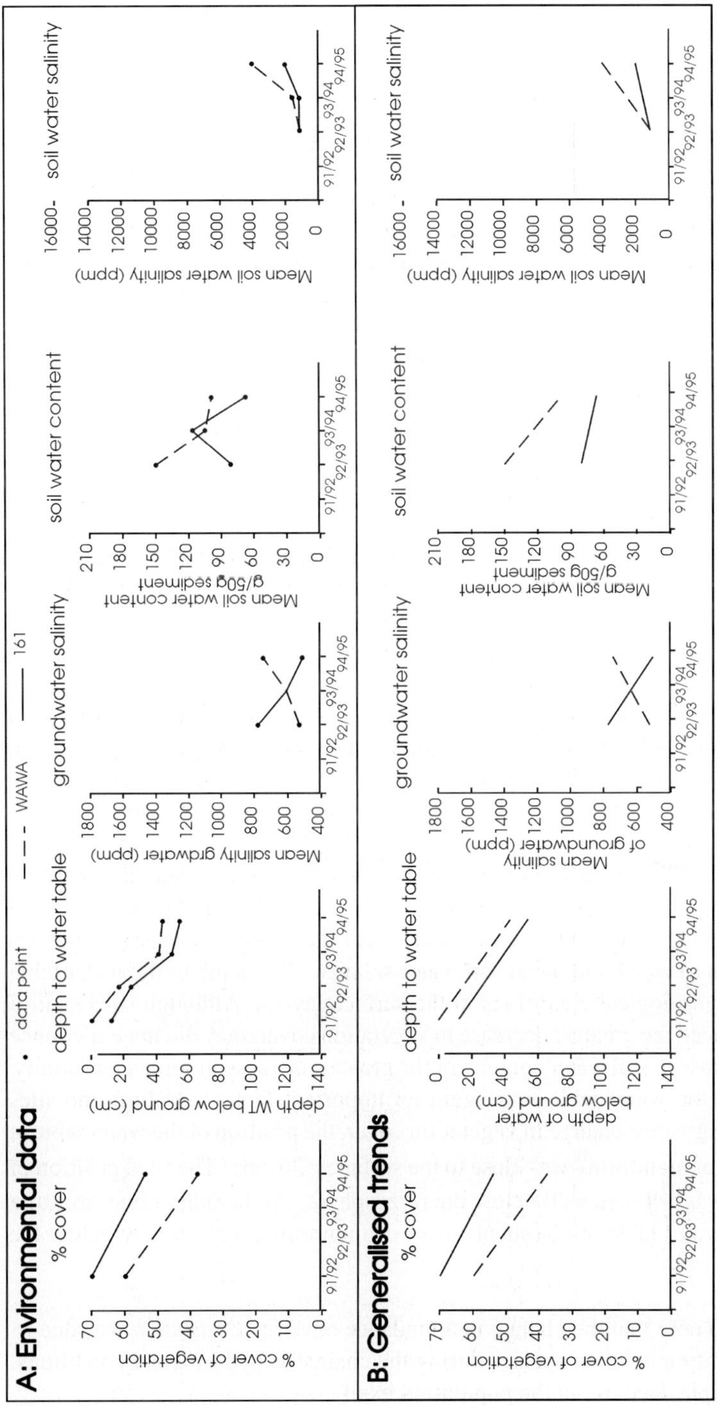

Figure 10-36. *Changes in % cover of B. articulata between 1991-1995 in relation to changes in water level, salinity, soil water content, and soil water salinity.*

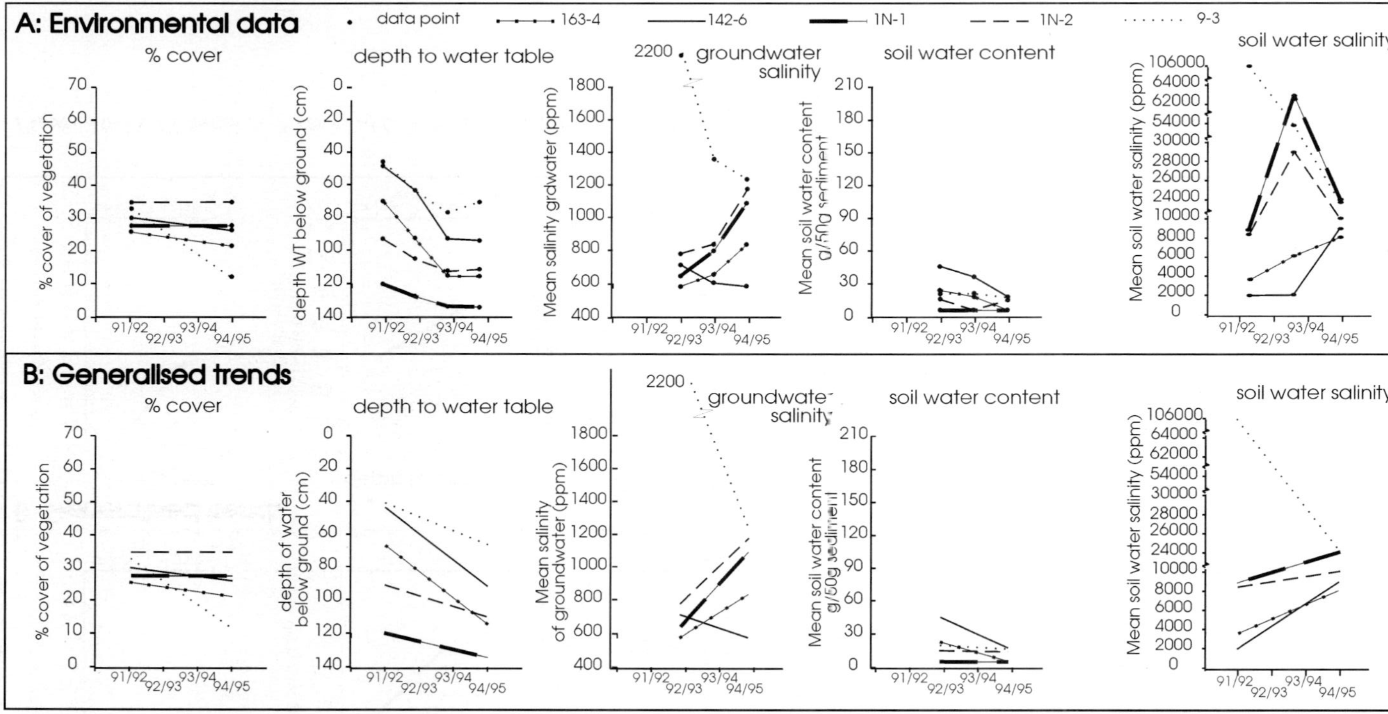

Figure 10-37. *Changes in % cover of B. juncea between 1991-1995 in relation to changes in water level, salinity, soil water content, and soil water salinity.*

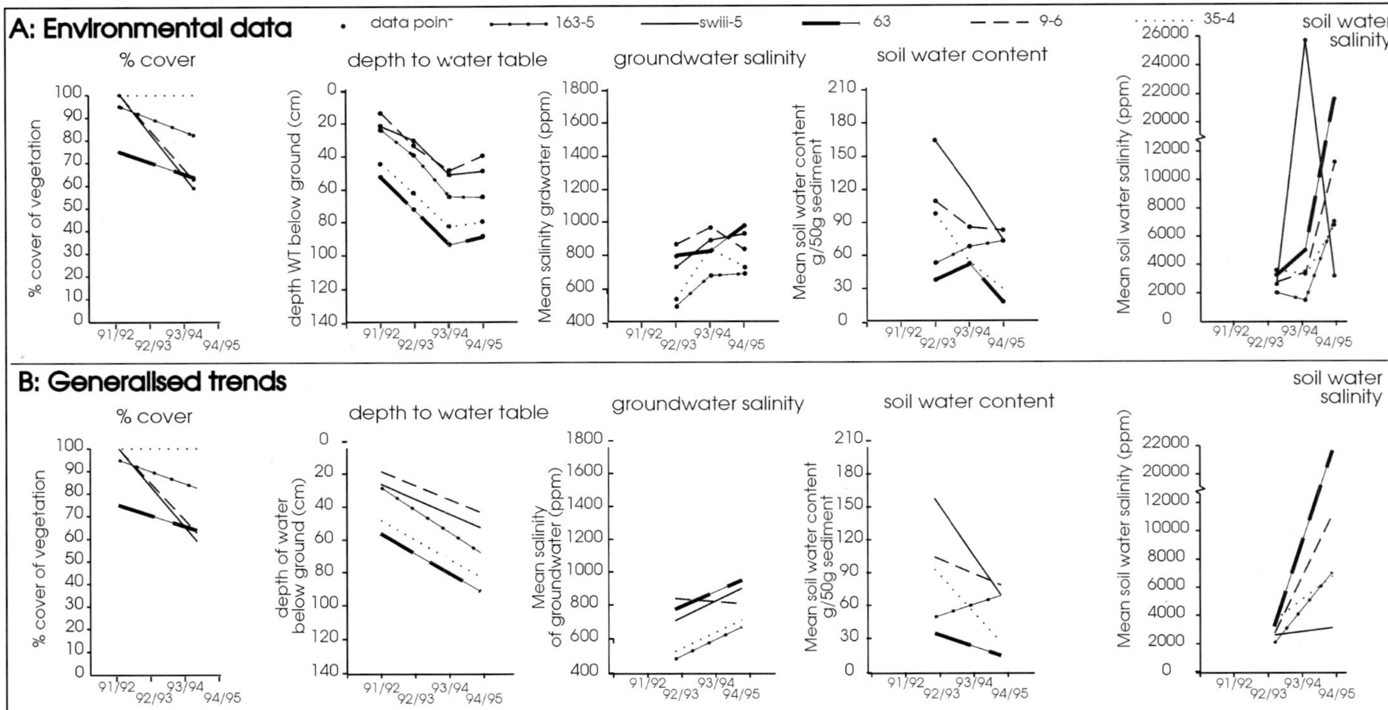

Figure 10-38. *Changes in % cover of C. asiatica between 1991-1995 in relation to changes in water level, salinity, soil water content, and soil water salinity.*

10.5 Plant adaptation to wetland hydrology

An analysis of the results of multivariate clustering and ordination combined with the additional data obtained from field site monitoring, showed that the availability and quality of the water was related to plant distribution and density. In this section, following an introduction about the location of structural vegetation types within the wetland basin, the presence of morphological features in individual species, possibly used for adaptation to water level fluctuations, are discussed.

10.5.1 Distribution of plant forms within a wetland basin

Groundwater availability is determined, in part, by the morphology of the land surface insofar as it affects the depth to water. The relationship between microtopography and wetland plant distribution, first described by Tansley (1949), expanded by Ivanov (1981) with respect to different levels of *Sphagnum*, and subsequently recognised by various authors (Andrus *et al.* 1983), is evident in the wetlands of the Becher area. The geomorphic elements of a wetland basin at Becher are: 1) central basin floor, 2) margin of basin floor at the base of ridge slope, 3) margin of basin floor within the swale (*i.e.*, north or south rim), 4) marginal slope of basin, 5) lateral bench within basin, and 6) mounds within the basin floor. In effect, these subdivisions within the wetland basin may be viewed as intra-wetland vegetation "microhabitats". Occurrence of species, as observed in the Becher Suite wetlands, in relation to these geomorphic elements (or "microhabitats"), is described below.

Melaleuca rhaphiophylla low forest to woodland occurs throughout the wetland basin but preferentially at the margins of the basin floor at the base of the ridge slope. *Melaleuca teretifolia* scrub occurs in clumps in the central portion of the wetland basin, whereas *M. viminea* heath to shrub colonises the lower slopes of basins and ridges within a wetland complex. *Baumea juncea* sedge colonises shallow, small scale wetland basins and flatter margins at the base of ridge slope of deeper basins, while *B. articulata* sedge colonises deep wetland basins. The sedge *L. gladiatum* colonises the lateral bench within a basin and the basin floor margin within the swale (*i.e.*, north or south rim). The herb *C. asiatica* colonises the central part of shallow wetlands and the margins of deeper wetlands. The rush *J. kraussii* colonises shallow wetland basins and occurs in small clumps in shallower parts of deeper wetland basins (mounds within the basin floor) and the basin floor margin within the swale (*i.e.*, north or south rim).

Generally, sedges and the shrub *M. teretifolia* occupy the deepest part of the wetland basin, while the basin floor margins are colonised by a variety of plant forms. The lower slopes are preferentially colonised by the shrub *M. viminea* and the upper slopes of the basin are colonised by the grass tree *X. preissii*.

Within this distribution of varying plant physiognomies, there was a range of densities. For density counts, each of the quadrats selected exhibited 100% cover abundance, however, it should be noted that a plant may have 100% canopy cover over the quadrat but not be rooted in it. Number of plants rooted in the quadrat varied from quadrat to quadrat and year to year, with changes in volume and permanence of water. The numbers of plants provided in Table 10.19 demonstrate the range for particular species within various common wetland vegetation assemblages as documented in 1995 survey.

Table 10.19 *Numbers of plants per m² showing range in densities: vegetation data 1995*

Species	*Range in plant population/m² (100% cover)*
M. rhaphiophylla	0-1
M. teretifolia	0-3
M. viminea	4-17
C. asiatica	356-1324
J. kraussii	58-388

10.5.2 Plant physiognomy

Multi-stemmed monocotyledons with reduced leaves, such as *B. articulata*, *B. juncea*, and *J. kraussii*, dominate the Becher wetlands. Trees and shrubs of *Melaleuca* are single and multi-stemmed, many are gnarled, but there is a surprising lack of epicormic development. At maturity, the shrubs develop a single trunk rather than multiple branches. Trees generally exhibited less height and girth than the same species in other wetlands. These latter features are attributed to the OME calcilutite sediment fill and its hydrological properties, (*e.g.*, lower pore water content) which differ from that of the 1-3 m of peat fill in other wetlands on the Swan Coastal Plain which are colonised by the same species.

Flowering spikes of monocotyledons are generally above the predicted water level maximum, ranging from 30 cm to 180 cm in height, but species of monocotyledons and the herb *C. asiatica* also use shoot elongation as a short term adaptive response to inundation. This response has been documented for other species in flood prone areas (Blom 1994; Blom and Voesenek 1996).

The extreme environmental conditions in the Becher wetlands means that all of the wetland plants exhibit structures and processes for reducing evapo-transpiration. Examples of features in the genus *Melaleuca* are schlerophyll leaves, leaves with reduced surface area (terete), and chartaceous bark covering to roots. Examples of processes include shedding of leaves, and orientation of leaves to reduce insolation. In the Cyperaceae family, numbers of leaves and their surface area are reduced.

The occurrence of xeromorphy in wetland plants was first recognised by Schimper (1898), cited in Wheeler (1999), and Yapp (1912), who, in a study of the marsh plant

Spiracea ulmaria, related changes in the frequency and length of hairs on the underside of leaves to changing evapo-transpiration, induced either by soil moisture deficits or atmospheric conditions. However, subsequent workers have related xeromorphy to nutrient deficiency (Small 1973), or reduction in uptake of soluble toxins associated with waterlogging and inundation (Jones and Etherington 1970).

10.5.3 Rhizome and root structures

Rhizome and root structures were examined more closely because it is in modifications of these structures that the seasonality of the water availability is likely to be expressed. Rhizomes comprise the non-photosynthetic perennial parts of the shoot system of monocotyledon plants, and are usually buried in the rooting substrate (Meney and Pate 1999). Rhizomes are typically robust, cylindrical stem like organs with smooth surfaces. For monocotyledons in the Becher wetlands, a study of the rhizome and root structures of *B. articulata*, *B. juncea*, and *J. kraussii* showed that their culms arise at regular intervals along their lengths (Figs. 10-39 to 10-44). Roots arise adventitiously and sequentially from the lower part of the rhizome system, but less regularly, and in most cases, at different frequency from the culms (Figs. 10-43, 44) (see Holttum 1955). In the text below, both the rhizomes and root structures are described together for the following species: shrubs and trees *M. rhaphiophylla*, *M. teretifolia*, *M. viminea*, the sedge *B. juncea*, the herb *C. asiatica*, and the rush *J. kraussii*.

Seven types of rhizomes were recognised amongst Restionaceae (Pate *et al.* 1991): stilt, tufted, and transitional varieties of ascending rhizomes, and the monomorphic, dimorphic, looping, and mostly unbranched varieties of horizontal rhizomes and these categories have been adopted herein for Cyperaceae and Juncaceae.

Generally in the literature, roots have been categorised as cable, lateral, tap, and root hairs. In this study the largest lateral roots (≥12 cm diameter), were categorised as cable roots, and were woody or covered in paperbark. Smaller, lateral roots (4-12 cm diameter) were categorised separately, simply as "lateral roots". The largest vertical roots (< 1 cm diameter) were termed "tap roots", and the fine short dendritic structures were termed "root hairs".

In the Becher wetlands, the roots and associated structures of the tree *M. rhaphiophylla*, at maturity, extend approximately 200 cm horizontally and up to 75 cm vertically, however much of the structure is often exposed above ground with the actual underground portion contained predominantly within 30 cm of the surface. Cable roots together with the root stock or boll, form the bulk of the structure, interlaced with lateral roots, all covered with root hairs (Fig. 10-39).

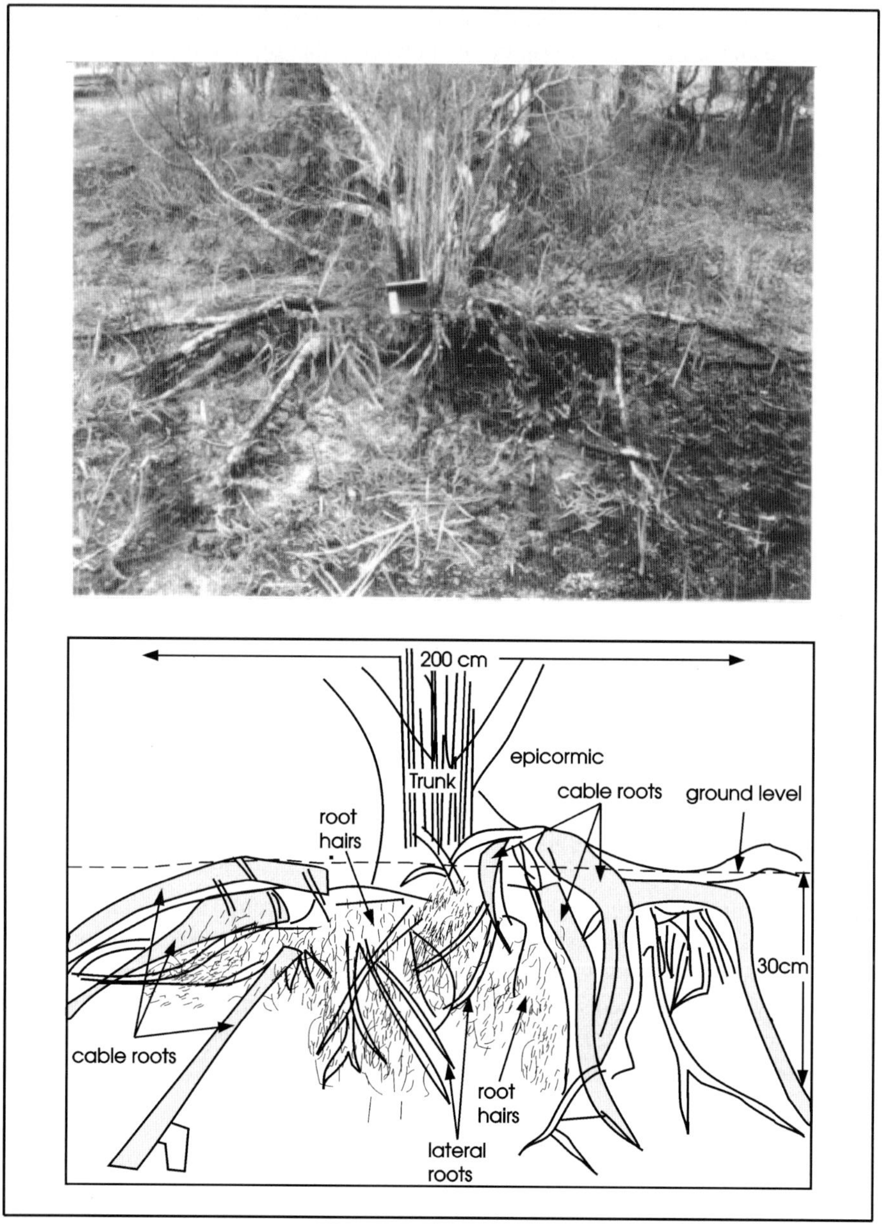

Figure 10-39. *Root structures in Melaleuca rhaphiophylla.*

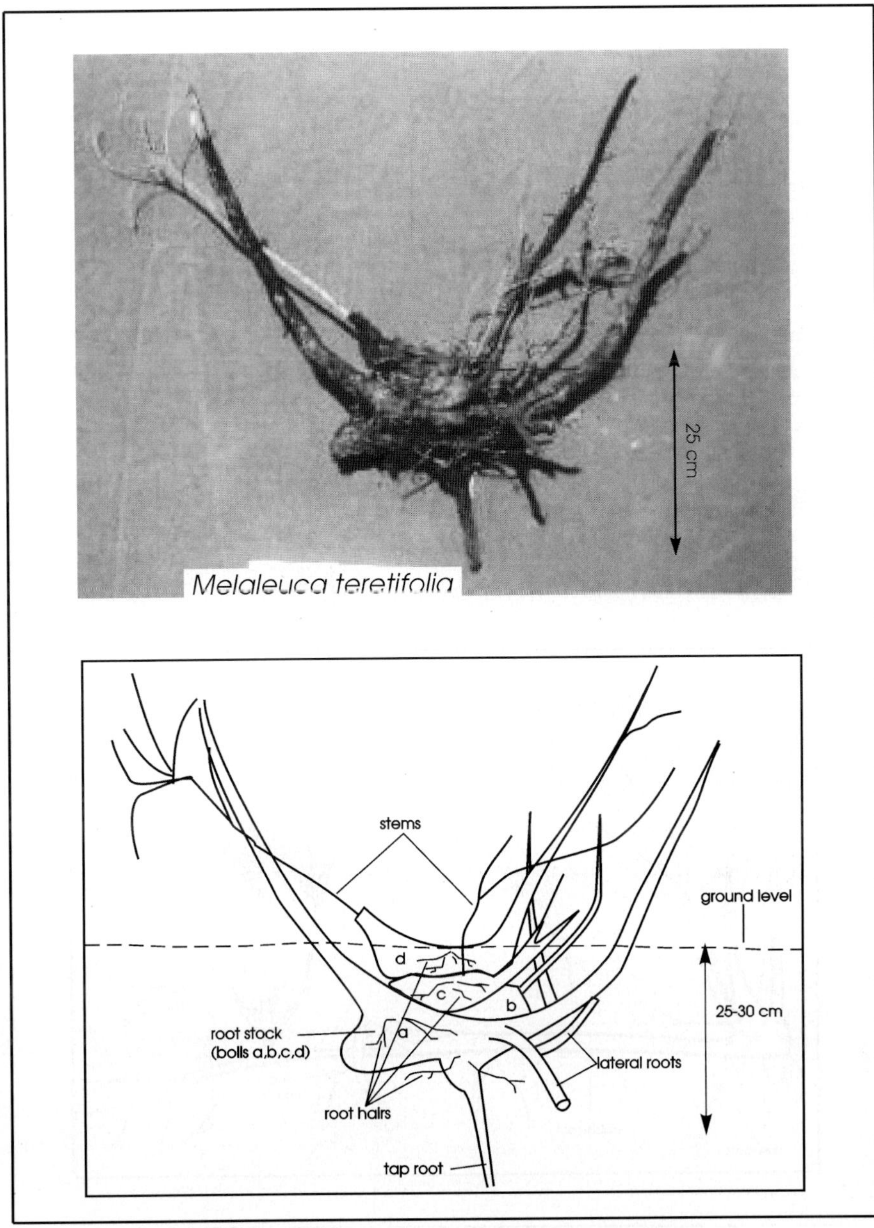

Figure 10-40. *Root structures in Melaleuca teretifolia.*

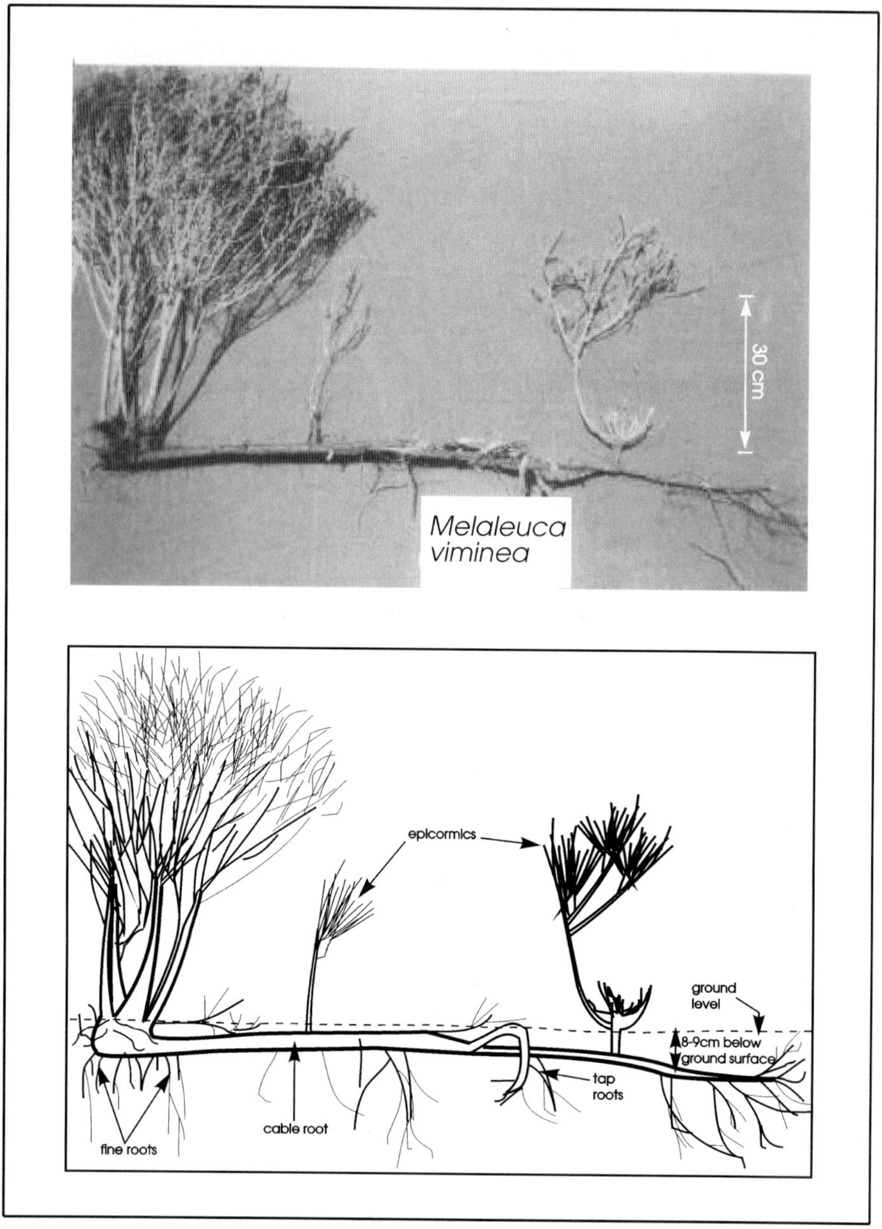

Figure 10-41. *Root structures in Melaleuca viminea.*

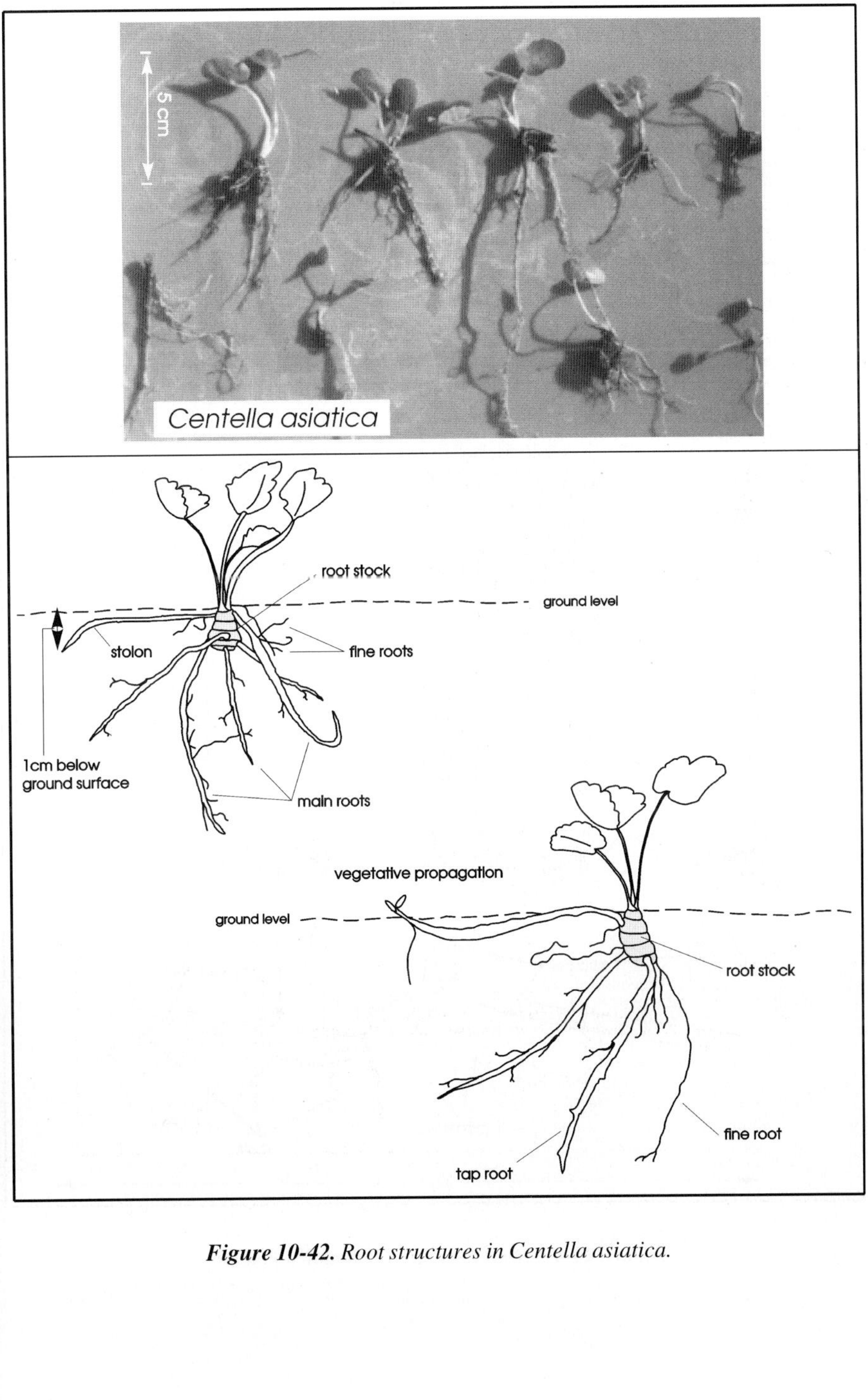

Figure 10-42. *Root structures in Centella asiatica.*

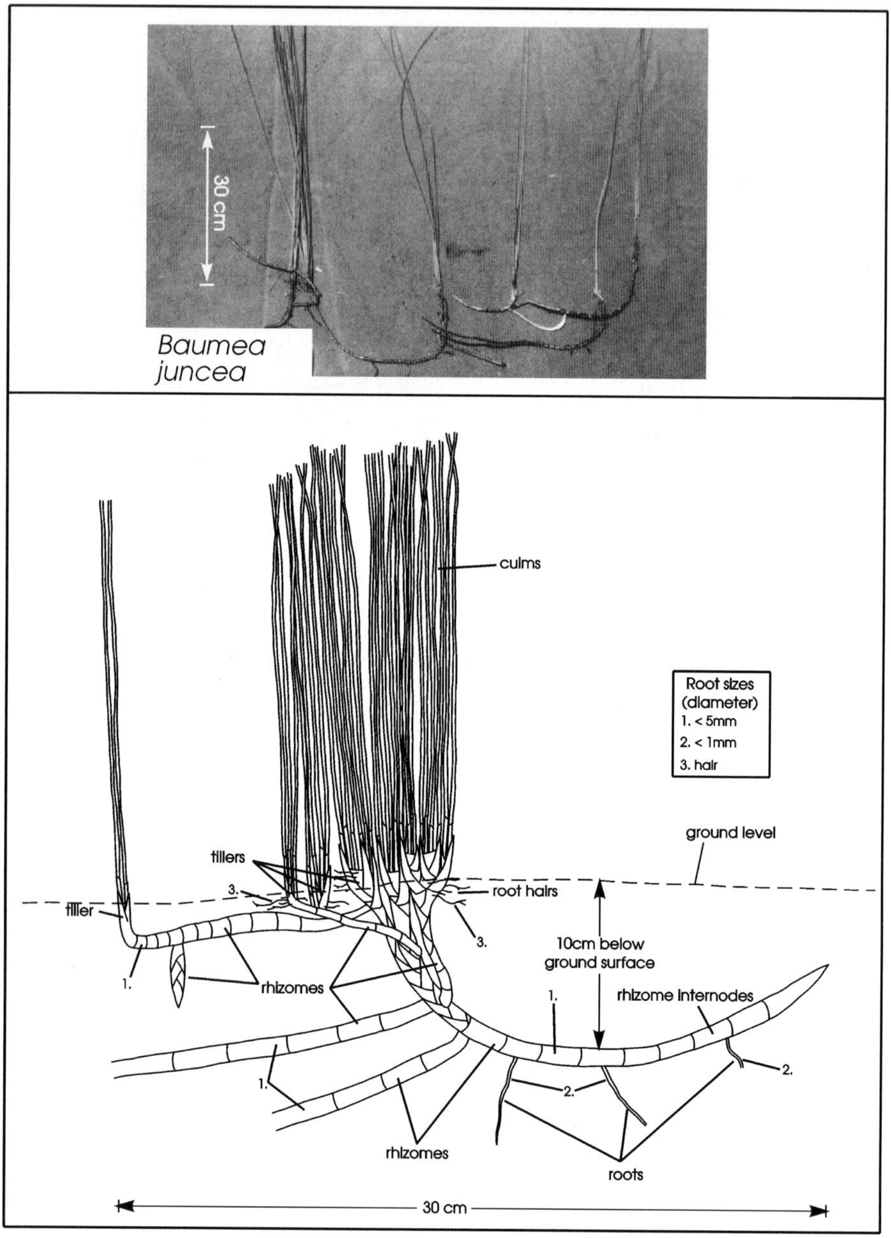

Figure 10-43. *Root structures in Baumea juncea.*

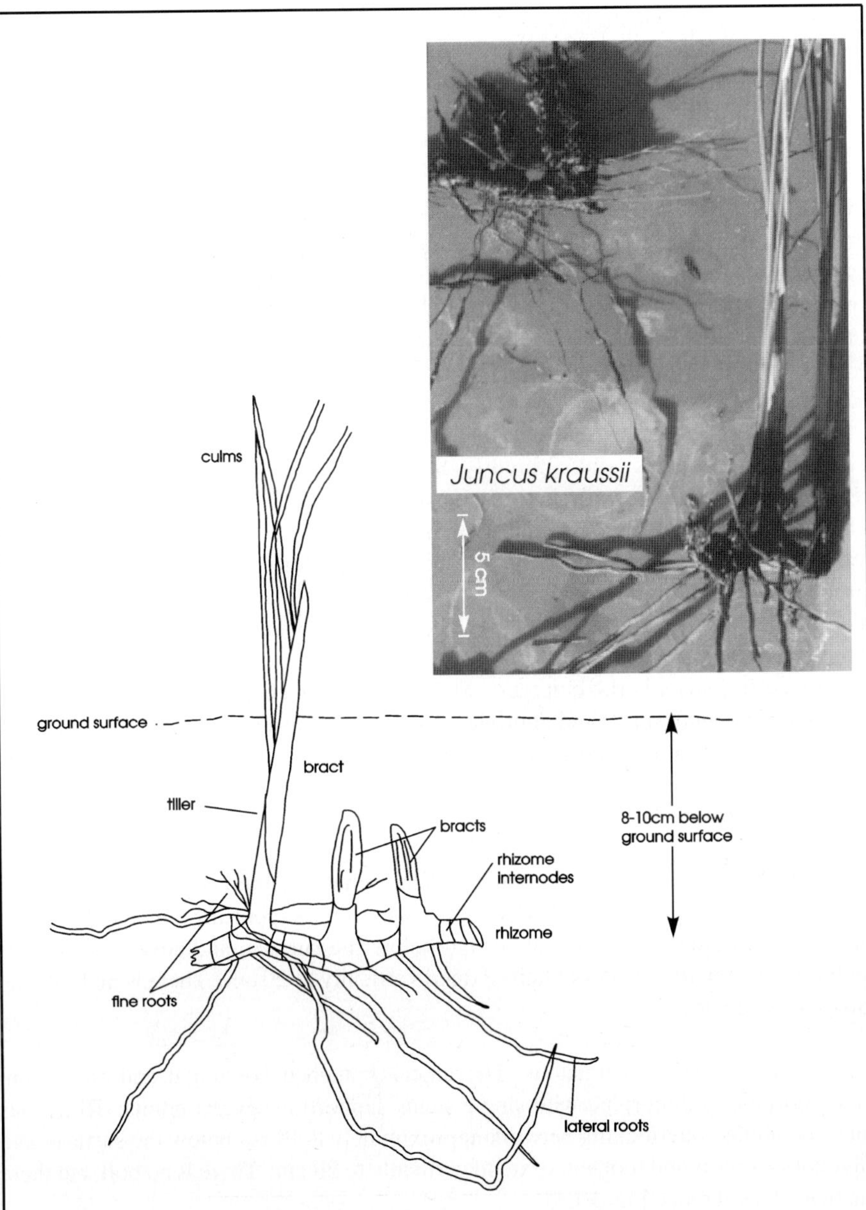

Figure 10-44. *Root structures in Juncus kraussii.*

The roots and associated structures of the shrub *M. teretifolia*, at maturity, extend approximately 120 cm horizontally and 35-60 cm vertically, however much of the structure is contained predominantly within 25-30 cm of the surface, with only the tap root extending further. Root stock or boll forms the bulk of the structure, interlaced with three or four major lateral roots, and root hairs (Fig. 10-40).

In the Becher wetlands, *M. viminea* exhibits a radial cable root system. The cable root or stem is unbranched, horizontal, and dimorphic with single stemmed shoots, which eventually forms separate individuals. The cable root is located approximately 8-9 cm below the surface, adventitious roots extend down to 30 cm vertically, with the tap root extending further. These cable roots act like an underground stem from which epicormics with root systems grow at intervals. There are no boll or lateral roots visible, and root hairs are less dense compared to other species (Fig. 10-41).

The herb *C. asiatica* is "stoloniferous". Stolons are unbranched, horizontal, smooth, white and monomorphic with single stems, which eventually form separate individuals. The stolon is located approximately 1 cm below the surface; there are no adventitious roots. Roots extend to 10 cm. The root stock or boll forms 50% of the underground structure, and the roots and multiple root hairs often form a mat comprising the other 50% of the root structure (Fig. 10-42). The longest root has notches along its length.

The sedge *B. juncea* is rhizomatous. The ramet is branched, horizontal, tufted, and monomorphic, with single and multiple stems supporting a variable number of culms. *R*hizomes form a complex interlocking network approximately 10 cm below the surface, and adventitious roots, including the tap root and root hairs, extend vertically to 20 cm. There is no boll, and lateral roots are few (Fig. 10-43).

The sedge *B. articulata* is rhizomatous. The ramet is unbranched, horizontal, and monomorphic with single and occasionally twin stems supporting single culms. The rhizome is located approximately 10 cm below the surface, and adventitious roots including the tap root, and root hairs extend vertically to 20 cm. There is no boll, and lateral roots are few.

The rush *J. kraussii* is rhizomatous. The ramet is branched, horizontal, and varies from monomorphic to dimorphic with single stems supporting several culms. Rhizomes form a complex interlocking network approximately 8-10 cm below the surface, and adventitious roots and root hairs extend vertically to 20 cm. There is no boll, but there are many lateral roots (Fig. 10-44).

An attempt was made to relate the pheno-morphology of rhizomes and roots of *B. articulata*, *B. juncea,* and *J. kraussii* to seasonal conditions exemplified by the hydrologic history of a given site over 4 years. No definitive linear pattern existed for stages of maturation with respect to rhizome length, diameter, internode number and length.

From this study, the following conclusions were drawn:

- in a growing season, culms can grow from any existing tiller;
- either a tiller or rhizomatous shoot can sprout from any leaf node along the rhizome segment between two existing tillers; and
- growth of rhizomes is only generally related to seasonal conditions.

However, the investigation deserves a more rigorously designed experimental and field monitoring programme than was possible for this study.

Rhizomes, roots and culm bases were also investigated for aerenchyma around the stem margin and aerenchimatic cells within the parenchyma. No adaptive cells were obvious in the plant material collected directly from the field except in the youngest roots. Minor development of aerenchyma in shallow rooted plants in wetlands has been observed in other studies (Metsavainio 1931, cited in Wheeler 1999), whereas with greater fractional root porosity, rooting depth increases under inundated conditions (Justin and Armstrong 1987; Armstrong *et al.* 1991).

10.6 The effects of vegetation on stratigraphy and hydrology

In contrast to studies concerned with the effect of environment on the growth of the plant, this section explores the pedogenic effects of wetland plants which subsequently influence the wetland hydrology. Plants create a variety of pedogenic, physical, and chemical effects at the bedding scale, in that they:

- directly contribute to the sediment through the addition of humus, peat, and root material *i.e.*, they contribute to sediment composition and texture;
- affect the structure of the sediments;
- are a factor in calcrete development;
- affect hydrology *e.g.*, alter the hydraulic conductivity of sediments by acting as conduits to the water table; affect discharge rates through evapo-transpiration; and
- affect soil water cation chemistry indirectly.

10.6.1 The relationship between plants and sediment

Plants directly contribute to the sediment by way of humus, peat and root material *i.e.*, composition and texture. The organic materials in the sediment derive from the remains of vascular plants such as sedges, rushes, grasses, and wood from shrubs and trees. Measurement of the percentage of organic matter in sediments was undertaken at replicate sites under selected vegetation assemblages. Variation of the density of the vegetation between sites (Table 10.19) resulted in a relatively large standard deviation from the mean. Measures (by weight) of total organic composition, including coarse plant debris such as roots, seeds, and stems, are contrasted with the percentage of mud sized organic material under various vegetation assemblages in the 0-10 cm layer (Fig. 10-45A and Table 10.20).

Table 10.20 *Total and mud sized organic composition in 0-10 cm layer under various plant assemblages*

Vegetation assemblage	*Percentage of total organic material in sediment*		*Percentage of mud in sediment*	*Percentage of mud sized organic matter in mud*	*Percentage of organic carbon in mud*
B. articulata	49 ± 26	n = 2	80	45	26
L. gladiatum	31	n = 1	49	23	13.3
M. teretifolia	19 ± 7	n = 5	73	15	8.7
J. kraussii	19 ± 4	n = 5	89	21	12.2
C. asiatica	18 ± 6	n = 5	61	16	9.3
M. rhaphiophylla	17 ± 6	n = 4	59	14	8.1
M. viminea	12 ± 5	n = 2	16	10	5.8
S. validus	12 ± 2	n = 3	23	5	2.9
B. juncea	12 ± 2	n = 4	76	3	1.7

Three of the sediment layers can be classed as organic, based on the criterion that organic carbon content exceeds 12% of the soil organic matter (Collins and Kuehl 2000). They are the surface sediments under sedges *B. articulata* and *L. gladiatum,* and under the rush *J. kraussii*. These species have high annual regeneration rates and high below ground biomass. Conversely, organic content is lowest under species *S. validus* and *B. juncea*, which are slender sedges which have lower above and below ground biomass. The measures of organic content in the surface layer under the three species of *Melaleuca* suggest similar rates of fall and decomposition. Macro-organic material is a minor component of the organic fraction under all assemblages except *B. juncea* and *L. gladiatum*. *B. juncea* has low above and below ground biomass and produces minor detritus. *L. gladiatum* produces substantial litter which is broken down relatively rapidly under seasonal waterlogging.

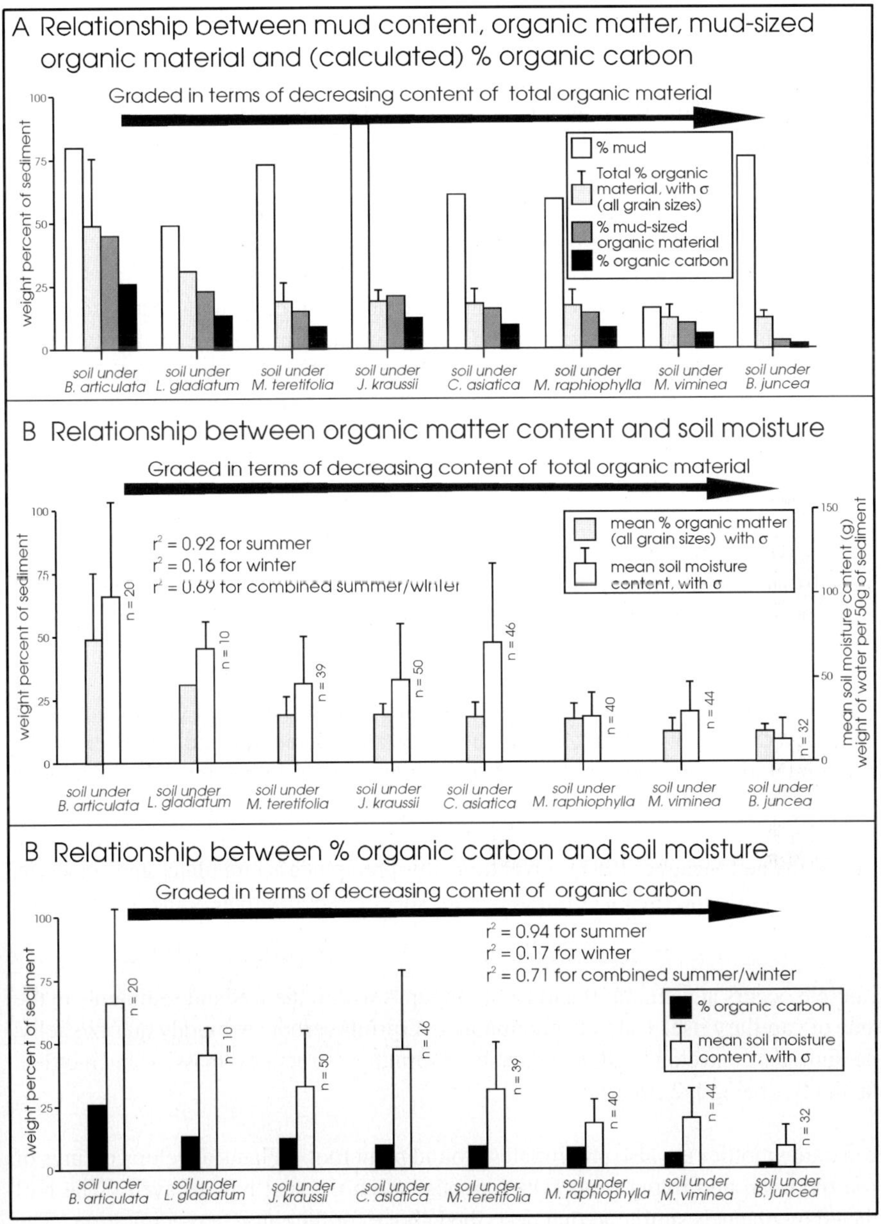

Figure 10-45. *Graphs showing relationship between mud content, % of organic carbon, and mean soil moisture under various vegetation formations.*

10.6.2 Plants affect the structure of the sediments

Interlocking roots, in the upper 20 cm of the majority of wetland sediments, disintegrate mud clods and partition homogeneous muddy sediments. In the Becher sediments many of the plants are rhizomatous, so that the sediments are dissected vertically and horizontally. In wet sediments, roots bind sediment in columnar peds. In hard, dry sediments, root penetration partitions the homogeneous mass. In soft, dry sediments, roots provide cohesiveness. Decaying roots may produce interstices within a homogeneous sediment, increasing porosity.

Plants, through development of bolls and high density root systems, can also create mounds within the sediments. Mounds under wetland plants are relatively small, *i.e.*, 10-20 cm high, and some 1-1.5 m in diameter. Mounds occur under the tree *M. rhaphiophylla* and the shrub *M. teretifolia*. It has been suggested that these mounds could be useful in reducing the period of waterlogging in a similar way to mounds resulting from differences in microtopography (Wheeler 1999).

10.6.3 Plants have chemical pedogenic effects

Apart from structural pedogenic effects such as root-structuring and the development of ped structures, plants have a chemical pedogenic effect in the development of calcrete and, infrequently, iron oxide mottles.

Calcrete is a 1-4 micron sized calcite occurring interstitially to a grain framework composed of sand sized quartz and carbonate grains (Semeniuk and Meagher 1981). In the interstices are roots and root hairs, often encased in calcrete. In calcareous sediments saturated by calcium carbonate rich groundwater, an interstitial precipitate of calcium carbonate results when plants extract water from the zone of capillary rise (Semeniuk and Meagher 1981). Over time, this precipitate accumulates and coalesces to form mottles and then a continuous thin sheet at approximately the level of root penetration.

Calcrete occurs at wetlands 9 and Cooloongup A within the wetland sediments in the zone of capillary rise. Calcrete precipitate overprints carbonate muddy quartz/shelly, medium sand, and clogs sediment interstices forming 2 structures: massive and mottled, the mottles being 1-2 cm in size.

Iron oxide mottles are also precipitates around plant roots. Plants develop coatings of iron oxide around living roots which persist when the root has decayed. The part played by plants is similar to that described above for the calcrete precipitation. The iron precipitate is a minor feature in the wetland sediment profile as iron is not particularly abundant in the groundwaters or in the sediment.

10.6.4 Plants affect hydrology

One of the major effects on wetland sediments is the penetration of relatively impermeable sediments by roots. Roots alter the characteristic porosity of the sediment by developing a network of vertical and horizontal conduits.

Plants also affect the sediment capacity for storage of interstitial water through the contribution of organic material. Although it is generally recognised that organic soils have high soil water holding capacity, it has been argued that much of this water is unavailable for plants because of the high concentration of pores with unfavourable dimensions within the matrix, *i.e.*, macropores facilitating rapid drainage and very small pores with high matric forces (Boelter and Blake 1964; Richardson *et al.* 2001). However, the characteristics of the sediments in the Becher wetlands are as follows: 1) there is a high proportion of organic carbon in the mud sized fraction (Table 10.21), 2) the sediment composition is an organic, mineral mix, *viz.*, peat and calcilutite, 3) the bulk densities of the sediments range from 0.2 to 0.75 g/cm^3, and 4) the porosity ranges from 46 to 60%. These physical properties are indicative not of organic soils (Collins and Kuehl 2001) but of organic enriched mineral soils.

Table 10.21 *Percentage of organic carbon in mud fraction of sediments in 0-10 cm layer calculated after the method of Soil Survey Staff (1996) cited in Collins and Kuehl (2001).*

Vegetation assemblage	*Percentage of organic carbon*	*Mean soil moisture content (weight water per 50 g sediment)*	
B. articulata	26	112	n = 20
L. gladiatum	13.3	66	n = 10
J. kraussii	12.2	52	n = 50
C. asiatica	9.3	69	n = 46
M. teretifolia	8.7	44	n = 39
M. rhaphiophylla	8.1	31	n = 30
M. viminea	5.8	29	n = 44
B. juncea	1.7	17	n = 32

When soil moisture content in the 0-10 cm layer (measured quarterly between 1992 and 1994) is compared with the organic carbon content in the mud fraction of the same sediment, there is a significant correlation between the mean soil water content for the summer period (0.94) and the amount of organic material. There is minimal correlation (0.17) between the two factors for the winter period (Fig. 10-45B).

10.6.5 Plants affect soil water and groundwater chemistry

Vegetation has an influence on the development of salinity of wetland basins through contribution of organic matter. As briefly described in Section 9.4.1, various plant species accumulate cations in their tissue, and through leaf litter, branch fall, bark shedding, and root death, directly contribute elevated concentrations of cations to the soil. The contribution of cations to the groundwater and interstitial water by rainwater is low, generally < 1 mM/L, but vegetation acts as a concentrating mechanism in the wetland. The vegetation within a wetland accumulates cations to the extent that leaves of wetland species generally contain ~ 200-700 mM/kg of Na, ~150-450 mM/kg of K, ~75-400 mM/kg of Ca, and ~40-250 mM/kg of Mg, while branches and roots generally contain ~75-300 mM/kg of Na, ~30-265 mM/kg of K, ~50-400 mM/kg of Ca, and ~35-100 mM/kg of Mg (Fig. 9.15). This cation rich organic matter, once accumulated into the sedimentary pile, results in peat, or OME calcilutite, humic calcilutite and humic sand. These sediments then have concentrations of cations elevated an order of magnitude above that contributed by rain. Leaching of these cation-enriched organic materials by infiltrating rainwater, or rising groundwater provides cations to the interstitial waters and to groundwater.

The leaching experiments described in Section 9.4.4. showed that after one day, rain can leach up to 38 mM/L of Na, 30 mM/L of K, 75 mM/L of Ca, and 90 mM/L of Mg, and that after a week of contact, rain can elute up to 80 mM/L of Na, 45 mM/L of K, 150 mM/L of Ca, and 205 mM/L of Mg (Fig. 9-10). These results have implications for the serial development of vegetation associations, since the occurrence of various wetland plant species are related to the salinity regime. For instance, *M. viminea*, *M. rhaphiophylla*, *S. validus* and *S. virginicus* can inhabit elevated to moderate hyposaline conditions, while *M. teretifolia* and *J. kraussii* can tolerate moderate to low hyposaline conditions, and species like *C. asiatica*, *B. articulata* and *B. juncea* inhabit conditions of freshwater (*i.e.* salinity ~500 ppm). Some of the salinity regimes encountered in the Becher wetlands will be limiting to species, in that certain species will be eliminated by low to moderate hyposaline conditions (*e.g.*, *C. asiatica* and *B. articulata*), which others will tolerate (*e.g.*, *M. rhaphiophylla, M. viminea*), while others actually thrive in such conditions (*M. cuticularis* and *S. virginicus*). Clearly, with the concentration of cations within vegetation, and its delivery to the sediment and soils, vegetation will play a role in the evolving salinity of the soil water, and hence in the development of vegetation associations (Fig. 9-43).

Of interest is the fact that cation content varies between species, as well as in different anatomical parts of a given species. A full analysis of the contribution of cations to the sedimentary store in a given wetland would involve many interrelated components of the wetlands and its flora. It would involve interrelating the floristic composition of vegetation associations in a specific wetland (*e.g.*, *M. rhaphiophylla* dominated, or *B. articulata* dominated, or mixed), the relative productivity of plant species and their contribution to the sedimentary pile, and the relative productivity, cation concentration and contribution of the various parts of the species (*e.g.*, high concentration of selected cations in bark vs lower concentration of cations in leaves, and their relative rates of production and fall). Such a study, however, is beyond the scope of this project, and the data obtained to date show that vegetation matter was a significant contributor of cations to the wetland sediment store.

Plants can also change the soil water chemistry through the liberation of other compounds by exudation from living roots, by secretion, and sloughing. Compounds include carbohydrates, amino acids, organic acids and lipids (Curl and Truelove 1986). Organic acids are good metal chelating compounds that play an important role in the absorption and translocation of nutrient elements. The principal factors that affect kinds and quantities of substances released by roots into the rhizosphere soil include the species and developmental stage of plants (Curl and Truelove 1986). Maximal quantities of root exudate have been obtained by partial desiccation, followed by wetting and continued growth, which is a common sequence within seasonal wetlands. Stress, created by low soil moisture, also increases exudation of various compounds (Curl and Truelove 1986).

10.7 Summary and discussion

As the setting and origin of the Becher Suite wetlands are uncommon, comparisons of vegetation dynamics are made with the most similar wetland type, the dune slack ecosystem. Corresponding vegetation characteristics and demonstrated links between environmental features and species distribution or community structure in dune slacks and the Becher Suite wetlands, are compared below.

- Very few species are confined to the slack habitat (Ranwell 1972) and this is also true of the inter-ridge wetlands on the Becher Cusp (Section 10.2). The flora of the Becher wetlands does not reflect the chemical composition of the sediments and waters. Of the 11 assemblages identified, only the sedgelands of *B. juncea*, *L. gladiatum*, *I. nodosa* and the rushland *J. kraussii* occur predominantly in calcareous sediments and these species are not restricted to these habitats.
- In many dune slacks, plant communities are arranged in mosaics (Sykes and Wilson 1987), whereas the arrangements in the Becher Suite wetlands grade from uniform to mosaic to concentric zonation, reflecting margin to centre environmental gradients within the wetland.
- Variable small scale differences in depth to water due to variable microtopography has been identified as one underlying determinant of species distribution in dune slacks (Crawford and Wishart 1967; Ranwell 1972; Van der Laan 1979), and this is corroborated by the multivariate analyses in this study.
- Sykes and Wilson (1987) argue that lack of clear vegetation community structure within New Zealand dune slacks indicates that there are a number of environmental factors which could be significant and point to soil cation and nutrient content as two potential factors, an idea supported by the ordination results in this study which separated sites with different vegetation assemblages into those

with consistent calcium and potassium content in pore waters and those with fluctuating content.

- Other environmental factors recognised as important are recharge by calcium rich groundwaters (Grootjans *et al.* 1991; Sival and Grootjans 1996); and soil moisture variability (Barbour *et al.* 1987). Distribution of a particular species may be controlled by sensitivity to moisture stress, *i.e.*, the tolerance of wetland plants to dry conditions (Keddy 1989, cited in Wheeler 1999). Several vegetation sites were distinguished on the basis of low water levels, *e.g.*, 1N, swii, 35-3, 142-7, and in supplementary work, soil moisture was found to be significant, demonstrating that plants were sensitive to both abundance and deficit.

In relation to plant species distribution and vegetation dynamics, a number of water related parameters were tested: inundation frequency, waterlogging frequency, hydroperiod over various years, mean water depths in wet and dry years, and soil moisture. This multi-pronged approach was used to ensure that the dynamic aspects of hydrology, and in particular, the longer term hydrological effects, were addressed in relation to vegetation. However, it was evident in the study of species dynamics between 1991-2001, that some species responded more quickly than others, resulting in a split between species in equilibrium with water conditions and those out of phase, a caveat given by van der Valk and Davis (1980). In retrospect, it appears that greater emphasis could have been placed on seasonality and further studies incorporating longer term rainfall fluctuations, and more detailed inter-annual and intra-annual changes could prove fruitful.

While it is possible to identify broad trends in relation to water table and plant distributions, the range of conditions occupied by individual species can be wide and inconsistent between sites (Wheeler 1999), therefore the work carried out pertaining to intra-basinal variation, on the effects on species density of a number of environmental factors, was valuable in establishing ranges for plants in the Becher wetland setting, and delineating potential species competition. Species occupied clear hydrological niches. Difficulties in defining clear vegetation-hydrochemical relationships stemmed from the considerable spatial and temporal variation in water composition, even within the same stand (Chapter 9), the effect of plant storage, the capacity of the sediment to supply nutrients through the mineralisation process, and the possibility that different nutrients were limiting in different habitats and for different species (Figs. 9-10, 9-36 to 9-40 and Wheeler *et al.* 1992).

However, the inter-relationships between plants and hydrochemical factors is an interesting one. It is clear that in changing the structure, texture, and composition of the sediments, plants change the hydrological properties of the sediments, and hence the small scale pathways of water movement throughout the profile and the wetland. Plants also have a long term affect on the chemistry of the interstitial and groundwaters as discussed above. These comments are expanded here using the example of cations, and some prognoses are presented.

Peat and organic enriched sediments derived from plants such as *B. articulata* will be generally lower in cation content than that derived from *J. kraussii*, or *M. cuticularis*. There exists the possibility that, as cations are contributed to the sedimentary pile from species with moderately elevated cation content, the increasing salinity can potentially eliminate those species intolerant of the hyposaline condition. In some instances, the contribution of cations by a given species may lead to a sufficient increase in salinity for the species eventually to be eliminated. For instance, *C. asiatica* preferentially inhabits wetlands of low salinity, viz., ~500 ppm, but its physiological concentrating mechanisms result in an excess of 2000 mM/Kg of cations in the plant matter. In a *C. asiatica*-dominated wetland, leaf production may result in a peat or organic-enriched soil that has elevated cation content, which, through the elution process, and in the absence of a strong groundwater throughflow to initiate regular flushing, will result, in time, in elevation of the cation content of the interstitial waters and groundwater in contact with the sediment. If there is insufficient elution and dilution balanced against the rate of plant productivity, solutes may progressively rise and this will result in the elimination of *C. asiatica*, and its replacement with *M. rhaphiophylla*, for example, which is more tolerant of hyposaline waters. However, if the accumulated cations formed under the former *C. asiatica* cover are continually leached out of the sedimentary stock, then the vegetation assemblage will persist. Or if the content of cations in the replacement species are continually leached out of the sedimentary stock such that cation content is reduced in the sediment, the solutes in interstitial water and groundwaters also reduce, and there may be a return of the *C. asiatica* assemblage. The elimination and recolonisation of freshwater and hyposaline species may seesaw, simulating a response to a fluctuating climate. Such a seesawing event also may be amplified by short term cyclic changes in climate.

In regard to the effect of hydrology on shallow *vs* deep rooted species, it is evident that in the seasonal wetlands of the Becher Suite, the prevailing hydrological condition is neither inundation nor waterlogging but seasonal fluctuation of the water table. At maximum water table position, plants experience inundation or waterlogging; at minimum water table position, plants experience stress due to water deficiency and hence undergo wilting. If inundation and waterlogging were to be a stress, plants whose root zone is located in the seasonally inundated layer 0-20 cm below surface would have an advantage over those with deeper roots located within the zone of permanent inundation and waterlogging. Generally, the inundation frequency for 0-20 cm zone was 1.6 months per annum compared to 7.8 months for the zone 50-60 cm.

If water deficiency is a stress, plants whose rooting zone is located in the seasonally inundated layer 0-20 cm below surface suffer a disadvantage during the period of minimum water table position compared to those with deeper rooting systems located within the zone of permanent inundation and waterlogging. However, the period of water deficiency for plants rooted in the layer 0-20 cm below surface is reduced by 1-2 months because the infiltration from the first rains occurs in this layer 1-2 months before recharge causes the groundwater to rise. Plants in this layer also have access to infiltration from aseasonal rain events. Soil moisture levels in this layer are really only low in March and April. The fact that such a large proportion of plants in the Becher wetlands are clonal suggests that physiological integration may be occurring, particularly between ramets with rhizomes at different depths below the ground surface. Physiological integration has been shown to benefit plants during times of fluctuating resource availability (de Kroon *et al.* 1996) such as occurs in seasonal wetlands. At intermediate positions of the water table, this surface 0-20 cm layer is almost continually in the zone of capillary rise.

Soil moisture content favoured shallow rooted species because in the 0-10 cm layer it was highest and showed a consistency between seasons for many of the sites (Figs. 8-31, 8-32). Results from the physical analysis showed that in most wetlands the 0-10 cm layer also was continually oxygenated and the source of plant nutrients (Figs. 9-36 to 9-40, 9-44). The average depth of decaying roots is located at 0-10 cm, although there were occurrences up to 40 cm (Figs. 6-3 to 6-23). Sand sized organic material was present at 0-5 cm, and mud sized organic material ranged from 0-70 cm but was most common at a depth of 20 cm (Figs. 6-44 to 6-56). Given that the major requirements of plants are water, oxygen, and nutrients, these results are consistent with the fact that most of the plant roots are located 0-30 cm below ground surface.

Given the annual pattern of seasonal fluctuations, as well as the longer term fluctuations associated with climatic cycles, *i.e.*, the 20-year cycle indicated by the 10-year moving average (Fig. 8-1), plants in these seasonal wetlands are continually having to adapt to changing hydrological conditions. Competition with adjacent upland species, annuals, and alien species is ongoing (Table 10.17), and evidence of a single remnant plant within a different assemblage (such as an *Acacia* sp. amongst *B. juncea* sedgeland) indicates that cycles of expansion and contraction of assemblages, as well as invasions, are an inherent part of the natural history of this vegetation. In this context, the most successful plants are those that can propagate quickly under favourable conditions (between October and February). It is not surprising that the majority of wetland plants in the Becher setting are rhizomatous, an advantage in rapid plant regeneration, and survival during unfavourable conditions.

11. VEGETATION HISTORY

11.1 Introduction

The purpose of the pollen study was to determine wetland habitat evolution during the last 4500 years, using vegetation as the indicator. Relationships between modern plant assemblages colonising the wetland basins and current water levels, aspects of hydrochemisty and sediment development were used to deduce ancient wetland habitats, based on the presence of certain plant species in the past, and their increases and decreases in population, as ascertained from pollen numbers. Because the current vegetation assemblages in the wetlands are variable, even in adjacent wetland basins of similar age and height above sea level, the question arose as to whether differences in vegetation cover and composition were the end result of independent evolutionary processes, or the product of fine scale hydrological/sedimentological variation in the modern wetland. Resolution of this question required the changes in vegetation through time to be traced for selected wetland basins within the stratigraphical framework.

The objectives of this pollen study were defined with respect to each wetland basin rather than the region. The specific objectives were:

- to determine how accurately pollen numbers and species in the surface sediments record extant vegetation in the wetlands today;
- using information from the surface pollen patterns, identify the constraints to interpreting ancient pollen accumulations;
- to document pollen down profile in selected wetlands;
- to examine the patterns of pollen abundance down profile and to interpret these distributions in order to construct the history of vegetation in selected wetlands;
- to relate the pollen distributions to habitat conditions and change within the wetland basins;
- to investigate the relationship between present day pollen accumulation and present day vegetation in the wetlands and surrounds of the Becher cuspate foreland; firstly; and
- to determine whether there is a climate history that may explain changes in wetlands and their vegetation through time.

The content of this chapter is drawn from two papers, dealing separately with the surface and down profile pollen within the wetlands of the Becher Suite, in which the methods, data and analyses are presented in finer detail (C. A. Semeniuk *et al.* 2006a,b). Methods of pollen extraction and techniques of analysis are described in Chapter 2.

11.1.1 Background

Species that may contribute to the pollen record, deriving from the wetlands and ridges on the Becher Cuspate foreland, and those from terrain further east, are briefly summarised in Tables 11.1-11.3. For a more comprehensive description of the vegetation on the cuspate foreland refer to Chapters 2 and 10.

Table 11.1 *Plant species within wetlands*

Family	*Species*
Apiaceae	*Centella asiatica*
Brassicaceae #	*Brassica tournefortii**
Chenopodiaceae	*Halosarcia halocnemoides*
Compositae #	*Sonchus asper**
Crassulaceae #	*Crassula* sp. *
Cyperaceae	*Baumea articulata*
	B. juncea
	Cyperus sp.*
	Isolepis nodosa
	I. cernua
	Lepidosperma gladiatum
	Schoenoplectus validus
	Typha domingensis
	T. orientalis *
Geraniaceae	*Pelargonium capitatum* *
Iridaceae #	*Romulea sp.**
Juncaceae	*Juncus kraussii*
Juncaginaceae	*Triglochin* sp. *striata or mucronata*
Lobeliaceae	*Lobelia alata*
Myrtaceae	*Melaleuca cuticularis*
	M. viminea
	M. rhaphiophylla
	M. teretifolia
Papilionaceae #	*Trifolium spp.* *
Poaceae	*Sporobolus virginicus*
Primulaceae	*Samolus repens*
Verbenaceae	*Phyla nodiflora* *
Xanthorrhoeaceae	*Xanthorrhoea preissii*

* denotes alien species

\# denotes annual species

Table 11.2 Plant species on beachridge/dunes of the cuspate foreland

Family	*Perennial species*
Chenopodiaceae	*Rhagodia baccata*
Compositae	*Olearia axillaris*
Cyperaceae	*Lepidosperma squamatum*
	Schoenus grandiflorus
Dasypogonaceae	*Acanthocarpus preissii*
	Lomandra maritima
Epacridaceae	*Leucopogon parviflorus*
Euphorbiaceae	*Adriana quadripartita*
	Phyllanthus calycinus
Geraniaceae	**Pelargonium capitatum*
Haemodoraceae	*Conostylis aculeata*
Mimoseae	*Acacia cyclops*
	A. lasiocarpa
	A. pulchella
	A. rostellifera
	A. saligna
Myrtaceae	*M. systena*
Poaceae	*Austrostipa flavescens*
Proteaceae	*Hakea prostrata*
	Jacksonia furcellata
Rhamnaceae	*Spyridium globulosum*
Santalaceae	*Exocarpos sparteus*

* denotes alien species

Table 11.3 Key plant species or families deriving from distal sources to the east

Family	*Perennial species*
Casuarinaceae	*Allocasuarina fraseriana* *Casuarina humilis*
Myrtaceae	*Eucalyptus gomphocephala* *Eucalyptus marginata* *Hypocalymma robusta*
Proteaceae	*Banksia attenuata* *Banksia menziesii* *Banksia grandis*

Of these species, it should be noted that pollen from the Casuarinaceae, Compositae and Poaceae families are well preserved and common. Pollen grains, often destroyed or badly damaged by the process of acetolysis, include Juncaceae, Myrtaceae and

Cyperaceae (Nilsson and Praglowski 1992), therefore, the absence of these pollen from surface and core samples cannot be interpreted as indicating absence of a species from the wetland basin or surrounds. The stratigraphic intervals over which pollen occurs in the basin fills generally are seasonally waterlogged, with the exception of deeper sections of wetland 161 (80-100 cm), which is permanently inundated, and thus pollen grains are subjected to alternate waterlogging and aeration, with variable hydroperiods. Pollen, which is highly susceptible to degradation under these conditions belong to the groups of *Baumea, Lepidosperma*, and *Xanthorrhoea*, and are expected to be in low numbers. Nonetheless, in this study, species of these genera were detected in the stratigraphic record.

Pollen dispersal and transport

Potential dispersal mechanisms and transporting agents for pollen include insects, avifauna, local processes of *in situ* generation, easterly and westerly winds, rain, sheet wash, and water transport (Figure 11-1).

At Becher, modern distribution of pollen by wind would be related to the prevailing northeasterly and southwesterly winds typical of spring, and the west to south westerly seabreezes and easterly quandrant landbreezes, both dominant in summer. The pattern of transport along these trajectories would result largely in a redistribution of pollen within the cuspate foreland, but pollen could also be transported from vegetation on the higher ridges in the Cooloongup Lake area and from the bordering Spearwood Dune ridge. Vegetation communities on the ridges at Cooloongup comprise similar species to the beachridges at Becher, with the important addition of tall woodland of *Eucalyptus gomphocephala*. The vegetation on the Spearwood Dune ridge is *E. marginata/Banksia* spp. low woodland. On the Becher cuspate foreland, pollen derived from local ridges and sub-regionally may be deposited in the wetlands directly by rain or by sheet wash. As the wetlands are located in swales, it is unlikely that such pollen deposited onto the litter or sediment of the vegetated wetland basins would be remobilised.

Perturbations within this general dissemination pattern may be caused by the higher than average ridges of the Becher cuspate foreland. By forming a partial obstruction to easterly and northeasterly wind flows, initially increasing wind velocity at the ridge crest and then decreasing velocity over the adjacent wetland, they are likely to have some impact on the amount and type of pollen deposited in any wetland on their lee side. In wetlands situated between the lower beachridges, deposition of allochthonous pollen into the centre of the basin, may be hampered by the vegetation structure at the wetland margin, *e.g.*, *Melaleuca* trees in wetlands 35 and 45, and the closed canopy of *X. preissii* in wetland 162.

Wetlands situated in swales between the adjacent beachridges may generally be regarded as pollen sinks for both *in situ* and imported pollen from local ridges. Surface pollen within the wetland basin may be redistributed by wind generated

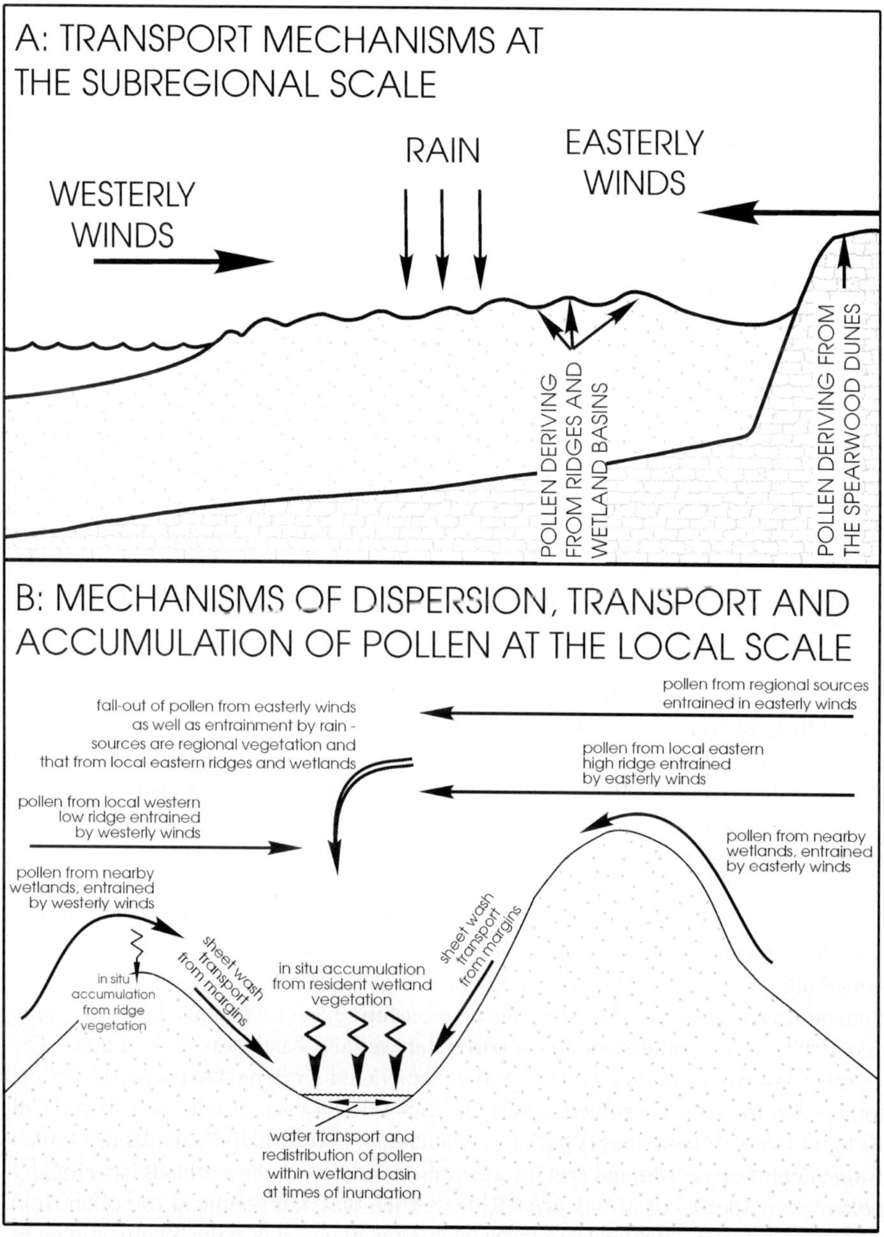

Figure 11-1. *Potential patterns of dispersal, transport and accumulation of pollen into wetland basins with in situ, local, and regional sources.*

water movement during inundation, and by bioturbation. Redistribution under the first process is likely to be minimal because of plant buffering. Redistribution under the second process is likely to be significant at the 10 cm scale.

11.2 Pollen in surface sediments

All wetlands were sampled for pollen at the surface in order to 1) encompass as complete a range of assemblages as possible, 2) determine the ratio of wetland/upland species contributing to the surface pollen in each wetland, 3) determine the various contributions to the wetland basin of local and regional pollen, 4) use the surface pollen assemblages as a baseline to interpret fossil sequences, and 5) see whether, by comparison of the proportion of wetland pollen between sites, the processes of distribution and preservation varied over the cuspate foreland. In each wetland, for the analysis of surface pollen, the species were separated into 6 categories (Fig. 11-2.):

1. pollen generated *in situ* from wetland basin vegetation;
2. pollen derived from *in situ* wetland margin vegetation;
3. pollen from allochthonous wetland vegetation;
4. pollen from local beachridge vegetation transported to the wetland basin;
5. pollen from distal vegetation in the region;
6. uncategorised pollen.

Allochthonous wetland pollen included those species currently colonising the Becher Suite wetlands elsewhere in the sub-region of the Becher cuspate foreland, but not within the basin from which the surface sample was obtained. "Ridge" pollen included species colonising the beachridges on the Becher cuspate foreland and "regional" pollen was categorised as deriving from beyond the cuspate foreland.

Overall, the majority of the surface pollen in the wetlands has been derived locally from the wetland and upland vegetation of the Becher cuspate foreland. The estimated cover and percent of total surface pollen for each species were summarised in C. A. Semeniuk *et al.* (2006a). From this, the relationship between extant wetland vegetation and surface pollen abundance in a particular wetland basin was determined. The results showed that *in situ* wetland pollen constituted a reasonable proportion of the total pollen found at the surface, varying from 5-44% (Fig. 11-2), with the exception of wetland swi. Wetland margin pollen was abundant in only half the wetlands. Pollen from allochthonous wetland species was present in most of the wetlands, specifically pollen from species of *Melaleuca*. Ridge pollen was sub-dominant to pollen from wetland species in 7 wetlands; in the remaining wetlands it was the dominant type of pollen. The contribution and significance of regional pollen varied (0-48%), from low numbers in most wetlands to relatively high numbers and significant proportions in wetlands 161 (17%), WAWA (48%), and 1N (20%). Uncategorised pollen ranged from 0-28%.

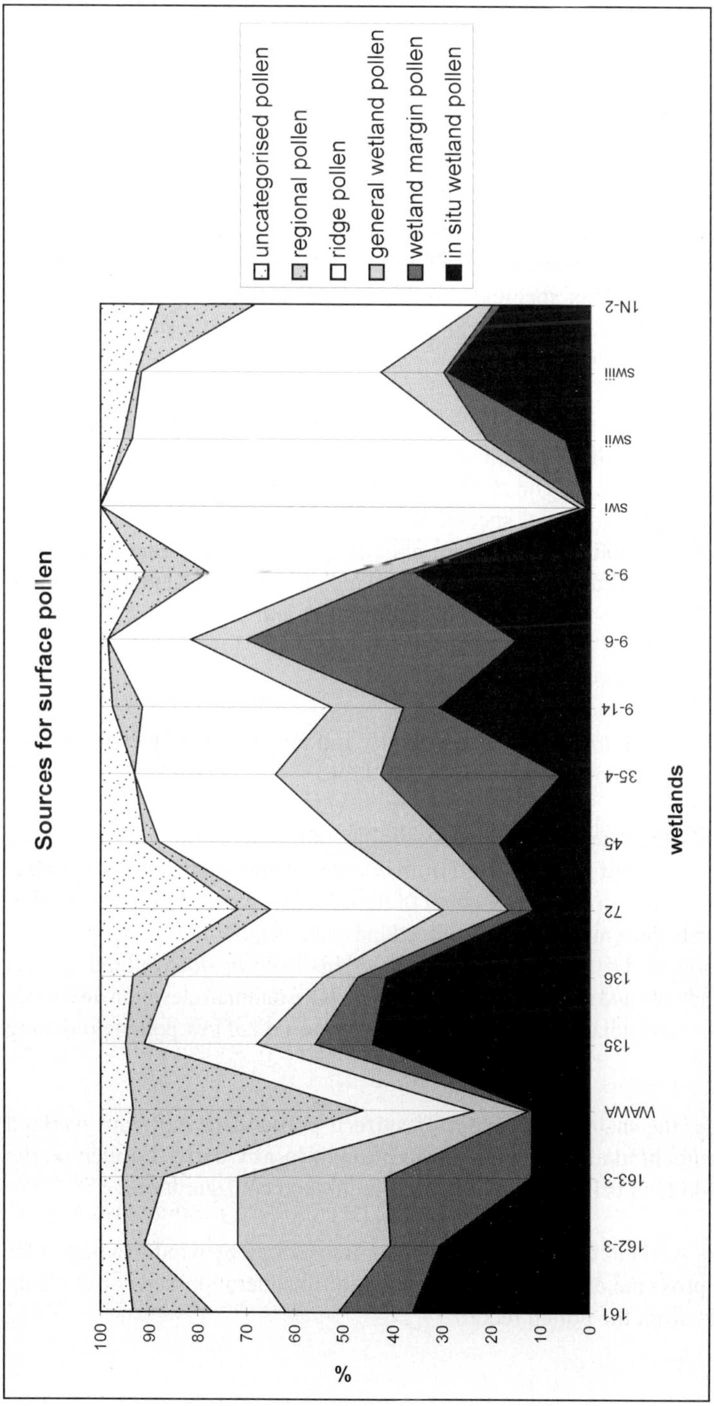

Figure11-2. *Composition of surface pollen for all sites.*

With the exception of *J. kraussii,* pollen for most species of plants colonising the wetland basin were present in the surface sediments (C. A. Semeniuk *et al.* 2006a). Pollen derived from *in situ* wetland plants was dominated by several species, *M. rhaphiophylla, S. virginicus* and *C. asiatica.* For the majority of species, the pollen numbers did not reflect the current abundance of plants in the wetlands. Pollen abundance for both species of *Baumea*, for two species of *Melaleuca* (*M. viminea* and *M. teretifolia*), and for *L. gladiatum* were consistently lower than the present vegetation cover would suggest. Pollen of *P. nodiflora* was also absent or rare even when cover abundance in a particular wetland was high, probably because of the short period of wetland colonisation by this species. Wetland plant species, *M. rhaphiophylla, S. virginicus, T. orientalis, I. nodosa,* and *C. asiatica*, were the most consistent contributors to the pollen assemblage.

Pollen from species in the wetland margins was also differentially represented, with an abundance of *I. nodosa* pollen and a deficit of pollen from *X. preissii* and *A. saligna* (C. A. Semeniuk *et al.* 2006a). The most consistent contributors to the surface pollen from ridges were *O. axillaris* and species belonging to Chenopodiaceae (probably *Rhagodia baccata*), the latter being more abundant in wetlands nearest the coast where the plants are more numerous. The regional pollen was dominated by species of Casuarinaceae (*Allocasuarina fraseriana, A. humilis)* and *Eucalyptus marginata,* important constituents of the *E. marginata/Banksia* spp. low woodland on the Spearwood Ridge to the east.

Percentage of pollen from *in situ* wetland plants, and percentage of wetland pollen which was not autochthonous is shown in Table 11.4.

Although many of the wetlands had very similar pollen densities in the surface sediments, the percentage of pollen derived from *in situ* wetland vegetation was variable, and depended on wetland size; ratio of cover of high pollen producing plants; number of adjacent wetlands; proximity of adjacent wetlands; and height of basin floor relative to surrounding ridges. Low numbers of pollen grains from *in situ* wetland species occurred in wetlands swi and swii which are very small, have minimal elevation difference between basin floor and adjacent ridges, and a high proportion of low pollen producing plants.

In most wetlands, the major component of surface pollen deriving from wetland vegetation was autochthonous. The high proportion of locally derived pollen on the Becher cuspate foreland is largely due to its geometry and configuration.

Whether pollen of wetland vegetation is *in situ*, or transported by wind to a particular basin from other proximal or distal basins, is a critical consideration in reconstructing vegetation history from the pollen record.

Table 11.4 *Wetlands listed in order of decreasing pollen abundance in surface sediments from wetland species*

(Key to species: Ac = *A. cyclops*, Ba = *B. articulata*, Bj = *B. juncea*, Ca = *C. asiatica*, In = *I. nodosa*, Lg = *L. gladiatum*, Mc = *M. cuticularis*, Mr = *M. rhaphiophylla*, Mt = *M. teretifolia*, Pc = *P. capitatum*, Scv = *S. validus*, Sv = *S. virginicus*, To = *T. orientalis*)

Wetlands	*In situ wetland pollen as percentage of total pollen*	*Main species in pollen assemblages*	*Percentage of wetland pollen which is not autochthonous*	*Total pollen counts/cm*^{3} *(rounded to 500)*
135	44	Sv, Mr	12	40,000
136	41.5	Mr, Sv	7	35,000
161	36	Ca, To, Ba	11	34,000
9-3	35.5	In, Bj, Mr	10.5	34,500
162	30.5	Ca, Mt, Sv	31.5	29,000
swiii	29	Scv, To, Lg	12.5	35,000
45	18.5	Mr, Ca	20	47,000
1N	18	In, Pc	3	6,000
9-14	15.5	Sv, Ac, Ca	14.5	31,000
9-6	15	Bj	11.5	81,000
163	12	Ca, Sv	30.5	29,000
72	11.5	Ca	13.5	29,000
WAWA	7.5	Ca	15.5	26,000
35	6	Mc	21.5	91,000
swii	5	Ca, Bj	4	88,000
swi	1	Bj	1.5	907,000

11.3 The use of surface and near surface pollen assemblages as a baseline for interpreting fossil wetland pollen sequences

Baumea articulata and *L. gladiatum* pollen appear to be reflecting *in situ* production and accumulation. The occurrences of pollen of *M. rhaphiophylla, M. viminea, M. teretifolia, M. cuticularis, Typha* spp., *S. virginicus*, and *C. asiatica* at first appearances would seem to be related in varying degrees to *in situ* production and contribution from wind. The occurrences of the four species of *Melaleuca* and the distribution of their pollen in surface sediments were compared (Fig. 11-3). The patterns show that where a species of *Melaleuca* is currently growing there is abundant pollen in the surface sediment, *e.g.*, the highest numbers of pollen of *M. rhaphiophylla* were in wetlands 135, 136, and 45, where the species is common. For *M. teretifolia* and *M. viminea,* the correlation was less pronounced.

There is evidence of local wind transport, particularly to wetlands nearby, *e.g.*, there is a high proportion of pollen of *M. cuticularis* in the wetlands near wetland 35, and pollen of *M. viminea* in the wetlands near wetland 9.

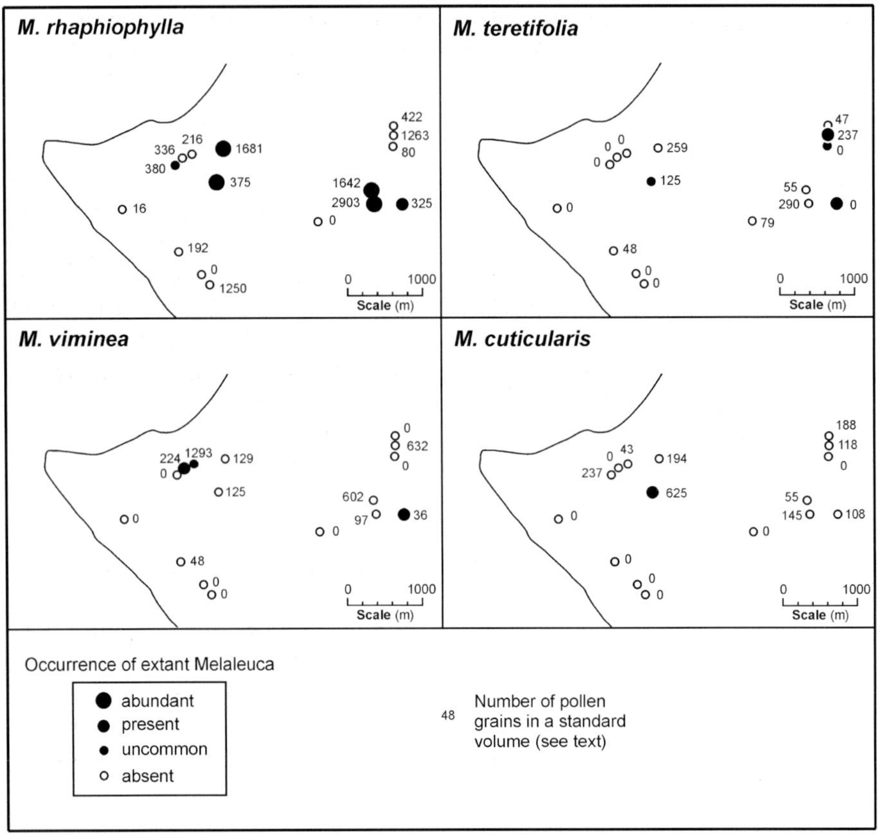

Figure 11-3. *Occurrence of extant species of Melaleuca in the wetland basins, and the relative abundance of their pollen spatially in the surface sediment.*

There are also occurrences of pollen species in wetlands, where the wetland species are absent today, which cannot easily be explained by wind transport, *e.g.*, *M. rhaphiophylla* in wetlands 161, 162, and 163, *M. teretifolia* in wetlands 136 and 45, and *M. viminea* in wetlands 162 and 135 (Fig. 11-3). This would suggest that these species have come and gone over the period of several decades or centuries, and that their pollen in the upper centimetre of sediment records these former populations. The recruitment and demise of such a population may be related to medium term climate changes. Drier conditions and increased wind activity may have taken place, increasing the salinity in wetland groundwaters, and favouring the more salt-tolerant species such as *M. cuticularis* and *S. virginicus* for a given wetland basin. In this context, the appearance of *M. cuticularis* and *S. virginicus* in response to a climate effect would be a local and not a sub-regional response.

A measure of the contribution of wind in supplying pollen from wetland species to a given basin was explored using the distribution of the pollen of upland species, as these species definitively indicate an origin outside of the wetland basin. Three species of regional and sub-regional pollen with distinct distal sources were used: *Olearia axillaris,* a primary dune species restricted to near coastal locations whose pollen would have to be delivered by westerly wind to the wetlands, *E. marginata* and Casuarinaceae spp. pollen which derive from the Spearwood Ridge and have to be delivered by easterly winds. The abundance of *O. axillaris,* Casuarinaceae spp., and *E. marginata* in the surface pollen varied from basin to basin, even for basins close to each other (Fig. 11-4). There was no clear gradient in abundance from source to distal wetland. For example, for Casuarinaceae spp., adjacent wetlands 161, 162, 163, registered 703, 79, and 40 pollen grains, respectively, and wetland 9 showed spatial variation from 0 to 129. Different wind fields and different grain fallout were evident for a single taxon, and between taxa, in an essentially isochronous layer, *i.e.*, the surface sediment of the wetland in a coastal climatic setting of today, and illustrated a wide variation in abundance across the receiving depositional surface.

The pollen record in the surface and near-surface sediment (*i.e.*, <1 cm depth, 3-5 cm depth, and 10 cm depth) in wetlands 161, 162, 163, 135 and 9-14 was noted against the extant wetland vegetation for a given wetland basin (C. A. Semeniuk *et al.* 2006b). The occurrence of pollen species in surface and near surface sediment when the plant species is absent in the wetland may be explained by 1) transportation into the basin by wind, 2) a sub-recent occurrence in the basin due either to hydrochemical changes or medium term climate changes , and 3) bioturbation into the surface layers from depths >10 cm.

It is concluded that the pollen of the main species of wetland plants in the stratigraphic profile mostly represents *in situ* accumulation. The following vegetation assemblages and their marginal components will be reflected in the pollen record, and can be used to interpret past vegetation patterns:

1. *B articulata* sedgeland
2. *Typha* spp. stands
3. mixed *B. articulata* and *Typha* spp. sedgeland
4. *M. teretifolia* scrub
5. *M. rhaphiophylla* forest/shrubland, with understorey of *C. asiatica*
6. *M. viminea* heath
7. *M. cuticularis* stands
8. wetland margins of *X. preissii*; understorey *I. nodosa* and *S. virginicus*

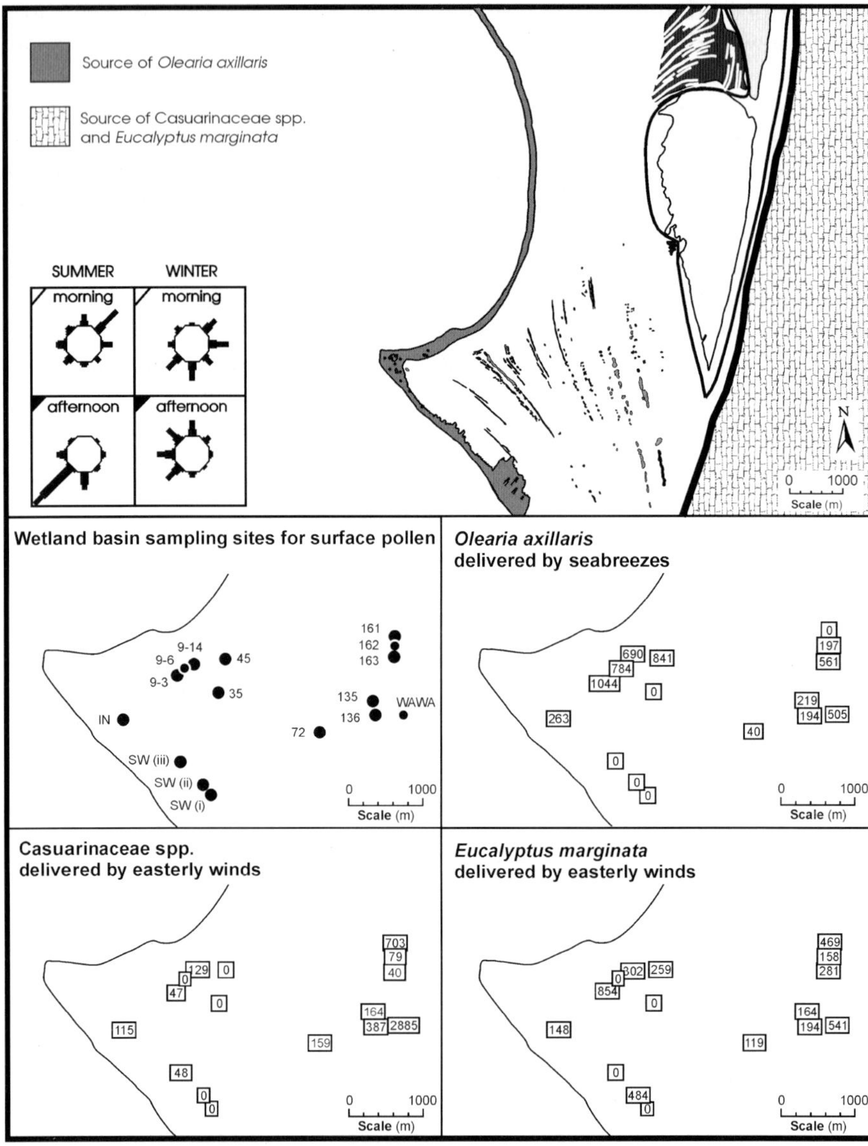

Figure 11-4. *Sources (coastal, inland regional) of the three types of upland pollen, summary of wind delivery systems, and the abundance of these pollen in surface sediments of wetlands.*

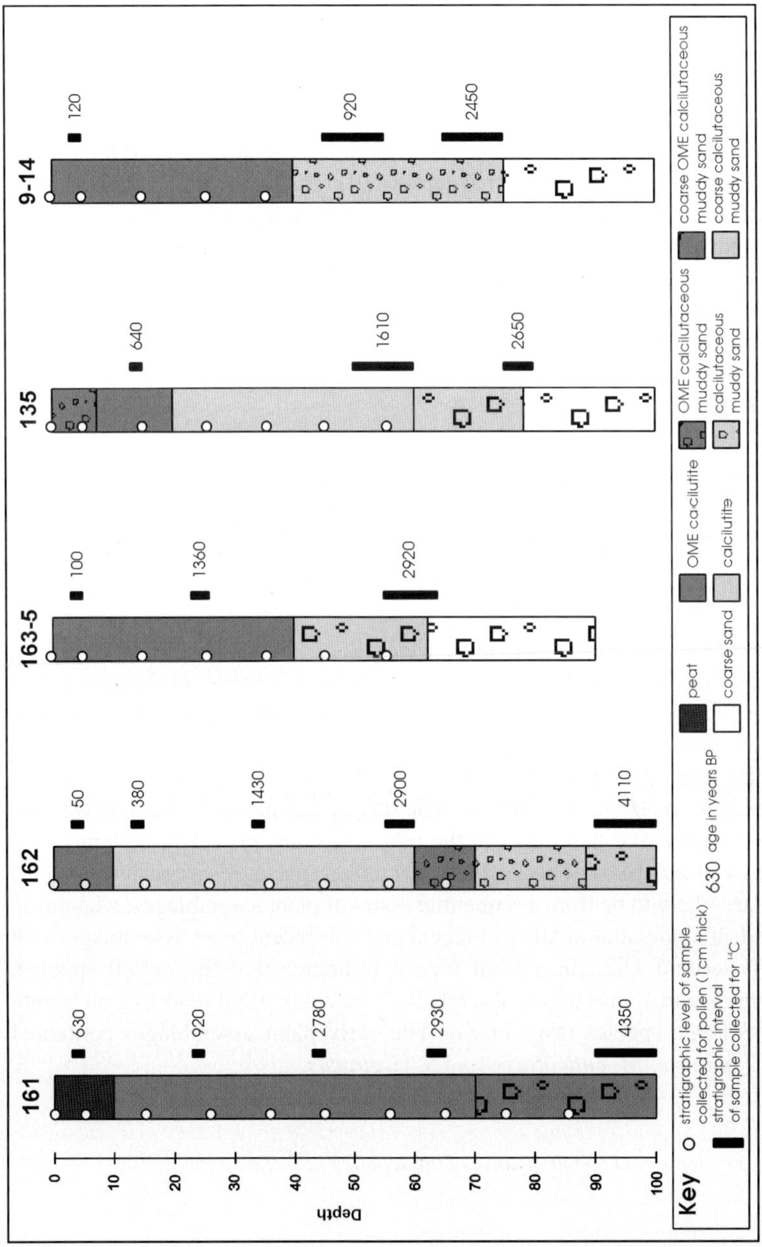

Figure 11-5. *Age structure of wetland fills in wetlands 161, 162, 163, 135, and 9-14, and sampling intervals for pollen.*

11.4 Pollen in selected cores

A core from each of the centres of five wetlands, selected to represent the range of ages, thickness of wetland fill, plant assemblages and geographic location on the Becher cuspate foreland (161, 162, 163, 135, and 9-14), was sampled for pollen abundance (C A Semeniuk *et al.* 2006b). The cores were sampled at the surface, at 3-5 cm, and thereafter at 10 cm intervals. To provide an age structure for the pollen, samples of mud were taken from selected depths and prepared for radiocarbon analysis (Fig. 11-5). The various lithologies sampled for pollen included carbonate mud, peat, and humic sand.

Pollen in the cores was categorised as follows (Fig. 11-6):

1. wetland pollen;
2. upland pollen;
3. regional pollen; and
4. undifferentiated pollen.

Pollen abundances down profile are presented and described in C A Semeniuk *et al.* (2006b).

11.4.1 History of vegetation in individual wetlands

Plots of pollen abundance against age structure in wetland 161 (Fig. 11-7) indicate that the extant species, *i.e.*, *B. articulata* and *C. asiatica,* were present in the early stage of wetland development (4350 ^{14}C yrs BP), but that *X. preissii* is a relatively recent arrival (920 ^{14}C yrs BP to the present). The extant assemblages appear to have replaced sub-recent assemblages of *Melaleuca* species. The surface sediments (0-5 cm) in wetland 161 span over 600 years in contrast to the other wetlands in which the equivalent interval spans 100-200 years. This suggests that pollen species recorded in the surface layer of 161 are likely to be from a sequential series of plant assemblages, whereas in other wetlands the potential mixture of recent and sub-recent plant assemblages will be less. In wetland 162, the pollen record indicates that the extant species, *i.e.*, *M. teretifolia* and *X. preissii* are also relatively recent (< 1000 years), again having replaced *Melaleuca* species (Fig. 11-8). The early plant assemblages contained *C. asiatica, Typha sp., M. cuticularis,* and *S. virginicus.*

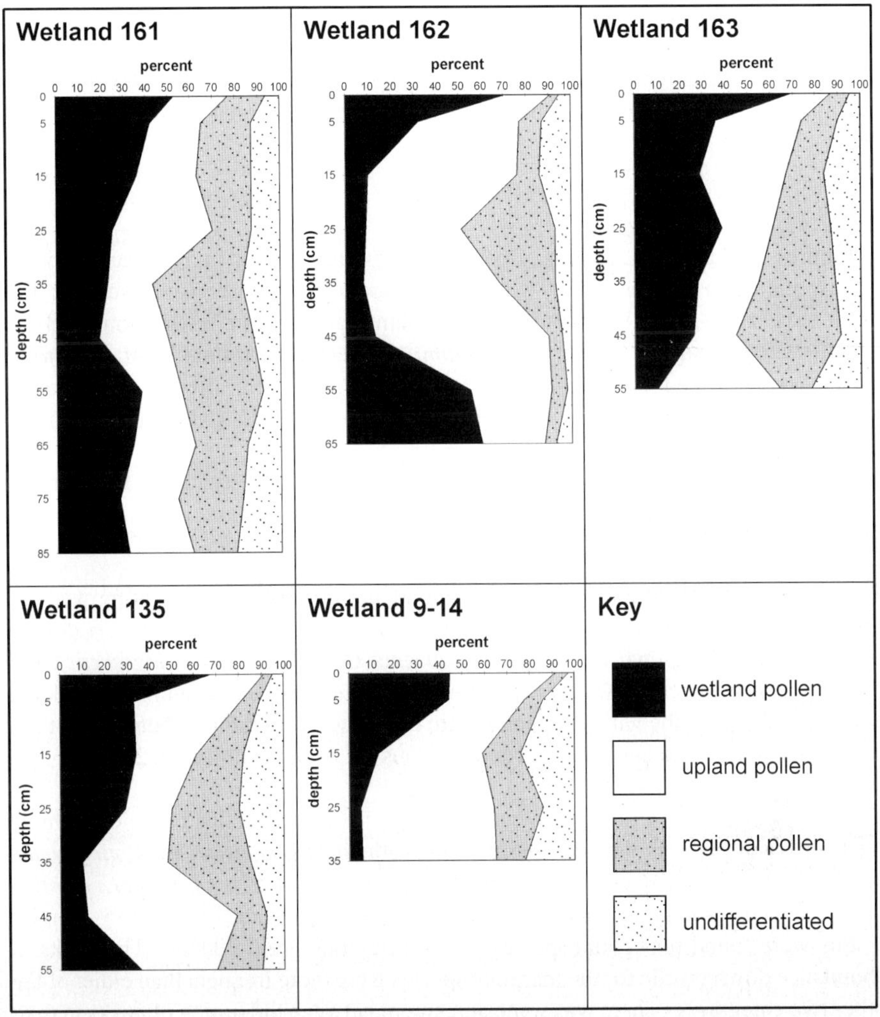

Figure 11-6. *Composition of pollen down profile for sites 161, 162, 163, 135 and 9-14.*

The first two species are likely to be related to the buried soil horizon and the latter two species to the commencement of carbonate mud deposition. In wetland 163, the pollen record indicates that extant species *C. asiatica* and *S. virginicus* were present from the time of wetland initiation up to the present, and *X. preissii* is a relatively recent arrival, (Fig. 11-9). *Melaleuca* pollen occurs in the sub-recent. In wetland 135, the pollen record indicates that the extant species *M. rhaphiophylla* and *C. asiatica* were present near the beginning of wetland development (Fig. 11-10), together with a number of other species. *Melaleuca* pollen occurs in the sub-recent. For wetland 9-14, it is not possible to determine the colonisation period for *J. kraussii* (Fig. 11-11), but *C. asiatica* was again an early coloniser. Although the pollen from plants in the current wetland assemblages is present throughout the stratigraphic cores, the earliest record of the two marginal species, *I. nodosa* and *X. preissii*, in the Becher area is *circa* 1500 ^{14}C yrs BP.

11.4.2 Species associations

In Figures 11-7 to 11-11, groups of pollen species are captured in 500 year intervals, corresponding to contructed 500-year spaced isochrons. These groups suggest that there have been repetitive occurrences of plants growing in association, such as 1) *B. articulata* and *C. asiatica*, 2) *M. rhaphiophylla* and *C. asiatica*, 3) *M. teretifolia* and *M. viminea*, and 4) *C. asiatica* and *S. virginicus*. There also are examples of species occurring alone, such as *M. cuticularis*, *C. asiatica* and *I. nodosa*; and examples of mixed groups.

11.4.3 Correlation of abundance patterns for selected pollen species between basins

The timing of peaks in the down profile abundances of pollen representing wetland species, (*C. asiatica, S. virginicus, M. rhaphiophylla, M. viminea* and *Triglochin* spp.), upland and regional species, (Casuarinaceae *O. axillaris* and *S. flavescens), and* wetland margin species, *(X. preissii* and *A. quadripartite),* were examined. Peaks in pollen numbers for the various species were correlated with radiocarbon dates of the sediments to facilitate inter-wetland comparisons (C. A. Semeniuk *et al.* 2006b).

There appeared to be some agreement in timing of peaks of wetland species within a wetland basin but less between basins. In wetland 161 *C. asiatica, S. virginicus, M. rhaphiophylla* and *M. viminea* exhibited peaks *circa* 1800 ^{14}C yrs BP. For *C. asiatica* there was a corresponding peak in wetland 163 but not in wetlands 162 or 135. Similar results were found for upland species (C. A. Semeniuk *et al.* 2006b). The peaks in abundance down profile for the marginal species were more frequent than either of the other two categories. There was some agreement between the timing of peaks in that, for many, there was a corresponding occurrence in at least one other wetland basin suggesting that the frequency of peaks could be greater if sampling was undertaken at a narrower interval.

11.4.4 Upland pollen

In terms of vegetation history, from the beginning of wetland development to the present, the species composition of both subregional and regional pollen has been similar, although there have been changes in abundance (Fig. 11-12).

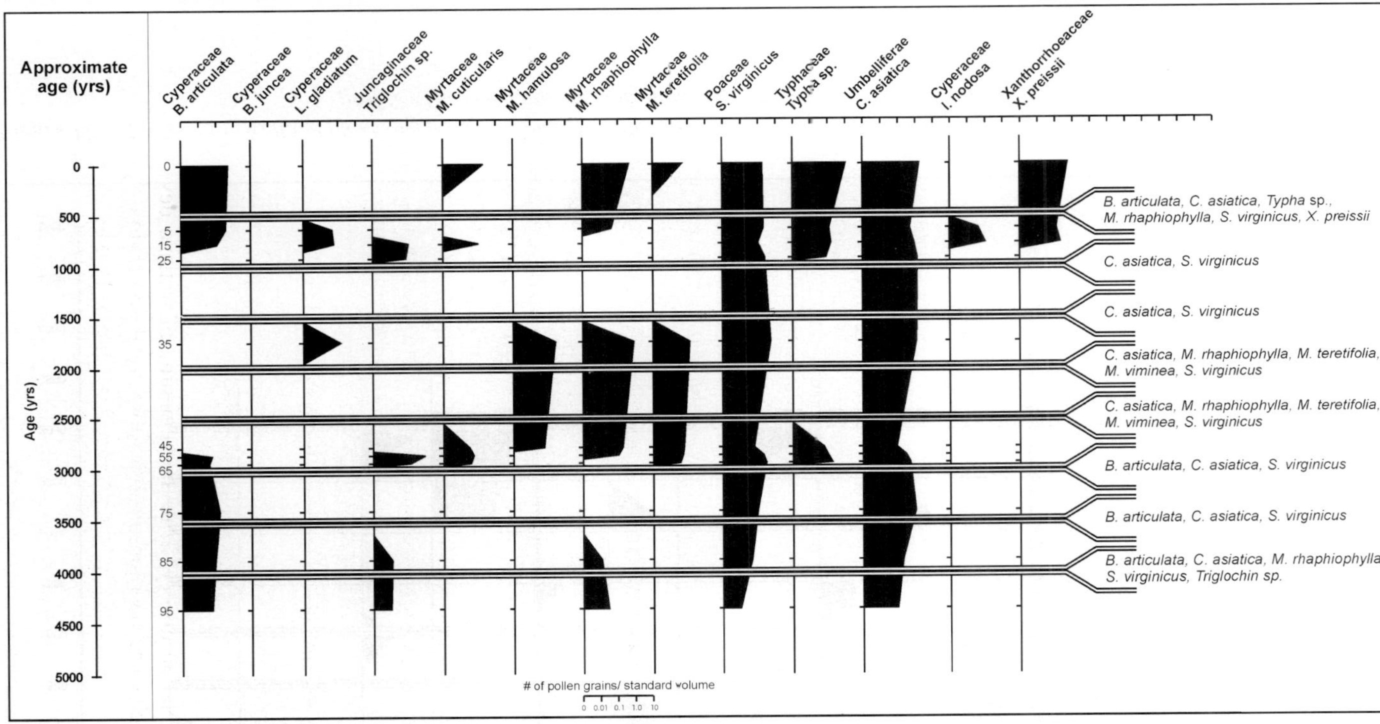

Figure 11-7. *Pollen from wetland plant species in wetland 161. Plant assemblages as determined by pollen composition down the stratigraphic profile. The vegetation composition is constructed at 500 year intervals.*

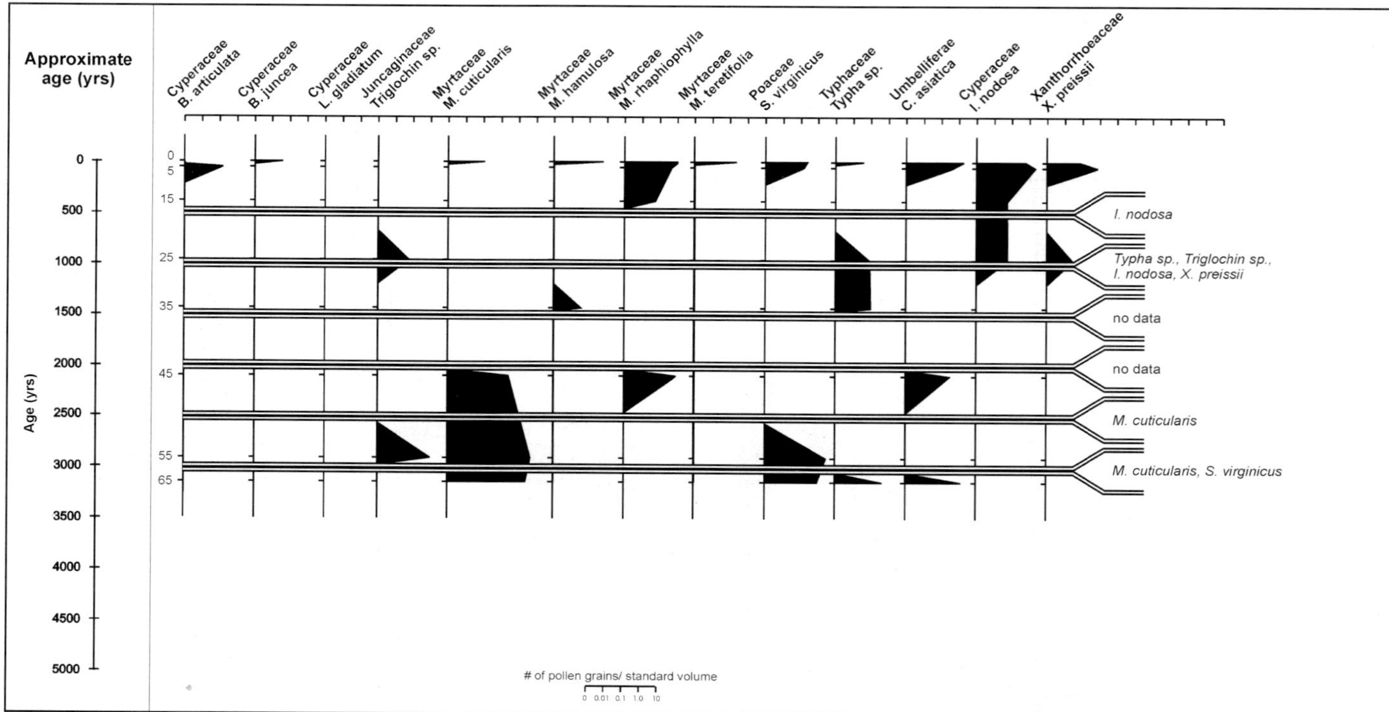

Figure 11-8. *Pollen from wetland plant species in wetland 162. Plant assemblages as determined by pollen composition down the stratigraphic profile. The vegetation composition is constructed at 500 year intervals.*

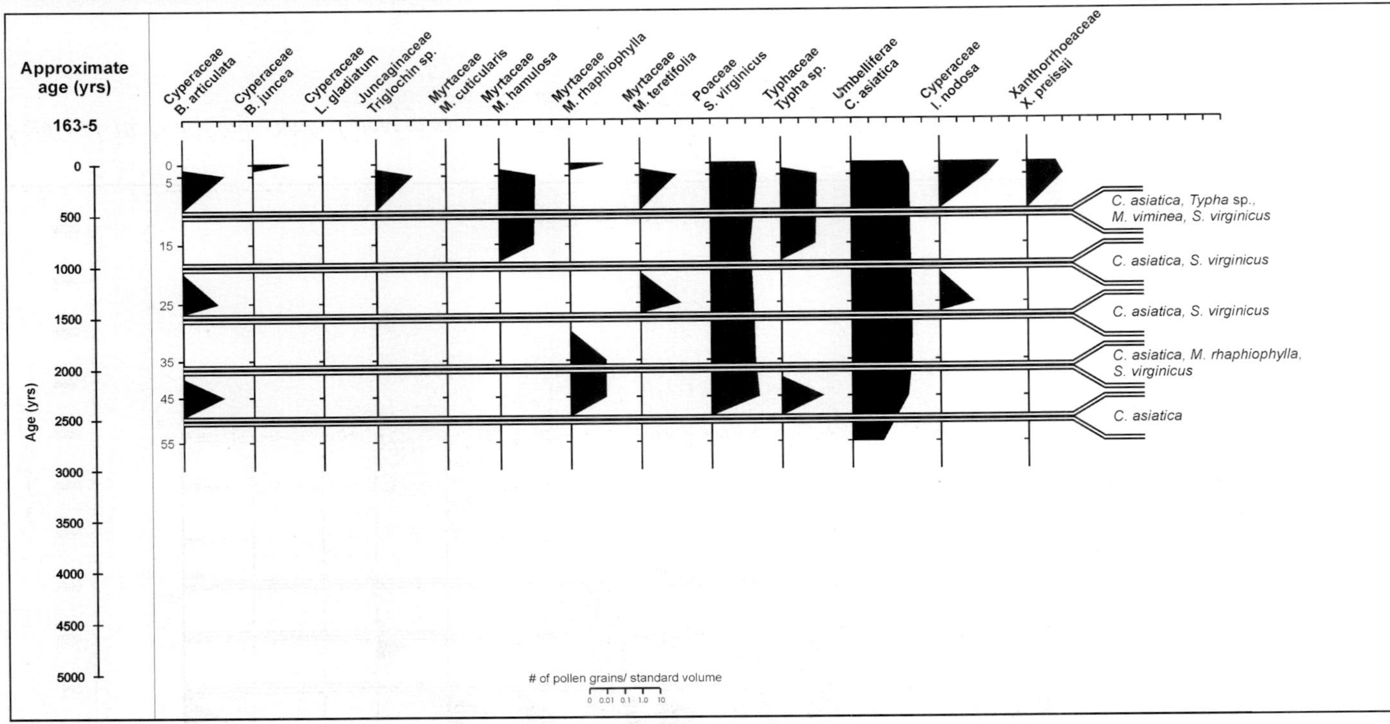

Figure 11-9. *Pollen from wetland plant species in wetland 163. Plant assemblages as determined by pollen composition down the stratigraphic profile. The vegetation composition is constructed at 500 year intervals.*

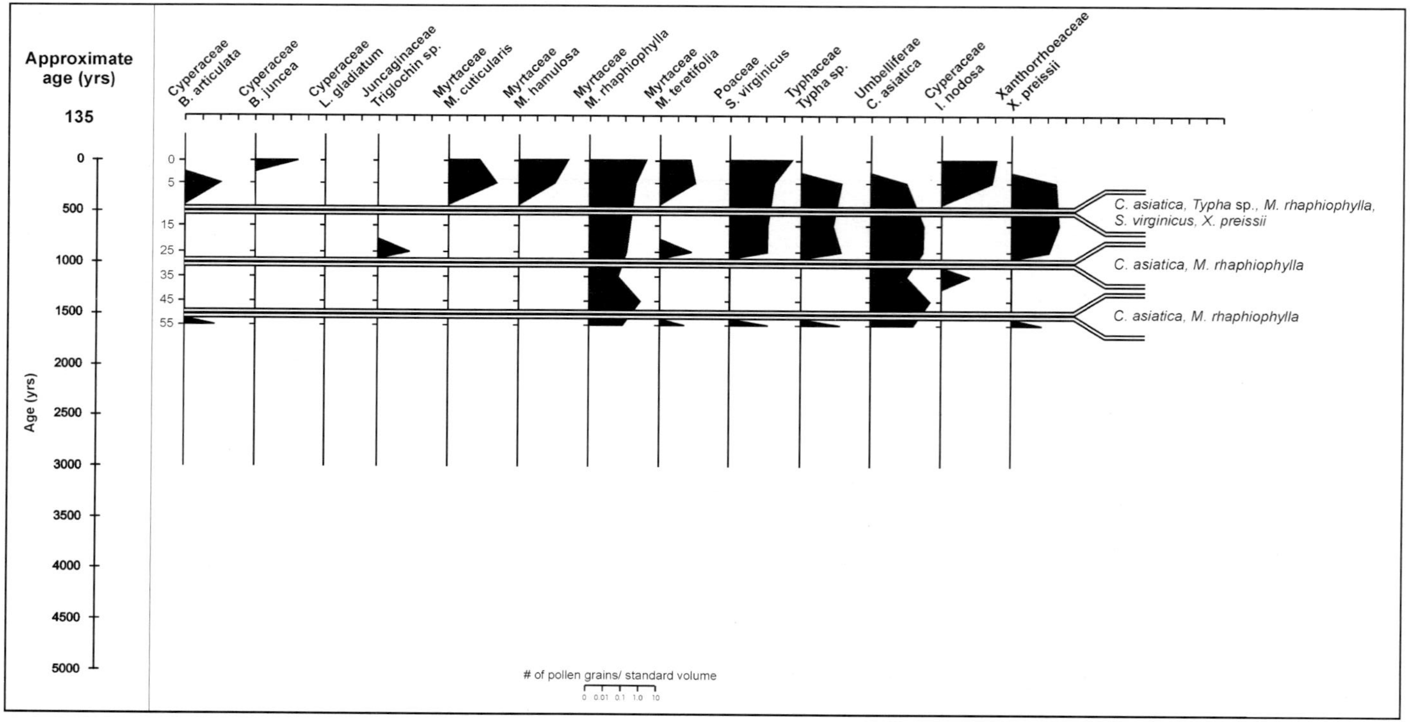

Figure 11-10. *Pollen from wetland plant species in wetland 135. Plant assemblages as determined by pollen composition down the stratigraphic profile. The vegetation composition is constructed at 500 year intervals.*

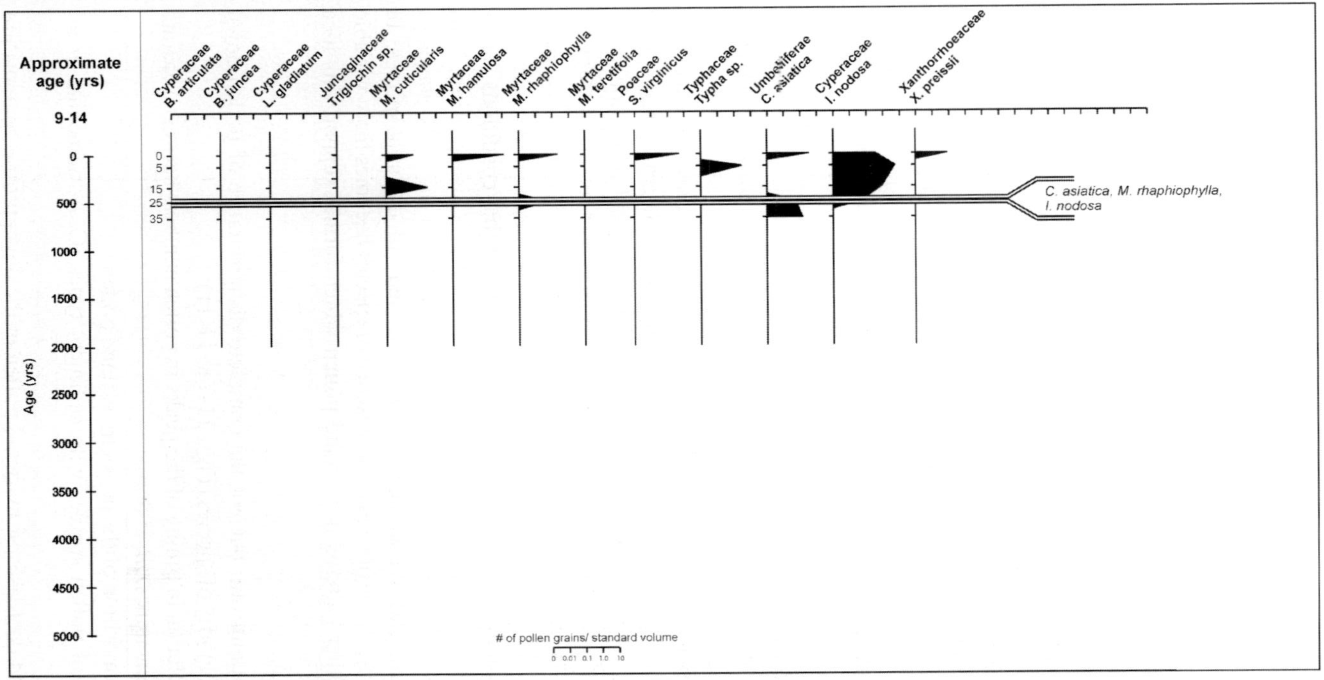

Figure 11-11. *Pollen from wetland plant species in wetland 9-14. Plant assemblages as determined by pollen composition down the stratigraphic profile. The vegetation composition is constructed at 500 year intervals.*

11.5 Interpretation of results

11.5.1 Pollen numbers in relation to sediment type

Ordering of pollen abundance with respect to sediment types showed that numbers in buried soil horizons were highest, followed by peat and OME carbonate mud, carbonate mud and muddy sand (C. A. Semeniuk *et al.* 2006b). Assemblages associated with distinct sediment types were examined. In wetland 161, the variation in numbers of *B. articulata, C. asiatica* and *Typha* spp. pollen matched the increases and decreases in organic matter content in the sediment layers, exemplified by several of these species reappearing in the buried soil horizon at the base of the carbonate mud (Fig. 11-7). In the carbonate muds, the numbers of *M. cuticularis* and *Triglochin* spp. increased. These relationships were clearer in the profile for wetland 162, but more ambiguous in wetland 135 (Figs. 11-8, 11-10). The relation between pollen numbers and sediment type was one of the factors used to deduce the importance of intra-basinal environments in determining wetland pollen assemblages compared to regional or climatic effects.

11.5.2 Relating wetland pollen to habitat type

Pollen derived from wetland species exhibits different composition, different abundances, and different patterns and timing in peaks in individual wetlands. This suggests that, even in instances where the same species occurred contemporaneously in more than one basin, distribution and population within separate basins were probably dissimilar, mirroring what is evident in the extant wetland vegetation today. It is concluded that the ancestral distribution and abundance of plant assemblages in the Becher wetlands were a function of intra-basin environmental changes caused by wetland evolution. Within a single wetland species, *e.g., C. asiatica,* the increases and decreases in abundance throughout the profile, suggest that conditions for its growth must have regularly fluctuated, without ever completely disappearing. This conclusion is supported by the fact that *C. asiatica* occurs in the early stages of wetland development in all five wetlands, but is associated with different species. In subsequent development, the presence and absence of species becomes more variable in the wetlands. Other patterns in wetland pollen which further support this idea include:

- the lack of continuity in, but not the complete disappearance of pollen down profile for the majority of species (Figs. 11-7 to 11-11);
- the lack of correlation in timing of the peaks in pollen numbers between separate basins (C. A. Semeniuk *et al.* 2006b);
- the variable composition of the peaks in wetland pollen ;
- the association of pollen species with sediment types (C. A. Semeniuk *et al.* 2006b); and
- the increases and decreases of marginal pollen in the down profile composition (Figs. 11-7 to 11-11).

Intra-wetland changes are likely to include responses to alternating periods of above and below average rainfall resulting in the expansion and contraction of plant assemblages and the invasion of wetland marginal species. The fluctuations in the pollen of marginal plant species are related to real presence and absence of a particular species as it migrates between the wetland centre and the margin in response to water availability. The regularity of these pollen peaks suggests that there is a cyclicity in the amount of rainfall which recharges the wetlands.

Given that the wetland plant species pool has been consistent over the last 4500 years, it is important to attempt to differentiate between the pollen record derived from one or two plants of a given species, and an assemblage. A list of pollen species occurring in sediments of similar age (C. A. Semeniuk *et al.* 2006b) shows that species may or may not be present in wetland basins at the same time, and that in most cases the abundance varies even when the composition is similar.

Plants growing in association can be used to interpret hydrological and hydrochemical changes, based on similar environmental attributes in their current habitats. For instance, *B. articulata* and *C. asiatica* indicate seasonal shallow inundation and fresh water, *M. rhaphiophylla* and *C. asiatica,* and *M. teretifolia* and *M. viminea* indicate seasonal waterlogging and fresh to hyposaline conditions, and possibly an expansion of species inhabiting the marginal zone, *M. cuticularis* indicates less frequent waterlogging and more saline conditions, and the occurrence of *C. asiatica* and *S. virginicus* together indicates short term changes between seasonal waterlogging and dry periods. The occurrences of a single pollen type suggest dominance within the wetland rather than a pure stand, *e.g.*, *M. cuticularis*, *C. asiatica*, and *I. nodosa*.

11.5.3 Relating upland pollen to habitat type

The pollen derived from upland vegetation exhibits continuity down profile, with fluctuations in abundance, however, the fluctuations cannot be rigorously correlated between the separate wetland basins (Fig. 11-12). Quantitative peaks down the stratigraphic profile in the three pollen types, illustrate total numbers of upland pollen, and the proportion of the three key species contributing to that total pollen number. Figure 11-12 shows that while total upland pollen decreased between 2000 and 500 ^{14}C yrs BP in wetland 162, the proportion of Casuarinaceae pollen increased at that time. In terms of interpreting the history of wind patterns that delivered the regional upland pollen to a given wetland basin, the graphs showing the proportions of contributing species are a better index of wind direction, while those illustrating total upland pollen numbers are a better index of wind intensity.

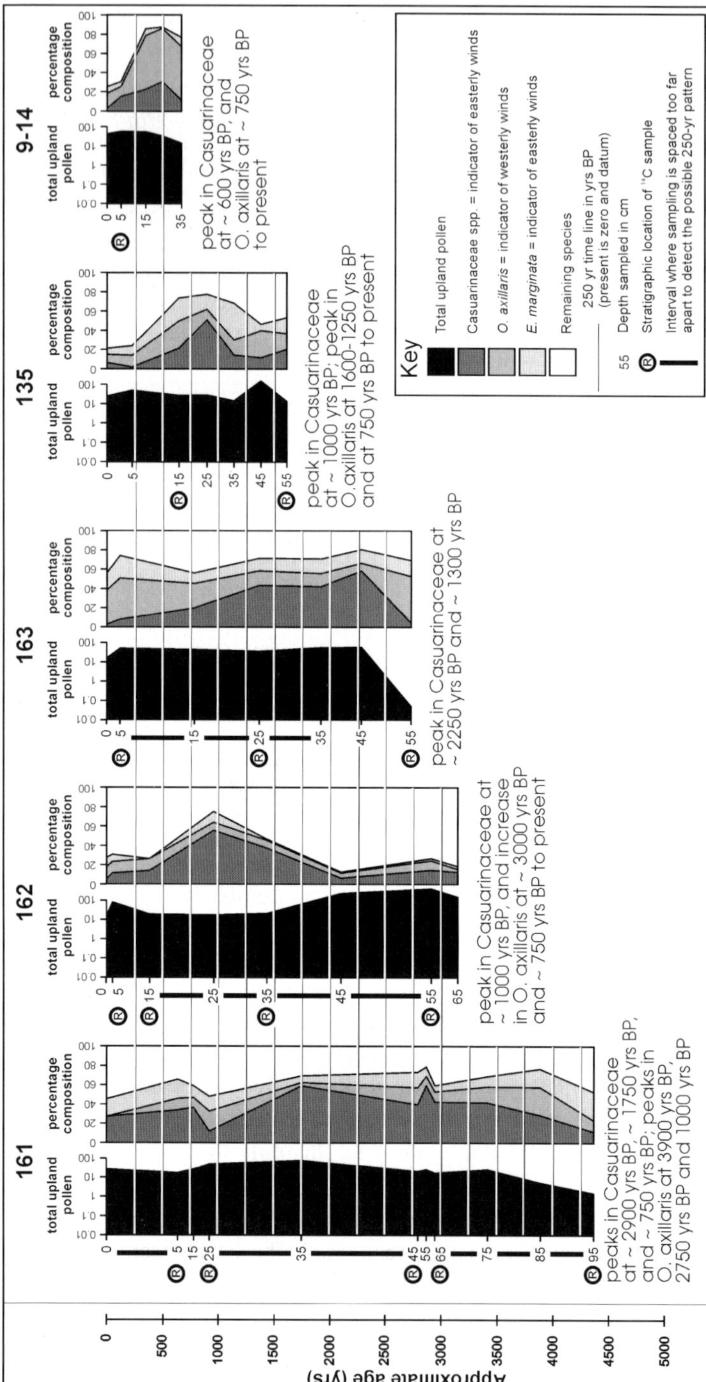

Figure 11-12. *Abundance and composition of upland pollen (selected species) down profile as an index of wind intensity and wind direction.*

Circa 2900 ^{14}C yrs BP, the numbers of Casuarinaceae and *E. marginata* pollen increased in the wetlands that were present at that time, *i.e.*, 161 and 162, suggesting that this period was associated with prevailing easterly winds. For the same period, in wetland 162, high numbers of pollen from Chenopodiaceae (probably *Rhagodia baccata*, a beachridge species), and increases in the pollen of *M. cuticularis, S. virginicus* and *Triglochin* spp. occurred, which could be interpreted as a change to more open conditions within the wetland basin with an increase in salt tolerant plants (Figs. 11-8, 11-12). The sediment accumulating during this period was carbonate mud, a depositional product associated with sub-regional groundwater rise (due to coastal progradation) rather than increased rainfall. In combination, these factors strongly suggest a wetland habitat adapting to drier climatic conditions.

In wetland 135 *circa* 1400 ^{14}C yrs BP, a similar increase in pollen from upland species occurred (Figs. 11-12). This increase, again coinciding with carbonate mud deposition, may also indicate a period during which vegetation cover in this basin became more open. A comparison with the other wetlands cannot be made because there were no samples for this period from wetlands 161 and 9-14, and in wetlands 162 and 163, although there are increases in the pollen of marginal wetland species (*M. viminea, I. nodosa*), and a decrease in *C. asiatica*, the composition of the wetland pollen does not suggest a sudden and intense change comparable to the arrival of *M. cuticularis* in the earlier period.

11.6 Serial development of wetland vegetation

Reconstruction of a wetland vegetation series was attempted on the following premise. As the origin of the wetlands was one of groundwater rise into an undulating interdune depression, there would potentially be a progression in water availability from seasonal wetness to seasonal waterlogging to seasonal inundation, and plant response is likely to reflect this gradation. The extant wetland vegetation assemblages can therefore be loosely categorised as sumpland and dampland types, and related to a rising water table. This single environmental factor of rising water table also embodies many other changes within the wetland basin such as organic content, hydrochemistry, and sediment accumulation. There is a general spatial chronosequence within the Becher Suite from east to west (4500-600 ^{14}C yrs BP) and this may be used to link wetland vegetation associations with stage of development.

The presence of pollen deriving from vegetation in the wetland margins is a key to determining the pathway of wetland vegetation changes because rising water levels in the Becher Suite wetlands cause a dynamic response in these water sensitive plant assemblages, resulting in a spatial shift within the wetland basin. Given the evidence of expansion and contraction of vegetation assemblages documented during the 10 years of variable rainfall during this study, the assumption is made that the present inhabitants of the wetland margins were present elsewhere in the basin in the past.

Beginning with the youngest of the selected wetlands, 9-14, a number of reconstructions are presented for vegetation succession in the Becher Suite wetlands, based on the pollen species in the cores as well as maps of current wetland vegetation assemblages within the study wetlands.

In the pre-wetland interdune depressions behind the younger beachridges in modern environments, the most common assemblage is sedgeland comprising *I. nodosa*. As wetness increases in the dune depression due to a rising water table, *B. juncea* begins to colonise the basin, *i.e.*, the assemblage in wetland 1N is the modern example, and pollen of Cyperaceae sp. and *I. nodosa* 0-15 cm in core 9-14 is the ancient example. If the depth to water is shallow, *J. kraussii* colonises the sandy substrate instead of *B. juncea*. This is demonstrated in the wetland vegetation in wetlands swi and swii and in the other two wetland suites in the Quindalup Dunes, Cooloongup and Peelhurst (C. A. Semeniuk 1988). Assemblages of *B. juncea* and *J. kraussii* produce a fine mud layer at the surface through plant decay and trapping of fine sediment grains (aeolian or suspended sediment), increasing the water holding capacity of the sediments. As a result of this, *C. asiatica* colonises the habitat (see pollen occurrences at the base of each of the stratigraphic cores, Figs. 11-7 to 11-11). This progressive series of assemblages is exemplified at various stages in the following wetlands in the Becher Suite, swi, swii, swiii, 1N.

In the pre-wetland interdune depressions behind the older beachridges in modern environments, the most common assemblage is a closed formation of grass trees comprising *X. preissii*. However, the pollen records show that *X. preissii* arrived in the Becher area *circa* 1500 ^{14}C yrs BP, so that pre-wetland interdune depressions prior to this were probably colonised by *L. gladiatum*. As wetness increases in the dune depression due to a rising water table and increasing organic content, *X. preissii* begins to migrate outwards and *B. juncea* begins to colonise the basin centre, as observed in many basins in the Becher Suite.

In wetlands which formed before the arrival of *I. nodosa*, swales were probably colonised by *J. kraussii* or *L. gladiatum*, although this cannot be verified by the pollen record because of their poor pollen preservation. If the water table rise was rapid, *L. gladiatum* would colonise the sandy substrate (*e.g.*, wetlands WAWA and 142). *L. gladiatum* produces abundant leaf litter at the surface through plant decay. As the water holding capacity of the sediments increased through organic detrital accumulation, *C. asiatica* would colonise the habitat *e.g.*, cores 135, 163 and 161. As the water table rises, and water longevity in the upper layers of the profile increases, flood tolerant plants such as *B. articulata* and *Schoenoplectus validus* would begin to compete with *C. asiatica* and it, too, would retract towards the wetland margins. This progressive series of assemblages is exemplified at various stages in the Becher Suite wetlands 161, 163, WAWA, 135, 136, 35, 72, 63, 45, 9-6, 9-14, swiii.

In the older wetlands, the down profile bands and mottles of OME carbonate mud and carbonate mud suggest successive humid and drier periods within the interval of wetland sediment deposition. Depending on the date of wetland initiation, these humid and dry climatic phases correspond with different stages of watertable rise in a particular basin. Where wetland initiation coincided with a wet period, the vegetation series described above are possible scenarios. Where wetland initiation coincided with a dry period, species tolerant of salinity such as *M. rhaphiophylla*, *M. viminea* and *M. cuticularis* are more likely to have colonised the wetland margins, with *S. virginicus* replacing *B. juncea* as the understorey and *Triglochin* sp. and Chara spp. colonising the centre, *e.g.*, the current vegetation assemblage in wetland 35 is the modern example, and pollen at 45 cm in core 162, pollen at 45-65 cm in core 161, and pollen in surface sediments in core 135 are the ancient examples. This progressive series of assemblages are exemplified at various stages in the Becher Suite wetlands 162, 142, 135, 136, 45, 35, 9-6.

It has been suggested elsewhere that the colonisation of dune slacks by sedges and rushes (such as *I. nodosa* and *J. kraussii)* during wet periods is the more common pathway in succession (Lubke and Avis 1982, Avis and Lubke 1996), the controlling environmental factors being rainfall and soil moisture. A similar conclusion is reached in this study. The production by *J. kraussii* and *L. gladiatum* of organic rich surface layers in dune slacks provides an increase in available nutrients and stimulates the growth of shrub and tree species (Grootjans *et al.* 1991). Avis and Lubke (1996) also suggest that in the dune environment, vegetation succession can either progress or regress under respective favourable and unfavourable climatic conditions, such that a vegetated dune slack can revert to plant species, cover and structure types indicative of pioneer slacks when sand mobility, moisture availability, and salt accumulation characteristics coincide with low rainfall. As this progression has been observed in the Becher wetlands during the period of this study, it is highly plausible that the same process has occurred earlier in the Holocene.

11.7 Discussion and conclusions

Interpreting any fossil pollen record or even its contemporary record is complex, as it is influenced by the relative rates of pollen production, taphonomic considerations, vectors of transport, and in regard to aeolian transport, wind directions and wind speeds in relation to flowering times. A full analysis of these factors in generating a fossil or contemporary record was beyond the scope of this study. In addition, there were a range of other factors contributing to the difficulty of interpreting the vegetation history of each specific pollen sequence in the Becher wetlands, as discussed below.

Firstly, if the vegetation had undergone fundamental changes in composition relatively quickly in response to 20-year and 250-year climatic fluctuations (Semeniuk & Semeniuk 2005), or hydrochemical changes in local groundwater, then to detect such changes, the pollen sequences within the rates of sedimentation determined by ^{14}C dating needed to have been sampled on a *circa* 1 mm interval. The monitoring of vegetation over the 10 years of study indicated that many species indeed fluctuated in abundance in response to the 20-year climatic variation.

Secondly, bioturbation mixes the sediments over at least a 10 cm depth, obliterating any potential fine scale sequencing of pollen. This means the pollen record from the sampling interval potentially records mixed wet and dry climate assemblages that may have alternated on a 20-year or 250-year turnaround. Even the surface pollen record from the wetlands may not in all cases be reflecting the extant assemblage compositions, but rather, given the accretion rates of sedimentary material within the wetlands, and bioturbation, a composite of up to several hundred years of record (*e.g.*, wetland 161).

Thirdly, the relative abundance of pollen is no indication of the relative abundance of plant cover. Without supporting autoecological and taphonomic information on the relative production rates of pollen from the various species, and the relative importance of flowering and seed production as a population maintenance strategy compared to cloning, the pollen record must be used with caution to interpret former relative proportions of species that contributed to the stratigraphic record.

Fourthly, in the absence of rigorous studies on the modern dispersion rates and dispersion patterns (involving identification of major wind flow paths, consistently generated eddies, and grain size fallout zones), it is not clear how far the pollen of a given species can be transported by wind and water. This means that it is difficult to differentiate wholly intra-basinal contributions from margin contributions or possible extra-basinal contributions. The discrimination between wind transported pollen and that generated within the basin in response to a medium term climate change to more arid conditions, is hampered by the fact that the species abundances and composition are likely to be the same in either situation. In addition, in the Becher wetlands, the species which are adapted to groundwater salinity are also the species which have efficient pollen dispersal and transport mechanisms.

Any medium term climate changes that involved modest shifts in wind direction and speeds, as governed by oceanic and/or interior arid hinterland effects, may not have been ubiquitous across the region or sub-region. Local topography of higher than normal beachridges, or of continuous swales acting like tunnels, may have influenced local deposition. Prominent marginal vegetation may have acted as interceptors to transported pollen in contrast to open basins with no barriers. Therefore the occurrences of upland pollen species within the wetland were used as coarse indicators of potential local wind deposition of pollen of wetland species.

Lastly, a major limitation to pollen studies in the wetlands is the lack of finely spaced dated material. While there have been a large number of ^{14}C dates determined in the area, in terms of sedimentary sequence, and for dating the base of the sequences, the sampling interval has not provided enough dates to correlate isochronous events across the various wetland basins. For example, while there is a marked pollen event in wetland 162, with an abundance of species *M. cuticularis* and *S. virginicus* at approximately ^{14}C age of 2900 yrs BP, the exact equivalent ^{14}C age interval to a precision of say, 250 years, has not been sampled in the other wetlands.

In spite of limitations described above, there are several conclusions which can be made from this pollen study and the vegetation history. Firstly, the species pool of wetland plants colonising the Becher Suite wetlands has remained fairly stable over the last 4500 years. Secondly, both the sediments and the pollen indicate that the history of the wetlands included alternate wetter and drier periods. The species adapted to one or other of the phases, came into dominance, and then faded again. Overall, many of the wetland plants are tolerant of periodic water salinity and water periodicity changes, and continually adapt to the annual cycle of wet and dry. Over a 20-year cycle, an assemblage would contract and expand, with the prevailing hydrological and sedimentological conditions determining the overall composition of the vegetation. During the course of the study, the wetlands of the Becher region experienced a transition from wet to dry, and while there were changes in the vegetation (Chapter 10), the major assemblages, floristically and structurally, remained essentially the same. For example, wetland 161 remained dominated by *B. articulata*, wetland 162 by *M. teretifolia* closed scrub, wetland 163 by *J. kraussii*, and wetlands 135 and 136 by low forest of *M. rhaphiophylla*. This would suggest that the 20-year climatic patterns may not effect enough change in vegetation to be detected in the pollen record.

Wetland plants are more likely to respond to intra-wetland environmental changes than regional changes, given that the factors which determine their distribution are small scale such as geohydrology, sediment chemistry and hydrochemistry, and that the species pool of wetland plants in the Becher Suite occurs throughout the entire southwest region of Western Australia, spanning humid to semi-arid climates. Other studies corroborate the findings herein that changes in wetland vegetation, particularly in seasonally inundated or waterlogged wetlands, are related to the localised fluctuations in the hydrological regime (Boyd 1990; Jenkins and Kershaw 1997). The exception to this general pattern has been the comparatively recent arrival of *X. preissii* and *I. nodosa* indicated by the occurrence of their pollen *circa* 1500 ^{14}C yrs BP. As their pollen is first recorded in each of the wetland cores around this time, and one or other of the species continues to occur in the record up to the present, it seems to suggest a response to changing climate.

The vegetation history of wetlands has been different from basin to basin (Figs. 11-7 to 11-11), suggesting that no defined successional series occurs after the colonisation by *C. asiatica*. The record of wetland vegetation postulated for the Becher wetlands is evident in the modern, spatially gradational sequence from young to old wetlands on the cuspate foreland. The series of wetland vegetation, postulated for the Becher wetlands historically, is based on vegetation in zones with dampland and sumpland characteristics, exhibiting a strong relationship with wetland hydroperiod. Moreover, changes in the hydrochemistry associated with sedimentological changes, and changes brought about by the vegetation itself, have probably played a role in the changing dominance of species abundances in the past similar to those documented in the wetlands today.

It was concluded that there was heterogeneity in distribution of upland pollen with respect to the present climatic conditions.

A combination of palynological and stratigraphic independently corrorborative information, within the *same* basin, may be used to pinpoint changes in climate. In the Becher area, examples of such changes occurred approximately 3000-2900, 1400 and 500 ^{14}C yrs BP. The 3000-2000 date records the commencement of carbonate mud accumulation in wetlands 161 and 162, and corresponds with the peaks in easterly wind transported pollen, and the occurrence of salt tolerant wetland species in the pollen record. The 1500 year date marks the arrival of *X. preissii* and *I. nodosa*, and the 500 year date marks the initiation of peat in surface sediments and the occurrence of a range of pollen from species adapted to freshwater and seasonal inundation.

12. SYNTHESIS

This chapter brings together the observations, measurements, and conclusions, drawn from the holistic study of wetland evolution on the Becher cuspate foreland. It should be noted that the intention of this final chapter is to combine the various elements of the study into a picture resembling the network of ecological interactions within the Becher Suite wetlands, *i.e.*, to some extent, identify causal and consequent processes. Therefore, discussion of findings in the light of international research presented at the end of each chapter is not revisited here.

As set forth in the introductory objectives, the thrust of this study has been the documentation of features and processes in the sedimentary wetland fills and the wetland hydrological systems, and in the extant and historic distribution of wetland plants in response to a changing habitat. While acknowledging the complexity and ramifying effects of internal processes and mechanisms within individual wetlands at the small scale, a simplified model of wetland development and functioning is presented here, based on processes reconstructed from buried stratigraphic features within the wetland basin, and extrapolation of short term, current, oscillations and fluctuations in wetland processes to the longer term climatic cycles in the mid to late Holocene. The synthesis has been drawn from information on the wetland processes occurring throughout the Becher Suite, representing the range of ages, stratigraphic types and hydrological settings.

12.1 Setting

The synthesis must begin at the regional scale because the setting in which the wetlands develop and function defines the fundamental properties of wetland geomorphology, geology, and hydrology. It is the background from which, and the context within which, evolution of the wetlands has taken place. The Becher Suite wetlands commenced their development within the swales of a prograding beachridge system on a cuspate foreland which set them apart from the majority of inter-ridge wetlands in other settings (Fig. 3-11). The stabilised ridges and basins contrast with deflation basins and flats within coastal dune terrain which often support dune slacks. The wetlands differ from barrier lagoons in that they were always land based, and from wetlands on many other beachridge plains in that they are not remnants or components of either an ancestral or modern fluvial system (Visser *et al.* 2000, Huh *et al.* 2001). Perhaps, the most important feature of the Becher beachridge plain setting is the convex shape of the water table, developed in response to the prograding beachridge plain, which has resulted in inundation of swales at different heights above the present sea level at different times (Fig. 12-1). In other cuspate forelands and chenier plains intersected by a horizontal planar water table surface, the type of wetland formed is usually a single large wetland (Logan and Rudolph 1997, Malcolm and Soulsby 2001).

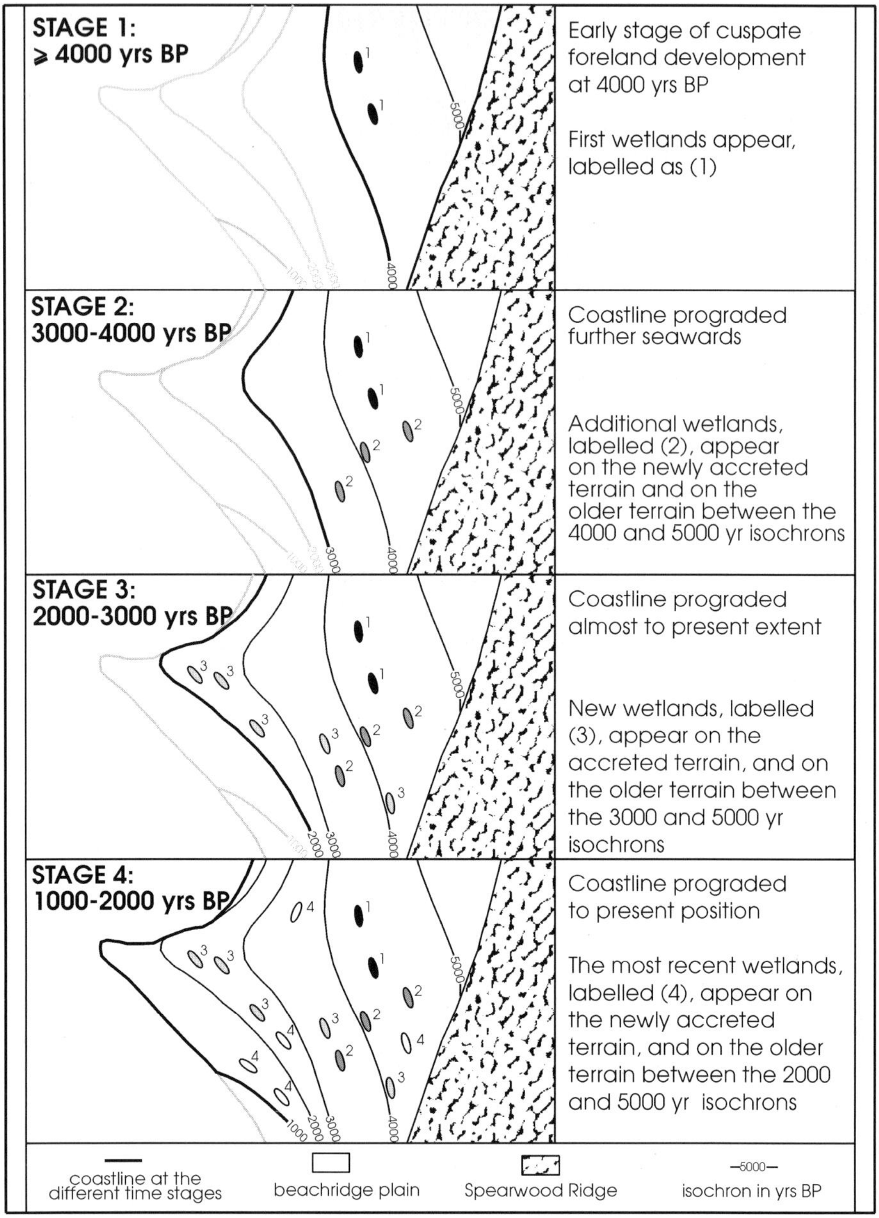

Fig 12-1. *Idealised diagram showing the progressive appearance of wetland basins on the landscape of the Becher cuspate foreland.*

From the perspective of individual wetland initiation and development within the Becher Suite, each wetland is slightly different. The environmental attributes of the different small scale settings determined the timing and style of inception, influenced their natural history, and continue to affect their development. Some of these factors are: 1) geographic position, 2) texture and composition of substrate and parent sand, 3) rate of water table rise, and 4) the period of initiation. Many of these factors are inter-related. The importance of geographic position is primarily in relation to the regional groundwater (Fig. 5-18). Along the central east/west axis of the cuspate foreland, where progradation was most rapid, vertical accretion and average depth to the water table were comparatively less than elsewhere along any swale, and wetland inception in this axial region was both more rapid and more frequent. Today, wetlands in this geographic location are still subject to more frequent inundation and more rapid accumulation of wetland sediments. In other areas of the cuspate foreland, where progradation was stalled by cycles of erosion and construction of higher ridges, the floors of the intervening swales also reached higher levels relative to AHD, and wetland initiation was delayed until later in into the Holocene when water tables had risen to sufficient height to intersect them. Geographic position today influences the hydraulic gradients within the regional groundwater body. Groundwater under inland areas of the cuspate foreland is dominated by low hydraulic gradients and slow throughflow, whereas at coastal margins groundwater throughflow is rapid.

Texture and composition, (in particular, the proportions of carbonate grains and coarse silica grains), of the inter-ridge substrate, and consequently the wetland basal sheet, vary according to the source from which the sediment was derived (*e.g.*, from the eroding Leschenault Barrier in the south, or local seagrass banks), and to the prevailing hydrodynamics involved in swale formation, that is, whether the swale host to the proto-wetland was built by swash or aeolian processes, and to what extent it has been modified by sheet wash. Older swales tend to have higher proportions of both carbonate and coarse silica grains.

The effects of various rates of water table rise may be seen in the number, thickness, structure, and composition of sediment types, and in the historical changes in vegetation. The rate of water table rise across the cuspate foreland was variable because it was linked to the rates of progradation. Evidence for variable rates of progradation include: 1) three styles of beachridge formation, and 2) the relic Becher Point geomorphic structures located near the apex of Becher Point itself (Semeniuk 1995). On the Becher cuspate foreland, the distribution of these geomorphic features suggests periods of rapid accretion (deduced from the number of low ridges and flat planar surfaces in the profile), associated with gentle nearshore gradients and an abundant supply of shoreward moving sand, regular accretion associated with low beachridges and continuous sand supply, and periods of stasis (cessation in progradation), or retreat, in shoreline development as a result of increased storminess, erosion, and construction of high beachridges (Fig. 5-2). A series of

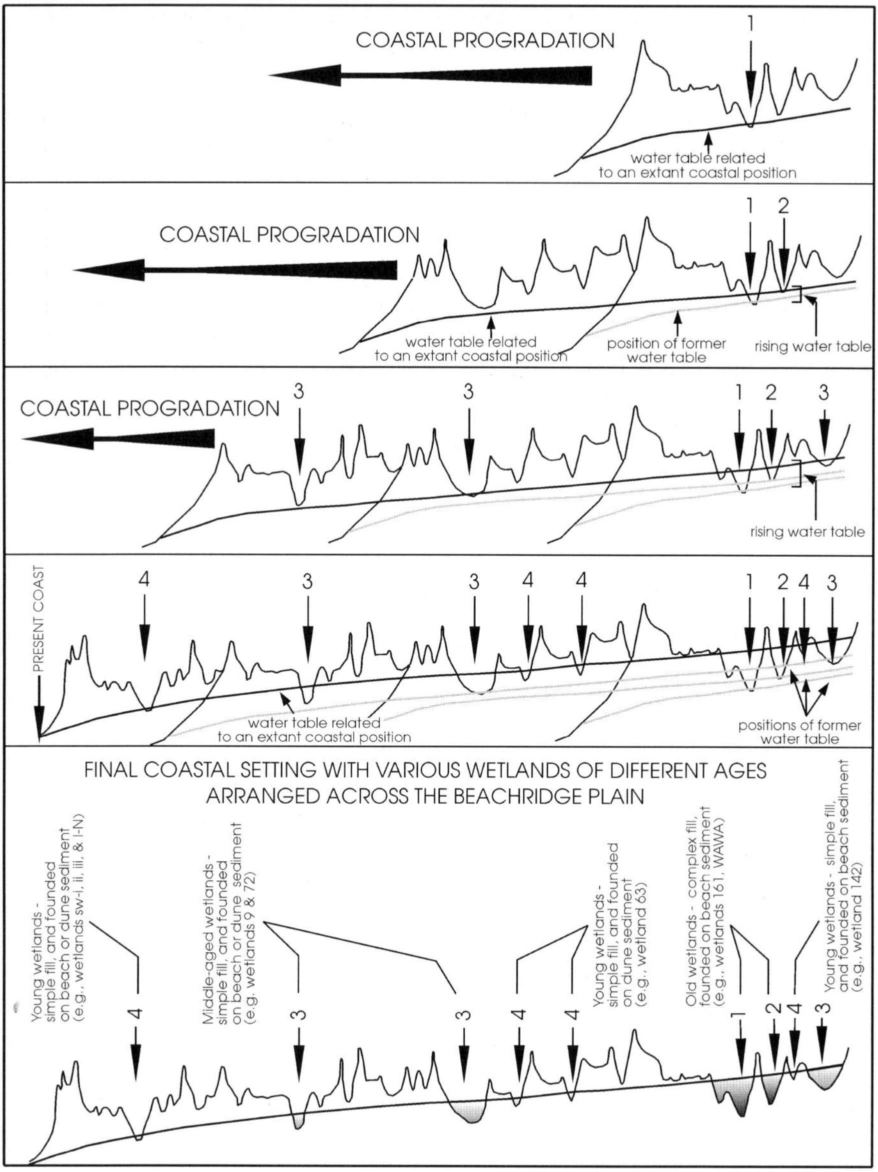

Figure 12-2. *Idealised diagram showing the progressive, scattered development of wetlands across the beachridge plain as the coast progrades, and as the water table inland from the coast rises and intersects the swales; the type of foundation sediment to the wetland; and the complexity of sedimentary fill. Numbers 1-4 show the order of appearance of the wetlands.*

relic curved ridge sections representing former apices of the cuspate foreland also were recognised as evidence of alternating erosion, accretion, and meandering progradation (Semeniuk 1995). During periods of rapid progradation, it is assumed that a corresponding response in water table rise would occur, *i.e.*, the water table would rise rapidly, while during slow coastal progradation, the rate of water table rise would also slow. The apparent rate of groundwater rise is also a function of the depth to water beneath any swale, shallow depths requiring less time for the water table to intersect the ground surface, other factors being equal.

Finally, the period in which wetland processes commenced would also influence wetland development. The coincidence of wetland initiation with either a relatively humid or relatively arid period would have significant effects on wetland sediment accumulation, hydrological and hydrochemical processes, and vegetation and faunal response. Wetland sediments and pollen in wetland 161, for example, indicate a comparatively wetter period of initiation than similar features in wetland 162 (Figs. 11-7, 11- 8).

12.2 Proto-wetland development

The development of small scale basins which were destined to become proto-wetlands, was the culmination of several regional coastal and aeolian processes which resulted in localised but important modifications to the regional beachridge swale geomorphology and hydrology (Figs. 5-8, 5-9), *e.g.*, segmentation of swales by linear sand ridges as a result of repetitive short term but intense storm activity, and progradation of the land surface seawards, concomitant with an increase in the height of the water table, at a fixed point, relative to AHD. Low topographic areas within the swales were intersected by rising groundwater (Fig. 12-2). Within the constraints of cuspate foreland development, the relative height between ground surface and water table in each basin was independent of distance from the sea, and therefore wetlands do not exhibit a progressive linear age, nutrient, or vegetation sequence from coast to inland. Unlike dune slacks which commence at the water table, the depth to water in the proto-wetland basins can range within the limits which define a wetland. This mode of wetland initiation has several implications for subsequent wetland development. Firstly, theoretically it is possible that amongst the 200 wetland basins of the Becher Suite, a wetland in the oldest swales furthest from the coast and a wetland in a swale nearest the prograding shoreline, commenced their development at the same time (Fig. 12-2). Secondly, even wetlands which commenced at the same time may have very different subsequent histories because of their position in the landscape. A wetland which commenced when the water table reached a maximum height of -0.5 m could have remained stable in this situation for a long period, or in contrast, this situation could have been of short duration due to the rapid rise of the water table to inundate the the surface. Examples of wetland pairs whichdemonstrate this principle are wetlands WAWA and 135 (which commenced as proto-wetlands *circa* 2700 years BP), wetlands 9-14 and 35 (which commenced as proto-wetlands *circa* 2450 years BP), and wetlands 72 and A9 (which commenced as proto-wetlands *circa* 2300 years BP), which exhibit different stratigraphy, hydrology and vegetation because of their different location on the cuspate foreland land surface. Along the east/west axis of the cusp where progradation was most rapid, there was a higher probability of intersection of the land surface by groundwater. As a consequence, there exists in the otherwise random spatial distribution of incipient wetlands, the general pattern of diminishing numbers of wetlands along any swale as distance to the north or south of the central axis of the cusp increases.

At the stage of the proto-wetland, regional environmental conditions played the dominant role in individual wetland development, the most important being regional water table configurations and climate. The rise of groundwater to the surface of low areas of the swales signified the commencement of physical, chemical, and biological processes that are recognised as wetland processes. Proto-wetland surfaces (marked by buried humic soils or carbonate mud infiltration into the sand floor) are a bench mark for measurement and comparison of subsequent intra-basinal changes (*e.g.*, Section 8.6). Pedogenic processes began to be influenced by hydrologic processes whose increasing influence is illustrated by the deposition of a carbonate mud sedimentary fill which was the product of inundation. Over time, pedogenic, diagenetic, and hydrological processes created stratified wetland sedimentary sequences which increasingly interacted with other features of the wetland such as water and vegetation in a way which continued to influence wetland evolution. The wetland sedimentary sequences exhibit imprints of these processes in compositional, structural and textural features. As such, they represent a shift in dominance from regional processes to intra-basinal processes.

12.3 Increase in stratigraphic heterogeneity

Generally, an increase in the heterogeneity of wetland fill has occurred in response to regional climatic processes, particularly the seasonality, and the long term variable frequency and intensity of rainfall. Variable long term rainfall effects on the wetland sedimentary deposits are evident in the alternating accumulation of the organic and calcareous mud, and the range of structures and textures throughout the sedimentary sequences. Carbonate mud in the Becher Suite wetlands would be regarded as palustrine carbonate (Platt and Wright 1992), Generally, an increase in the heterogeneity of wetland fill has occurred in response to regional climatic processes, particularly the seasonality, and the long term variable frequency and intensity of rainfall. Variable long term rainfall effects on the wetland sedimentary deposits are evident in the alternating accumulation of the organic and calcareous mud, and the range of structures and textures throughout the sedimentary sequences. Carbonate mud in the Becher wetlands would be regarded as palustrine carbonate (Platt and Wright 1992), which is the product of a strongly seasonal climate. Palustrine carbonate deposits typically are shallow fresh water and have pedogenic features, indicating emergence and water saturation. Palustrine carbonates, both modern and ancient, typically occur in metre carbonate mud, coarse calcareous sand, and medium/fine calcareous sand (Fig. 12-3). There is a difference in height between the top of the beach sediments under the western ridge and under the wetland, the result of carbonate dissolution. As a result of the complex stratigraphy, the hydrologic processes in wetland 161 are more numerous and multi-directional. They include: 1) recharge to the groundwater by meteoric infiltration, 2) infiltration of rain into peat layer where it is ponded above the carbonate mud, 3) rapid infiltration to groundwater via cliffed muddy sand margin, 4) seasonal upwelling along western margin, 5) lateral flows from ridge to wetland margin from both east and west, 6) westward discharge of acidic water from wetland organic horizons below the level of the wetland fill, and 7) evapo-transpiration (Fig. 12-4).

Flow paths are not only affected by stratigraphy within the wetland, but by local hydraulic gradients arising from the disparity in water table recharge rates under beachridges and wetlands, which is a function of both beachridge and wetland stratigraphy and topography.

At the time of the most comprehensive water level survey (September 1994), the regional groundwater contours in the main body of the Becher cuspate foreland exhibited three patterns: widely spaced furthest from the coast; closer together in the central part; and wider again in the vicinity of the protruding point (Fig. 8-6A). These patterns suggest that under conditions when water tables are located within the wetland fills, wetlands situated in the most landward part of the Becher cuspate foreland or near the coast are likely to experience slow but generally unimpeded flow, and wetlands in the central part of the area are likely to experience unimpeded flow out from but not into this subregion. At times of higher water levels, groundwater contours are generally closer spaced corresponding to more rapid throughflow, except furthest from the coast where minimal lateral flow is indicated. At times of lower water levels, the spatial variation is no longer observable, and the groundwater contours are consistent, indicating more slow but unimpeded throughflow subregionally (Fig. 8-6B, C).

In contrast to the uncomplicated pattern of subregional hydrology, at the basin scale the wetlands exhibited perceptible differences in hydrological response. Near the coast, where the wetlands are young, the wetland fill is shallow and the period of dominance of meteoric input over throughflow is short, and depends on the frequency and intensity of rainfall events. In the wetlands further from the coast, waterlogging of wetland sediments can occur for one to six months, *i.e.*, half way through winter to halfway through summer (Fig. 8-15). Analysis of subregional, local, and ridge to wetland hydraulic gradients for the latter wetlands showed that the east and west ridge to wetland gradients were consistent, with flows to and from the wetlands within the period that wetland sediments were waterlogged (Figs. 8-19 to 25, 8-26 to 29). The local east/west hydraulic gradients were more persistent in wetlands with no perching, *e.g.*, wetland WAWA where maximum slopes were reached during maximum rain input. In other wetlands with less permeable sediments, the gradients during the period of

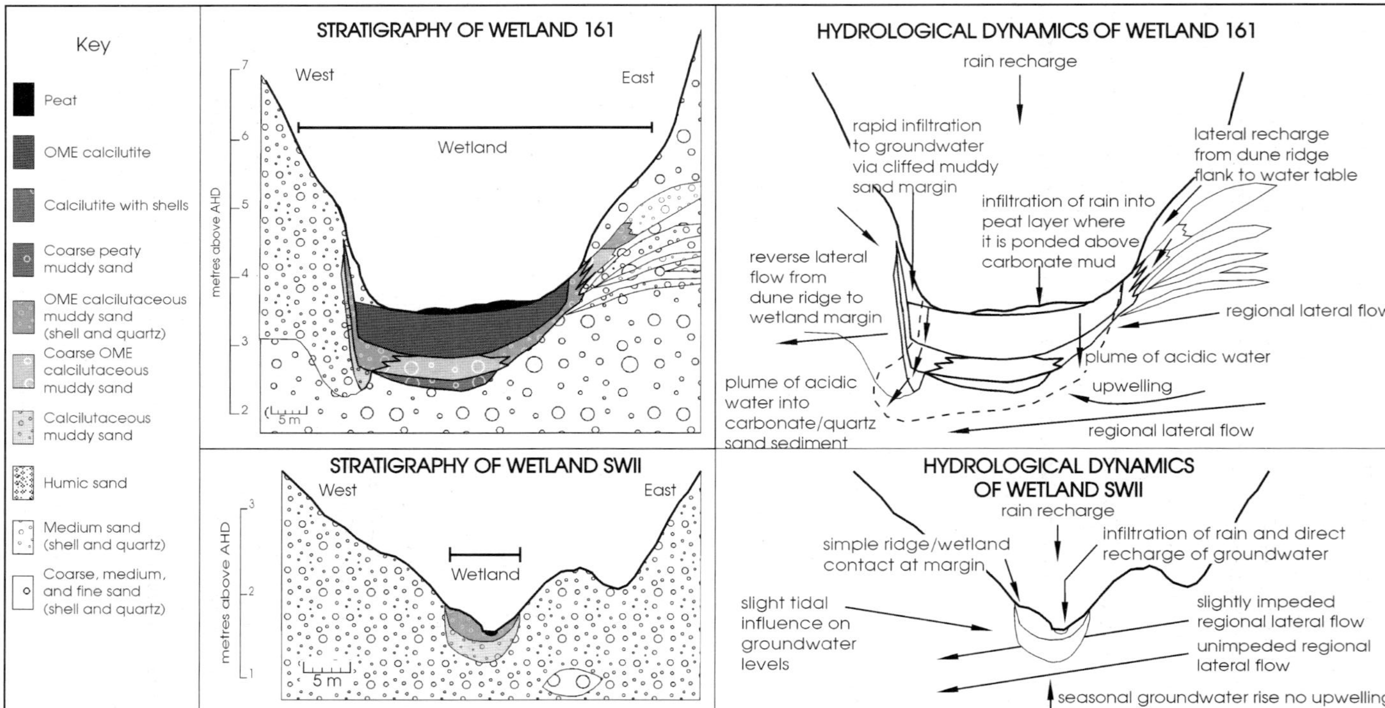

Figure 12-4. *Summary of stratigraphic patterns and interpretation of hydrologic reponses, comparing the oldest and youngest wetlands, 161 and swii, respectively.*

scale cycles (Platt and Wright 1992). They are bioturbated and comprise a low-salinity biota of charophytes, ostracods and gastropods. Climate is a critical control on the nature and development of palustrine depositional sequences. Platt and Wright (1992) recognised three characteristic types of palustrine cycle, each relating to a specific climate: semi-arid type; intermediate type; and sub-humid type. The semi-arid type is characterised by surface evaporite layers or extensive dessication, the intermediate type is characterised by root structures, rhizoliths, gastropods and calcretes, and the sub-humid type is characterised by organic matter. The Becher wetland carbonate mud sequences are characterised by nearly all of the features listed. They most closely resemble the intermediate group, but have elements of both the semi-arid and the sub-humid type. For example, in the Cooloongup basin, a dessication layer was described at site A2, (Fig. 6-21) but rhizolites and calcrete fabrics were present in the same location. Mottled calcretes are present in wetland 9 (Figs. 6-14, 6-15). All wetlands exhibit extensive root structuring, predominantly in the surface layers but also at depth (50 cm or more), and all wetlands exhibit various degrees of organic matter enrichment, again, predominantly in the surface layers but locally, also in deeper layers (Figs. 6-3, 6-4, 6-5, 6-6). With reference to the ancient palustrine facies described by Platt and Wright (1992), the carbonate mud sequences of the Becher Suite wetlands indicate a change from predominantly dry conditions through intermediate to wetter conditions, retaining marked seasonality throughout. Clearly, in this context, the heterogeneity of any wetland fill is a function of the number of climatic cycles experienced by the wetland, and is therefore related to wetland age. The older wetlands, 161, 162, WAWA, Cooloongup, exhibit the greater number of lithologies and sedimentary cycles (Figs. 6-3, 6-4, 6-6, 6-21, 6-22, 6-23).

In individual wetlands, a range of hydroperiods and water quality, generated by the variable patterns of long term rainfall frequency and intensity, have produced different thicknesses of lithologies, numbers of layers, and dominance of different compositional types (lithology) within the profile. Seasonal variations have resulted in a greater or lesser degree of bioturbation, root development, and colour mottling. An increase in the heterogeneous nature of the wetland fill can be observed from younger to older wetlands, in the increasing number and types of structures, fabrics, textures, and modes of sediment composition (Figs. 6-3 to 6-20). In the surface layers, structure is dominated by root morphology, in the intermediate layers, structure is variably texture mottled in younger wetlands (swi, swii, swiii), and either texture mottled, layered, homogeneous, or a combination of these in older wetlands (wetlands 161, 162, 135). In lower sediment layers, younger wetlands exhibit structural homogeneity while older wetlands exhibit root structures, texture mottles or homogeneous structure. Sediment fabrics and textures generally reflect prevailing hydrological conditions in that muds are generated in shallow, inundated basins, and sands are dominant in drier basins due to the cumulative effects of imported or *in situ* detritus disintegration and the lower mud content. Thus, in the wetland fill sequence in younger wetlands (63, swi, swii, Figs. 6-51, 6-55), the fabric is commonly packstone which changes down profile to

grainstone. In older wetlands, the fabric is commonly mudstone which changes down profile to packstone, then grainstone (Figs. 6-44, 6-45, 6-49, 6-52). Similarly, the texture of the wetland fill changes from sand dominated to mud dominated with time. This simple pattern is sometimes modified by the input of sheet wash sand into the surface layers, and there are important differences between the wetlands in the rate of change down profile of textural types and in the respective ratios of mud to sand in any single layer (Figs. 6-44 to 6-56).

Compositional differences are a most obvious contributor to stratigraphic heterogeneity. The end members of the wetland fill are: 1) peat 2) carbonate mud and 3) calcareous sand, which are the products of different hydrological, hydrochemical and biological processes, and wetland history. The three compositional types occur most commonly as layers of mixed composition in the younger wetlands and as separate or mixed layers in older wetlands. Stratigraphic heterogeneity is increased through diagenetic overprints on the sediments such as cementation and dissolution. Cementation, present in both the middle (wetland 9) and oldest (Cooloongup) of the wetlands, is a product of transpiration from the vadose zone, and is unrelated to stage of development. In contrast, dissolution of carbonate grains is strongly related to wetland developmental stage through its dependence on formation of both peat and carbonate mud, and through the cumulative effect of time.

12.4 Effect of stratigraphy on hydrology

The effect of stratigraphy on hydrological functions, in and adjacent to, a wetland basin increases with its heterogeneity. In the wetland basins, meteoric input is affected by the lenses and ribbons of wetland fill, through which it must pass to reach the water table. For example, the fabric of the wetland fill sequence generally changes down profile from mudstone through packstone to grainstone, a trend which reflects increasing permeability and lower water holding capacity. Depending on the prevalence of fabric type in the wetland fill, sedimentary bodies range from being highly impermeable, to absorbant, to relatively free draining. A corollary of increasing permeability, also influenced by stratigraphic heterogeneity, is decreasing pore water content. In all wetland sites, the water content fluctuated over a relatively shallow sequence of wetland fill under the influence of grain size (texture) and matrix type (fabric). Within any wetland, soil moisture content was highest in the muds, then muddy sand, then sand, thus generally decreasing down the profile following the stratigraphic sequence. Since the mud content increases in the wetland fills of older wetlands, because of the increasing thickness of sediment and the prevalence of mudstone over packstone fabric, the length of time that the wetland sediments are waterlogged and the degree of waterlogging both increase with the age of the wetland (Figs. 7-2 and 8-15, 8-33).

Rates of groundwater flow within and through wetland fill became more varied with time as a result of increasing thickness and complexity of sedimentary stratigraphic sequences. Rate and volume of recharge to the groundwater from a given volume of rain were related to textural and compositional characteristics of the sediment, *i.e.*, pore size and configuration, and to initial water content (Bouwer 1978). These factors, as they occurred in peaty mud, carbonate mud, and carbonate muddy sands, varied throughout a profile, with all three sediment types influencing the height and rate of groundwater rise and fall (Fig. 8-18). Rate of flow through peaty mud was greater than through carbonate mud, so that wetlands with a lower ratio of carbonate mud to peaty mud, *i.e.*, younger wetlands, exhibited shorter periods of inundation and waterlogging in the surface layers.

As a given wetland progressed in its stratigraphic development, flow paths became more numerous as a result of structural and microstructural differentiation in the stratigraphic sequence of the wetlands due to small scale features such as layering, roots and texture mottling (due to bioturbation), and, locally, led to dominance of vertical flow within the surface sedimentary layers, or lateral flow in interlayered sediments (Chapter 8). Other macropores, including burrows and lenses or layers of shell gravel, were common in the intermediate layers of wetland fill in the middle and older Becher wetlands (Figs. 6-3 to 6-5, 6-8, 6-11, 6-14, 6-19). In this study, stratigraphic interlayering between wetland and beachridge sediments was found to affect groundwater levels at the western wetland margins (Chapter 7). Stratigraphic interlayering also occurred in the middle layers of the wetland fill in wetlands 161 and 163 (Figs. 6-3, 6-5), and may have facilitated water movement and/or retention here also. Stratigraphic layering becomes more complex in older wetlands, as will be demonstrated with examples below.

The effect of stratigraphy on hydrological functions can best be illustrated by comparing examples of young wetlands (1N, swi, swii) and old wetlands (161). In the youngest wetland, 1N, the sedimentary sequence comprises two weakly differentiated calcareous sand layers, a root structured humic surface layer and a homogeneous basal sand layer. In the second youngest wetlands, swi and swii, the sequence comprises a thin layer of root structured peaty/carbonate mud, burrow mottled carbonate muddy sand and a homogeneous basal sand layer, with a "simple" contact between the wetland fill and beachridge sediment (Fig. 12-3). The hydrologic processes in these wetlands are: 1) recharge to the groundwater by meteoric infiltration, 2) westward throughflow, and 3) evapo-transpiration (Fig. 12-4). In contrast, in wetland 161, the oldest of the study wetlands, the sequence comprises root structured peaty mud, colour mottled and layered peaty/carbonate mud, colour mottled and burrow mottled carbonate mud and carbonate muddy sand, a buried soil horizon, and gravel lag deposits of marine and freshwater shells (Fig. 12-3). The wetland sediments at the margin are higher than the current wetland surface and indicate the former level of inundation. Sheet wash deposits have buried the wetland sediments at the margin. The western contact between wetland and ridge sediments is cliffed and exhibits interdigitating

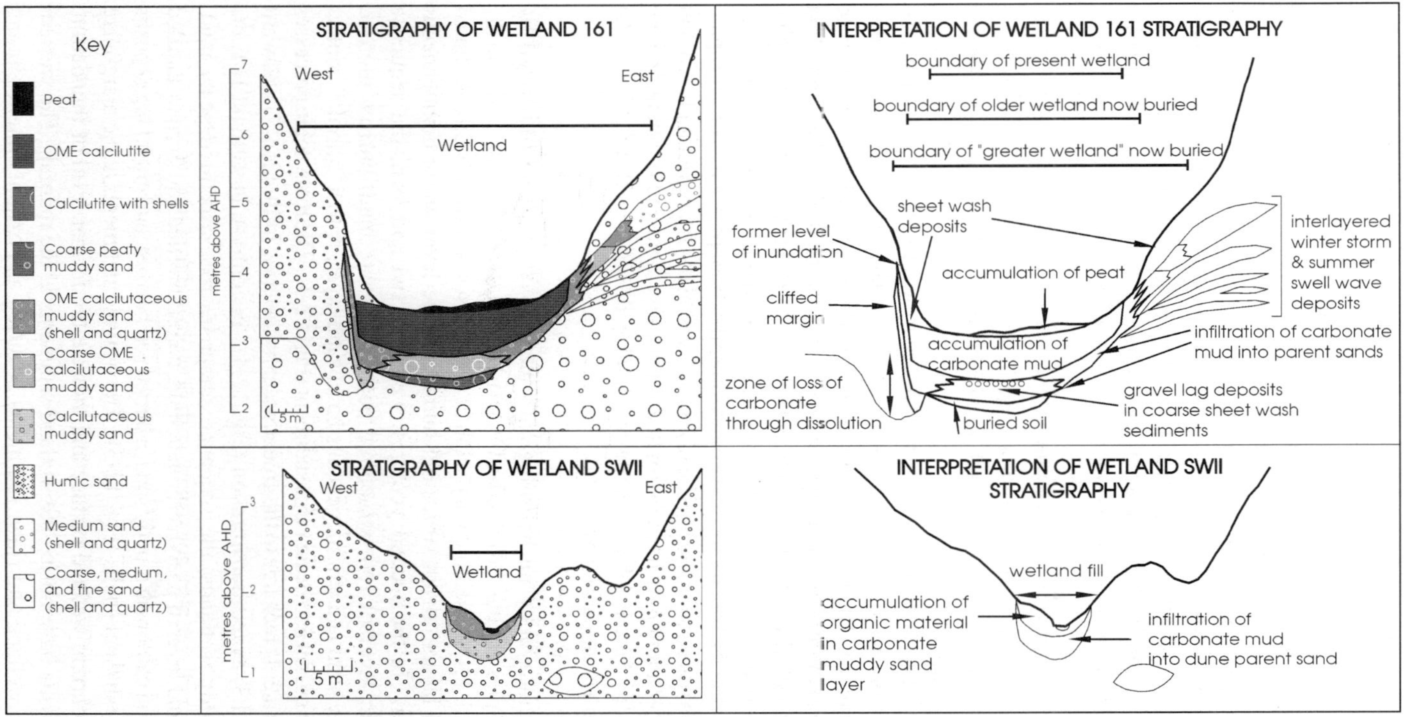

Figure 12-3. *Summary of stratigraphic patterns and interpretation of wetland history, comparing the oldest and youngest wetlands, 161 and swii, respectively.*

carbonate mud, coarse calcareous sand, and medium/fine calcareous sand (Fig. 12-3). There is a difference in height between the top of the beach sediments under the western ridge and under the wetland, the result of carbonate dissolution. As a result of the complex stratigraphy, the hydrologic processes in wetland 161 are more numerous and multi-directional. They include: 1) recharge to the groundwater by meteoric infiltration, 2) infiltration of rain into peat layer where it is ponded above the carbonate mud, 3) rapid infiltration to groundwater via cliffed muddy sand margin, 4) seasonal upwelling along western margin, 5) lateral flows from ridge to wetland margin from both east and west, 6) westward discharge of acidic water from wetland organic horizons below the level of the wetland fill, and 7) evapo-transpiration (Fig. 12-4).

Flow paths are not only affected by stratigraphy within the wetland, but by local hydraulic gradients arising from the disparity in water table recharge rates under beachridges and wetlands, which is a function of both beachridge and wetland stratigraphy and topography.

At the time of the most comprehensive water level survey (September 1994), the regional groundwater contours in the main body of the Becher cuspate foreland exhibited three patterns: widely spaced furthest from the coast; closer together in the central part; and wider again in the vicinity of the protruding point (Fig. 8-6A). These patterns suggest that under conditions when water tables are located within the wetland fills, wetlands situated in the most landward part of the Becher cuspate foreland or near the coast are likely to experience slow but generally unimpeded flow, and wetlands in the central part of the area are likely to experience unimpeded flow out from but not into this subregion. At times of higher water levels, groundwater contours are generally closer spaced corresponding to more rapid throughflow, except furthest from the coast where minimal lateral flow is indicated. At times of lower water levels, the spatial variation is no longer observable, and the groundwater contours are consistent, indicating more slow but unimpeded throughflow subregionally (Fig. 8-6B, C).

In contrast to the uncomplicated pattern of subregional hydrology, at the basin scale the wetlands exhibited perceptible differences in hydrological response. Near the coast, where the wetlands are young, the wetland fill is shallow and the period of dominance of meteoric input over throughflow is short, and depends on the frequency and intensity of rainfall events. In the wetlands further from the coast, waterlogging of wetland sediments can occur for one to six months, *i.e.*, half way through winter to halfway through summer (Fig. 8-15). Analysis of subregional, local, and ridge to wetland hydraulic gradients for the latter wetlands showed that the east and west ridge to wetland gradients were consistent, with flows to and from the wetlands within the period that wetland sediments were waterlogged (Figs. 8-19 to 25, 8-26 to 29). The local east/west hydraulic gradients were more persistent in wetlands with no perching, *e.g.*, wetland WAWA where maximum slopes were reached during maximum rain input. In other wetlands with less permeable sediments, the gradients during the period of

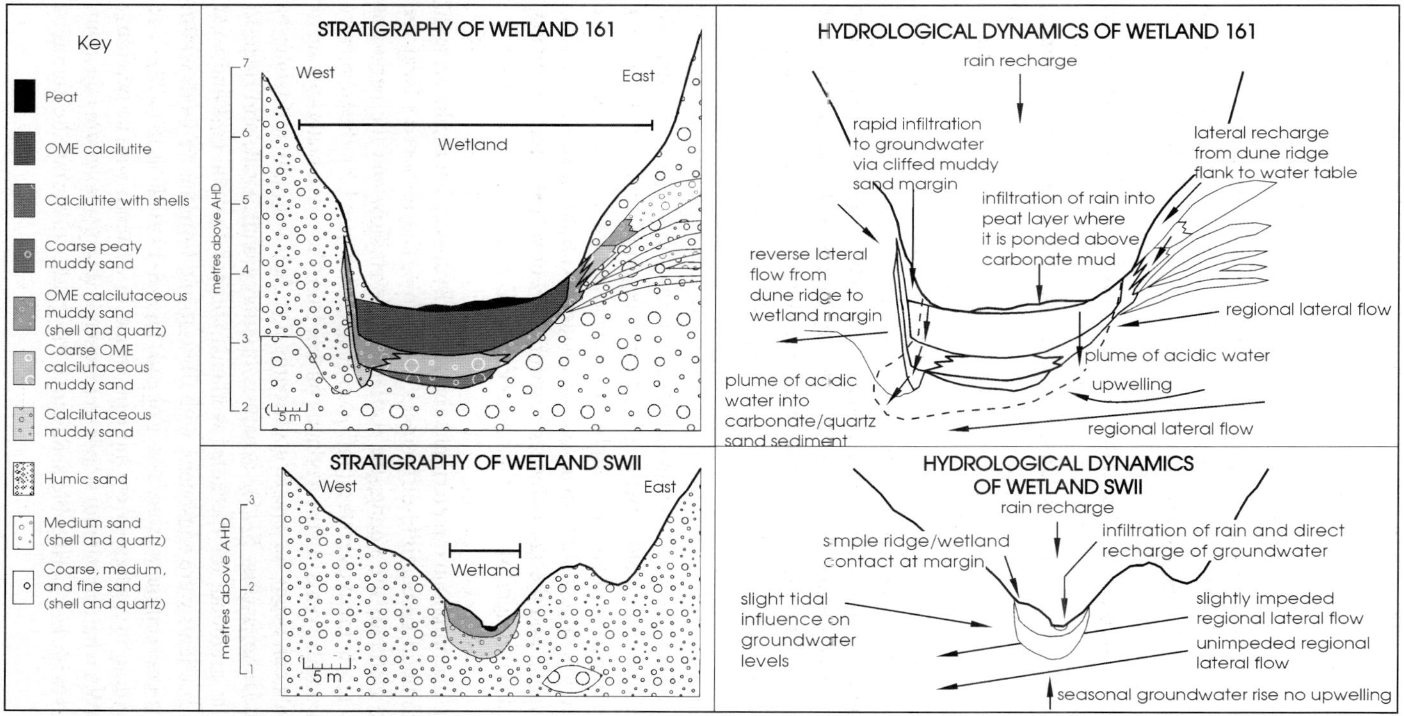

Figure 12-4. *Summary of stratigraphic patterns and interpretation of hydrologic reponses, comparing the oldest and youngest wetlands, 161 and swii, respectively.*

saturation were less pronounced and their presence was more intermittent, *e.g.*, wetland 35. Reversals of flow occurred down gradient of wetlands, where the adjacent western ridge was comparatively low (wetlands 135, 35 and 9-6), where the wetland margin was steep (wetlands 161, 162), where wetland sediments at the margin had been buried (wetlands 161, 163, 142), and where the wetland fill was thickest (wetland 161). These reversals of flow were augmented by seasonal upwelling during early winter, spring and early summer (Figs. 8-13, 8-14). During the winter, frequency of rain events was often sufficient to maintain this situation, *e.g.*, wetland 135.

The patterns described above indicate a trend away from the hydrodynamics of persistent throughflow in younger wetlands to dominant recharge by rainfall in older wetlands, at least during the winter and spring seasons. The sequence from young to old wetlands can be viewed as the progression from unimpeded lateral flow to increasingly impeded flow with variations, *e.g.*, unimpeded throughflow (wetland 1N), impeded throughflow (wetlands swi, swii, 63, 72, 142), impeded throughflow and early winter perching (wetlands 162, 135, 136, 9-3, 9-6), impeded throughflow with seasonal upwelling down gradient of wetland (wetland 163, 35), and impeded throughflow and perching with intermittent reversals of flow (wetlands 135, 161). Locally, where there is a sufficient plug of relatively impermeable wetland sediment, wetlands can also perform a discharge function for short periods, transporting water from the wetland centre to the margins (wetlands 161, 162, 163, WAWA, 9-6), or to basal sheets in the western perimeter.

The subregional contour configuration indicates that higher rainfall volumes and extended rainfall seasons would increase the importance of lateral flow. At present, the maximum subregional lateral flow occurs in March/April when water levels within the calcareous sands beneath the wetlands are aligned with the regional gradient. However, this alignment coincides with minimum water levels so that regional gradients are comparatively flatter. As the configuration and height of the water table surface varies with the volume of water in the aquifer, lateral flow at maximum water table heights during periods of average or above average rainfall could be expected to be significant, and wetland throughflow may be dominant. This is shown by the increased hydraulic gradients during October to December, the period of maximum water table height in wetter years. In this situation, for all wetland basins except those nearest the coast, local and ridge to wetland margin flows would occur from a groundwater level under the eastern ridge higher than the wetland surface, generating flows from groundwater to surface water. Under present climatic conditions, this process has been comparatively infrequent. However, on the western side of the same wetlands, any higher than normal rainfall could result in a mound at the wetland margin, creating local flow back into the wetland. Depending on the nature of the wetland/ridge contact and the frequency of rainfall, the local flow may continue to prevent discharge from the wetland.

12.5 Effect of stratigraphy on hydrochemistry

Processes involved in pedogenesis of the calcareous quartzose Safety Bay Sand underlying the beachridges were: dissolution of carbonates resulting in a decrease of Mg-calcite, calcium, and magnesium down profile, and eastward across the chronosequence; accumulation of plant material resulting in an increase of total organic matter at the surface; and chemical breakdown in organic material resulting in a constant acid soluble fraction in the organic layer (Woods 1984). Within the wetlands, in addition to these pedogenic effects, waterlogging and inundation had two important additional impacts: 1) they increased the period under which the sediments were in contact with water and 2) they decreased the period under which the sediments were oxygenated.

In conjunction with the evolution of wetland geomorphology, stratigraphy, hydrologic functioning, and vegetation in relation to the age of the wetlands and their complexity of basin fill, there has been a related evolution in their hydrochemistry. It was beyond the scope of this study to fully explore the effect of layered sediments on hydrochemistry, however, from the investigations undertaken, the following patterns emerged. Salinity was only affected by sediment composition insofar as it affected water content. Cation concentrations were affected by sediment type because waters interstitial to different layers within the stratigraphy leached cations from the sediments in various amounts depending on the sediment composition and texture. Simple wetlands and simple stratigraphic fills result in a relatively simple hydrochemistry. As a consequence of the development of complex sedimentary fills in the wetlands (*viz.*, calcareous quartzose sand, muddy sands, carbonate muds, and peat), the hydrochemistry has become more complex as percolating meteoric water and throughflowing groundwater interact with, and leach, the various sedimentary units in the sedimentary pile. Extant vegetation adds to this hydrochemical complexity. As the plant associations evolve, the various species with their cation requirements and cation storage, alter the cation pool at the surface and at depth via metabolic extraction of cations, and contribution at death and leaf fall. Thus, with the various ages of the wetlands, and their various sedimentary fills, there is a chemical layering of sediment types which results in some degree of hydrochemical layering. Rising and falling groundwater, and percolating meteoric water interact with this sedimentary chemical and hydrochemical layering, resulting in hydrochemical complexity related to the residency of water at a given soil chemical level, plant abstraction, and physicochemical processes such as evaporation and solution (dissolution and leaching) as expressed in groundwaters and interstitial waters. As a result, the combined effects of a local and variable wetland stratigraphy, and plant uptake and release on cation concentrations down profile in sediments, interstitial waters and groundwater, result in very localised hydrochemical signatures specific to type of sedimentary fills, their evolutionary stage, and to their extant vegetation type.

The vertical fluctuation of groundwater through the layered stratigraphy also affected its cation concentration at each level when rates of rise and fall varied. Under conditions when water levels rose rapidly, the effect was usually a decrease in cation concentrations due to dilution by rain. When water levels fell rapidly, the effect was usually an increase in cation concentration, but when water levels changed slowly, the effect could be in either direction, depending on ionic saturation, the composition and solubility of the sediment, and the hydrologic and/or hydrochemical requirement of individual plant species. Nutrient levels were related to sediment type (peat, carbonate mud, and sand), but the effect of a varied stratigraphy on nutrient distribution and levels was largely unexplored.

12.6 Stratigraphy as a record of hydrochemical processes

In the wetland basins, dissolution of carbonate grains in the underlying sands has resulted in wetland subsidence, in the order of 15-90 cm (90 cm in wetland 161; 70 cm in 162; 90 cm in WAWA; Figs. 6-26, 6-27, 6-29). The zone of dissolution is located in the basement sands below the carbonate mud and extends westward from the middle of the wetland to the basal slopes of the western ridge effectively mirroring the direction of discharge by throughflow. The thickness of a dissolution zone varies with the ratio of organic material to carbonate mud, and the age of the wetland. The width of the zone depends on the hydraulic gradient during infiltration. The zone is denoted by a sagging beach sand horizon and a wedge of infiltrated medium sand between the coarse sand underlying the ridge and wetland (Fig. 12-5). Dissolution has been shown also to occur in the basal sheet under the centre of several wetlands, but here, thinning of the muddy sand layer is partly compensated by carbonate mud accumulation *e.g.*, wetland 161. Buried cliffs also testify to the vertical displacement of the wetland fills (Fig. 12-5). In the Becher setting, carbonate dissolution has become sufficiently significant to rank as a factor in the formation of wetlands and their subsequent progression to increasing wetness. One of the factors influencing the magnitude of the wetland deepening through dissolution is the percentage of carbonate grains in the calcareous sands. At the lower end of the composition spectrum in the Becher calcareous sands, *i.e.*, 30% carbonate content, the effect of dissolution may be fully compensated by sediment input at the surface. If the rate of dissolution and hence subsidence is equal to, or less than the rate of sediment accumulation, the frequency of inundation will continue to be a function of climate variability and stratigraphic heterogeneity. If the rate of dissolution and subsidence is greater than the rate of sediment accumulation, the frequency of inundation is likely to increase, *e.g.*, wetlands 162, WAWA. With this predicted increase in hydroperiod, the mode of sedimentation may alter.

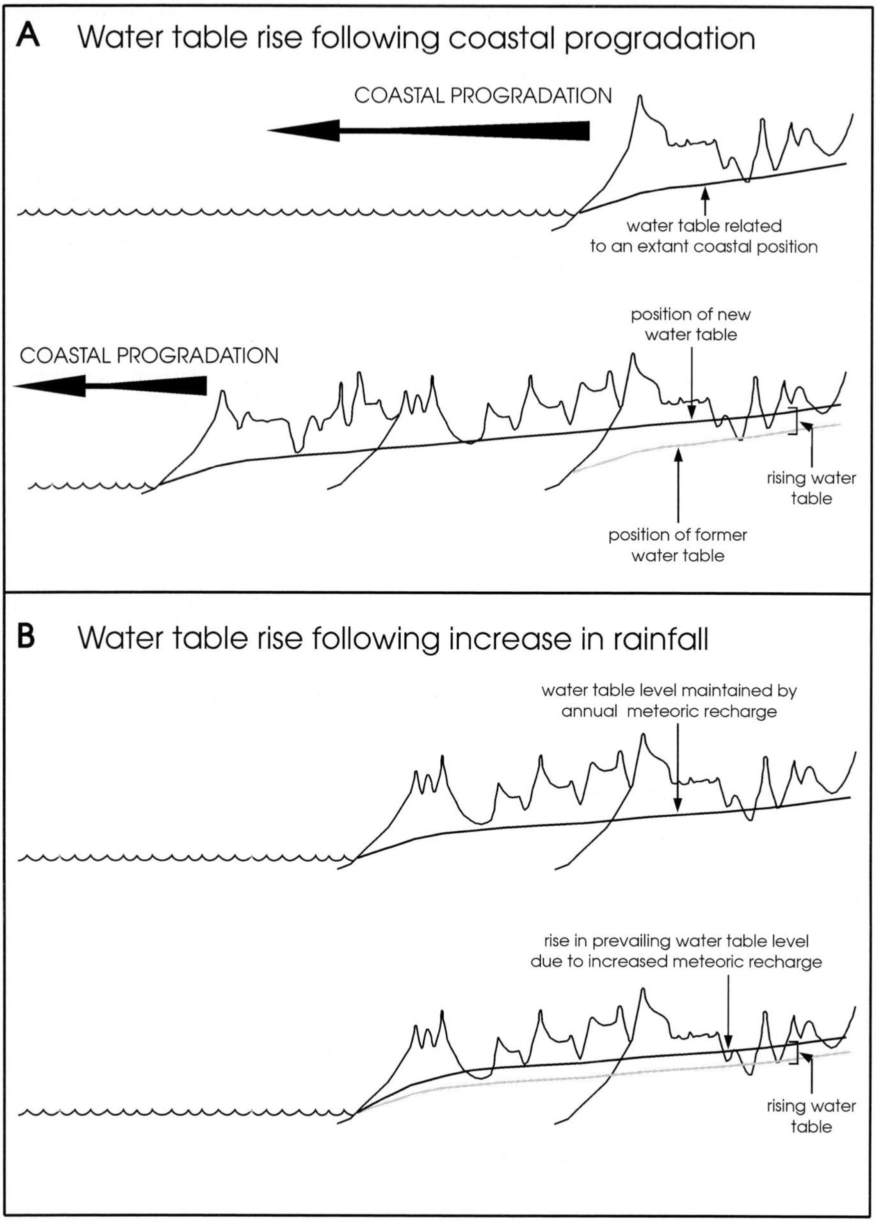

Figure 12-5. *Two processes by which prevailing water tables rise at the local scale. A. General permanent rise due to coastal progradation B. Permanent, or short to medium term rise as a result of increased meteoric recharge due to climate changes.*

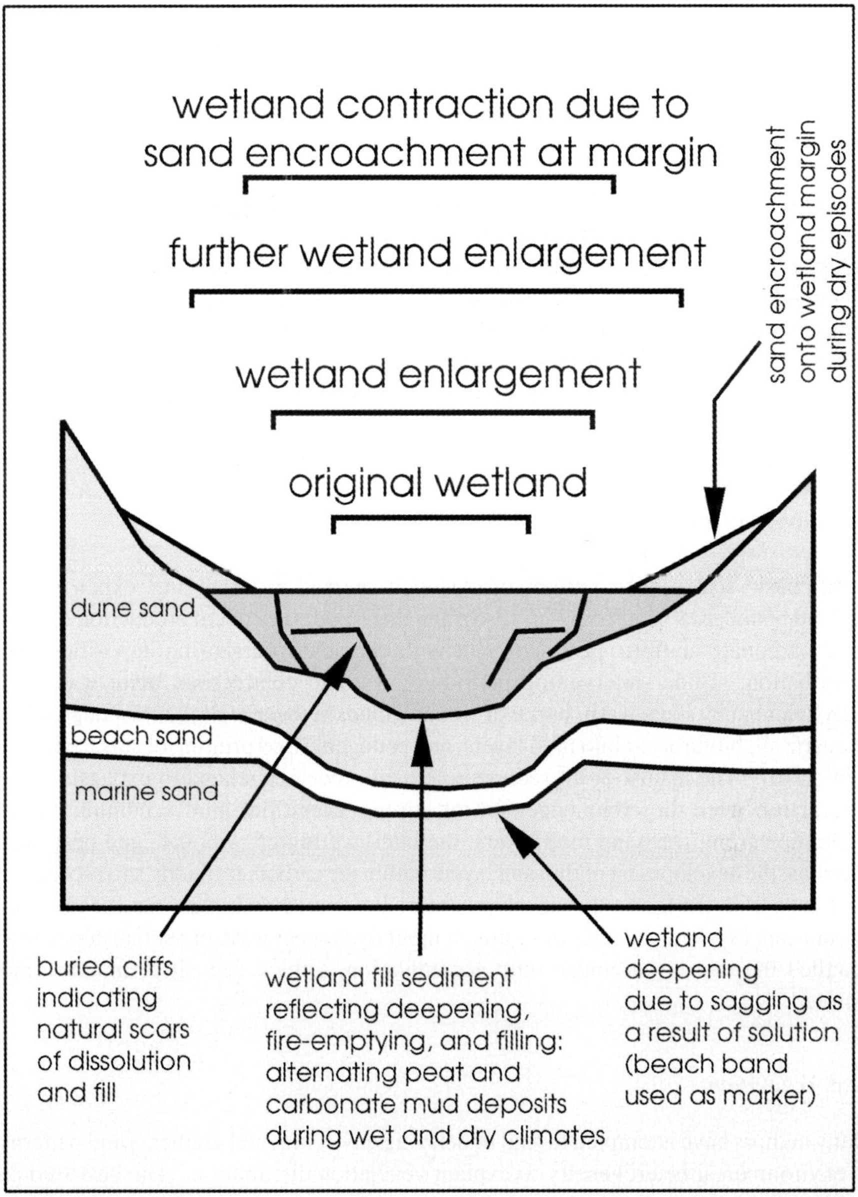

Figure 12-6. *Features and processes of the Becher wetland fills and their history - wetland filling, emptying, deepening, enlargement and contractions.*

12.7 Stratigraphy as a record of sedimentological and climatic processes

The wetland fills also reveal a range of sedimentological processes related to wetland developmental history and fluctuating rain cycles. The most obvious example is the contrast in the type of sediment generated under conditions of groundwater inundation resulting from sub-regional groundwater rise, and the type of sediment generated under inundation by groundwater rise in response to increased meteoric recharge (Fig. 12-6). The sub-regional rise in groundwater following coastal progradation occurs independently of any increase in rainfall. In small inundated basins in swales Charophytes were able to extract calcium carbonate in solution from waters residing in carbonate rich sands to produce a carbonate mud deposit over time. With a gradual change towards more humid conditions beginning around 2000 years ago and continuing to the present, the role of rainfall has become increasingly important in the hydrological functioning of the wetlands and in the development of intra-wetland basinal deposits. Under present conditions of inundation, meteoric recharge increasingly determines the chemistry of surface and near surface waters, creating conditions suitable for peat formation.

At the basin scale, under periods of extended wetland waterlogging, expansion of wetland boundaries occurred (Fig. 12-5), and there was consequent production of mud sized carbonate or fibric peat sediment with characteristic settling deposition and bioturbation. Under intervening periods of wetland contraction, there was sand encroachment by sheet wash, burial of wetland muds at the margin, aeolian deposition of quartz silt bioturbated into mud layers, and pedogenic overprinting of surface layers (Fig. 12-6). The results of all sedimentological processes related to expansion and contraction were the commencement of simple interstitial mud accumulation or infiltration from overlying mud layers, the interlayering of carbonate and peat mud horizons, the development of thin soil layers within the carbonate muddy sand sequence (wetlands 162, 9-6), and the development of brecciated carbonate mud layers. In circumstances where the rates of sediment input from sheet wash or aeolian deposition equalled the rate of carbonate mud accumulation, a thick deposit of muddy sand resulted.

12.8 Vegetation

Many authors have attempted to find underlying environmental gradients and patterns of environmental heterogeneity to explain vegetation distribution. The best studies are those which examine a single attribute for one or two species of a single genus in a variety of controlled settings, or several species within a single wetland, because it is at this scale that understanding of population distribution is best expressed (Justin and Armstrong 1987: Glenn *et al.* 1995; de Kroon *et al.* 1996; Warwick and Bailey 1997; Visser *et al.* 2000). Even in this setting within the Becher wetlands, which lends itself to analysis of vegetation patterns because the vegetation pool is limited, and there is a gradation in age of wetland matching the complexity of stratigraphy, hydrology, hydrochemistry and their interactions, it has become apparent that vegetation distribution is not to be understood by studies at the spatial scale of the wetland basin and the temporal scale of monthly sampling. However, two interesting trends and indicators were identified which could be incorporated into the general model of wetland evolution, and these are discussed below.

Firstly, vegetation across the range of wetlands in the study area exhibited an increase in complexity of pattern and form from latiform to maculiform to concentriform with increased age of wetland. These patterns reflect variation in topography, concomitant wetland sediment deposition, hydrological and hydrochemical variation, as well as an increase in the complexity of their interactions. Topographical variation is not only inherent in the basin geometry, but is modified by sedimentological processes such as wetland deepening, variable rates of sediment accumulation, sheet wash into wetland margins, and by sediment and root mounds developed under some plant species (*e.g.*, *M. rhaphiophylla* and *M. teretifolia*). Topographic micro-relief results in patchiness of vegetation assemblages. Subtle to sharp vegetation zonation occurs in response to 1) the interplay between basin morphology and water levels, 2) buried wetland sediments at the margins (wetlands WAWA, 163), and 3) zones of irregular and often extreme hydrological and hydrochemical effects at wetland margins (wetlands 161, 162, WAWA, 9-6). A small component of the patchiness also may be attributed to the history of vegetation in the wetland basin. Residual stands of species more representative of wetter or drier phases in the climate history may be mixed with species adapted to the current range of climatic parameters. For instance, there are individual occurrences of one or two shrubs or trees within sedgeland, or woodland composed of a different species, indicating that these plants are remnants of a former assemblage, *e.g.*, *M. teretifolia* in wetland 163.

In the youngest wetlands, the vegetation pattern tends to be latiform sedgeland comprising two species, *I. nodosa* and *B. juncea*. In the next stage, the pattern changes to maculiform sedgelands comprising *B. juncea, L. gladiatum, S. validus*, herblands comprising *C. asiatica,* and rushlands comprising *J. kraussii*. With increasing age, the pattern becomes concentricform with two, three and then four zones. Wetlands with two zones comprise an inner zone of mixed sedgeland and herbland (*B. juncea/ C. asiatica*), and an outer zone of closed grass tree (*X. preissii*). Wetlands with three zones are similar, but the inner zone contains only herbland and the intermediate zone contains only sedgeland. Wetlands with four zones exhibit more diverse composition. While the outer and intermediate zones are similar to the two and three zoned wetlands, the outer zone may also contain *A. saligna*, and the intermediate zone may contain *M. hamulosa*. The central zone exhibits the greatest variability in that it may comprise sedgeland (*B. articulata*), shrubland (*M. teretifolia*), or low forest (*M. rhaphiophylla*).

A second characteristic of the vegetation in the Becher Suite wetlands is its dynamic nature in response to progressive, and/or decadal variation in water levels as well as inter-annual rainfall variation. The dynamic nature of the vegetation is expressed in constantly changing geometry of zones and patches as a result of changing species abundance. Changes in species abundance, evident from density counts and from quadrat analysis, may be rapid, usually short term, and often reversible when it involves annuals, aggressively opportunistic, or highly adaptive species. Changes in vegetation can occur between seasons for some species (*C. asiatica, S. virginicus*), between years for others (*M. hamulosa, B. articulata, A. cyclops*), and over a time span greater than 11 years for others (*M. rhaphiophylla*). Changes in species abundance appears to be one of the most frequently used adaptations to variable environmental conditions within the wetland, and they overprint and confound vegetation changes responding to evolving habitat.

There appears to be no correlation between plant species and sediment type, and the ranges of plant tolerance to water depth, and groundwater and soil water salinity, overlap. This is not a surprising result given that experimental studies elsewhere have demonstrated that plants, even within the same genus, have different mechanisms for coping with environmental conditions at the margins of their tolerance level (Justin and Armstrong 1987, Blom *et al.* 1994, de Kroon *et al.* 1996, Visser *et al.* 2000), and in some investigations, initial hypotheses concerning plant responses to controlled experimental conditions have been proven to be only partly correct (Zedler *et al.* 1990). Plant sustaining mechanisms range from interactive dependence (rhizomes), to individual cell control (aerenchyma), to selective storage in different parts of their anatomy, to rapid or slow response to pathogenic conditions.

12.9 Vegetation history

Pollen preserved in wetland sediments provides a record of past vegetation and environmental change. The pollen records contained in the Becher wetlands included the vegetation history of several basins as well as that of the sub-region, however, this discussion concentrates primarily on the pollen deriving from *in situ* wetland species, the reconstruction of the internal history of each basin, and the arguments as to whether that history demonstrates sequential succession or merely vegetation dynamics.

Pollen abundances in cores from several wetlands of decreasing age show no single uni-directional trend in vegetation. Similar taxa were present throughout the cores, indicating that the species pool was consistent, and there was no recognisable sequence in vegetation life forms. Pollen from trees was present from the base to the surface of the cores, however, analysis of this aspect of the pollen is confounded by the difficulty of separating allochthonous tree pollen and that derived *in situ*. Several observations from the study of dune vegetation succession by Avis and Lubke (1996) have a bearing on this study. Firstly, Avis and Lubke (1996) observed that the dune slack habitat was subject to invasion by species from communities with both more and less complex structural diversity and species richness, under respective favourable and unfavourable climatic conditions. Observations of wetland vegetation in the Becher Suite corroborate these findings. It explains why no uni-directional trend in vegetation was apparent, why species of pollen not only fluctuated down profile, but appeared and disappeared, and why the pollen abundance graphs of the Becher wetlands show an increase in *in situ* species diversity. Secondly, Avis and Lubke (1996) noted that a vegetated dune slack could revert to plant species, cover and structure types indicative of pioneer slacks when important environmental conditions became more variable. In the Becher wetlands, this factor would explain why there are patches of vegetation indicated by the pollen record and observed in the extant vegetation, which more typically represent incipient wetland stages. Thirdly, Avis and Lubke (1996) concluded that the succession terminated before reaching the most complex plant community because of environmental constraints. Environmental constraints and inter-species competition explain the diversity of wetland assemblages present in the Becher Suite today. Serial colonisation, as described in Chapter 11, occurs in the early stages of wetland development in response to a gradual change from parent sand to a wetland filling with sediment, and the associated increase in water availability through increasing soil moisture and decreasing depth to water. Once water dependent plants colonise the wetland, successional changes become blurred. The complexity of wetland functioning and the dynamic changes emphasised by seasonality and longer term climatic variation can mean that no set of environmental conditions exist long enough for a particular species to prevail. As this process has been described in the Becher wetlands during the period of this study, it is plausible that the same process occurred in the past.

However, the pollen abundance graphs of the Becher wetlands (Figs. 11-7 to 11-11) do show an increase in *in situ* species diversity and an increase in pollen from marginal species, which could signify a trend towards greater complexity in vegetation structure and arrangement. Such factors have been identified as influencing successional pathways (Glenn-Lewin 1980, McCook 1994).

Succession has been best demonstrated in studies, model simulations, or experiments, of either resource competition, usually nitrogen or light (Tilman 1990, Ernst *et al.* 1994), or strategic resource acquisition by one species (van der Valk 1981, Huston and Smith 1987). The focus of studies of resource competition is the dominance which arises from the interaction between a community of plants with different life history traits, and an environment with different resource levels. In practice, this approach requires that the particular resource should change in one direction, *e.g.*, that nutrients should increase with the buildup of organic matter. However, of the environmental resources studied in the Becher wetlands, the most important is water, which is characterised by changing availability and quality. In this latter context, the progress of succession in the Becher wetlands more closely resembles that of Johnstone (1986), who proposed a stochastic interaction between species invasion, maintenance, and decline.

The study of strategic resource acquisition by one species is based on the premise that species interactions occur at the level of the individual. A plant's particular life history traits determine its success at a particular region of any environmental gradient (McCook 1994) and can be measured as rates of growth, mortality, and photosynthesis. Five patterns of species abundances resulted from the simulations of interaction between two species with different life history traits (Huston and Smith 1987): sequential succession; divergence; total suppression; convergence; and pseudo-cyclic replacements. The patterns in the wetland pollen taxa at Becher most closely resemble that of the pseudo-cyclic replacements. Theoretically, sequential succession *sensu* Huston and Smith (1987) can be generated from correlation of traits where species are otherwise not sequential (McCook 1994), but in practice it is difficult to distinguish population dynamics from this type of succession.

Pollen of two species, *C. asiatica* and *S. virginicus*, representing two distinct assemblages (herbland and grassland) is present in all basins and throughout cores (Figs. 11-7 to 11-11). The fluctuations in the pollen abundances indicate expansion and contraction of these vegetation assemblages in response to beneficial or environmentally adverse conditions within the wetland basins. Although it is possible that the life history traits of these plants could have changed (McCook 1994), *C. asiatica* currently thrives under freshwater conditions in organic/carbonate mud, with high soil moisture content, but can survive in any of the wetland sediment types and in a range of hydroperiods. *S. virginicus* currently preferentially inhabitats carbonate mud with low soil moisture content, and high pore water salinity, but can survive increased moisture and lower salinity although it is not a strong competitor in these conditions. Perhaps the most adaptable species in the Becher wetlands is *J. kraussii* which exhibited weak correlation with most of the environmental factors tested, indicating a high tolerance to a range of habitats. It has been argued that species such as *C. asiatica* and *J. kraussii,* that are successful because of their tolerance to a broad range of conditions, or *S. virginicus,* which is successful because of its tolerance to conditions unfavourable to others, are nonetheless likely to do better in absolute terms at higher resource levels (McCook 1994), and this is borne out by increased density and luxuriance of plants, as well as in terms of surface area colonised in the Becher wetlands. Other attributes that are common to species which prevail are superior size (*Melaleuca* spp.), more rapid growth rates through vegetative propogation (sedges), and inhibitory mechanisms such as chemical leaf toxicity (*Melaleuca* spp.). These attributes probably contribute to the dominance of a species in individual wetlands, *e.g.*, wetlands 162, 142, 135, 136, 45. While it is important to consider the nature of successional interactions as species interactions (McCook 1994), in few of these cases of superior competitive ability has a species been able to develop and maintain dominance throughout the Becher suite, with the possible exceptions of *M. cuticularis* in the core of wetland 162,extant *M. teretifolia* in wetland 162, and *J. kraussii* in wetland 163. *M. cuticularis* is well known as a species adapted to

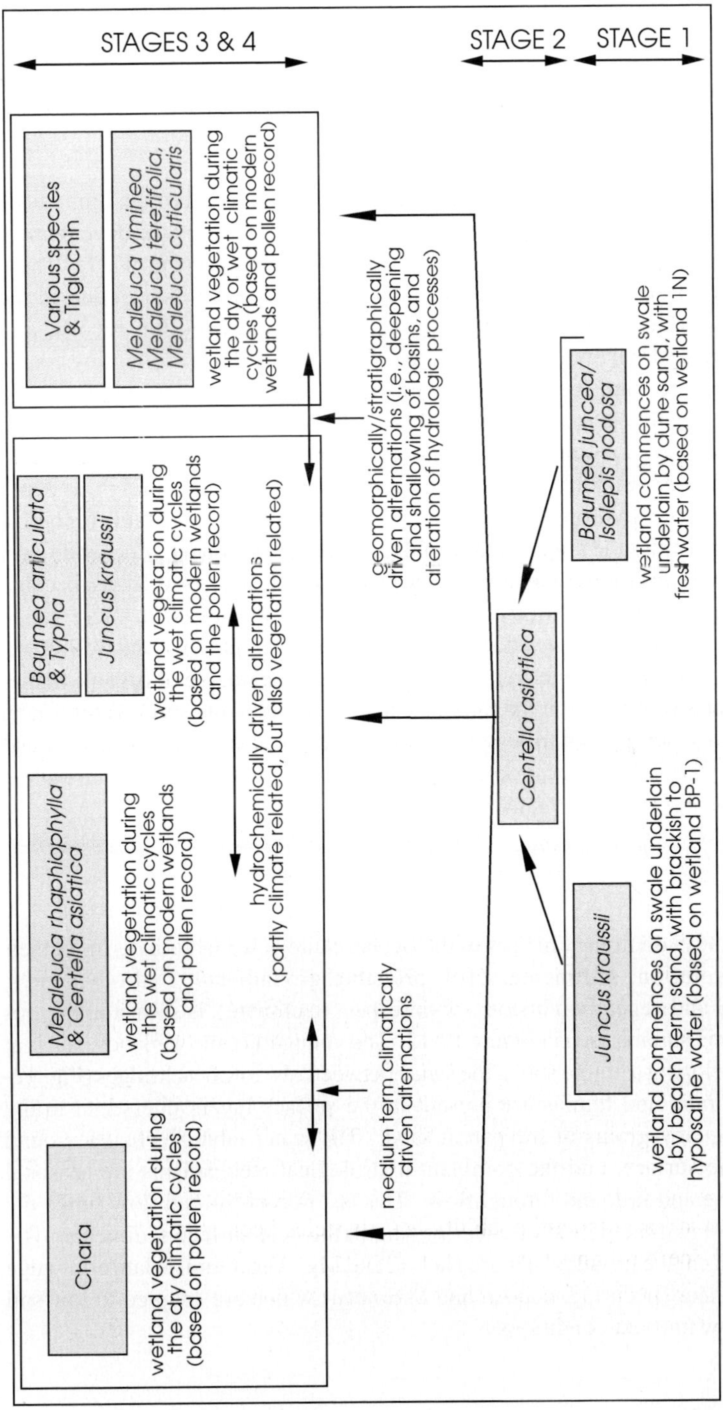

Figure 12-7. *Summary of stages in the development of wetland vegetation, showing changes in relation to overall wetland evolution, and responses to smaller-scale climatic effects (wetland margins omitted).*

relatively saline conditions (Marchant *et al.* 1987). The corresponding occurrence of a dramatic increase in the pollen abundance of this species with the formation of carbonate mud could indicate a natural environmental disturbance of the type required to generate a shift in the vegetation which would satisfy the definition of successional change.

The occurrence of plant species and development of assemblages through time can be related to habitat, and to changes in habitat which correspond to wetland development stages in response to small to medium term climatic fluctuations (Fig. 12-7). These findings are based on observations of the current plant species distribution in the Becher Suite wetlands, together with the evidence of the pollen record, and stratigraphic and hydrochemical information. Today, in swales underlain by coarse calcareous/quartzose sand (beach berm) with fresh to hyposaline groundwater, the most common assemblage is *J. kraussii* rushland. In swales underlain by medium to fine calcareous/quartzose sand (dune) with fresh groundwater, the most common assemblage is *I. nodosa* and *B. juncea* sedgeland. In both of these settings, as depth to water decreases and hydroperiod lengthens, the habitat becomes suitable for herbs, and *C. asiatica* becomes the dominant assemblage. Under subsequent drier conditions, on carbonate mud substrates, with more saline groundwater, species of *Chara* become important. Under drier conditions, on organic mud substrates, with more saline groundwater, species of *Melaleuca* dominate. Under wetter conditions, and freshwater, various species will colonise the wetland, depending on the degree of organic matter which accumulates within or over the carbonate mud. The following species correspond to the gradation from greater to lesser amounts of organic mud: *B. articulata* and *T. orientalis* or *T. domingensis*; *J. kraussii;* and *M. rhaphiophylla* with *C. asiatica* understorey (Fig. 12-7).

12.10 Evolution of wetlands

A summary of the evolutionary history of the Becher Suite of wetlands in terms of their geomorphic evolution, sedimentary fill, probable groundwater hydrodynamics, hydrochemistry and vegetation history (with respect to climate), is shown in Figures 12-8 to 12-11, and summarised in Figure 12-12. The youngest of the wetlands, 1N, is at the stage of developing a humic soil in the swale between two low beachridges (Fig. 12-8). Carbonate mud and humus are present in the surface layers interstitial to the carbonate and quartz grains of the parent sand. The water table fluctuates around 1.2 m below the surface, and the dominant hydrological mechanisms are seasonal groundwater rise and fall, and throughflow. The fast rate of throughflow limits the groundwater rise and the period of waterlogging in the near surface sediments. The dominant cations in the groundwater are Na, Ca, and Mg. Vegetation is latiform sedge comprising pioneer species (*I. nodosa* and *B. juncea*) which are adapted to low soil moisture and low nutrient conditions.

Wetlands swi and swii represent the next stage of wetland development on the Becher plain (Fig. 12-9). The accumulation of organic matter and carbonate mud in the surface layers of these basins is sufficient to have formed a semi-permeable muddy sediment. The water table fluctuates around 0.6-0.8 m below the surface and the dominant hydrological mechanisms are seasonal groundwater rise and fall, and throughflow. The fast rate of throughflow limits the groundwater rise, but meteoric percolation results in a degree of waterlogging in the near surface sediments. The dominant cations in the groundwater are again Na, Ca and Mg. Vegetation is maculiform sedge and herbland comprising pioneer and secondary successional species (*B. juncea, J. kraussii, L. gladiatum* and *C. asiatica*), which are adapted to more consistent soil moisture, and low nutrient conditions.

Wetland swiii represents the next stage of wetland development. There is no wetland currently displaying all the attributes shown in Figure 12-10, however, most of the wetlands, including swiii, exhibit buried carbonate mud fills composed of disintegrated particles of *Chara*, and this fill extends beyond the current boundaries of the wetland, as illustrated in this diagram. The wetland fill is semi-permeable, carbonate sandy mud and mud. The carbonate mud has been slightly infiltrated and/or overlain by organic mud, and bioturbation has also contributed to the mixing of the two compositional types in the surface layers. The water table fluctuates around -0.4 m and inundates the surface under average rainfall conditions. There are two hydrological mechanisms, seasonal groundwater rise and fall, and throughflow. The fast rate of throughflow limits the groundwater rise, but temporary perching of rain and retardation of percolation results in waterlogging in the near surface sediments. The dominant cations in the groundwater are Na, Ca and Mg, with increases in concentration in the latter two cations. Vegetation is maculiform sedge and herbland comprising mainly secondary successional species (*S. validus, B. juncea* and *C. asiatica*) with patches of pioneer species (*L. gladiatum, J. kraussii*). The successional species are adapted to more consistent soil moisture, waterlogging, and low nutrient conditions.

In other wetlands, which are older than wetland swiii, but at a similar stage of development (wetlands 9, 35, 45), there are additional features which are reminders of the fact that individual wetlands have independent histories. Wetland 9 contains a sheet of calcrete, evidence for a drier period in the wetland's history. In wetlands 9 and 35, there are remnants of species of wetland shrubs and trees which also testify to earlier drier and more saline periods (*M. hamulosa* and *M. cuticularis*). In contrast, in wetland 45, the degree of organic development suggests that this wetland has experienced extensive wet periods.

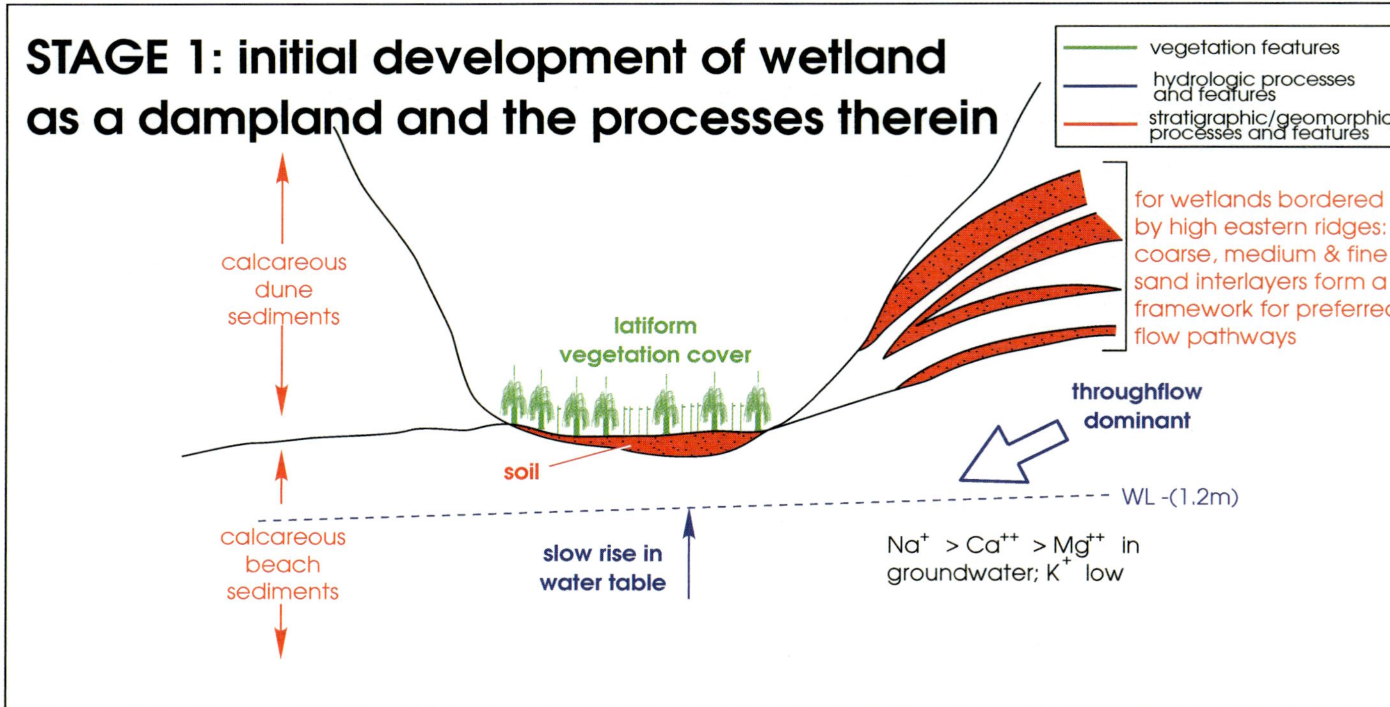

Figure 12-8. *Early stage of wetland development (e.g., wetland 1N)*
(various geomorphic. stratigraphic, hydrologic, and hydrochemical features and processes, and vegetation responses are colour coded).

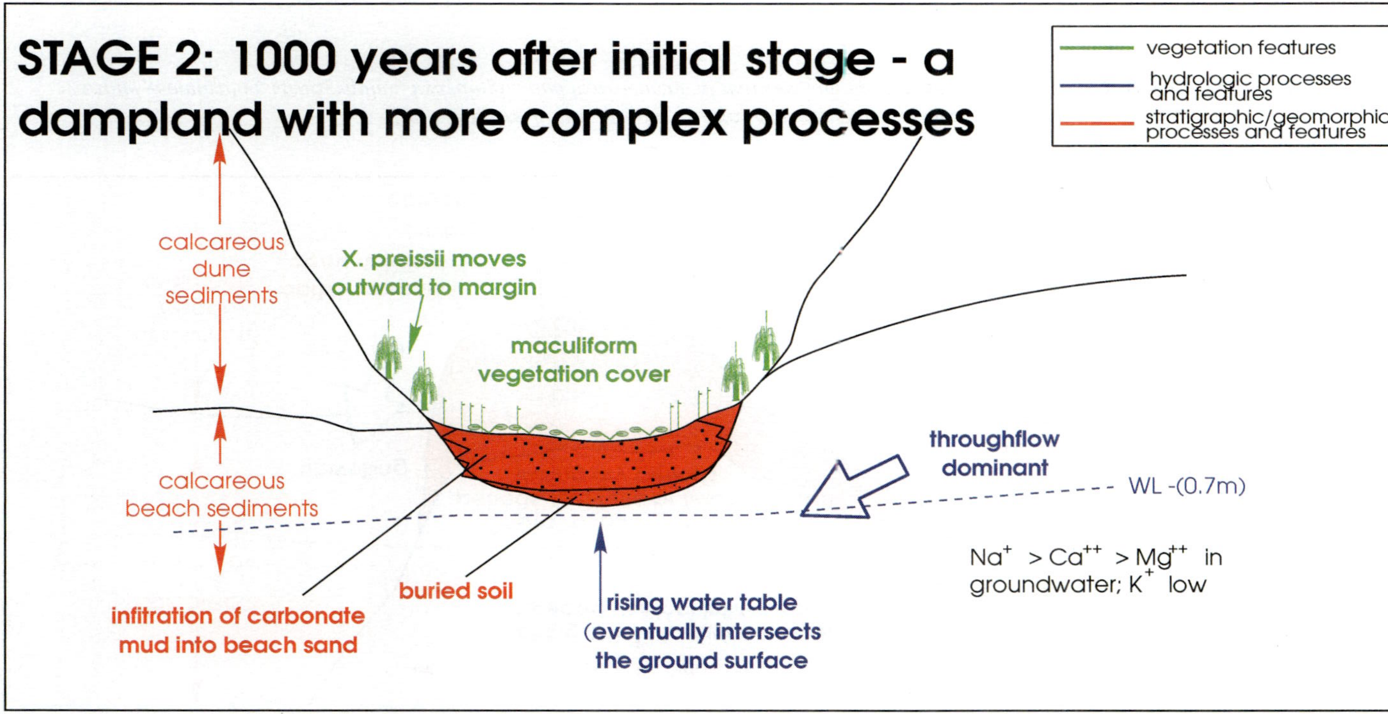

Figure 12-9. *Development of a dampland 1000 years after Stage 1 (e.g., wetland swi) (various geomorphic. stratigraphic, hydrologic, and hydrochemical features and processes, and vegetation responses are colour coded).*

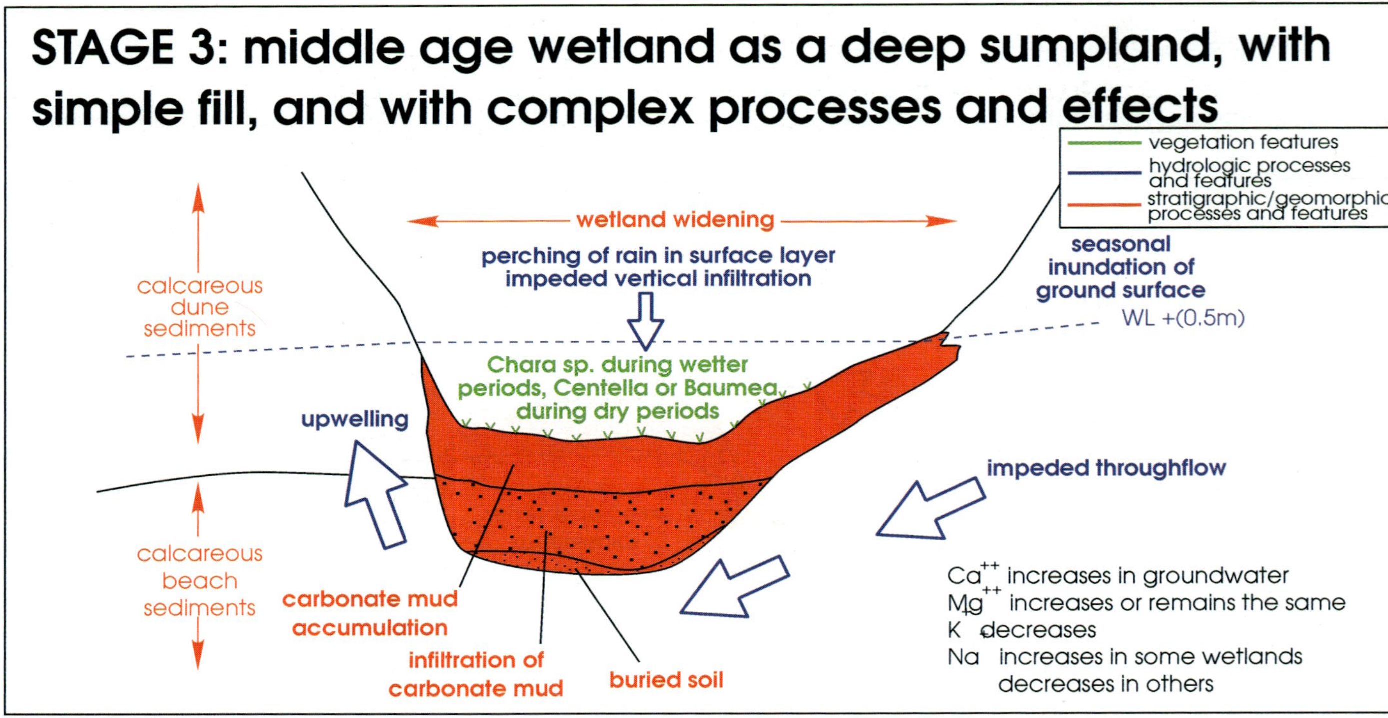

Figure 12-10. *Middle age stage of wetland development (various geomorphic. stratigraphic, hydrologic, and hydrochemical features and processes, and vegetation responses are colour coded).*

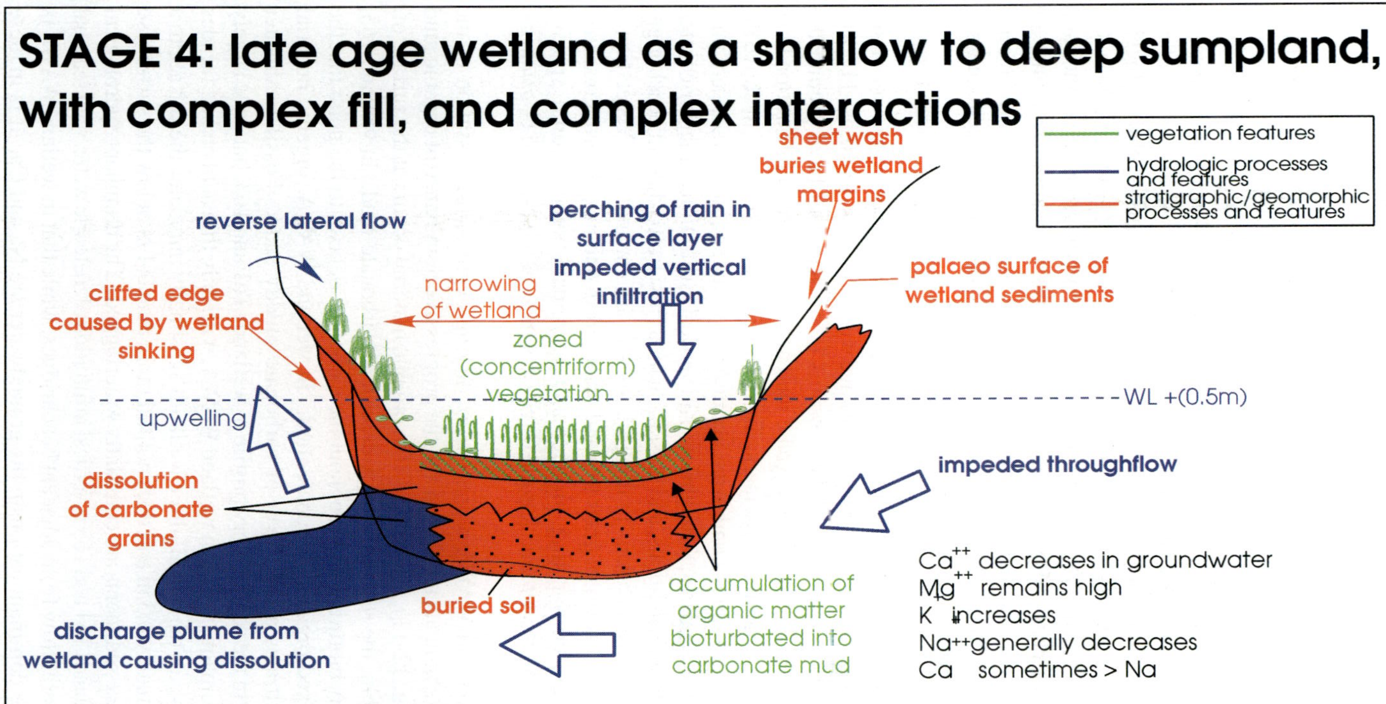

Figure 12-11. *Late stage of wetland development (e.g., wetland 161)*
(various geomorphic. stratigraphic, hydrologic, and hydrochemical features and processes, and vegetation responses are colour coded).

It is approximately at this central location on the cuspate foreland that the pioneering species in the vegetation succession change. The incipient wetlands here exhibit humic soils and closed structure of *X. preissii* (grass trees). Wetlands 63 and 72 represent the earliest stage of wetland development on this part of the Becher plain (Fig. 12-9). Accumulation and infiltration of carbonate mud into the humic soils produced a gradational muddy sand to mud fill. The two hydrological mechanisms, seasonal groundwater rise and fall, and throughflow, result in a water table fluctuating around 0.6-0.8 m below the surface. There is a slower rate of throughflow under these basins which does not limit the groundwater rise, but reduces the period of waterlogging in the near surface sediments. The dominant cations in the groundwater are Na, Ca, and, to a lesser extent, Mg. Vegetation is concentriform but the zones are in the form of discontinuous patches. From periphery to centre the vegetation structures are grass tree, heath, sedge, and herbland, comprising pioneer and secondary successional species (*I. nodosa, B. juncea, X. preissii* and *C. asiatica*), which reflect fluctuating soil moisture, and low nutrient conditions.

Wetlands 135, 136, 142, and 162 represent the stage of wetland development on the inland Becher plain between the middle and late age wetlands (Figs. 12-10, 12-11). The accumulation of carbonate mud in these basins is sufficient to have produced a relatively thick deposit, but minimal organic mud accumulation and infiltration have occurred. The water table fluctuates markedly from -1.2 m below ground to above the surface, under average rainfall conditions, because of surface and subsurface perching. The hydrological mechanisms, throughflow, and vertical infiltration of meteoric water, alternate depending on conditions, and seasonal lateral flow from adjacent beachridges is evident. The dominant cations in the groundwater are Na and Mg. Vegetation is variable. It includes gradiform grass tree and open forest, and maculiform grass tree, heath, and sedges, comprising pioneer and secondary successional species (*M. rhaphiophylla, M. teretifolia, X. preissii, S. virginicus, B. juncea,* and *C. asiatica*).

Wetlands 161, 163 and WAWA represent, to varying degrees, the cessation of the phase of carbonate mud accumulation and the progressive dominance of organic mud accumulation (Fig. 12-11). In wetland 161, organic material occurs at the surface of, and in bands within, the relatively thick deposit of carbonate mud. In wetland 163, organic material is bioturbated throughout the carbonate deposit to the point that the sediment is a mixed organic/carbonate mud. In wetland WAWA, organic material overlies and infiltrates a thin layer of carbonate muddy sand. Fossil freshwater gastropods occur in the sedimentary profile in wetlands 161 and 163, but have largely been dissolved from the profile in wetland WAWA, with only thin acid etched facades remaining. The water table in all three wetlands seasonally inundates the surface to a depth of approximately 0.3 m and does not fall more than 0.4 m below the surface, so that the near surface sediments are constantly waterlogged. The dominant hydrological mechanism in wetlands 161 and 163 is vertical infiltration of meteoric water, with seasonal upwelling and lateral flow from adjacent beachridges, while that in wetland WAWA is throughflow. The dominant cationsin the groundwater are Na and Ca. It is in these

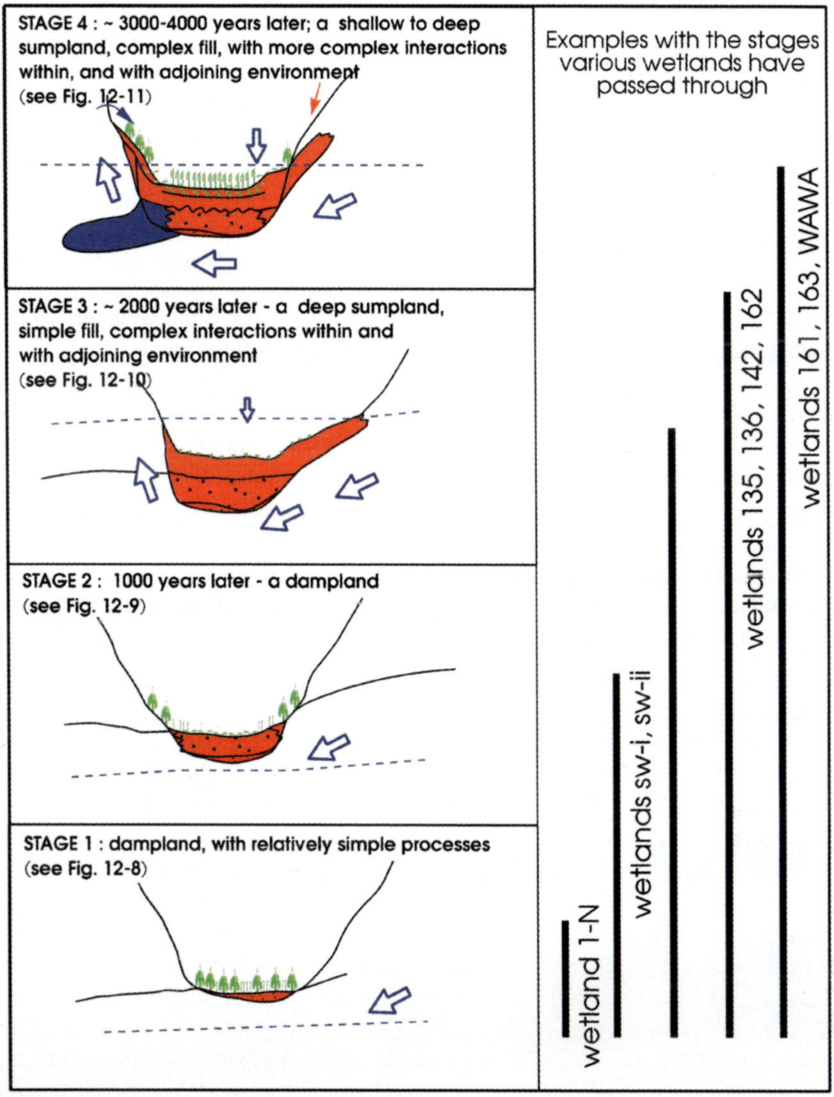

Figure 12-12. *Summary of stages in the development of wetlands in Figs 12-8 to 12-11, showing progressive complexity of stratigraphic fill, hydrologic, hydrochemical, and vegetation responses, and examples of wetlands representative of each stage. It is interpreted that the more complex, older wetlands had passed through the earlier simpler stages.*

three wetlands that wetland deepening through dissolution of carbonate grains has been the most pronounced. Vegetation is variable. It includes gradiform grass tree and sedge, and maculiform grass tree, forest, heath, sedge, rush and herblands, comprising pioneer and secondary successional species (*J. kraussii, X. preissii, B. juncea, B. articulata, S. validus, M. rhaphiophylla, M. teretifolia, M. hamulosa, S. virginicus*, and *C. asiatica*).

12.11 Conclusion

Over 4,500 years of wetland history shows that the Becher Suite wetlands contain valuable information about wetland processes, wetland evolution, and climate change within the late Holocene period, and that less than 1 metre of wetland sediment can initiate a response within a wetland basin which not only sets it apart from larger scale regional processes, but establishes a physical, chemical and biological system which continues to evolve independently. The study has shown that the climatic variability during this period was of sufficient magnitude to cause a current evolutionary trend in the wetlands towards increasingly complex stratigraphy, hydrology and hydrochemistry, which ironically, if taken to completion, could result in the return to a simpler, larger peat filled wetland with complex hydrological and hydrochemical interactions along the margins, a type of wetland which is evident in many settings around the world.

REFERENCES

Adams, M. A., & Pate, J. S. (1992). Availability of organic and inorganic forms of phosphorous to lupins (*Lupinus* spp.). *Plant and Soil, 145*, 107-113.

Allen, A. D. (1976). The Swan Coastal Plain - Hydrogeology of superficial formations, In B. A. Carbon (Ed.), *Groundwater Resources of the Swan Coastal Plain.* Australia: C.S.I.R.O. Division of Land Resources Management, Symposium Proceedings Perth 1975.

Almendinger, J. E., & Leete, J. H. (1998). Regional and local hydrogeology of calcareous fens in the Minnesota River Basin, USA. *Wetlands, 18*(2), 184-202.

Anadon, P., Cabrera, L., & Kelts, K. (Eds.), (1991). *Lacustrine Facies Analysis.* Special Publication No 13 International Association of Sedimentologists Blackwell Scientific Publications.

Anderson, F. O. (1976). Primary production in a shallow water lake with special reference to a reed swamp. *Oikos, 27*, 243-250.

Anderson, M. P., & Munter, J. A. (1981). Seasonal reversals of groundwater flow around lakes and the relevance to stagnation points and lake budgets. *Water Resources Research, 17*, 1138-1150.

Anderson, W. P. Jr, Evans, D. J., & Snyder, S. W. (2000). The effects of Holocene barrier island evolution on water-table elevations, Hatteras Island, North Carolina, USA. *Hydrogeology Journal, 8*, 390- 404.

Andrus, R. E., Wagner, D. J., & Titus, J. E. (1983). Vertical zonation of *Sphagnum* mosses along hummock-hollow gradients. *Canadian Journal of Botany, 61*, 3128-3139.

Anthony, E. J. (1989). Chenier plain development in northern Sierra Leone, West Africa. *Marine Geology, 90*, 297-309.

Anthony, E. J. (1991). Beachridge plain development: Sherbro Island, Sierra Leone. *Zeitschrift fuer Geomorphologie,* N F Suppl Bd *81,* 85-98.

Anthony, E. J. (1995). Beachridge development and sediment supply: examples from West Africa. *Marine Geology, 129*, 175-186.

Armstrong, W., Justin, S. H. F. W., Beckett, P. M., & Lythe, S. (1991). Root adaptation to soil waterlogging. *Aquatic Botany, 39*, 57-73.

Arndt, J. L., & Richardson, J. L. (1989). Geochemistry of hydric soil salinity in a rechargethroughflow-discharge prairie-pothole wetland system. *Soil Science Society of America Journal, 53*, 848-855.

Arndt, J. L., & Richardson, J. L. (1993). Temporal variations in the salinity of shallow groundwater from the periphery of some North Dakota wetlands (USA). *Journal of Hydrology, 141*, 75-105.

Attiwill, P. M., & Leeper, G. W. (1987). *Forest Soils and Nutrient Cycles*. Melbourne: Melbourne University Press.

Austin, M. P. (1976). Performance of four ordination techniques assuming three different nonlinear species response models. *Vegetatio, 33*(1), 43- 49.

Austin, M. P. (1981). Permanent quadrats: an interface for theory and practice. *Vegetatio, 46*, 1-10.

Avis, A. M., & Lubke, R. A. (1996). Dynamics and succession of coastal dune vegetation in the Eastern Cape, South Africa. *Landscape and Urban Planning, 34*, 237-254.

Backhouse, J. (1993). Holocene vegetation and climate record from Barker Swamp, Rottnest Island, Western Australia. *Journal of the Royal Society of Western Australia, 76*(2), 53-61.

Barbour, M. G., Burk, J. H., & Pitts, W. D. (1987). *Terrestrial Plant Ecology*. (pp. 182-207) California: Benjamin Cummings Publishing Co.

Bathurst, R. G. C. (1975). Carbonate sediments and their diagenesis. In *Developments in Sedimentology* No 12 (2nd ed.) London: Elsevier.

Baxter, J. L. (1977). *Heavy mineral sand deposits of Western Australia*. Mineral Resources Bulletin 10. Geological Survey of Western Australia.

Belbin, L. (1991). Semi strong hybrid scaling, a new ordination algorithm. *Journal of Vegetation Science 2*, 491-496.

Belbin, L. (1995). *PATN Pattern Analysis Package*. Technical Reference. CSIRO Division of Wildlife and Ecology Australia.

Bhattacharyya, G. K., & Johnson, R. A. (1977). *Statistical concepts and methods*. New York: John Wiley and Sons.

Bianchi, W. C., & Haskell, E. E. Jr. (1966). Air in the vadose zone as it affects water movements beneath a recharge basin. *Water Resources Research, 2*, 315-322.

Bigarella, J. J. (1965). Sand ridge structures from Parana coastal plain. *Marine Geology, 3*, 269-278.

Bird, E. C. F. (1960). The formation of sand beach-ridges. *Australian Journal of Science, 22*, 349-350.

Bird, E. C. F. (1985). The origin of foredunes and sandy beachridges. *Search, 16*(3-4), 109-111.

Blom, C. W. P. M., & Voesenek, L. A. C. J. (1996). Flooding: the survival strategies of plants. *Trends in Ecology and Evolution, 11*(7), 290-295.

Blom, C. W. P. M., Voesenek, L. A. C. J., Banga, M., Engelaar, W. M. H. G., Rijnders, J. H. G. M., van de Steeg, H. M., & Visser, E. J. W. (1994). Physiological ecology of riverside species: adaptive responses of plants to submergence. *Annals of Botany, 74*, 253-263.

Boelter, D. H., & Blake, G. R. (1964). Importance of volumetric expression of water contents in organic soils. *Soil Science Society of America Proceedings, 28*, 176-178.

Bonorino, G. G., Bujalesky, G., Colombo, F., & Ferrero, M. (1999). Holocene coastal palaeoenvironments in Atlantic Patagonia, Argentina. *Journal of South American Earth Sciences, 12*, 325-331.

Bouma, J., & Wosten, J. H. M. (1979). Flow patterns during extended saturated flow in two undisturbed swelling clay soils with different macropores. *Soil Science Society of America Journal, 3*, 16-22.

Bouwer, H. (1978). *Groundwater hydrology*. New York: Mc Graw-Hill Book Co.

Bowman, G., & Harvey, N. (1986). Geomorphic evolution of a Holocene beach-ridge complex, Le Fevre Peninsula, South Australia. *Journal of Coastal Research, 2*, 345-362.

Boyd, C. E. (1978). Chemical composition of wetland plants. In R. E. Good, D. F. Whigham, & R. L. Simpson (Eds.), *Freshwater wetlands ecological processes and management potential* (pp. 155-167). New York: Academic Press.

Boyd, W. E. (1990). Quaternary pollen analysis in the arid zone of Australia: Dalhousie Springs, Central Australia. *Reviews of Palaeobotany and Palynology, 64*, 331-341.

Boyer, M. L. H., & Wheeler, B. D. (1989). Vegetation patterns in spring-fed calcareous fens: calcite precipitation and constraints on fertility. *Journal of Ecology, 77*, 597-609.

Briggs, S. V., Maher, M. T., & Carpenter, S. M. (1985). Limnological studies of waterfowl habitat in south-western New South Wales. 1. Water chemistry. *Australian Journal of Marine* and *Freshwater Research, 36*, 59-67.

Brinson, M. M. (1977). Decomposition and nutrient exchange of litter in an alluvial swamp forest. *Ecology, 58*, 601-609.

Broberg, O., & Pettersson, K. (1988). Analytical determination of orthophosphate in water. *Hydrobiologia, 170*, 45-59.

Brown, E. M., Burbidge, A. H., Dell, J., Edinger, J., Hopper, S. D., & Wills, R. T. (1997). *Pollination in Western Australia: a database of animals visiting flowers*. Handbook No. 15 WA Naturalists Club Perth.

Bureau of Meteorology (1975). *Climatic Averages Western Australia*. Canberra: Australian Government Publishing Service.

Butler, J. J. (Jr) (1998). *The design, performance and analysis of slug tests*. Boca Raton: Lewis Publishers.

Campbell, K. L., & Capece, J. C. (1999). Hydrologic processes influencing phosphorous transport. In K. R. Reddy, G. A. O'Connor, & C. L. Schelske (Eds.) *Phosphorous biogeochemistry in subtropical systems* (pp. 343-354). Boca Raton: Lewis Publishers.

Carter, M. R. (Ed.) (1993). *Soil Sampling and Methods of Analysis*. Canadian Society of Soil Science Lewis Publishers.

Carter, V. (1986). An overview of the hydrologic concerns related to wetlands in the United States. *Canadian Journal of Botany, 64*, 364-374.

Carter, V., Bedinger, M. S., Novitzki, R. P., & Wilen, W. O. (1979). Water resources and wetlands. In P. E. Greeson, J. R. Clark, & J. E. Clark (Eds.) *Wetland functions and values* (pp. 344376). American Water Research Association MN.

Chen, C., & Wagenet, R. J. (1992). Simulation of water and chemicals in macropore soils. Part 1: Representation of the equivalent macropore influence and its effect on soilwater flow. *Journal of Hydrology, 130*, 105-126.

Cherkauer, D. S., & Zager, J. P. (1989). Groundwater interaction with a kettle hole lake: relation of observations to digital simulations. *Journal of Hydrology, 109*, 167-184.

Churchward, H. M., & McArthur, W. M. (1978). Landforms and soils of the Darling System, Western Australia. In Department of Conservation and Environment (1980). *Atlas of Natural Resources Darling System Western Australia* (pp. 25-33) Perth.

Clarke, D. J., & Eliot, I. G. (1987). Groundwater-level changes in a coastal dune, sea-level fluctuations and shoreline movement on a sandy beach. *Marine Geology, 77*, 319-326.

Clementson, L. A., & Wayte, S. E. (1992). The effect of frozen storage on open-ocean seawater samples on the concentration of dissolved phosphate and nitrate. *Water Resources Research, 26*, 1171-1176.

Clemmensen, L. B. (1986). Storm generated eolian sand shadows and their sedimentary structures, Vejers Strand, Denmark. *Journal of Sedimentary Petrology, 56*(4), 520-527.

Clymo, R. S. (1983). Peat. In A. J. P. Gore (Ed.), *Mires: Swamp, Bog, Fen, and Moor, Ecosystems of the world* 4A (pp. 159-224). Amsterdam: Elsevier.

Coakley, J. P. (1989). The origin and evolution of a complex cuspate foreland: Pointe-aux-Pins, Lake Eerie, Ontario. *Geographe Physique Quaternaire, 43*(1), 68-76.

Cockbain, A. E., & Playford, P. E. (1973). *Stratigraphic nomenclature of Cretaceous rocks in the Perth Basin.* Western Australian Geological Survey Annual Report (1972), 26-31. Perth.

Cole, J. J. & Fisher, S. G. (1979). Nutrient budgets of a temporary pond ecosystem. *Hydrobiologia, 63*, 213-222.

Collins, L. B. (1983). *Post glacial sediments and history, southern Rottnest Shelf, Western Australia.* PhD thesis Department Geology University of WA, Perth. Unpublished.

Collins, M. E., & Kuehl, R. J. (2001). Organic matter accumulation and organic soils. In J. L. Richardson & M. J. Vepraskas (Eds.) *Wetland Soils - Genesis, Hydrology, Landscapes, and Classification.* (pp. 137-162). Boca Raton: Lewis Publishers.

Congdon, R. A., & McComb, A. J. (1976). The nutrients and plants of Lake Joondalup, a mildly eutrophic lake experiencing large seasonal changes in volume. *Journal of the Royal Society of Western Australia, 59*, 14-23.

Cowgill, U. M. (1973). Biogeochemical cycles for the chemical elements in *Nymphaea odorata* AIT. and the aphid *Rhopalosiphum nymphaeae* (L.) living in Linsley Pond. *Science of the Total Environment, 2*, 259-303.

Crawford, R. M. & Wishart, D. (1966). A multivariate analysis of the development of dune slack vegetation in relation to coastal accretion at Tentsmuir, Fife. *Journal of Ecology, 54*, 729-743.

Cressie, N. A. C. (1991). *Statistics for spatial data.* NewYork: Wiley Interscience.

Crisp, D. T. (1966). Input and output of minerals for an area of pennine moorland: the importance of precipitation, drainage, peat erosion and animals. *Journal of Applied Ecology, 3*, 327-348.

Crow, E. L., Davis, F. A., & Maxfield, M. W. (1961). *Statistics Manual.* Research Dept US Naval Ordnance Test Station New York: Dover Publications.

Curl, E. A. & Truelove, B. (1986). *The rhizosphere.* Advanced Series in Agricultural Sciences 15. Berlin: Springer-Verlag.

Currie, R. G., & Fairbridge, R. W. (1985). Periodic 18.6 year and cyclic 11 year induced drought and flood in northeastern China and some global implications. *Quaternary Science Reviews, 4*, 109-134.

D'Amore, D. V., Stewart, S. R., Huddleston, H., & Glasmann, J. R. (2000). Stratigraphy and hydrology of the Jackson-Frazier wetland, Oregon. *Soil Science Society of America Journal, 64*, 1535-1543.

Davies, J. L. (1957). The significance of cut and fill in the development of sand beach ridges. *Australian Journal of Science, 20*, 105-111.

Davies, J. L. (1958). Analysis of height variation of sand beach ridges. *Australian Journal of Science, 21*, 51-52.

Day, F. P. (1982). Litter decomposition rates in the seasonally flooded Great Dismal Swamp. *Ecology, 63*, 670-678.

Deer, W. A., Howie, R. A., & Zussman, J. (1966). *An introduction to the rock forming minerals*. London: Longmans Green & Co Ltd.

de Kroon, H., Fransen, B., van Rheenen, J. W. A., van Dijk, A., & Kreulen, R. (1996). High levels of inter-ramet water translocation in two rhizomatous *Carex* species, as quantified by deuterium labelling. *Oecologia, 106*, 73-84.

de Raeve, F. (1987). Sand dune vegetation and management dynamics. In F. van der Meulen, P. D. Jungerius, & J. H. Visser (Eds.), *Perspectives in coastal dune management* Proceedings of European Symposium 1987 Leiden, Netherlands.

Dickinson, K. J. M., & Mark, A. F. (1994). Forest wetland vegetation patterns associated with a Holocene dune-slack sequence Haast Ecological District, south western New Zealand. *Journal of Biogeography, 21*(3), 259-281.

Dietvorst, P., van der Maarel, E., & van der Putten, H. (1982). A new approach to the minimal area of a plant community. *Vegetatio, 50*, 77-91.

Doeglas, D. L., (1946). Origin and destruction of beach-ridges. *Leidse Geologische Melededeligen, 20*, 34-37.

Doss, P. K. (1993). The nature of a dynamic water table in a system of non-tidal, freshwater coastal wetlands. *Journal of Hydrology, 141*, 107-126.

Edwards, W. M., van der Ploeg, R. R. & Ehlers, W. (1979). A numerical study of the effects of non-capillary sized pores upon infiltration. *Soil Science Society Of America Journal, 43*, 851-856.

Eghbal, M. K., Southard, R. J., & Whittig, L. D. (1989). Dynamics of evaporite distribution in soils on a fan-playa transect in the Carrizo Plain, California. *Soil Science Society of America Journal, 53*, 898-903.

Erdtman, G. (1960). The acetolysis method: a revised description. *Svensk Botanisk Tidskrift, 54*, 561-564.

Ernst, W. H. O., Slings, Q. L., & Nelissen, H. J. M. (1996). Pedogenesis in coastal wet dune slacks after sod-cutting in relation to revegetation. *Plant and Soil, 180*, 219-230.

Eser, P. & Rosen, M. R. (1999). The influence of groundwater hydrology and stratigraphy on hydrochemistry of Stump bay, South Taupo wetland, New Zealand. *Journal of Hydrology, 220*, 27-46.

Eshleman, K. N., Pollard, J. S., & O'Brien, A. K., (1994). Interactions between groundwater and surface water in a Virginia coastal plain watershed. 1. Hydrological flowpaths. *Hydrological Processes, 8*, 389-410.

Fairbridge, R. W., & Hillaire-Marcel, C. (1977). An 8,000-yr palaeoclimatic record of the "Double-Hale" 45-yr solar cycle. *Nature, 268*, 413-416.

Fetter, C. W. (1994). *Applied hydrogeology*. 3rd ed. New Jersey: Prentice Hall.

Fisher, S. G., Grimm, N. B., Marti, E., & Gomez, R. (1998). Hierarchy, spatial configuration, and nutrient cycling in a desert stream. *Australian Journal of Ecology, 23*, 41-52.

Flugel, E. (1982). Microfacies analysis of limestones (pp.366-382) (K. Christenson, Trans.). Berlin: Springer-Verlag. (Original work published 1978).

Folk, R. L. (1974). *Petrology of sedimentary rocks*. Texas: Hemphill Publishing Co.

Fontolan, G., & Simeoni, U. (1999). Holocene cuspate forelands in the Strait of Magellan, southern Chile. *Revista Geologica de Chile, 26*(2), 175-186.

Franson, M. A. H. (Ed.), (1985). *Standard methods for the examination of water and wastewater. (*16th ed.) Washington: American Public Health Association, American Waterworks Association, Water Pollution Control.

Gahoonia, T. S., Claassen, N., & Jungk, A. (1992). Mobilisation of phosphate in different soils by ryegrass supplied with ammonium or nitrate. *Plant and Soil, 140*, 241-248.

Garcia, C. M., Garcia-Ruiz, R., Rendon, M., Niell, F. X., & Lucena, J. (1997). Hydrological cycle and inter annual variability of the aquatic community in a temporary saline lake (Fuente de Piedra, Southern Spain). *Hydrobiologia, 345*, 131-141.

Gasse, F., Fontes, J. C., Plaziat, J. C., Carbonel, P., Kaczmarska, I., de Deckker, P., Soulie-Marsche, I., Callot, Y., & Dupeuble, P. A. (1987). Biological remains, geochemistry and stable isotopes for the reconstruction of environmental changes in the Holocene lakes from North Sahara. *Palaeogeoraphy Palaeoclimatology and Palaeoecology, 60*, 1-46.

Gehrels, J., & Mulamoottil, G. (1990). Hydrologic processes in a southern Ontario wetland. *Hydrobiologia, 208*, 221-234.

Gentilli, J. (1972). *Australian Climate Patterns*. Melbourne: Nelson.

Gentilli, J. & Fairbridge, R. W. (1951). The physiographic regions of Australia. In J. Gentilli (Ed.), *Western Landscapes*. (1979) (pp. 49-87). Perth: UWA Press.

Gerla, P. J. (1992). The relationship of water table changes to the capillary fringe, evapotranspiration, and precipitation in intermittent wetlands. *Wetlands, 12*(2), 91-98.

Germann, P. F., & Bevan, K. (1981). Water flow in soil macropores 1. an experimental approach. *Journal of Soil Science, 32*, 1-13.

Germann, P. F. & Bevan, K. (1985). Kinematic wave approximation to infiltration into soils with sorbing macropores. *Water Resources Research, 21*, 990-996.

Gibson, N., Keighery, B. J., Keighery, G. J., Burbidge, A. H., & Lyons, M. N. (1994). *A floristic survey of the southern Swan Coastal Plain*. Unpublished Report to Australian Heritage Commission by Department of Conservation and Land Management and Conservation Council of WA.

Gillham, R. W. (1984). The capillary fringe and its effect on water-table response. *Journal of Hydrology, 67*, 307-324.

Glassford, D. K. (1980). *Late Cainozoic desert eolian sedimentation in Western Australia*. PhD thesis University of Western Australia, Perth. Unpublished.

Glen, E., Thompson, T. L., Frye, R., Riley, J., & Baumgartner, D. (1995). Effects of salinity and evapotranspiration of *Typha domingensis* Pers. *Aquatic Botany, 52*, 75-91.

Glenn-Lewin, D. C. (1980). The individualistic nature of plant community development. *Vegetatio, 43*, 141-146.

Godwin, H., & Bharucha, F. R. (1932). Studies in the ecology of Wicken Fen. II the fen water table and its control of plant communities. *Journal of Ecology, 20*, 158-191.

Good, R. E., Whigham, D. F., & Simpson, R. L, (Eds.) (1978). *Freshwater wetlands: ecological processes and management potential*. New York: Academic Press.

Gore, A. J. P. (Ed.) (1983). *Ecosystems of the world. Mires: Swamp, bog, fen and moor.* Vol 4A and 4B. Amsterdam: Elsevier.

Grootjans, A. P., Ernst, W. H. O., & Stuyfzand, P. J. (1998). European dune slacks: strong interactions of biology, pedogenesis and hydrology. *Trends in Ecology and Evolution, 13*(3), 96-100.

Grootjans, A. P., Hartog, P. S., Fresco, L. F. M., & Esselink, H. (1991). Succession and fluctuation in a wet dune slack in relation to hydrological changes. *Journal of Vegetation Science, 2*, 545-554.

Grootjans, A. P., Sival, F. P., & Stuyfzand, P. J. (1996). Hydro-geochemical analysis of a degraded dune slack. *Vegetatio, 126*, 27-38.

Gulliver, F. P. (1896). Cuspate forelands. *Bulletin of the Geological Society of America, 7*, 399-422.

Gulliver, F. P. (1897). Dungeness Foreland. *Geographical Journal,* Vol *ix*, 536-546.

Gurnell, A. (1997). The hydrological and geomorphological significance of forested floodplains. *Global Ecology and Biogeographical Letters, 6*, 219-229.

Hails, J. R. (1969). The origin and development of the Umina-WoyWoy beachridge system, Broken Bay, NSW. *The Australian Geographer, 11*, 1-12.

Hammer, U. T. (1986). *Saline lake ecosystems of the world.* Monographiae Biologicae Vol. 59. Series Editor H. J. Dumont. Dordrecht: Dr W Junk Publishers.

Harvey, J. W., & Nuttle, W. K. (1995). Fluxes of water and solute in a coastal wetland sediment. 2. Effect of macropores on solute exchange with surface water. *Journal of Hydrology, 164*, 109-125.

Hassel, C. W., & Kneebone, E. W. S. (1960). *The geology of Rottnest Island.* Honours thesis Department Geology University of Western Australia, Perth. Unpublished.

Hayashi, M., van der Kamp, G., & Rudolph, D. L. (1998). Water and solute transfer between a prairie wetland and adjacent uplands, 1. Water balance. *Journal of Hydrology, 207*, 42-55.

Hayden, B. P., Santos, M. C. F. V., Shao, G., & Kochel, R. C. (1995). Geomorphological controls on coastal vegetation at the Virginia Coast Reserve. *Geomorphology, 13*, 283-300.

Heddle, E. M., Loneragan, O. W., & Havel, J. J. (1978). Vegetation complexes of the Darling System, Western Australia. In Department of Conservation and Environment (1980). *Atlas of Natural Resources Darling System Western Australia.* Perth.

Hellemaa, P. (1998). The development of coastal dunes and their vegetation in Finland. *Fennia, 176*(1), 111-221.

Hemond, H. F. (1980). Biogeochemistry of Thoreau's Bog, Concord, Massachusetts. *Ecological Monographs, 50*, 507-526.

Hesp, P. (1984). The formation of sand "beach ridges" and foredunes. *Search, 15,* (9-10), 289.

Hite, C. D., & Cheng, S. (1996). Spatial characterisation of hydrochemistry within a constructed fen, Greene County, Ohio. *Ground Water, 34*(3), 415-424.

Hobbs, J. E. (1988). Recent climatic change in Australasia. In S. Gregory (Ed.), *Recent climatic change, a regional approach* (pp. 285-297). London: Belhaven Press.

Hodgkin, E. P., & Di Lollo, V. (1957). The tides of south-western Australia. *Proceedings of the Royal Society of Western Australia, 41*, 42-54.

Holtan, H., Kamp-Nielsen, L., & Stuanes, A. O. (1988). Phosphorous in soil, water and sediment: an overview. *Hydrobiologia, 170*, 19-34.

Holttum, R. E. (1955). Growth habits of monocotyledons - variations on a theme. *Phytomorphology, 5*, 399-413.

Howard-Williams, C. (1985). Cycling and retention of nitrogen and phosphorus in wetlands; a theoretical and applied perspective. *Freshwater Biology, 15*, 391-431.

Howard-Williams, C., Davies, B. R., & Cross, R. H. M. (1978). The influence of periphyton on the surface structure of a *Potamogeton pectinatus* L. leaf (an hypothesis). *Aquatic Botany, 5*, 87-91.

Huh, O. K., Walker, N. D. & Moeller, C. (2001). Sedimentation along the eastern chenier plain coast: down drift impact of a delta complex shift. *Journal of Coastal Research, 17*(1), 72-81.

Hunt, R. J., Walker, J. F., & Krabbenhoft, D. P. (1999). Characterising hydrology and the importance of groundwater discharge in natural and constructed wetlands. *Wetlands, 19*(2).

Huston, M., & Smith, T. (1987). Plant succession: life history and competition. *American Naturalist, 130*, 168-198.

Hutchinson, G. E. (1975). *A treatise on limnology. Vol 1* (Pt 2) *Chemistry of lakes*. New York: John Wiley and Sons.

Ingram, H. A. P. (1983). Hydrology. In A. J. P. Gore (Ed.) *Ecosystems of the world. Mires: Swamp, Bog, Fen, and Moor* Vol. 4A (pp. 67-158). Amsterdam: Elsevier.

Isla F I, Cortizo L. C., & Schnack, E. J, (1996). Pleistocene and Holocene beaches and estuaries along the southern barrier of Buenos Aires, Argentina. *Quaternary Science Reviews, 15*, 833-841.

Ivanov, K. E. (1981). *Water movement in mirelands* (A. Thomson & H. A. P. Ingram, Trans.) London: Academic Press. (Original work published in 1975).

Jackson, J. A. (1997). *Glossary of geology*. 4th ed. Alexandria, Virginia: American Geological Institute.

Jenkins, M. A., & Kershaw, A. P. (1997). A mid-late Holocene vegetation record from an interdunal swamp, Mornington Peninsula, Victoria. *Proceedings of the Royal Society of Victoria, 109*(2), 133-148.

Johnson, D. W. (1919). *Shore processes and shoreline development.* New York: John Wiley & Sons.

Johnstone, I. M. (1986). Plant invasion windows: a time based classification of invasion potential. *Biological Reviews, 61*, 369-394.

Jones, H. E., & Etherington, J. R. (1970). Comparative studies of plant growth and distribution in relation to waterlogging. I. The survival of *Erica cinerea L.* and *Erica tetralix L.* and its apparent relationship to iron and manganese uptake in waterlogged soils. *Journal of Ecology, 58*, 487-496.

Jones, R., & Etherington, J. R. (1971). Comparative studies of plant growth and distribution in relation to waterlogging. Part IV. The growth of dune and dune slack plants. *Journal of Ecology, 59*, 793-801.

Jones, P. S., & Etherington, J. R. (1987). Ecological and physiological studies of sand dune slack vegetation, Kenfig Pool and Dune local nature reserve, Mid-Glamorgan, Wales, UK. In F. van der Meulen, P. D. Jungerius, & J. H. Visser (Eds.) *Perspectives in coastal dune management. Proceedings of European Symposium 1987,* Leiden Netherlands.

Jungk, A., Seeling, B., & Gerke, J. (1993). Mobilisation of different phosphate fractions in the rhizosphere. *Plant and Soil, 155/156*, 91-94.

Justin, S. H. F. W., & Armstrong, W. (1987). The anatomical characteristics of roots and plant response to soil flooding. *New Phytologist, 106*, 465-495.

Kadlec, R. H. (1999). Transport of phosphorous in wetlands. In K. R. Reddy, G. A. O'Connor, & C. L. Schelske (Eds.), *Phosphorous biogeochemistry in subtropical systems* (pp. 355-375). Boca Raton: Lewis Publishers.

Kehew, A. E., Passero, R. N., Krishnamurthy, R. V., Lovett, C. K., Betts, M. A., & Dayharsh, B. A. (1998). Hydrochemical interaction between a wetland and an unconfined glacial drift aquifer, southwestern Michigan. *Ground Water, 36*(5), 849-856.

Kemmers, R. H., & Jansen, P. C. (1988). Hydrochemistry of rich fen and water management. *Agricultural Water Management, 14*, 399-412.

Kent, M., & Coker, P. (1992). *Vegetation description and analysis A practical approach.* London: CRC Press.

Kieft, T. L., Soroker, E., & Firestone, M. K. (1987). Microbial biomass response to a rapid increase in water potential when dry soil is wetted. *Soil Biology and Biochemistry, 19*, 119-126.

King, C. A. M. & Barnes, F. A. (1964). Changes in the configuration of the intertidal beach zone of part of the Lincolnshire coast since 1951. *Zeitschrift fuer Geomorphologie* Suppl, *8*, 105-126.

Klenowski, G. (1976). Geotechnical properties of the Coastal Limestone in the Perth Metropolitan area. In *Geological Survey of Western Australia Annual Report 1995* (pp. 42-46). Perth.

Kling, G. W. (1986). The physicochemistry of some dune ponds on the Outer Banks, North Carolina. *Hydrobiologia, 134*, 3-10.

Klopatek, J. M. (1978). Nutrient dynamics of freshwater riverine marshes and the role of emergent macrophytes. In R. E. Good, D. F. Whigham, & R. L. Simpson, (Eds.), *Freshwater wetlands ecological processes and management potential* (pp. 195-216). New York: Academic Press.

Koerselman, W. (1989). Groundwater and surface water hydrology of a small groundwater-fed fen. *Wetlands Ecology and Management, 1*(1), 31-43.

Koerselman, W. & Verhoeven, J. T. A. (1992). Nutrient dynamics in mires of various trophic status: nutrient inputs and outputs and the internal nutrient cycle. In J. T. A. Verhoeven (Ed.), *Fens and bogs in the Netherlands: Vegetation, history, nutrient dynamics and conservation* (pp. 397-432). Netherlands: Kluwer Academic Publishers.

Kutilek, M. & Nielsen, D. R. (1994). *Soil hydrology*. Cremlingen-Destedt (Germany): Catena.

LaBaugh, J. W. (1986). Limnological characteristics of selected lakes in the Nebraska sandhills, USA and their relation to chemical characteristics of adjacent ground water. *Journal of Hydrology, 86*, 279-298.

LaBaugh, J. W., Winter, T. C., Swanson, G. A., Rosenberry, D. O., Nelson, R. D., & Euliss, N. H. Jr (1996). Changes in atmospheric circulation patterns affect midcontinent wetlands sensitive to climate. *Limnology and Oceanography, 41*, 864-870.

Lammerts, E. J., & Grootjans, A. P. (1997). Nutrient deficiency in dune slack pioneer vegetation: a review. *Journal of Coastal Conservation, 3*, 87-94.

Lammerts, E. J., Maas, C., & Grootjans, A. P. (2001). Groundwater variables and vegetation in dune slacks. *Ecological Engineer, 17*, 33-47.

Leeper, G. W., & Uren, N. C. (1993). *Soil science: An introduction.* Melbourne: C.S.I.R.O. Australia and Melbourne University Press.

Lees, B. G. (1987). Age structure of the Point Stuart chenier plain: a reassessment. *Search, 18*(5), 257-259.

Lewis, W. V. (1932). The formation of Dungeness Foreland. *Geographyr Journal, 80*, 390-424.

Legendre, P., & Fortin, M. J. (1989). Spatial pattern and ecological analysis. *Vegetatio, 80*, 107 138.

Logan, B. W. (1974). Inventory of diagenesis in Holocene-Recent carbonate sediments, Shark Bay, Western Australia. In B. W. Logan, J. F. Read, G. M. Hagan, P. Hoffman, R. G. Brown, P. J. Woods & C. D. Gebelein (Eds.), *Evolution and diagenesis of Quaternary carbonate sequences, Shark Bay, Western Australia* (pp. 195-249). Oklahoma USA: American Association of Petroleum Geologists.

Logan, W. S., & Rudolph, D. L. (1997). Microdepression-focused recharge in a coastal wetland, La Plata, Argentina. *Journal of Hydrology, 194*, 221-238.

Lubke, R. A., & Avis, A. M. (1982). Factors affecting the distribution of *Scirpus nodosus* plants in a dune slack community. *South African Journal of Botany, 1*, 97-103.

MacNally, R., & Quinn, G. P. (1998). Symposium introduction: The importance of scale in ecology. *Australian Journal of Ecology, 23*, 1-7.

Maine, M. A., Hammerly, M. S., Leguizamon, M. S., & Pizarro, M. J. (1992). Influence of the pH and redox potential on phosphate activity in the Parana Medio system. *Hydrobiologia, 228*, 83-90.

Major, G. A., Dal Pont, G., Klye, J. & Newell, B. (1972). *Laboratory techniques in marine chemistry - a manual.* Report No 51, 18. C.S.I.R.O. Division of Fisheries and Oceanography.

Malcolm, R., & Soulsby, C. (2001). Hydrochemistry of groundwater in coastal wetlands: implications for coastal conservation in Scotland. *Science of the Total Environment, 265*, 269-280.

Malmer, N. (1986). Vegetational gradients in relation to environmental conditions in northwestern European mires. *Canadian Journal of Botany, 64*, 375-383.

Mann, C. J., & Wetzel, R. G. (2000a). Hydrology of an impounded lotic wetland - wetland sediment characteristics. *Wetlands, 20*(1), 23-32.

Mann, C. J. & Wetzel, R. G. (2000b). Hydrology of an impounded lotic wetland - subsurface hydrology. *Wetlands, 20*(1), 33-47.

Marchant, N. G. & Abbott, I. (1981). Historical and recent observations of the flora of Garden Island, Western Australia. Western Australian *Herbarium Research Notes No. 5*, 49-62.

Marchant, N. G., Wheeler, J. R., Rye, B. L., Bennett, E. M., Lander, N. S., & McFarlane, T. D. (1987). *Flora of the Perth Region Parts 1 and 2.* Perth: Western Australian Herbarium, Department of Agriculture, Western Australia.

Martinova, M. V. (1993). Nitrogen and phosphor compounds in bottom sediments: mechanisms of accumulation, transformation and release. *Hydrobiologia, 252*, 1-22.

Mason, C. F., & Bryant, R. J. (1975). Production, nutrient content and decomposition of *Phragmites communis* Trin and *Typha angustifolia* L. *Journal of Ecology, 63*, 71-95.

McArthur, W. M., & Bartle, G. A., (1979). *Landforms and soils of the Perth Metropolitan southwest corridor, Western Australia*. C.S.I.R.O. Land Resources Management Series No. 5. Perth: CSIRO Division of Land Resources Management.

McArthur, W. M., & Bettenay, E. (1960). *The development and distribution of soils of the Swan Coastal Plain, Western Australia*. C.S.I.R.O. Soil Publication No. 16. Perth: C.S.I.R.O. Western Australia

McCook, L. J. (1994). Understanding ecological community succession: causal models and theories, a review. *Vegetatio, 110*, 115-147.

McKnight, D., Thurman, E. M., Wershaw, R. L., & Hemond, H. (1985). Biogeochemistry of aquatic humic substances in Thoreau's Bog, Concord, Massachusetts. *Ecology, 66*(4), 1339-1352.

McLachlan, A., Kerley, G., & Rickard, C. (1996). Ecology and energetics of slacks in the Alexandria coastal dunefield. *Landscape and Urban Planning, 34*, 267-276.

Meney, K. A., & Pate, J. S. (Eds.) (1999). *Australian rushes biology: Identification and conservation of Restionaceae and allied families*. Perth: UWA Press.

Mertes, L. A. K., Daniel, D. L., Melack, J. M., Nelson, B., Martinelli, L. A., & Forsberg, B. R. (1995). Spatial patterns of hydrology, geomorphology, and vegetation on the floodplain of the Amazon River in Brazil from a remote sensing perspective. *Geomorphology, 13*, 215-232.

Miller, J. J., Acton, D. F., & St Arnaud, R. J. (1985). The effect of groundwater on soil formation in a morainal landscape in Saskatchewan. *Canadian Journal of Soil Science, 65*, 293-307.

Mitsch, W. J., & Gosselink, J. G. (1986). *Wetlands*. New York: Van Nostrand Reinhold.

Moore, P. A., & Reddy, K. R. (1994). Role of pH and eH on phosphorous geochemistry in sediments of Lake Okeechobee, Florida. *Journal of Environmental Quality, 23*, 955-964.

Moreno-Casasola, P. (1988). Patterns of plant species distribution on coastal dunes along the Gulf of Mexico. *Journal of Biogeography, 15*, 787-806.

Moreno-Casasola, P., & Espejel, I. (1986). Classification and ordination of coastal sand dune vegetation along the Gulf and Caribbean Sea of Mexico. *Vegetatio, 66*, 147-182. Moslow, T. F., & Heron, S. D. Jr (1981). Holocene depositional history of a microtidal cuspate foreland cape: Cape Lookout, North Carolina. *Marine Geology, 41*, 251-270.

Mudroch, A., Azcue, J. M., & Mudroch, P. (Eds.), (1997). *Manual of physico-chemical analysis of aquatic sediments* (pp. 7-67). CRC Lewis Publishers.

Munoz Reinoso, J. C., (2001). Sequential pattern in the stabilised dunes of Donana Biological Reserve (SW Spain). *Journal of Coastal Research, 17*(1), 90-94.

Murphy, J., & Riley, J. P. (1962). A modified single solution method for the determination of phosphate in natural waters. *Analytica Chimera Acta, 27*, 31-36.

Ndifon, G. T. & Ukoli, F. M. A. (1989). Ecology of freshwater snails in south-western Nigeria. 1.Distribution and habitat preferences. *Hydrobiologia, 171*, 231-253.

Nichols, D. S., & Keeney, D. R. (1973). Nitrogen and phosphorus release from decaying water milfoil. *Hydrobiologia, 42*, 509-525.

Nilsson, S., & Praglowski, J. (1992). *Erdtman's Handbook of Palynology* 2nd ed. Copenhagen: Munksgaard.
Nixdorf, B. & Deneke, R. (1997). Why 'very shallow' lakes are more successful opposing reduced nutrient loads. *Hydrobiologia, 342/343*, 269-284.

Noest, V. (1994). A hydrology-vegetation interaction model for predicting the occurrence of plant species in dune slacks. *Journal of Environmental Management, 40*, 119-128.

Novakowski, K. S., & Gillham, R. W. (1988). Field investigations of the nature of water-table response to precipitation in shallow water-table environments. *Journal of Hydrology, 97*, 23-32.

Olff, H., Huisman, J., & van Tooren, B. F. (1993). Species dynamics and nutrient accumulation during early orimary succession in coastal sand dunes. *Journal of Ecology, 81*, 693-706.

Otvos, E. G. (2000). Beach ridges - definitions and significance. *Geomorphology, 32*, 83-108.

Ovington, J. D. (1951). The afforestation of Tentsmuir Sands. *Journal of Ecology, 39*, 363-375.

Palmer, M. W. (1995). How should one count species? *Natural Areas Journal, 15*(2), 124-135.

Passmore, J. R. (1967). The geology, hydrology and contamination of shallow coastal aquifers in the Rockingham district, Western Australia. PhD thesis University of Western Australia. Unpublished.

Passmore, J. R. (1970). Shallow coastal aquifers in the Rockingham district, Western Australia. Water Resesearch Foundation of Australia Bulletin *18*, 83.

Pate, J. S., Meney, K. A., & Dixon, K. W. (1991). Contrasting growth and morphological characteristics of fire sensitive (obligate seeder) and fire resistant (resprouter) species of Restionaceae (S. Hemisphere Rushes) from south-western Western Australia. *Australian Journal of Botany, 39*, 505-525.

Penland, S., & Suter, J. R. (1989). The geomorphology of the Mississippi River chenier plain. *Marine Geology, 90*, 259-267.

Perez Corona, M. E., van der Klundert, I., & Verhoeven, J. T. A. (1996). Availability of organic and inorganic phosphorous compounds as phosphorous sources for *Carex* species. *New Phytologis,t 133*, 225-231.

Phillips, P. J., & Shedlock, R. J. (1993). Hydrology and chemistry of groundwater and seasonal ponds in the Atlantic Coastal Plain in Delaware, USA. *Journal of Hydrology, 141*, 157
178.

Phipps, D., & Playford, G. (1984). *Laboratory techniques for extraction of palynomorphs from sediments.* Department of Geology, University of Queensland No. 11, 1-23.

Platt, N. H. (1989). Lacustrine carbonates and pedogenesis: sedimentology and origin of palustrine deposits from the Early Cretaceous Rupelo Formation, W. Cameros Basin, N. Spain. *Sedimentology, 36*, 665-684.

Platt, N. H., & Wright, V. P. (1992). Palustrine carbonates and the Florida Everglades: towards an exposure index for the freshwater environment. *Journal of Sedimentary Petrology, 52*, 1058-1071.

Playford, P. E. (1988). *Guidebook to the geology of Rottnest Island. Excursion Guidebook No 2.* Perth: Geology Society of Australia and Geological Survey of Western Australia.

Playford, P. E., Cockbain, A. E., & Low, G. H. (1976). *Geology of the Perth Basin Western Australia*. Bulletin 124 Perth: Geological Survey of Western Australia.

Ponnamperuna, F. N., (1972). The chemistry of submerged soils. *Advances in Agronomy, 24*, 29-96.

Price, A. G., & Watters, R. J. (1989). The influence of the overstorey, understorey and upper soil horizons on the fluxes of some ions in a mixed deciduous forest. *Journal of Hydrology, 109*, 185-197.

Price, W. A. (1982). In M. L. Schwartz (Ed.) *The encyclopaedia of beaches and coastal environments*. In R. W. Fairbridge (Series editor) Encyclopaedia of Earth Sciences, Vol 15 Pennsylvania: Hutchinson Ross Publishing Co.

Proctor, M. C. F. (1995). Hydrochemistry of the raised bog and fens at Malham Tarn National nature Reserve, Yorkshire, UK. In J. Hughes & L. Heathwaite (Eds.), *Hydrology and hydrochemistry of British wetlands* (pp. 273-289). London: John Wiley & Sons.

Psuty, N. P. (1965). Beachridge development in Tabasco, Mexico. *Annals of the Association of American Geographers, 55*, 112-124.

Purser, B. H., & Schroeder, J. H. (1986). The diagenesis of reefs: A brief review of our present understanding. In J. H. Schroeder & B. H. Purser (Eds.), *Reef diagenesis* (pp. 424-446). Berlin: Springer-Verlag.

Ramm, K., & Scheps, V. (1997). Phosphorus balance of a polytrophic shallow lake with the consideration of phosphorus release. *Hydrobiologia, 342/343*, 43-53. Ranwell, D. (1959). Newborough Warren, Anglesey 1. The dune system and dune slack habitat. *Journal of Ecology, 47*, 571-601.

Ranwell, D. (1972). *Ecology of salt marshes and sand dunes*. London: Chapman and Hall.

Rao, C. P. (1996). *Modern carbonates: tropical temperate and polar. Introduction to sedimentology and geochemistry* (pp. 11-20). Hobart: University of Tasmania.

Rasch, M., Jacobsen, B. H., & Nielsen, N. (1997). Geomorphology and sedimentary record of three cuspate forelands as indicators of late Holocene relative sea level changes, Disko, West Greenland. *Geografisk Tidsskrift, 97*, 33-46.

Rayment, G. E., & Higginson, F. R. (1992). Australian laboratory handbook of soil and water chemical methods (pp. 10-209). Melbourne: Inkata Press.

Read, J. F. (1974). Calcrete deposits and Quaternary sediments, Edel Province, Shark Bay, Western Australia. In B. W. Logan, J. F. Read, G. M. Hagan, P. Hoffman, R. G. Brown, P. J. Woods & C. D. Gebelein (Eds.), *Evolution and diagenesis of Quaternary carbonate sequences, Shark Bay, Western Australia* (pp. 250-282). Oklahoma USA: American Association of Petroleum Geologists.

Rhue, R. D., & Harris, W. G. (1999). Phosphorous sorption/desorption reactions in soils and sediments. In K. R. Reddy, G. A. O'Connor, & C. L. Schelske (Eds.), *Phosphorous biogeochemistry in subtropical systems* (pp. 187-204). Boca Raton: Lewis Publishers.

Richardson, C. J. (1985). Mechanisms controlling phosphorus retention capacity in freshwater wetlands. *Science, 228*, 1424-1427.

Richardson, C. J., Tilton, D. L., Kadlec, J. A., Chamie, J. P. M., & Wentz, W. A., (1978). Nutrient dynamics of northern wetland ecosystems. In R. E. Good, D. F. Whigham, & R. L. Simpson (Eds.), *Freshwater wetlands ecological processes and management potential* (pp. 217-241). New York: Academic Press.

Richardson, J. L., Arndt, J. L., & Freeland, J. (1994). Wetland soils of the prairie potholes. *Advanced Agronomy, 52*, 121-171.

Richardson, J. L., Arndt, J. L., & Montgomery, J. A. (2001). Hydrology of wetland and related soils. In J. L. Richardson, & M. J. Vepraskas (Eds.), *Wetland soils: Genesis, hydrology, landscapes and classification* (pp. 35-84). Bocca Raton: Lewis Publishers.

Robertson, A. W., Mark, A. F., & Wilson, J. B. (1991). Ecology of a coastal lagoon to dune forest sequence, south Westland, New Zealand. *New Zealand Journal of Botany, 29*, 17-30.

Ross, S. M. (1995). Overview of the hydrochemistry and solute processes in British wetlands. In J. Hughes & L. Heathwaite (Eds.), *Hydrology and hydrochemistry of British wetlands* (pp. 133-181). John Wiley & Sons.

Rosengren, N. (1981). Dune systems on cuspate forelands, east Gippsland, Victoria. *Proceedings of the Royal Society of Victoria, 92*(1-2), 137-147.

Roy, P. S., Thom, B. G., & Wright, L. D. (1980). Holocene sequences on an embayed high energy coast: an evolutionary model. *Sedimentary Geology, 26*, 1-19.

Russell, M. A., & Maltby, E. (1995). The role of hydrologic regime on phosphorus dynamics in a seasonally waterlogged soil. In J. Hughes & L. Heathwaite (Eds.), *Hydrology and hydrochemistry of British wetlands* (pp. 245-260). John Wiley & Sons.

Saito, Y., Nishimura, A., Sato, Y., Yokota, S., Wei, H., & Zhou, Y. (2000). Delta progradation and chenier formation in the Huanghe (Yellow River) Delta, China. *Journal of Asian Earth Sciences, 18*(4), 489-497.

Sanderson, P. G. (2000). A comparison of reef protected environments in Western Australia: the central west and Ningaloo coasts. *Earth Surface Processes and Landforms, 25*(4), 397
419.

Sanderson, P. G., Eliot, I., Hegge, B., & Maxwell, S. (2000). Regional variation of coastal morphology in southwestern Australia: a synthesis. *Geomorphology, 34*, 73-88.

Schlumberger (1985). *Log interpretation charts*. Schlumberger Well Services USA.

Scholle, P. A., Bebout, D. G., & Moore, C. H. (Eds.) (1983). *Carbonate depositional environments*. Memoir 33. Oklahoma: American Association of Petroleum Geologists.

Schot, P. P., & Wassen, M. J. (1993). Calcium concentrations in wetland groundwater in relation to water sources and soil conditions in the recharge area. *Journal of Hydrology, 141*, 197-217.

Schwartz, M. L. (Ed.) (1986). *The Encyclopaedia of beaches and coastal environments*. In R. W. Fairbridge (Series Ed.), Encyclopaedia of Earth Sciences, Vol 15. Pennsylvania: Hutchinson Ross Publishing Company.

Searle, D. J. (1984). A sedimentation model of the Cape Bouvard to Trigg Island Sector of the Rottnest Shelf, Western Australia. PhD thesis University of Western Australia Unpublished.

Searle, D. J., & Semeniuk, V. (1985). The natural sectors of the inner Rottnest Shelf coast adjoining the Swan Coastal Plain. *Journal of the Royal Society of Western Australia, 67*, 115-135.

Searle, D. J., & Semeniuk, V. (1988). Petrology and origin of beach sand along the Rottnest Shelf coast, Southwestern Australia. *Journal of the Royal Society of Western Australia, 70*(4), 119-128.

Searle, D. J., Semeniuk, V. & Woods, P. J. (1988). Geomorphology, stratigraphy and Holocene history of the Rockingham-Becher Plain, South-western Australia. *Journal of the Royal Society of Western Australia, 70*(4), 89-109.

Searle, D. J., & Woods, P. J. (1986). Detailed documentation of Holocene sea-level record in the Perth region, South Western Australia. *Quaternary Research, 26*, 299-308.

Semeniuk, C. A. (1987). Wetlands of the Darling System - a geomorphic approach to habitat classification. *Journal of the Royal Society of Western Australia, 69*(3), 95-112.

Semeniuk, C. A. (1988). Consanguineous wetlands and their distribution in the Darling System, Southwestern Australia. *Journal of the Royal Society of Western Australia, 70*(3), 69-87.

Semeniuk, C. A., & Semeniuk, V. (1995). A geomorphic approach to global classification for inland wetlands. *Vegetatio, 118*, 103-124.

Semeniuk, C. A., Milne, L. A., Ladd, P., & Semeniuk, V. (2006a). Pollen in the surface sediments of wetlands in the Becher Point area, southwestern Australia: a baseline for use in interpreting Holocene sequences. *Journal of the Royal Society Western Australia, (in press).*

Semeniuk, C. A., Milne, L. A., Ladd, P., & Semeniuk, V. (2006b). Holocene palynology of five wetland basins in the Becher Point area, southwestern Australia: *Journal of the Royal Society Western Australia, (in press).*

Semeniuk, C. A., Semeniuk, V., Cresswell, I. D., & Marchant, N. G. (1990). Wetlands of the Darling System, southwestern Australia: a descriptive classification using vegetation pattern and form. *Journal of the Royal Society of Western Australia, 69*(3), 95-112.

Semeniuk, V. (1983). The Quaternary stratigraphy and geological history of the Australind-Leschenault area. *Journal of the Royal Society of Western Australia, 66*(3), 71-83.

Semeniuk, V. (1985). The age structure of a Holocene barrier dune system and its implication for sealevel history reconstructions in southwestern Australia. *Marine Geology, 67*, 197-212.

Semeniuk, V. (1986). Holocene climate history of coastal southwestern Australia using calcrete as an indicator. *Palaeogeography Palaeoclimatology Palaeoecology, 53*, 289-308.

Semeniuk, V. (1995a). The Holocene record of climatic, eustatic and tectonic events along the coastal zone of Western Australia - a review. In C. W. Jr Finkl (Ed.), *Holocene Cycles: Climate, sea levels and sedimentation. Journal of Coastal Research,* Coastal Education and Research Foundation Jubilee Volume Special Issue No 17, pp. 247-259.

Semeniuk, V. (1995b). New Pleistocene and Holocene stratigraphic units in the Yalgorup Plain area, southern Swan Coastal Plain. *Journal of the Royal Society of Western Australia, 78*, 67-79.

Semeniuk, V., & Glassford, D. K. (1989). Bassendean and Spearwood Dunes: their geomorphology, stratigraphy and soils as a basis for habitats of *Banksia* woodlands. *Journal of the Royal Society of Western Australia, 71*(4), 87-88.

Semeniuk, V., & Johnson, D. P. (1982). recent and Pleistocene beach/dune sequences, Western Australia. *Sedimentary Geology, 32*, 301-328.

Semeniuk, V., & Meagher, T. D. (1981). The geomorphology and surface processes of the Australind-Leschenault Inlet coastal area. *Journal of the Royal Society of Western Australia, 64*, 33-51.

Semeniuk, V. & Searle, D. J. (1985a). Distribution of calcrete in Holocene coastal sands in relationship to climate, southwestern Australia. *Journal of Sedimentary Petrology, 55*(1), 86-95.

Semeniuk, V., & Searle, D. J. (1985b). The Becher Sand, a new stratigraphic unit for the Holocene of the Perth Basin. *Journal of the Royal Society of Western Australia, 67*, 109-115.

Semeniuk, V., & Searle, D. J. (1986a). The Whitfords Cusp-its geomorphology, stratigraphy and age structure. *Journal of the Royal Society of Western Australia, 68*, 29-36.

Semeniuk, V., & Searle, D. J. (1986b). Variability of Holocene sealevel history along the southwestern coast of Australia - evidence for the effect of significant local tectonism. *Marine Geology, 72*, 47-58.

Semeniuk, V., & Searle, D. J. (1987). The Bridport Calcilutite. *Journal of the Royal Society of Western Australia,* 70(1), 25-27.

Semeniuk, V., & Searle, D. J. (1987). Beach rock ridges/bands along a high-energy coast in Southwestern Australia - their significance and use in coastal history. *Journal of Coastal Research, 3*(3), 331-342.

Semeniuk, V., Searle, D. J., & Woods, P. J. (1988). The sedimentology and stratigraphy of a cuspate foreland, Southwestern Australia. *Journal of Coastal Research, 4*(4), 551-564.

Semeniuk, V., Cresswell, I. D., & Wurm, P. A. S. (1989). The Quindalup Dunes: the regional system, physical framework and vegetation habitats. *Journal of the Royal Society of Western Australia, 71*(2 and 3), 23-47.

Semeniuk, V., Milne, L. A., & Waterhouse, H. (2000). Contemporary and sub-recent pollen record for the Leschenault Inlet Estuary: towards a palynological baseline. *Journal of the Royal Society of Western Australia, 83*(4), 335-348.

Semeniuk, V., & Semeniuk, C. A. (2001). Human impacts on globally to regionally significant geoheritage features of the Swan Coastal Plain and adjoining coastal zone, southwestern Australia. In V. A. Gostin (Ed.), *Gondwana to Greenhouse*: *Australian Environmental Geoscience* (pp. 181-199). Geology Society of Australia Special Publication 21.

Semeniuk, V., & Semeniuk, C. A. (2004). Sedimentary fill of basin wetlands, central Swan Coastal Plain, southwestern Australia. Part I: sediment particles, typical sediments, and classification of depositional systems. *Journal of the Royal Society of Western Australia,* 87, 139-186.

Semeniuk, V., & Semeniuk, C. A. (2005). Wetland sediments and soils on the Swan Coastal Plain, south-southwestern Australia: types, distribution, susceptibility to combustion, and implications for fire management. *Journal of the Royal Society of Western Australia*, 88: 91-120.

Shaw, J. N., & Rabenhorst, M. C. (1997). The geomorphology, characteristics, and origin of the freshwater marl sediments in the Great Limestone Valley, Maryland, USA. *Catena, 30*, 41-59.

Shedlock, R. J., Wilcox, D. A., Thompson, T. A., & Cohen, D. A. (1993). Interactions between groundwater and wetlands, southern shore of Lake Michigan, USA. *Journal of Hydrology, 141*, 127-155.

Shinn, E. A. (1983). Tidal flat environment. In P. A. Scholle, D. G. Bebout & C. H. Moore (Eds.), *Carbonate depositional environments* Chapter 4. Memoir 33. Oklahoma: American Association of Petroleum Geologists.

Short, A. D., & Hesp, P. A. (1982). Wave, beach and dune interactions in southeastern Australia. *Marine Geology, 48*, 259-284.

Short, A. D., Buckley, R. C., & Fotheringham, D. G. (1989). Preliminary investigations of beachridge progradation on Eyre Peninsula and Kangaroo Island. *Transactions of the Royal Society of SA, 113*(3), 145-161.

Shotyk, W. (1988). Review of the inorganic geochemistry of peats and peatland waters. *Earth Science Reviews, 25*, 95-176.

Siegel, D. I., & Glaser, P. H. (1987). Groundwater flow in a bog-fen complex, Lost River peatland, northern Minnesota. *Journal of Ecology, 75*, 743-754.

Sival, F. P. (1996). Mesotrophic basiphilious communities affected by changes in soil properties in two dune slack chronosequences. *Acta Botanica Neerlandia, 45*(1), 95-106.

Sival, F. P., & Grootjans, A. P. (1996). Dynamics of seasonal bicarbonate supply in a dune slack: effects on organic matter, nitrogen pool and vegetation succession. *Vegetatio, 126*, 1-50.

Sjors, H. (1952). On the relation between vegetation and electrolytes in North Swedish mire waters. *Oikos, 2*, 239-258. Sjors, H., & Gunnarsson, U., (2002). Calcium and pH in north and central Swedish mire waters. *Journal of Ecology, 90*, 650-657.

Small, E. (1973). Water relations of plants in raised *Sphagnum* peat bogs. *Ecology, 53*, 726-728.

Smith, S. M., Allen, R. B., & Daly, B. K. (1985). Soil-vegetation relationships on a sequence of sand dunes, Tautuku Beach, South-east Otago, New Zealand. *Journal of the Royal Society of New Zealand, 15*(3), 295-312.

Smith, S. V., Chambers, R. M., & Hollibaugh, J. T. (1996). Dissolved and particulate nutrient transport through a coastal watershed-estuary system. *Journal of Hydrology, 176*, 181
203.

Snead, R. E. (1982). *Coastal Landforms and surface features: A photographic atlas and glossary*. Pennsylvania: Hutchinson Ross Publishing Company.

Soil Science Society of America (1997). *Glossary of soil science terms 1996*. Madison WI: Soil Science Society of America Inc.

Specht, R. L. (1981). Foliage projective cover and standing biomass. In A. N. Gillison & D. J. Anderson (Eds.), *Vegetation Classification in Australia*. Proceedings of a workshop by C.S.I.R.O. Division of Land Use Research Canberra.

Stapor, F. W. (1975). Holocene beachridge plain development, northwest Florida. *Zeitschrift fuer Geomorphologie, 22*, 116-144.

Stapor, F. W. (1982). *Beachridges and beachridge coasts*. In M. L. Schwartz (Ed.), *The encyclopaedia of beaches and coastal environments*. In R. W. Fairbridge (Series editor) Encyclopaedia of Earth Sciences, Vol. 15. Pennsylvania: Hutchinson Ross Publishing Co.

Stapor, F. W., & Tanner, W. F. (1975). Hydrodynamic implications of beach, beachridge and dune grain size studies. *Journal of Sedimentary Petrology, 45*(4), 926-931.

Steedman, R. K., & Craig, P. D. (1979). *Numerical model study of circulation and other oceanographic aspects of Cockburn Sound*. Report to Cockburn Sound Study Group. Western Australia: Dept Conservation and Environment.

Steedman, R. K., & Craig, P. D. (1983). Wind driven circulation of Cockburn Sound. *Australian Journal of Marine and Freshwater Research, 34*, 187-212.

Steinwand, A. L., & Fenton, T. E. (1995). Landscape evolution and shallow groundwater hydrology of a till landscape in Central Iowa. *Soil Science Society of America Journal, 59*, 1370-1377.

Stuyfzand, P. J. (1993). *Hydrochemistry and hydrology of the coastal dune area of the Western Netherlands*. Nieuwegein Netherlands: KIWA N V Research & Consultancy.

Sykes, M. T., & Wilson, J. B. (1987). The vegetation of a New Zealand dune slack. *Vegetatio, 71*, 13-19.

Syrinx Environmental (2000). Chaot and beachridge vegetation of the Point Becher region. Perth: Unpublished report.

Tanner, W. F. (1993). An 8000 year record of sea-level change from grain size parameters: data from beachridges in Denmark. *The Holocene, 3*, 220-231.

Tanner, W. F. (1995). Origin of beachridges and swales. *Marine Geology, 129*, 149-161.

Tanner, W. F., & Stapor, F. W. (1971). Tabasco beachridge Plain: An eroding coast. *Transactions of the Gulf Coast Association Geology Society, 21*, 231-232.

Tansley, A, G. (1949). *The British Islands and their vegetation.* 2nd ed. Cambridge: University Press Cambridge.

Taylor, M., & Stone, G. W. (1996). Beachridges: a review. *Journal of Coastal Research, 12* (3), 612-621.

Thom, B. G. (1963). Origin of sand beach ridges. *Australian Journal of Science, 26*(1), 351.

Thom, B. G. (Ed.) (1984). *Coastal Geomorphology in Australia.* Sydney: Academic Press.

Tilman, D. (1990). Constraints and tradeoffs: toward a predictive theory of competition and succession. *Oikos, 58*, 3-15.

Treloar, J. M. (1978). *Sediments, depositional environments and history of sedimentation in Peel Inlet, Western Australia.* MSc thesis Department Geology University of Western Australia Unpublished.

Trudgen, M. (1988). *A report on the flora and vegetation of the Port Kennedy area.* Unpublished report to Bowman, Bishaw and Associates. Perth, WA.

Tucker, M. E., & Wright, V. P. (1990). *Carbonate Sedimentology.* Oxford: Blackwell Scientific Press.

V & C Semeniuk Research Group (1991). *Wetlands of the City of Rockingham: Their classification, significance and management.* Report to City of Rockingham and Heritage Council of Western Australia Perth, WA.

van den Berg, J. A., & Ullersma, P. (1994). Ponded infiltration as a cause of the instability of continuous macropores. *Journal of Hydrology, 159*, 169-186.

van der Laan, D. (1979). Spatial and temporal variation in the vegetation of dune slacks in relation to the groundwater regime. *Vegetatio, 39*(1), 43-51.

van der Maarel, E. (1988). Vegetation dynamics: Patterns in time and space. *Vegetatio, 77*, 7-19.

van der Valk, A. G. (1981). Succession in wetlands: A Gleasonian approach. *Ecology, 62*, 688-696.

van der Valk, A. G. & Davis, C. B. (1980). The impact of a natural drawdown on the growth of four emergent species in a prairie glacial marsh. *Aquatic Botany, 9*, 301-322.

van Dijk, H. W. J. & Grootjans, A. P. (1993). Wet dune slacks: decline and new opportunities. *Hydrobiologia, 265*, 281-304.

Vazquez, G., Moreno-Casasola, P., & Barrera, O. (1998). Interaction between algae and seed germination in tropical dune slack species: A facilitation process. *Aquatic Botany, 60*(4), 409-416.

Verhoeven, J. T. A., Whigham, D. F., van Kerkhoven M., O'Neill, J., & Maltby, E. (1994). Comparative study of nutrient-related processes in geographically separated wetlands: towards a science base for functional assessment procedures. In W. J. Mitsch (Ed.), *Global Wetlands: Old World and New.* Amsterdam: Elsevier Science.

Verry, E. S., & Timmons, D. R. (1982). Waterborne nutrient flow through an upland-peatland watershed in Minnesota. *Ecology, 63*, 1456-1467.

Visser, E. J. W., Bogemann, G. M., van de Steeg, H. M., Pierik, R., & Blom, C. W. P. M. (2000). Flooding tolerance of *Carex* species in relation to field distribution and aerenchyma formation. *New Phytologist, 148*, 93-103.

Vitt, D. H., & Chee, W. L. (1990). The relationships of vegetation to surface water chemistry and peat chemistry in fens of Alberta, Canada. *Vegetatio, 89*, 87-106.

Warner, R. F. (1977). Hydrology. In D. N. Jeans (Ed.) *Australia: A geography*. Sydney University Press.

Warwick, N. W. M., & Bailey, P. C. E. (1997). The effect of increasing salinity on the growth and ion content of three non-halophytic wetland macrophytes. *Aquatic Botany, 58*, 73-88.

Water Authority of Western Australia (1988). *Stakehill groundwater area management plan.* Perth: Unpublished report.

Wells, M. R., Oma, V. P. M., & Richards, N. B. L. (1985). *Shire of Rockingham-A study of land resources and planning considerations.* Technical Report 44. Division of Resource Management Department of Agriculture WA.

Wetzel, R. G. (1983). *Limnology.* 2nd ed. USA: Saunders College Publishing.

Wetzel, R. G. (1999). Organic phosphorous mineralisation in soils and sediments. In K. R. Reddy, G. A. O'Connor, & C. L. Schelske (Eds.), *Phosphorous biogeochemistry in subtropical systems* (pp. 225-245). Boca Raton: Lewis Publishers.

Wheeler, B. D. (1999). Water and plants in freshwater wetlands. In A. J. Baird & R. L. Wilby (Eds.), *Eco-hydrolog -Plants and water in terrestrial and aquatic environments* (pp. 127180). Routledge.

Wheeler, B. D., Shaw, S. C., & Cook, R. E. D. (1992). Phytometric assessment of the fertility of undrained rich-fen soils. *Journal of Applied Ecology, 29*, 466-475.

White, A. F., & Brantley, S. L. (1995). Chemical weathering rates of silicate minerals. *Reviews in Mineralogy,* 31, 583.

White, E. M. (1971). Grass cycling of calcium, magnesium, potassium and sodium in solodization. *Soil Science Society of America Proceedings, 35*, 309-311.

Wilcox, D. A., Shedlock, R. J., & Hendrickson, W. H. (1986). Hydrology, water chemistry and ecological relations in the raised mound of Cowles Bog. *Journal of Ecology, 74*, 11031117.

Willard, H. H., Merritt, L. L., Dean, J. A., & Settle, F. A. (1988). *Instrumental methods of analysis* 7th ed. California: Wadsworth Publishing Company.

Willis, A. J., Folkes, J. F., Hope-Simpson, J. F., & Yemm, E. W. (1959). Braunton Burrows: the dune system and its vegetation. Part I. *Journal of Ecology, 47*(1), 11-24.

Willis, A. J., Folkes, J. F., Hope-Simpson, J. F., & Yemm, E. W. (1959). Braunton Burrows: The dune system and its vegetation. Part II. *Journal of Ecology, 47*(2), 249-288.

Winter, T. C. (1983). The interaction of lakes with variably saturated porous media. *Water Resources Research, 19*, 1203-1218.

Winter, T. C. (1986). Effect of groundwater recharge on configuration of the water table beneath sand dunes and on seepage in lakes in the sandhills of Nebraska, USA. *Journal of Hydrology, 86*, 221-237.

Winter, T. C., & Pfannkuch, H. O. (1985). Affect of anisotropy and groundwater system geometry on seepage through lakebeds. *Journal of Hydrology, 75*, 239-253.

Winter, T. C., & Rosenberry, D. O. (1998). Hydrology of prairie pothole wetlands during drought and deluge: A 17 year study of the Cottonwood Lake wetland complex in North Dakota in the perspective of longer term measures and proxy hydrological records. *Climatic Change, 40*, 189-289.

Woods, P. J. (1984). *Evolution of, and soil development on Holocene beachridge sequencs, west coast Western Australia*. PhD thesis University of Western Australia Unpublished.

Woods, P. J., & Searle, D. J. (1983). Radiocarbon dating and Holocene history of the Becher/ Rockingham beachridge plain, west coast Western Australia. *Search, 14* (1-2), 44-46.

Wright, L. D. (1970). The influence of sediment availability on patterns of beachridge development in the vicinity of the Shoalhaven River Delta, NSW. *Australian Geographer, 11* (3), 336-348.

Wright, V. P., & Platt, N. H. (1995). Seasonal wetland carbonate sequences and dynamic catenas: A re-appraisal of palustraine limestones. *Sedimentary Geology, 99*, 65-71.

Yapp, R. H. (1912). *Spirea ulmaria L.* and its bearing on the problem of xeromorphy in marsh plants. *Annals of Botany, 26*, 815-870.

Ying Wang & Xiankun Ke (1989). Cheniers on the east coastal plain of China. *Marine Geology, 90*, 321-335.

Yu, B., & Neil, D. T. (1993). Long term variations in regional rainfall in the south-west of Western Australia and the difference between average and high intensity rainfalls. *International Journal of Climatology, 13*, 77-88.

Zedler, J. B., Paling, E., & McComb, A. (1990). Differential responses to salinity to help explain the replacement of native *Juncus kraussii* by *Typha orientalis* in Western Australian salt marshes. *Australian Journal of Ecology, 15*, 57-72.

Zeeb, P. J., & Hemond, H. F. (1998). Hydrologic response of a wetland to changing moisture conditions: modelling effects of soil heterogeneity. *Climatic Change, 40*, 211-227.

Zenkovich, V. P. (1959). On the genesis of cuspate spits along lagoon shores. *Journal of Geology, 67*, 269-277.

SUBJECT INDEX

Wetlands: Ecology, Conservation and Management

1. C. Semeniuk: *The Becher Wetlands – A Ramsar Site*. Evolution of Wetland Habitats and Vegetation Associations on a Holocene Coastal Plain, South-Western Australia. 2007 ISBN 1-4020-4671-5